Probiotics in Agroecosystem

Vivek Kumar • Manoj Kumar
Shivesh Sharma • Ram Prasad
Editors

Probiotics in Agroecosystem

Springer

Editors
Vivek Kumar
Himalayan School of Biosciences
Swami Rama Himalayan University
Jolly Grant
Dehradun
Uttarakhand
India

Manoj Kumar
Amity Institute of Microbial Technology
Amity University
Noida
Uttar Pradesh
India

Shivesh Sharma
Department of Biotechnology
Motilal Nehru National Institute of Technology
Allahabad
Uttar Pradesh
India

Ram Prasad
Amity Institute of Microbial Technology
Amity University
Noida
Uttar Pradesh
India

ISBN 978-981-10-4058-0 ISBN 978-981-10-4059-7 (eBook)
DOI 10.1007/978-981-10-4059-7

Library of Congress Control Number: 2017953188

Printed on acid-free paper

This Springer imprint is published by Springer Nature
The registered company is Springer Nature Singapore Pte Ltd.
The registered company address is: 152 Beach Road, #21-01/04 Gateway East, Singapore 189721, Singapore

Preface

Probiotics in Agro-Ecosystems

As a general notion, probiotics are beneficial microbes for human health, and are, by definition, living microbes, which when administered appropriately confer a benefit to the host. Advertisements and recent research claim that probiotic products are good for our health, resulting in improved digestion, immunity, and management of allergies and colds. However, the probiotic prospective applications in nondairy-food products and agriculture have not received proper recognition. Presently there is increased interest in food and agricultural applications of probiotics, selection of new probiotic strains, and the development of new applications. The agricultural applications of probiotics with regard to animal, fish, and crop plants have increased steadily, yet a number of uncertainties concerning technological, microbiological, regulatory, and ignored aspects do exist.

Human systems obtain benefits from the beneficial bacteria of probiotics. Likewise, plants also reflect a dependency on certain eco-friendly microbes that act in symbiosis, i.e. plant strengtheners, bioinoculants, phytostimulators, and biopesticides, which eventually benefit human health and agro-ecosystems. The way these microbes are associated with or inhabit plant systems and the fate of their interaction are still poorly understood at a metabolic level. It most likely differs according to microbial plethora, age, and species of the plant, although numerous environmental factors do influence this association.

Scientists have known for decades that legume plants harbor beneficial bacteria in nodules, which fix unavailable nitrogen into a form the plant can easily use. On the other hand, the plant root surface, especially the rhizosphere region, harbors diverse beneficial bacteria and fungi along with various types of endophytes, which are present in the host tissue. This endophytic plant relationship is a matter of adaptation during the process of evolution. Plants have a restricted capacity to genetically adapt to rapidly changing environmental conditions such as temperature, water stress, pathogens, or limited nutrient resources. Therefore, plants may use microbes that have the potential to evolve rapidly owing to their short life cycles and simple genetic material, and help the plant to overcome unfavorable conditions. During the process of selection, the host plant chooses or favors the right microbes for particular conditions, which helps the plants to be healthier and competitive. In this way, it is comparable to humans taking probiotics to improve their health.

The increasing interest in the preservation of the environment and the health of consumers is demanding change in production methods and food consumption habits. Consumers demand functional foods because they contain bioactive compounds in bioavailable forms that are involved in health protection. To fulfill consumers' demands, plants are inoculated with biofertilizers, which are linked to the roots or move inside them, thus acting as plant probiotics and to some extent they become reliable substitutes for chemical fertilizers.

These beneficial microbes are plant probiotics, which promote plant growth through diverse mechanisms such as phosphate solubilization, nitrogen fixation, phytohormone and siderophore production, and by mitigating abiotic and biotic stress. They act as vector carriers that take up unavailable nutrients, move them through the soil, and mobilize them to the root. This concept posits that less is wasted, whatever is available is utilized, and that less needs to be applied. The plant thrives in an environment of lower pollution, and nutrients taken up by the organisms are not available to be leached into the ground and surface waters. In addition, this concept creates a healthy soil that produces superior and healthy plants and involves much more than using only chemical inputs. Regular application of only synthetic inputs leads to reduced soil quality and fortifies plants with chemicals, which results in unhealthy produce, imparting negative effects on human health.

Health-conscious society has encouraged farmers and organic growers to adopt these microbial-based probiotic technologies to inoculate seeds/soils/roots to provide nutrients like phosphate, nitrogen, and other phytostimulatory compounds. In addition, microorganisms have also attracted worldwide consideration owing to their role in disease management, drought tolerance, and remediation of polluted soils. Accordingly, selected and potentially selected microbial communities are possible tools for sustainable crop production and can set a trend for a healthy future. Scientific researchers draw on multidisciplinary approaches to understanding the complexity and practical utility of a wide spectrum of microbes for the benefit of crops. The success of crop improvement, however, largely depends on the performance of microbes and the willingness and acceptance by growers to cooperate. A substantial amount of research has been carried out to highlight the role of microbes in the improvement of crops, but very little attempt is made to organize such findings in a way that can significantly help students, academics, researchers, and farmers.

"Plant Probiotics in Agro-Ecosystems" is conceptualized by experts providing a broad source of information on strategies and theories of probiotic microbes with sustainable crop improvement in diverse agro-ecosystems. The book presents strategies for nutrient fortification, adaptation of plants in contaminated soils, and mitigating pathogenesis, and explores ways of integrating diverse approaches to accomplish anticipated levels of crop production under outdated and conventional agro-ecosystems. It is believed that the enthusiasm and noteworthy opportunities presented in this work regarding our recent understanding of the challenges and relationships that bring about learning plant probiotic and synergistic approaches towards plant and human health will inspire readers to push the field forward to new frontiers.

Dehradun, Uttarakhand, India Vivek Kumar
Noida, Uttar Pradesh, India Manoj Kumar
Allahabad, Uttar Pradesh, India Shivesh Sharma
Noida, Uttar Pradesh, India Ram Prasad

Contents

About the Editors

Vivek Kumar, PhD is Associate Professor, involved in teaching, research and guidance, with a pledge to enduring knowledge. Dr. Kumar works at Himalayan School of Biosciences, Swami Rama Himalayan University, Dehradun, India. He obtained his Masters and Ph.D degrees from CCS Haryana Agricultural University, Hisar, Haryana, India. He serves as an editor and reviewer of several reputed international journals. He has published 61 research papers, 30 book chapters, six review articles, and four books. Dr. Kumar has also served as a microbiologist for 8 years in the Department of Soil and Water Research, Public Authority of Agricultural Affairs and Fish Resources, Kuwait. He has been credited with first-time reporting and identification of pink rot inflorescence disease of the date palm in Kuwait caused by *Serratia marcescens.* He has also organized a number of conferences/workshops as convener/organizing secretary.

Dr. Kumar's research areas are plant-microbe-interactions, environmental microbiology, and bioremediation. He was awarded the 'Young Scientist Award' for the year 2002 in 'Agricultural Microbiology' by the Association of Microbiologists of India (AMI).

Manoj Kumar, PhD is a positive-minded scientist who has a passion for research and development, with a commitment to lifelong learning. He is devoted to high quality science that contributes broadly to both increasing the intellectual knowledge of plant development and to increasing the ecological niche. He has a high level of professional desire and intellectual curiosity, and the potential to fulfil the dream of his high impact publications and the future recognition of these by academic peers.

Dr. Kumar completed his PhD in plant biotechnology at the prestigious Jawaharlal Nehru University and was then awarded two post-doctoral fellowships consecutively: i) DBT-PDF from IISc Bangalore in 2005 and then NRF-PDF from University of Pretoria.

Dr. Manoj Kumar is a researcher of plant biotechnology in the Amity University Uttar Pradesh, India. His present research goal is to understand the metabolic fate of microbial-mediated precursors in whole plant physiology and genetics through processes occurring at the level of metabolism, particularly through processes of rhizosphere communication under *in situ* and *in vitro* plant conditions. Many graduate and undergraduate students who have worked with him have been placed in prestigious organizations worldwide.

Dr. Kumar has published his research work in many prestigious journals and played the role of reviewer for many of his peers.

Shivesh Sharma, PhD is Associate Professor and Head of the Department of Biotechnology at Motilal Nehru National Institute of Technology (MNNIT), Allahabad, Uttar Pradesh, India. Dr. Shivesh Sharma has completed his Masters and Ph.D. in the field of Microbiology. His research interests include environmental microbiology/biotechnology, plant-microbe interaction and bioformulations. Before joining MNNIT Allahabad, Dr. Shivesh Sharma worked at S.B.S (PG) Institute of Biomedical Sciences and Research, Balawala, Dehradun Uttarakhand from 2002–2009. He has been involved in a number of research projects funded both externally (DBT, UGC, DST, MHRD) and internally in the fields of his research interests. His teaching interests include microbiology, environmental microbiology/biotechnology, food biotechnology, bacteriology, and IPR.

He has successfully supervised seven Ph.D., eight M. Tech. and 28 M.Sc. students. He has more than 80 publications in different research journals and various book chapters to his credit. He has also organized various outreach activities.

Ram Prasad, PhD is Assistant Professor at the Amity University, Uttar Pradesh, India. Dr. Prasad completed his Ph.D. from the Department of Microbiology, Chaudhary Charan Singh University, Meerut, UP, India, in collaboration with the School of Life Sciences, Jawaharlal Nehru University (JNU), New Delhi, India. Dr. Prasad received his M.Sc. in Life Sciences at JNU and also qualified CSIR-NET, ASRB-NET, and GATE. His research interest includes plant-microbe interactions, sustainable agriculture, and microbial nanobiotechnology. Dr. Prasad has 93 publications to his credit including research papers and book chapters, five patents issued or pending, and has edited or authored several books. Dr. Prasad has 11 years of teaching experience and he has been awarded the Young Scientist Award (2007) and Prof. J.S. Datta Munshi Gold Medal (2009) by the International Society for Ecological Communications; FSAB fellowship (2010) by the Society for Applied Biotechnology; Outstanding Scientist Award (2015) in the field of Microbiology by Venus International Foundation, and the American Cancer Society UICC International Fellowship for Beginning Investigators (USA, 2014). In 2014–2015, Dr. Prasad served as Visiting Assistant Professor in the Department of Mechanical Engineering at Johns Hopkins University, USA.

Contributors

Saida Aarab Département de Biologie, Faculté des Sciences et Techniques d'Al Hoceima, Al Hoceima, Morocco

Equipe de Recherche de Biotechnologies et Génie des Biomolécules (ERBGB), Faculté des Sciences et Techniques de Tangier, Tangier, Morocco

Adebayo Adeyemo Department of Crop, Soil and Pest Management, Federal University of Technology, Akungba-Akoko, Ondo State, Nigeria

Department of Agronomy, Adekunle Ajasin University, Akungba-Akoko, Ondo State, Nigeria

Ees Ahmad Department of Agricultural Microbiology, Faculty of Agricultural Sciences, Aligarh Muslim University, Aligarh, Uttar Pradesh, India

Iqbal Ahmad Department of Agricultural Microbiology, Aligarh Muslim University, Aligarh, India

Shafaqat Ali Department of Environmental Sciences and Engineering, Government College University, Faisalabad, Pakistan

Mohd Musheer Altaf Department of Agricultural Microbiology, Aligarh Muslim University, Aligarh, India

Abdullah Safar Al-Thubiani Department of Biology, Umm Al-Qura University, Makkah, Saudi Arabia

Rizwan Ali Ansari Section of Plant Pathology and Nematology, Department of Botany, Aligarh Muslim University, Aligarh, Uttar Pradesh, India

Abdelhay Arakrak Equipe de Recherche de Biotechnologies et Génie des Biomolécules (ERBGB), Faculté des Sciences et Techniques de Tangier, Tangier, Morocco

Priyanka Arora School of Sciences, Noida International University, Greater Noida, India

Moses Awodun Department of Crop, Soil and Pest Management, Federal University of TechnologyAkure, OndoState, Nigeria

Department of Agronomy, Adekunle Ajasin University, Akungba-Akoko, Ondo State, Nigeria

Mohammed Bakkali Equipe de Recherche de Biotechnologies et Génie des Biomolécules (ERBGB), Faculté des Sciences et Techniques de Tangier, Tangier, Morocco

Danas Baniulis Institute of Horticulture, Lithuanian Research Centre for Agriculture and Forestry, Babtai, Lithuania

Driss Bouhraoua Equipe de Recherche de Biotechnologies et Génie des Biomolécules (ERBGB), Faculté des Sciences et Techniques de Tangier, Tangier, Morocco

Lorena Carro School of Biology, Newcastle University, Tyne, UK

Deepika Chaudhary Department of Microbiology, CCS Haryana Agricultural University Hisar, Haryana, India

Kumud Dubey Centre for Social Forestry and Eco-Rehabilitation, Allahabad, Uttar Pradesh, India

K.P. Dubey CCF/General Manager, (East), Uttar Pradesh Forest Corporation Allahabad, Allahabad, Uttar Pradesh, India

Renato Fernandes Galdiano Júnior Department of Technology, State Paulista University – UNESP, São Paulo, Brazil

Paula García-Fraile Institute of Microbiology of the CAS, v. v. i, Vestec, Czech Republic

Satwant Kaur Gosal Department of Microbiology, Punjab Agricultural University, Ludhiana, India

Zaid ul Hassan Department of Environmental Sciences and Engineering, Government College University, Faisalabad, Pakistan

Muhammad Ibrahim Department of Environmental Sciences and Engineering, Government College University, Faisalabad, Pakistan

Rosangela Naomi Inui Kishi Department of Technology, State Paulista University – UNESP, São Paulo, Brazil

Veluswamy Karthikeyan Centre for Research, Department of Botany and Biotechnology, Thiagarajar College, Madurai, Tamil Nadu, India

Jupinder Kaur Department of Microbiology, Punjab Agricultural University, Ludhiana, India

Mohd Saghir Khan Department of Agricultural Microbiology, Aligarh Muslim University, Aligarh, Uttar Pradesh, India

Mohammad Saghir Khan Department of Agricultural Microbiology, Faculty of Agricultural Sciences, Aligarh Muslim University, Aligarh, Uttar Pradesh, India

Luciano Takeshi Kishi Department of Technology, State Paulista University – UNESP, São Paulo, Brazil

Martin Kostovčík Institute of Microbiology of the CAS, v. v. i, Vestec, Czech Republic

Manoj Kumar Amity Institute of Microbial Technology, Amity University, Noida, Uttar Pradesh, India

Nitin Kumar Department of Biotechnology, Motilal Nehru National Institute of Technology Allahabad, Allahabad, Uttar Pradesh, India

Rakesh Kumar Department of Microbiology, CCS Haryana Agricultural University Hisar, Hisar, Haryana, India

Vivek Kumar Himalayan School of Biosciences, Swami Rama Himalayan University, Jolly Grant, Dehradun, Uttarakhand, India

Anju Kumari Centre of Food Science and Technology CCS Haryana Agricultural University Hisar, Hisar, Haryana, India

Amin Laglaoui Equipe de Recherche de Biotechnologies et Génie des Biomolécules (ERBGB), Faculté des Sciences et Techniques de Tangier, Tangier, Morocco

Salvador Lladó Institute of Microbiology of the CAS, v. v. i, Prague, Czech Republic

Rafiq A. Lone School of Studies in Botany, Jiwaji University, Gwalior, Madhya Pradesh, India

Department of Natural Sciences, SBBS University, Khiala (Padhiana), Jalandhar, Punjab, India

Rubén López-Mondéjar Institute of Microbiology of the CAS, v. v. i, Vestec, Czech Republic

Mamta Department of Environmental Science, Jiwaji University, Gwalior, Madhya Pradesh, India

Irshad Mahmood Section of Plant Pathology and Nematology, Department of Botany, Aligarh Muslim University, Aligarh, Uttar Pradesh, India

Manuel Megias Departamento de Microbiología y Parasitología, Facultad de Farmacia, Universidad de Sevilla, Sevilla, Spain

Inga Tamosiune Institute of Horticulture, Lithuanian Research Centre for Agriculture and Forestry, Babtai, Lithuania

M. Miransari Department of Book & Article, AbtinBerkeh Scientific Ltd. Company, Isfahan, Iran

Rohit Mishra Centre for Medical Diagnostic and Research MNNIT, Allahabad, Uttar Pradesh, India

Átila Francisco Mógor Federal University of Paraná, Department of Plant Science and Crop Protection, Curitiba, Paraná, Brazil

S. Muthulakshmi P.G and Research Centre, Department of Botany, Thiagarajar College, Madurai, Tamil Nadu, India

Muhammad Nafees Institute of Soil & Environmental Sciences. University of Agriculture Faisalabad, Faisalabad, Pakistan

Segun Oladele Department of Crop, Soil and Pest Management, Federal University of Technology, Akungba-Akoko, Ondo State, Nigeria

Department of Agronomy, Adekunle Ajasin University, Akungba-Akoko, Ondo State, Nigeria

Francisco Javier Ollero Departamento de Microbiología, Facultad de Biología, Universidad de Sevilla, Sevilla, Spain

Alok Pandey Centre for Social Forestry and Eco-Rehabilitation, Allahabad, Uttar Pradesh, India

Ram Prasad Amity Institute of Microbial Technology, Amity University, Noida, Uttar Pradesh, India

Kuppu Rajendran Center for Research, Department of Botany and Biotechnology, Thiagarajar College, Madurai, Tamil Nadu, India

Raman Jangra Department of Microbiology, CCS Haryana Agricultural University Hisar, Haryana, India

Rashmi Department of Microbiology, CCS Haryana Agricultural University Hisar, Haryana, India

Asfa Rizvi Department of Agricultural Microbiology, Faculty of Agricultural Sciences, Aligarh Muslim University, Aligarh, Uttar Pradesh, India

Rose Rizvi Section of Plant Pathology and Nematology, Department of Botany, Aligarh Muslim University, Aligarh, Uttar Pradesh, India

Muhammad Rizwan Department of Environmental Sciences and Engineering, Government College University, Faisalabad, Pakistan

B. Sadhana P.G and Research Centre, Department of Botany, Thiagarajar College, Madurai, Tamil Nadu, India

Safiuddin Section of Plant Pathology and Nematology, Department of Botany, Aligarh Muslim University, Aligarh, Uttar Pradesh, India

Saima Saif Department of Agricultural Microbiology, Faculty of Agricultural Sciences, Aligarh Muslim University, Aligarh, Uttar Pradesh, India

Kulliyan Sathiyadash Center for Research, Department of Botany and Biotechnology, Thiagarajar College, Madurai, Tamil Nadu, India

Seema B. Sharma Department of Earth and Environmental Science, KSKV Kachchh University, Gujarat, India

Shivesh Sharma Department of Biotechnology, Motilal Nehru National Institute of Technology, Allahabad, Uttar Pradesh, India

Centre for Medical Diagnostic and Research MNNIT, Allahabad, Uttar Pradesh, India

Razia Shuab School of Studies in Botany, Jiwaji University, Gwalior, Madhya Pradesh, India

Khushboo V. Singh Department of Microbiology, Gujarat University, Ahmedabad, Gujarat, India

Vidmantas Stanys Institute of Horticulture, Lithuanian Research Centre for Agriculture and Forestry, Babtai, Lithuania

Aisha Sumbul Section of Plant Pathology and Nematology, Department of Botany, Aligarh Muslim University, Aligarh, Uttar Pradesh, India

Vivian Jaskiw Szilagyi Zecchin Federal University of Paraná, Department of Plant Science and Crop Protection, Curitiba, Paraná, Brazil

Priyanku Teotia Department of Botany, CCS University, Meerut, India

Nitika Thakur Shoolini University of Biotechnology and Management Sciences, Solan, HP, India

Archana Tiwari School of Sciences, Noida International University, Greater Noida, India

Praveen Tripathi Centre for Social Forestry and Eco-Rehabilitation, Allahabad, Uttar Pradesh, India

Neha Upadhyay Department of Biotechnology, Motilal Nehru National Institute of Technology Allahabad, Allahabad, Uttar Pradesh, India

Silvana Pompéia Val-Moraes Department of Technology, State Paulista University – UNESP, São Paulo, Brazil

Sunita J. Varjani School of Biological Sciences and Biotechnology, University and Institute of Advanced Research, Gandhinagar, Gujarat, India

Rishi Kumar Verma Department of Biotechnology, Motilal Nehru National Institute of Technology Allahabad, Allahabad, Uttar Pradesh, India

Kanchan Vishwakarma Department of Biotechnology, Motilal Nehru National Institute of Technology Allahabad, Allahabad, Uttar Pradesh, India

Khursheed Ahmad Wani Department of Environmental Science, ITM University, Gwalior, Madhya Pradesh, India

Muhammad Waseem Department of Microbiology, Government College University, Faisalabad, Pakistan

Gaurav Yadav Department of Biotechnology, Motilal Nehru National Institute of Technology Allahabad, Allahabad, Uttar Pradesh, India

Centre for Medical Diagnostic and Research MNNIT, Allahabad, Uttar Pradesh, India

Almas Zaidi Department of Agricultural Microbiology, Faculty of Agricultural Sciences, Aligarh Muslim University, Aligarh, Uttar Pradesh, India

Mohammad Javad Zarea Faculty of Agriculture, University of Ilam, Ilam, Iran

Faculty of Agriculture, University of Kurdistan, Sanandaj, Iran

Role of Endophytic Bacteria in Stress Tolerance of Agricultural Plants: Diversity of Microorganisms and Molecular Mechanisms

1

Inga Tamosiune, Danas Baniulis, and Vidmantas Stanys

Abstract

Bacterial endophytes are a group of endosymbiotic microorganisms widespread among plants. An association of plants with endophytic bacteria includes a vast diversity of bacterial taxa and host plants. In this review we present an overview of taxonomic composition of the bacterial endophytes identified in common agricultural crops with special emphasis on the most recent results obtained using metagenomic analysis. Endophytic microbiome constitutes a part of larger soil microbial community and is susceptible to direct or indirect effect of agricultural practices: soil tillage, irrigation, use of pesticides and fertilizers has a major effect on function and structure of soil and endophytic microbial populations. Therefore, the use of agricultural practices that maintain natural diversity of plant endophytic bacteria becomes important element of sustainable agriculture that ensures plant productivity and quality of agricultural production. On the other hand, the endophytic microbiome itself have been shown to have multiple effects on their host plant, including modulation of phytohormone signaling, metabolic activity, and plant defense response pathways. It has been demonstrated that these effects could be helpful for plant adaptation to abiotic or biotic stresses. Therefore, application of endophytic bacteria to improve crop performance under cold, drought, salinity, and heavy metal contamination stress conditions or to enhance disease resistance presents an important potential for sustainable agricultural production.

I. Tamosiune • D. Baniulis • V. Stanys (✉)
Institute of Horticulture, Lithuanian Research Centre for Agriculture and Forestry, Kaunas str. 30, Babtai, Kaunas reg, Lithuania
e-mail: i.miliute@lsdi.lt; d.baniulis@lsdi.lt; v.stanys@lsdi.lt

V. Kumar et al. (eds.), *Probiotics in Agroecosystem*,
DOI 10.1007/978-981-10-4059-7_1

1.1 Introduction

An intensification of agricultural production has been crucial in sustaining population growth throughout civilization history (Ellis et al. 2013). During the last century, the agricultural intensification has been largely achieved through improvement in crop productivity and the use of farm equipment, irrigation, intensive tillage, fertilizers, pesticides, and other manufactured inputs (Foley et al. 2005; 2011). However, these agricultural practices often lead to detrimental effects on environment as well as human health. Therefore, new environmentally benign pathways have to be employed to maintain increase in agricultural production while greatly reducing unsustainable uses of water, nutrients, and agricultural chemicals. This requires new means to overcome threats that cause loss of crop yield, including plant stresses associated with unfavorable environmental conditions, such as drought, temperature extremes, or soil salinity, as well as biotic stress induced by plant pathogens and pests. Therefore, the attention is drawn to exploitation of mutualistic and antagonistic biotic interactions within agroecosystems that would increase crop productivity and improve sustainability of pest control technologies (Gaba et al. 2014).

Plants live in intimate association with microorganisms that fulfill important functions in agricultural ecosystems and represent an important resource for improvement of plant performance through enhancing crop nutrition or reducing damages caused by pathogens or environmental stress (Jha et al. 2013; Singh et al. 2011). Bacteria constitute the most numerous group of microorganisms in soil (Whitman et al. 1998). They exist as free-living organisms, attached to the surface of roots or phyllosphere, and establish interactions with plants. The extreme forms of plant–microbe interactions could be categorized into commensal (acquire nutrients from the plant without damaging), mutualistic (positively influence plant health), and pathogenic (damage plant) type, yet many microorganisms exploit different forms of relationship with plants during their life cycles (Newton et al. 2010). Endophytic bacteria are a group of endosymbiotic microorganisms that live in internal plant tissues of apparently healthy host plants and do not normally cause any substantial disease symptoms (Schulz and Boyle 2006).

Endophytic bacteria colonize intercellular spaces of the cell walls and xylem vessels of plant roots, stems, and leaves, and they are also found in tissues of flowers (Compant et al. 2011), fruits (de Melo Pereira et al. 2012), and seeds (Cankar et al. 2005; Johnston-Monje and Raizada 2011; Trognitz et al. 2014). Meanwhile it is generally believed that endophytic bacteria reside in apoplast of plant cells, several studies of intracellular colonization of cytosol have been published (Cocking et al. 2006; Koskimaki et al. 2015; Thomas and Sekhar 2014; White et al. 2014). Plant roots have been established as the main entry point of the potential endophytes from soil and provide a base camp for colonization of other plant organs. Higher density of endophyte populations is characteristic to plant roots and other belowground tissues as compared to phyllosphere, and an ascending migration of endophytic bacteria from roots to leaves of rice plants has been demonstrated (Chi et al. 2005). It has been also shown that plant roots are capable to take up bacteria from surrounding environment (Paungfoo-Lonhienne et al. 2010). Isolation of endophytic bacteria from seeds suggests an alternative transmission route (Cankar et al. 2005; Johnston-Monje and Raizada 2011; Trognitz et al. 2014). Structure of the endophytic community is defined by abiotic and biotic factors such as environmental conditions, microbe–microbe interactions, and plant–microbe interactions (Ryan et al. 2008).

Diverse effects of endophytic bacteria on plant health and growth have been well documented. The endophytes aid nutrient availability and uptake, enhance stress tolerance, and provide disease resistance (Hamilton et al. 2012; Ryan et al. 2008). The plant growth-promoting capability of endophytes is established through activity that increases accessibility of nutrients, such as nitrogen and phosphorus, or is mediated by compounds produced by the microorganisms and the host cells, such as plant growth hormones (Brader et al. 2014; Glick 2012; Reinhold-Hurek and Hurek 2011). Disease protection properties are associated with ability of endophytic bacteria to produce compounds, such as antibiotics and fungal cell-wall lytic enzymes, which can inhibit growth of plant pathogens (Brader et al. 2014; Christina et al. 2013; Raaijmakers and Mazzola 2012; Wang et al. 2014) or priming plant response to pathogens by induced systemic resistance (ISR) mechanism (Pieterse et al. 2014). Owing to their plant growth-promoting and disease control properties, endophytes can be used in the form of bioinoculants in agriculture to benefit development of sustainable agricultural production practices (Mei and Flinn 2010).

The aim of this review is to outline the understanding about diversity of endophytic bacterial communities of agricultural crops and their implication in plant adaptation to stress and disease resistance. We provide a summary of the extensive information on taxonomic composition of bacterial endophytes identified in major agricultural crop plants that has been remarkably expanded due to application of advanced metagenomic analysis methods. Effect of different agricultural practices on the diversity of endophytic bacterial communities is assessed. Further, an implication of endophytes in plant adaptation to stress and disease resistance through modulation of phytohormone balance or induction of stress-related metabolites or systemic resistance signaling pathway is presented.

1.2 Assessment of Diversity of Bacterial Endophytes Using Cultivation Techniques and Metagenomic Analysis

Plants are naturally associated with continuum of other organisms, the majority of which are bacterial endophytes. Population densities of endophytic bacteria are extremely variable in different plants and tissues and have been shown to vary from hundreds to reaching as high as 9×10^9 of bacteria per gram of plant tissue (Chi et al. 2005; Jacobs et al. 1985; Misaghi and Donndelinger 1990). Initial studies of diversity of endophyte community were mostly based on the classic microbial culture techniques; therefore, bacterial endophytes isolated using surface sterilization methods have been reported for most species of agricultural plants (Rakotoniriana et al. 2013). One of the early reviews by Hallman et al. (1997) presented the list of isolated bacterial endophytes from various plant parts of different agricultural crops.

The list was supplemented by latter studies on endophyte diversity (Bacon and Hinton 2007; Lodewyckx et al. 2002; Rosenblueth and Martinez-Romero 2006; Ryan et al. 2008; Sturz et al. 2003).

Innovative culture-independent sequencing technologies allow much deeper assessment of microbial communities and improve our understanding about diverse microbiomes occupying plants. In recent years, extensive information about diversity of endophytic microbiota has been gathered using metagenomic sequencing platforms. Application of hypervariable regions from small subunit ribosomal RNA gene (16S rRNA) for the metagenomic sequencing allows precise taxonomic identification (Turner et al. 2013). Direct amplification of microbial DNA from plant tissue samples and application of modern bioinformatics tools allow analysis of growing numbers of plant material samples, and such studies have revealed rarely reported endophyte species of *δ*- and *ε-Proteobacteria* (Sun et al. 2008). In addition, culture-independent high-throughput sequencing technologies reflect variations of total microbial diversity and their physiological potential and ecological functions (Akinsanya et al. 2015; Turner et al. 2013; van Overbeek and van Elsas 2008). For example, Tian and associates (Tian et al. 2015) used second-generation sequencing technology to assess diversity of bacterial endophytes before and after nematode attack, and the study revealed that nematode infection was associated with variation and differentiation of the endophyte bacterial populations.

Studies of microbial diversity using culture-independent molecular techniques could be limited by relatively low abundance of endophytic bacteria that results in underrepresentation in metagenomic library. This problem is associated with difficulties in separation and high sequence homology of endophytic bacteria, plant nuclei, plastids, mitochondria, and plant-associated microbial DNA (Govindasamy et al. 2014). In recent years, gene enrichment strategies have been broadly used. Bacterial DNA extraction from host plant tissues and enrichment is the key step in preparation of the metagenomic library harboring representative sample of microbial diversity. In order to recover target genes of metagenome, a suitable enrichment method should be used before DNA amplification (Mutondo et al. 2010). Jiao et al. (2006) enriched target genes from a metagenome by optimized hydrolysis of the plant cell walls, followed by differential centrifugation. Wang et al. (2008) efficiently enriched bacterial DNA from medicinal plant by specific enzymatic treatment. The same method increased representation of less abundant grapevine-associated bacteria (Bulgari et al. 2009). Series of differential centrifugation steps followed by a density gradient centrifugation efficiently enriched proportion of microbial DNA in stems of soybean (Ikeda et al. 2009). Maropola and colleagues (2015) analyzed the impact of metagenomic DNA extraction procedures on the endophytic bacterial diversity in sorghum and found that different DNA extraction methods introduce significant biases in community diversities. The authors stated that despite the differences in results of extraction of DNA, the agriculturally important genera such as *Microbacterium*, *Agrobacterium*, *Sphingobacterium*, *Herbaspirillum*, *Erwinia*, *Pseudomonas*, and *Stenotrophomonas* were predominant. An enrichment method useful for extraction of plant-associated bacteria of potato tubers was developed by Nikolic et al. (2011) and involved overnight shaking of small pieces of potato tubers in sodium chloride solution.

Although 16S rRNA gene clone library technique provides efficient means to study different agricultural plant microbiota in detail (genetics and physiology), however, not all endophytes are easily amenable using this method as well (Sessitsch et al. 2012). The methods for microbe enrichment in plant tissues may lead to overrepresentation of high-abundance bacterial species and reduced representation of low-abundance species. Therefore, a combination of microbial cultivation and culture-independent metagenomic analysis methods provides broader perspective of the diversity of endophytes.

A summary of the most widespread bacterial isolates identified in common agricultural crop plants is presented in Table 1.1. Due to a vast diversity of bacterial species and host plants described to this day, the list is not complete and presents a sample of important agricultural crops and overview of associated endophytic bacterial species identified using both, cultivation and metagenomic, analysis methods.

A study of direct comparison of culture-dependent and culture-independent approaches for assessing bacterial communities in the phyllosphere of apple has been published by Yashiro et al. (2011). Among the cultivated isolates only order of *Actinomycetales* has been found, while metagenomic approach has revealed the presence of *Bacteroidales*, *Enterobacteriales*, *Myxococcales*, and *Sphingobacteriales*. Differences between plant-associated microbial phyla are revealed when comparing the niches of rhizosphere, endosphere, and phyllosphere. The largest diversity is found in the roots, as it is the primary site of interaction between plants and soil microorganisms (Hardoim et al. 2011). Maropolla and colleagues (2015) found that diversity of sorghum-associated endophytic bacteria is lower in stems than that of rhizospheric communities. Rhizospheric endophytic species mostly belong to *α*-, *β*-, and *γ-Proteobacteria* subgroups and are closely related to epiphytic species (Kuklinsky-Sobral et al. 2004). The group of *γ-Proteobacteria* is found to be the most diverse. Culture-dependent methods revealed bacteria species that belong to the *Proteobacteria*, meanwhile *Firmicutes*, *Actinobacteria*, and also *Bacteroides* are less common (Reinhold-Hurek and Hurek 2011).

Culture-independent approach suggests a 100–1000-fold higher diversity of the bacterial communities in economically important crops (Suman et al. 2016; Turner et al. 2013). Sessitsch with associates (Chi et al. 2005) investigated genomic characteristics of the most abundant bacterial endophytes colonizing rice roots under field conditions without cultivation bias. In this study, the members of *γ-Proteobacteria*, comprising mostly *Enterobacter*-related endophytes, were predominant. Metagenomic analyses demonstrated that rhizobia (and other *α-Proteobacteria*) were the most abundant plant-associated endophytes, including *β-Proteobacteria*, *γ-Proteobacteria*, and *Firmicutes* (Turner et al. 2013). However, it was found that only culture-independent techniques were able to identify endophytic archaea (*Euryarchaeota*) (Suman et al. 2016). In general, the species of *Pseudomonas*, *Bacillus*, *Enterobacter*, *Klebsiella*, *Rhizobium*, *Sphingomonas*, *Pantoea*, *Microbacterium*, *Acinetobacter*, *Erwinia*, and *Arthrobacter* were defined as the most dominant using both methods.

Table 1.1 Endophytic bacteria isolated from common agricultural crop species plants

Phylum	Subgroup	Genus	Onion	Sorghum	Wheat	Rice	Sugarcane	Maize	Potato	Tomato	Soybean	Strawberry	Sugar beet	Bean	Pepper	Carrot	Grapevine
Proteobacteria	α	*Agrobacterium* sp.		■		■		■	■		■						
		Bradyrhizobium sp.				■		■					■	■			
		Brevundimonas sp.			■	■		■	■		■			■			
		Caulobacter sp.				■		■									■
		Devosia sp.						■	■					■			
		Mesorhizobium sp.						■						■			■
		Methylobacterium sp.			■	■		■	■		■			■			
		Novosphingobium sp.				■	■										
		Novosphingobium sp.									■			■			
		Rhizobium sp.			■	■		■	■		■			■		■	
		Rhodopseudomonas sp.				■		■									
		Sinorhizobium sp.			■	■					■			■			
		Sphingomonas sp.				■	■	■	■		■		■	■	■		
		Sphingopyxis sp.				■						■					
	β	*Achromobacter* sp.			■	■	■	■									
		Acidovorax sp.				■			■								
		Burkholderia sp.	■		■	■	■	■							■		■
		Comamonas sp.				■			■		■						■
		Delftia sp.			■	■					■			■			
		Duganella sp.			■	■		■									
		Gallionella sp.			■	■											

		Herbaspirillum sp.		■	■	■	■	■									
		Ralstonia sp.						■	■		■						■
		Variovorax sp.			■	■		■	■		■						
	γ	*Acinetobacter* sp.			■	■		■	■		■			■			
		Azotobacter sp.			■			■									
		Enterobacter sp.			■	■	■	■	■		■	■		■			■
		Erwinia sp.		■			■	■	■		■		■				
		Klebsiella sp.			■	■	■	■	■		■					■	■
		Kluyvera sp.			■			■									
		Pantoea sp.			■	■	■	■	■		■	■					
		Pseudomonas sp.	■	■	■	■	■	■	■	■	■	■	■	■	■	■	■
		Psychrobacter sp.			■				■								
		Serratia sp.			■	■		■	■						■		
		Stenotrophomonas sp.		■	■	■		■			■			■	■		■
		Xanthomonas sp.			■			■	■				■				■
		Geobacter sp.			■	■								■			
		Syntrophus sp.						■						■			
Firmicutes		*Bacillus* sp.			■	■	■	■	■	■	■	■	■	■	■	■	■
		Brevibacillus sp.								■				■	■		
		Clostridium sp.			■			■						■			
		Exiguobacterium sp.			■						■						
		Lactobacillus sp.										■	■				
		Lysinibacillus sp.			■			■	■					■	■		
		Paenibacillus sp.			■			■	■		■			■	■	■	■
		Staphylococcus sp.			■			■	■		■			■			■

(continued)

Table 1.1 (continued)

Phylum	Subgroup	Genus	Onion	Sorghum	Wheat	Rice	Sugarcane	Maize	Potato	Tomato	Soybean	Strawberry	Sugar beet	Bean	Pepper	Carrot	Grapevine
Actinobacteria		*Actinomyces* sp.							■								■
		Amycolatopsis sp.				■		■									
		Arthrobacter sp.			■			■	■		■				■		■
		Aureobacterium sp.				■		■									
		Brachybacterium sp.						■									■
		Clavibacter sp.						■			■						■
		Corynebacterium sp.			■	■		■	■				■				
		Curtobacterium sp.				■			■			■			■		■
		Dactylosporangium sp.				■		■									
		Frankia sp.				■		■									
		Kocuria sp.			■			■						■			
		Microbacterium sp.		■		■	■	■	■		■			■	■		■
		Micrococcus sp.			■	■			■					■	■		■
		Micromonospora sp.			■	■		■									
		Mycobacterium sp.			■	■											
		Nocardioides sp.			■	■											
		Rhodococcus sp.				■		■	■					■			■
		Rothia sp.						■									
		Streptomyces sp.			■	■		■					■				■
		Chryseobacterium sp.						■			■				■		

	Cytophagales sp.				■					■						
	Flavobacterium sp.			■	■		■	■		■				■		■
	Pedobacter sp.				■		■	■		■						
	Sphingobacterium sp.		■		■					■			■			

Onion (*Allium cepa* L.) (Frommel et al. 1991; Weilharter et al. 2011), sorghum (*Sorghum bicolor*) (James et al. 1997; Maropola et al. 2015), wheat (*Triticum aestivum*) (Balandreau et al. 2001; Coombs and Franco 2003; Iniguez et al. 2004; Jha and Kumar 2009; Larran et al. 2002; Mavingui et al. 1992; Velazquez-Sepilveda et al. 2012; Verma et al. 2013, 2014, 2015; Zinniel et al. 2002), rice (*Oryza sativa*) (Chaintreuil et al. 2000; Elbeltagy et al. 2001; Engelhard et al. 2000; Govindarajan et al. 2008; Kaneko et al. 2010; Mbai et al. 2013; Naik et al. 2009; Piromyou et al. 2015; Rangjaroen et al. 2014; Sandhiya et al. 2005; Sessitsch et al. 2012; Stoltzfus et al. 1997; Sun et al. 2008; Tian et al. 2007; Yan et al. 2008; Yanni et al. 1997; You and Zhou 1988), sugar cane (*Saccharum officinarum*) (Govindarajan et al. 2007; Mendes et al. 2007; Nutaratat et al. 2014; Quecine et al. 2012; Suman et al. 2001; Tam and Diep 2014), maize (*Zea mays*) (Araujo et al. 2000; Chelius and Triplett 2001; Fisher et al. 1992; Fouts et al. 2008; Hallman et al. 1997; Lalande et al. 1989; Matsumura et al. 2015; Mcinroy and Kloepper 1995; Montanez et al. 2012; Palus et al. 1996; Rai et al. 2007; Rosenblueth and Martinez-Romero 2006; Tam and Diep 2014; Zinniel et al. 2002), potato (*Solanum tuberosum*) (Berg et al. 2005; De Boer and Copeman 1974; Hallman et al. 1997; Han et al. 2011; Hollis 1951; Manter et al. 2010; Pageni et al. 2013; Pavlo et al. 2011; Rado et al. 2015; Reiter et al. 2002; Sessitsch et al. 2004; Sturz et al. 1988, 2003), tomato (*Solanum lycopersicum*) (Hallman et al. 1997; Li et al. 2014; Patel et al. 2012; Pillay and Nowak 1997; Samish et al. 1961; Yang et al. 2011), soybean (*Glycine max*) (Hung and Annapurna 2004; Kuklinsky-Sobral et al. 2004; Mingma et al. 2014; Okubo et al. 2009; Pimentel et al. 2006; Zinniel et al. 2002), strawberry (*Fragaria ananassa*) (de Melo Pereira et al. 2012; Dias et al. 2009; Hardoim et al. 2011, 2012), sugar beet (*Beta vulgaris*) (Dent et al. 2004; Jacobs et al. 1985; Tsurumaru et al. 2015), common bean (*Phaseolus vulgaris*) (de Oliveira Costa et al. 2012; Suyal et al. 2015), pepper (*Piper* sp.) (Aravind et al. 2009; Paul et al. 2013; Sziderics et al. 2007; Xia et al. 2015), carrot (*Daucus carota* L.) (Surette et al. 2003; Wu et al. 2015), grapevine (*Vitis vinifera*) (Baldan et al. 2014; Bell et al. 1995; Bulgari et al. 2009; Campisano et al. 2014a; Compant et al. 2011; Westet al. 2010).

1.3 Effect of Agricultural Practices on Diversity of Endophytic Bacterial Communities

Bacteria constitute the most numerous group of microorganisms in soil (Whitman et al. 1998), and many endophytic bacteria originate from the population of plant-associated microorganisms in rhizosphere (Hardoim et al. 2008). Microbial diversity of the plant rhizosphere itself is defined by overall composition of microbial pool of soil and further refined by specific plant–microbe interactions that are largely mediated by root exudates (Sorensen and Sessitsch 2006). It has been demonstrated that endophytic community represents a plant genotype-specific subset of the wider microbial population of soil (Bulgarelli et al. 2012; Lundberg et al. 2012). Agricultural land management, such as tillage or irrigation, greatly alters soil characteristics that may lead to reduction in soil microbial diversity due to mechanical destruction, soil compaction, reduced pore volume, desiccation, and disruption of access to food resources (Garcia-Orenes et al. 2013; Jangid et al. 2008). Several studies have established the effect of tillage systems on soil microbial communities in different soils and cropping systems (Balota et al. 2003; Dorr de Quadros et al. 2012; Mathew et al. 2012). The effect of excessive use of pesticides can induce significant changes in the function and structure of soil microbial populations due to direct inhibition of microbial growth or overall changes in the structure of agricultural ecosystems (Pampulha and Oliveira 2006). Balanced mineral or organic fertilizers have been shown to have positive effect on diversity and metabolic activity of the soil microbial community (Zhong et al. 2010).

The effect of the agronomic practices on the overall soil microbial community could be expected to reflect differences in endophyte populations of agricultural crop plants. However, the research aimed to elicit effect of agricultural practices on composition of the endophytic bacteria populations is limited to several studies. An early study by Fuentes-Ramirez et al. (1999) demonstrated that colonization ability of nitrogen-fixing endophytic bacterium *Acetobacter diazotrophicus* was largely decreased in the sugarcane plants fertilized with high levels of nitrogen. A recent study using automated ribosomal intergenic spacer analysis showed that structure of rice root endophytic community was affected by the nitrogen fertilization level (Sasaki et al. 2013). Another study assessed root bacterial endophyte diversity in maize grown using different fertilizer application conditions. Application of PCR-based group-specific markers revealed that type I methanotroph patterns were different for plants cultivated using mineral and organic fertilizer (Seghers et al. 2004).

Recently, culture-based and metagenomic analyses were employed to assess bacterial endophyte diversity of plants grown using conventional and organic practices. An extensive study by Xia et al. (2015) evaluated diversity of culturable bacterial endophytes in different tissues of corn, tomato, melon, and pepper grown using organic or conventional practices. The endophyte diversity was significantly higher among all the crops grown organically versus those grown using conventional practices. There were 32 species isolated from organically grown plants and 28 species from plants grown using conventional practices.

No significant effect of herbicide treatment on composition of the maize root endophyte population was detected using the PCR-based group-specific markers (Seghers et al. 2004). However, recent study using automated ribosomal intergenic spacer fingerprinting and metagenomic analysis using 16S rDNA pyrosequencing identified differences in the composition of endophytic communities in grapevines cultivated using organic and integrated pest management conditions (Campisano et al. 2014a). While a different outcome of the two studies might be a consequence of improvement in the capability of the analysis methods, it could as well be related to differences specific to the plant species or pesticide treatment conditions.

The studies described in this section showed that agricultural conditions could alter diversity of endophytic bacteria populations; however, further insight would be required to elucidate the mechanisms that mediate such changes. The variation in bacterial diversity could be a consequence of changes in overall soil microbial population upon the fertilizer treatment or application of other agronomic practices. On the other hand, the agronomical conditions potentially had a direct effect on the root endophytic bacterial community as was suggested by Xia et al. (2015). In addition, an important role might be attributed to differences in plant physiological state and changes in composition of the plant root exudates that influence growth of endophytic bacteria (Paungfoo-Lonhienne et al. 2010). This notion that factors related to plant biochemistry regulate endophyte diversity was supported by the study demonstrating that application of chitin resulted in changes in bacterial communities in soil, rhizosphere, and cotton roots, and the organic amendment supported the endophytic species in cotton roots that otherwise did not occur (Hallman et al. 1999). Intriguingly, it was shown that composition of the endophytic community was largely different from that of the rhizosphere; therefore, the amendment of chitin, which enhanced chitinase and peroxidase concentrations, might have changed preference of the plants for certain bacterial endophytes.

Another aspect related to the effect of agricultural practices on soil and plant microbiome is reflected by disease-suppressive soil phenomenon that is associated with the capability of soils to suppress or reduce plant disease of susceptible host plants in the presence of virulent pathogen (Weller et al. 2002). It was shown several decades ago that disease-suppressive properties of soil were largely induced by long-term cultivation of wheat and potato monoculture leading to buildup of host-specific microbial community (Lorang et al. 1989; Scher and Baker 1980; Whipps 1997). Further studies elucidated possible mechanisms of disease suppression that include competition for space and nutrients, antagonism due to production of secondary metabolites, and elicitation of ISR by soil microbiota (Philippot et al. 2013; Pieterse et al. 2014). Specific role of the endophytic bacteria in the development of the disease-suppressive traits was rarely addressed in the studies on disease-suppressive soil communities; however, bacteria of genus *Streptomyces*, *Bacillus*, *Actinomyces*, and *Pseudomonas* that are known to lead endophytic lifestyle were shown to contribute to the disease-suppressive traits of soils (Haas and Defago 2005; Kinkel et al. 2012; Mendes et al. 2011; Siddiqui and Ehteshamul-Haque 2001; Weller et al. 2002).

The importance of agricultural practices that maintain natural diversity of plant endophytic bacteria is emphasized by the observations that agricultural plants may become a niche for human pathogens and a source for outbreaks of food-borne illness (Brandl 2006). Use of manures contaminated with virulent bacteria was identified as a main source of human pathogens (Brandl 2006; Holden et al. 2009; van Overbeek et al. 2014). Other routes included irrigation water (Erickson et al. 2010) or flies (Talley et al. 2009). Meanwhile a decline of species antagonistic to the pathogenic bacteria in soil and endosphere was associated with plant colonization by human pathogen species (Latz et al. 2012); it was also demonstrated that the presence of certain plant pathogens and other species living in soil plays an important role in colonization of plants by human pathogens (Barak and Liang 2008; Brandl 2008; Brandl et al. 2013). On the other hand, typical plant-associated bacteria species belonging to the genera of *Enterobacter*, *Serratia*, and *Klebsiella* could become virulent to humans by acquisition of mobile genetic elements from human pathogens through horizontal gene transfer (van Overbeek et al. 2014). Pathogenic bacteria of the family *Enterobacteriaceae*, including pathogenic *Salmonella* genus strains, *E. coli*, *Klebsiella pneumoniae*, and *Vibrio cholerae* strains, and the human opportunistic pathogens *Pseudomonas aeruginosa* and *Propionibacterium acnes* were described as endophytic colonizers of plants (Campisano et al. 2014b; Deering et al. 2012; El-Awady et al. 2015; Kumar et al. 2013; Kutter et al. 2006; Schikora et al. 2008).

1.4 Role of Endophytic Bacteria in Adaptation of Agriculture Crops to Biotic and Abiotic Environmental Stress

1.4.1 Induction of Accumulation of Stress-Related Metabolites and Enzymes

Plants are capable to acclimate to environmental stresses by altering physiology to attain state adopted to overcome stress factors such as dehydration, mechanical injury, nutrient deficiency, high solar radiation, or stress-induced increase in concentration of reactive oxygen species. This acclimation is associated with enhanced production of compounds that mediate osmotic adjustment, stabilize cell components, and act as free radical scavengers. It has been observed that plant inoculation with endophytic bacteria leads to accumulation of such compounds, including proline, phenolic compounds, carbohydrates, and antioxidants.

It was shown that bacterial endophyte *Burkholderia phytofirmans* PsJN enhances cold tolerance of grapevine plants by altering photosynthetic activity and metabolism of carbohydrates involved in cold stress tolerance (Ait Barka et al. 2006; Fernandez et al. 2012). The presence of the bacterium in the plant promoted acclimation to chilling temperatures resulting in lower cell damage, higher photosynthetic activity, and accumulation of cold-stress-related metabolites such as starch, proline, and phenolic compounds (Ait Barka et al. 2006). Fernandez et al. (2012)

demonstrated that bacterization of grapevine plants resulted in a twofold increase in soluble sugar content, and the plantlets inoculated with the bacterium displayed higher concentrations of the sugars known to be involved in low-temperature tolerance, such glucose, sucrose, and raffinose with its precursor, galactinol.

Positive effect of the *B. phytofirmans* PsJN strain on metabolic balance and reduced effect of drought stress was demonstrated in wheat plants grown under reduced irrigation conditions (Naveed et al. 2014). Inoculation with the bacterium resulted in higher antioxidant activity of plants compared to control under drought stress. However, in contrast to the grapevine plants in the study by Fernandez et al. (2012), the bacterium had no effect on sugar contents of the wheat, and phenolic contents decreased in the bacterized plants as compared to control.

Another endophytic bacterium, *Bacillus subtilis* B26, reduced a phenotypic effect of drought stress in *Brachypodium distachyon* grass compared to plants not harboring the bacterium (Gagne-Bourque et al. 2015). The protection from drought stress was associated with increase in total soluble sugars, glucose, fructose, and starch contents. However, no accumulation of stress response-related raffinose family carbohydrates was observed in either inoculated or control plants.

Pandey et al. (2012) evaluated cross-species stress reducing effect of wheat endophytic bacterium *Pseudomonas aeruginosa* PW09 in cucumber. Application of the PW09 strain induced increase in accumulation of proline and total phenolics under NaCl stress and pathogen *Sclerotium rolfsii* inoculation. Also, increase in activities of the enzymes involved biosynthesis of phenolic compounds, polyphenol oxidase, and phenylalanine ammonia lyase, as well as the antioxidative enzyme superoxide dismutase (SOD) was observed under biotic and abiotic stress conditions. Similarly, effect of six bacterial strains on stress-related biochemical traits of gladiolus plants was assessed in another study (Damodaran et al. 2014). The bacteria strains were shown to induce increase in activities of SOD, phenylalanine lyase, catalase, peroxidase enzymes, and accumulation of higher concentrations of proline and phenolic compounds in gladiolus plants grown in soil with high concentration of sodium. However, the capability of the different bacterial strains, isolated from soil, roots, culms, and leaves of grasses, to colonize endophytic niche was not explicitly confirmed.

A proline accumulation stimulating effect by endophytic strains of *Arthrobacter* sp. and *Bacillus* sp. was reported in pepper (*Capsicum annuum* L.) plants in vitro (Sziderics et al. 2007). Osmotic stress caused a similar increase in the content of free proline in the leaves of both inoculated and non-inoculated plants. However, higher concentration of proline was accumulated in leaves of unstressed plants inoculated with either of the two strains compared with unstressed non-inoculated plants. The bacterization resulted in a significantly reduced upregulation or downregulation of the stress-inducible genes suggesting that both strains reduced abiotic stress in pepper under osmotic stress conditions.

Endophytic bacteria *Pseudomonas pseudoalcaligenes* was shown to induce accumulation of higher concentrations of glycine betaine-like compounds leading to improved salinity stress tolerance in rice (Jha et al. 2011). At higher salinity levels, bacterization with mixture of both *P. pseudoalcaligenes* and rhizospheric

Bacillus pumilus showed better response against the adverse effects of salinity. In this study, bacterization with either *P. pseudoalcaligenes* or both *P. pseudoalcaligenes* and *B. pumilus* resulted in lower levels of proline accumulation under the stress conditions, suggesting that different strategies of accumulation of osmoprotectant proteins in endophyte-inoculated plants were either plant or bacterium genotype-specific phenomena. Related study demonstrated that both of the bacterial strains induced production of defense-related enzymes, chitinase, peroxidase, and polyphenol oxidase, under biotic stress conditions in the presence of *Magnaporthe grisea* pathogen (Jha and Subramanian 2009).

Chen et al. (2014) demonstrated that endophytic bacteria *Sphingomonas SaMR12* influenced the contents of root exudates, which were important for chelating cadmium ions and resulted in alleviation of the toxic metal stress in *Sedum alfredii*. Exudation of oxalic acid, malic acid, and tartaric acid was significantly affected by the inoculation of the endophytic bacterium in a manner dependent on cadmium treatment levels.

1.4.2 Effect on Phytohormone Balance

Ethylene (ET) is important for plant growth and development and has been extensively studied as mediator of plant stress response signaling (Gamalero and Glick 2015). Stress-induced accumulation of ET is usually deleterious to plant growth and health. ET is formed from methionine via S-adenosyl-L-methionine, which is converted into 1-aminocyclopropane-1-carboxylic acid (ACC) by the enzyme ACC oxidase (Bleecker and Kende 2000). ET is a key mediator of the plant defense response pathways that regulate colonization of plant tissue by endophytic bacteria (Iniguez et al. 2005). Endophytes may produce the enzyme ACC deaminase that has no function in bacteria but contributes to plant growth promotion and improved stress tolerance by cleaving the ET precursor ACC (Campbell and Thompson 1996). There are numerous reports on ACC deaminase-containing plant-associated bacteria and their role in improved plant growth and stress tolerance that has been recently reviewed by (Glick 2014).

Qin et al. (2014) isolated 13 ACC deaminase-producing putative endophytic bacteria of genera *Bacillus*, *Pseudomonas*, *Klebsiella*, *Serratia*, *Arthrobacter*, *Streptomyces*, *Isoptericola*, and *Microbacterium* from the halophyte plant *Limonium sinense*. It was suggested that the bacteria might play an important role in higher salinity tolerance of the plant as four of the selected ACC deaminase-producing strains were shown to stimulate growth of the host plants. An improved growth of *Catharanthus roseus* plant in 150 mM NaCl-containing soils was demonstrated for the plants inoculated with the *Achromobacter xylosoxidans* AUM54 strain (Karthikeyan et al. 2012). The bacterium was one of the four isolates isolated from *C. roseus* grown in saline soil and was shown to produce ACC deaminase.

In another study, tomato plants bacterized with ACC deaminase-containing endophytic *Pseudomonas fluorescens* YsS6 and *Pseudomonas migulae* 8R6 strains exhibited higher gain of biomass and a greater number of flowers and buds when grown under 165 mM and 185 mM NaCl levels as compared to the plants treated with ACC deaminase-deficient mutants of the bacteria or control with no bacterial treatment (Ali et al. 2014). Intriguingly, the study suggested the presence of different mechanisms of salt tolerance that might be plant genotype specific or stimulated by ACC deaminase-producing bacteria. It was shown that endophytic *Pseudomonas* sp. used in the study limited the concentration of sodium in tomato plant shoots (Ali et al. 2014). This was in contrast to previously reported rhizospheric *Pseudomonas putida* UW4 strain that was shown to be able to reduce ET levels in canola plants due to ACC deaminase activity (Cheng et al. 2007). In this case, sodium accumulated in root tissues and presumably partitioned into the vacuole.

In addition to salt stress tolerance, ACC deaminase-producing *P. agglomerans* Jp3-3 and *Achromobacter xylosoxidans* strain Ax10 were shown to alleviate stress of *Brassica* sp. plants grown in copper-contaminated soils and improved copper uptake by the plants (Ma et al. 2009; Zhang et al. 2011a). ACC deaminase-producing isolates from *Commelina communis* plants grown on lead and zinc mine soils were shown to improve growth of rape plants in the lead-contaminated soil (Zhang et al. 2011b).

Abscisic acid (ABA) is another phytohormone involved in plant stress response and is important for regulation of plant water balance and osmotic stress tolerance (Tuteja 2007). Information about role of ABA in endophytic bacteria-mediated stress tolerance is limited. It was described that endophytic bacteria *Bacillus licheniformis* Rt4M10 and *Pseudomonas fluorescens* Rt6M10 had drought stress reducing activity on plants grown in vitro that was associated with accumulation of high ABA levels in leaves of bacterized plants (Salomon et al. 2014). Cohen et al. (2009) showed that bacterization with *Azospirillum lipoferum* enhanced ABA accumulation and drought stress tolerance in maize plants. It was also suggested that plant performance under stress conditions was further enhanced by *A. lipoferum*-produced gibberellins (GAs). It is an intriguing observation as it is commonly believed that response to abiotic stress is associated with reduced plant growth-promoting hormone GA levels. However, GA signaling is closely integrated to ABA and ET signaling during the response to abiotic stress (Colebrook et al. 2014), and the interaction of the stress response pathways and exogenous hormone produced by plant growth-promoting bacteria remains ambiguous.

1.4.3 Induced Systemic Resistance and Priming of Response to Biotic Stress

Pathogen defense response priming, termed as induced systemic resistance, is activated by nonpathogenic plant-associated microorganisms. The ISR primes plant defense mechanisms and protects non-exposed plant parts against a future attack by pathogenic microbes and herbivorous insects. Plant hormones jasmonic acid (JA) and ET play a major regulatory role in the network of interconnected signaling pathways involved in ISR induction; however, the details of mechanism of the defense priming during ISR remain vague (Pieterse et al. 2014). There is an evidence for the role of transcription co-regulator NPR1 in the JA-/ET-dependent ISR and its cytosol-specific function that is different from the function involved in pathogen-induced systemic acquired resistance (SAR) (Spoel et al. 2003; Stein et al. 2008). Further, the role of transcription factors MYB72 and MYC2 in establishment of the ISR induced by rhizobacteria and priming of JA-/ET-dependent defense genes has been demonstrated (Pozo et al. 2008; Van Der Ent et al. 2008). In addition to the JA-/ET-mediated ISR activation pathway, an evidence that salicylic acid produced by plant growth-promoting bacteria could elicit ISR response has been discussed (Bakker et al. 2014).

Many studies have been dedicated to the ISR mediated by free-living rhizobacterial strains (Choudhary and Johri 2009); however, a number of endophytic bacteria have been reported to have the ISR-inducing activity as well. The first study demonstrating that endophytic bacteria could elicit ISR in plants was published in 1991 and showed that inoculation of cucumber roots with endophytic *Pseudomonas fluorescens* strain 89B-61 could induce resistance against cucumber anthracnose in the plant leaves (Kloepper and Ryu 2006; Wei et al. 1991). Subsequently attention was drawn to ISR mediated by several other endophytic species of genus *Pseudomonas* and the effect was characterized in different plant–pathogen systems. *Pseudomonas* sp. strain PsJN isolated from onion roots (Frommel et al. 1991) was shown to suppress verticillium wilt (*Verticillium dahliae*) on tomato seedlings and tissue culture plantlets grown in vitro, and it was proposed that the protection was mediated through the ISR activation (Sharma and Nowak 1998). *Pseudomonas* sp. strain 63–28 was shown to induce systemic resistance in tomato and pea plants leading to reduced damage by *Fusarium oxysporum* root pathogen (Benhamou et al. 1996; M'Piga et al. 1997). *Pseudomonas putida* MGY2 was isolated from papaya fruits and had reducing effect on postharvest decay of papaya fruit caused by *Colletotrichum gloeosporioides* (Shi et al. 2011). It was established that bacterization with the endophytic pseudomonad upregulated expression of enzymes involved in plant defense response, phenylalanine ammonia lyase, catalase, and peroxidase. A study by Ardanov et al. (2011) demonstrated that *Pseudomonas* sp. IMBG294 reduced symptoms of soft rot disease caused by bacterial pathogen *Pectobacterium atrosepticum* in potato plants. The assessment of expression of PR2 and PDF1.2, the molecular markers of the SAR and ISR, respectively, in *Arabidopsis–Pseudomonas syringae* model revealed that the endophytic bacterium was able to induce disease resistance via defense priming.

The asporogenous pseudomonads demonstrated poor performance when used in commercial plant protection products due to lack of long-term viability; therefore, subsequently much attention was drawn by plant growth-promoting strains of *Bacillus* sp. (Kloepper et al. 2004). ISR mediated by endophytic *Bacillus pumilus* strain SE34 was described by Benhamou et al. (1996; 1998). The bacterial strain reduced symptoms of root-rotting fungus *Fusarium oxysporum* infection in pea through induction of plant defense mechanism leading to accumulation of callose and phenolic compounds in the root epidermal and cortical cell walls and formation of the barriers beyond the infection sites (Benhamou et al. 1996). The same *B. pumilus* SE34 strain induced resistance to *Fusarium oxysporum* infection in tomato plants (Benhamou et al. 1998).

In addition, ISR mediated by endophytic *Serratia* sp. (Benhamou et al. 2000), *Methylobacterium* sp. (Ardanov et al. 2011), and actinobacteria *Streptomyces* sp. (Conn et al. 2008) was described. The early study by Benhamou et al. (2000) demonstrated that *Serratia plymuthica* strain R1GC4 sensitized susceptible cucumber seedlings to react more rapidly and more efficiently to infection by soilborne pathogen *Pythium ultimum* (Benhamou et al. 2000). The defense reaction was associated with deposition of enlarged callose-enriched wall appositions, also containing pectin, cellulose, and phenolic compounds.

The capability of endophytic actinobacteria *Streptomyces* sp. strains, isolated from wheat tissues, to activate the SAR or ISR pathways was assessed using *Arabidopsis thaliana* (Conn et al. 2008). It was demonstrated that the *Streptomyces* sp. EN27 was able to prime both pathways depending on the infecting pathogen. Resistance to *Erwinia carotovora* subsp. *carotovora* occurred via an NPR1-independent pathway and required salicylic acid, whereas the JA/ET signaling molecules were not essential. In contrast, induction of resistance to *Fusarium oxysporum* was mediated by NPR1-dependent pathway but also required salicylic acid and it was JA/ET independent.

Intriguingly, the study on induction of disease resistance to soft rot pathogen *Pectobacterium atrosepticum* in potato plants demonstrated that priming capacities of *Methylobacterium* sp. IMBG290 was inversely proportional to bacterial inoculants size (Ardanov et al. 2011). The difference in plant response mechanisms was associated with different patterns of activity of reactive oxygen species scavenging enzymes SOD and catalase. Plants treated with a low titer of *Methylobacterium* sp. showed higher SOD activity and unchanged catalase activity resulting in the development of ISR; meanwhile higher *Methylobacterium* sp. density caused SOD inactivation and catalase activation after inoculation with the pathogen and was followed by hypersensitive response.

1.5 Concluding Remarks

Several decades of research on endophytes in agricultural plants have revealed an immense taxonomic diversity of the endophytic bacteria. The endophytic species have been mostly reported throughout α-, β-, and γ-subgroups of phylum *Proteobacteria*, the latter being the most diverse and dominant group that includes common soil and endophytic bacteria of *Pseudomonas* sp. Next to the pseudomonads, much attention has been dedicated to members of *Bacillus* sp. that belongs to phylum *Firmicutes*. A number of other species of phyla *Firmicutes* and *Actinobacteria* have been identified as endophytic bacteria as well. During the last decade, development of metagenomic analysis techniques has brought to light new aspects of the diversity of endophytic bacteria including identification of new unculturable species and establishment of the dynamics of endophyte diversity that provide hints about physiological significance and ecological functions of the complex host plant and endophytic bacteria interactions.

Composition of the endophytic microbiome depends on plant genotype as well as environmental factors. Evidence has been presented that agricultural practices play an important role in shaping structure of the endophytic microbial community of agricultural crop plants. Therefore, assessment of the capability of modern agronomical techniques to maintain natural diversity of plant endophytic bacteria should become an important element in the development of sustainable agricultural practices. In addition, numerous studies have demonstrated beneficial effects of the endophytic bacteria on plant growth and adaptability to biotic or abiotic stresses through modulation of phytohormone signaling, production of metabolites involved in stress response, and priming of plant defense response pathways. The endophytes play an integral role in balancing plant physiology and functioning of agroecosystems; thus, understanding of composition and functioning of the plant-associated microbial communities has a large potential for improvement of performance of agricultural crops and development of integrated plant disease management systems.

Acknowledgments This research was funded by the long-term research program "Genetics and directed genotype development of agricultural and forestry plants" implemented by the Lithuanian Research Centre for Agriculture and Forestry.

References

Ait Barka E, Nowak J, Clement C (2006) Enhancement of chilling resistance of inoculated grapevine plantlets with a plant growth-promoting rhizobacterium, *Burkholderia phytofirmans* strain PsJN. Appl Environ Microbiol 72:7246–7252

Akinsanya MA, Goh JK, Lim SP, Ting AS (2015) Metagenomics study of endophytic bacteria in Aloe Vera using next-generation technology. Genom Data 6:159–163

Ali S, Charles TC, Glick BR (2014) Amelioration of high salinity stress damage by plant growth-promoting bacterial endophytes that contain ACC deaminase. Plant Physiol Biochem 80:160–167

Araujo JM, Silva AC, Azevedo JL (2000) Isolation of endophytic actinomycetes from roots and leaves of maize *(Zea mays* L.*)* Braz Arch Biol Technol 43:447–451

Aravind R, Kumar A, Eapen SJ, Ramana KV (2009) Endophytic bacterial flora in root and stem tissues of black pepper (*Piper nigrum* L.) genotype: isolation, identification and evaluation against *Phytophthora capsici*. Lett Appl Microbiol 48:58–64

Ardanov P, Ovcharenko L, Zaets I, Kozyrovska N, Pirttila AM (2011) Endophytic bacteria enhancing growth and disease resistance of potato (*Solanum tuberosum* L.) Biol Control 56:43–49

Bacon CW, Hinton DM (2007) Bacterial endophytes: the endophytic niche, its occupants, and its utility. In: Gnanamanickam SS (ed) Plant-associated bacteria. Springer, Dordrecht, pp 155–164

Bakker PAHM, Ran LX, Mercado-Blanco J (2014) Rhizobacterial salicylate production provokes headaches! Plant Soil 382:1–16

Balandreau J, Viallard V, Cournoyer B, Coenye T, Laevens S, Vandamme P (2001) *Burkholderia cepacia* genomovar III is a common plant-associated bacterium. Appl Environ Microbiol 67:982–985

Baldan E, Nigris S, Populin F, Zottini M, Squartini A, Baldan B (2014) Identification of culturable bacterial endophyte community isolated from tissues of *Vitis vinifera* "Glera". Plant Biosyst 148:508–516

Balota EL, Colozzi-Filho A, Andrade DS, Dick RP (2003) Microbial biomass in soils under different tillage and crop rotation systems. Biol Fertil Soils 38:15–20
Barak JD, Liang AS (2008) Role of soil, crop debris, and a plant pathogen in *Salmonella enterica* contamination of tomato plants. PLoS One 3:e1657
Bell CR, Dickie GA, Harvey WLG, Chan JWYF (1995) Endophytic bacteria in grapevine. Can J Microbiol 41:46–53
Benhamou N, Gagne S, Le QD, Dehbi L (2000) Bacterial-mediated induced resistance in cucumber: beneficial effect of the endophytic bacterium *Serratia plymuthica* on the protection against infection by *Pythium ultimum*. Phytopathology 90:45–56
Benhamou N, Kloepper JW, Quadt-Hallman A, Tuzun S (1996) Induction of defense-related ultrastructural modifications in pea root tissues inoculated with endophytic bacteria. Plant Physiol 112:919–929
Benhamou N, Kloepper JW, Tuzun S (1998) Induction of resistance against *Fusarium* wilt of tomato by combination of chitosan with an endophytic bacterial strain: ultrastructure and cytochemistry of the host response. Planta 204:153–168
Berg G, Krechel A, Ditz M, Sikora RA, Ulrich A, Hallmann J (2005) Endophytic and ectophytic potato-associated bacterial communities differ in structure and antagonistic function against plant pathogenic fungi. FEMS Microbiol Ecol 51:215–229
Bleecker AB, Kende H (2000) Ethylene: a gaseous signal molecule in plants. Annu Rev Cell Dev Biol 16:1–18
Brader G, Compant S, Mitter B, Trognitz F, Sessitsch A (2014) Metabolic potential of endophytic bacteria. Curr Opin Biotechnol 27:30–37
Brandl MT (2006) Fitness of human enteric pathogens on plants and implications for food safety. Annu Rev Phytopathol 44:367–392
Brandl MT (2008) Plant lesions promote the rapid multiplication of *Escherichia coli* O157:H7 on postharvest lettuce. Appl Environ Microbiol 74:5285–5289
Brandl MT, Cox CE, Teplitski M (2013) *Salmonella* interactions with plants and their associated microbiota. Phytopathology 103:316–325
Bulgarelli D, Rott M, Schlaeppi K, van TE VL, Ahmadinejad N, Assenza F, Rauf P, Huettel B, Reinhardt R, Schmelzer E, Peplies J, Gloeckner FO, Amann R, Eickhorst T, Schulze-Lefert P (2012) Revealing structure and assembly cues for *Arabidopsis* root-inhabiting bacterial microbiota. Nature 488:91–95
Bulgari D, Casati P, Brusetti L, Quaglino F, Brasca M, Daffonchio D, Bianco PA (2009) Endophytic bacterial diversity in grapevine (*Vitis vinifera* L.) leaves described by 16S rRNA gene sequence analysis and length heterogeneity-PCR. J Microbiol 47:393–401
Campbell BG, Thompson JA (1996) 1-aminocyclopropane-1-carboxylate deaminase genes from *Pseudomonas* strains. FEMS Microbiol Lett 138:207–210
Campisano A, Antonielli L, Pancher M, Yousaf S, Pindo M, Pertot I (2014a) Bacterial endophytic communities in the grapevine depend on pest management. PLoS One 9:e112763
Campisano A, Ometto L, Compant S, Pancher M, Antonielli L, Yousaf S, Varotto C, Anfora G, Pertot I, Sessitsch A, Rota-Stabelli O (2014b) Interkingdom transfer of the acne-causing agent, *Propionibacterium acnes*, from human to grapevine. Mol Biol Evol 31:1059–1065
Cankar K, Kraigher H, Ravnikar M, Rupnik M (2005) Bacterial endophytes from seeds of Norway spruce (*Picea abies* L. karst). FEMS Microbiol Lett 244:341–345
Chaintreuil C, Giraud E, Prin Y, Lorquin J, Ba A, Gillis M, de LP, Dreyfus B (2000) Photosynthetic bradyrhizobia are natural endophytes of the African wild rice *Oryza breviligulata*. Appl Environ Microbiol 66:5437–5447
Chelius MK, Triplett EW (2001) The diversity of archaea and bacteria in association with the roots of *Zea mays* L. Microb Ecol 41:252–263
Chen B, Zhang Y, Rafiq MT, Khan KY, Pan F, Yang X, Feng Y (2014) Improvement of cadmium uptake and accumulation in *Sedum alfredii* by endophytic bacteria *Sphingomonas* SaMR12: effects on plant growth and root exudates. Chemosphere 117:367–373

Cheng Z, Park E, Glick BR (2007) 1-aminocyclopropane-1-carboxylate deaminase from *Pseudomonas putida* UW4 facilitates the growth of canola in the presence of salt. Can J Microbiol 53:912–918

Chi F, Shen SH, Cheng HP, Jing YX, Yanni YG, Dazzo FB (2005) Ascending migration of endophytic rhizobia, from roots to leaves, inside rice plants and assessment of benefits to rice growth physiology. Appl Environ Microbiol 71:7271–7278

Choudhary DK, Johri BN (2009) Interactions of *Bacillus* spp. and plants - with special reference to induced systemic resistance (ISR). Microbiol Res 164:493–513

Christina A, Christapher V, Bhore SJ (2013) Endophytic bacteria as a source of novel antibiotics: an overview. Pharmacogn Rev 7:11–16

Cocking EC, Stone PJ, Davey MR (2006) Intracellular colonization of roots of *Arabidopsis* and crop plants by *Gluconacetobacter diazotrophicus*. In Vitro Cell Dev - Pl 42:74–82

Cohen AC, Travaglia CN, Bottini R, Piccoli PN (2009) Participation of abscisic acid and gibberellins produced by endophytic *Azospirillum* in the alleviation of drought effects in maize. Botany 87:455–462

Colebrook EH, Thomas SG, Phillips AL, Hedden P (2014) The role of gibberellin signalling in plant responses to abiotic stress. J Exp Biol 217:67–75

Compant S, Mitter B, Colli-Mull JG, Gangl H, Sessitsch A (2011) Endophytes of grapevine flowers, berries, and seeds: identification of cultivable bacteria, comparison with other plant parts, and visualization of niches of colonization. Microb Ecol 62:188–197

Conn VM, Walker AR, Franco CM (2008) Endophytic actinobacteria induce defense pathways in *Arabidopsis Thaliana*. Mol Plant-Microbe Interact 21:208–218

Coombs JT, Franco CM (2003) Isolation and identification of actinobacteria from surface-sterilized wheat roots. Appl Environ Microbiol 69:5603–5608

Damodaran T, Rai RB, Jha SK, Kannan R, Pandey BK, Sah V, Mishra VK, Sharma DK (2014) Rhizosphere and endophytic bacteria for induction of salt tolerance in gladiolus grown in sodic soils. J Plant Interact 9:577–584

De Boer SH, Copeman RJ (1974) Endophytic bacterial flora in *Solanum tuberosum* and its significance in bacterial ring rot disease. Can J Microbiol 54:115–122

de Melo Pereira GV, Magalhaes KT, Lorenzetii ER, Souza TP, Schwan RF (2012) A multiphasic approach for the identification of endophytic bacterial in strawberry fruit and their potential for plant growth promotion. Microb Ecol 63:405–417

de Oliveira Costa LE, de Queiroz MV, Borges AC, de Moraes CA, de Araujo EF (2012) Isolation and characterization of endophytic bacteria isolated from the leaves of the common bean (*Phaseolus vulgaris*). Braz J Microbiol 43:1562–1575

Deering AJ, Mauer LJ, Pruitt RE (2012) Internalization of *E. coli* O157:H7 and *Salmonella* spp. in plants: a review. Food Res Int 45:567–575

Dent KC, Stephen JR, Finch-Savage WE (2004) Molecular profiling of microbial communities associated with seeds of *Beta vulgaris* subsp. *vulgaris* (sugar beet). J Microbiol Methods 56:17–26

Dias ACF, Costa FEC, Andreote FD, Lacava PT, Teixeira MA, Assumpcao LC (2009) Isolation of micropropagated strawberry endophytic bacteria and assessment of their potential for plant growth promotion. World J Microbiol Biotechnol 25:189–195

Dorr de Quadros P, Zhalnina K, Davis-Richardson A, Fagen JR, Drew J, Bayer C, Camargo FAO, Triplett EW (2012) The effect of tillage system and crop rotation on soil microbial diversity and composition in a subtropical acrisol. Diversity 4:375–395

El-Awady MAM, Hassan MM, Al-Sodany YM (2015) Isolation and characterization of salt tolerant endophytic and rhizospheric plant growth-promoting bacteria (PGPB) associated with the halophyte plant (*Sesuvium verrucosum*) grown in KSA. Int J Appl Sci Biotechnol 3:552–560

Elbeltagy A, Nishioka K, Sato T, Suzuki H, Ye B, Hamada T, Isawa T, Mitsui H, Minamisava K (2001) Isolation and characterization of endophytic bacteria from wild and traditionally cultivated rice varieties. Soil Sci Plant Nutr 46:617–629

Ellis EC, Kaplan JO, Fuller DQ, Vavrus S, Klein GK, Verburg PH (2013) Used planet: a global history. Proc Natl Acad Sci U S A 110:7978–7985

Engelhard M, Hurek T, Reinhold-Hurek B (2000) Preferential occurrence of diazotrophic endophytes, *Azoarcus* spp., in wild rice species and land races of *Oryza sativa* in comparison with modern races. Environ Microbiol 2:131–141

Erickson MC, Webb CC, Diaz-Perez JC, Phatak SC, Silvoy JJ, Davey L, Payton AS, Liao J, Ma L, Doyle MP (2010) Surface and internalized *Escherichia coli* O157:H7 on field-grown spinach and lettuce treated with spray-contaminated irrigation water. J Food Prot 73:1023–1029

Fernandez O, Theocharis A, Bordiec S, Feil R, Jacquens L, Clement C, Fontaine F, Barka EA (2012) *Burkholderia phytofirmans* PsJN acclimates grapevine to cold by modulating carbohydrate metabolism. Mol Plant-Microbe Interact 25:496–504

Fisher PJ, Petrini O, Scott HML (1992) The distribution of some fungal and bacterial endophytes in maize (*Zea mays* L.) New Phytol 122:299–305

Foley JA, Defries R, Asner GP, Barford C, Bonan G, Carpenter SR, Chapin FS, Coe MT, Daily GC, Gibbs HK, Helkowski JH, Holloway T, Howard EA, Kucharik CJ, Monfreda C, Patz JA, Prentice IC, Ramankutty N, Snyder PK (2005) Global consequences of land use. Science 309:570–574

Foley JA, Ramankutty N, Brauman KA, Cassidy ES, Gerber JS, Johnston M, Mueller ND, O'Connell C, Ray DK, West PC, Balzer C, Bennett EM, Carpenter SR, Hill J, Monfreda C, Polasky S, Rockstrom J, Sheehan J, Siebert S, Tilman D, Zaks DP (2011) Solutions for a cultivated planet. Nature 478:337–342

Fouts DE, Tyler HL, DeBoy RT, Daugherty S, Ren Q, Badger JH, Durkin AS, Huot H, Shrivastava S, Kothari S, Dodson RJ, Mohamoud Y, Khouri H, Roesch LF, Krogfelt KA, Struve C, Triplett EW, Methe BA (2008) Complete genome sequence of the N_2-fixing broad host range endophyte *Klebsiella pneumoniae* 342 and virulence predictions verified in mice. PLoS Genet 4:e1000141

Frommel MI, Nowak J, Lazarovits G (1991) Growth enhancement and developmental modifications of in vitro grown potato (*Solanum tuberosum* spp. *tuberosum*) as affected by a nonfluorescent *Pseudomonas* sp. Plant Physiol 96:928–936

Fuentes-Ramirez LE, Caballero-Mellado J, Sepulveda J, Martinez-Romero E (1999) Colonization of sugarcane by *Acetobacter diazotrophicus* is inhibited by high N-fertilization. FEMS Microbiol Ecol 29:117–128

Gaba S, Bretagnolle F, Rigaud T, Philippot L (2014) Managing biotic interactions for ecological intensification of agroecosystems. Front Ecol Evol 2:1–9

Gagne-Bourque F, Mayer BF, Charron JB, Vali H, Bertrand A, Jabaji S (2015) Accelerated growth rate and increased drought stress resilience of the model grass *Brachypodium distachyon* colonized by *Bacillus subtilis* B26. PLoS One 10:e0130456

Gamalero E, Glick BR (2015) Bacterial modulation of plant ethylene levels. Plant Physiol 169:13–22

Garcia-Orenes F, Morugan-Coronado A, Zornoza R, Scow K (2013) Changes in soil microbial community structure influenced by agricultural management practices in a mediterranean agroecosystem. PLoS One 8:e80522

Glick BR (2012) Plant growth-promoting bacteria: mechanisms and applications. Scientifica 2012:1–15

Glick BR (2014) Bacteria with ACC deaminase can promote plant growth and help to feed the world. Microbiol Res 169:30–39

Govindarajan M, Balandreau J, Kwon SW, Weon HY, Lakshminarasimhan C (2008) Effects of the inoculation of *Burkholderia vietnamensis* and related endophytic diazotrophic bacteria on grain yield of rice. Microb Ecol 55:21–37

Govindarajan M, Kwoon SW, Weon HY (2007) Isolation, molecular characterization and growth-promoting activities of endophytic sugarcane diazotroph *Klebsiella* sp. GR9. World J Microbiol Biotechnol 23:997–1006

Govindasamy V, Franco CMM, Gupta VVSR (2014) Endophytic actinobacteria: diversity and ecology. In: Verma VC, Gange AC (eds) Advances in endophytic research. Springer, New Dehli, pp 27–59

Haas D, Defago G (2005) Biological control of soil-borne pathogens by fluorescent pseudomonads. Nat Rev Microbiol 3:307–319

Hallman J, Quadt-Hallman A, Mahafee WF, Kloepper JW (1997) Bacterial endophytes in agricultural crops. Can J Microbiol 43:895–914

Hallman J, Rodriguez-Kabana R, Kloepper JW (1999) Chitin-mediated changes in bacterial communities of the soil, rhizosphere and within roots of cotton in relation to nematode control. Soil Biol Biochem 31:551–560

Hamilton CE, Gundel PE, Helander M, Saikkonen K (2012) Endophytic mediation of reactive oxygen species and antioxidant activity in plants: a review. Fungal Div 54:1–10

Han JI, Choi HK, Lee SW, Orwin PM, Kim J, Laroe SL, Kim TG, O'Neil J, Leadbetter JR, Lee SY, Hur CG, Spain JC, Ovchinnikova G, Goodwin L, Han C (2011) Complete genome sequence of the metabolically versatile plant growth-promoting endophyte *Variovorax paradoxus* S110. J Bacteriol 193:1183–1190

Hardoim P, Nissinen R, van Elas JD (2012) Ecology of bacterial endophytes in sustainable agriculture. In: Maheshwari DK (ed) Bacteria in agrobiology: plant probiotics. Springer, Berlin/Heidelberg, pp 97–126

Hardoim PR, Andreote FD, Reinhold-Hurek B, Sessitsch A, van Overbeek LS, van Elsas JD (2011) Rice root-associated bacteria: insights into community structures across 10 cultivars. FEMS Microbiol Ecol 77:154–164

Hardoim PR, van Overbeek LS, Elsas JD (2008) Properties of bacterial endophytes and their proposed role in plant growth. Trends Microbiol 16:463–471

Holden N, Pritchard L, Toth I (2009) Colonization outwith the colon: plants as an alternative environmental reservoir for human pathogenic enterobacteria. FEMS Microbiol Rev 33:689–703

Hollis JP (1951) Bacteria in healthy potato tissue. Phytopathology 41:350–367

Hung PQ, Annapurna K (2004) Isolation and characterization of endophytic bacteria in soybean (*Glycine* sp.) Omonrice 12:92–101

Ikeda S, Kaneko T, Okubo T, Rallos LE, Eda S, Mitsui H, Sato S, Nakamura Y, Tabata S, Minamisawa K (2009) Development of a bacterial cell enrichment method and its application to the community analysis in soybean stems. Microb Ecol 58:703–714

Iniguez AL, Dong Y, Carter HD, Ahmer BM, Stone JM, Triplett EW (2005) Regulation of enteric endophytic bacterial colonization by plant defenses. Mol Plant-Microbe Interact 18:169–178

Iniguez AL, Dong Y, Triplett EW (2004) Nitrogen fixation in wheat provided by *Klebsiella pneumoniae* 342. Mol Plant-Microbe Interact 17:1078–1085

Jacobs MJ, Bugbee WM, Gabrielson DA (1985) Enumeration, location, and characterization of endophytic bacteria within sugar beet roots. Can J Bot 63:1262–1265

James EK, Olivares FL, Baldani JI, Dobereiner J (1997) *Herbaspirillum*, an endophytic diazotroph colonizing vascular tissue of *Sorghum bicolor* L. Moench J Exp Bot 48:785–798

Jangid K, Williams MA, Franzluebbers AJ, Sanderlin JS, Reeves JH, Jenkins MB, Endale DM, Coleman DC, Whitman WB (2008) Relative impacts of land-use, management intensity and fertilization upon soil microbial community structure in agricultural systems. Soil Biol Biochem 40:2843–2853

Jha P, Kumar A (2009) Characterization of novel plant growth promoting endophytic bacterium *Achromobacter xylosoxidans* from wheat plant. Microb Ecol 58:179–188

Jha PN, Gupta G, Jha P, Mehrotra R (2013) Association of rhizospheric/endophytic bacteria with plants: a potential gateway to sustainable agriculture. Greener J Agricult Sci 3:73–84

Jha Y, Subramanian RB (2009) Endophytic *Pseudomonas pseudoalcaligenes* shows better response against the *Magnaporthe grisea* than a rhizospheric *Bacillus pumilus* in *Oryza sativa* (Rice). Arch Phytopathol Plant Protect 44:592–604

Jha Y, Subramanian RB, Patel S (2011) Combination of endophytic and rhizospheric plant growth promoting rhizobacteria in *Oryza sativa* shows higher accumulation of osmoprotectant against saline stress. Acta Physiol Plant 33:797–802

Jiao JY, Wang HX, Zeng Y, Shen YM (2006) Enrichment for microbes living in association with plant tissues. J Appl Microbiol 100:830–837

Johnston-Monje D, Raizada MN (2011) Conservation and diversity of seed associated endophytes in *Zea* across boundaries of evolution, ethnography and ecology. PLoS One 6:e20396

Kaneko T, Minamisawa K, Isawa T, Nakatsukasa H, Mitsui H, Kawaharada Y, Nakamura Y, Watanabe A, Kawashima K, Ono A, Shimizu Y, Takahashi C, Minami C, Fujishiro T, Kohara M, Katoh M, Nakazaki N, Nakayama S, Yamada M, Tabata S, Sato S (2010) Complete genomic structure of the cultivated rice endophyte *Azospirillum* sp. B510. DNA Res 17:37–50

Karthikeyan B, Joe MM, Islam R, Sa T (2012) ACC deaminase containing diazotrophic endophytic bacteria ameliorate salt stress in *Catharanthus roseus* through reduced ethylene levels and induction of antioxidative defense systems. Symbiosis 56:77–86

Kinkel LL, Schlatter DC, Bakker MG, Arenz BE (2012) Streptomyces competition and coevolution in relation to plant disease suppression. Res Microbiol 163:490–499

Kloepper JW, Ryu CM, Zhang S (2004) Induced systemic resistance and promotion of plant growth by *Bacillus* spp. Phytopathology 94:1259–1266

Kloepper JW, Ryu C-M (2006) Bacterial endophytes as elicitors of induced systemic resistance. In: Schulz B, Boyle C, Sieber TN (eds) Microbial root endophytes. Springer, Berlin/Heidelberg, pp 33–52

Koskimaki JJ, Pirttila AM, Ihantola EL, Halonen O, Frank AC (2015) The intracellular scots pine shoot symbiont *Methylobacterium extorquens* DSM13060 aggregates around the host nucleus and encodes eukaryote-like proteins. MBio 6:1–12

Kuklinsky-Sobral J, Araujo WL, Mendes R, Geraldi IO, Pizzirani-Kleiner AA, Azevedo JL (2004) Isolation and characterization of soybean-associated bacteria and their potential for plant growth promotion. Environ Microbiol 6:1244–1251

Kumar A, Munder A, Aravind R, Eapen SJ, Tummler B, Raaijmakers JM (2013) Friend or foe: genetic and functional characterization of plant endophytic *Pseudomonas aeruginosa*. Environ Microbiol 15:764–779

Kutter S, Hartmann A, Schmid M (2006) Colonization of barley (*Hordeum vulgare*) with *Salmonella enterica* and *Listeria* spp. FEMS Microbiol Ecol 56:262–271

Lalande RN, Bissonnette N, Coutlee D, Antoun H (1989) Identification of rhizobacteria from maize and determination of their plant-growth promoting potential. Plant Soil 115:11

Larran S, Perello A, Simon MR, Moreno V (2002) Isolation and analysis of endophytic microorganisms in wheat (*Triticum aestivum* L.) leaves. World J Microbiol Biotechnol 18:683–686

Latz E, Eisenhauer N, Rall BC, Allan E, Roscher C, Scheu S, Jousset A (2012) Plant diversity improves protection against soil-borne pathogens by fostering antagonistic bacterial communities. J Ecol 100:597–604

Li GJ, Dong QE, Ma L, Huang Y, Zhu ML, Ji YP, Wang QH, Mo MH, Zhang KQ (2014) Management of *Meloidogyne incognita* on tomato with endophytic bacteria and fresh residue of *Wasabia japonica*. J Appl Microbiol 117:1159–1167

Lodewyckx C, Mergeay M, Vangronsveld J, Clijsters H, Van der Lelie D (2002) Isolation, characterization, and identification of bacteria associated with the zinc hyperaccumulator Thlaspi Caerulescens subsp. calaminaria. Int J Phytoremediation 4:101–115

Lorang JM, Anderson NA, Lauer FI, Wildung DK (1989) Disease decline in a Minnesota potato scab plot. Am Potato J 66:531

Lundberg DS, Lebeis SL, Paredes SH, Yourstone S, Gehring J, Malfatti S, Tremblay J, Engelbrektson A, Kunin V, del Rio TG, Edgar RC, Eickhorst T, Ley RE, Hugenholtz P, Tringe SG, Dangl JL (2012) Defining the core *Arabidopsis thaliana* root microbiome. Nature 488:86–90

M'Piga P, Belanger RR, Paulitz TC, Benhamou N (1997) Increased resistance to *Fusarium oxysporum* f. Sp. *radicis-lycopersici* in tomato plants treated with the endophytic bacterium *Pseudomonas fluorescens* strain 63-28. Physiol Mol Plant Pathol 50:301–320

Ma Y, Rajkumar M, Freitas H (2009) Inoculation of plant growth promoting bacterium *Achromobacter xylosoxidans* strain Ax10 for the improvement of copper phytoextraction by *Brassica juncea*. J Environ Manag 90:831–837

Manter DK, Delgado JA, Holm DG, Stong RA (2010) Pyrosequencing reveals a highly diverse and cultivar-specific bacterial endophyte community in potato roots. Microb Ecol 60:157–166

Maropola MKA, Ramond JB, Trindade M (2015) Impact of metagenomic DNA extraction procedures on the identifiable endophytic bacterial diversity in *Sorghum bicolor* (L. Moench). J Microbiol Methods 112:104–117

Mathew RP, Feng Y, Githinji L, Ankumah R, Balkcom KS (2012) Impact of no-tillage and conventional tillage systems on soil microbial communities. Appl Environ Soil Sci 2012:1–10

Matsumura EE, Secco VA, Moreira RS, dos Santos OJAP, Hungria M, de Olivera ALM (2015) Composition and activity of endophytic bacterial communities in field grown maize plants inoculated with *Azospirillum brasilense*. Ann Microbiol 65:2187–2200

Mavingui P, Laguerre G, Berge O, Heulin T (1992) Genetic and phenotypic diversity of *Bacillus polymyxa* in soil and in the wheat rhizosphere. Appl Environ Microbiol 58:1894–1903

Mbai FN, Magiri EN, Matiru VN, Nganga J, Nyambati VCS (2013) Isolation and characterization of bacterial root endophytes with potential to enhance plant growth from Kenyan basmati rice. Am Int J Contemp Res 3:25–40

Mcinroy JA, Kloepper JW (1995) Survey of indigenous bacterial endophytes from cotton and sweet corn. Plant Soil 173:337–342

Mei C, Flinn BS (2010) The use of beneficial microbial endophytes for plant biomass and stress tolerance improvement. Recent Pat Biotechnol 4:81–95

Mendes R, Kruijt M, de Bruijn I, Dekkers E, van der Voort M, Schneider JH, Piceno YM, DeSantis TZ, Andersen GL, Bakker PA, Raaijmakers JM (2011) Deciphering the rhizosphere microbiome for disease-suppressive bacteria. Science 332:1097–1100

Mendes R, Pizzirani-Kleiner AA, Araujo WL, Raaijmakers JM (2007) Diversity of cultivated endophytic bacteria from sugarcane: genetic and biochemical characterization of *Burkholderia cepacia* complex isolates. Appl Environ Microbiol 73:7259–7267

Mingma R, Pathom-aree W, Trakulnaleamsai S, Thamchaipenet A, Duangmal K (2014) Isolation of rhizospheric and roots endophytic actinomycetes from *Leguminosae* plant and their activities to inhibit soybean pathogen, *Xanthomonas campestris* pv. *Glycine*. World J Microbiol Biotechnol 30:271–280

Misaghi IJ, Donndelinger CR (1990) Endophytic bacteria in symptom-free cotton plants. Phytopathology 80:808–811

Montanez A, Blanco AR, Barlocco C, Beracochea M, Sicardi M (2012) Characterization of cultivable putative endophytic plant growth promoting bacteria associated with maize cultivars (*Zea mays* L.) and their inoculation effects in vitro. Appl Soil Ecol 58:21–28

Mutondo MS, Huddy RJ, Bauer R, Tuffin MI, Cowan DA (2010) Metagenomic gene discovery. In: Li RW (ed) Metagenomics and its applications in agriculture, biomedicine, and environmental studies. Nova Science Publishers, New York, pp 287–320

Naik BS, Shashikala J, Krishnamurthy YL (2009) Study on the diversity of endophytic communities from rice (*Oryza sativa* L.) and their antagonistic activities in vitro. Microbiol Res 164:290–296

Naveed M, Hussain MB, Zahir ZA, Mitter B, Sessitsch A (2014) Drought stress amelioration in wheat through inoculation with *Burkholderia phytofirmans* strain PsJN. Plant Growth Regul 73:121–131

Newton AC, Fitt BD, Atkins SD, Walters DR, Daniell TJ (2010) Pathogenesis, parasitism and mutualism in the trophic space of microbe-plant interactions. Trends Microbiol 18:365–373

Nikolic B, Schwab H, Sessitsch A (2011) Metagenomic analysis of the 1-aminocyclopropane-1-carboxylate deaminase gene (acdS) operon of an uncultured bacterial endophyte colonizing *Solanum tuberosum* L. Arch Microbiol 193:665–676

Nutaratat P, Srisuk N, Arunrattiyakorn P, Limtong S (2014) Plant growth-promoting traits of epiphytic and endophytic yeasts isolated from rice and sugar cane leaves in Thailand. Fungal Biol 118:683–694

Okubo T, Ikeda S, Kaneko T, Eda S, Mitsui H, Sato S, Tabata S, Minamisawa K (2009) Nodulation-dependent communities of culturable bacterial endophytes from stems of field-grown soybeans. Microbes Environ 24:253–258

Pageni BB, Lupwayi NZ, Larney FJ, Kawchuk LM, Gan Y (2013) Populations, diversity and identities of bacterial endophytes in potato (*Solanum tuberosum* L.) cropping systems. Can J Plant Sci 93:1125–1142

Palus JA, Bonrneman J, Ludden PW, Triplett EW (1996) A diazotrophic bacterial endophyte isolated from stems of *Zea mays* L. and *Zea luxurians* Iltis and Doebley. Plant Soil 186:135–142

Pampulha ME, Oliveira A (2006) Impact of an herbicide combination of bromoxynil and prosulfuron on soil microorganisms. Curr Microbiol 53:238–243

Pandey PK, Yadav SK, Singh A, Sarma BK, Mishra A, Singh HB (2012) Cross-species alleviation of biotic and abiotic stresses by the endophyte *Pseudomonas aeruginosa* PW09. J Phytopathol 160:532–539

Patel HA, Patel RK, Khristi SM, Parikh K, Rajendran G (2012) Isolation and characterization of bacterial endophytes from *Lycopersicon esculentum* plant and their plant growth promoting characteristics. Nepal J Biotechnol 2:37–52

Paul NC, Ji SH, Deng JX, Yu SH (2013) Assemblages of endophytic bacteria in chili pepper (*Capsicum Annuum* L.) and their antifungal activity against phytopathogens in vitro. Plant Omics J 6:441–448

Paungfoo-Lonhienne C, Rentsch D, Robatzek S, Webb RI, Sagulenko E, Nasholm T, Schmidt S, Lonhienne TG (2010) Turning the table: plants consume microbes as a source of nutrients. PLoS One 5:e11915

Pavlo A, Leonid O, Iryna Z, Natalia K, Maria PA (2011) Endophytic bacteria enhancing growth and disease resistance of potato (*Solanum tuberosum* L.) Biol Control 56:43–49

Philippot L, Raaijmakers JM, Lemanceau P, van der Putten WH (2013) Going back to the roots: the microbial ecology of the rhizosphere. Nat Rev Microbiol 11:789–799

Pieterse CM, Zamioudis C, Berendsen RL, Weller DM, Van Wees SC, Bakker PA (2014) Induced systemic resistance by beneficial microbes. Annu Rev Phytopathol 52:347–375

Pillay VK, Nowak J (1997) Inoculum density, temperature, and genotype effects on *in vitro* growth promotion and epiphytic and endophytic colonization of tomato (*Lycopersicon esculentum* L.) seedlings inoculated with a pseudomonad bacterium. Can J Microbiol 43:354–361

Pimentel IC, Glienke-Blanco C, Gabardo J, Stuart RM, Azevedo JL (2006) Identification and colonization of endophytic fungi from soybean (*Glycine max* (L.) Merril) under different environmental conditions. Braz Arch Biol Technol 49:705–711

Piromyou P, Greetatorn T, Teamtisong K, Okubo T, Shinoda R, Nuntakij A, Tittabutr P, Boonkerd N, Minamisawa K, Teaumroong N (2015) Preferential association of endophytic bradyrhizobia with different rice cultivars and its implications for rice endophyte evolution. Appl Environ Microbiol 81:3049–3061

Pozo MJ, Van Der Ent S, Van Loon LC, Pieterse CM (2008) Transcription factor MYC2 is involved in priming for enhanced defense during rhizobacteria-induced systemic resistance in *Arabidopsis thaliana*. New Phytol 180:511–523

Qin S, Zhang Y-J, Yuan B, Xu P-Y, Xing K, Wang J, Jiang J-H (2014) Isolation of ACC deaminase-producing habitat-adapted symbiotic bacteria associated with halophyte *Limonium sinense* (Girard) Kuntze and evaluating their plant growth-promoting activity under salt stress. Plant Soil 374:753–766

Quecine M, Araujo W, Rossetto P, Ferreira A, Tsui S, Lacava P, Mondin M, Azevedo J, Pizirani-Kleiner A (2012) Sugarcane growth promotion by the endophytic bacterium *Pantoea agglomerans* 33.1. Appl Environ Microbiol 78:7511–7518

Raaijmakers JM, Mazzola M (2012) Diversity and natural functions of antibiotics produced by beneficial and plant pathogenic bacteria. Annu Rev Phytopathol 50:403–424

Rado R, Andrianarisoa B, Ravelomanantsoa S, Rakotoarimanga S, Rahetlah V, Fienena FR, Andriambeloson O (2015) Biocontrol of potato wilt by selective rhizospheric and endophytic bacteria associated with potato plant. Afr J Food Agric Nutr Dev 15:9762–9226

Rai R, Dash PK, Prasanna BM, Singh A (2007) Endophytic bacterial flora in the stem tissue of a tropical maize (*Zea mays* L.) genotype: isolation, identification and enumeration. World J Microbiol Biotechnol 23:853–858

Rakotoniriana EF, Rafamantanana M, Randriamampionona D, Rabemanantsoa C, Urveg-Ratsimamanga S, El JM, Munaut F, Corbisier AM, Quetin-Leclercq J, Declerck S (2013) Study in vitro of the impact of endophytic bacteria isolated from *Centella asiatica* on the disease incidence caused by the hemibiotrophic fungus *Colletotrichum higginsianum*. Antonie Van Leeuwenhoek 103:121–133

Rangjaroen C, Rerkasem B, Teaumroong N, Sungthong R, Lumyong S (2014) Comparative study of endophytic and endophytic diazotrophic bacterial communities across rice landraces grown in the highlands of northern Thailand. Arch Microbiol 196:35–49
Reinhold-Hurek B, Hurek T (2011) Living inside plants: bacterial endophytes. Curr Opin Plant Biol 14:435–443
Reiter B, Pfeifer U, Schwab H, Sessitsch A (2002) Response of endophytic bacterial communities in potato plants to infection with *Erwinia carotovora* subsp. atroseptica. Appl Environ Microbiol 68:2261–2268
Rosenblueth M, Martinez-Romero E (2006) Bacterial endophytes and their interactions with hosts. Mol Plant-Microbe Interact 19:827–837
Ryan RP, Germaine K, Franks A, Ryan DJ, Dowling DN (2008) Bacterial endophytes: recent developments and applications. FEMS Microbiol Lett 278:1–9
Salomon MV, Bottini R, de Souza Filho GA, Cohen AC, Moreno D, Gil M, Piccoli P (2014) Bacteria isolated from roots and rhizosphere of *Vitis vinifera* retard water losses, induce abscisic acid accumulation and synthesis of defense-related terpenes in in vitro cultured grapevine. Physiol Plant 151:359–374
Samish Z, Etinger-Tulczynska R, Bick M (1961) Microflora within healthy tomatoes. Appl Microbiol 9:20–25
Sandhiya GS, Sugitha TKC, Balachandar D, Kumar K (2005) Endophytic colonization and in planta nitrogen fixation by a diazotrophic *Serratia* sp. in rice. Indian J Exp Biol 43:802–807
Sasaki K, Ikeda S, Ohkubo T, Kisara C, Sato T, Minamisawa K (2013) Effects of plant genotype and nitrogen level on bacterial communities in rice shoots and roots. Microbes Environ 28:391–395
Scher FM, Baker R (1980) Mechanism of biological control in a Fusarium-suppressive soil. Phytopathology 70:412–417
Schikora A, Carreri A, Charpentier E, Hirt H (2008) The dark side of the salad: *Salmonella typhimurium* overcomes the innate immune response of *Arabidopsis thaliana* and shows an endopathogenic lifestyle. PLoS One 3:e2279
Schulz B, Boyle C (2006) What are endophytes? In: Schulz B, Boyle C, Sieber T (eds) Microbial root endophytes. Springer, Berlin, pp 1–13
Seghers D, Wittebolle L, Top EM, Verstraete W, Siciliano SD (2004) Impact of agricultural practices on the *Zea mays* L. endophytic community. Appl Environ Microbiol 70:1475–1482
Sessitsch A, Hardoim P, Doring J, Weilharter A, Krause A, Woyke T, Mitter B, Hauberg-Lotte L, Friedrich F, Rahalkar M, Hurek T, Sarkar A, Bodrossy L, Brar D, van Elsas JD, Reinhold-Hurek B (2012) Functional characteristics of an endophyte community colonizing rice roots as revealed by metagenomic analysis. Mol Plant-Microbe Interact 25:28–36
Sessitsch A, Reiter B, Berg G (2004) Endophytic bacterial communities of field-grown potato plants and their plant-growth-promoting and antagonistic abilities. Can J Microbiol 50:239–249
Sharma VK, Nowak J (1998) Enhancement of verticillium wilt resistance in tomato transplants by in vitro coculture of seedlings with a plant growth-promoting rhizobacterium (*Pseudomonas* sp. strain PsJN). Can J Microbiol 44:528–536
Shi J, Liu A, Li X, Feng S, Chen W (2011) Inhibitory mechanisms induced by the endophytic bacterium MGY2 in controlling anthracnose of papaya. Biol Control 56:2–8
Siddiqui IA, Ehteshamul-Haque S (2001) Suppression of the root rot–root knot disease complex by *Pseudomonas aeruginosa* in tomato: the influence of inoculum density, nematode populations, moisture and other plant-associated bacteria. Plant Soil 237:81–89
Singh JS, Pandey VC, Singh DP (2011) Efficient soil microorganisms: a new dimension for sustainable agriculture and environmental development. Agric Ecosyst Environ 140:339–353
Sorensen J, Sessitsch A (2006) Plant-associated bacteria lifestyle and molecular interactions. In: van Elsas JD, Jansson JK, Trevors JT (eds) Modern soil microbiology. CRC Press, Boca Raton, pp 211–236

Spoel SH, Koornneef A, Claessens SM, Korzelius JP, Van Pelt JA, Mueller MJ, Buchala AJ, Metraux JP, Brown R, Kazan K, Van Loon LC, Dong X, Pieterse CM (2003) NPR1 modulates cross-talk between salicylate- and jasmonate-dependent defense pathways through a novel function in the cytosol. Plant Cell 15:760–770

Stein E, Molitor A, Kogel KH, Waller F (2008) Systemic resistance in *Arabidopsis* conferred by the mycorrhizal fungus *Piriformospora indica* requires jasmonic acid signaling and the cytoplasmic function of NPR1. Plant Cell Physiol 49:1747–1751

Stoltzfus JR, So R, Malarvithi PP, de Bruijn FJ (1997) Isolation of endophytic bacteria from rice and assessment of their potential for supplying rice with biologically fixed nitrogen. Plant Soil 194:25–36

Sturz AV, Chriestie BR, Matheson BG (1988) Association of bacterial endophyte populations from red clover and potato crops with potential for beneficial allelopathy. Can J Microbiol 44:162–167

Sturz AV, Christie BR, Nowak J (2003) Bacterial endophytes: potential role in developing sustainable systems of crop production. Crit Rev Plant Sci 19:1–30

Suman A, Shasany AK, Singh M, Shahi HN, Gaur A, Khanuja SPS (2001) Molecular assessment of diversity among endophytic diazotrophs isolated from subtropical Indian sugarcane. World J Microbiol Biotechnol 17:39–45

Suman A, Yadav AN, Verma P (2016) Endophytic microbes in crops: diversity and beneficial impact for sustainable agriculture. In: Singh DP, Singh AB, Prabha R (eds) Microbial inoculants in sustainable agricultural productivity. Springer India, New Delhi, pp 117–143

Sun L, Qiu F, Zhang X, Dai X, Dong X, Song W (2008) Endophytic bacterial diversity in rice (*Oryza sativa* L.) roots estimated by 16S rDNA sequence analysis. Microb Ecol 55:415–424

Surette MA, Sturz AV, Lada RR, Nowak J (2003) Bacterial endophytes in processing carrots (*Daucus carota* L. Var. *sativus*): their localization, population density, biodiversity and their effects on plant growth. Plant Soil 253:381–390

Suyal DC, Yadav A, Shouche Y, Goel R (2015) Bacterial diversity and community structure of western Indian Himalayan red kidney bean (*Phaseolus vulgaris*) rhizosphere as revealed by 16S rRNA gene sequences. Biologia 70:303–313

Sziderics AH, Rasche F, Trognitz F, Sessitsch A, Wilhelm E (2007) Bacterial endophytes contribute to abiotic stress adaptation in pepper plants (*Capsicum annuum* L.) Can J Microbiol 53:1195–1202

Talley JL, Wayadande AC, Wasala LP, Gerry AC, Fletcher J, DeSilva U, Gilliland SE (2009) Association of *Escherichia coli* O157:H7 with filth flies (Muscidae and Calliphoridae) captured in leafy greens fields and experimental transmission of *E. coli* O157:H7 to spinach leaves by house flies (Diptera: Muscidae). J Food Prot 72:1547–1552

Tam HM, Diep CN (2014) Isolation, characterization and identification of endophytic bacteria in sugarcane *(Saccharum* sp. L.*)* cultivated on soils of the Dong Nai province, southeast of Vietnam. Am J Life Sci 2:361–368

Thomas P, Sekhar AC (2014) Live cell imaging reveals extensive intracellular cytoplasmic colonization of banana by normally non-cultivable endophytic bacteria. AoB Plants 6:1–12

Tian BY, Cao Y, Zhang KQ (2015) Metagenomic insights into communities, functions of endophytes, and their associates with infection by root-knot nematode, *Meloidogyne incognita,* in tomato roots. Sci Rep 5:1–15

Tian X, Cao L, Tan H, Han W, Chen M, Liu Y, Zhou S (2007) Diversity of cultivated and uncultivated actinobacterial endophytes in the stems and roots of rice. Microb Ecol 53:700–707

Trognitz F, Piller K, Nagel M, Borner A, Bacher C-F, Rechlik M, Mayrhofer H, Sessitsch A (2014) Isolation and characterization of endophytes isolated from seeds of different plants and the application to increase juvenile development. In: Brandstetter A, Geppner M, Grausgruber H, Buchgraber K (eds) Tagung Zukünftiges Saatgut - Produktion, Vermarktung, Nutzung und Konzervierung (future seed - production, marketing, use and conservation). Hohere Bundeslehr und Forschungsanstalt, Raumberg-Gumpenstein, pp 25–28

Tsurumaru H, Okubo T, Okazaki K, Hashimoto M, Kakizaki K, Hanzawa E, Takahashi H, Asanome N, Tanaka F, Sekiyama Y, Ikeda S, Minamisawa K (2015) Metagenomic analysis of the bacterial community associated with the taproot of sugar beet. Microbes Environ 30:63–69

Turner TR, James EK, Poole PS (2013) The plant microbiome. Genome Biol 14:209

Tuteja N (2007) Abscisic acid and abiotic stress signaling. Plant Signal Behav 2:135–138

Van Der Ent S, Verhagen BW, Van Doorn R, Bakker D, Verlaan MG, Pel MJ, Joosten RG, Proveniers MC, Van Loon LC, Ton J, Pieterse CM (2008) MYB72 is required in early signaling steps of rhizobacteria-induced systemic resistance in *Arabidopsis*. Plant Physiol 146:1293–1304

van Overbeek L, van Elsas JD (2008) Effects of plant genotype and growth stage on the structure of bacterial communities associated with potato (*Solanum Tuberosum* L.) FEMS Microbiol Ecol 64:283–296

van Overbeek LS, van DJ, Wichers JH, van AA, van Roermund HJ, Willemsen PT (2014) The arable ecosystem as battleground for emergence of new human pathogens. Front Microbiol 5:104

Velazquez-Sepilveda I, Orozco-Mosqueda MC, Prieto-Barajas CM, Santoyo G (2012) Bacterial diversity associated with the rhizosphere of wheat plants (*Triticum aestivum*): toward a metagenomic analysis. Int J Exp Bot 81:81–87

Verma P, Yadav AJ, Khannamkazy S, Saxera AK, Suman A (2013) Elucidating the diversity and plant growth promoting attributes of wheat (*Triticum aestivum*) associated acidotolerant bacteria from southern hills zone of India. Natl J Life Sci 10:219–227

Verma P, Yadav AN, Kazy SK, Saxena AK, Suman A (2014) Evaluating the diversity and phylogeny of plant growth promoting bacteria associated with wheat (*Triticum aestivum*) growing in central zone of India. Int J Curr Microbiol App Sci 3:432–447

Verma P, Yadav AN, Khannam S, Panjiar N, Kumar S, Saxena AK, Suman A (2015) Assessment of genetic diversity and plant growth promoting attributes of psychrotolerant bacteria allied with wheat *(Triticum aestivum)* from the northern hills zone of India. Appl Environ Microbiol 65:1885–1899

Wang M, Xing Y, Wang J, Xu Y, Wang G (2014) The role of the *chi1* gene from the endophytic bacteria *Serratia proteamaculans* 336x in the biological control of wheat take-all. Can J Microbiol 60:533–540

Wang M, Xing Y, Xu Y, Wang G (2008) The role of the chi1 gene from the endophytic bacteria *Serratia proteamaculans* 336x in the biological control of wheat take-all. Can J Microbiol 60:533–540

Wei G, Kloepper JW, Tuzun S (1991) Induction of systemic resistance of cucumber to *Colletotrichum orbiculare* by select strains of plant growth-promoting rhizobacteria. Phytopathology 81:1508–1512

Weilharter A, Mitter B, Shin MV, Chain PS, Nowak J, Sessitsch A (2011) Complete genome sequence of the plant growth-promoting endophyte *Burkholderia phytofirmans* strain PsJN. J Bacteriol 193:3383–3384

Weller DM, Raaijmakers JM, Gardener BB, Thomashow LS (2002) Microbial populations responsible for specific soil suppressiveness to plant pathogens. Annu Rev Phytopathol 40:309–348

West ER, Cother EJ, Steel CC, Ash GJ (2010) The characterization and diversity of bacterial endophytes of grapevine. Can J Microbiol 56:209–216

Whipps JM (1997) Developments in the biological control of soil-borne plant pathogens. Adv Bot Res 26:1–134

White JF Jr, Torres MS, Somu MP, Johnson H, Irizarry I, Chen Q, Zhang N, Walsh E, Tadych M, Bergen M (2014) Hydrogen peroxide staining to visualize intracellular bacterial infections of seedling root cells. Microsc Res Tech 77:566–573

Whitman WB, Coleman DC, Wiebe WJ (1998) Prokaryotes: the unseen majority. Proc Natl Acad Sci U S A 95:6578–6583

Wu Q, Zhu L, Jiang L, Xu X, Xu Q, Zhang Z, Huang H (2015) Draft genome sequence of *Paenibacillus dauci* sp. nov., a carrot-associated endophytic actinobacteria. Genom Data 5:241–253

Xia Y, DeBolt S, Dreyer J, Scott D, Williams MA (2015) Characterization of culturable bacterial endophytes and their capacity to promote plant growth from plants grown using organic or conventional practices. Front Plant Sci 6:490

Yan Y, Yang J, Dou Y, Chen M, Ping S, Peng J, Lu W, Zhang W, Yao Z, Li H, Liu W, He S, Geng L, Zhang X, Yang F, Yu H, Zhan Y, Li D, Lin Z, Wang Y, Elmerich C, Lin M, Jin Q (2008) Nitrogen fixation island and rhizosphere competence traits in the genome of root-associated *Pseudomonas stutzeri* A1501. Proc Natl Acad Sci U S A 105:7564–7569

Yang CJ, Zhang XG, Shi GY, Zhao HY, Chen L, Tao K, Hou TP (2011) Isolation and identification of endophytic bacterium W4 against tomato *Botrytis cinerea* and antagonistic activity stability. Afr J Microbiol 5:131–136

Yanni YG, Rizk RY, Corich V, Squartini A, Ninke K, Philip-Hollingsworth S, Orgambide G, de Bruijn F, Stoltzfus J, Buckley D, Schmidt TM, Mateos PF, Ladha JK, Dazzo FB (1997) Natural endophytic associations between *Rhizobium leguminosarum* bv. *trifolii* and rice roots and assessment of its potential to promote rice growth. Plant Soil 194:114

Yashiro E, Spear RN, McManus PS (2011) Culture-dependent and culture-independent assessment of bacteria in the apple phyllosphere. J Appl Microbiol 110:1284–1296

You C, Zhou F (1988) Non nodular endophytic nitrogen fixation in wetland rice. Can J Microbiol 35:408

Zhang YF, He LY, Chen ZJ, Wang QY, Qian M, Sheng XF (2011a) Characterization of ACC deaminase-producing endophytic bacteria isolated from copper-tolerant plants and their potential in promoting the growth and copper accumulation of *Brassica napus*. Chemosphere 83:57–62

Zhang YF, He LY, Chen ZJ, Zhang W-H, Wang QY, Qian M, Sheng XF (2011b) Characterization of lead-resistant and ACC deaminase-producing endophytic bacteria and their potential in promoting lead accumulation of rape. J Hazard Mater 186:1720–1725

Zhong W, Gu T, Wang W, Zhang B, Lin X, Huang Q, Shen W (2010) The effects of mineral fertilizer and organic manure on soil microbial community and diversity. Plant Soil 326:511–522

Zinniel DK, Lambrecht P, Harris NB, Feng Z, Kuczmarski D, Higley P, Ishimaru CA, Arunakumari A, Barletta RG, Vidaver AK (2002) Isolation and characterization of endophytic colonizing bacteria from agronomic crops and prairie plants. Appl Environ Microbiol 68:2198–2208

The Interactions of Soil Microbes Affecting Stress Alleviation in Agroecosystems

2

M. Miransari

Abstract

Crop plants are subjected to different kinds of stresses, and as a result, their growth is adversely affected. Different mechanisms may be used by crop plants to tolerate the stress including the morphological and physiological ones. However, the efficiency of such mechanisms differs in sensitive and tolerant crop species, and the tolerant species can utilize such mechanisms more efficiently. The other important aspect of stress tolerance in crop plants is related to their interactions with the soil microbes. A wide range of soil microbes including arbuscular mycorrhizal (AM) fungi, plant growth-promoting rhizobacteria (PGPR), and endophytic bacteria as well as their interactions can affect stress tolerance in crop plants. Such a topic is among the most important research subjects and can greatly affect the efficiency of crop plants under stress. Mycorrhizal fungi are soil fungi, developing a symbiotic association with their nonspecific host plants, and increase their growth by enhancing the uptake of water and nutrients. PGPR are soil bacteria, which can enhance the growth of their host plant by different mechanisms through developing a nonsymbiotic association. The endophytic microbes are able to colonize the inner parts of their host plant and affect its growth under different conditions including stress. The interactions of soil microbes in most cases can positively affect the growth of the host plant under different conditions including stress. The important point, which deserves investigation, is the interaction of mycorrhizal fungi, PGPR, and the endophytic bacteria, which reside in plant roots affecting plant growth and yield production. Such details will be useful for the production of more tolerant microbial inoculums, which are more efficient under different conditions including stress. Some

M. Miransari
Department of Book & Article, AbtinBerkeh Scientific Ltd. Company,
#8, Fourth Floor, Sadaf Building, Hakim Nezami Ave, Isfahan 8175676486, Iran
e-mail: Miransari1@gmail.com; Info@abtinberkeh.com

V. Kumar et al. (eds.), *Probiotics in Agroecosystem*,
DOI 10.1007/978-981-10-4059-7_2

of the most important and recent findings related to the growth of crop plants under stress, as affected by the interactions of soil microbes, along with the future perspectives are presented, reviewed, and analyzed.

2.1 Introduction

Crop plants are subjected to different kinds of stresses, and as a result, their growth is adversely affected. Under stress, crop plants use different morphological and physiological mechanisms to alleviate the stress. However, the level of stress tolerance is different in sensitive and tolerant crop plants. Depending on the level of stress, crop response in sensitive and tolerant species is different, and accordingly they can tolerate the stress up to some level (Ramegowda and Senthil-Kumar 2015).

Crop plants interact with a wide range of microbes in the soil; however, the combination of soil microbes in the bulk soil and in the rhizosphere is different indicating that the crop plants are able to determine their combination of microbes in the microbiome (Bisseling et al. 2009; Zamioudis and Pieterse 2012). A wide range of beneficial soil microbes including mycorrhizal fungi, plant growth-promoting rhizobacteria (PGPR), and the endophytic bacteria and fungi are able to develop a symbiotic or nonsymbiotic association with their host plants and enhance the growth of their host plant. High rate of recognition and coordination between the host plant and the soil microbes are essential for the development of an intimate symbiosis association (Giovannetti et al. 2006; Berg 2009; Newton et al. 2010).

The important issue related to the growth of crop plants under stress is to provide them with their essential nutrients. Although chemical fertilizer can quickly provide crop plants with the nutrients necessary for their growth, due to the environmental and economic consequences, the single use of such method of fertilization is not recommendable. Accordingly, a suitable method of fertilization is the integrated use of chemical and biological fertilization (Miransari 2011a, b; Hoseinzade et al. 2016). According to FAO (2015), the world demand for N fertilizer has been equal to 141 682000 T, P_2O_5 at 51940000 T, and K_2O at 36367000 T, and the total value of fertilizer demand in 2015 has been equal to 223,064,000 T. The highest fertilizer use has been related to China, India, and the USA.

Interestingly, the soil microbes are able to determine the structure of plant community and plant traits. Accordingly, plant growth and yield production, nutrient uptake, and the functioning of ecosystem are affected by soil microbes (Degens 1998; Marschner and Rumberger 2004; Bell et al. 2005; Bonkowski and Roy 2005; Lau and Lennon 2011; Huang et al. 2014). This is an important approach toward the alleviation of soil stresses by soil microbes and development of tolerant plant and microbial species.

There has been a great and growing research on the use of biological fertilization including mycorrhizal fungi and PGPR integrated with chemical fertilization during the recent years, mainly for increasing the efficiency use of fertilization. In the past time, the use of microbial inoculants has been mostly for plant growth promotion and biological control. However, it has been just recently that biological fertilization has been used for the increased uptake of crop plants. More than 50% of N fertilizer is lost due to leaching and volatilization with long-time environmental consequences including the production of greenhouse gases, depletion of ozone, global warming, and acid rain (Flessa et al. 2002; Ma et al. 2007; Miransari and Mackenzie 2015).

The important role of soil microbes in the alleviation of stresses has been indicated by different research works. For example, the biochemical effects of soil microbes on the alleviation of soil stresses are by the production of (1) biofilm and exopolysaccharides affecting the properties of soil and (2) different organic products such as osmolytes, stress proteins, etc. A set of interactions between the soil, the microbes, and the plant affects the biological, the physical, and the chemical properties of soil (Flemming and Wingender 2010; Singh et al. 2011).

The production of microbial polysaccharides by the soil microbes results in the binding of soil particles and as a result improves the structure of the soil and plant tolerance under stress. The growth of the mycrorrhizal hyphae into the soil pores can stabilize the structure of the soil and increase the uptake of water and nutrients by the host plant under different conditions including stress (Sandhya et al. 2009a, b). The soil microbes can be used as models for understanding how soil stresses can affect crop growth and hence can be genetically modified for a more efficient use under stress conditions (Mantelin and Touraine 2004; Grover et al. 2011; Schenk et al. 2012; Bashan et al. 2014).

PGPR can affect crop growth and the environment by the following mechanisms: (1) the production of plant hormones, phosphorus (P)-solubilizing products, and siderophores; (2) biological N fixation; and (3) controlling pathogens. The PGPR can hence decrease the use of chemical fertilization, herbicides and pesticides, which is of environmental and economic significance (Yasmin et al. 2004; Yu et al. 2011). Such beneficial effects have resulted in the wide use of soil microbes including PGPR and arbuscular mycorrhizal fungi as important sources of fertilization, namely, biological fertilization (Adesemoye et al. 2009; Berg 2009; Miransari 2011a, b).

PGPR are also able to alleviate stress by the production of plant hormones such as cytokinin and antioxidants, which can scavenge the production of reactive oxygen species under stress. The PGPR are also able to alleviate the stress in the host plant by the induction of the stress genes and the production of the enzyme 1-aminocyclopropane-1-carboxylate (ACC) deaminase, which is able to degrade the prerequisite for ethylene production and decrease the adverse effects of the stress hormone, ethylene, which its level increases under stress. The enzyme is able to increase root growth under drought stress and increases water uptake and water efficiency by the host plant (Berg 2009; Grover et al. 2011).

Although mycorrhizal fungi are able to develop an intimate symbiotic association with their host plant, at high concentration of P (greater than 100 mg/kg) such a symbiosis decreases significantly (Amijee et al. 1989; Koide 1991). Accordingly, at high fertile soils, the use of mycorrhizal fungi may not be beneficial, as the host plant is able to receive its essential P from the soil (Koide and Li 1990; Stewart et al. 2005).

The use of plant biostimulants such as soil microbes is an effective method to alleviate the adverse effect of stress on plant growth. Microbial inoculums including arbuscular mycorrhizal fungi, plant growth-promoting rhizobacteria, and N-fixing rhizobium can significantly alleviate the negative effects of stress on plant growth. The interactions between the soil microbes and between the plant and the soil microbes affect the efficiency of stress alleviation by the soil microbes (Calvo et al. 2014; Kalita et al. 2015; Mabrouk and Belhadj 2016).

Among the most studied PGPR strains is *Azospirillum* spp., which is able to develop a nonsymbiotic association with their host plant and reside on the root surface or inside the plant tissues (Hartmann and Bashan 2009; Bashan et al. 2014). The N-fixing ability of *Azospirillum* spp. has been indicated by different research works. For example, the rate of fixed N in wheat by *Azospirillum lipoferum* was in the range of 7–12%; however, in sugarcane, *Azospirillum diazotrophicus* was able to fix 60–80% of the plant total N (Saubidet et al. 2000; Calvo et al. 2014).

Different PGPR strains including *Bacillus subtilis* (Arkhipova et al. 2005), *Pseudomonas* spp. (Park et al. 2005), *Streptomyces* spp. (Glick 2015), *Micrococcus* spp. (Dastager et al. 2010), *Achromobacter* spp. (Pereg and McMillan 2015), *Flavobacterium* spp. (Glick 2015), *Azospirillum* spp. (Arzanesh et al. 2011; Miransari 2014), and *Erwinia* spp. (Hsieh et al. 2010) are able to increase the availability of P in the soil.

The PGPR are able to enhance the availability of P by the production of phosphatases and organic acids. The hydroxyl and carboxyl present in the organic acids are able to enhance the solubility of P by the following mechanisms: (1) chelating the P anions and (2) decreasing the pH of rhizosphere resulting in the release of P anion (Kpomblekou and Tabatabai 1994; Singh et al. 2011). The PGPR strains determine the types of organic acids, produced in the rhizosphere. For example, the organic acids including acetic, lactic, isobutyric, and isovaleric acids are produced by *Bacillus amyloliquefaciens* and *Bacillus licheniformis*. However, *Azospirillum* spp., *Pseudomonas* spp., and *Erwinia* spp. are able to produce gluconic acid (Martínez et al. 2011; Zhang et al. 2014).

The Gram-negative PGPR including *Enterobacter*, *Pseudomonas*, *Citrobacter*, etc. can increase P solubility by producing acid phosphatases. Using the organic source of phytate and by the production of phytase, the PGPR including *Bacillus* spp. and *Pseudomonas* spp. are able to increase the availability of P, for plant use (Martínez et al. 2011; Singh and Satyanarayana 2011).

The soil beneficial microbes are able to increase the growth of the host plant by the following mechanisms: (1) increased uptake of water and nutrients, (2) the fixation of nutrients such as nitrogen (N) by PGPR including rhizobium (symbiosis) and other microbes (nonsymbiosis), (3) production of different plant hormones, (4) controlling pathogens, (5) alleviating stress, and (6) interaction among soil microbes (Pozo and Azcon-Aguilar 2007; De Vleesschauwer and Höfte 2009; Lugtenberg and Kamilova 2009; Zamioudis and Pieterse 2012; Miransari et al. 2013).

The important point about the performance of soil microbes under stress is how the soil microbes are affected by the stress. Under stress different microbial genes are activated making the microbes tolerate the stress. However, if the soil microbes are isolated from stressed soils, they can be used more efficiently for the alleviation of stresses. Different osmoregulators are produced by PGPR under stress including K+, proline glutamate, trehalose, betaine, glycine, etc., which are able to regulate the osmotic potential of cytoplasm. The important role of trehalose, in rhizobium during the signaling with the host plant, and its effects on plant growth and yield under stress has been indicated. It has also been shown that the genetic modification of trehalose-signaling pathway in rhizobium can significantly affect the performance of legumes (Suarez et al. 2008).

The availability of potassium (K^+) as the other important nutrient is also affected by the activity of PGPR. Such PGPR are able to solubilize the K^+ in the minerals including illite, micas, and orthoclases by the production of organic acids. Accordingly, by dissolving K^+ or chelating silicon, the solubility of K+ increases (Miransari 2013; Ahmed et al. 2014). Singh et al. (2010) indicated that the strains of PGPR such as *Rhizobium* spp. and *Azotobacter chroococcum* are able to increase the availability of potassium (increased solubility) from mica for maize and wheat use. Accordingly, plant growth, K content, chlorophyll, and crude protein rate in plant increased. The authors accordingly indicated that it is possible to provide the plant with K using PGPR and mica.

Sheng and He (2006) found that the increased solubility of K by *Bacillus edaphicus* is due to the production of organic acids including oxalic, citric, α-ketogluconic, tartaric, and succinic by PGPR resulting in dissolving K and chelating silicon ions. The bacterial inoculants are able to enhance the availability and the uptake of different nutrients by the host plant, although more research work will have to be conducted to illustrate the related details.

With respect to the abovementioned details and importance of interactions among different soil microbes mainly PGPR, mycorrhizal fungi, and endophytic bacteria, in this chapter, the effects of soil microbes and their interactions on the growth of crop plants are presented, reviewed, and analyzed. Such kind of analyses can be useful for the development of methods and techniques, which can be used for the production of tolerant crop plants and efficient microbial inoculums under stress.

2.2 Crop and Stress

Stress significantly decreases plant growth by adversely affecting plant morphology and physiology. Plant by itself can tolerate the stress through the modification of its morphology and physiology. Plant roots are among the most important parts affecting plant response under stress. Several important functions, indicated in the following, are fulfilled by the roots, making the plant grow under different conditions including stress: (1) maintaining the plant in the soil, (2) absorbing water and nutrients for plant growth and yield production, (3) affecting the properties of the soil specifically the rhizosphere, (4) interacting with the other plant roots, (5) interacting with the soil microbes, and (6) production of different biochemicals.

Accordingly, if the root traits including root architecture and functioning are modified using molecular techniques, it is possible to enhance the root potentials including its interaction with the soil microbes affecting plant tolerance under stress. The following indicates how the modification of plant roots may affect root properties including its interaction with the soil microbial activities such as the process of symbiosis: (1) modifying rhizosphere pH affecting root functioning; (2) enhancing root interactions with soil microbes, which affects root functionality; (3) proliferation of roots enhancing nutrient uptake; (4) modifying the root exudates, which results in the increased uptake of nutrients (Haichar et al. 2014); and (5) the increased number and length of lateral roots improving nutrient uptake and its symbiosis with soil microbes (Meister et al. 2014).

The following mechanisms, which are the results of signaling communications between the host plant and the soil microbes, resulting in the subsequent colonization of plant roots by the microbes, indicate how the two symbionts may interact under stress: (1) the production of volatiles by bacteria affects the translocation of Na + and its uptake by plant; (2) the production of ACC deaminase by bacteria decreases the level of ethylene in plant; (3) the bacteria produces cytokinin, resulting in the increased production of ABA in plant; (4) bacterial production of antioxidants scavenges the production of reactive oxygen species in plant; (5) the production of exopolysaccharides by bacteria improves the properties of soil; and (6) the bacteria are able to produce IAA and some unknown growth substances, which increase root growth under different conditions including stress (Grover et al. 2011).

Production of tolerant crop plants under stress can increase crop growth and yield. Different methods have been used so far including the breeding techniques and the use of soil microbes, both of which have been indicated to be effective on the growth of crop plants under stress. The use of soil microbes has also been indicated to be useful for the growth of crop plants under stress. A wide range of soil microbes have been tested under stress; however, more research is essential on the production and use of microbial inoculums under different conditions including stress.

2.3 Arbuscular Mycorrhizal Fungi and Stress

The symbiotic mycorrhizal fungi are able to increase plant growth under different conditions including stress. In such a symbiotic association, the fungi provide the host plant with water and nutrients for carbon, which is utilized by the fungi for their growth and activities. The fungi can increase plant growth under stress by the following mechanisms: (1) the extensive hyphal network, (2) production of plant hormones, (3) interaction with the other soil microbes, (4) improving the root growth of the host plant, and (5) increased uptake of different nutrients mainly P.

Different research works have indicated the alleviating effects of mycorrhizal fungi under salt stress in different crop plants mainly due to the improved production of proline, resulting in osmoregulation. However, other processes such as the increased uptake of P and higher concentration of sugar can also improve the host plant tolerance under salinity stress (Ben Khaled et al. 2003; Daei et al. 2009; Talaat and Shawky 2014; López-Ráez 2015; Garg and Singla 2016).

Numerous research works have indicated the positive effects of mycorrhizal fungi on plant growth and yield production under different types of stresses including salinity, drought, acidity, compaction, flooding, heavy metals, cool temperature, etc. according to the following details. Among the most important potentials of mycorrhizal fungi under stress is their extensive network of hyphal, which is able to enhance plant host tolerance by significantly increasing the uptake of water and nutrients. The fungal hyphae are able to grow even in the finest soil micropores where the root hairs are not able to grow and absorb water and nutrients for the host plant use. Such ability is important for affecting plant growth under compaction stress (Miransari 2010; Garg and Chandel 2010).

The alleviating effects of mycorrhizal fungi on different stresses including salinity (Al-Karaki 2000; Colla et al. 2008; Daei et al. 2009), drought (Subramanian et al. 2006; Wu et al. 2008), acidity (Raju et al. 1988; Vosatka et al. 1999; Muthukumar et al. 2014), compaction (Miransari et al. 2008, 2009), heavy metals (Audet and Charest 2007; Miransari 2011c), flooding (Rutto et al. 2002; Carvalho et al. 2003), temperature (Bunn et al. 2009; Zhu et al. 2010), and nutrient deficiency (Miransari 2010; Smith et al. 2010) have been indicated.

For example, the improving effects of mycorrhizal fungi on the growth of plant under osmotic and drought stress have been indicated by the following mechanisms: (1) enhanced activities of antioxidants such as catalase, peroxidase, and superoxide dismutases; (2) decreased production of soluble protein and malondialdehyde; (3) the increased levels of nonstructural carbohydrate; and (4) the increased rate of K^+, $Ca2^+$, and Mg^{2+} (Wu and Xia 2006).

2.4 Plant Growth-Promoting Rhizobacteria and Stress

Research work has indicated the enhancing effects of PGPR on plant growth under stress. The mechanisms, which PGPR use under stress to alleviate the stress and increase plant growth, are (1) the modification of plant morphology and physiology; (2) the production of plant hormones; (3) the increased rate of proline; (4) the increased uptake of nutrients by the host plant; (5) the decreased production of the stress hormone, ethylene, by the production of ACC deaminase; and (6) the interactions with the other soil microbes (Kaymak 2010; Miransari 2014). The promoting effects of *Rhizobium* spp. and *Azospirillum* spp. on the growth of plant under drought and saline conditions have been indicated (Hamaoui et al. 2001, Arzanesh et al. 2011). The rhizobium and PGPR strains have been isolated from stress conditions. The adaptation of soil microbes under stress is a function of gene regulation, resulting in the survival of the microbes (Ali et al. 2009). For example, the use of *Pseudomonas* spp. strain AMK-P6 increased the thermotolerance of sorghum seedlings under heat stress by the following: (1) the production of high molecular weight protein in plant leaf, (2) increased plant biomass, and (3) enhanced production of amino acid, sugar, proline, and chlorophyll II content (Ali et al. 2009).

Under stress PGPR use different mechanisms to handle the stress. For example, in a saline environment, the bacterial cells lose their water and as a result the dehydration of the cytoplasm decreases. Microbes also utilize the following mechanisms to alleviate the adverse effects of stress on their growth and activities including:

1. The increased ionic strength, which increases the salt concentration of cytoplasm equal to the surrounding environment.
2. The increased uptake of K^+ as well as the enhanced accumulation of the compatible solutes including amino acids, sugars, polyols, and betaines by the microbial cells. Such solutes are synthesized by the bacterial cells or taken up from the environment (Street et al. 2006; Paul and Nair 2008). The authors indicated that under salt stress the PGPR *Pseudomonas fluorescens* MSP-393 was able to tolerate the stress by the production of different solutes including glycine, alanine, serine, glutamic acid, asparatic acid, and threonine. Such solutes can also stabilize the structure of proteins under stress (Street et al. 2006; Paul and Nair 2008).
3. The alteration of cellular composition under stress, which modifies the structure of proteins, saccharides, and glucans, is also another mechanism used by the microbes to alleviate stress. The production of exopolysaccharides, which are able to stabilize the water content and regulate the carbon diffusion into the surrounding environment of the cell by *Pseudomonas* under stress, increased the bacterial tolerance to survive the stress (Sandhya et al. 2009a).
4. The expression of salt-responsive genes in the microbes under salinity stress can regulate the microbial response under stress (Paul and Lade 2014; Chakraborty et al. 2015).

PGPR are also able to increase the growth of the host plant by increasing the availability of iron (Fe III) in the rhizosphere where the concentration of Fe III is little. The Fe-binding chelators (siderophores) can bind FE III under little concentration of FE and transfer it to the cell (Sayyed et al. 2005; Dimkpa et al. 2009; Marschner et al. 2011). For example, the enhancing effects of PGPR such as *Streptomyces* and *Bacillus* on the growth of different plants, due to the production of siderophores, have been indicated (Imbert et al. 1995; Fiedler et al. 2001; Temirov et al. 2003). The production of siderophores by PGPR is also of environmental significance as they can chelate different heavy metals and hence can be used for the bioremediation of contaminated sites.

2.5 PGPR and Plant Hormones

The soil microbes not only develop the suitable mechanisms for their survival under stress, they are also able to enhance the host plant tolerance under stress. For example, the production of different plant hormones by PGPR increases the growth of plant roots including root length, the number of root tips, and root surface area. Accordingly, the uptake of water and nutrients and hence plant growth increases by the host plant (Adesemoye et al. 2009; Egamberdieva and Kucharova 2009).

Production of plant hormones such as auxins, cytokinins, ethylene, and gibberellins is another important mechanism by PGPR affecting the growth of the host plant. PGPR are able to modify the production of hormones by the host plant. The growth of different plant parts including the roots (root length, root initiation, and formation of root hairs) is regulated by plant hormones. A high number of research works have indicated the production of auxin by PGPR; the hormone is able to affect plant growth by (1) affecting cellular growth and division, (2) root growth, (3) differentiation of vesicular bundles, (4) apical dominance, (5) ethylene production, and (6) expression of different plant genes (Döbbelaere et al. 1999; Spaepen et al. 2008).

The production of plant hormones by PGPR is regulated by different mechanisms. For example, the root exudates can modify the production of plant hormones by *A. brasilense*; if the production of root exudates decreases and is not at a suitable rate for bacterial growth, the production of IAA by PGPR increases resulting in the production of root hairs and lateral roots. The production of IAA by *A. brasilense* Sp245 increased leaf length and the growth of the aerial part in spring wheat, related to the control treatment (Spaepen et al. 2008).

The other important plant hormone, which is produced by PGPR, is cytokinin inducing plant cellular activities specifically cell division, as well as leaf growth and senescence. The production of cytokinin by *Bacillus subtilis* increased the rate of cytokinin production and the subsequent plant growth in lettuce. The cytokinin-producing PGPR can also increase plant growth under drought stress (Arkhipova et al. 2005).

Gibberellins can affect plant growth by affecting the cellular activities and growth (Eichmann and Schäfer 2015). The hormone can affect different stages of plant growth including the floral, the fruit, the aerial part, and the root growth as well as the seed germination. Similar to the other plant hormones, such as auxin and cytokinins, gibberellins can also act in combination and cross talk with the other plant hormones. Different species of *Azospirillum* can produce gibberellins affecting plant growth and yield production (Piccoli et al. 1997; Spaepen et al. 2009). The production of gibberellins has also been indicated in the other bacterial strains such as *Bacillus* spp., *Herbaspirillum seropedicae*, and *Acetobacter diazotrophicus*. Although the exact mechanism, which promotes plant growth by gibberellins, has not been indicated, the increased root growth, specifically the density of root hairs by the hormone, enhances the uptake of water and nutrients by plant (Gutiérrez-Mañero et al. 2001; Kang et al. 2012; Cassán et al. 2014).

The other plant hormone, ethylene, regulates different plant activities such as fruit ripening, cell growth, the germination of seeds, and the senescence of leaf and flower. The production of the hormone also increases under stress, and as a result, it is called the stress hormone (Spaepen et al. 2009) adversely affecting plant growth, specifically the root growth. The production of the hormone is regulated by ACC synthase. Research work has indicated that PGPR are able to decrease the production of the stress hormone ethylene in plant by the production of 1-aminocycloprop ane-1-carboxylate (ACC) deaminase, which is able to degrade ACC synthase into α-ketobutyrate and ammonia. Accordingly, the level of stress hormone in plant decreases, and as a result, plant growth increases. The production of ACC synthase by plant roots results in the degradation of the enzyme by soil PGPR and due to the decreased concentration of enzyme in the rhizosphere plant exudates more enzyme into the soil, and as a result, the level of enzyme deceases in plant, and hence plant biomass increases (Jalili et al. 2009; Glick 2015).

2.6 Endophytic Microbes and Stress

The endophytic microbes colonize the endosphere and the rhizosphere. Such soil microbes are able to enhance plant growth under different conditions including stress. The endophytic microbes including bacteria and fungi are able to enter plant tissues, establish a systemic nonpathogenic and intercellular association with the host plant, and complete their life cycle. A wide range of endophytic microbes are found in a single plant affecting plant growth and yield production (Naveed et al. 2014a).

The endophytic microbes are able to alleviate the adverse effects of stress on plant growth by morphological, biochemical, and physiological adaptation. The microbes are also able to affect the host plant growth by affecting the availability of different soil nutrients by the following mechanisms: (1) the reduction of the root diameter, (2) enhancing the length of root hairs, and (3) the production of different biochemicals including the phenolic products. Such compounds are also able to affect root growth under acidic conditions by the sequestration of aluminum on the root surfaces (Malinowski and Belesky 2000; Barac et al. 2004; Bauer and Mathesius 2004). The alleviating effects of endophytic fungi on plant growth are by the following activities: (1) inducing systemic resistance in the host plant, (2) activation of stress enzymes, and (3) production of different metabolites (Yuan et al. 2010).

Naveed et al. (2014a) investigated the microbial strains, which may enhance maize growth under biotic and abiotic stresses. The five endophytic strains of bacteria were isolated from the maize roots. All strains were able to increase maize growth under nonstressed and stressed conditions; however, *Enterobacter* sp. was among the most efficient strain under control and stress conditions. The bacteria were able to enhance maize morphology and physiology and were able to colonize plant roots and stems as well as the rhizosphere.

In another experiment Naveed et al. (2014b) evaluated the effects of *Burkholderia phytofirmans* and *Enterobacter* sp. on the growth, photosynthetic activity, and water content of two maize genotypes grown under drought stress. The plants were subjected to drought stress 45 days after planting during the vegetative growth stage by withholding irrigation. The bacterial inoculants were able to inoculate maize seedlings efficiently and be isolated from different plant parts under control and stress conditions. The bacteria were able to enhance plant growth under stress by improving plant morphology, physiology, and water content compared with uninoculated plants. Strain *B. phytofirmans* was the more efficient strain than *Enterobacter* sp. under the stress. The authors accordingly indicated that it is possible to alleviate drought stress on maize growth although such a potential is a function of plant genotype and bacterial strains.

Waqas et al. (2014) investigated how soybean growth and yield are affected by the combined effects of biochar and endophytic fungi producing plant hormones under heavy metal stress. The endophytic fungi were isolated from a wetland area polluted with zinc. According to the results, the combined or the single use of the endophytic fungi and biochar (15% w/w) significantly increased soybean growth under nonstressed and stressed conditions (Zn at 5253.6 mg/kg). The treatments decreased the uptake of Zn by plant tissues. The fungi were able to inoculate the plant roots in the presence of biochar under nonstressed and stressed conditions. The treatments induced plant systemic resistance by significantly increasing the production of jasmonic acid. The authors accordingly indicated that the combined use of the endophytic fungi and biochar is recommendable under the stress of heavy metal for soybean production.

The alleviating effects of *Enterobacter* sp. EJ01 isolated from sea china pink (*Dianthus japonicus* Thunb) in salty areas of South Korea were shown by Kim et al. (2014). The bacteria were used for the inoculation of tomato and *Arabidopsis* under salty conditions and were able to alleviate the stress by the production of ACC deaminase and indole-3-acetic acid (IAA). The growth parameters of plants including fresh and dry weight and plant height were increased by the bacteria under the stress conditions. The effects of the bacteria on plant growth at the molecular level were by the enhanced expression of the *Arabidopsis* genes, which are responsive under salt stress including RAB18, DREB2b, RD29A, and RD29B.

The bacteria were also able to upregulate the expression of genes, which results in the production of proline such as P5CS1 and P5CS2 and in priming such as MPK3 and MPK6 under stress. The bacteria also increased the scavenging process of reactive oxygen species in tomato plants subjected to the stress conditions. In conclusion, the authors indicated that the newly isolated bacteria are able to alleviate the stress of salinity on tomato and *Arabidopsis* growth by activating a set of different mechanisms, most importantly the related salt stress signaling pathway.

2.7 Interactions of Soil Microbes Affecting Soil Stresses

Although the single use of soil microbes has been proved to be effective on the alleviation of soil stresses, their combined use has also indicated to be a great tool for plant growth under stress. However, more research work in this respect is essential. This is because, if the possible interactions between the soil microbes are illustrated under different conditions including stress, it is possible to recognize and select the most efficient microbial consortium for inoculum production. However, it has also to be mentioned that the interactions between the soil microbes may be positive or negative, and for the production of effective inoculums under stress, the microbes with positive interactions must be selected (Miransari 2011a, b). The following examples show how the soil microbes may interact under different conditions including stress.

The coinoculation of corn plants with *Rhizobium* and *Pseudomonas* enhanced corn tolerance under salt stress by the following mechanisms: (1) decreased electrolyte leakage, (2) increased production of proline, (3) maintaining leaf water content, and (4) selective uptake of K^+ (Bano and Fatima 2009; Paul and Lade 2014). Alami et al. (2000) found that rhizobium enhanced sunflower tolerance under drought stress by improving the structure of soil due to the production of exopolysaccharide.

According to Chen et al. (2007), there is a correlation between the production of proline under drought and salinity stresses. Accordingly, the authors inserted the *pro*BA gene from *Bacillus subtilis* into *Arabidopsis thaliana*, which resulted in the increased production of proline in the plant and enhanced plant growth under stress. Proline is also able to neutralize the cellular redox potential. The coinoculation of corn (*Zea mays* L.) plants with *Rhizobium* and *Pseudomonas* enhanced the salt tolerance of the plant by (1) the increased production of proline, (2) decreased electrolyte leakage, and (3) increased uptake of K^+. Under salinity and temperature stresses, the positive effects of proline on the cell growth are by the protection of cellular membranes and the proteins resulted from the adverse effects of the stress. Proline is also able to act as a protein like molecule and scavenge the hydroxyl radical molecules produced under stress (Bano and Fatima 2009).

The coinoculation of lettuce plants with PGPR (*Pseudomonas* spp.) and mycorrhizal fungi (*Glomus* sp.) increased the production of catalase under severe drought stress indicating that the combined use of such soil microbes can be used for the alleviation of drought stress. However, interestingly the alleviating effects of mycorrhizal fungi under salinity stress have been more evident and constant than under drought stress (Kohler et al. 2008).

Research work has indicated that a wide range of PGPR strains in combination with mycorrhizal fungi are able to increase plant growth and yield production significantly. The related PGPR are *Azospirillum* spp., *Pseudomonas* spp., *Acinetobacter* spp., and *Bacillus* spp., which in combination with AM fungi increased the uptake of different nutrients including Ca, Mg, S, Mn, Fe, Zn, and Cu in different crop plants (Liu et al. 2000; Khan 2005; Kohler et al. 2008).

Kohler et al. (2010) investigated the single and the combined use of *Glomus mosseae* and *Pseudomonas mendocina* on the structural properties of lettuce (*Lactuca sativa* L.) soil under salt stress. The PGPR significantly increased the plant growth compared with the control treatment under control and saline conditions; however, mycorrhizal fungi just increased plant growth under the moderate level of salinity. With increasing the level of salinity, even in the presence of microbial inoculation, the aggregate stability of soil decreased, compared with the control treatment. The high level of salinity decreased the concentration of glomalin-related soil protein, although the highest level was related to the inoculated soil.

The authors accordingly indicated that the use of mycorrhizal fungi and PGPR may be restricted under salinity stress due to the adverse effects of salinity on the structure of the soil, which is due to the increased concentration of sodium and the less concentration of glomalin-related soil protein, compared with control conditions. Under adverse conditions such as drought, the production of exopolysaccharides by the soil bacteria enhances the aggregate stability of soil, and as a result, the water retention of soil increases (Kohler et al. 2010).

Franzini et al. (2010) investigated the tripartite symbiosis of mycorrhizal fungi and rhizobium in four different genotypes of bean (*Phaseolus vulgaris*) under moderate drought stress. Surprisingly, most of the microbial treatments including one species of fungi and one strain of rhizobium adversely affected the growth of bean plants under moderate drought conditions. However, such a deleterious effect was a function of plant genotype and microbial species. Mycorrhizal fungi decreased plant growth by inhibiting nodule development and N_2-fixation. At the moderate level of drought stress, the combined use of AM fungi and rhizobium negatively affected bean growth, indicating the importance of selecting the right combination of bean genotype and microbial species.

The interactions between soil microbes have been reviewed by different researchers. For example, Gopal et al. (2012) reviewed the interactions between mycorrhizal fungi and soil bacteria and the interactions between mycorrhizal fungi and spore-associated bacteria. The authors accordingly indicated that a more clear understating on the interaction among mycorrhizal fungi, spore-associated bacteria, and PGPR can enhance the quality of inoculums. Franzini et al. (2013) suggested that although the interactions between mycorrhizal fungi and PGPR are usually positive enhancing the legume growth and yield, under drought stress such kind of interactions may negatively affect legume growth and yield. Accordingly, in their experiment they evaluated the combined effects of six rhizobium strains and two mycorrhizal species on the growth of *Phaseolus vulgaris* and *Zea mays* under moderate drought conditions. Their results indicated that the combined combinations of rhizobium and mycorrhizal fungi in some cases decreased the growth of bean and corn under moderate drought conditions.

Abdel-Rahman et al. (2011) investigated the effects of single or combined inoculation with mycorrhizal fungi and *Bacillus subtilis* on the growth, oil (yield and percentage), and nutrient uptake of three different genotypes of sweet basil under the salinity levels of 0, 1000, 2000, and 4000 mg/l. The results indicated that high salinity level (4000 mg/l) significantly decreased plant growth, oil, and the nutrient uptake of N, P, and K of all the genotypes. The salinity treatment also highly increased the concentration of sodium in the plants. The effects of mycorrhizal fungi on the growth of plant under salinity stress were more than the bacteria and the combined inoculation intensified such a positive response. The response of genotypes was different under the stress. The authors accordingly indicated that it is possible to enhance the response of sweet basil to salinity stress using the single or combined use of mycorrhizal fungi and *Bacillus subtilis*.

Saia et al. (2015) investigated the single and the combined effects of mycorrhizal fungi and PGPR on the metabolic activities of durum wheat roots under N-limited and P-high conditions. The soil natural conditions were used as the control treatment. Inoculation with AM fungi highly colonized crop roots and reduced the concentrations of metabolic compounds including amino acids and saturated fatty acids in the roots. However, the combined inoculation with the fungi and PGPR increased the concentration of such compounds compared with the single inoculation with AM fungi. The authors attributed such a difference to the mineralizing role of PGPR on the organic matter and hence the release of N by PGPR for the use of the host plant compared with the single inoculation with mycorrhizal fungi.

The above details indicate how the soil microbes including mycorrhizal fungi and PGPR may interact under stress affecting plant growth and yield production. However, the other important point, which must be researched in great details, is how such microbes may interact with endophytic bacteria residing in plant roots. If such a case is indicated, it will be possible to use the combined use of soil microbes, including the endophytic microbes for inoculum production. Accordingly, the production of inoculums will be more beneficial under different conditions including stress.

2.8 Conclusion and Future Perspectives

Some important details related to the effects of soil microbes on plant growth and yield production have been indicated. The role of the most important soil microbes including mycorrhizal fungi, PGPR, and endophytic microbes on the growth of crop plants under stress has been presented. Accordingly, research work has indicated that the single use of soil microbes can positively affect plant growth. Some details are also available on the combined use of soil microbes, specifically mycorrhizal fungi, PGPR, and rhizobium affecting the growth of crop plants. However, the other important point, which must be researched in greater details, is the interactions of soil microbes with the endophytic bacteria residing in plant roots. If such details are illustrated, it will be possible to produce microbial inoculums, which are more effective on plant growth and yield production under different conditions including stress.

References

Abdel-Rahman S, Abdel-Kader A, Khalil S (2011) Response of three sweet basil cultivars to inoculation with Bacillus subtilis and arbuscular mycorrhizal fungi under salt stress conditions. Nat Sci 9:93–111

Adesemoye A, Torbert H, Kloepper J (2009) Plant growth-promoting rhizobacteria allow reduced application rates of chemical fertilizers. Microb Ecol 58:921–929

Ahmed E, Hassan E, El Tobgy K, Ramadan E (2014) Evaluation of rhizobacteria of some medicinal plants for plant growth promotion and biological control. Ann Agric Sci 59:273–280

Alami Y, Achouak W, Marol C, Heulin T (2000) Rhizosphere soil aggregation and plant growth promotion of sunflowers by exopolysaccharide producing Rhizobium sp. strain isolated from sunflower roots. Appl Environ Microbiol 66:3393–3398

Ali S, Sandhya V, Grover M, Kishore N, Rao L, Venkateswarlu B (2009) *Pseudomonas* sp. strain AKM-P6 enhances tolerance of sorghum seedlings to elevated temperatures. Biol Fertil Soils 46:45–55

Al-Karaki GN (2000) Growth of mycorrhizal tomato and mineral acquisition under salt stress. Mycorrhiza 10:51–54

Amijee F, Tinker P, Stribley D (1989) The development of endomycorrhizal root systems. VII. A detailed study of effects of soil phosphorus on colonization. New Phytol 111:435–446

Arkhipova T, Veselov S, Melentiev A, Martynenko E, Kudoyarova G (2005) Ability of bacterium *Bacillus subtilis* to produce cytokinins and to influence the growth and endogenous hormone content of lettuce plants. Plant Soil 272:201–209

Arzanesh MH, Alikhani HA, Khavazi K, Rahimian HA, Miransari M (2011) Wheat (*Triticum aestivum* L.) growth enhancement by Azospirillum spp. under drought stress. World J Microbiol Biotechnol 27:197–205

Audet P, Charest C (2007) Dynamics of arbuscular mycorrhizal symbiosis in heavy metal phytoremediation: meta-analytical and conceptual perspectives. Environ Pollut 147:609–614

Bano A, Fatima M (2009) Salt tolerance in *Zea mays* (L.) following inoculation with *Rhizobium* and *Pseudomonas*. Biol Fertil Soils 45:405–413

Barac T, Taghavi S, Borremans B, Provoost A, Oeyen L, Colpaert JV, Vangronsveld J, van der Lelie D (2004) Engineered endophytic bacteria improve phytoremediation of water-soluble, volatile, organic pollutants. Nat Biotechnol 22:583–588

Bashan Y, de Bashan L, Prabhu R, Hernandez J (2014) Advances in plant growth-promoting bacterial inoculant technology: formulations and practical perspectives (1998–2013). Plant Soil 378:1–33

Bauer W, Mathesius U (2004) Plant responses to bacterial quorum sensing signals. Curr Opin Plant Biol 7:429–433

Bell T, Newman J, Silverman B, Turner S, Lilley A (2005) The contribution of species richness and composition to bacterial services. Nature 436:1157–1160

Ben Khaled L, Gomez AM, Ourraqi EM, Oihabi A (2003) Physiological and biochemical responses to salt stress of mycorrhized and/or nodulated clover seedlings (*Trifolium alexandrinum* L.) Agronomie 23:571–580

Berg G (2009) Plant–microbe interactions promoting plant growth and health: perspectives for controlled use of microorganisms in agriculture. Appl Microbiol Biotechnol 84:11–18

Bisseling T, Dangl JL, Schulze-Lefert P (2009) Next-generation communication. Science 324:691–691

Bonkowski M, Roy J (2005) Soil microbial diversity and soil functioning affect competition among grasses in experimental microcosms. Oecologia 143:232–240

Bunn R, Lekberg Y, Zabinski C (2009) Arbuscular mycorrhizal fungi ameliorate temperature stress in thermophilic plants. Ecology 90:1378–1388

Calvo P, Nelson L, Kloepper J (2014) Agricultural uses of plant biostimulants. Plant Soil 383:3–41

Carvalho L, Correia P, Caçador I, Martins-Loução M (2003) Effects of salinity and flooding on the infectivity of salt marsh arbuscular mycorrhizal fungi in *Aster tripolium* L. Biol Fertil Soils 38:137–143

Cassan F, Vanderleyden J, Spaepen S (2014) Physiological and agronomical aspects of phytohormone production by model plant-growth-promoting rhizobacteria (PGPR) belonging to the genus Azospirillum. J Plant Growth Regul 33:440–459

Chakraborty U, Chakraborty B, Dey P Chakraborty AP (2015) Role of microorganisms in alleviation of abiotic stresses for sustainable agriculture. In: Chakraborty U, Chakraborty B (eds.) Abiotic stresses in crop plants, ISBN: 9781780643731, CAB International, Wallingford, pp 232

Chen M, Wei H, Cao J, Liu R, Wang Y, Zheng C (2007) Expression of *Bacillus subtilis pro*AB genes and reduction of feedback inhibition of proline synthesis increases proline production and confers osmotolerance in transgenic *Arabdopsis*. J Biochem Mol Biol 40:396–403

Colla G, Rouphael Y, Cardarelli M, Tullio M, Rivera CM, Rea E (2008) Alleviation of salt stress by arbuscular mycorrhizal in zucchini plants grown at low and high phosphorus concentration. Biol Fertil Soils 44:501–509

Daei G, Ardekani MR, Rejali F, Teimuri S, Miransari M (2009) Alleviation of salinity stress on wheat yield, yield components, and nutrient uptake using arbuscular mycorrhizal fungi under field conditions. J Plant Physiol 166:617–625

Dastager G, Deepa C, Pandey A (2010) Isolation and characterization of novel plant growth promoting micrococcus sp NII-0909 and its interaction with cowpea. Plant Physiol Biochem 48:987–992

Degens BP (1998) Decreases in microbial functional diversity do not result in corresponding changes in decomposition under different moisture conditions. Soil Biol Biochem 30:1989–2000

De Vleesschauwer D, Höfte M (2009) Rhizobacteria-induced systemic resistance. In: Van Loon L (ed) Plant innate immunity. Academic Press Ltd./Elsevier Science Ltd., London, pp 223–281

Dimkpa C, Weinand T, Asch F (2009) Plant–rhizobacteria interactions alleviate abiotic stress conditions. Plant Cell Environ 32:1682–1694

Döbbelaere S, Croonenborghs A, Thys A, Vande Broek A, Vanderleyden J (1999) Phytostimulatory effect of *Azospirillum brasilense* wild type and mutant strains altered in IAA production on wheat. Plant Soil 212:153–162

Egamberdieva D, Kucharova Z (2009) Selection for root colonising bacteria stimulating wheat growth in saline soils. Biol Fertil Soils 45:563–571

Eichmann R, Schäfer P (2015) Growth versus immunity-a redirection of the cell cycle? Curr Opin Plant Biol 26:106–112

FAO, 2015. www.fao.org

Fiedler H, Krastel P, Müller J, Gebhardt K, Zeeck A (2001) Enterobactin: the characteristic catecholate siderophore of Enterobacteriaceae is produced by Streptomyces species. FEMS Microbiol Lett 196:147–151

Flemming H, Wingender J (2010) The biofilm matrix. Nat Rev Microbiol 8:623–633

Flessa H, Ruser R, Dörsch P, Kamp T, Jimenez MA, Munch JC, Beese F (2002) Integrated evaluation of greenhouse gas emissions (CO_2, CH_4, N_2O) from two farming systems in southern Germany. Agric Ecosyst Environ 91:175–189

Franzini VI, Azcon R, Mendes FL, Aroca R (2010) Interactions between Glomus species and Rhizobium strains affect the nutritional physiology of drought-stressed legume hosts. J Plant Physiol 167:614–619

Franzini VI, Azcon R, Mendes FL, Aroca R (2013) Different interaction among Glomus and Rhizobium species on *Phaseolus vulgaris* and *Zea mays* plant growth, physiology and symbiotic development under moderate drought stress conditions. Plant Growth Regul 70:265–273

Garg N, Chandel S (2010) Arbuscular mycorrhizal networks: process and functions. A review. Agron Sustain Dev 30:581–599

Garg N, Singla P (2016) Stimulation of nitrogen fixation and trehalose biosynthesis by naringenin (Nar) and arbuscular mycorrhiza (AM) in chickpea under salinity stress. Plant Growth Regul. doi:10.1007/s10725-016-0174-y

Giovannetti M, Avio L, Fortuna P, Pellegrino E, Sbrana C, Strani P (2006) At the root of the wood wide web. Self-recognition and nonself incompatibility in mycorrhizal networks. Plant Signal Behav 1:1–5

Glick BR (ed.) (2015) Beneficial plant-bacterial interactions. Springer International Publishing, Switzerland

Gopal S, Chandrasekaran M, Shagol C, Kim KY, Sa TM (2012) Spore associated bacteria (sab) of arbuscular mycorrhizal fungi (amf) and plant growth promoting rhizobacteria (pgpr) increase nutrient uptake and plant growth under stress conditions. Korean J Soil Sci Fert 45:582–592

Grover M, Ali S, Sandhya V, Rasul A, Venkateswarlu B (2011) Role of microorganisms in adaptation of agriculture crops to abiotic stresses. World J Microbiol Biotechnol 27:1231–1240

Gutierrez-Manero F, Ramos-Solano B, Probanza A, Mehouachi J, Tadeo F, Talon M (2001) The plant-growth promoting rhizobacteria *Bacillus pumilus* and *Bacillus licheniformis* produce high amounts of physiologically active gibberellins. Physiol Plant 111:206–211

Haichar E-ZF, Santaella C, Heulin T, Achouak W (2014) Root exudates mediated interactions belowground. Soil Biol Biochem 77:69–80

Hamaoui B, Abbadi J, Burdman S, Rashid A, Sarig S, Okon Y (2001) Effects of inoculation with *Azospirillum brasilense* on chickpeas (*Cicer arietinum*) and faba beans (*Vicia faba*) under different growth conditions. Agronomie 21:553–560

Hartmann A, Bashan Y (2009) Ecology and application of *Azospirillum* and other plant growth-promoting bacteria (PGPB)-special issue. Eur J Soil Biol 45:1–2

Hoseinzade H, Ardakani MR, Shahdi A, Rahmani HA, Noormohammadi G, Miransari M (2016) Rice (*Oryza sativa* L.) nutrient management using mycorrhizal fungi and endophytic *Herbaspirillum seropedicae*. J Integr Agric 5:1385–1394

Hsieh T, Huang H, Erickson R (2010) Spread of seed-borne *Erwinia rhapontici* in bean, pea and wheat. Eur J Plant Pathol 127:579–584

Huang X, Chaparro J, Reardon K, Zhang R, Shen Q, Vivanco J (2014) Rhizosphere interactions: root exudates, microbes, and microbial communities. Botany 92:267–275

Imbert M, Béchet M, Blondeau R (1995) Comparison of the main siderophores produced by some species of Streptomyces. Curr Microbiol 31:129–133

Jalili F, Khavazi K, Pazira E, Nejati A, Rahmani HA, Sadaghiani HR, Miransari M (2009) Isolation and characterization of ACC deaminase-producing fluorescent pseudomonads, to alleviate salinity stress on canola (*Brassica napus* L.) growth. J Plant Physiol 166:667–674

Kalita M, Bharadwaz M, Dey T, Gogoi K, Dowarah P, Unni BG, Ozah D, Saikia I (2015) Developing novel bacterial based bioformulation having PGPR properties for enhanced production of agricultural crops. Indian J Exp Biol 53:56–60

Kang S, Khan A, Hamayun M, Hussain J, Joo G, You Y, Kim J, Lee I (2012) Gibberellin-producing Promicromonospora sp. SE188 improves *Solanum lycopersicum* plant growth and influences endogenous plant hormones. J Microbiol 50:902–909

Kaymak H (2010) Potential of PGPR in agricultural innovations. In: Maheshwari, Dinesh K (eds) Plant growth and health promoting bacteria. Springer, Berlin/Heidelberg, pp 45–79

Khan AG (2005) Role of soil microbes in the rhizospheres of plants growing on trace metal contaminated soils in phytoremediation. J Trace Elem Med Biol 18:355–364

Kim K, Jang YJ, Lee SM, Oh BT, Chae JC, Lee KJ (2014) Alleviation of salt stress by *Enterobacter* sp. EJ01 in tomato and *Arabidopsis* is accompanied by up-regulation of conserved salinity responsive factors in plants. Mol Cells 37:109–117

Kohler J, Hernández JA, Caravaca F, Roldán A (2008) Plant-growth-promoting rhizobacteria and arbuscular mycorrhizal fungi modify alleviation biochemical mechanisms in water stressed plants. Funct Plant Biol 35:141–151

Kohler J, Caravaca F, Roldán A (2010) An AM fungus and a PGPR intensify the adverse effects of salinity on the stability of rhizosphere soil aggregates of *Lactuca sativa*. Soil Biol Biochem 42:429–434

Koide R, Li M (1990) On host regulation of the vesicular-arbuscular mycorrhizal symbiosis. New Phytol 114:59–74

Koide R (1991) Tansley review no. 29: nutrient supply, nutrient demand, and plant response to mycorrhizal infection. New Phytol 117:365–386

Kpomblekou K, Tabatabai M (1994) Effect of organic acids on release of phosphorus from phosphate rocks. Soil Sci Soc Am J 158:442–453

Lau J, Lennon J (2011) Evolutionary ecology of plant–microbe interactions: soil microbial structure alters selection on plant traits. New Phytol 192:215–224

Liu A, Hamel C, Hamilton RI, Ma BL, Smith DL (2000) Acquisition of Cu, Zn, Mn and Fe by mycorrhizal maize (*Zea mays* L.) grown in soil at different P and micronutrient levels. Mycorrhiza 9:331–336

Lopez-Raez JA (2015) How drought and salinity affect arbuscular mycorrhizal symbiosis and strigolactone biosynthesis? Planta 243(6):1375–1385

Lugtenberg B, Kamilova F (2009) Plant-growth-promoting rhizobacteria. Annu Rev Microbiol 63:541–556

Ma J, Li XL, Xu H, Han Y, Cai Z, Yagi K (2007) Effects of nitrogen fertilizer and wheat straw application on CH_4 and N_2O emissions from a paddy rice field. Aust J Soil Res 45:359–367

Mabrouk Y, Belhadj O (2016) Enhancing the biological nitrogen fixation of leguminous crops grown under stressed environments. Afr J Biotechnol 11:10809–10815

Malinowski D, Belesky D (2000) Adaptations of endophyte-infected cool-season grasses to environmental stresses: mechanisms of drought and mineral stress tolerance. Crop Sci 40:923–940

Mantelin S, Touraine B (2004) Plant growth-promoting bacteria and nitrate availability: impacts on root development and nitrate uptake. J Exp Bot 55:27–34

Marschner P, Rumberger A (2004) Rapid changes in the rhizosphere bacterial community structure during re-colonization of sterilized soil. Biol Fertil Soils 40:1–6

Marschner P, Crowley D, Rengel Z (2011) Rhizosphere interactions between microorganisms and plants govern iron and phosphorus acquisition along the root axis model and research methods. Soil Biol Biochem 43:883–894

Martínez O, Jorquera M, Crowley D, de la Luz MM (2011) Influence of nitrogen fertilisation on pasture culturable rhizobacteria occurrence and the role of environmental factors on their potential PGPR activities. Biol Fertil Soils 47:875–885

Meister R, Rajani MS, Ruzicka D, Schachtman DP (2014) Challenges of modifying root traits in crops for agriculture. Trends Plant Sci 19:779–788

Miransari M, Bahrami HA, Rejali F, Malakouti MJ (2008) Using arbuscular mycorrhiza to alleviate the stress of soil compaction on wheat (*Triticum aestivum* L.) growth. Soil Biol Biochem 40:1197–1206

Miransari M, Bahrami HA, Rejali F, Malakouti MJ (2009) Effects of soil compaction and arbuscular mycorrhiza on corn (*Zea mays* L.) nutrient uptake. Soil Tillage Res 103:282–290

Miransari M (2010) Contribution of arbuscular mycorrhizal symbiosis to plant growth under different types of soil stress. Plant Biol 12:563–569

Miransari M (2011a) Interactions between arbuscular mycorrhizal fungi and soil bacteria. Appl Microbiol Biotechnol 89:917–930

Miransari M (2011b) Soil microbes and plant fertilization. Appl Microbiol Biotechnol 92:875–885

Miransari M (2011c) Hyperaccumulators, arbuscular mycorrhizal fungi and stress of heavy metals. Biotechnol Adv 29:645–653

Miransari M (2013) Soil microbes and the availability of soil nutrients. Acta Physiologiae Plan 35:3075–3084

Miransari M, Riahi H, Eftekhar F, Minaie A, Smith DL (2013) Improving soybean (*Glycine max* L.) N_2 fixation under stress. J Plant Growth Regul 32:909–921

Miransari M (2014) Plant growth promoting rhizobacteria. J Plant Nutr 37:2227–2235

Miransari M, Mackenzie AF (2015) Development of soil N testing for wheat production using soil residual mineral N. J Plant Nutr 38:1995–2005

Muthukumar T, Priyadharsini P, Uma E, Jaison S, Pandey R (2014) Role of arbuscular mycorrhizal fungi in alleviation of acidity stress on plant growth. In: Miransari M (ed) Use of microbes for the alleviation of soil stresses, vol 1. Springer, New York, pp 43–71

Naveed M, Mitter B, Yousaf S, Pastar M, Afzal M, Sessitsch A (2014a) The endophyte *Enterobacter* sp. FD17: a maize enhancer selected based on rigorous testing of plant beneficial traits and colonization characteristics. Biol Fertil Soils 50:249–262

Naveed M, Mitter B, Reichenauer TG, Wieczorek K, Sessitsch A (2014b) Increased drought stress resilience of maize through endophytic colonization by *Burkholderia phytofirmans* PsJN and Enterobacter sp. FD17. Environ Exp Bot 97:30–39

Newton A, Fitt B, Atkins S, Walters D, Daniell T (2010) Pathogenesis, parasitism and mutualism in the trophic space of microbe–plant interactions. Trends Microbiol 18:365–373

Park M, Kim C, Yang J, Lee H, Shin W, Kim S, Sa T (2005) Isolation and characterization of diazotrophic growth promoting bacteria from rhizosphere of agricultural crops of Korea. Microbiol Res 160:127–133

Paul D, Nair S (2008) Stress adaptations in a plant growth promoting rhizobacterium (PGPR) with increasing salinity in the coastal agricultural soils. J Basic Microbiol 48:378–384

Paul D, Lade H (2014) Plant-growth-promoting rhizobacteria to improve crop growth in saline soils: a review. Agron Sustain Dev 34:737–752

Pereg L, McMillan M (2015) Scoping the potential uses of beneficial microorganisms for increasing productivity in cotton cropping systems. Soil Biol Biochem 80:349–358

Piccoli P, Lucangeli D, Schneider G, Bottini R (1997) Hydrolysis of [17,17-2H2]Gibberellin A20-Glucoside and [17, 17-2H2]Gibberellin A20-glucosyl ester by *Azospirillum lipoferum* cultured in a nitrogen free biotin based chemically-defined medium. Plant Growth Regul 23:179–182

Pozo M, Azcon-Aguilar C (2007) Unraveling mycorrhiza-induced resistance. Curr Opin Plant Biol 10:393–398

Raju P, Clark R, Ellis J, Maranville J (1988) Effects of VA mycorrhizae on growth and mineral uptake of sorghum grown at varied levels of soil acidity. Commun Soil Sci Plant Analysis 19:919–931

Ramegowda V, Senthil-Kumar M (2015) The interactive effects of simultaneous biotic and abiotic stresses on plants: mechanistic understanding from drought and pathogen combination. J Plant Physiol 176:47–54

Rutto K, Mizutani F, Kadoya K (2002) Effect of root-zone flooding on mycorrhizal and non-mycorrhizal peach (*Prunus persica* Batsch) seedlings. Sci Hortic 94:285–295

Sandhya V, Ask Z, Grover M, Reddy G, Venkateswarlu B (2009a) Alleviation of drought stress effects in sunflower seedlings by the exopolysaccharides producing *Pseudomonas putida* strain GAPP45. Biol Fertil Soils 46:17–26

Sandhya V, Ali S, Grover M, Kishore N, Venkateswarlu B (2009b) *Pseudomonas* sp. strain P45 protects sunflowers seedlings from drought stress through improved soil structure. J Oilseed Res 26:600–601

Saia S, Ruisi P, Fileccia V, Di Miceli G, Amato G, Martinelli F (2015) Metabolomics suggests that soil inoculation with arbuscular mycorrhizal fungi decreased free amino acid content in roots of durum wheat grown under N-limited, P-rich field conditions. PLoS One 10:e0129591

Saubidet M, Fatta N, Barneix A (2000) The effects of inoculation with *Azospirillum brasilense* on growth and nitrogen utilization by wheat plants. Plant Soil 245:215–222

Sayyed R, Badgujar M, Sonawane H, Mhaske M, Chincholkar S (2005) Production of microbial iron chelators (siderophores) by fluorescent pseudomonads. Indian J Biotechnol 4: 484–490

Schenk P, Carvalhais L, Kazan K (2012) Unraveling plant–microbe interactions: can multi-species transcriptomics help? Trends Biotechnol 30:177–184

Sheng X, He L (2006) Solubilization of potassium-bearing minerals by a wild-type strain of *Bacillus edaphicus* and its mutants and increased potassium uptake by wheat. Can J Microbiol 52:66–72

Singh G, Biswas D, Marwaha T (2010) Mobilization of potassium from waste mica by plant growth promoting rhizobacteria and its assimilation by maize (*Zea mays*) and wheat (*Triticum aestivum* L.): a hydroponics study under phytotron growth chamber. J Plant Nutr 33:1236–1251

Singh B, Satyanarayana T (2011) Microbial phytases in phosphorus acquisition and plant growth promotion. Physiol Mol Biol Plants 17:93–103

Singh J, Pandey V, Singh D (2011) Efficient soil microorganisms: a new dimension for sustainable agriculture and environmental development. Agric Ecosyst Environ 140:339–353

Smith S, Facelli E, Pope S, Smith F (2010) Plant performance in stressful environments: interpreting new and established knowledge of the roles of arbuscular mycorrhizas. Plant Soil 326:3–20

Spaepen S, Dobbelaere S, Croonenborghs A, Vanderleyden J (2008) Effects of *Azospirillum brasilense* indole-3-acetic acid production on inoculated wheat plants. Plant Soil 312:15–23

Spaepen S, Vanderleyden J, Okon Y (2009) Plant growth promoting actions of rhizobacteria. Adv Bot Res 51:283–320

Stewart L, Hamel C, Hogue R, Moutoglis P (2005) Response of strawberry to inoculation with arbuscular mycorrhizal fungi under very high soil phosphorus conditions. Mycorrhiza 15:612–619

Suarez R, Wong A, Ramirez M, Barraza A, OrozcoMdel C, Cevallos MA, Lara M, Hernandez G, Iturriaga G (2008) Improvement of drought tolerance and grain yield in common bean by over-expressing trehalose-6-phosphate synthase in rhizobia. Mol Plant-Microbe Interact 21:958–996

Subramanian K, Santhanakrishnan P, Balasubramanian P (2006) Responses of field grown tomato plants to arbuscular mycorrhizal fungal colonization under varying intensities of drought stress. Sci Hortic 107:245–253

Street T, Bolen D, Rose G (2006) A molecular mechanism for osmolyte-induced protein stability. Proc Natl Acad Sci U S A 103:13997–14002

Talaat N, Shawky B (2014) Protective effects of arbuscular mycorrhizal fungi on wheat (*Triticum aestivum* L.) plants exposed to salinity. Environ Exp Bot 98:20–31

Temirov Y, Esikova T, Kashparov IA (2003) A catecholic siderophore produced by the thermoresistant Bacillus licheniformis VK21 strain. Russ J Bioorg Chem 29:542–549

Vosatka M, Batkhuugyin E, Albrechtova J (1999) Response of three arbuscular mycorrhizal fungi to simulated acid rain and aluminium stress. Biol Plant 42:289–296

Waqas M, Khan AL, Kang SM, Kim YH, Lee IJ (2014) Phytohormone-producing fungal endophytes and hardwood-derived biochar interact to ameliorate heavy metal stress in soybeans. Biol Fertil Soils 50:1155–1167

Wu Q, Xia R (2006) Arbuscular mycorrhizal fungi influence growth, osmotic adjustment and photosynthesis of citrus under well-watered and water stress conditions. J Plant Physiol 163:417–425

Wu Q, Xia R, Zou Y (2008) Improved soil structure and citrus growth after inoculation with three arbuscular mycorrhizal fungi under drought stress. Eur J Soil Biol 44:122–128

Yasmin S, Bakar MAR, Malik KA, Hafeez FY (2004) Isolation, characterization and beneficial effects of rice associated plant growth promoting bacteria from Zanzibar soils. J Basic Microbiol 3:241–252

Yu XM, Ai CX, Xin L, Zhou GF (2011) The siderophore-producing bacterium, *Bacillus subtilis* CAS15, has a biocontrol effect on Fusarium wilt and promotes the growth of pepper. Eur J Soil Biol 47:138–145

Yuan ZL, Zhang CL, Lin FC (2010) Role of diverse non-systemic fungal endophytes in plant performance and response to stress: progress and approaches. J Plant Growth Regul 29:116–126

Zamioudis C, Pieterse C (2012) Modulation of host immunity by beneficial microbes. Mol Plant-Microbe Interact 25:139–150

Zhang N, Wang D, Liu Y, Li S, Shen Q, Zhang R (2014) Effects of different plant root exudates and their organic acid components on chemotaxis, biofilm formation and colonization by beneficial rhizosphere-associated bacterial strains. Plant Soil 374:689–700

Zhu X, Song F, Xu H (2010) Arbuscular mycorrhizae improves low temperature stress in maize via alterations in host water status and photosynthesis. Plant Soil 331:129–137

Phosphate-Solubilizing Microorganisms in Sustainable Production of Wheat: Current Perspective

3

Mohammed Saghir Khan, Asfa Rizvi, Saima Saif, and Almas Zaidi

Abstract

In terms of global production, wheat among cereals ranks third after rice and maize, contributing about 35% of the total food grain production. Wheat due to high nutritional value is considered one of the important dietary constituents and, hence, has become one of the better food choices around the world. For growth and development, wheat requires large amounts of major plant nutrients especially phosphorus (P). Application of sufficient amounts of P has many beneficial impacts on wheat including its role in growth, grain formation, and development, and in straw yield. Phosphorus deficiency, however, may adversely affect the growth and, therefore, hampers the physiological processes leading eventually to overall stunting of the plant. In order to circumvent the phosphorus problems and hence to achieve optimum yields, wheat growers usually apply excessive amounts of chemical phosphatic fertilizer which is both expensive and destructive to soil fertility. To overcome these problems, a physiologically versatile array of microorganisms especially belonging to phosphate-solubilizing group has been introduced into the agricultural system for improving wheat production. The P-solubilizing microorganisms (PSM) solubilize unavailable soil P and make it available for uptake by plants. The use of microbial phosphatic fertilizer (microphos) in wheat production system is considered an eco-friendly strategy without adversely affecting the soil health. Despite numerous informations available on the impact of P-solubilizing microorganisms on various plants, literature suggesting the use of PSM in wheat production is limited. Realizing the importance of PSM in enhancing the overall performance of wheat, attempt has been made to better understand as to how the PSM affects wheat production in variable agricultural practices. Also, efforts will be made to find PSM which could be

M.S. Khan (✉) • A. Rizvi • S. Saif • A. Zaidi
Department of Agricultural Microbiology, Faculty of Agricultural Sciences, Aligarh Muslim University, Aligarh 202002, Uttar Pradesh, India
e-mail: khanms17@rediffmail.com

V. Kumar et al. (eds.), *Probiotics in Agroecosystem*,
DOI 10.1007/978-981-10-4059-7_3

applied to facilitate the growth and development of wheat grown in different agroecological niches. Constant and sustainable application of PSM is expected to decrease the use of fertilizers in wheat production strategies.

3.1 Introduction

Wheat (*Triticum aestivum*) is one of the important cereal crops which is extensively cultivated and used as food worldwide. Wheat is cultivated in variable climatic conditions that range between 47°S and 57°N latitudes on different soils including sandy and clayey soil and has the highest adaptation among all the crop species (Naresh et al. 2014). Globally, about 680 million tons of wheat was produced during 2008–2012, while in 2011, the production rate reached to almost 700 million tons (FAO Stat http://faostat.fao.org/site/291/default.aspx). Moreover, the statistics reveal that more than 600 million tons of wheat is harvested annually. Among various wheat-producing countries, India is the second largest wheat-producing country with 11.9% production from approximately 12% of total area (USDA 2010). In India, wheat is cultivated in about 30 million hectares land resulting in 93 million tons yield. The national productivity is estimated as 2.98 tons ha^{-1}. However, despite being an important food crop of the country, the average production of wheat is slowly dwindling due to several reasons (Ray et al. 2012). Chief among them has been the declining cultivable lands, fluctuating environmental conditions (global warming) and excessive usage of chemical fertilizers in order to obtain maximum yields.

The consistently increasing cost of chemical fertilizers and their deleterious impact on soil fertility and human health (via food chain) are some of the vital problems of farmers growing wheat across different regions (Eman et al. 2008; Singh et al. 2008). In order to reduce the use of fertilizers in wheat production, scientists require searching for some other inexpensive alternatives. In this regard, biofertilizers, defined as a biological product containing living microorganisms which, when applied to seed, plant surfaces, or soil, colonize the rhizosphere or the interior of the plant and facilitate growth by enhancing the supply or availability of primary nutrients to the host plant (Vessey 2003), involving PSM that have offered a better alternative to the expensive and environmentally unfriendly fertilizers in wheat production (Sarker et al. 2014; Sharma et al. 2013). The chief benefit of applying PSM in wheat cultivation is their ability to better colonize and establish in the rhizosphere and to make P available to plants constantly (Sharma 2002). Apart from P, PSM also facilitates the growth of plants by providing gibberellins, cytokinins, and IAA (Priya et al. 2013; Sharma et al. 2012), improves uptake of water and nutrients (Abbasniayzare et al. 2012; Khan et al. 2007), secretes antibiotics and other toxic products (Shanmugam et al. 2011; Lipping et al. 2008), and supplies vitamins belonging to β-group (Revillas et al. 2000). Also, PSM improves the wheat growth by inhibiting plant pathogens (Salma et al. 2014). For instance, P-solubilizing bacteria like *Pseudomonas* and *Bacillus* species when used alone or in combination profoundly increased grain yield, tiller formation, and seed P of wheat (Afzal et al.

2005). Similarly, the pre-sowing application of PSM inoculated wheat resulted in a considerable increase in yield relative to uninoculated wheat (Dwivedi et al. 2004). These and other similar studies suggest that the sole or combined application of PSM could be used to enhance the overall performance of wheat.

3.2 Nutritional Importance of Wheat

Wheat is one of the main staple food crops for majority of the world population and is the major staple food in many Asian countries. Wheat is an important cereal crop of human dietary systems all over the world. Nutritionally, it is highly rich in carbohydrate and serves as a good source of energy. Interestingly, proteins and fibers are also the major constituents of wheat composition. In addition, wheat also contains significant amounts of vitamins, minerals, lipids, and a few phytochemicals. The nutrient composition of wheat is represented in Table 3.1.

Table 3.1 Dietary components of winter wheat

Energy	1368 KJ (327 k Cal)
Carbohydrates	71.18 g
Sugars	0.41
Dietary fiber	12.2 g
Fat	1.54 g
Proteins	12.6 g
Vitamins	
Thiamine (B1)	33% (0.383 mg)
Riboflavin (B2)	10% (0.115 mg)
Niacin (B3)	36% (5.464 mg)
Pantothenic acid (B5)	19% (0.954 mg)
Vitamin B6	23% (0.3 mg)
Folate (B9)	10% (38 μg)
Choline	6% (31.2 mg)
Vit. E	7% (1.01 mg)
Vit. K	2% (1.9 mg)
Minerals	
Calcium	3% (29 mg)
Iron	25% (3.19 mg)
Magnesium	35% (126 mg)
Manganese	190% (3.985 mg)
Phosphorus	41% (288 mg)
Potassium	8% (363 mg)
Sodium	0% (2 mg)
Zinc	28% (2.65 mg)
Other constituents	
Selenium	70.7 μg

Source: USDA nutrient database

3.3 Role of Phosphorus in Promoting Wheat Growth

Plants in general require variety of nutrient elements for survival and growth. These elements are categorized as macro- and micronutrients depending upon the requirement of various crop plants. Phosphorus among plant nutrients is required in larger quantities by plants, the deficiency of which severely affects the whole metabolism of many plants including wheat. Phosphorus plays an important part in photosynthesis, energy transfer, utilization of sugar and starch, nucleus formation and cell division, signal transduction, macromolecular biosynthesis and respiration (Khan et al. 2010), and N_2 fixation (Wibisono et al. 2015). It also initiates root growth and development and maintains the overall health of the plants.

In wheat, P plays a prime role in strengthening of the straw and results in better fruit production (Anonymous 1988). Presence of P also accounts for a better tillering in wheat, promotes early maturation of plant, and assists seed formation. For example, in a study, the combined effect of irrigation and phosphorus demonstrated a positive impact on the developmental stages of wheat. Furthermore, an increase in the number and weight of the grains was recorded. Topical application of P further increased the size of wheat grains (Hossain et al. 1996; Turk and Tawaha 2002). Phosphorus fertilization is therefore a major input in crop production across different regions, because some soils lack sufficient P to optimize crop quality and yields (Griffith 2009). Effective nutrient management, hence, requires that P be available in adequate amounts when needed by the plant (Sarfraz et al. 2009). In order to supply available and soluble form of P to wheat plants, a group of soil microorganisms collectively called as phosphate-solubilizing microorganisms (PSM) have been applied. To substantiate this further, Islam and Hossain (2012) proposed that the use of synthetic phosphatic fertilizer could be reduced if the insoluble soil P is solubilized naturally by PSM and is made available to plants. In some other experiments, P-solubilizing bacteria (PSB) when applied as bioinoculants have been shown to solubilize the fixed soil P, thereby improving the crop yield (Gull et al. 2004) including those of wheat (Afzal and Asghari 2008). In addition to the sole application of PSM, there has been considerable reports where PSM in synergism with other organisms, for example, AM fungi, has been found to enhance uptake of solubilized soil P (Barea et al. 2002) and concurrently increased nutrient uptake and yield of wheat and maize (Raja et al. 2002).

3.4 Rationale for Using Phosphate-Solubilizing Microorganisms in Wheat Production

Phosphorus, one of the major plant nutrients, affects many stages of plant growth and enhances grain yield and yield components. On the contrary, in many agricultural production systems, P has been identified as the most deficient essential nutrient after N. And hence, nutrient supply to agronomic production systems has increased to achieve optimum yields in order to sustain the growing populations demand around the world. When soil is deficient in available P, phosphatic fertilizers are applied to soils. Although inorganic fertilizers are readily available, they are slowly converted to unavailable forms due to precipitation/complexation. Among cereals, wheat requires more nutrients than other crops. Worldwide, the commercial production of fertilizers needs substantial amounts of energy, and, hence, it becomes costly. Moreover, phosphatic fertilizers are consistently used in wheat production to achieve higher yields. The excessive and continued application of phosphatic fertilizers, however, destruct the soil fertility (Younis et al. 2013). High dose of P fertilizers causes abrupt shoot growth, while it limits root growth. Also, following accumulation within soils, P can pollute the ground water resources. Protein digestion inhibitors deposited in plant cell vacuoles were not taken up by sucking herbivores but hampered the chewing herbivores (Mattson 1980). Moreover, the uptake of P by plants is considerably low due to its rapid fixation with Fe and Al oxides in acidic soils (Goldstein 1986; Norrish and Rosser 1983) and calcium in neutral or calcareous soils (Lindsay et al. 1989). Due to these, approximately 75–90% of the P fertilizers applied to soil are lost, and, hence, plants generally suffer from P deficiency. The combined use of the phosphatic fertilizers to maximize the wheat production without experiencing any toxicological hazards is therefore urgently required. In order to overcome the cost of production, abolish the toxic effect of fertilizers, and fulfill the P requirements of wheat, it has become imperative to search for some newer and inexpensive option that could solve such difficulties. To address such problems, the focus is shifted toward the use of PSM both singly (Kumar et al. 2014a) and as mixture with fertilizers (Babana et al. 2013) or as coculture (Upadhyay et al. 2012) to improve soil fertility leading to increase in wheat yields (Zaidi and Khan 2005). When applied properly, PSM in agricultural practices has been found to decrease the use of costly phosphatic fertilizers (Ali et al. 2014; Dalve et al. 2009). For example, Ramlakshmi and Bhrathiraja (2015) in a study conducted for marigold production have suggested that the mixture of *Paenibacillus polymyxa* (phospho-bacterium) and *Glomus fasciculatum* (AM fungi) could decrease the application of P fertilizer by 25%. The integrated nutrient management (Chaitra and Patil 2007) which involves the use of PSM carrying variable characteristics has, therefore, motivated wheat growers worldwide (Naqvi and Ahmad 2012; Goes et al. 2012). Phosphate-solubilizing microorganisms also act as biological control agents (Zaidi et al. 2014) and by limiting the phytopathogens increase the performance of plants (Basharat et al. 2011).

3.5 PSM Improves Wheat Production

3.5.1 PSM: Definition, Origin, and Selection of Phosphate-Solubilizing Microorganisms

Phosphate-solubilizing microorganisms are a group of useful microorganisms which hydrolyze organic and inorganic P (Chen et al. 2006b). Numerous PSM have been recovered from non-rhizosphere (Onyia and Anyanwu 2013) and rhizosphere soils (Qiao et al. 2013), rhizoplane (Sarkar et al. 2012), phyllosphere (Mwajita et al.

2013), rock phosphate (RP) deposit area soil (Mardad et al. 2013), marine environment (Mujahid et al. 2014), and polluted soils (Susilowati and Syekhfani 2014). Some of the important P-solubilizing bacteria belongs to genera *Achromobacter* (Ma et al. 2009), *Acinetobacter* (Gulati et al. 2010), *Sphingomonas* and *Burkholderia* (Panhwar et al. 2014; Song et al. 2008), *Bacillus* (Tallapragada and Usha 2012), *Serratia* (Selvakumar et al. 2008), *Enterobacter* (Frank and Julius 2012), *Micrococcus* (Reena et al. 2013), *Pseudomonas* (Mehnaz et al. 2010), rhizobia (Kumar et al. 2014; Kenasa et al. 2014), and actinomycetes (Saif et al. 2014). The important P-solubilizing fungi belong to genera *Penicillium* (Reena et al. 2013), *Aspergillus* (Coutinho et al. 2012), and *Trichoderma* (Yasser et al. 2014). However, among various phosphate solubilizers, P-solubilizing fungi (PSF) in general have been found superior P solubilizer compared to PSB (Venkateswarlu et al. 1984). Like any other plant, wheat too represents a habitat for diverse PSM, which colonize the (i) rhizosphere (Majeed et al. 2015; Kundu et al. 2009), (ii) the phyllosphere (Verma et al. 2016a), (iii) PSM living inside tissues (endophytes) (Verma et al. 2016a), and (iv) stressed environment.

Rhizosphere PSM The region of soil that is directly influenced by root exudates and associated soil microbiota is generally termed as rhizosphere. The term rhizosphere (Greek word "rhizo" meaning root and "sphere" is one field of action, influence, or existence) was introduced by Hiltner in 1904. The rhizosphere is generally rich in rhizodeposition (sloughed-off plant cells), proteins, and sugars released by roots. These exudates support the growth of various microbial communities including PSM. Like many other microbial communities, PSB have been recovered from many crop rhizospheres including those of wheat. Some of them have been identified as *Pseudomonas aeruginosa* (Kumar et al. 2015), *P. fluorescens*, *P. putida* (Zabihi et al. 2011), *P. stutzeri* (Venieraki et al. 2011), *Bacillus* (Ogut and Er 2016; Ogut et al. 2011), *Lysinibacillus sphaericus*, *Paenibacillus polymyxa*, *Staphylococcus succinus*, *Sporosarcina* sp. (Verma et al. 2016b), *Azotobacter chrococcum* (Kumar and Narula 1999), *Thiobacillus* sp. (Babana et al. 2016), *Vibrio splendidus* (Babana et al. 2013), *Proteus* sp. (Billah and Bano 2015), *Azospirillum brasilense* (Venieraki et al. 2011), *Acinetobacter* (Ogut et al. 2010), *Stenotrophomonas* sp. AJK3 (Majeed et al. 2015), *Enterobacter* sp., *Arthrobacter chlorophenolicus* (Kumar et al. 2014a), and *Serratia marcescens* (Lavania and Nautiyal 2013). Of the filamentous fungi involved in solubilization of insoluble P, *Aspergillus niger* (Shrivastava and D'Souza 2014), *Penicillium bilaii* (Ram et al. 2015), *Penicillium oxalicum* (Xiao et al. 2013), and *Mucor ramosissimus* (Xiao et al. 2009) are the most important PSF, while strains of *Candida krissi* (Xiao et al. 2009) have also been identified as P solubilizer which solubilized insoluble P by secreting organic acids.

Phyllosphere PSM The term phyllosphere refers to the total aboveground portions of plants inhabited by microorganisms (Last 1955; Ruinen 1956). Phosphate-solubilizing microbes in the wheat phyllospheres have been reported and identified using PCR technique. For instance, Verma et al. (2014) isolated wheat-associated epiphytic bacteria from five locations in central zone (one of the wheat agroecological zones) in India. The phosphate-solubilizing bacteria isolated from phyllosphere (N = 89) belonged to genera *Arthrobacter*, *Bacillus*, *Corynebacterium*, *Methylobacterium*, *Paenibacillus*, *Pseudomonas*, and *Psychrobacter*. Of these, *Arthrobacter humicola* showed the highest P-solubilizing activity. Other genera identified to species level using 16S rRNA gene sequencing and subsequent molecular phylogeny analysis included *Paenibacillus amylolyticus*, *Bacillus aryabhattai*, *Methylobacterium extorquens*, *Methylobacterium mesophilicum*, *Methylobacterium radiotolerans*, *Psychrobacter fozii*, and *Pseudomonas fuscovaginae*.

Endophyte PSM An endosymbiont (bacterium or fungus) often called an endophyte resides inside plant tissues (Hardoim et al. 2008) for longer periods of its life cycle but causes no diseases to plants (Vijayabharathi et al. 2016; Puri et al. 2015; Hardoim et al. 2015). Also, the endophytic bacteria improve plant growth and nutrition more efficiently compared to rhizospheric bacteria because they show more intimate relationship with plant tissues. The endophytes have an ecological advantage over epiphyte microbes because they are shielded from unfavorable environmental conditions such as high temperature, salinity, drought, pH, osmotic potentials, and ultraviolet radiation (Seghers et al. 2004). The endophytes generally adhere to root hair zone of apical roots and enter through a crack or damage. Following entry inside, they colonize the differentiation zone and intercellular spaces in epidermis (Raven et al. 2009). After crossing the exodermal barrier, they colonize different regions such as point of entry, deep inside cortex, and the cortical intercellular spaces. The plant tissue type, plant growth stage, and soil fertilizer treatment all contribute to composition of endophyte bacterial community in wheat (Robinson et al. 2015). Like rhizosphere/phyllosphere microbial communities, endophytes also facilitate the growth of plants by various mechanisms (Gaiero et al. 2013) including P-solubilization (Wakelin et al. 2004). There are other studies also which suggest that soil inoculation with P-solubilizing *Bacillus* spp. can solubilize unavailable soil P and applied P, leading to a better plant development and greater yields (Canbolat et al. 2006). The root endophytes Bacillus, *Enterobacter*, *Micrococcus*, and *Pseudomonas* genera identified by Mbai et al. (2013) have also shown to have potential to solubilize P. In other study, Jha and Kumar (2009) isolated a diazotrophic bacterium identified as *Achromobacter xylosoxidans* WM234C–3 from surface-sterilized roots and culms of wheat variety Malviya 234 which had a significant P-solubilizing activity. Zinniel et al. (2002) also isolated diazotrophic endophytic bacteria from wheat, whereas the filamentous *Actinobacteria* and some fungi were observed in wheat plants by Coombs and Franco (2003). Recently, Oteino et al. (2015) reported that majority of the endophytic *Pseudomonas* strains produced gluconic acid (GA) (14-169 mM) and demonstrated moderate to high P-solubilizing activity (400–1300 mgl^{-1}). Thus, the study of endophytes is important primarily for two reasons – (i) it helps to better understand its ecology and (ii) the bioactive molecules secreted by endophytes facilitate the growth of plants in sustainable agricultural practices. Therefore, concerted efforts should be directed to find some new and potentially exciting endophytes for ultimate use in different agricultural production systems across different ecological niches.

Stress-Tolerant PSM Phosphate-solubilizing microorganisms in general have regularly been isolated from conventional environment. However, the isolation of PSM from derelict/stressed environment would be advantageous because such stress-tolerant PSM could be beneficial for crops growing in stressed/polluted soils. In this context, various P-solubilizing bacteria have been recovered from wheat growing in disturbed environments, for example, low temperature (Mishra et al. 2011; Verma et al. 2015a), drought (Verma et al. 2014), acidic soil (Verma et al. 2013), and salinity (Egamberdieva et al. 2008; Tiwari et al. 2011) using culture-dependent techniques. Verma et al. (2016a) in a recent investigation assessed the diversity and functional attributes of thermotolerant bacteria hosting leaves, shoots, roots, and rhizospheric soils of wheat growing in the peninsular zone of India. Majority of the isolated genera demonstrated P-solubilizing activity and belonged to genera *Achromobacter*, *Alcaligenes*, *Arthrobacter*, *Bacillus*, *Methylobacterium*, *Pseudomonas*, *Rhodobacter*, *Salmonella*, and *Staphylococcus*. Verma et al. (2016b) in a follow-up experiment recovered the epiphytic bacterial strains identified as *B. amyloliquefaciens*, *A. faecalis*, and *P. poae* from the wheat phyllosphere. In a similar study, *A. faecalis* and *P. poae* were isolated from wheat growing at arid land and high-temperature regions and were identified for the first time as epiphytic PGPB (Joo et al. 2005). Similar studies were carried out by Verma et al. (2013) to reveal acidotolerant P-solubilizing bacteria exhibiting other PGP activities from phyllosphere of two varieties of wheat growing in acidic soil in the southern hills zone of India. Among these PSB, *Variovorax soli* (21.52 ± 1.3 μg P mg^{-1} day^{-1}) was isolated from wheat var. HD2833 and *Methylobacterium* sp. (36.35 ± 1 μg P mg^{-1} day^{-1}) and *M. radiotolerans* (21.35 ± 1 μg P mg^{-1} day^{-1}) from wheat var. HW3094. Apart from bacteria and fungi, the P-solubilizing actinomycetes can also survive in extreme environments (e.g., drought, fire), and through their ability to secrete antibiotics and phytohormone-like compounds, they can enhance plant growth. *Micromonospora aurantiaca* and *Streptomyces griseus*, for example, have shown the greatest stimulatory effect on wheat due to their P-solubilizing efficiency and plant growth-promoting activities (Hamdali et al. 2008, Jog et al. 2012). This ability of actinomycetes of surviving under extremes of environmental conditions has therefore attracted greater attention toward their use as biological agents in stressful conditions. Stress-tolerant microbial inoculants are required for inoculation under extreme environments like high temperature so that such organisms could survive under such adverse environment while maintaining their plant growth-promoting activities. And hence, the selection of thermotolerant P-solubilizing microorganisms carrying numerous PGP traits could be used to produce inoculants for crops grown in the arid, subarid, high-plateau, and high-temperature zones. Furthermore, considering the variety of PSM's widely spread in different habitats, there is ample scope to find many more prospective microorganisms from variable environments for eventual transfer to end users/farmers.

3.6 Selection of Phosphate Solubilizers

P-solubilizing microorganisms have been recovered from conventional (Surapat et al. 2013) to derelict soils (Susilowati and Syekhfani 2014) and from rhizosphere (Ranjan et al. 2013) to endophytic (Resende et al. 2014) environment. They have subsequently been used in various agronomic practices with greater positive impact on different crops (Sonmez and Tufenkci 2015) including wheat (Sial et al. 2015) under different production systems. And hence, the use of P-solubilizing organisms in crop production is increasing which is likely to substitute or/decrease the use of phosphatic fertilizers considerably (Adesemoye et al. 2009). Considering the importance of PSM in sustainable crop production, many workers have isolated such beneficial microbes (Ahemad and Khan 2012) employing serial plate dilution technique or enrichment culture method. Generally, the PSM are isolated using media containing insoluble tricalcium phosphate (TCP), and the best suitable and most widely used medium for this purpose has been the Pikovskaya medium (Pikovskaya 1948) (g/l: glucose 10, $Ca_3 (PO_4)_2$ 5, $(NH_4)_2 SO_4$ 0.5, NaCl 0.2, $MgSO_4.7H_2O$ 0.1, KCl 0.1, yeast extract 0.5, $MnSO_4$ and $FeSO_4$ trace, and pH 7). Rhizospheric or non-rhizospheric soil samples are diluted serially and spread plated (100 μl) or streaked or spot (10 μl) inoculated on Pikovskaya agar plates or any plates having insoluble P and properly incubated for 5–7 days (bacteria) and 3–5 days (fungi and actinomycetes) at 28 ± 2 °C. Organisms showing PS activities are identified by the appearance of zone of solubilization (clear halo) near microbial growth (Plate 3.1) on TCP/insoluble P supplemented plates. The consistency of this technique is, however, not

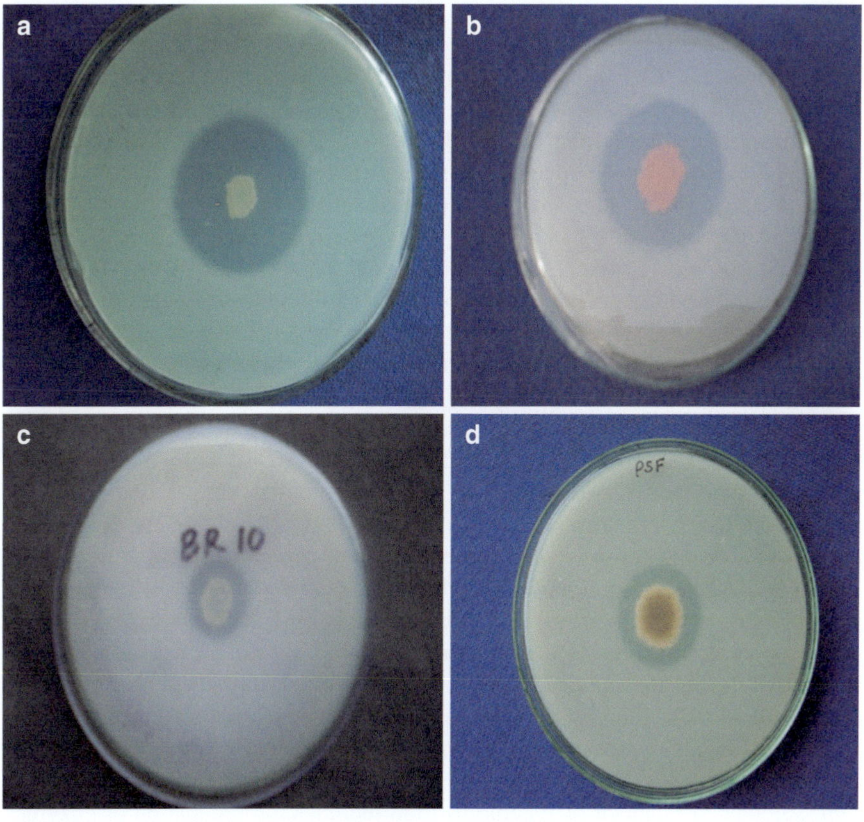

Plate 3.1 Tricalcium phosphate solubilization on Pikovskaya plates by (**a**) *Pseudomonas* sp., (**b**) *Serratia* sp., (**c**) *Bacillus* sp., and (**d**) fungal species

accepted by many workers since numerous bacterial isolates in other studies have even though failed to yield any visible zone of P-solubilization on agar plates but instead solubilized insoluble P in culture medium. However, considering halo formation as a positive indicator of P-solubilizers, the organisms (colonies) showing halo on plates are picked up and used to determine their capability to solubilize insoluble P under liquid medium. After evaluating their P-solubilizing ability on agar plates and in liquid medium, the PSM are assessed for their in vitro potential to secrete plant growth-promoting bioactive molecules. The most putative strains are then identified to species level employing molecular techniques, for example, 16S rDNA sequencing and phylogenetic method. The organisms identified by biochemical (identified up to genus level only) or molecular methods (identified up to species level) and showing single or many plant growth-promoting activities apart from their intrinsic P-solubilizing activity are selected and checked in pots and fields using seed treatment, seedling dipping, or soil application methods for their final transfer to growers for consequent application in agronomic practices as an inexpensive phosphatic option.

3.7 Phosphate Solubilization: How It Occurs?

The heterogenous microbial populations in different habitat include PSM also. Out of the total P-solubilizing microorganisms, 1–50% is contributed by PSB, whereas only 0.1–0.5% populations account for PSF (Chen et al. 2006a). Both phosphate-solubilizing bacteria and fungi convert the insoluble form of phosphorus (inorganic and organic) into soluble and available form of P which is taken up as a source of P by plants. Broadly, the solubilization/mineralization of inorganic/organic P occurs by one of the three mechanisms: (a) production of organic acids, (b) excretion of H^+ions, and (c) synthesis of enzymes (Kapri and Tewari 2010; Arcand and Schneider 2006). Of these, solubilization of inorganic P through the organic acids secreted by PSM (Khan et al. 2014) is one of the most extensively accepted theories of P-solubilization. The organic acids produced by PSM largely include gluconic acid, oxalic acid, ketogluconic acid, citric acid etc. (Table 3.2). The release of organic acids is directly associated with the lowering of pH of the medium (Mardad et al. 2013; Whitelaw 2000; Maliha et al. 2004). The effectiveness of solubilization/mineralization, however, depends on the types and amounts of organic acids/enzymes secreted into the liquid medium. Also, the inherent properties of the organic acids are vital than the whole amount of acids released by P-solubilizers (Scervino et al. 2010). Moreover, the insoluble P is also converted into soluble P without the secretion of OA by microbes (Illmer and Schinner 1992). For example, ammonia assimilation resulting in proton extrusion has been found as an alternative mechanism for P-solubilization (Parks et al. 1990; Ogut et al. 2011). Phosphate solubilization by PSM also occurs by enhancing the process of chelation of cations that are bound to the soil P or by generating soluble compounds with cations linked with insoluble soil P so as to discharge the P into the soil system. Another mechanism of P-solubilization includes the production of enzymes

Table 3.2 Organic acids produced by phosphate-solubilizing microorganisms

P-solubilizing microorganisms	Organic acids produced	Initial pH of medium	Final pH of medium	Amount of P solubilized (mg/l)	Reference
Aspergillus sp.; *Penicillium* sp.	Citric, gluconic, oxalic, succinic, glycolic, malic	7.0; 7.0	3.2; 3.3	392 ; 381	Sane and Mehta (2015)
Aspergillus niger	Oxalic, lactic			3640	Padmavathi (2015)
Trichoderma harzianum	Lactic, succinic, tartaric, citric	5.4	4.3	ND	Li et al. (2015)
Rhizobium tropici, *Paenibacillus kribbensis*, *Acinetobacter* sp.	Malic, 2-ketoglutaric, lactic, succinic, tartaric, propionic, gluconic	5.0; 7.0	5.0; 4.0	70 ; 75; 75 (approx.)	Marra et al. (2015)
Pseudomonas sp. *R7*	Lactic, isocitric, tartaric, pyruvic	7.03	4.92	19.5	Mihalache et al. (2015)
Pseudomonas fluorescens L228	Gluconic	7.0	4.06	1312	Oteino et al. (2015)
Rhizobium sp. *VM-2*	Organic acids	7.0	4.93	799	Satyanandam et al. (2014)
Rhizobium sp. *RASH6*	Succinic, gluconic	7.0	3.4	275	Singh et al. (2014)
Bacillus megaterium	Malic, quinic	7.0	4.0	ND	Kang et al. (2014)
Trichoderma harzianum	Citric, lactic, succinic	7.2	4.68	9.31	Promwee et al. (2014)
Pantoea agglomerans, *Burkholderia anthina*, *Enterobacter ludwigii*	Gluconic, oxalic, citric	7.0	3.2; 3.5; 4.0	575.16; 384.28; 600	Walpola and Yoon (2013a, b)
Enterobacter hormaechei sub sp. *steigerwaltii*	Gluconic, succinic, malic, glutamic	7.0	3.5	505	Mardad et al. 2013
Azospirillum, *Bacillus*, *Enterobacter*	Acetic, citric, gluconic	7.0	Reduced pH	218.1; 298.3; 258.6	Tahir et al. (2013)
Burkholderia ambifaria KS 01, *B. Tropica* KS 04	Acetic, citric, gluconic, lactic, succinic, propionic	6.6	4.86; 4.05	433.81; 499.85	Surapat et al. (2013)
Penicillium sp.	Gluconic, citric	6.25	3.22	39.2–86.1	Nath et al. (2012)
Acinetobacter sp. *WR 1222*	Gluconic	7.0	4.21	414	Ogut et al. (2010)

ND not determined

like (a) phosphatases (meant for dephosphorylation of phospho-ester bonds), (b) phytases (responsible for the release of P from phytic acid), and (c) phosphatases (enzymes that cleave the C-P linkage in organophosphonates). The phosphatases and phytases together mediate the P mineralization process. The phosphatases play an important role in releasing the inorganic phosphates through scavenging of phospho-ester bonds, whereas most of the phytases are involved in the cleavage of C-P bonds. Another interesting role attributed to these enzymes is the degradation of phytate, thereby leading to mineralization of organic P present within the soil (Behera et al. 2014).

3.8 How Phosphate Solubilizers Facilitate Wheat Growth

Indeed, the plant growth enhancement by P-solubilizing microorganisms in P-deficient soil occurs greatly through solubilization of insoluble P. The soluble P is then taken up as a major nutrient element by plants. Apart from making soluble P available to plants, the PSM also secretes some important active biomolecules (Table 3.3) that directly or indirectly enhance the growth and productivity of crop plants (Fig. 3.1). Chief among them is the release of siderophores (Ghosh et al. 2015), indole acetic acid (Chitraselvi et al. 2015), and gibberellic acid (Jha and Subramanian 2014). Synthesis of siderophores, an iron-chelating substance by PSB, for instance, *Pantoea agglomerans* and *Burkholderia anthina*, may indirectly affect the growth of plants (Datta and Chakrabartty 2014; Walpola and Yoon 2013b, c). Siderophores released by PSB form a complex with iron (Fe^{3+}) in the rhizosphere and limit its availability to the phytopathogens and concomitantly prevent phytopathogens from causing damage to plants. Thus, PSB due their ability to secrete siderophores could be developed as biocontrol agent as well. Secretion of IAA by phosphate solubilizers, for example, *Azospirillum*, *Bacillus*, and *Enterobacter*, (Tahir et al. 2013) is yet another microbiological trait that has shown greater positive impact on overall performance of wheat plants. IAA secreted as a secondary metabolite due to rich supply of substrates by PSB control cell elongation and division, phototropism, and apical dominance in plants (Remnas et al. 2008; Ali et al. 2009). Also, IAA helps in the expansion of roots and increases the number of root hairs and lateral roots which participate in uptake of nutrients from soil (Datta and Basu 2000). Indole acetic acid also inhibits or impedes the abscission of leaves inducing flowering and fruiting (Zhao 2010). Interestingly, phosphate-solubilizing organism, for example, *Bacillus*, also secretes cyanogenic compounds (Agrawal and Agrawal 2013), and *Burkholderia tropica* (Tenorio-Salgado et al. 2013) and phosphate-solubilizing actinomycetes *Streptomyces* spp. (Jog et al. 2014) exhibited antifungal activity which suppress the fungal phytopathogens and indirectly promote the growth of plants (Singh et al. 2014). Among various P-solubilizing bacteria, some bacterial strains, like *Stenotrophomonas rhizophila*, *Enterobacter cloacae* etc., have been

Table 3.3 Examples of plant growth regulators synthesized by phosphate-solubilizing microorganisms with reference to wheat

Phosphate-solubilizing microorganisms	Source	PGP activities	Reference
Pantoea sp.	Wheat seeds	IAA, siderophore, N_2 fixation	Herrera et al. (2016)
Burkholderia sp. *Enterobacter* sp.	Wheat rhizosphere	IAA, siderophore	Moriera et al. (2016)
Pseudomonas fluorescens	Wheat rhizosphere	Siderophore, IAA	Safari et al. (2016)
Serratia marcescens *P. aeruginosa*	Vegetables rhizosphere	IAA NH_4 and HCN	Kumar et al. (2015)
Psychrobacter maritimus *Serratia proteamaculans* *Bacillus anthracis*	Wheat rhizosphere	IAA, siderophore	Amara et al. (2015)
Serratia grimesii *Serratia marcescens*	Wheat rhizosphere	N_2 fixation, zinc solubilization, EPS activity, ACC deaminase, biocontrol activity, IAA production	Abaid-Ullah et al. (2015)
Stenotrophomonas rhizophila	Wheat rhizosphere	N_2 fixation, IAA, catalase and cytochrome oxidase	Majeed et al. (2015)
Bacillus sp.	Wheat rhizosphere	Catalase and cytochrome oxidase production	
Enterobacter cloacae subsp. Dissolvens	Soyabean rhizosphere	IAA production, siderophore production, ammonia production, potassium and zinc solubilization	Ramesh et al. (2014)
Pseudomonas fuscovaginae	Wheat phyllosphere	N_2 fixation, biocontrol activity, IAA, siderophore production, and NH_4 production	Verma et al. (2014)
Psychrobacter fozii	Wheat phyllosphere	ACC deaminase activity, biocontrol activity, IAA, siderophore production, Gibberellic acid production, HCN and NH_4 production	
Streptomyces sp.	Wheat rhizosphere	Chitinase, phytase, siderophore production, IAA production	Jog et al. (2014)
Planococcus rifietoensis	Wheat rhizosphere	IAA production, ACC deaminase activity	Rajput et al. (2013)

(continued)

Table 3.3 (continued)

Phosphate-solubilizing microorganisms	Source	PGP activities	Reference
Providencia sp.	Wheat rhizosphere	NH_4 production, siderophore, HCN, indolic compound, antifungal activities, Zn solubilization, antibacterial activity	Rana et al. (2012)
Azospirillum isolates	Wheat rhizosphere	N_2 fixation, IAA production	Venieraki et al. (2011)
Pseudomonas aeruginosa	Wheat endorhizosphere	Siderophore, ACC Deaminase, IAA production, NH_4 production, antifungal enzyme production as cellulase, protease, pectinase	Sharma et al. (2011)

Abbreviations used in this table are: *IAA* indole acetic acid, *HCN* hydrogen cyanide, *NH4* ammonia, A*CC* 1-aminocyclopropane-1-carboxylate, and *EPS* exopolysaccharide

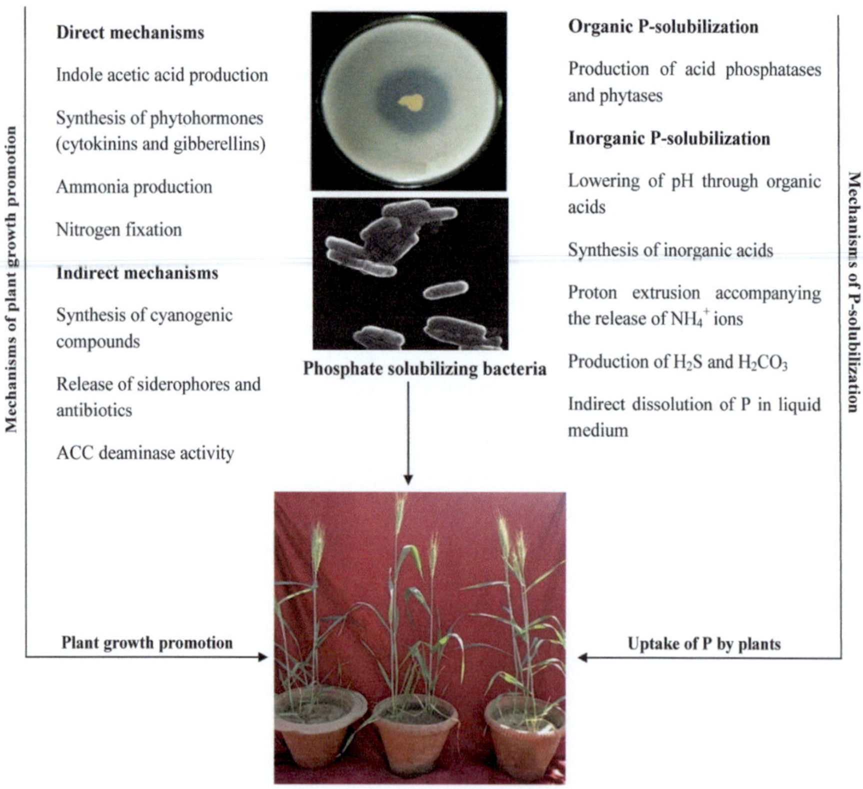

Fig. 3.1 Mechanism of phosphate solubilization and wheat growth enhancement by phosphate-solubilizing microorganisms

reported to fix atmospheric nitrogen (Majeed et al. 2015; Singh and Jha 2015). This property of N_2 fixation by PSB could be of special interest for growers for soils deficient in both N and P since application of single PSB expressing dual activity of P-solubilization and N_2 fixation together are likely to overcome the N and P deficiency in P- and N-deficient soils. Among plant growth regulators, ACC deaminase is an important plant growth regulator that induces metabolic changes and, hence, increases the growth of plants indirectly by hindering/reducing ethylene secretion (Glick et al. 2007; Bal et al. 2013; Magnucka and Pietr 2015). In a recent study, Singh and Jha (2015) isolated a phosphate-solubilizing bacterium *Enterobacter cloacae* from rhizospheric roots of *Aerva javanica* carrying various plant growth-promoting activities such as the isolate was able to produce ACC deaminase, display nitrogen fixation ability, synthesize IAA, and secrete ammonia. Considering the multifarious traits of this bacterium, it was suggested that such bacteria could be used as biofertilizers for increasing the production of crops including those growing in salt-stressed environment.

3.9 Examples of Sole and Composite Effects of PSM on Wheat

With ever increasing human populations, there is greater pressure on agriculture to satisfy human food demands across different regions. Indeed, modern agriculture especially green revolution has reduced some of the human problems by producing more and more foods. To achieve optimum yields, practitioners have, however, extensively used agrochemicals including fertilizers and pesticides in agronomic practices. The expensive and injudicious applications of agrochemicals have undoubtedly increased food production, but their use over the years has backfired as well. The excessive application of agrichemicals has resulted in destruction of microbial diversity and consequently the loss of soil fertility. To obviate such threatening problems of soil pollution and to restore soil fertility, the use of inexpensive natural resources, for example, PSM, has been practiced in different production systems in recent times. Wheat is a high P-demanding cereal crop, and at global scale, wheat production suffers from certain problems; one of the key limitations in enhancing wheat production is the inappropriate use of plant nutrients, especially P and K, and the mean P uptake of wheat is about 3.8 kg P/t of grains (Timsina and Connor 2001). The recovery of P by wheat from fertilizers is quite low, and it is estimated that about 15–20% of the applied P is recovered, while the 80–85% P is fixed as insoluble soil P (Rodríguez and Fraga 1999). It is reported that only 0.1% of the total P remains in soluble form which is available for uptake by plants. Constantly increasing costs of synthetic phosphatic fertilizers together with its high complexation ability in soil have warranted the search for alternative and viable means of P nutrition of wheat. In this context, the sole (Agrawal and Pathak 2010) or composite (Saxena et al. 2014) application of PSM have been considered as a suitable and practicable choice for providing soluble P to wheat while reducing dependence on chemical fertilizers (Table 3.4).

Table 3.4 Influence of single and multiple applications of phosphate-solubilizing microorganisms on biological and chemical characteristics of wheat

PSM inoculants	Growth parameters of wheat	References
Single microorganism		
Penicillium bilaii	Grain yield, spike density	Ram et al. (2015)
Pseudomonas fluorescens	Growth traits and yield	Zia-ul-Hassan et al. (2015)
Pseudomonas sp.	Nutrient uptake and seedling growth	Sarker et al. (2014)
Bacillus megaterium var. phosphaticum	No. of kernels/spike, grain yield, grain protein ratio	Bulut (2013)
PSB strain MR1	Grain and straw yield	Haque et al. (2013)
Aspergillus awamori	Dry matter accumulation at tillering and ear emergence, grain and straw yield	Sharma et al. (2012)
Pseudomonas sp.	Straw yield and P uptake	Babana et al. (2012)
Penicillium oxalicum	Shoot and root dry weight, grain yield, P accumulation	Singh and Reddy (2011)
Azotobacter chroococcum	Grain and straw yield	Narula et al. (2005)
Composite culture		
Azotobacter sp. +mycorrhiza	Seed protein, NPK in seeds and root colonization	Amraei et al. (2015)
Bacillus megaterium BHU1+ *Arthrobacter chlorophenolicus* BHU3	Plant height, grain yield, straw yield and nutrient acquisition	Kumar et al. (2014)
P. fluorescens + *B. cepacia* + *G. etunicatum*	Growth, yield and nutrient uptake	Minaxi et al. (2013)
B. lentus + *P. putida*	Tiller number, grain yield, total biomass	Saber et al. (2012)

3.10 Inoculation Impact of Phosphate-Solubilizing Bacteria on Wheat

Numerous bacterial species identified as PS bacteria have been used as biofertilizer (microbial inoculants) in agricultural practices largely because of their immense ability to improve the availability of applied and soil P (Vessey 2003). In general, the valuable effects of PSB on crop production have been extensively reported (Khan et al. 2007; Zaidi et al. 2009), but the application of PSB as microbial inoculants (biofertilizer) in wheat cultivation is limited. Considering the importance of PSB and lack of sufficient information on the role of PSB in wheat productivity, an attempt is made here to highlight the impact of single or dual culture of PSB in the improvement of wheat grown under different agroecological niches.

Application of PS bacterium *Bacillus megaterium* var. phosphaticum [M-13] in the presence of P fertilizers greatly increased the grain and straw yield of wheat when grown in pots. Also, an increase of 27.3%–53.3% in number of spikes per square meter was observed in the presence of inoculated PSB strain over control treatment (Bulut 2013). In a follow-up study, Hossain and Sattar (2014) investigated the effect of mineral P fertilizer and phosphate-solubilizing bacteria (*Pseudomonas* sp. and *Klebsiella* sp.) used singly or as mixture on the growth, yield, nutrient uptake, and P use efficiency of wheat grown in field soils treated with varying levels of inorganic phosphorus (triple super phosphate) fertilizer. The highest grain and straw yield (2.13 and 2.84 t ha^{-1}) were observed when *Pseudomonas* sp. and *Klebsiella* sp. were applied with 15 kg P ha^{-1} at Pabna and Rajshahi, respectively. Inoculation of *Pseudomonas* sp. for Pabna and *Klebsiella* sp. for Rajshahi in the presence of triple super phosphate resulted in better yield and nutrient uptake of wheat and quality of soil compared to other treatments. When used alone, PS bacteria increased the efficiency of P during crop production, and a positive significant correlation was found between yield contributing characters and grain yield of wheat. This study clearly indicated that PSB could solubilize unavailable P to available form and made it available to crops resulting in greater nutrient uptake and yield of wheat.

Similarly, Afzal et al. (2005) found a significant enhancement in grain and biological yield of wheat grown in presence of PSB (*Pseudomonas* and *Bacillus* species) used either alone or in combinations. Moreover, a statistically significant improvement in seed P content and tillers per m^2 over control was recorded. It was concluded from this study that P-solubilizing microorganisms when used singly or jointly with other organisms showed a significant impact on grain and biological yield, tillers per m^2, and seed P content. An increase in straw and grain yields of wheat following interaction between levels of phosphatic fertilizers and PSB inoculations have been reported (Dwivedi et al. 2004). In a similar study, a synergistic relationship between P-solubilizing microorganisms, for example, *Pseudomonas striata* and *Penicillium*, and asymbiotic N_2 fixer *A. chroococcum* facilitated a better uptake of poorly soluble P and, consequently, enhanced dry biomass, grain yield, and P uptake of wheat plants (Zaidi and Khan 2005). Later on, Sarker et al. (2014) observed a considerable increment in growth and nutrient uptake of *Pseudomonas* inoculated wheat plants. Following inoculation with *Pseudomonas* sp., the dry biomass of shoots increased significantly over uninoculated control. Additionally, the concentration of macronutrients like, N, P, and K in root and shoot tissues were found maximum in inoculated wheat plants. Kumar et al. (2001), in a pot experiment carried out in greenhouse, assayed the survival of P-solubilizing strains of *A. chroococcum*, including soil isolates and their mutants, in the rhizosphere, and their influence on biological characteristics (growth and root biomass) of three genetically diverse wheat cultivars. Wheat plants inoculated with/without microbial cultures were grown in soils treated with different dose rates of N and P fertilizers. Seeds of wheat bacterized with P-solubilizing and plant hormone producing *A. chroococcum* displayed superior response relative to uninoculated controls. Furthermore, grain and straw yields were increased significantly by 12.6% and

11.4%, respectively, following inoculation of mutant strains of *A. chroococcum* over control. The survival of mutant strain of *A. chroococcum* in the rhizosphere was enhanced by 12–14% as compared to parent soil isolate. Of the mutant strain, strain M37 was found superior for all three varieties and significantly increased grain yield and root biomass by 14% and 11.4%, respectively, over control. In an experiment, the application of P-solubilizing bacteria (*Thiobacillus thiooxidans*) in combination with fertilizers (Tilemsi rock phosphate) has resulted in a tremendous increase in wheat yields. Formulation of RP fertilizers along with *T. thiooxidans* AHB411 and *T. thiooxidans* AHB417 increased the yield up to 33.3% and 11.9%, respectively. Other biological parameters like number of tillers per plant and length of panicle and seed characteristics such as grains per panicle and 1000 grain weight were significantly improved. Mixed inoculation of *T. thiooxidans* and Bio TRP1 increased the grain yield of wheat by 46%, whereas straw yield was enhanced by 74% relative to control (Babana et al. 2016).

3.11 Response of Wheat to PSF Inoculations

Apart from phosphate-solubilizing bacteria, fungi have also been found as a better P-solubilizing organism (Khan et al. 2010; Yasser et al. 2014), and upon inoculation, they have shown considerable improvement in wheat production (Wahid and Mehana 2000). For instance, Ram et al. (2015) in a recent field experiment determined the effect of seed treatment with PSF, *Penicillium bilaii* at varying levels of P on growth, P concentration in leaves, and production of wheat. In the absence of P, the single application of PSF profoundly enhanced grain yield by 12.6% over uninoculated control. On the contrary, PSF in the presence of 50% P fertilizer augmented wheat yield which was equal to the application of 100% P but without PSF inoculation. The interaction between PSF inoculation and P levels affected the spike density significantly. When the P levels were 0 and 50%, the spike density increased significantly to about 7% as compared to control, without PSF application. The PS fungus *Penicillium bilaii* was capable of enhancing the number of grains per spike and grain yield of wheat remarkably when compared with uninoculated treatments. A 3.7% increase in the 1000 grain weight was recorded following PSF application in wheat relative to control. Also, the application of *P. bilaii* and phosphatic fertilizer together increased the concentration of P both in grains and straw of wheat plants. When measured at 30 DAS, the P content in the leaves of *P. bilaii* inoculated wheat plants was found to increase. The study in general reflected a growth enhancement in wheat as a result of PSF inoculation as well as application of phosphatic fertilizer (Ram et al. 2015). In a similar experiment performed by Singh and Reddy (2011), the growth of wheat plants was enhanced due to inoculation with *Penicillium oxalicum*. *Penicillium oxalicum* in combination with rock phosphate (RP) increased the shoot length by 1.5 times compared to uninoculated plants. Moreover, the dry biomass of shoots and roots of inoculated plants grown in soil treated with rock phosphate was comparatively higher than control. The mixture of *P. oxalicum* and RP, however, also increased the yield by 42%. The total P content of wheat plants also increased in the presence of *P. oxalicum*. Generally, the P accumulation within various plant organs like shoot, root, and grains of wheat plants inoculated with mixture of *P. oxalicum* and RP was almost three times higher than the P accumulated in untreated plants. Apart from increase in P, the phosphatase and phytase activity were also enhanced following *P. oxalicum.* The overall improvement in the performance of wheat plants was, therefore, attributed to the inoculation of phosphate-solubilizing fungus *P. oxalicum* which increased the soil available P and concurrently facilitated the growth of wheat plants.

3.12 Influence of Composite Inoculations on Wheat Production

Wheat crop requires a larger quantity of some essential plant nutrients, such as N and P. The deficiency of such nutrient elements restricts the growth of plants severely. And hence, to supply such plant nutrients, inoculation of inexpensive and favorably interacting microorganisms have been found effective and viable. Also, where P is limited, it has been reported that plants inoculated with AM-fungi, either singly or as co-culture with PSM enhanced the uptake of P by wheat plants (Raja et al. 2002). In view of this, Saxena et al. (2014) studied the interactive effect of an AMF, for instance, *Glomus etunicatum*, and a PSB, *Burkholderia cepacia* BAM-6, on wheat plants sown in pots having low available P in order to find bio-inoculants for semiarid regions. The composite application of *G. etunicatum* and *B. cepacia* increased the growth and yield in comparison to the single application of *G. etunicatum* and *B. cepacia.* Crop yield was increased by more than 50%, while N concentration was enhanced by 90%, due to the co-inoculation. The root colonization infected by AMF and population of PSB in rhizosphere also increased with time in soil. This study suggested that *B. cepacia* BAM-6 interacted synergistically with AMF and enhanced the growth and nutrient uptake of wheat plants. Therefore, the mixture of two unrelated organisms could be used as biofertilizer for wheat crop grown in arid to semiarid regions. In other study, Tomar et al. (1998) used various combinations of *Azotobacter*, AM fungi, PSB, and NPK fertilizers in wheat production practices. The highest (3.80 tons ha^{-1}) grain yield was recorded with dual inoculation of AM fungi and P-solubilizing bacteria in the presence of NPK which was followed by 3.41 tons ha^{-1} with NPK only and 2.63 tons ha^{-1} for control. In a similar experiment, the synergistic effects of plant growth-promoting rhizobacteria and an AM fungus *G. fasciculatum* on plant growth, yield, and nutrient uptake of wheat plants grown under field conditions were assayed by Khan and Zaidi (2007). The tripartite combination of asymbiotic nitrogen fixer *A. chroococcum* with PS bacterium *Bacillus* and *G. fasciculatum* significantly augmented the dry biomass by 2.6-fold compared to control. At 135 days after sowing (DAS), the grain yield of wheat plants bacterized with *A. chroococcum*, *Bacillus* sp., and *G. fasciculatum* was twofold greater in comparison to non-inoculated plants. Grain protein (GP) was maximum (255.2 mg g^{-1}) in plants treated with four cultures namely, *A. chroococcum*, *Bacillus* sp., *G. fasciculatum*, and *Penicillium variabile* (PSF), while the minimum GP (113.7 mg g–1) was obtained with sole application of *G. fasciculatum.* The N and P contents were maximum (33.6 and 67.8 mg plant^{-1}, respectively) in wheat plants co-inoculated with *A. chroococcum*, *Bacillus* sp., and *G. fasciculatum*. However, the N and P contents of soil measured at 135 DAS varied among treatments. Use of *P. variabile* along with single or dual cultures had negative impact on the measured parameters. Percentage root infection, spore density of the AM fungus, populations of *A. chroococcum*, and P-solubilizing microorganisms were enhanced at 80 DAS. This result demonstrated that the various combinations of PGPR constantly amplified the growth and yield, N and P contents, and grain quality of wheat. In a field study conducted consecutively for 2 years, Kaur and Reddy (2015) used two phosphate-solubilizing bacteria, *Pantoea cypripedii* (PSB-3) and *Pseudomonas plecoglossicida* (PSB-5), which were applied singly or as mixture with RP against maize and wheat crops, and their impact was compared with chemical P fertilizer (diammonium phosphate, DAP). Application of PSB along with RP improved the shoot height, shoot and root dry matter, grain yield, and total P concentration in both maize and wheat crops in comparison to the other treatments. Available soil P, enzyme activities, and PSB populations in both maize and wheat rhizosphere were significantly increased due to inoculation of PSB X RP fertilization relative to DAP application. The mixed application of PSB and RP was found more economical, and, hence, it was suggested that the composite application of PSB and RP would be a suitable alternative to phosphatic fertilizer in sustainable production of wheat. The composite culture of phosphate-solubilizing bacterial strains *Pseudomonas fluorescens* (BAM-4) and *B. cepacia* (BAM-12) and *Glomus etunicatum* enhanced the shoot and root dry biomass and grain yields of wheat plants relative to the uninoculated plants (Saxena et al. 2013). The solubilization of insoluble P is carried out effectively by PSB, whereas the process of P uptake by plant roots is attributed to AM fungi which assist the transportation of solubilized P through plant roots. A composite inoculation of PSB and AM fungus showed better growth and yield of wheat plants in comparison to the plants inoculated with single microbial culture (Minaxi et al. 2013). Also, the AM fungi enhance the P uptake of plants by enhancing the contact surface and volume of soil (Clark and Zeto 2000).

3.13 Inoculation Effects of Immobilized Culture on Wheat Production

Immobilization of bacterial cells has commonly been used in agriculture, pharmaceutical, food, and other industries to obtain a defensive structure or a capsule that could allow immobilization, protection, release, and function of active ingredients. And hence, bacterial cells face little challenge from adverse environmental conditions since encapsulation helps bacterial cells to stabilize and enhance their viability and stability during production, storage, and handling of cultures. Also, encapsulation provides extra protection to bacterial cells during rehydration and lyophilization. In addition, the use of microbial cultures into soil has shown that some microbial inoculants can augment plant uptake of nutrients and consequently increase the use efficiency of applied chemical fertilizers (Adesemoye and Kloepper 2009). In this context, rhizobacteria can play an important role in creating a suitable system for crop production. However, the application of free-living PSB into soil is difficult because it is not easy to maintain the survivability of cells around plant roots largely because they are highly susceptible to environmental variables, for example, temperature, humidity, and stressor molecules. The variation in PSB effect on plants is chiefly due to the differences in the quality of microbial inoculants. Due to these, the efforts should be directed to find an adequate formulation that could be developed as a commercial inoculants product. Considering these, Schoebitz et al. (2013) evaluated the P-solubilizing ability of rhizobacteria using RP as insoluble P and the assimilation of soluble P by wheat plants in quartz sand potted experiments. For this, two P-solubilizing bacteria such as *P. fluorescens* and *Serratia* sp. were encapsulated in sodium alginate and potato starch beads. They were further tested for enzyme activity (alkaline and acid phosphatase) and P-solubilization in Pikovskaya liquid medium. A considerable decrease in pH was obtained following P-solubilization. A total of 89 and 93 μg P ml^{-1} was solubilized by immobilized P-solubilizing bacteria, which was significantly greater than those observed for autoclaved alginate-starch beads. An appreciable increase of 64% in P uptake by wheat plants was observed after 60 days of growth when wheat plants were treated with immobilized *P. fluorescens* + 3.25 ppm of P. This finding suggests that use of the immobilized rhizobacteria could be a viable option for increasing the P level in wheat grown in different agroecological niches.

Conclusion

The phosphate-solubilizing microorganisms are a boon to the agricultural system. It is indeed an inexpensive and an environmentally friendly strategy to reduce the use of chemical fertilizers in farming practices. The enhancement in biological and chemical properties of wheat plants has been reported due to inoculation with variety of phosphate solubilizers including bacteria, fungi, actinomycetes, and mycorrhizae etc. The yield and other growth parameters of wheat have been enhanced, in general, following inoculation with PSM when used singly or as mixture with other free-living PGPR/AM fungi. Another positive aspect of using these microorganisms is that the health of soil is not compromised at any stage of plant growth. Thus, the phosphate-solubilizing microorganisms in general are considered a useful soil microflora which could be developed at commercial scale as bioinoculants for enhancing the production of wheat while reducing dependence on chemical fertilizer.

References

Abaid-Ullah M, Hassan MN, Jamil M, Brader G, Shah MKN, Sessitsch A, Hafeez FY (2015) Plant growth promoting rhizobacteria: an alternate way to improve yield and quality of wheat (*Triticum aestivum*). Int J Agric Biol 17:51–60

Abbasniayzare SK, Sedaghathoor S, Dahkaei MNP (2012) Effect of biofertilizer application on growth parameters of *Spathiphyllum illusion*. Am Eurasian J Agric Environ Sci 12:669–673

Adesemoye AO, Torbert HA, Kloepper JW (2009) Plant growth-promoting rhizobacteria allow reduced application rates of chemical fertilizers. Microb Ecol 58:921–929

Adesemoye AO, Kloepper JW (2009) Plant–microbes interactions in enhanced fertilizer-use efficiency. Appl Microbiol Biotechnol 85:1–12
Afzal A, Asghari B (2008) *Rhizobium* and phosphate solubilizing bacteria improve the yield and phosphorus uptake in wheat *(Triticum aestivum)*. Int J Agric Biol 10:85–88
Afzal A, Ashraf M, Asad SA, Farooq M (2005) Effect of phosphate solubilizing microorganisms on phosphorus uptake, yield and yield traits of wheat (*Triticum aestivum* L.) in rainfed area. Int J Agric Biol 7:207–209
Agrawal S, Pathak RK (2010) Response of phosphate solubilizing microorganism on quality of wheat (*Triticum aestivum* L.) plant grown conventionally in temperate climate. J Biofertil Biopestici 2:2
Agrawal DPK, Agrawal S (2013) Characterization of *Bacillus* sp. strains isolated from rhizosphere of tomato plants (*Lycopersicon esculentum*) for their use as potential plant growth promoting rhizobacteria. Int J Curr Microbiol App Sci 2:406–417
Ahemad M, Khan MS (2012) Effect of fungicides on plant growth promoting activities of phosphate solubilizing *Pseudomonas putida* isolated from mustard (*Brassica campestris*) rhizosphere. Chemosphere 86(9):945–950
Ali A, Tahir M, Rashid H, Ajmal B, Sajjad RN, Adeel A (2014) Investigation of biofertilizers influence on vegetative growth, flower quality, bulb yield and nutrient uptake in gladiolus (*Gladiolus grandiflorus* L.) Intern J Plant Anim Environ Sci 4:94–99
Ali B, Sabri AN, Ljung K, Hasnain S (2009) Quantification of indole-3-acetic acid from plant associated Bacillus spp. and their phytostimulatory effect on *Vigna radiata* (L.). World J Microbiol Biotechnol 25:519–526
Amara U, Wang YX, Cui XL, Khalid R, Ali S, Shabbir G, Hayat R (2015) Screening and identification of soil bacteria for growth promotion of wheat (*Triticum aestivum* L.) J Bio Env Sci 7(3):87–99
Amraei B, Ardakani MR, Rafiei M, Paknejad F, Rejali F (2015) Effect of Mycorrhiza and *Azotobacter* on concentration of macroelements and root colonization percentage in different cultivars of wheat (Triticum aestivum L). Biol Forum An Int J 7:895–900
Anonymous (1988) Better crops with plant food. PPI, Atlanta, pp 26
Arcand MM, Schneider KD (2006) Plant- and microbial-based mechanisms to improve the agronomic effectiveness of phosphate rock: a review. An Acad Bras Cienc 78:791–807
Babana AH, Antoun H, Dicko AH, Maïga K, Traoré D (2012) Effect of *Pseudomonas* sp. on wheat roots colonization by mycorrhizal fungi and phosphate-solubilizing microorganisms, wheat growth and P-uptake. Int J Microbiol 1(1):1–7
Babana AH, Dicko AH, Maïga K, Traoré D (2013) Characterization of rock phosphate-solubilizing microorganisms isolated from wheat (*Triticum aestivum* L.) rhizosphere in Mali. J Microbiol Res 1(1):1–6
Babana AH, Kassogué A, Dicko AH, Maîga K, Samaké F, Traoré D, Fané R, Faradji FA (2016) Development of a biological phosphate fertilizer to improve wheat (*Triticum aestivum* L.) production in Mali. Procedia Eng 138:319–324
Bal HB, Das S, Dangar TK, Adhya TK (2013) ACC deaminase and IAA producing growth promoting bacteria from the rhizosphere soil of tropical rice plants. J Basic Microbiol 53:972–984
Barea JM, Toro M, Orozco MO, Campos E, Azcón R (2002) The application of isotopic 32P and 15N-dilution techniques to evaluate the interactive effect of phosphate-solubilizing rhizobacteria, mycorrhizal fungi and *Rhizobium* to improve the agronomic efficiency of rock phosphate for legume crops. Nutr Cycl Agroecosyst 63:35–42
Basharat N, Sobita S, Vijay SR (2011) Effect of *Pseudomonas fluorescens* on *Fusarium oxysporum* f. sp. gladioli causing cormrot disease of gladiolus. J Plant Dis Sci 6:51–53
Behera BC, Singdevsachan SK, Mishra RR, Dutta SK, Thatoi HN (2014) Diversity, mechanism and biotechnology of phosphate solubilizing microorganism in mangrove- a review. Biocatal Agric Biotechnol 3(2):97–110
Billah M, Bano A (2015) Role of plant growth promoting rhizobacteria in modulating the efficiency of poultry litter composting with rock phosphate and its effect on growth and yield of wheat. Waste Mgmt Res 33:63–72

Bulut S (2013) Evaluation of yield and quality parameters of phosphorous-solubilizing and N-fixing bacteria inoculated in wheat (*Triticum aestivum* L.) Turk J Agric For 37:545–554
Canbolat MY, Bilen S, Cakmakci RS, Ahin F, Aydın A (2006) Effect of plant growth-promoting bacteria and soil compaction on barley seedling growth, nutrient uptake, soil properties and rhizosphere microflora. Biol Fertil Soils 42:350–357
Chaitra R, Patil VS (2007) Integrated nutrient management studies in China aster (*Callistephus chinensis* Nees) cv. Kamini Karnataka. J Agric Sci 20:689–690
Chen YP, Rekha PD, Arunshen AB, Lai WA, Young CC (2006a) Phosphate solubilizing bacteria from subtropical soil and their tricalcium phosphate solubilizing abilities. Appl Soil Ecol 34:33–41
Chen BD, Zhu YG, Smith FA (2006b) Effect of *arbuscular mycorrhizal* inoculation on uranium and arsenic accumulation by Chinese brake fern (*Pteris vittata* L.) from uranium mining impacted soil. Chemosphere 62:1464–1473
Chitraselvi RPE, Kalidass S, Kant R (2015) Efficiency of rhizosphere bacteria in production of indole acetic acid, siderophore and phosphate solubilization. Int J Chem Tech Res 7(6):2557–2564
Clark RB, Zeto SK (2000) Mineral acquisition by arbuscular mycorrhizal plants. J Pl Nutr 23:867–902
Coombs JT, Franco CMM (2003) Isolation and identification of actinobacteria from surface-sterilized wheat roots. Appl Environ Microbiol 69:5603–5608
Coutinho FP, Felix WP, Yano-Melo AM (2012) Solubilization of phosphates *in vitro* by *Aspergillus* spp. and *Penicillium* spp. Ecol Eng 42:85–89
Dalve PD, Mane SV, Nimbalkar RR (2009) Effect of biofertilizers on growth, flowering and yield of gladiolus. Asian J Hortic 4:227–229
Datta C, Basu P (2000) Lndole acetic acid production by a *Rhizobium* species from root nodules of a leguminous shrub *Cajanus cajan*. Microbiol Res 155:123–127
Datta B, Chakrabartty PK (2014) Siderophore biosynthesis genes of *Rhizobium* sp. isolated from *Cicer arietinum* L. 3 Biotech 4:391–401
Dwivedi BS, Singh VK, Dwivedi V (2004) Application of phosphate rock, with or without *Aspergillus awamori* inoculation, to meet phosphorus demands of rice–wheat systems in the indo–Gangetic Plains of India. Austr J Exptl Agric 44:1041–1050
Egamberdieva D, Kamilova F, Validov S, Gafurova L, Kucharova Z, Lugtenberg B (2008) High incidence of plant growth-stimulating bacteria associated with the rhizosphere of wheat grown on salinated soil in Uzbekistan. Environ Microbiol 10:1–9
Eman AA, Monem AE, Saleh MMS, Mostafa EAM (2008) Minimizing the quantity of mineral nitrogen fertilization grapevine by using humic acid organic and biofertilizers. Res J Agric Sci 4:46–50
Frank O, Julius O (2012) Some characteristics of a plant growth promoting *Enterobacter* sp. isolated from the roots of maize. Adv Microbiol 2:368–374
Gaiero JR, Mc Call CA, Thompson KA, Day NJ, Best AS, Dunfield KE (2013) Inside the root microbiome: bacterial root endophytes and plant growth promotion. Amer J Bot 100:1738–1750
Ghosh P, Rathinasabapathi B, Ma LQ (2015) Phosphorus solubilization and plant growth enhancement by arsenic-resistant bacteria. Chemosphere 134:1–6
Glick BR, Todorovic B, Czarny J, Cheng Z, Duan J, McConkey B (2007) Promotion of plant growth by bacterial ACC deaminase. Crit Rev Plant Sci 26:227–242
Goes KCGPD, Cattelan AJ, De Carvalho CGP (2012) Biochemical and molecular characterization of high population density bacteria isolated from sunflower. J Microbiol Biotechnol 22:437–447
Goldstein AH (1986) Bacterial mineral phosphate solubilization: historical perspective and future prospects. Am J Alternat Agric 1:57–65
Griffith DB (2009) Efficient fertilizer use – phosphorus. Sci Agric 433:23–67
Gulati A, Sharma N, Vyas P, Sood S, Rahi P, Pathania V, Prasad R (2010) Organic acid production and plant growth promotion as a function of phosphate solubilization by *Acinetobacter rhizosphaerae* strain BIHB 723 isolated from the cold deserts of the trans-Himalayas. Arch Microbiol 192:975–983

Gull M, Hafeez FY, Saleem M, Malik KA (2004) Phosphorus uptake and growth promotion of chickpea by co-inoculation of mineral phosphate solubilising bacteria and a mixed rhizobial culture. Aust J Exp Agric 44:623–628

Hamdali H, Hafidi M, Virolle MJ, Ouhdouch Y (2008) Growth promotion and protection against damping-off of wheat by two rock phosphate solubilizing actinomycetes in a P-deficient soil under greenhouse conditions. Appl Soil Ecol 40(3):510–517

Haque MA, Sattar MA, Islam MR, Hashem MA, Khan MK (2013) Performance of phosphate solubilizing bacteria with various phosphorus levels on wheat in pot culture. J Environ Sci Nat Res 6:221–226

Hardoim PR, van Overbeek LS, Berg G, Pirttilä AM, Compant S, Campisano A, Döring M, Sessitsch A (2015) The hidden world within plants: ecological and evolutionary considerations for defining functioning of microbial endophytes. Microbiol Mol Biol Rev 79:293–320

Hardoim PR, Van Overbeek LS, Van Elsas JD (2008) Properties of bacterial endophytes and their proposed role in plant growth. Trends Microbiol 16:463–471

Herrera SD, Grossi C, Zawoznik M, Groppa MD (2016) Wheat seeds harbour bacterial endophytes with potential as plant growth promoters and biocontrol agents of *Fusarium graminearum*. Microbiol Res 186:37–34

Hossain MA, Begum S, Rahman AKMM, Arabinda S, Salahuddin ABM (1996) Growth analysis of mustard and rapeseed in relation to grain filling period and yield potential. J Agric Res 34:59–369

Hossain MB, Sattar MA (2014) Effect of inorganic phosphorus fertilizer and inoculants on yield and phosphorus use efficiency of wheat. J Environ Sci Natural Res 7:75–79

Illmer P, Schinner F (1992) Solubilization of inorganic phosphates by microorganisms isolated from forest soil. Soil Biol Biochem 24:389–395

Islam MT, Hossain MM (2012) Plant probiotics in phosphorus nutrition in crops, with special reference to Rice. In: Maheshwari DK (ed) Bacteria in agrobiology: plant probiotics. Springer, Berlin/Heidelberg, pp 325–363

Jha P, Kumar A (2009) Characterization of novel plant growth promoting endophytic bacterium *Achromobacter xylosoxidans* from wheat plant. Microb Ecol 58:179–188

Jha Y, Subramanian RB (2014) Characterization of root-associated bacteria from paddy and its growth promotion efficacy. 3 Biotech 4:325–330

Jog R, Nareshkumar G, Rajkumar S (2012) Plant growth promoting potential and soil enzyme production of the most abundant *Streptomyces* spp. from wheat rhizosphere. J Appl Microbiol 113:1154–1164

Jog R, Pandya M, Nareshkumar G, Rajkumar S (2014) Mechanism of phosphate solubilization and antifungal activity of *Streptomyces* spp. isolated from wheat roots and rhizosphere and their application in improving plant growth. Microbiol 160:778–788

Joo HS, Hirai M, Shoda M (2005) Characteristics of ammonium removal by heterotrophic nitrification-aerobic denitrification by *Alcaligenes faecalis* no. 4. J Biosci Bioeng 100:184–191

Kang SM, Radhakrishnan R, You YH, Joo GJ, Lee IJ, Lee KE, Kim JH (2014) Phosphate solubilizing *Bacillus megaterium* mj 1212 regulates endogenous plant carbohydrates and amino acid contents to promote mustard plant growth. Indian J Microbiol 54:427–433

Kapri A, Tewari L (2010) Phosphate solubilization potential and phosphatase activity of rhizospheric *Trichoderma* spp. Braz J Microbiol 41:787–795

Kaur G, Reddy MS (2015) Effects of phosphate-solubilizing bacteria, rock phosphate and chemical fertilizers on maize-wheat cropping cycle and economics. Pedosphere 25:428–437

Kenasa G, Jida M, Assefa F (2014) Characterization of phosphate solubilizing faba bean (*Vicia faba L.*) nodulating rhizobia isolated from acidic soils of Wollega, Ethiopia. Sci Technol Arts Res J 3:11–17

Khan MS, Zaidi A, Ahemad M, Oves M, Wani PA (2010) Plant growth promotion by phosphate solubilizing fungi – current perspective. Arch Agron Soil Sci 56:73–98

Khan MS, Zaidi A (2007) Synergistic effects of the inoculation with plant growth-promoting rhizobacteria and an arbuscular mycorrhizal fungus on the performance of wheat. Turk J Agric For 31:355–362

Khan MS, Zaidi A, Wani PA (2007) Role of phosphate solubilizing microorganisms in sustainable agriculture: a review. AgronSustain Dev 27:29–43

Khan MS, Zaidi A, Ahmad E (2014) Mechanism of phosphate solubilization and physiological functions of phosphate solubilizing microorganisms. In: Phosphate solubilising microorganisms: principles and application of microphos technology. Springer, Switzerland; p 31–62

Kumar A, Maurya BR, Raghuwanshi R (2014a) Isolation and characterization of PGPR and their effect on growth, yield and nutrient content in wheat (*Triticum aestivum* L.) Biocatal Agric Biotechnol 3:121–128

Kumar A, Shukla UK, Singh A, Poonam AK, Prasad S, Singh SK, Kumar D (2014) Evaluation of *Pseudomonas* isolates from wheat for some important plant growth promoting traits. Afr J Microbiol Res 8:2604–2608

Kumar A, Maurya BR, Raghuwanshi R (2015) Characterization of bacterial strains and their impact on plant growth promotion and yield of wheat and microbial populations of soil. Afr J Agri Res 10:1367–1375

Kumar V, Narula N (1999) Solubilization of inorganic phosphates and growth emergence of wheat as affected by *Azotobacter chroococcum* mutants. Biol Fertil Soils 28(3):301–305

Kumar V, Behl RK, Narula N (2001) Establishment of phosphate-solubilizing strains of *Azotobacter chroococcum* in the rhizosphere and their effect on wheat cultivars under green house conditions. Microbiol Res 156:87–93

Kundu BS, Nehra K, Yadav R, Tomar M (2009) Biodiversity of phosphate solubilizing bacteria in rhizosphere of chickpea, mustard and wheat grown in different regions of Haryana. Indian J Microbiol 49:120–127

Last FT (1955) Seasonal incidence of *Sporobolomyces* on cereal leaves. Trans Br Mycol Soc 38:221–239

Lavania M, Nautiyal CS (2013) Solubilization of tricalcium phosphate by temperature and salt tolerant *Serratia marcescens* NBRI1213 isolated from alkaline soils. Afr J Microbiol Res 7:4403–4413

Li RX, Cai F, Pang G, Shen QR, Li R, Chen W (2015) Solubilisation of phosphate and micronutrients by *Trichoderma harzianum* and its relationship with the promotion of tomato plant growth. PLoS One 10(6):e0130081. doi:10.1371/journal.pone.0130081

Lindsay WL, Vlek PLG, Chien SH (1989) Phosphate minerals. In: Dixon JB, Weed SB, (eds) Soil environment, 2nd ed. Soil Science Society of America, Madison, pp 1089–1130

Lipping Y, Jiatao X, Daohong J, Yanping F, Guoqing L, Fangcan L (2008) Antifungal substances produced by *Penicillium oxalicum* strain PY-1 potential antibiotics against plant pathogenic fungi. World J Microbiol Biotechnol 24:909–915

Ma Y, Rajkumar M, Freitas H (2009) Inoculation of plant growth promoting bacterium *Achromobacter xylosoxidans* strain Ax10 for the improvement of copper phytoextraction by *Brassica juncea*. J Environ Manag 90:831–837

Magnucka EG, Pietr SJ (2015) Various effects of fluorescent bacteria of the genus *Pseudomonas* containing ACC deaminase on wheat seedling growth. Microbiol Res 181:112–119

Majeed A, Abbasi MK, Hameed S, Imran A, Rahim N (2015) Isolation and characterization of plant growth-promoting rhizobacteria from wheat rhizosphere and their effect on plant growth promotion. Front Microbiol 6:1–10

Maliha R, Samina K, Najma A, Sadia A, Farooq L (2004) Organic acids production and phosphate solubilization by phosphate solubilizing microorganisms under in vitro conditions. Pak J Biol Sci 7:187–196

Mardad I, Serrano A, Soukri A (2013) Solubilization of inorganic phosphate and production of organic acids by bacteria isolated from a Moroccan mineral phosphate deposit. Afr J Microbiol Res 7:626–635

Marra LM, Longatti SMO, Soares CRFS, Lima JM, Olivares FL, Moreira FMS (2015) Initial pH of medium affects organic acid production but do not affect phosphate solubilization. Braz J Microbiol 46:367–375

Mattson WJ Jr (1980) Herbivory in relation to plant nitrogen content. Annu Rev Ecol Syst 11:119–161

Mbai FN, Magiri EN, Matiru VN, Nganga J, Nyambati VCS (2013) Isolation and characterisation of bacterial root endophytes with potential to enhance plant growth from kenyan basmati rice. Amer Intern J Contem Res 3:25–40

Mehnaz S, Baig DN, Lazarovits G (2010) Genetic and phenotypic diversity of plant growth promoting rhizobacteria isolated from sugarcane plants growing in Pakistan. J Microbiol Biotechnol 20:1614–1623

Mihalache G, Zamfirache MM, Mihasan M, Ivanov I, Stefan M, Raus L (2015) Phosphate solubilizing bacteria associated with runner bean rhizosphere. Arch Biol Sci 67:793–800

Minaxi SJ, Chandra S, Nain L (2013) Synergistic effect of phosphate solubilizing rhizobacteria and arbuscular mycorrhiza on growth and yield of wheat plants. J Soil Sci Pl Nutr 13:511–525

Mishra PK, Bisht SC, Ruwari P, Selvakumar G, Joshi GK, Bisht JK, Bhatt JC, Gupta HS (2011) Alleviation of cold stress in inoculated wheat (*Triticum aestivum* L.) seedlings with psychrotolerant pseudomonads from NW Himalayas. Arch Microbiol 193:497–513

Moreira FS, Costa PB, Rd S, Beneduzi A, Lisboa BB, Vargas LK, Passaglia LM (2016) Functional abilities of cultivable plant growth promoting bacteria associated with wheat (*Triticum aestivum*L.) crops. Genet Mol Biol 39:111–121

Mujahid TY, Siddiqui K, Ahmed R, Kazmi SU, Ahmed N (2014) Isolation and partial characterization of phosphate solubilizing bacteria isolated from soil and marine samples. Pak J Pharm Sci 27:1483–1490

Mwajita MR, Murage H, Tani A, Kahangi EM (2013) Evaluation of rhizosphere, rhizoplane and phyllosphere bacteria and fungi isolated from rice in Kenya for plant growth promoters. SpringerPlus 2:606

Naqvi SDY, Ahmad S (2012) Effect of *Pseudomonas fluorescens* on *Fusarium oxysporum* f.sp. *gladioli* causing corm rot disease of gladiolus. J Stored Prod Postharvest Res 3:49–51

Naresh R, Tomar S, Purushottam S, Kumar D, Pratap B, Kumar V, Nanher A (2014) Testing and evaluation of planting methods on wheat grain yield and yield contributing parameters in irrigated agro-ecosystem of western Uttar Pradesh, India. Afr J Agric Res 9:176–182

Narula N, Kumar V, Behl RK (2005) Effect of phosphate solubilizing strains of *Azotobacter chroococcum* on yield traits and their survival in the rhizosphere of wheat genotypes under field conditions. Acta Agronomica Hungarica 49:141–149

Nath R, Sharma GD, Barooah M (2012) Efficiency of tricalcium phosphate solubilization by two different Endophytic *Penicillium sp.* isolated from tea (Camellia sinensis L.) Eur J Exp Biol 2:1354–1358

Norrish K, Rosser H (1983) Mineral phosphate soils: an Australianviewpoint. Sponsored by the Division of Soils, Commonwealth Scientific and Industrial Research Organization. Academic Press/CSIRO, Melbourne /London . pp 335–361

Ogut M, Er F (2016) Mineral composition of field grown winter wheat inoculated with phosphorus solubilizing bacteria at different plant growth stages. J Plant Nutr 39:479–490

Ogut M, Er F, Neumann G (2011) Increased proton extrusion of wheat roots by inoculation with phosphorus solubilising microorganims. Plant Soil 339:285–297

Ogut M, Er F, Kandemir N (2010) Phosphate solubilization potentials of soil *Acinetobacter* strains. Biol Fertil Soils 46:707–715

Onyia CE, Anyanwu CU (2013) Comparative study on solubilization of tri-calcium phosphate (TCP) by phosphate solubilizing fungi (PSF) isolated from Nsukka pepper plant rhizosphere and root free soil. J Yeast Fungal Res 4:52–57

Oteino N, Lally RD, Kiwanuka S, Lloyd A, Ryan D, Germaine KJ, Dowling DN (2015) Plant growth promotion induced by phosphate solubilizing endophytic *Pseudomonas* isolates. Front Microbiol 6:745. doi:10.3389/fmicb.2015.00745

Padmavathi T (2015) Optimization of phosphate solubilization by *Aspergillus niger* using plackett-burman and response surface methodology. J Soil Sci Pl Nutr 15(3):781–793

Panhwar QA, Naher UA, Jusop S, Othman R, Latif MA, Ismail MR (2014) Biochemical and molecular characterization of potential phosphate solubilizing bacteria in acid sulphate soils and their beneficial effects on rice growth. PLoS One 9 : PMC4186749; e97241

Parks EJ, Olson GJ, Brinckman FE, Baldi F (1990) Characterization by high performance liquid chromatography (HPLC) of the solubilization of phosphorus in iron ore by a fungus. J Ind Microbiol Biotechnol 5:183–189

Pikovskaya RI (1948) Mobilization of phosphorus in soil in connection with vital activity of some microbial species. Microbiologiya 17:362–370

Priya S, Panneerselvam T, Sivakumar T (2013) Evaluation of indole-3-acetic acid in phosphate solubilizing microbes isolated from rhizosphere soil. Int J Curr Microbiol App Sci 2:29–36

Promwee A, Issarakraisila M, Intana W, Chamswarng C, Yenjit P (2014) Phosphate solubilization and growth promotion of rubber tree (*Hevea brasiliensis* Muell. Arg.) by *Trichoderma* strains. J Agric Sci 6:8. doi:10.5539/jas.v6n9p8

Puri A, Padda KP, Chanway CP (2015) Can a diazotrophic endophyte originally isolated from lodgepole pine colonize an agricultural crop (corn) and promote its growth? Soil Biol Biochem 89:210–216

Qiao ZW, Hong JP, Xie YH, Li LX (2013) Screening, identification and phosphate-solubilizing characteristics of *Rahnella* sp. phosphate-solubilizing bacteria in calcareous soil. Ying Yong Sheng Tai Xue Bao 24:2294–2300

Raja AK, Shah KH, Aslam M, Memon MY (2002) Response of phosphobacterial and mycorrhizal inoculation in wheat. Asian J Plant Sc 1:322–323

Rajput L, Imran A, Mubeen F, Hafeez FY (2013) Salt tolerant PGPR strain *Planococcus rifietoensis* promotes the growth and yield of wheat (*Triticum aestivum* L.) cultivated in saline soil. Pak J Bot 45:1955–1962

Ram H, Malik SS, Dhaliwal SS, Kumar B, Singh Y (2015) Growth and productivity of wheat affected by phosphorus-solubilizing fungi and phosphorus levels. Plant Soil Environ 61:122–126

Ramesh A, Sharma SK, Sharma MP, Yadav N, Joshi OP (2014) Plant growth-promoting traits in *Enterobacter cloacae subsp. dissolvens* MDSR9 isolated from soybean rhizosphere and its impact on growth and nutrition of soybean and wheat upon inoculation. Agri Res 3:53–66

Ramlakshmi R, Bharathiraja S (2015) AM fungi and phosphate solubilizing bacteria (*Paenibacillus polymyxa*) as a potential bioinoculant for marigold. Intern J Curr Res 7:12264–12266

Rana A, Joshi M, Prasanna R, Shivay YS, Nain L (2012) Biofortification of wheat through inoculation of plant growth promoting rhizobacteria and cyanobacteria. Eur J Soil Biol 50:118–126

Ranjan A, Mahalakshmi M, Sridevi M (2013) Isolation and characterization of phosphate solubilizing bacterial species from different crop fields of Salem, Tamil Nadu, India. Intern J Nutr, Pharmacol, Neurol dis 3:29–33

Raven JA, Beardall J, Flynn KJ, Maberly SC (2009) Phagotrophy in the origins of photosynthesis in eukaryotes and as complementary mode of nutrition in phototrophs: relation to Darwin's insectivorous plants. J Exp Bot 60:3975–3987

Ray DK, Ramankutty N, Mueller ND, West PC, Foley JA (2012) Recent patterns of crop yield growth and stagnation. Nat Commun 3:1293

Reena TD, Deepthi H, Pravitha MS, Lecturer D (2013) Isolation of phosphate solubilizing bacteria and fungi from rhizospheres soil from banana plants and its effect on the growth of *Amaranthus cruentus* L. IOSR J Pharmacy Biol Sci 5:06–11

Remans R, Beebe S, Blair M, Manrique G, Tovar E, Rao I, Croonenborghs A, Torres-Gutierrez R, El-Howeity M, Michiels J, Vanderleyden J (2008) Physiological and genetic analysis of root responsiveness to auxin-producing plant growth-promoting bacteria in common bean (*Phaseolus vulgaris* L.) Plant Soil 302:149–161

Resende MP, Jakoby ICMC, dos Santos LCR, Soares MA, Pereira FD, Souchie EL, Silva FG (2014) Phosphate solubilization and phytohormone production by endophytic and rhizosphere *Trichoderma* isolates of guanandi (*Calophyllum brasiliense* Cambess). Afr J Microbiol Res 8(27):2616–2623

Revillas JJ, Rodelas B, Pozo C, Martı'nez-Toledo V, Gonza'lez-Lo´ pez J (2000) Production of B-group vitamins by two *Azotobacter* strains with phenolic compounds as sole carbon source under diazotrophic and adiazotrophic conditions. J Appl Microbiol 89:486–493

Robinson RJ, Fraaije BA, Clark IM, Jackson RW, Hirsch PR, Mauchline TH (2015) Endophytic bacterial community composition in wheat (*Triticum aestivum*) is determined by plant tissue type, developmental stage and soil nutrient availability. Plant Soil 405:381–396

Rodríguez H, Fraga R (1999) Phosphate solubilizing bacteria and their role in plant growth promotion. Biotechnol Adv 17:319–339

Ruinen J (1956) Occurrence of *Beijerinckia* species in the phyllosphere. Nature 178:220–221

Saber Z, Pirdashti H, Esmaeili M, Abbasian A, Heidarzadeh A (2012) Response of wheat growth parameters to co-inoculationof plant growth promoting Rhizobacteria (PGPR) and different levels of inorganic nitrogen and phosphorus. World Appl Sci J 16:213–219

Safari D, Jamali F, Nooryazdan HR, Bayat F (2016) Screening fluorescent pseudomonads isolated from wheat rhizosphere for plant growth- promoting and salt tolerance properties. Biol Forum – An Int Journal 8:35–42

Saif S, Khan MS, Zaidi A, Ahmad E (2014)Role of phosphate solubilizing actinomycetes in plant growth promotion: current perspective. In: Khan MS, Zaidi A, Musarrat J (eds) Phosphate solubilising microorganisms: principles and application of microphos technology. Springer, Switzerland; pp 137–156

Salma Z, Sindhu SS, Ahlawat VP (2014) Suppression of *Fusarium* wilt disease in gladiolus by using rhizobacterial strains. J Crop Weed 10:466–471

Sane SA, Mehta SK (2015) Isolation and evaluation of rock phosphate solubilizing fungi as potential biofertilizer. J Fertil Pestic 6(2):156–160

Sarfraz M, Abid M, Mehdi SM (2009) External and internal phosphorus requirements of wheat in Rasulpur soil series of Pakistan. Soil Environ 28:38–44

Sarkar A, Islam T, Biswas GC, Alam S, Hossain M, Talukder NM (2012) Screening for phosphate solubilizing bacteria inhabiting the rhizoplane of rice grown in acidic soil in Bangladesh. ActaMicrobiol Immunol Hung 59:199–213

Sarker A, Talukder NM, Islam MT (2014) Phosphate solubilizing bacteria promote growth and enhance nutrient uptake by wheat. Plant Sci Today 1:86–93

Satyanandam T, Babu K, Rosaiah G, Vijayalakshmi M (2014) Screening of *Rhizobium* strains isolated from the root nodules of *Vigna mungo* cultivated in rice fallows for their phosphate solubilizing ability and enzymatic activities. Brit Microbiol Res J4:996–1006

Saxena J, Minaxi, Jha A (2014) Impact of a phosphate solubilizing bacterium and an arbuscular mycorrhizal fungus (*Glomus etunicatum*) on growth, yield and P concentration in wheat plants. Clean (Weinh) 42:1248–1252

Saxena J, Chandra S, Nain L (2013) Synergistic effect of phosphate solubilizing rhizobacteria and arbuscular mycorrhiza on growth and yield of wheat plants. J Soil Sci Pl Nutr 13: 511–525

Scervino JM, Mesa MP, Mónica ID, Recchi M, Moreno NS, Godeas A (2010) Soil fungal isolates produce different organic acid patterns involved in phosphate salts solubilization. Biol Fertil Soils 46:755–763

Schoebitz M, Ceballos C, Ciampi L (2013) Effect of immobilized phosphate solubilizing bacteria on wheat growth and phosphate uptake. J Soil Sci Plant Nutr 13(1):1–10

Selvakumar G, Mohan M, Kundu S, Gupta AD, Joshi P, Nazim S, Gupta HS (2008) Cold tolerance and plant growth promotion potential of *Serratia marcescens* strain SRM (MTCC 8708) isolated from flowers of summer squash (Cucurbita Pepo). Lett Appl Microbiol 46: 171–175

Seghers D, Wittebolle L, Top EM, Verstraete W, Siciliano SD (2004) Impact of agricultural practices on the *Zea mays* L. endophytic community. Appl Environ Microbiol 70:1475–1482

Shanmugam V, Kanoujia N, Singh M, Singh S, Prasad R (2011) Biocontrol of vascular wilt and corm rot of gladiolus caused by *Fusarium oxysporum* f. Sp. gladioli using plant growth promoting rhizobacterial mixture. Crop Prot 30:807–813

Sharma AK (2002) Bio-fertilizers for sustainable agriculture. AgrobiosIndia, Jodhpur, p 407

Sharma SK, Johri BN, Ramesh A, Joshi OP, Prasad SS (2011) Selection of plant growth-promoting *Pseudomonas* spp. that enhanced productivity of soybean-wheat cropping system in Central India. J Microbiol Biotechnol 21:1127–1142

Sharma A, Rawat US, Yadav BK (2012) Influence of phosphorus levels and phosphorus solubilizing fungi on yield and nutrient uptake by wheat under sub-humid region of Rajasthan, India. ISRN Agronomy. Article ID 234656, 9 pages. doi:10.5402/2012/234656

Sharma SB, Sayyed RZ, Trivedi MH, Gobi TA (2013) Phosphate solubilizing microbes: sustainable approach for managing phosphorus deficiency in agricultural soils. SpringerPlus 2:587

Shrivastave M, D'Souza SF (2014) Bio-solubilization of rock phosphate and plant growth promotion by *Aspergillus niger* TMPS1 in ultisol and vertisol. International symposium on managing soils for food security and climate change adaptation and mitigation. Vienna, Austria

Sial NA, Memon MY, Abro SA, Shah JA, Depar ND, Abbas M (2015) Effect of phosphate solubilizing bacteria (*Bacillus megaterium*) and phosphate fertilizer on yield and yield components of wheat. Pak J Biotechnol 12:35–40

Singh H, Reddy MS (2011) Effect of inoculation with phosphate solubilizing fungus on growth and nutrient uptake of wheat and maize plants fertilized with rock phosphate in alkaline soils. Eur J Soil Biol 47:30–34

Singh P, Kumar V, Agrawal S (2014) Evaluation of phytase producing bacteria for their plant growth promoting activities. Int J Microbiol 2014; Article eID 426483, 7 pages, doi:10.1155/2014/426483

Singh RP, Jha PN (2015) Plant growth promoting potential of ACC deaminase rhizospheric bacteria isolated from *Aerva javanica*: a plant adapted to saline environments. Int J Curr Microbiol App Sci 4(7):142–152

Singh N, Pandey P, Dubey RC, Maheshwari DK (2008) Biological control of root rot fungus *Macrophomina phaseolina* and growth enhancement of *Pinus roxburghii* (Sarg.) by rhizosphere competent *Bacillus subtilis* BN1. World J Microbiol Biotechnol 24:1669–1679

Song OR, Lee SJ, Lee YS, Lee SC, Kim KK, Choi YL (2008) Solubilization of insoluble inorganic phosphate by *Burkholderia cepacia* DA 23 isolated from cultivated soil. Braz J Microbiol 39:151–156

Sönmez F, Tüfenkçi Ş (2015) Investigation the effects of different doses organic fertilizers and phosphate solubilizing bacterias on yield and nutrient contents in chickpea (*Cicer arietinum* L.) Int J Second Metab (IJSM) 2:43–53

Surapat W, Pakahuta C, Rattanachaikunsopon P, Aimi T, Boonlue S (2013) Characteristics of phosphate solubilization by phosphate solubilizing bacteria isolated from agricultural chilli soil and their efficiency on the growth of chilli (*Capsicum frutescens* L. Cv. Hua Rua). Chiang Mai J Sci 40:11–25

Susilowati LE, Syekhfani (2014) Characterization of phosphate solubilizing bacteria isolated from Pb contaminated soils and their potential for dissolving tricalcium phosphate. J Degrad Min Lands Manag 1:57–62

Tahir M, Mirza MS, Zaheer A, Dimitrov MR, Smidt H, Hameed S (2013) Isolation and identification of phosphate solubilizer *Azospirillum, Bacillus* and *Enterobacter* strains by 16SrRNA sequence analysis and their effect on growth of wheat (*Triticum aestivum* L.) Aus J Crop Sci 7:1284

Tallapragada P, Usha S (2012) Phosphate-solubilizing microbes and their occurrence in the rhizospheres of *Piper betel* in Karnataka, India. Turk J Biol 36:25–35

Timsina J, Connor DJ (2001) Productivity and management of rice-wheat cropping systems: issues and challenges. Field Crop Res 59:93–132

Tiwari S, Singh P, Tiwari R, Meena KK, Yandigeri M, Singh DP, Arora DK (2011) Salt-tolerant rhizobacteria-mediated induced tolerance in wheat (*Triticum aestivum*) and chemical diversity in rhizosphere enhance plant growth. Biol Fertil Soils 47:907–916

Tenorio-Salgado S, Tinoco R, Vazquez-Duhalt R, Caballero-Mellado J, Perez-Rueda E (2013) Identification of volatile compounds produced by the bacterium *Burkholderia tropica* that inhibit the growth of fungal pathogens. Bioengineered 4:236–243

Tomar US, Tomar IS, Badaya AK (1998) Response of chemical and biofertilizer on some matric traits in wheat. Crop Res 16:408–410

Turk MA, Tawaha AM (2002) Impact of seedling rate, seeding date, rate and method of phosphorus application in faba bean (*Vicia faba* L. Minor) in the absence of moisture stress. Biotechnol Agron Soc Environ 6:171–178

Upadhyay SK, Singh JS, Saxena AK, Singh DP (2012) Impact of PGPR inoculation on growth and antioxidant status of wheat under saline conditions. Plant Biol 14:605–611

USDA (2010) Grain Report No. IN 1011. Washington, Global Agricultural Information Network

Venkateswarlu B, Rao AV, Raina P (1984) Evaluation of phosphorous solubilization by microorganisms isolated from aridisols. J Ind Soc Soil Sc 32:273–277

Venieraki A, Dimou M, Pergalis P, Kefalogianni I, Chatzipavlidis I, Katinakis P (2011) The genetic diversity of culturable nitrogen-fixing bacteria in the rhizosphere of wheat. Microb Ecol 61:277–285

Verma P, Yadav AN, Kazy SK, Saxena AK, Suman A (2013) Elucidating the diversity and plant growth promoting attributes of wheat (*Triticum aestivum*) associated acidotolerant bacteria from southern hills zone of India. Natl J Life Sci 10:219–226

Verma P, Yadav AN, Kazy SK, Saxena AK, Suman A (2014) Evaluating the diversity and phylogeny of plant growth promoting bacteria associated with wheat (*Triticum aestivum*) growing in central zone of India. Int J Curr Microbiol App Sci 3:432–447

Verma P, Yadav AN, Khannam KS, Mishra S, Kumar S, Saxena AK, Suman A (2016a) Appraisal of diversity and functional attributes of thermotolerant wheat associated bacteria from the peninsular zone of India. Saudi J Biol Sci. doi:10.1016/j.sjbs.2016.01.042

Verma P, Yadav AN, Khannam KS, Kumar S, Saxena AK, Suman A (2016b) Molecular diversity and multifarious plant growth promoting attributes of bacilli associated with wheat (*Triticum aestivum* L.) rhizosphere from six diverse agro-ecological zones of India. J Basic Microbiol 56:44–58

Verma P, Yadav AN, Khannam KS, Panjiar N, Kumar S, Saxena AK, Suman A (2015a) Assessment of genetic diversity and plant growth promoting attributes of psychrotolerant bacteria allied with wheat (*Triticum aestivum*) from the northern hills zone of India. Ann Microbiol 65:1885–1899

Vessey KJ (2003) Plant growth promoting rhizobacteria as biofertilizers. Plant Soil 255:123–112

Vijayabharathi R, Sathya A, Gopalakrishnan S (2016) A renaissance in plant growth-promoting and biocontrol agents by endophytes. In: Microbial inoculants in sustainable agricultural productivity. Springer, India; pp 37–60.

Wahid OA, Mehana TA (2000) Impact of phosphate-solubilizing fungi on the yield and phosphorus-uptake by wheat and faba bean plants. Microbiol Res 5(3):221–227

Wakelin S, Warren R, Harvey P, Ryder M (2004) Phosphate solubilization by Penicillium spp. closely associated with wheat roots. Biol Fertil Soils 40:36–43

Walpola BC, Yoon MH (2013a) *In vitro* solubilization of inorganic phosphates by phosphate solubilizing microorganisms. Afr J Microbiol Res 7:3534–3541

Walpola BC, Yoon MH (2013b) Phosphate solubilizing bacteria: assessment of their effect on growth promotion and phosphorus uptake of mung bean (*Vigna radiata* [L.]R. Wilzcek). Chilean J Agric Res 73:275–281

Walpola C, Yoon MH (2013c) Isolation and characterization of phosphate solubilizing bacteria and their co-inoculation efficiency on tomato plant growth and phosphorous uptake. Afr J Microbiol Res 7:266–275

Whitelaw MA (2000) Growth promotion of plants inoculated with phosphate solubilizing fungi. Adv Agron 69:99–151

Wibisono MG, Veneklaas E, Mendham DS, Hardiyanto EB (2015) Nitrogen fixation of *Acacia mangium* Willd. From two seed sources grown at different levels of phosphorus in an ultisol, South Sumatra, Indonesia. South Forests A J Forest Sci 77:59–64

Xiao C, Chi R, He H, Qiu G, Wang D, Zhang W (2009) Isolation of phosphate-solubilizing fungi from phosphate mines and their effect on wheat seedling growth. Appl Biochem Biotechnol 159(2):330–342

Xiao CQ, Chi RA, Hu LH (2013) Solubilization of aluminum phosphate by specific *Penicillium* spp. J Cent South Univ 20:2109–2114

Yasser MM, Mousa ASM, Massoud ON, Nasr SH (2014) Solubilization of inorganic phosphate by phosphate solubilizing fungi isolated from Egyptian soils. J Biol Earth Sci 4:B83–B90

Younis A, Riaz A, Ikram S, Nawaz T, Hameed M, Fatima S, Batool R, Ahmad F (2013) Salinity-induced structural and functional changes in three cultivars of *Alternanthera bettzickiana* (Regel) G. Nicholson. Turk J Agri For 37:674–687

Zabihi HR, Savaghebi GR, Khavazi K, Ganjali A, Miransari M (2011) Pseudomonas bacteria and phosphorous fertilization, affecting wheat (*Triticum aestivum* L.) yield and P uptake under greenhouse and field conditions. Acta Physiol Planta 33:145–152

Zaidi A, Ahmad E, Khan MS (2014) Role of phosphate solubilising microbes in the management of plant diseases. In: Khan MS,Zaidi A, Musarrat J (eds) Phosphate solubilising microorganisms: principles and application of Microphos technology. Springer, Switzerland; pp 225–256

Zaidi A, Khan MS, Ahemad M, Oves M (2009) Plant growth promotion by phosphate solubilizing bacteria. Acta Microbiol Immunol Hung 56:263–284

Zaidi A, Khan MS (2005) Interactive effect of rhizospheric microorganisms on growth yield and nutrient uptake of wheat. J Plant Nutr 28:2079–2092

Zhao Y (2010) Auxin biosynthesis and its role in plant development. Annu Rev Plant Biol 61:49–64

Zia-ul-Hassan ATS, Shah AN, Jamro GM, Rajpar I (2015) Biopriming of wheat seeds with rhizobacteria containing ACC deaminase and phosphate solubilizing activities increases wheat growth and yield under phosphorus deficiency. Pak J Agri Agril Engg Vet Sci 31:24–32

Zinniel DK, Lambrecht P, Harris NB, Feng Z, Kuezmarski D, Higley P, Ishimaru CA, Arunakumari A, Barletta RG, Vidaver AK (2002) Isolation and characterization of endophytic colonizing bacteria from agronomic crops and prairie plants. Appl Environ Microbiol 68:2198–2208

Arbuscular Mycorrhization and Growth Promotion of Peanut (*Arachis hypogaea* L.) After Inoculation with PGPR

4

Driss Bouhraoua, Saida Aarab, Amin Laglaoui, Mohammed Bakkali, and Abdelhay Arakrak

Abstract

Because of their potential to increase plant nutrition and yield, the use of some microorganisms in low-input agriculture and forestry has been addressed for successful agroecological investigations. A pot experiment was conducted to study the effects of some *plant growth-promoting rhizobacteria* (PGPR) on arbuscular mycorrhizal fungi (AMF) and growth of KP29's peanut variety grown in the northwest of Morocco. Seeds were inoculated with three *Pseudomonas* (PP22, GP70, and GR1) and two *Aeromonas* strains (PR29 and GR70). Then, they were grown in two unsterilized soils collected from subsistence farmers' fields of Laaouamra and Moulay Bousselham. Plant harvesting was made after 60 days of cultivation under growth chamber conditions, and the roots were removed and rinsed carefully. Results showed positive and negative effects of these rhizobacteria on growth and mycorrhization of peanut. Pseudomonad strains gave the greatest plant nutrient content (N, P, and K) and growth parameters. Also, bacterial inoculation had a positive impact on peanut mycorrhization by enhancing arbuscular abundance. Highest stimulation was noticed with pseudomonad strains on both soils. In addition, PR29 exhibited maximum values of mycorrhizal colonization on the soil of Laaouamra. However, the magnitude effect of inoculation on plant growth and mycorrhizal infection varied according to the origin of soils. On the other hand, only PP22 stimulated nodules formation on the

D. Bouhraoua • A. Laglaoui • M. Bakkali • A. Arakrak (✉)
Equipe de Recherche de Biotechnologies et Génie des Biomolécules (ERBGB), Faculté des Sciences et Techniques de Tangier, Tangier, Morocco
e-mail: arakrak_abdelhay@yahoo.fr

S. Aarab
Equipe de Recherche de Biotechnologies et Génie des Biomolécules (ERBGB), Faculté des Sciences et Techniques de Tangier, Tangier, Morocco

Département de Biologie, Faculté des Sciences et Techniques d'Al Hoceima, Al Hoceima, Morocco

V. Kumar et al. (eds.), *Probiotics in Agroecosystem*,
DOI 10.1007/978-981-10-4059-7_4

soil of Laaouamra. In conclusion, this study reveals that GP70, GR1, and PP22 can enhance growth, yield, and nutrient uptake of peanut. They can also enhance biological nitrogen fixation and mineral uptake in combination with AMF.

4.1 Introduction

An intensive farming practice that warrants high yield and quality requires extensive use of chemical fertilizers, which are costly and can create serious environmental problems. Large amounts of chemical fertilizers are used to replace soil nitrogen and phosphorus. Despite the deleterious environmental effects, the total amount of inorganic fertilizers used worldwide is expected to produce more food via intensive agriculture for the increasing world population (Adesemoye et al. 2009). The challenge therefore is to continue more agricultural productivity in a way that minimizes harmful environmental effects of fertilizers. Current efforts have been focused on the decreased use of inorganic fertilizers in agriculture, prompting the search for alternative ways to improve soil fertility and crop production. Beneficial plant-microbe interactions in the rhizosphere are the determinants of plant health and soil fertility (Jeffries et al. 2003). Soil microorganisms are very important in the biogeochemical cycles of both inorganic and organic nutrients in the soil and in the maintenance of soil health and quality (Jeffries et al. 2003). The growth-promoting activities of some microbiota on plants can be explained in various ways, including through biocontrol and induction of disease resistance in the inoculated plant, biological N_2 fixation, phosphorus solubilization, and/or production of phytohormones (Mia et al. 2012). The symbiotic association between arbuscular mycorrhizae (AM) fungi and root provides a significant contribution to plant nutrition and growth; they are frequently associated with species from around 90% of plant families. AMF provide many benefits to plants and the environmental stability which includes nutrient uptake enhancement, drought tolerance, root pathogens, and soil aggregation improvement (Smith and Read 2008). In the last decades, the interest concerning elucidation of the mechanisms involved in establishing the complex interactions of microorganisms in the rhizosphere and their role in protecting and stimulating plant development had increased. The present study is designed to evaluate the effect of some PGPR inoculations on mycorrhization, yield, and nutrient uptake of peanut grown in the northwest region of Morocco.

4.2 Materials and Methods

4.2.1 Plant Material and Soils Used

The commercial KP29's variety of peanut (*Arachis hypogaea* L.), currently cultivated by farmers in various parts of Moulay Bousselham and Laaouamra, was used as plant material. This peanut variety is a legume that belongs to the botanical group of "Valencia."

Sampling of soils, cultivated previously by peanut, is performed in the first 20 cm deep on both sites of Laaouamra (clay 10.10%, silt 6.11%, sand 80.81%, pH (H_2O) 6.1, total organic matter 1.1%, P_2O_5 97.47 ppm, total nitrogen 50 ppm) and Moulay Bousselham (clay 5.03%, silt 8%, sand 85.43%, pH (H_2O) 6.5, total organic matter 0.71%, P_2O_5 62.91 ppm, total nitrogen 35 ppm). The soil was air dried, sieved on 2 mm mesh sieves, and placed in favorable conditions throughout the duration of the study.

4.2.2 Inoculation of Seedlings with PGPR

The bacterial strains used as inocula are isolated from the rhizosphere of three varieties of rice (Puntal, Elio, and Guadiamar). There are three *Pseudomonas* (PP22, GP70, and GR1) and two *Aeromonas* (PR29 and GR70) (Aarab et al. 2015a, b). These bacterial strains were chosen because of their ability to solubilize the tricalcium phosphate and to secrete indole acetic acid (IAA). The peanut's seeds were surface sterilized with 0.5% sodium hypochlorite for 5 min, rinsed, and left to soak for 6 h in sterile distilled water. Seeds were then transferred aseptically to Petri dishes filled with 1% (w:v) water agar medium, and plates were incubated at 28 °C in the dark for 72 h. After germination, seedlings were planted in both soils of Laaouamra and Moulay Bousselham. In fact, two groups of plastic pots (18 cm diameter, 20 cm height) were filled with 3 kg of non-sterilized soil. One germinated seed was sown in each pot, of both soils, and inoculated directly with 1.5 ml of bacterial culture (10^8 cfu ml^{-1}) grown in TSB. All pots were placed in a growth chamber with a 16:8 h light:dark photoperiod at 28 ± 2 °C and a photosynthetically active radiation (PAR) of 400 μE m^{-2} s^{-1}. Four replications were maintained for each treatment.

4.2.3 Mycorrhizal and Growth Parameters

Sixty days after sowing, plants were harvested. The leaf area was calculated by using the equation described by Ahmed and Morsy (1999): Leaf area (cm^2) = 0.70 (length × width) – 1.06. Then, plants were uprooted carefully from the soil and washed with water. The presence of root nodules was checked visually on each plant replicate. Plant growth was evaluated by measuring the dry mass of shoots (62 °C for 72 h) for each of the four plant replicates per treatment. A part of the root of each plant was collected, cleared, and stained as described by Phillips and Hayman (1970) and finally mounted on slides. Quantification of arbuscular mycorrhizal infection and colonization was evaluated using the notation scale described by Trouvelot et al. (1986). Parameters of mycorrhization were calculated with MYCOCALC software, available at: http://www.dijon.inra.fr/mychintec/Mycocalc-prg/download.html.

4.2.4 Plant Mineral Analysis (N, P, K)

Shoot samples were oven-dried at 62 °C for 72 h, ground and passed through a 1 mm sieve. Then, the Kjeldahl method was used to determine total nitrogen (N) after wet digestion with concentrated sulfuric acid. Also, phosphorus (P) and potassium (K) were determined using the method "ICP: inductively coupled plasma spectrophotometer" at the National Center for Scientific and Technical Research (CNRST) in Rabat, Morocco.

4.2.5 Statistical Analysis

Statistical analyses of the experimental data were carried out using ANOVA test; p-value ≤ 0.05 was considered statistically significant. Data analysis was performed on mycorrhizal infection, vegetative growth, and mineral nutrition.

4.3 Results

Most published works on study of PGPR and AMF have been focused as alternative fertilizers to improve yield and productivity of legume plants. The soil microorganisms may play a decisive role on nutrient uptake and ecological growth of peanut in the northwest of Morocco. As plants are valuable sources of nutrients for many categories of soil microorganisms, they represent the center of different types of interrelations, competition, or cooperation, in order to gain access to mineral nutrient.

4.3.1 Nodulation and Mycorrhization of Peanut After Inoculation with PGPR

The results of nodules number and mycorrhizal root colonization are shown in Table 4.1. No nodules were observed on hairy roots of uninoculated control and inoculated plants with PGPR on both soils, except with PP22 (three nodules/plant)

Table 4.1 Effect of PGPR inoculations on nutrient uptake (N, P, K) of peanut plants

Parameters*	Soils	Control	GP70	GT1	PR29	GR70	PP22
N (mg plant^{-1})	MB	7.59 ± 1.90^{bc}	9.50 ± 1.91^{ab}	9.86 ± 1.57^{ab}	7.89 ± 1.56^{bc}	5.76 ± 1.19^{c}	11.29 ± 2.40^{a}
	Laa	7.89 ± 2.09^{b}	8.87 ± 2.36^{ab}	9.52 ± 1.48^{ab}	8.63 ± 1.13^{ab}	7.66 ± 1.97^{b}	11.28 ± 3.65^{a}
P (mg plant^{-1})	MB	1.58 ± 0.51^{bc}	2.58 ± 0.41^{a}	2.43 ± 0.64^{a}	2.23 ± 0.66^{ab}	1.07 ± 0.31^{c}	2.19 ± 0.64^{ab}
	Laa	1.86 ± 0.12^{c}	2.07 ± 0.43^{bc}	2.48 ± 0.14^{ab}	1.99 ± 0.15^{c}	2.01 ± 0.38^{c}	2.66 ± 0.39^{a}
K (mg plant^{-1})	MB	12.10 ± 0.89^{ab}	14.39 ± 5.95^{a}	11.47 ± 1.99^{ab}	10.28 ± 3.40^{ab}	09.37 ± 0.90^{b}	12.04 ± 0.90^{ab}
	Laa	8.06 ± 0.83^{b}	10.37 ± 2.14^{ab}	12.51 ± 1.82^{a}	9.51 ± 1.50^{b}	8.77 ± 1.90^{b}	12.38 ± 2.57^{a}

*Values in lines followed by letter a, b and c differ significantly according to Fisher-protected LSD test ($p < 0.05$)

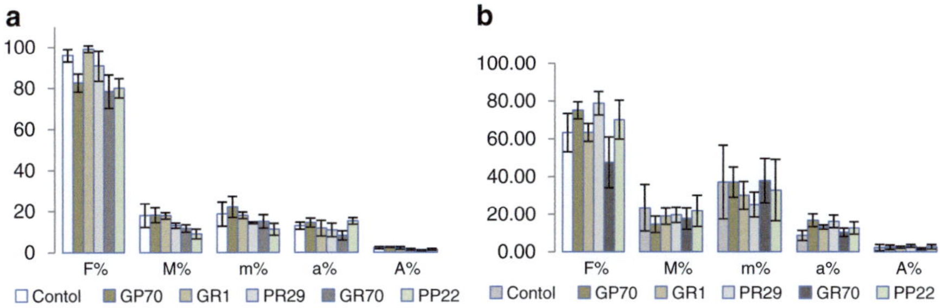

Fig. 4.1 Mycorrhizal parameters of peanut (KP29 variety) on the soil of Moulay Bousselham (**a**) and Laaouamra (**b**)

on the soil of Laaouamra. The promotion of mycorrhizal infection of peanut plants by native AM fungi after inoculation with PGPR was expressed by the frequency of root colonization, mycorrhizal intensity of the root cortex colonization, and abundance of arbuscules. On the soil of Moulay Bousselham, plants inoculation with GP70, GR70, and PP22 reduced frequency of root colonization (F%) compared to control (Fig. 4.1). Also, mycorrhizal intensity (M%) was decreased significantly by PR29, GR70, and PP22 strains. On the soil of Laaouamra, PR29 increased significantly the frequency of root colonization, and only GR70 had a negative impact on this mycorrhizal parameter. On the other hand, plant inoculation with GP70, GR1, PR29, and PP22 improved (sometimes not significantly) arbuscular abundance in the mycorrhizal root cortex (A% and a%), especially on the soil of Laaouamra.

4.3.2 Response of Peanut Plants to Inoculation with PGPR

In this experiment, inoculation with PGPR had a positive effect on yield and peanut growth, except with GR70 which presents sometimes negative effects. Generally, the significant improvements were registered after inoculation with *Pseudomonas* strains.

4.3.3 Shoot and Root Length

The results show significant differences due to bacterial inoculation. Significant increases in shoot length were observed after inoculation with GP70 and PP22 on the soil of Moulay Bousselham. On the soil of Laaouamra, all bacterial inoculants increased shoot length, but only PR29 had significant effect. On the other hand, a significant root shortening due to GR70 inoculation was detected on the soil of Moulay Bousselham. The roots of plants treated with GR1 and PR29 on the soil of Laaouamra and GP70 on the soil of Moulay Bousselham were significantly longer than roots of control plants.

Table 4.2 Effect of PGPR inoculations on growth and yield of peanut plants

Parameters*	Soils	Control	GP70	GR1	PR29	GR70	PP22
Root length (cm)	MB	17.00 ± 2.94^{ab}	20.85 ± 4.57^{a}	17.30 ± 4.42^{ab}	15.21 ± 1.29^{b}	7.83 ± 1.65^{c}	14.93 ± 1.65^{b}
	Laa	11.12 ± 2.65^{b}	13.25 ± 3.30^{ab}	18.66 ± 3.09^{a}	20.00 ± 8.64^{a}	14.42 ± 2.38^{ab}	15.00 ± 2.16^{ab}
Shoot height (cm)	MB	28.75 ± 3.20^{c}	40.38 ± 1.25^{a}	33.75 ± 3.77^{b}	33.50 ± 2.08^{b}	27.63 ± 2.13^{c}	38.25 ± 2.06^{a}
	Laa	34.50 ± 4.2^{b}	37.50 ± 5.50^{b}	43.25 ± 2.75^{a}	36.13 ± 3.42^{b}	37.63 ± 1.70^{b}	39.75 ± 3.30^{ab}
Leaf area (cm^2)	MB	1.01 ± 0.47^{b}	1.37 ± 0.17^{ab}	1.01 ± 0.22^{b}	1.51 ± 0.13^{a}	1.39 ± 0.17^{ab}	1.47 ± 0.23^{a}
	Laa	0.73 ± 0.2^{d}	1.07 ± 0.108^{c}	1.49 ± 0.342^{a}	1.12 ± 0.116^{bc}	1.36 ± 0.139^{ab}	1.45 ± 0.035^{a}
Nodules number	MB	00 ± 00^{a}	00 ± 00^{a}	00 ± 00^{a}	00 ± 00^{a}	00 ± 00^{a}	00 ± 00^{a}
	Laa	00 ± 00^{b}	00 ± 00^{b}	00 ± 00^{b}	00 ± 00^{b}	00 ± 00^{b}	03 ± 1.29^{a}
Fresh root weight (g)	MB	0.30 ± 0.02^{ab}	0.35 ± 0.05^{a}	0.21 ± 0.04^{c}	0.23 ± 0.06^{bc}	0.13 ± 0.03^{d}	0.25 ± 0.05^{bc}
	Laa	0.19 ± 0.09^{b}	0.20 ± 0.04^{b}	0.27 ± 0.05^{ab}	0.34 ± 0.09^{a}	0.22 ± 0.03^{b}	0.27 ± 0.04^{ab}
Fresh shoot weight (g)	MB	2.06 ± 0.69^{bc}	3.23 ± 0.71^{a}	2.80 ± 0.37^{ab}	2.24 ± 0.37^{bc}	1.78 ± 0.59^{c}	3.13 ± 0.55^{a}
	Laa	2.12 ± 0.36^{c}	2.71 ± 0.24^{abc}	3.24 ± 0.59^{a}	2.98 ± 0.48^{ab}	2.58 ± 0.35^{bc}	3.18 ± 0.41^{ab}
Dry root weight (g)	MB	0.05 ± 0.008^{ab}	0.07 ± 0.017^{a}	0.046 ± 0.012^{b}	0.048 ± 0.012^{b}	0.02 ± 0.008^{c}	0.045 ± 0.012^{b}
	Laa	0.042 ± 0.017^{a}	0.042 ± 0.012^{a}	0.045 ± 0.008^{a}	0.046 ± 0.015^{a}	0.042 ± 0.010^{a}	0.045 ± 0.023^{a}
Dry shoot weight (g)	MB	0.30 ± 0.089^{b}	0.46 ± 0.054^{a}	0.36 ± 0.064^{a}	0.30 ± 0.073^{b}	0.20 ± 0.041^{c}	0.37 ± 0.09^{a}
	Laa	0.30 ± 0.036^{c}	0.43 ± 0.055^{a}	0.45 ± 0.109^{a}	0.36 ± 0.091^{ab}	0.36 ± 0.051^{ab}	0.46 ± 0.084^{a}

*Values in lines followed by letter a, b, c and d differ significantly according to Fisher-protected LSD test ($p < 0.05$)

4.3.4 Leaf Area

On the soil of Laaouamra, all bacterial treatments increased significantly peanut's leaf area as compared to the control (Table 4.2), and maximum values were obtained with the GR1 and PP22 inoculants. However, only PR29 and PP22 increased this parameter growth on the soil of Moulay Bousselham.

4.3.5 Fresh and Dry Shoot Weight

Inoculation of plants with GP70 and PP22 increased significantly fresh shoot weight on the soil of Moulay Bousselham. In addition to these two bacterial strains, GR1 had the same effect on peanut's dry shoot weight. On the soil of Laaouamra, bacterial strains had a positive impact on fresh shoot weight of peanut's plants. Indeed, they all had a significant effect on dry shoot weight.

4.3.6 Macroelement Content

Concerning mineral content, the studied macroelements varied in a similar way depending on treatments. The P and K contents were significantly favored by GR1 and PP22 on the soil of Laaouamra. On the soil of Moulay Bousselham, inoculation had no significant effect on K content of plants. Nevertheless, GR1 had a positive significant impact on P content, compared with the control plants. Results showed also that only PP22 inoculants contributed to the significant increase of mineral N on both soils of Moulay Bousselham and Laaouamra.

4.4 Discussion

Our results demonstrate that soils origin influences magnitude and inoculum impact degree on plant growth and symbiosis establishment between peanut and microorganisms. In fact, the response of peanut yield and nutrient uptake to bacterial inoculation is of great relationship with soils' physicochemical characteristics. The inferred data suggest that inoculations might determine which bacterial strain can be exploited to increase such parameter of peanut growth on both soils of Moulay Bousselham and Laaouamra. Reduction of mycorrhizal frequency and intensity on the soil of Moulay Bousselham could be explained by negative impact of bacteria. They can establish biofilms on the surface of peanut root after inoculation. Thus, Frey-Klett et al. (2007) proposed that the bacteria exert two opposite effects on plant growth: the obvious beneficial ones independent of cell density and some detrimental ones toward the plant or fungus when at high densities. Then, bacteria may have physical effects on peanut mycorrhization through biofilm formation, as well as chemical effects through the release of compounds in the exudates. A harmful impact of bacterially derived volatiles (acids, alcohols, methane, ethyl acetate, acetaldehyde, acetoin, and diacetyl) on mycorrhiza formation has been previously demonstrated (Mackie and Wheatley 1999; Bruce et al. 2003). Moreover, significant negative effects on mycorrhiza formation were observed in the case of the spent broth experiment, particularly for spent broth produced during exponential phase growth (Aspray et al. 2005). On the other hand, GP70, GR1, PR29, and PP22 can positively influence the efficiency of mycorrhization by increasing arbuscular exchange area. The so-called mycorrhiza helper bacteria (MHB) have been shown to promote mycelial growth and mycorrhiza formation (Frey-Klett et al. 2007; Garbaye 1994). These bacteria can facilitate hyphal penetration through the soil, and when hyphae colonize plant tissues, they can continue their functions. The bacteria located on hyphae can be released to the intercellular spaces after hyphae penetration in roots. In the intercellular spaces, this bacteria cause dilatation of the cortical root cells and establish highly branched hyphae that develop between the fungal cell wall and the plasma membrane of plant cells.

In response to bacterial treatments, N, P, and K contents were favored by all bacterial inoculations. Among bacterial strains, which might be estimated as MHB, pseudomonads bacteria (PP22, GP70, and GR1) had higher positive impact on plant N and P content. Solubilizing phosphate is one of the most important actions of these three bacterial strains (Aarab et al. 2015b). Thus, they were also estimated as phosphate-solubilizing bacteria (PSB). Moreover, AMF provide more necessary mineral nutrients to plant in easily assimilated via improvement of arbuscular exchange area. Therefore, they could explain the increase of plant mineral content. Enhancement of leaf area, shoot height, and biomass might be attributed to P-solubilizing ability and the adequate uptake of other nutrients to roots. By increasing P availability and its uptake, PP22 stimulated nodule formation and nitrogen fixation on the soil of Laaouamra. Improvement of plant growth parameters could be also a result of the increased synthesis of IAA (Aarab et al. 2015b), which is a plant growth promoter. The production of IAA has been reported by many species, and *Pseudomonas spp.* has always been to produce more auxin, thus increasing plant height, shoot weight, and more biomass production. Glick (1995) viewed that the mechanism most commonly invoked to explain the various effect of PGPR on plants is the production of phytohormones, and IAA may play the most role in growth promotion. Also, there is evidence that mycorrhizal plants contain higher concentration of growth hormones than their non-mycorrhizal equivalents. Effective nutrient acquisition by AMF is generally attributed to the extensive hyphal growth beyond the nutrient depletion zone surrounding the root and principal avoidance.

Conclusion

Our results confirm the suitability of PGPR inoculation to improve plant peanut growth. The management of the application of these rhizobacteria represents considerable extent especially with pseudomonads strains. The positive interactions with other beneficial microorganisms such as native AMF can also be taken into account as a method of enhancing peanut yield and growth. However, due to the high specificity involved in these types of bacterial inoculations, a previous screening to select the best microbe-host plant combination should be done in order to optimize results.

References

Aarab S, Ollero J, Megías M, Laglaoui A, Bakkali M, Arakrak A (2015a) Isolation and screening of bacteria from rhizospheric soils of rice fields in Northwestern Morocco for different plant growth promotion (PGP) activities: an *in vitro* study. Int J Curr Microbiol App Sci 4(1):260–269

Aarab S, Ollero J, Megías M, Laglaoui A, Bakkali M, Arakrak A (2015b) Isolation and screening of inorganic phosphate solubilizing *Pseudomonas* strains from rice rhizosphere soil from Northwestern Morocco. Am J Res Commun 3(4):29–39

Adesemoye AO, Torbert HA, Kloepper JW (2009) Plant growth-promoting rhizobacteria allow reduced application rates of chemical fertilizers. Microb Ecol 58:921–929

Ahmed FF, Morsy MH (1999) A new methods for measuring leaf area in different fruit species. Minia J Agric Res Dev 19:97–105

Aspray TJ, Eirian Jones E, Whipps JM, Bending GD (2005) Importance of mycorrhization helper bacteria cell density and metabolite localization for the *Pinus sylvestris-Lactarius rufus* symbiosis. FEMS Microbiol Ecol 56:25–33

Bruce A, Stewart D, Verrall S, Wheatley RE (2003) Effect of volatiles from bacteria and yeast on the growth and pigmentation of sapstain fungi. Int Biodeter Biodegr 51:101–108

Frey-Klett P, Garbaye J, Tarkka M (2007) The mycorrhiza helper bacteria revisited. New Phytol 176:22–36

Garbaye J (1994) Helper bacteria: a new dimension to the mycorrhizal symbiosis. New Phytol 128:197–210

Glick BR (1995) The enhancement of plant growth by free living bacteria. Can J Microbiol 41:1376–1381

Jeffries P, Gianinazzi S, Perotto S, Turnau K, Barea JM (2003) The contribution of arbuscular mycorrhizal fungi in sustainable maintenance of plant health and soil fertility. Biol Fertil Soils 37:1–16

Mackie AE, Wheatley RE (1999) Effects and incidence of volatile organic compound interactions between soil bacterial and fungal isolates. Soil Biol Biochem 31:375–385

Mia MAB, Shamsuddin ZH, Mahmood M (2012) Effects of rhizobia and plant growth promoting bacteria inoculation on germination and seedling vigor of lowland rice. Afr J Biotechnol 11:3758–3765

Phillips JM, Hayman DS (1970) Improved procedures for clearing roots and staining parasitic and vesicular–arbuscular mycorrhizal fungi for rapid assessment of infection. Trans Br Mycol Soc 55:158–161

Smith SE, Read DJ (2008) Mycorrhizal symbiosis, 3rd edn. London, UK: Academic Press

Trouvelot A, Kough JL, Gianinazzi-Pearson V (1986) Mesure du taux de mycorhization VA d'un système radiculaire. Recherche de méthodes d'estimation ayant une signification fonctionnelle. In: Gianinazzi-Pearson V, Gianinazzi S (eds) Physiological and genetical aspects of mycorrhizae. INRA edition, Paris

Biosynthesis of Nanoparticles by Microorganisms and Their Significance in Sustainable Agriculture

5

Deepika Chaudhary, Rakesh Kumar, Anju Kumari, Rashmi, and Raman Jangra

Abstract

An assuring novel technique in the field of micro-biotechnology is the application of potential microbes for producing inorganic nanoparticles. These biological agents are safe, eco-friendly, and good source of producing green nanoparticles. Since large species of microbial agents with different metabolic complexities have potential of producing metal nanoparticles (NPs), the exact means or procedure of nanoparticle synthesis is not well understood. The interest in the nanotechnology field is triggered by exclusive assets of nanoparticles and their possible and probable fields of application including electronics, medication, and agriculture. With the ever enhancement in global populace and demand for food, the nanotechnology technique may be the most hopeful and reassuring way to improve the overall agricultural production. The potential applications of nanoparticles in agriculture sector include biosensors; gradual, time-consuming, and controlled delivery of chemical fertilizers and pesticides; detection and control of plant diseases; soil and water remediation; etc. At present however the use and employment of nanotechnology in the agricultural field is in the infant stage; however, if discovered gradually and used in a sustainable way, this new technology can help in the orientation of our agriculture and society today to new heights in the future.

D. Chaudhary • R. Kumar (✉) • Rashmi • Raman Jangra
Department of Microbiology, CCS Haryana Agricultural University Hisar,
Hisar, Haryana, India
e-mail: sehrawatrk@gmail.com

A. Kumari
Centre of Food Science and Technology CCS Haryana Agricultural University Hisar,
Hisar, Haryana, India

V. Kumar et al. (eds.), *Probiotics in Agroecosystem*,
DOI 10.1007/978-981-10-4059-7_5

5.1 Introduction

Agriculture is one of the significant area that gives food to human, in a roundabout way or straightforwardly notwithstanding bolster, fiber, fire, and fuels. The perpetually expanding population results in expansion of interest of all above immensely regardless of constant availability of natural resources. Nowadays, the agriculture system in Indian scenario has guaranteed far-reaching acceptance and implementation of genetically superior novel seeds, synthetic fertilizers, pesticides, escalated watering system, ideal agronomic conditions, and advanced hardware. These processes plainly take out unique genotype from the nature. Right away the agricultural production system faces the confront of enhancing the crop production and providing adequate diets for the expanding population from nutritional points of view, under indeterminate climatic extremes, water lack, in restricted (and debased at numerous spots) land zone, and generally with poor quality water and air, associated with quick changes in natural biodiversity. In this context, nanotechnology might be an expectation for better agrarian advancement (Anonymous 2009).

The father of nanotechnology, Richard Feynman, a physicist, and Nobel Laureate begat the term and clarified the roll of nanotechnology at the yearly meeting of the American Physical Society in 1959 as "There is plenty of room at the bottom," in which he alluded to the boundless number of potential outcomes of manipulating and controlling things on the small scale (Drexler 2009). As of now the above remark holds impeccably and execute in all areas under the umbrella of nanotechnology (Anonymous 2009; Marchiol 2012). Indeed, "nano" is produced from the Greek word signifying "dwarf." In more specialized terms, "nano" implies 10^{-9}, or one billionth of meter (for instance, a virus is approximately 100 nm in size). Actually, the word nanotechnology developed because of the use of nanometer-size particles (size of 1–100 nm), nanoparticles (Lyons 2010). Nanotechnology is a developing and promising field of interdisciplinary research and opens up a wide cluster of chances in various area including drug, pharmaceuticals, gadgets, and agriculture. However, the application of this technology possesses a great deal of difficulty in agriculture sector, but still it is worthwhile at large due to its immense importance in agriculture. This novel technology is helpful in the nano-DNA crystals process, recycling of agricultural waste, and management of insect and pest through potential formulations of nanomaterial-based chemical pesticides and insecticides. The augmentation of agricultural products may be achieved by employing bio-conjugated nanoparticles (encapsulation) especially for sluggish liberation of macro and micronutrients and also water (Bhattacharyya et al. 2014).

There are several techniques related to physical, chemical, and fusion methods which are available to manufacture diverse types of nanoparticles (Liu et al. 2011; Luechinger et al. 2010; Tiwari et al. 2008; Mohanpuria et al. 2008). Though the chemical and physical approaches are more common and prevalent in the synthesis of nanoparticles, due to these techniques we are able to get good amount of nanoparticles with defined size and morphology in a comparatively short duration. Furthermore, these nanoparticles are complicated, multifaceted, expensive, and inefficient and produce dangerous and risky toxic wastes that are injurious, not only to the environment but also to the human health which momentously limits their uses. Consequently, the development of dependable, harmless, and environmentally friendly or green methods for the synthesis of nanoparticles is very important to diversify their applications. To achieve this goal, one option could be the application of potential microbes to prepare or manufacture unique nanoparticles. The nanoparticles generated by biogenic enzymatic process are distantly excellent, in numerous ways, compared to those nanoparticles prepared by other methods. Employing an enzymatic procedure, the use of costly chemicals will not be required, and therefore, more suitable and acceptable, green method is not energy and cost intensive compared to the chemical method employed; moreover, it is also environment friendly. The application of "biogenic" tactic is further maintained by the point that the bulk of microbes dwell ambient conditions of varying environmental factors such as temperature, pH, and pressure (Li 2011). The nanoparticles produced by these methods have greater catalytic reactivity, better specific surface area, and an enhanced, updated contact between enzyme and the metal salt in question due to the microbial carrier matrix (Bhattacharya and Mukherjee 2008; Simkiss and Wilbur 1989). For the synthesis of nanoparticles, among the biological agents exploited, microbes like fungi, bacteria, Actinomycetes, and yeasts are majorly utilized (Mohanpuria et al. 2008; Rai et al. 2008; Thakkar et al. 2010; Sharma et al. 2009). Additionally, there are reports that some algae and blue green algae have also been efficaciously used for nanoparticle synthesis (Govindraju et al. 2008). Thus, it can be suggested that different microbes prove and validate enormous biodiversity for the nanoparticles synthesis which leads to the development of eco-friendly nanotechnology.

5.2 Why Microbes Synthesize Nanoparticles?

Microorganisms are present almost everywhere in nature. They can survive even in extreme conditions. In order to adapt in different environmental conditions, organisms have developed special cell functions, and nanoparticle synthesis is one of them. There are three main motives for them to manufacture nanoparticles, which are as follows:

1. Chemolithotrophy for the production of energy
2. Employment of nanoparticles for distinct purposes
3. Detoxification for survival in toxic the environments

In the chemolithotrophy technique, inorganic substrates (generally of mineral origin) are metabolized by organisms to get reducing equivalents for biosynthesis or energy storage through an aerobic or anaerobic respiration approach (Krumov et al. 2009). One of the best examples is sulfate-reducing bacteria (SRB) which reduce sulfate as oxidizing agent to sulfide. Most SRB can also take up other oxidized sulfur compounds, such as sulfite and thiosulfate, or the elemental sulfur. Frequently, the metal sulfides are produced as a side product, e.g., in the case of yeast *Torulopsis* sp. where nanoparticles PbS are created (Kowshik et al. 2002). In order to begin special cell functions, some microbial species manufacture inorganic materials at nanoscale and assimilate them as the functional components (Slocik et al. 2004), one of the best examples is synthesis of magnetosomes by magnetotactic bacteria. Several microaerophilic microbes employ intracellular chains of magnetic crystals of Fe_3O_4 or Fe_3S_4 to position them according to the earth's geomagnetic field. Microbes such as *Aquaspirillum* sp. live in aquatic systems ca. 2–6 m below the sea level (Slocik et al. 2004; Balkwill et al. 1980). In order to survive under the toxic environments, microbes have developed several cleansing or decontamination methods and techniques. The metal ions such as Cd and Hg are unnecessary and prove toxic; other metals like Cu and Zn are vital and indispensable for normal physiological functions in living organisms (Mehra and Winge 1991). Nevertheless, augmented levels of crucial metal ions can also become fatal and toxic. Consequently, the production of nanoparticles from these metal ions is one of the means to control the intracellular concentration of ions. Insoluble nontoxic nanoclusters of Ag^0, Au^0, ZnS, CdS, and Ag_2S have been found in dissimilar microbes (Slocik et al. 2004).

5.3 Microbial Groups Producing Nanoparticles

5.3.1 Bacteria

The study has focused deeply on prokaryotic organisms as methods of manufacturing the nanoparticles. Potential bacteria are an excellent pick for study owing to their plenty in environment and their capability to acclimatize to the acute and extreme conditions. Moreover, these microbes are also fast growing, easily cultivable, and simple to handle and manipulate. Most of the bacteria are also known to manufacture either intracellularly or extracellularly inorganic materials. The common examples of bacteria manufacturing inorganic metal nanoparticles include the magnetotactic bacteria (also known as magnetite nanoparticles) (Bazylinski and Frankel 2004), S layer bacteria, and many others (Table 5.1) (Pum and Sleytr 1999). Different bacterial groups synthesize nanoparticles having different size and morphology. In the last few years, the reports on synthesis of silver nanoparticles have been increased extensively due to its immense applications as antimicrobial agents.

5.3.2 Fungi

Synthesis of nanoparticles using fungal systems is referred to as mycosynthesis. The expression "mycosynthesis" was first time expressed by Ingle and his coworkers in 2008, to describe the manufacture of nanoparticles by fungal strain *Fusarium acuminatum*. Rai et al. (2009) suggested the expression "myconanotechnology" with reference to research carried out on fungal nanoparticles. This is a novel field which includes cohesive subject of mycology (study of fungi) and nanotechnology. There are several fungal species which have been used for synthesis of diverse metal

Table 5.1 Synthesis of nanoparticles by different bacteria

Bacteria	Nanoparticles	Size (nm)	Shape	References
Brevibacterium casei	Ag Au	10–50 10–50	Spherical	Kalishwaralal et al. (2010)
Bacillus indicus (MTCC 4374)	Ag	2.5–13.3	Spherical	Shivaji et al. (2011)
Pseudomonas aeruginosa	Ag	13	Spherical	Kumar and Mamidyala (2011)
Klebsiella pneumoniae	Ag	1–6	Spherical	Mokhtari et al. (2009)
Lactobacillus casei subsp. casei	Ag	25–50	Spherical	Korbekandi et al. (2012)
Serratia nematodiphila	Ag	10–31	Spherical, crystalline	Malarkodi et al. (2013)
Rhodopseudomonas capsulata	Au	10–20	Spherical	Shiying et al. (2007)
Desulfovibrio desulfuricans	Au	20–50	Triangular, hexagonal, rods	Deplanche and Macaskie (2008)
Pseudomonas fluorescens	Au	50–70	Spherical	Rajasree and Suman (2012)
Escherichia coli K12	Au	50	Circular	Srivastava et al. (2013)
Geobacillus sp.	Au	5–50	Quasi-hexagonal	Correa-Llantén et al. (2013)
Stenotrophomonas maltophilia	Au	40	Spherical	Sharma et al. (2012)
Escherichia coli	CdS	2–5	Spherical, elliptical	Sweeney et al. (2004)
Streptomyces sp. HBUM 171191	$MnSO_4$, $ZnSO_4$	10–20	Polymorphic	Waghmare et al. (2011)
Selenium respiring bacteria	Se	200–400	Spherical	Oremland et al. (2004) Ajayan et al. (2004)
Rhodobacter sphaeroides	ZnS	8	Spherical	Bai et al. (2006)
Lactobacillus strains	Ag–Au alloys	100–300	Crystalline, cluster	Nair and Pradeep (2002)
Magnetospirillum magnetotacticum	Magnetite	–	Cluster	Philipse and Maas (2002)

nanoparticles, which are of many shapes and sizes (Table 5.2). Application of fungi in manufacturing metallic nanoparticles has obtained noteworthy interest as these offer some benefits over employment of bacteria in synthesis of the nanoparticles. The comfort and simplicity of scaling up and downstream processing, the monetary possibility, and the increased surface area owing to large size of mycelium are some

Table 5.2 Synthesis of nanoparticles by different fungi

Fungi	Nanoparticles	Size (nm)	Shape	References
Alternaria alternata	Ag	20–60	Spherical	Gajbhiye et al. (2009)
Fusarium oxysporum	Ag	5–15	Spherical	Duran et al. (2005)
Aspergillus niger	Ag	20	Spherical	Gade et al. (2008)
Phanerochaete chrysosporium	Ag	50–200	Pyramidal, hexagonal	Vigneshwaran et al. (2006)
Penicillium fellutanum	Ag	5–25	Spherical	Kathiresan et al. (2009)
Colletotrichum spp.	Au	20–40	Spherical	Shankar et al. (2003)
Trichothecium spp.	Au	5–200	Triangle, hexagonal	Ahmad et al. (2005)
Verticillium luteoalbum	Au	10	Spherical	Gericke and Pinches (2006)
Fusarium oxysporum	Au	20–40	Triangular, spherical	Mukherjee et al. (2002)
Fusarium oxysporum	Au–Ag alloy	8–14	Spherical, ellipsoidal	Senapati et al. (2005)
Coriolus versicolor	CdS	25–75	Spherical	Sanghi and Verma (2009)
Fusarium oxysporum	Silica	5–15	Quasi-spherical	Bansal et al. (2005)
Fusarium oxysporum	Ti	6–13	Spherical	Bansal et al. (2005)
Aspergillus terreus	ZnO	54.8–82.6	Spherical	Baskar et al. (2013)

significant benefits that can be considered (Nasreen et al. 2014). Mohanpuria et al. (2007) also recommended that since fungi excrete meaningfully higher quantities of proteins compared to bacteria, this will increase the nanoparticle production. Biosynthesis of the nanoparticles by employing fungal strains is widely accepted because monodisperse particles with distinct and definite sizes and with dissimilar chemical compositions and sizes can also be obtained. Fungi have the probability to provide comparatively rapid and ecologically “green” metallic nanoparticles and, therefore, can be treated as eco-friendly bio-factories (Rai et al. 2009). Therefore, the fungal strains can be considered as extremely good nominees for synthesis of the metal nanoparticles.

5.3.3 Yeasts

The numerous publications have revealed that all yeast genera can accumulate different heavy metals as well as a majority of them are capable of synthesizing intracellular nanomaterials (Table 5.3), with few exceptions. They have the ability to accumulate significant amounts of highly toxic metals. Yeasts synthesize nanoparticles as one of the mechanisms for overcoming the toxic effects of heavy metals (Breierova et al. 2002). Stringent control of intracellular metal ions is required by

Table 5.3 Synthesis of nanoparticles by yeasts, algae, and actinomycetes

Type of organism	Name of organism	Nanoparticles	Size (nm)	Shape	References
Yeasts	*Pichia jadinii*	Au	<100	Spherical	Gericke and Pinches (2006)
	Candida glabrata	CdS	0.20 ± 0.03	Spherical	Dameron et al. (1989)
	Sachharomyces cerevisiae	CdS	2.5–5.5	Spherical	Prasad and Jha (2010)
	Yarrowia lipolytica	Au	15	Hexagonal, triangular	Agnihotri et al. (2009)
	Schizosaccharomyces pombe	CdS	~0.18	Spherical	Dameron et al. (1989)
Algae	*Plectonema boryanum*	Au	10–6000	Octahedral platelets	Lengke et al. (2006)
	Spirulina platensis	Ag Au Au core and Ag shell	7–16 6–10 17–25	All spherical	Govindraju et al. (2008), Tsibakhashvili et al. (2010)
Cyanobacteria	Cyanobacteria	Au Ag Pd Pt	5.4 ± 0.6 to 25.0 ± 3.2 10.0 ± 1.2 to 40.0 ± 4.2 35.0 ± 0.3 and 10.0 ± 1.2 3.2 ± 0.3	Triangular, hexagonal Spherical Spherical Not mentioned	Brayner et al. (2007)
Actinomycetes	*Rhodococcus sp.*	Au	5–15	Spherical	Ahmad et al. (2003)
	Thermospora spp.	Au	9–10	Spherical	Sastry et al. (2003)

yeast cells to avoid negative or lethal effects. Toxicity to the cells results due to over storage of essential metal ions or by the exposure of the cell to metals which do not have any biological significance such as mercury, lead, or cadmium. Detoxification mechanisms in yeast cells are brought about by glutathione and two groups of metal-binding ligands—metallothioneins and phytochelatins. In most of the yeasts species studied, these molecules determine the mechanism for the formation of nanoparticles and stabilize the complexes. The considerable variations in size, particle location, monodispersity, and properties are due to different mechanisms employed by yeast strains of different genera for nanoparticle synthesis (Nasreen et al. 2014). The yeasts are usually known as "semiconductor crystals" or "quantum semiconductor crystals" (Dameron et al. 1989). Yeasts are mainly known for their ability to synthesize semiconductor nanoparticles, particularly cadmium sulfide. Among the eukaryotes, yeast species are the most studied and used in bioprocesses. This aspect selects them as a prominent object for nanoparticle synthesis.

5.3.4 Algae and Actinomycetes

A limited number of scientific reports describe the biosynthesis of nanoparticles by actinomycetes and algae. In algae, most of the reports are on the synthesis of gold and silver nanoparticles and in actinomycetes on gold (Table 5.3). Shape and size of the nanoparticles varies depending upon the target organism and experimental conditions (Rai et al. 2013).

5.4 Synthesis of Nanoparticles by Different Microorganisms

The precise mechanism for the synthesis of nanoparticles employing microorganisms has not been conceived yet. This is because different microbes react differently with metal ions leading to the formation of nanoparticles. Many microorganisms produce inorganic materials either intra- or extracellularly. In the intracellular synthesis of nanoparticles, the cell wall of the microorganisms plays an important role. The mechanism involves electrostatic interaction of the positive charge of the metal ions with negative charge of the cell wall. A special ion transportation system in the cell wall is employed for the transfer of ions in the cell. The enzymes which are present in the cytoplasm reduce the ions to nanoparticles. After synthesis, nanoparticles are capped by different molecules of microbes to make them stable (Fig. 5.1). Later, these nanoparticles either get diffused off through the cell wall or get accumulated in the cell (Nasreen et al. 2014). In the synthesis and stabilization process of nanoparticles, many enzymes such as reductases, synthases, hydrolases, and hydrogenases are involved in different microbial species. Among reductases NADH

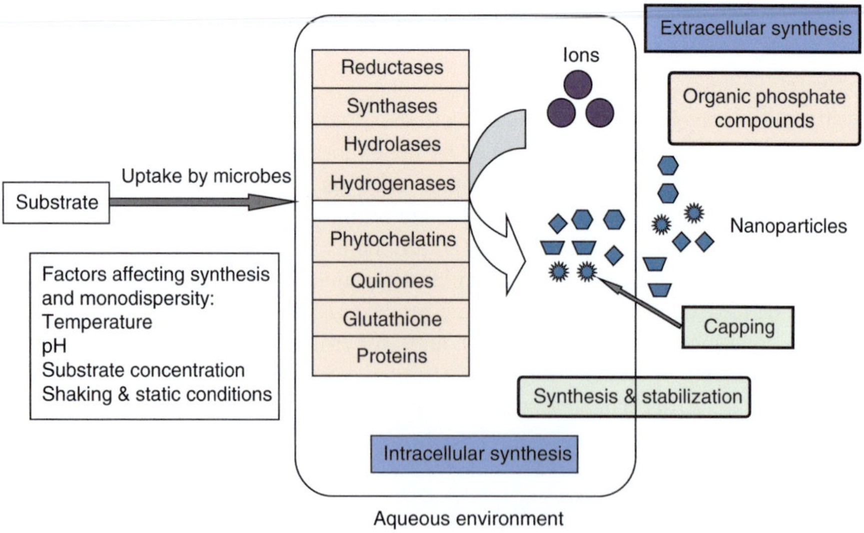

Fig. 5.1 Generalized mechanism of synthesis of nanoparticles by different microbes

and NADPH, dependent enzymes are most commonly employed. In addition to this, many other compounds such as phytochelatins, quinones, glutathione, and some proteins are also important (Ramezani et al. 2010).

The mechanism of extracellular synthesis of nanoparticles involves secretion of microbial enzymes in the medium and subsequent reduction of metal ions to their respective nanoparticles by these enzymes. In extracellular synthesis, usually the enzyme which synthesizes nanoparticles is also responsible for their capping.

5.5 Control of Size and Morphology of Nanoparticles

It is well known that the electronic and optical properties of nanoparticles are heavily dependent on their size and shape. Thus, there has been tremendous interest in controlling the size and shape of nanoparticles. Particular emphasis has recently been placed on the control of shape, because in many cases it allows properties to be fine-tuned with a great versatility that gives the particles a unique nature. Microbes, which are regarded as potent eco-friendly green nanofactories, have the potential to control the size and shape of biological nanoparticles. The rate of particle formation, particle size, and morphology could, to an extent, be manipulated by controlling parameters such as pH, temperature, substrate concentration, and exposure time to substrate (Li et al. 2011). In some cases some proteins or peptides of host microbe also affect shape and size of nanoparticles, e.g., *Magnetospirillum magneticum* AMB-1 Mms6 protein has a strong effect in regulating the size and shape of magnetite particles during the synthesis process (Amemiya et al. 2007). Once the reaction conditions are standardized for a specific microbe, particles of uniform size and morphology can be obtained.

5.6 Mechanism of Synthesis of Gold Nanoparticles by Different Microorganisms

Gold NPs are formed by a various potential microbes; however, the usual underlying means involved in synthesis remains the same. There are number of biomolecules which are involved in the manufacture and maintenance of the nanoparticles. In spite of huge quantity of scientific reports on the microbial mediated gold nanoparticle synthesis, the exact mechanistic facets have not been deciphered fully, and there is great need to find out the real mechanism.

5.6.1 Mechanism of Au Reduction

In the biosynthetic process of nanoparticles, there are two chief precursors of gold nanoparticles: (i) $HAuCl_4$ which dissociates to (gold) Au^{3+} ions (Khan et al. 2013) and (ii) AuCl which further dissociates to Au^+ (Zeng et al. 2010). This gold (Au^+)

precursor is less studied, owing to the greater solubility of the Au^{3+} ions compared to the Au^{+} ions. Das et al. (2012) studied the single electron reduction of Au^{+} to Au^{0} is completed in a single step; the three e^{-} (electron) reduction process of Au^{3+} to Au^{0} is the arrangement of number of the chemical transformations. Gold nanoparticle formation could happen either in extracellular or intracellular space. Extracellular gold nanoparticle creation is informed when the Au^{3+} ions are imprisoned and reduced by the cell wall proteins or by excreted enzymes in medium. In case of the intracellular gold nanoparticle formation, the Au^{3+} ions dispersed through cell membrane and are also reduced by the cystolic redox intermediaries. Conversely, it is not very clear whether the dissemination of Au^{3+} ions through the cell membrane occurs via active bioaccumulation process or by the passive biosorption methods in various microorganisms (Das et al. 2012). On the other hand, the responsible enzymes for the biosynthesis of gold nanoparticles may be the same or may differ in diverse microbes (Xiaorong et al. 2011).

In addition to the enzymes and proteins, some other biomolecules are also responsible for the nanoparticle biosynthesis process. In fungus *Fusarium oxysporum*, there is NADH-dependent reductase enzyme involved in the bioreduction process (He et al. 2007; Xiaorong et al. 2011). However, the definite protein(s) involved in Au (gold) reduction has not been identified hitherto. The probable participation of aromatic amino acids such as tyrosine and tryptophan in reduction of Au^{3+} ions has been detected in fungus *Phanerochaete chrysosporium* (Sanghi et al. 2011). Other means for fungal gold nanoparticle synthesis have also been proposed, though the fungal pathogen *Candida albicans* is also proficient of producing phytochelatins. This also helps in biosynthesis of gold nanoparticles, which are capped by the antioxidant glutathione (Chauhan et al. 2011). Organic phosphatic compounds have also been reported to play a significant role in nanoparticle biosynthesis in bacterium *Bacillus subtilis* (Beveridge and Murray 1980). In bacteria *Escherichia coli* and *Desulfovibrio desulfuricans*, the hydrogen gas (H_2) acts as an electron donor, and periplasmic hydrogenase enzymes are proposed to contribute in Au^{3+} ion reductions and bioaccumulation of gold nanoparticles (Deplanche and Macaskie 2008). In fungus *Phanerochaete chrysosporium*, the enzymes such as ligninase and laccase have been reported to biosynthesize the intracellular and extracellular gold nanoparticles correspondingly (Sanghi et al. 2011). In the actinomycete *Thermomonospora* sp., enzymes are known to play a significant role in the reduction of the metal ions along with the stability of nanoparticles. This results in proficient production of monodispersed gold nanoparticles (Ahmad et al. 2003). The algae *Shewanella* could also reduce the Au^{3+} ions in an anaerobic environment by employing hydrogen gas (Konishi et al. 2004).

5.6.2 Capping Process

Little gold nanocrystals are unsteady, and owing to this reason, microbes employ proteins and certain other combinations as capping agents to lessen the gold nanoparticles accumulation and, therefore, stabilize the nanocrystals. This procedure is similar to dispersion of AuNPs through the chemical agents and results in development of dispersed AuNPs with a bigger particle size distribution. The full complete control of capping step in AuNPs synthesis procedure may result in the production of AuNPs with fine size and shape distribution that could be promptly employed in the industrial, agricultural, and biomedical applications. The morphology of Au^0 nanoparticles is also influenced by various abiotic factors such as pH, temperature, humidity, and substrate concentration, shaky and stationary conditions in all microorganisms (Gericke and Pinches 2006). Subsequently a diversity of microbial and also enzymes and proteins are involved in the biosynthesis and maintenance process; therefore, the precise carbon-based molecule that acts as capping agent cannot be noticed. The proteins and amino acid residues such as cysteine, tyrosine, and tryptophan are also stated to play a significant role in the stabilization of AuNPs. In some of the cases, the amino acid tyrosine can bind to gold surface through amine groups and reduce silver ions at high pH, thus producing gold core silver shell nanoparticles. Antioxidant glutathione also plays a vital role in steadiness of AuNPs in some microbes (Sastry et al. 2003; Gole et al. 2001; Selvakannan et al. 2004; Si and Mandal 2007).

5.7 Importance of Nanoparticles in Agriculture

Nowadays, nanotechnology plays a very important role in apparently all areas of research and development. The role of nanoparticles in agriculture is an emerging area, and till now, it is mostly theoretical. In spite of this, using this technology, one can deliver insecticides encapsulated in nanomaterials for measured and restrained release, steadiness of biopesticides, slow release of nanomaterial-assisted fertilizers, bioinoculants, and including the vital micronutrients for well-organized use along with the field applications of agrochemicals. One can also apply this technology in transfer of genetic material for crop development (Bhattacharyya et al. 2014). Recognition of potential plant pathogens based on nanosensor technology helps in controlling plant diseases. It is also reported that silver nanoparticles as antifungal and antibacterial agents have role in agricultural crop protection where these particles also regulate proper nutrition to plants (Ghormade et al. 2011). Consequently, the application of nanotechnology in agriculture segment can overawed several problems associated with traditional farming, and agricultural production can be increased in an eco-friendly way. Certain uses of nanotechnology in agriculture sector are mentioned below.

5.7.1 Nanobiosensors

Nanoparticles possess interesting electronic and optical properties which provide an opportunity to use them as biosensors. These biosensors basically generate signals depending on the concentration level of target, i.e., pathogens, herbicides, pesticides, nutrients, etc. (Bhattacharyya et al. 2014). Factors like plant growth and important field conditions such as crop disease, moisture level, soil potency, temperature, crop nutrient status, insects, type of weeds, etc. can be supervised using the bionanosensors. The application of these sensors delivers a significant and essential data for agronomic practices, viz., ideal time of crop planting and harvesting of the produce. It is also beneficial in monitoring the time and quantity of irrigation, fertilizer application, pesticide use (herbicides), and other important treatments. This has moved meticulous agriculture system to much higher level of control, such as in water usage, leading ultimately to water conservation (Agrawal and Rathore 2014). It is known that under nutrient limitation, crops secrete carbonaceous compounds into rhizosphere to enable biotic mineralization of N and/or P from soil organic matter. So, these root exudates can be considered as environmental signals and be selected to prepare nanobiosensors that will be incorporated into novel nanofertilizers.

Characteristics of an ideal nanobiosensor (Rai et al. 2012):

- Very explicit for the analysis purpose means thereby a sensor must be able to differentiate between analyte and any "other" material.
- Should be unchanging under normal storing conditions.
- Sensor and analyte should be independent of physical parameters such as temperature, stirring, and pH.
- Responses achieved should be exact, specific, and actual, results can be reproduced, and there is also no electrical noise.
- Minimal reaction time.
- Should be inexpensive, transportable, and easy to use.
- Bionanosensor should be miniature, biocompatible, non-antigenic, and nontoxic.

5.7.2 Nanofertilizers

In Indian scenario, chemical fertilizers, seed quantity and quality, and time of irrigation are primarily responsible for higher production of food grain. It has been established conclusively that chemical fertilizers contribute about 35–40% of the productivity of majority of crops. Excessive application of chemical fertilizers leads to unfair and unnecessary fertilization, vital and micronutrient deficiencies, and reduction in soil organic matter. Therefore, it is imperative to develop a nano-based fertilizer preparation having manifold functions. The use of nanofertilizer knowhow is very state of the art, but it is meagerly reported in the literature.

Presently, research is under process to create nanocomposites to supply all the essential important and vital nutrients in suitable proportion through the smart delivery system. Keeping the above in view, the chemical fertilizers are encapsulated within nanoparticles, which helps to coordinate the discharge of fertilizers with their uptake and transportation by crops, which prevent the unwanted nutrient losses. Furthermore, noteworthy upsurge in yields has also been observed due to the foliar application of nanoparticles as chemical fertilizers (Tarafdar 2012; Tarafdar et al. 2012). It is a significant point that nanotechnology can be employed to enhance the fertilizer preparations and manufacture of more eco-friendly fertilizers. It can enhance the performing and working of fertilizers in other ways also. Such as owing to its photocatalytic characteristics, nanosize titanium dioxide has also been included into chemical fertilizers as a bactericidal adjunct. Additionally, nanosilica particles up taken by plant roots have been demonstrated to form films at the cell wall level, which can significantly enhance plant's resistance to abiotic and biotic stresses and, therefore, increases the crop production.

5.7.3 Nano-herbicide

Weeds are great threats to the agriculture. The common herbicides that are available in market are designed in such a way to suppress or destroy the aboveground part of the weeds. None of the prevalent herbicides prevent the function or activity of viable belowground plant parts such as rhizomes or tubers or roots, which perform as a main source of new weeds in the subsequent season. Enhancements and upgradation in the effectiveness of herbicides by applying nanotechnology can result in superior crop production. Encapsulated nano-herbicides are pertinent and keeping in view the necessity to plan and make a nano-herbicide that can be protected under natural environmental conditions and should act only when it is required such as in a spell of rainfall, which can actually mimic the rain-fed system. Objective-specific herbicide particle which is encapsulated within a nanoparticle is designed in a way that it aims at precise receptors in the target weeds' roots. The nanoparticle enters into the weed root system and is transported to other parts of the plant which inhibits the growth of weeds (Chinnamuthu and Kokiladevi 2007). Several adjuvants for herbicide application are presently available in market that claims to contain nanomaterials.

5.7.4 Nano-pesticide

The application of pesticides in the early stage of crop growth aids in lowering the pest population below the economic threshold level, which guarantees an effective method of control for a longer duration. Henceforth, the application of effective ingredients applied on the surface remains effective in controlling insect pests and also as one of the most cost-effective and multipurpose means. The nano-encapsulation method can be effectively employed to protect the active ingredient from the hostile environmental conditions. This approach also promotes persistence, which is an effective nanotechnology approach which is used to progress the insecticidal value.

The nano-encapsulation of pesticides such as insecticides, fungicides, or nematicides will help in manufacturing a combination which will offer actual control of pests while preventing or reducing the buildup of residues in the soil. The application of nano-pesticides will also lessen the rate of application since the amount of product actually being effective is usually at least 10–15 times reduced than that applied with classical formulations, henceforth a much smaller compared to the normal amount which is required to have much better and prolonged management. There are many pesticide manufacturers which are developing pesticides which are encapsulated in nanoparticles (OECD and Allianz 2008). These encapsulated pesticides may be time released or discharged upon occurrence of required environmental trigger (abiotic factors such as light, temperature, and humidity). It is still not well known whether these pesticide products will be commercially available for short term.

5.7.5 Detection and Control of Plant Diseases

Any stage of plant growth may be attached by plant microbial pathogens, which results in vast damage of food and produce. If we can detect the plant diseases at an early stage, then we can save million tons of food and produce from the probable outbreak of microbial diseases. Using the help of precise nanotechnology microbial plant diseases can not only be detected but also can be easily monitored and controlled. The combination of micro-biotechnology and nanotechnology in sensor development will create equipment having increased sensitivity, allowing an earlier response to ecological changes and infections. Linking of autonomous nanobiosensors into a global positioning system (GPS) will help to monitor the field, soil conditions, and crop conditions; certainly it would be of great help to farmers and scientists. Certain nanoparticles which have entered into the field to control plant diseases are mainly nanoforms of silver, silica, and aluminosilicates (Biswal et al. 2012). Among the biosynthesized nanoparticles, nano-silver is the most studied particle and has been utilized for controlling the plant diseases. It has long been understood that nanoparticles have tough inhibitory and bactericidal effects as well as a wide-ranging range of antimicrobial activities. Even though the Ag ions do have excellent antimicrobial function, the silver nanoparticles, owing to its high surface area and high fraction of surface atoms, are more potential compared to the bulk silver. It eradicates unsolicited microbes in planter soils and also in hydroponics systems. It is being employed as foliar spray to stop pathogens like fungi, molds, rot, and numerous other potent plant pathogens. In addition in controlling the diseases, Ag is also an exceptional plant growth stimulator. The silver nanoparticles were found to effectively control the powdery mildew disease of rose caused by fungus *Sphaerotheca pannosa* var rosae. Furthermore, nanosilica-silver composite "silicon" (Si) is also known to be absorbed by the plants to upsurge the disease resistance along with stress resistance (Brecht et al. 2003). The aqueous silicate solution, which is used to treat plants, has been reported to display brilliant preventive effects on pathogenic microbes causing powdery mildew disease or downy mildew disease in plants. Additionally, it also promotes the physiological activities and plant growth and induces the disease resistance in crop plants (Garver et al. 1998; Kanto et al. 2004). Since the silica has no direct disinfection effects on pathogenic microbes in plants, it does not display any effect on well-known diseases. Moreover, the effects of silica meaningfully differ with the physiological environmental conditions, and consequently, they are less employed compared to Ag nanoparticles in agriculture sector.

5.7.6 Soil Remediation

The nanoparticles can also be applied in dissimilar ways in soil remediation owing to its sorption capability. These can also be employed to interact with polymorphs, contaminants, minerals, and macro and micronutrients. Particularly, they can be applied to absorb and adsorb metal and anionic soil contaminants such as arsenic (As), chromium, lead (Pb), mercury, selenium, copper (Cu), and uranium, natural organic matter, organic acids, and also heavy metals (Changa and Chen 2005; Mercier and Pinnavaia 1997; Yang et al. 2006). Lü et al. (2007) reported that contaminant sequestration is attained by surface complexation or by encapsulation technology in interior interfaces of the nanoparticle aggregates; Tungittiplakorn et al. (2005) also supported this concept. Particularly, the copper oxide (CuO) nanoparticles have also been applied for As III and V adsorption (Martinson and Reddy 2009) and zero valent iron for remediation of organic pollutants (Ponder et al. 2001).

5.7.7 Water Remediation

The nanotechnology proposes the probable of innovative and unusual nanomaterials for treatment of the ground water, surface water, and wastewater which has been contaminated by toxic metal ions, inorganic and organic solutes, and various microbes. Owing to their exceptional activity toward recalcitrant contaminants, many nanomaterials are under studies and research for their application in water purification. Performance of nanoparticles, in combination with some other methods, improves significantly. Waterborne pathogens are traditionally removed by chlorination or ozonation which results in the formation of unsafe and hazardous byproducts (Shannon et al. 2008; Zhang et al. 2010). To overcome this, some new methods have been developed; one of these is the use of ultraviolet light merged with the nanotechnology to increase the photon effect by means of photocatalytic nanostructures. Tayade et al. (2006) reported that some nanoparticles such as transition metal oxide (TiO_2) nanoparticles are able to eradicate various bacteria and viruses by producing oxygen-based radicals after irradiation. Mura et al. (2011) also supported this notion. These nanoparticles are not spent or used up during the process and are, consequently, referred as “Green technology” for water purification and disinfection. By applying the principle of photocatalysis, the other metal oxide nanoparticles such as SnO_2 and ZnO could also be employed for decomposition of lethal herbicides and pesticides, which usually under normal climatic conditions take a long period to be degraded (Malato et al. 2002). Positive and fruitful results have been reported for degradation of some of the herbicides, such as Dicamba (Prevot et al. 2001), 2,4-D (Herrmann and Guillard 2000), atrazine (Zhanqi et al.

2007), and some other common pesticides, such as cyproconazole (Lhomme et al. 2007), 2,4-dichlorophenoxyacetic, also known as the DPA (Shankar et al. 2004), carbofuran (Mahalakshmi et al. 2007), and dichlorvos (Evgenidou et al. 2005).

5.8 Disadvantages and Challenges

The very minute size makes the nanoparticles of huge and enormous usefulness; sadly the same feature also causes numerous contrary and opposing effects and may signify noteworthy risks to the animals, microbes, environment, human beings, and plants when used non-sensibly (Mueller and Nowack 2010; Zhan 2009). The researchers are developing novel ways to observe probable and potential nanoparticle toxicity to macro and microorganisms of interest such as bacteria, fungi, plants, human, and animals (Grieger et al. 2010; Karn et al. 2009). A decrease in the size to the nanoscale level has resulted in huge upsurge in the surface to volume ratio, so comparatively more particles of the chemical nature are present on surface, consequently increasing the intrinsic toxicity. This could be one of the reasons why nanoparticles are commonly more lethal compared to larger particles of the same material while comparing on a mass dose base ratio. In the soil, the eco-toxicity is likely to be changed by several environmental issues that affect the colloid behavior of the particles. These factors include temperature, pH, humidity, ionic strength, divalent ions such as Ca^{2+}, and the presence of organic matter (Handy et al. 2008; Mühlfeld et al. 2008).

The nanoparticles employed as pesticides, chemical fertilizers, or in other preparations may deposit on aboveground parts of crop plants, if they are airborne. These NPs may clog stomata and create a very fine corporeal and toxic barrier layer on the stigma, which prevents the pollen tube penetration. These particles may also go in the vascular tissue and harm transportation of minerals, water, and other photosynthates. Harmfulness of ZnO nanoparticles was established on rye grass and in corn crop where it inhibited the germination process (Lin and Xing 2007). Ma et al. (2010) conducted an experiment and showed that in plants aluminum oxide NPs could be lethal to a variety of crops such as corn, soybeans, carrots, cabbage, and cucumbers as inhibitors of root elongation. Nanoparticles can also be absorbed and adsorbed from soil into the plants, as established in a study conducted by Gardea-Torresdey et al. (2002) on gold nanoparticles. Effect of these tiny materials on a living organism can occur once absorbed, which is not understood properly, and one may ask a question "whether the nanoparticles are taken up by plants could be toxic or nontoxic"? To assess the environmental risk of NPs, a clear understanding of these nanoparticles' stability, mobility, bioavailability, reactivity, eco-toxicity, and above all persistency is also required. Presently, we do not know much about the biodegradation aspects of nanoparticles, though this could be an area of potential research.

The animals may also inhale nanoparticles which may result into various negative effects and illness. These nanoparticles may enter into the bloodstream, which may cause serious ailments. The airborne nanoparticles present may cause precise threats for human health; these may enter the body through the respiratory system and may settle down deep in alveoli. Due to the entry of these nanoparticles into the lungs and then in the blood stream, there is probability of ailments like inflammation, protein fibrillation, and induction of genotoxicity. The scientists noted that nanoparticles breathed in by rats settled in the brain and lungs, which may led to noteworthy upsurges in biomarkers for swelling and stress response.

On the other hand, the nanoparticles do have harmful effect on valuable and beneficial microorganisms. Particularly, it was observed that nanoparticles such as TiO_2, SiO_2, and ZnO suspended in water for bacteria *Bacillus subtilis* and *Escherichia coli* are lethal for other Gram-positive and Gram-negative bacteria. Negative aspects of nanoparticles have also been exhibited on soil bacterial communities as application of ZnO nanoparticles in soil was found to lessen the soil enzyme activities, microbial biomass, and their diversity (Ge et al. 2011). Additionally, the process of immobilization and aggregation of these nanoparticles in soil displays phytotoxic effects and results in reduced biomass and plant root length (Kim et al. 2011).

Owing to these risk factors, the application of nanotechnology in agriculture sector has to be addressed very carefully and thoughtfully, and this permits obligatory need to critically examine and observe the risks involved with nanoparticle formulations.

5.9 Conclusion and Future Perspectives

Even though the nanotechnology is a developing and an emergent field of science greatly manipulated in different sectors mainly electronics, energy, and medical science, its application in agriculture is rarely exploited in India and abroad though it has tremendous potential as a new effective tool in the agricultural field to gain solutions to unresolved field problems. This technology enhances agricultural productivity in many ways such as by remediating soil and water, enabling smart delivery systems for fertilizers and pesticides, and detecting and controlling plant diseases at early stage. Moreover, synthesis of nanoparticles using microorganisms is an eco-friendly approach. Since many prokaryotic and eukaryotic groups of microorganisms can synthesize nanoparticles and they possess a well-regulated cellular system, they could be a beneficial substitute to common physical and chemical approaches which release many harmful chemicals and radiations in the environment during nanoparticle synthesis. So, in future, the use of nanotechnology in agriculture can not only increase production but also solve many problems associated with traditional farming. However, the toxic effects of nanoparticles on living organisms and environment must be of concern.

References

Agnihotri M, Joshi S, Kumar AR, Zinjarde S, Kulkarni S (2009) Biosynthesis of gold nanoparticles by the tropical marine yeast *Yarrowia lipolytica* NCIM 3589. Mater Lett 63:1231–1234

Agrawal S, Rathore P (2014) Nanotechnology pros and cons to agriculture: a review. Int J Curr Microbiol App Sci 3:43–55

Ahmad A, Khan MI, Senapati S, Kumar R, Sastry M (2003) Langmuir 19:3550–3553

Ahmad A, Senapati S, Khan I, Kumar R, Sastry M (2005) Extra-/intracellular biosynthesis of gold nanoparticles by an alkalotolerant fungus *Trichothecium* sp. J Biomed Nanotechnol 1:47–53

Ajayan PM, Sutto T, Ellis AV, Curran S (2004) Structural and spectral features of selenium nanospheres produced by Se-respiring bacteria. Appl Environ Microbiol 70:52–60
Amemiya Y, Arakaki A, Staniland SS, Tanaka T, Matsunaga T (2007) Controlled formation of magnetite crystal by partial oxidation of ferrous hydroxide in the presence of recombinant magnetotactic bacterial protein Mms6. Biomaterials 28:5381–5389
Anonymous N (2009) Nanotechnology and nanoscience applications: revolution in India and beyond. Strateg Appl Integr Nano Sci
Bai H, Zhang Z, Gong J (2006) Biological synthesis of semiconductor zinc sulfide nanoparticles by immobilized *Rhodobacter sphaeroides*. Biotechnol Lett 28:1135–1139
Balkwill DL, Maratea D, Blakemore RP (1980) Ultrastructure of a magnetotactic *spirillum*. J Bacteriol 141:1399–1408
Bansal V, Rautaray D, Bharde A, Ahire K, Sanyal A, Ahmad A, Sastry M, Mater J (2005) Fungus-mediated biosynthesis of silica and titania particles. J Mater 15:2583–2589
Baskar G, Chandhuru J, Fahad SK, Praveen AS (2013) Mycological synthesis, characterization and antifungal activity of zinc oxide nanoparticles. Asian J Pharm Tech 3:142–146
Bazylinski DA, Frankel RB (2004) Magnetosome formation in prokaryotes. Nat Rev Microbiol 2:217–230
Beveridge TJ, Murray RGE (1980) Sites of metal deposition in the cell wall of *Bacillus subtilis*. J Bacteriol 141:876–887
Bhattacharya R, Mukherjee P (2008) Biological properties of "naked" metal nanoparticles. Adv Drug Deliv Rev 60:1289–1306
Bhattacharyya A, Chandrasekar R, Chandra AK, Epidi TT, Prakasham RS (2014) Application of nanoparticles in sustainable agriculture: its current status. Short views on insect. Biochem Mol Biol 2:429–448
Biswal SK, Nayak AK, Parida UK, Nayak PL (2012) Applications of nanotechnology in agriculture and food sciences. Int J Sci Innov Discoveries 2:21–36
Brayner R, Barberousse H, Hemadi M, Djedjat C, Yéprémian C, Coradin T, Livage J, Fiévet F, Couté A (2007) Cyanobacteria as bioreactors for the synthesis of Au, Ag, Pd, and Pt nanoparticles via an enzyme-mediated route. J Nanosci Nanotechnol 7:2696–2708
Brecht M, Datnoff L, Nagata R, Kucharek T (2003) The role of silicon in suppressing tray leaf spot development in St. Augustine grass. Publication in University of Florida, University Report, 1–4
Breierova E, Vajczikova I, Sasinkova V, Stratilova E, Fisera M, Gregor T, Sajbidor J (2002) Biosorption of cadmium ions by different species. Z Naturforsch 57c:634–639
Changa YC, Chen DH (2005) Preparation and adsorption properties of monodisperse chitosan-bound Fe_3O_4 magnetic nanoparticles for removal of Cu(II) ions. J Colloid Interface Sci 283:446–456
Chauhan A, Zubair S, Tufail S, Sherwani A, Sajid M, Raman S et al (2011) Fungus-mediated biological synthesis of gold nanoparticles: potential in detection of liver cancer. Int J Nanomedicine 6:2305–2319
Chinnamuthu CR, Kokiladevi E (2007) Weed management through nanoherbicides. In: Chinnamuthu CR, Chandrasekaran B, Ramasamy C (eds) Application of nanotechnology in agriculture. Tamil Nadu Agricultural University, Coimbatore
Correa-Llantén DN, Muñoz-Ibacache SA, Castro ME, Muñoz PA, Blamey JM (2013) Gold nanoparticles synthesized by *Geobacillus* sp. strain ID17 a thermophilic bacterium isolated from Deception Island, Antarctica. Microb Cell Factories 12:75
Dameron CT, Reese RN, Mehra RK, Kortan AR, Carroll PJ, Steigerwald ML, Brus LE, Winge DR (1989) Biosynthesis of cadmium sulfide quantum semiconductor crystallites. Nature 338:596–597
Das SK, Liang J, Schmidt M, Laffir F, Marsili E (2012) Biomineralization mechanism of gold by zygomycete fungi *Rhizopus oryzae*. ACS Nano 6:6165–6173
Deplanche K, Macaskie LE (2008) Biorecovery of gold by *E. coli* and *Desulfovibrio desulfuricans*. Biotechnol Bioeng 99:1055–1064
Drexler E (2009) There's plenty of room at the bottom (Richard Feynman, Pasadena, 29 December 1959). Metamodern, The Trajectory of Technol 12:29

Duran N, Marcato PD, Alves OL, DSouza G, Esposito E (2005) Mechanistic aspects of biosynthesis of silver nanoparticles by several *Fusarium oxysporum* strains. J Nanobiotechnol 3:8–14

Evgenidou E, Fytianos K, Poulios I (2005) Semiconductor-sensitized photodegradation of dichlorvos in water using TiO_2 and ZnO as catalysts. Appl Catal B Environ 59:81–89

Gade AK, Bonde P, Ingle AP, Marcato PD, Duran N, Rai MK (2008) Exploitation of *Aspergillus niger* for synthesis of silver nanoparticles. J Biobaased Mater Bioenergy 2:243

Gajbhiye M, Kesharwani J, Ingle A, Gade A, Rai M (2009) Fungus-mediated synthesis of silver nanoparticles and their activity against pathogenic fungi in combination with fluconazole. Nanomed Nanotechnol Biol Med 5:382

Gardea-Torresdey JL, Parsons JG, Gomez E, Peralta-Videa J, Troiani HE, Santiago P, Yacaman MJ (2002) Formation and growth of Au nanoparticles inside live alfalfa plants. Nano Lett 2:397–401

Garver TLW, Thomas BJ, Robbins MP, Zeyen RJ (1998) Phenylalanine ammonia-lyase inhibition, auto fluorescence, and localized accumulation of silicon, calcium and manganese in oat epidermis attacked by the powdery mildew fungus *Blumeria graminis* (DC) speer. Physiol Mol Plant Pathol 52:223–243

Ge Y, Schimel JP, Holden PA (2011) Evidence for negative effects of TiO_2 and ZnO nanoparticles on soil bacterial communities. Environ Sci Technol 45:1659–1664

Gericke M, Pinches A (2006) Biological synthesis of metal nanoparticles. Hydrometal 83:132

Ghormade V, Deshpande MV, Paknikar KM (2011) Perspectives for nano-biotechnology enabled protection and nutrition of plants. Biotechnol Adv 29:792–803

Gole A, Dash C, Ramakrishnan V, Sainkar SR, Mandale AB, Rao M et al (2001) Pepsin–gold colloid conjugates: preparation, characterization and enzymatic activity. Langmuir 17:1674–1679

Govindraju K, Basha SK, Kumar VG, Singaravelu G (2008) Silver, gold and bimetallic nanoparticles production using single-cell protein (*Spirulina platensis*) Geitler. J Mat Sci 43:5115–5122

Grieger KD, Fjordbøge A, Hartmann NB, Eriksson E, Bjerg PL, Baun A (2010) Environmental benefits and risks of zero-valent iron nanoparticles (nZVI) for in situ remediation: risk mitigation or trade-off. J Contam Hydrol 118:165–183

Handy RD, Henry TB, Scown TM, Johnstone BD, Tyler CR (2008) Manufactured nanoparticles: their uptake and effects on fish: a mechanistic analysis. Ecotoxico 17:396–409

He S, Guo Z, Zhang Y, Zhang S, Wang J, Gu N (2007) Biosynthesis of gold nanoparticles using the bacteria *Rhodopseudomonas capsulate*. Mater Lett 61:3984–3987

Herrmann JM, Guillard C (2000) Photocatalytic degradation of pesticides in agricultural used waters. Surface Chem Cat 23:417

Ingle A, Gade A, Pierrat S, Sonnichsen C, Rai M (2008) Mycosynthesis of silver nanoparticle using the fungus *Fusarium acuminatum* and its activity against some human pathogenic bacteria. Cur Nanosci 4:141–144

Kalishwaralal K, Deepaka V, Pandiana SR, Kottaisamy M, ManiKantha BS, Kartikeyana B, Gurunathana S (2010) Biosynthesis of silver and gold nanoparticles using *Brevibacterium casei*. Coll Surf B Bioint 77:257–262

Kanto T, Miyoshi A, Ogawa T, Maekawa K, Aino M (2004) Suppressive effect of potassium silicate on powdery mildew of strawberry in hydroponics. J Gen Plant Pathol 70:207–211

Karn B, Kuiken T, Otto M (2009) Nanotechnology and in situ remediation: a review of the benefits and potential risks. Environ Health Perspect 117:1813–1831

Kathiresan K, Manivannan S, Nabeel MA, Dhivya B (2009) Studies on silver nanoparticles synthesized by a marine fungus, *Penicillium fellutanum* isolated from coastal mangrove sediment. Colloids Surf B Biointerfaces 71:33

Khan MM, Kalathil S, Han TH, Lee J, Cho MH (2013) Positively charged gold nanoparticles synthesized by electrochemically active biofilm-a biogenic approach. Nanosci Nanotechnol 13:6079–6085

Kim S, Kim J, Lee I (2011) Effects of Zn and ZnO nanoparticles and Zn^{2+} on soil enzyme activity and bioaccumulation of Zn in *Cucumis sativus*. Chem Ecol 27:49–55

Konishi Y, Nomura T, Tsukiyama T, Saitoh N (2004) Microbial preparation of gold nanoparticles by anaerobic bacterium. J Trans Mater Res Soc Jpn 29:2341

Korbekandi H, Iravani S, Abbasi S (2012) Optimization of biological synthesis of silver nanoparticles using *Lactobacillus casei* subsp. Casei J Chemi Technol Biotechnol 87:932–937

Kowshik M, Deshmukh N, Vogel W, Urban J, Kulkarni SK, Paknikar KM (2002) Microbial synthesis of semiconductor CdS nanoparticles, their characterization and their use in the fabrication of an ideal diode. Biotechnol Bioeng 78:583–588

Krumov N, Nochta IP, Oder S, Gotcheva V, Angelov A, Posten C (2009) Production of inorganic nanoparticles by microorganisms. Chem Eng Technol 32:1026–1035

Kumar GC, Mamidyala SK (2011) Extracellular synthesis of silver nanoparticles using culture supernatant of *Pseudomonas aeruginosa*. Colloids Surf B: Biointerfaces 84:462–466

Lengke M, Ravel B, Feet ME, Wanger G, Gordon RA, Southam G (2006) Mechanisms of gold bioaccumulation by filamentous cynobacteria from gold (III)-chloride complex. Environ Sci Technol 40:6304–6309

Lhomme L, Brossilon S, Woolbert D (2007) Photocatalytic degradation of a triazole pesticide, cyproconazole, in water. J Photochem Photobiol 188:34–42

Li X, Xu H, Chen ZS, Chen G (2011) Biosynthesis of nanoparticles by microorganisms and their applications. J Nanomaterials. doi:10.1155/2011/270974

Lin D, Xing B (2007) Phytotoxicity of nanoparticles: inhibition of seed germination and root growth. Environ Pollut 150:243–250

Liu J, Qiao SZ, Hu QH, Lu GQ (2011) Magnetic nanocomposites with mesoporous structures: synthesis and applications. Small 7:425–443

Lü QF, Huang MR, Li XG (2007) Synthesis and heavy metal ion sorption of pure sulfophenylenediamine copolymer nanoparticles with intrinsic conductivity and stability. Chem Eur J 13: 6009–6018

Luechinger NA, Grass RN, Athanassiou EK, Stark WJ (2010) Bottom-up fabrication of metal/metal nanocomposites from nanoparticles of immiscible metals. Chem Mater 22:155–160

Lyons K (2010) Nanotechnology: transforming food and the environment. Food First Backgrounder 16:1–4

Ma Y, Kuang L, HeX BW, Ding Y, Zhang Z, Zhao Y, Chai Z (2010) Effects of rare earth oxide nanoparticles on root elongation of plants. Chemosphere 78:273–279

Mahalakshmi M, Arabindoo B, Palanichamy M, Murugesan V (2007) Photocatalytic degradation of carbofuran using semiconductor oxides. J Hazard Mater 143:240–245

Malarkodi C, Rajeshkumar S, Paulkumar K, Vanaja M, Jobitha GDG, Annadurai G (2013) Bactericidal activity of biomediated silver nanoparticles synthesized by *Serratia nematodiphila*. Drug Invention Today 5:119–125

Malato A, Blanco J, Vidal A, Richter C (2002) Photocatalysis with solar energy at a pilot-plant scale: an overview. Appl Catal B-Environ 37:1–15

Marchiol L (2012) Synthesis of metal nanoparticles in living plants. Ita J Agro 37:274–282

Martinson CA, Reddy KJ (2009) Adsorption of arsenic (III) and arsenic (V) by cupric oxide nanoparticles. J Colloid Interf Sci 336:406–411

Mehra RK, Winge DR (1991) Metal ion resistance in fungi: molecular mechanisms and their regulated expression. J Cell Biochem 45:30–40

Mercier L, Pinnavaia TJ (1997) Access in mesoporous materials: advantages of a uniform pore structure in the design of a heavy metal ion adsorbent for environmental remediation. Adv Mater 9:500–503

Mohanpuria P, Rana NK, Yadav SK (2007) Cadmium induced oxidative stress influence on glutathione metabolic genes of *Camellia sinensis* (L.) O. Kuntze. Environ Toxicol 22:368–374

Mohanpuria P, Rana NK, Yadav SK (2008) Biosynthesis of nanoparticles: technological concepts and future applications. J Nanopart Res 10:507–517

Mokhtari N, Daneshpajouh S, Seyedbagheri S, Atashdehghan R, Ab di K, Sarkar S, Minaian S, Shahverdi HR, Shahverdi AR (2009) Biological synthesis of very small silver nanoparticles by culture supernatant of *Klebsiella pneumoniae*: the effects of visible-light irradiation and the liquid mixing process. Mat Res Bull 44:1415–1421

Mueller N, Nowack B (2010) Nanoparticles for remediation: solving big problems with little particles. Elements 6:395–400

Mühlfeld C, Gehr P, Rothen-Rutishauser B (2008) Translocation and cellular entering mechanisms of nanoparticles in the respiratory tract. Swiss Med Wkly 138:387–391

Mukherjee P, Senapati S, Mandal D, Ahmad A, Khan MI, Kumar R, Sastry M (2002) Extracellular synthesis of gold nanoparticles by the fungus *Fusarium oxysporum*. Chem Bio Chem 3:461

Mura S, Greppi GF, Roggio A, Malfatti L, Innocenzi P (2011) Polypeptide binding to mesostructured titania films. Microporous Mesoporous Mat 142:1–6

Nair B, Pradeep T (2002) Coalescence of nanoclusters and formation of submicron crystallites assisted by *Lactobacillus* strains. Crystal Growth Design 2:293–298

Nasreen I, Hulkoti TC, Taranath TC (2014) Biosynthesis of nanoparticles using microbes: a review. Colloids Surf B: Biointerfaces 121:474–483

OECD and Allianz (2008) Sizes that matter: opportunities and risks of nanotechnologies. Report in cooperation with the OECD International Futures Programme. http://www.oecd.org/dataoecd/32/1/44108334.pdf

Oremland RS, Herbel MJ, Blum JS, Langley S, Beveridge TJ, Ajayan PM, Sutto T, Ellis AV, Curran S (2004) Structural and spectral features of selenium nanospheres produced by Se-respiring bacteria. Appl Environ Microbiol 70:52–60

Philipse AP, Maas D (2002) Magnetic colloids from magnetotactic bacteria: chain formation and colloidal stability. Langmuir 18:9977–9984

Ponder SM, Darab JG, Bucher J, Caulder D, Craig I, Davis L, Edelstein N, Lukens W, Nitsche H, Rao L, Shuh DK, Mallouk TE (2001) Surface chemistry and electrochemistry of supported zerovalent iron nanoparticles in the remediation of aqueous metal contaminants. Chem Mater 13:479–486

Prasad K, Jha AK (2010) Biosynthesis of CdS nanoparticles: an improved green and rapid procedure. J Coll Interf Sci 342:68–72

Prevot AB, Fabbri D, Pramauro E, Rubio AM, de la Guardia M (2001) Continuous monitoring of photocatalytic treatments by flow injection. Degradation of dicamba in aqueous TiO_2 dispersions. Chemosphere 44:249–255

Pum D, Sleytr UB (1999) The application of bacterial S-layers in molecular nanotechnology. Trends Biotechnol 17:8–12

Rai M, Yadav A, Gade A (2008) Current trends in phytosynthesis of metal nanoparticles. Crit Rev Biotech 28:277–284

Rai M, Yadav A, Bridge P, Gade A (2009) Myconanotechnology: a new and emerging science. In: Rai M, Bridge PD (eds) Applied mycology, 1st edn. CAB International, New York, pp 258–267

Rai V, Acharya S, Dey N (2012) Implications of nanobiosensors in agriculture. J Biomat Nanobiotechnol 3:15–324

Rai M, Ingle AP, Gupta IR, Birla SS, Yadav AP, Kamel A (2013) Potential role of biological systems in formation of nanoparticles: mechanism of synthesis and biomedical applications. Curr Nanosci 9:576–587

Rajasree SR, Suman TY (2012) Extracellular biosynthesis of gold nanoparticles using a gram negative bacterium *Pseudomonas fluorescens*. Asian Pac J Trop Dis 2:796–799

Ramezani H, Holm S, Allard A, Ståhl G (2010) Monitoring landscape metrics by point sampling: accuracy in estimating Shannon's diversity and edge density. Environ Monit Assess 164:403–421

Sanghi R, Verma PA (2009) Facile green extracellular biosynthesis of CdS nanoparticles by immobilized fungus. Chem Eng J 155:886–891

Sanghi R, Verma P, Puri S (2011) Enzymatic formation of gold nanoparticles using *Phanerochaete Chrysosporium*. J Adv Chem Eng Sci 1:154–162

Sastry M, Ahmad A, Khan MI, Kumar R (2003) Biosynthesis of metal nanoparticles using fungi and actinomycetes. Curr Sci 85:162–170

Selvakannan PR, Swami A, Srisathiyanarayanan D, Shirude PS, Pasricha R, Mandale AB et al (2004) Synthesis of aqueous Au core-Ag shell nanoparticles using tyrosine as a pH-dependent reducing agent and assembling phase-transferred silver nanoparticles at the air-water interface. Langmuir 20:7825–7836

Senapati S, Ahmad A, Khan MI, Sastry M, Kumar R (2005) Extracellular biosynthesis of bimetallic Au-Ag alloy nanoparticles. Small 1:517–520

Shankar SS, Ahmad A, Pasricha R, Sastry M (2003) Bioreduction of chloroaurate ions by geranium leaves and its endophytic fungus yields gold nanoparticles of different shapes. J Mater Chem 13:1822–1826

Shankar MV, Anandan S, Venkatachalam N, Arabindoo B, Murugesan V (2004) Novel thin-film reactor for photocatalytic degradation of pesticides in an aqueous solution. J Chem Technol Biot 79:1279–1285

Shannon MA, Bohn PW, Elimelech M, Georgiadis JG, Mariñas BJ, Mayes AM (2008) Science and technology for water purification in the coming decades. Nat 452:301–310

Sharma VK, Yngard RA, Lin Y (2009) Silver nanoparticles: green synthesis and their antimicrobial activities. Adv Coll Interf Sci 145:83–96

Sharma N, Pinnaka AK, Raje MA, Fnu Bhattacharyya MS, Choudhury AR (2012) Exploitation of marine bacteria for production of gold nanoparticles. Microb Cell Factories 11:86

Shivaji S, Madhu S, Singh S (2011) Extracellular synthesis of antibacterial silver nanoparticles using psychrophilic bacteria. Process Biochem 49:830–837

Shiying H, Zhirui G, Yu Z, Song Z, Jing W, Ning G (2007) Biosynthesis of gold nanoparticles using the bacteria *Rhodopseudomonas capsulate*. Mater Lett 61:3984–3987. doi:10.1016/j.matlet.2007.01.018

Si S, Mandal TK (2007) Tryptophan-based peptides to synthesize gold and silver nanoparticles: a mechanistic and kinetic study. Chem Eur J 13:3160–3168

Simkiss K, Wilbur KM (1989) Biomineralization. Academic Press, San Diego

Slocik JM, Knecht MR, Wright DW (2004) Biogenic nanoparticles. In: Nalwa HS (ed) The encyclopedia of nanoscience and nanotechnology. American Scientific Publishers, Stevenson Ranch, pp 293–308

Srivastava SK, Yamada R, Ogino C, Kondo A (2013) Biogenic synthesis and characterization of gold nanoparticles by Escherichia coli K12 and its heterogeneous catalysis in degradation of 4-nitrophenol. Nanoscale Res Lett 8:70

Sweeney RY, Mao CX, Gao JL, Burt AM, Belcher G, Georgiou BL, Iverson BL (2004) Bacterial biosynthesis of cadmium sulfide nanocrystals. Chem Boil 11:1553

Tarafdar JC (2012) Perspectives of nanotechnological applications for crop production. NAAS News 12:8–11

Tarafdar JC, Raliya R, Rathor I (2012) Microbial synthesis of phosphorus nanoparticles from Tricalcium phosphate using Aspergillus tubingensis TFR-5. J Bionanosci 6:84–89

Tayade RJ, Kulkarni RG, Jasra RV (2006) Transition metal ion impregnated mesoporous TiO_2 for photocatalytic degradation of organic contaminants in water. Ind Eng Chem Res 45:5231–5238

Thakkar KN, Mhatre SS, Parikh RY (2010) Biological synthesis of metallic nanoparticles. Nanomed 6:257–262

Tiwari DK, Behari J, Sen P (2008) Time and dose-dependent antimicrobial potential of Ag nanoparticles synthesized by top-down approach. Curr Sci 95:647–655

Tsibakhashvili N, Kalabegishvili T, Gabunia V, Gintury E, Kuchava N, Bagdavadze N, Pataraya D, Gurielidzse M, Gvarjaladze D, Lomidze L (2010) Synthesis of silver nanoparticles using bacteria. Nano Studies 2:179–182

Tungittiplakorn W, Cohen C, Lion LV (2005) Engineered polymeric nanoparticles for bioremediation of hydrophobic contaminants. Environ Sci Technol 39:1354–1358
Vigneshwaran N, Kathe AA, Varadarajan PV, Nachane RP, Balasubramanya RH (2006) Biomimetics of silver nanoparticles by white rot fungus, *Phanerochaete chrysosporium*. Coll Surf B Biointerg 53:55–59
Waghmare SS, Deshmukh AM, Kulkarni SW, Oswaldo LA (2011) Biosynthesis and characterization of manganese and zinc nanoparticles. Univ J Environ Res Technol 1:64
Xiaorong Z, Xiaoxiao H, Kemin W, Xiaohai Y (2011) Different active biomolecules involved in biosynthesis of gold nanoparticles by three fungus species. J Biomed Nanotech 7:245–254
Yang K, Zhu L, Xing B (2006) Adsorption of polycyclic aromatic hydrocarbons by carbon nanomaterials. Environ Sci Technol 40:1855–1861
Zeng J, Ma Y, Jeong U, Xia Y (2010) Au(I): an alternative and potentially better precursor than Au(III) for the synthesis of Au nanostructures. J Mater Chem 20:2290–2301
Zhan J (2009) Multifunctional colloidal particles for in situ remediation of chlorinated hydrocarbons. Environ Sci Technol 43:8616–8621
Zhang D, Li G, Yu JC (2010) Inorganic materials for photocatalytic water disinfection. J Mater Chem 20:4529–4536
Zhanqi G, Shaogui Y, Na T, Cheng S (2007) Microwave-assisted rapid and complete degradation of atrazine using TiO_2 nanotube photocatalyst suspensions. J Hazard Mater 145:424–430

Soil Microbiome and Their Effects on Nutrient Management for Plants

6

Rosangela Naomi Inui Kishi, Renato Fernandes Galdiano Júnior, Silvana Pompéia Val-Moraes, and Luciano Takeshi Kishi

Abstract

The soil microbiome is a diverse system composed of microorganisms with different functions. Microorganisms known as plant growth-promoting microorganisms (PGPMs) can help plants with nutrient uptake and consequently with crop yields. From this class of microorganisms, we can isolate nitrogen-fixing bacteria (NFB), phosphorus-solubilizing microorganisms (PSMs), and the microbes that are able to produce phytohormones. The use of these microorganisms in improving nutrient uptake by plants has been acceptable because of reduced costs and the safety of application for humans and the environment. It is for this reason that inoculant products have been developed. During the process of inoculant development, it is possible to use molecular biology techniques, such as 16S rRNA gene sequencing. This technique helps with the identification of potential microorganisms adapted for different conditions and crops. Moreover, these microorganisms can be used in degradable areas or as pathogen controls. It is also important to consider the siderophore, which is a biological molecule produced by various bacteria, and which has an immense application in agriculture. Another important symbiosis that occurs is realized by mycorrhizas, which are essential for transferring nutrients and water from the soil to plants.

6.1 Introduction

The association between plants, soil, and soil microbes is like a system, which influences plant health and productivity. To illustrate this, recent advances in "omics" research can provide a common understanding and management of these interactions (Chaparro et al. 2012).

R.N. Inui Kishi (✉) • R.F. Galdiano Júnior • S.P. Val-Moraes • L.T. Kishi
Departament of Technology, State Paulista University – UNESP,
Via de Acesso Prof. Paulo Donato Castellane s/n, Jaboticabal, São Paulo, Brazil, 14884-900
e-mail: rosangela.inui@gmail.com

V. Kumar et al. (eds.), *Probiotics in Agroecosystem*,
DOI 10.1007/978-981-10-4059-7_6

The phytomicrobiome is characterized by microbial communities related to plants (Smith and Zhou 2014; Smith et al. 2015). This phytomicrobiome can be divided into the rhizomicrobiome, which is located in and around the roots or the rhizosphere (Lundberg et al. 2012), the phyllomicrobiome, which refers to the microbiome present in aerial parts of the plant (Rastogi et al. 2012; Kembel et al. 2014), and the endosphere, which is inside the plant (Berg et al. 2014). The structure of these microbial communities can vary according to the interactions between plant – microorganism and/or microorganism-microorganism. These connections are mediated by compounds that are released by plants or microorganisms as exudates (East 2013). Understanding the form and function of these compounds is essential for the possibility of using these microbes to develop new technologies for crop growth promotion, industrial process optimization (e.g., fermentation), and biocontrol mechanism development (East 2013).

Of all the microbial communities, the rhizomicrobiome shows most relevance for field crops. This group of microorganisms is very diverse and dynamic in response to environmental conditions and the interactions between plants and microbes, which in some cases are specific. Plant growth–promoting microorganisms (PGPMs) are one group from the rhizomicrobiome which live in soil or close to plant roots (Gray and Smith 2005; Mabood et al. 2014). These groups of bacteria are distinguished by some inherent characters: (i) able to be established at the root surface; (ii) remain and compete with other microbes; (iii) promote plant growth (Kloepper 1994).

PGPMs have acquired relevance in agriculture because they are considered an alternative to the traditional management of crops (Bhattacharyya and Jha 2012), and an environmentally friendly practice. A huge miscellanea of microbes have been used as PGPMs (Ahemad and Kibret 2014); one of the most used is Rhizobia, due to the high yield increases that result when inoculated in plants (Rathore 2014). Moreover, PGPMs are classified according to their functionality (Gray and Smith 2005; Mabood et al. 2014). As biofertilizers, they can intensify the acquisition of nutrients by plants, through nitrogen fixation (Vessey 2003; Bhattacharyya and Jha 2012) and phosphate solubilization (Inui-Kishi et al. 2012; Trabelsi and Mhamdi 2013). Other uses of PGPMs are as phytostimulators (inducing plant growth through phytohormes), rhizomediators (used for restoration of degraded environments) (Antoun and Prévost 2005), and for the production of metal chelators and siderophores (Vessey 2003; Bhattacharyya and Jha 2012).

Other mechanisms of PGPMs include the production of 1-aminocycloropropane-1-carboxyla deaminase (maintaining ethylene levels in plant tissues under stress situations) (Penrose and Glick 2003), the induction of innate resistance or suppression of disease by antibiotics produced by fungi or bacteria (Antoun and Prévost 2005; Bhattacharyya and Jha 2012), the production of cell wall lytic enzymes (Haas and Defago 2005; Rathore 2014), and quorum sensing and interference on biofilm formation (Bhattacharyya and Jha 2012).

PGPMs can be divided into two groups according to the proximity of bacteria to the root (Gray and Smith 2005). One group is called ePGPM (extracellular) and is present in the rhizosphere or on the rhizoplane. *Agrobacterium*, *Arthobacter*, *Azospirillum*, *Pseudomonas*, and *Serratia* are just some examples of ePGPMs

(Bhattacharyya and Jha 2012). Another group is iPGRM (intracellular), which are present inside roots cells and are represented by *Allorhizobium*, *Azorhizobium*, *Bradyrhizobium*, *Mesorhizobium*, and *Rhizobium* (Gray and Smith 2005).

In summary, microorganisms can act in coordination with the soil microbiome for the purpose of benefiting plant health and development. Much evidence shows that plants are able to determine microbial communities through root exudates. These exudates maintain a molecular conversation according to the plants' phenological stage, interaction with others species, and management techniques (Chaparro et al. 2012). This chapter discusses the detail and actual knowledge about the importance of the soil microbiome on nutrient acquisition together with other applications of these microorganisms.

6.2 Soil Microbiome Diversity and Function

The rhizosphere of plants is a huge source of soil microbes (Köberl et al. 2013), which are captivated by root secretions and/or rhizodeposits (Compant et al. 2010). Plant species can co-ordinate the rhizosphere microbiome, which is dependent on the soil microbial community (Smalla et al. 2001). Furthermore, microbial communities are contingent on soil type, pedoclimatic conditions, plant health, phenological stage, and edaphoclimatic factors (Singh and Mukerji 2006).

The higher activity observed by microbes in the rhizosphere brings several biological and ecological benefits to the environment and improves plant yield. The rhizosphere contains a large number of microorganisms with the ability to fix nitrogen, solubilize phosphorus (P), enhance plant pathogen resistance (Arjun and Harikrishnan 2011), and help recover degraded environments (Antoun and Prévost 2005). These microorganisms assume an important aspect from an agronomic point of view, as we can outline below.

Knowledge of the diversity of the rhizosphere is very limited. It has been estimated that less than 1% of the soil microbiome has been isolated in pure culture. In order to understand the soil microbiome, metagenomics can help in analyzing complex genomes of microbial communities through culture-independent molecular approaches (Peix et al. 2007). Moreover, with molecular approaches it is possible to verify the existence and determine the quantity of microorganisms (Oliveira et al. 2009). For bacterial diversity analysis, the molecular marker 16S rRNA gene is utilized (Richardson et al. 2011). Arjun and Harikrishnan (2011) investigated the microbial diversity present in the rice rhizosphere from a paddy field ecosystem in Kerala, India. They used culture-independent molecular techniques, 16S rRNA clone library generation obtained by RFLP, sequencing, and phylogenetic analysis. Through sequence analysis of 16S rRNA genes, they observed major diversity in the bacterial community, with the majority of microbes being related to Proteobacteria. Just a small portion of the 16S rRNA sequences were highly similar to rRNA from the Acidobacteria, Firmicutes, and Bacterioides groups. Knowledge of the less known microbial community is very useful for the comprehension of their individual roles as related to plant health, yield, and metabolic capabilities. Moreover, metagenomics promises to bring many more questions regarding the uncultivable fraction of the rhizosphere community.

Köberl et al. (2013) present a study performed with the purpose of analyzing the microbiome of medicinal plants (*Matricaria chamomilla* L., *Calendula officinalis* L., and *Solanum distichum* Schumach. and Thonn.) planted in an organic desert farm in Egypt. These plants have a distinguished microbiome due to their particular and structurally divergent active secondary metabolites. These secondary metabolites present the major reason for their high specificity for related microorganisms (Qi et al. 2012). Soil microbiomes of desert environments are more abundant in Gram-positive bacteria related to pathogen suppression. These authors observed an evident selection of specific microbes by plants, as well as highly specific diazotrophic communities that demonstrated the importance of plant species on microbial diversity. Moreover, they found *Bacillus* spec. div. strains were able to promote plant growth and improve flavonoid production. These results emphasize the numerous connections between the plant microbiome and the plant metabolome.

Several surveys have demonstrated that the soil microbiome diversity has been reduced due to the intensification of land use in in the agriculture (Maeder et al. 2002; de Vries et al. 2013), This demonstrates some of the negative effects of agriculture on the environment and the unsustainability of agricultural production (Sala et al. 2000). The decline in soil biodiversity is sometimes discussed in terms of functional redundancy. Functional redundancy suggests that different species can have the same function in an ecosystem, and therefore declines in species diversity do not necessarily affect ecosystem functioning. Research by Philippot et al. (2013) counters this viewpoint, suggesting that microbial diversity loss can affect ecosystem processes.

Mendes et al. (2015) hypothesize that the microbial community diversity and functional diversity are much lower in undisturbed than disturbed soils, with consequences for functional redundancy in the soil microbiome. To explain this hypothesis in detail, they used soil DNA shotgun metagenomics to assess the soil microbiome in a chronological sequence of land use with native forest, followed by deforestation and cultivation of soybean and pasture in different seasons. The results obtained by these authors demonstrated that an agriculture and pasture soil shows the most diversity and higher functional redundancy. Conversely, the equilibrium in forest ecosystems was maintained with a lower diversity and higher abundance of microorganisms. These results indicate that land use is an important factor in the composition of the soil microbiome. Knowledge of the diversity of the soil microbial community could help in the identification of microbial candidates to act as PGPMs and for development of inoculant products.

6.3 Biological Nitrogen Fixation

Nitrogen (**N**) is one of the most important elements for plant development because it is an essential part of nucleic acids, enzymes, and proteins. Seventy-eight percent of N is in gaseous form. Despite this, N is unavailable to plants and is thus considered one of the most growth-limiting nutrients (Dalton and Krammer 2006). To become available to plants, atmospheric nitrogen (N_2) needs to be modified or fixed to ammonia (NH_3) by nitrogen-fixing microorganisms (Kim and Rees 1994). Biological nitrogen fixation (BNF) contributes to two-thirds of the nitrogen fixed worldwide. Mild temperature is generally one of the conditions that promotes BNF by diazotrophic microorganisms, which are dispersed in nature (Raymond et al. 2004).

Microorganisms with BNF activity are separated into groups as being (a) symbiotic N_2-fixing bacteria, including the rhizobiaceae family (Ahemad and Khan, 2012b); (b) non-leguminous trees (e.g. *Frankia*); and (c) non-symbiotic (free living and endophytes) nitrogen-fixing forms like cyanobacteria *Anabaena*, *Nostoc*, *Azospirrilum*, *Azoarcus*, and others (Battacharrya and Jha 2012). Non-symbiotic microorganisms provide a small amount of the fixed nitrogen that is required by plants (Glick 2012) due to the limitation of carbon and the inhibition of nitrogenase by oxygen, which is the enzyme responsible by nitrogen fixation (Oldroyd and Downie 2004).

For the purpose of acquiring nutrients, plants have formed symbiotic interactions with microorganisms such as legumes and rhizobia. In this symbiosis, the bacteria penetrate the plant root and remain restricted in intracellular space, such as nodules, where N_2 is transformed into ammonia, which is then absorbed by plants (Oldroyd and Downie 2008).

6.3.1 Molecular Signaling in Nitrogen-Fixing Bacteria

The molecular communication between plant and bacteria occurs due to the detection of flavonoids and related molecules that are secreted from legume roots by rhizobia (Perret et al. 2000). Flavonoids are transcriptional regulators recognized by NodD proteins that bind to a signaling molecule and have the possibility to activate gene expression (Long 1996).

The formation of nodules in legume roots is activated by nodulation (Nod) components (NC), which are signaling molecules that induce developmental changes in plants. Nod components have in their structure a chitin backbone with an N-linked fatty acid moiety connected with a non-reducing terminal sugar. Some modifications can occur in NC structure among species of rhizobia. These modifications can define the specificity between the rhizobia and the host plant (Oldroyd and Downie 2008).

Bacterial invasion normally happens by root hair cells or disruption of root cells. The entry via root hair cells begins by bacterial attachment to root hairs (Oldroyd and Downie 2004). Concurrently, cortical cells activate cell partitioning to establish the nodule meristem. The epidermal response to the entry of roots is correlated with the recognition of nodulation components that govern calcium spiking-subordinate signaling pathway, and with root hair distortion through a signaling pathway autonomous to calcium spiking. Cortical cell division is correlated with an increase in the concentration of cytokinin and auxin. Transcription factors (TFs) from nodulating signaling pathway, NSP1 and NSP2, are essential for nodulation and are obligatory in the epidermis with induction of initial nodulation genes (INODs), cortical cell division, and nodule inception (NIN). These TFs are responsible for activating NC gene expression. In disruption invasion, violation occurs through the epidermis and the microbes receive

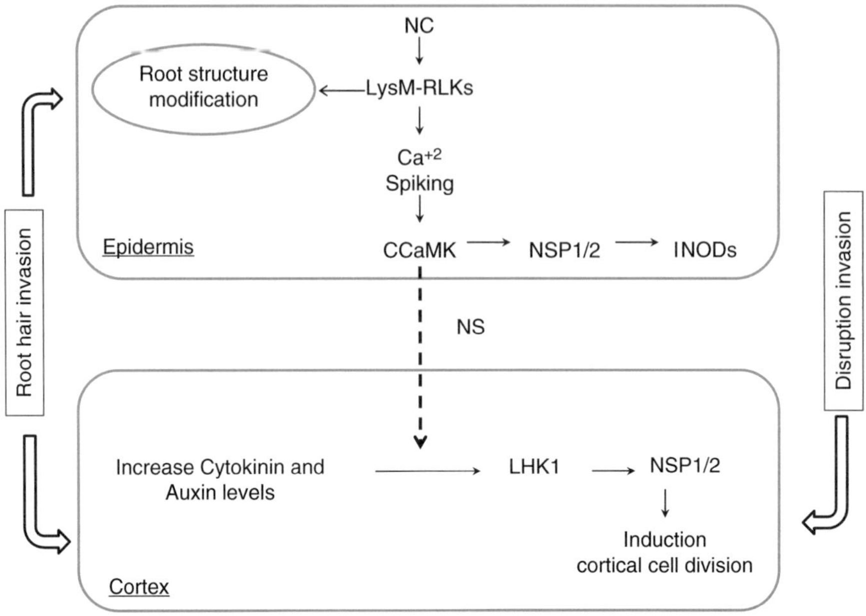

Fig. 6.1 Changes in the roots during the nodulation process. *NC* nodulation components, *INODs* initial nodulation genes, *NS* nodule start, *LysM-RLK* lysin motif receptor kinase, *LHK1* Lothushistidine kinase (Adapted from Oldroyd and Downie 2008)

access to cortical cells. The relation with Nod components is considered important in some species, which endure disruption entry, but NC-independent of disruption invasion also exists and may be associated with rhizobial modifications of cytokinins (Fig. 6.1) (Oldroyd and Downie 2008; Oldroyd et al. 2011).

After the nodule formation occurs, the bacteria enlarge and differentiate into nitrogen-fixing forms, and are denominated as bacteroids. These bacteroids are surrounded by plant membrane, known as a symbiosome that is a kind of organelle, which has the function of reducing nitrogen (Oldroyd and Downie 2004).

The enzyme responsible for nitrogen fixation in diazotrophs is generally molybdenum nitrogenase. As an alternative to molybdenum nitrogenase, some diazotrophic microorganims have vanadium and/or iron nitrogenases. The structure of the molibidenum nitrogenase enzyme is composed of *nif*DK and *nif*H genes (Rubbio and Ludden 2008).

6.3.2 Optimization of Elements of Biological Nitrogen-Fixing System

One of the main topics discussed in the Edinburg Declaration on Reactive Nitrogen (2011) was related to improving nitrogen availability to plants (Gutierrez 2012). Here we can consider biological nitrogen fixation (BNF) (Galloway et al. 2004).

Due to the importance of BNF in decreasing the necessity for chemical nitrogen fertilizers, it will be interesting to understand how to improve the biological process. For this, we can verify what happens during the optimal use of a nitrogen-fixing system, search for new plant-microorganism fixing associations, and transfer the ability of BNF to non-fixing microbes (Olivares et al. 2013).

There are many ways to improve the BNF through legumes, like legume adoption by farmers during combined cultivation and rotations between crops (Sessitsch et al. 2002). With the purpose of maximizing BNF by legumes, nodulation with the appropriate rhizobia should be evaluated. However, there are soils with low numbers of compatible rhizobia, thus it is necessary to carry out inoculation and to select an inoculant strain. For the selection of this strain it is necessary to verify the bacterial compatibility and nitrogen-fixing efficiency with the plant. Moreover, environmental conditions need to be analyzed because they may limit BNF activity and periodical inoculation must also be adopted (Hungria et al. 2005).

The election of inoculants is usually based on existing microbe diversity. Therefore, there are some genetic modifications that can be done with the purpose of enhancing the BNF of a given strain, by reiteration or overexpression of genes related to nitrogenase enzyme activity. Peralta et al. (2004) investigated improving the symbiotic efficiency in *Rhizobium etli – Phaseolus vulagris*. With the purpose of improving nitrogenase production, these authors built a chimeric nifHDK operon regulated by a strong nifHc promoter and verifed it in symbiosis with *P. vulgaris*. Bacterial strains with overexpression of nitrogenase had increased nitrogenase activity, plant weight (improved around 32%), concentration of nitrogen in plants and seed, and seed yield. Moreover, the overexpression of the chimeric *nif*HDK operon contributed to increased symbiosis. In another study, Wang et al. (2013) recognized a cluster consisting of nine *nif* genes in the genome of *Paenibacillus* sp. WLY78. With the purpose of analyzing the genetic requirements for fixing nitrogen, they inserted a *Paenibacillus nif* gene cluster in *Escherichia coli.* A minimum *nif* gene cluster allows the production of active nitrogenase in *E. coli*. However, on deletion analysis it was verified that in addition to the core *nif* genes, *hes*A (one of the genes of *nif* clusters) participates in an important aspect on nitrogen fixation and is sensitive to molybdenum. Wang et al. (2013) wanted to demonstrate the possibility of transferring the BNF activity with a short set of *nif* gene cluster. Breeding for enhanced nitrogen fixation is not an easy task, but you can analyze characteristics such as plant and seed yield, and others, to quantify the efficiency of BNF (Olivares et al. 2013).

6.4 Phosphorus in Soil

Phosphorus and nitrogen are important to keeping a healthy nutritional life for plants, but, unlike nitrogen, phosphorus is not present in large amounts that can become available to plants.

Many peculiarities are associated with phosphorus nutrition and this element shows an important role in metabolic processes (Khan et al. 2010). Microbial associations are associated with P-fixation as with N-fixation in legumes. Phenological development, crop yield, and resistance to plant diseases are also described in relation to P nutrition (Hao et al. 2002).

A considerable amount of inorganic P is quickly retained in insoluble mineral complexes after frequent application of phosphate fertilizers (Rodriguez and Fraga 1999; Igual et al. 2001). The retention of P is around 75% of the total amount applied. Organic forms (20–80% of P in soils) are the other important storage of immobilized P (Richardson et al. 1994). Phosphorus is present in soil on average at around 0.05% (w/w); however, only 0.1% of it is usable by plants (Zhu et al. 2011). Physicochemical (adsorption – desorption) and biological (immobilization-mineralization) means are responsible for soil P dynamics and consequently for P fixation. Most P that is administrated as fertilizer becomes static through a condensation reaction with Al^{3+}, Fe^{+3}, and Ca^{+2} (Hao et al. 2002).

The application of chemical fertilizers represents an extra cost to agricultural production and moreover causes negative impacts on different environments (Tilman et al. 2001). Reduction of fertility by lost microbial diversity, which consequently reduces the crop yields, is also observed (Gyaneshwar et al. 2002). Beyond that, P has limited sources in rock phosphate and there are estimations that the reserve of P will be exhausted in the current century (Cordell et al. 2009). The high cost of P chemical fertilizers and problems with P availability make it necessary to search for an environmental alternative and a production strategy for improving crop yields without environmental problems. One option available to achieve this purpose is the use of microbe inoculated fertilizer with P-solubilizing properties in agriculture (Sharma et al. 2013).

6.4.1 Phosphorus-Solubilizing Microorganisms

In 1903 the natural occurrence of phosphorus-solubilizing microbes (PSMs) was observed. These have the ability to use diverse means to solubilize and mineralize P, with the purpose of converting inorganic and organic soil P into available forms for plants (Khan et al. 2007).

Bacteria represent the majority of the microbial community with 1–50%, with fungi constituting only 0.1–0.5% of the total respective population. Several species of bacteria and fungi have been tested in relation to their potential as a PSM. Generally, these microbes are found in the phytomicrobiome and soil areas with P deposits. The isolation can be done by serial dilution methods or through enhancement culture techniques (Zaidi et al. 2009). Another area that a PSM could be isolated in is in stressful environments, such as the moderately halophic bacterium, *Kushneria* sp., found by Zhu et al. (2011), in the solid residue of Daqiao saltern on the east coast of China, YCWA18.

Many factors could affect the potential of these PSMs, such as the amount of iron ore, temperature, carbon, and nitrogen origin. One of the biggest problems that generates controversy is the source of insoluble P used to isolate PSMs: tricalcium phosphate (TCP). The TCP is able to select a huge number of isolates. However, when these isolates are evaluated in relation to the available P provided to plants, many of them fail as PSMs because there are often other sources of P that are less soluble than TCP such as iron/aluminum/calcium phosphate (Bashan et al. 2013).

Soil is complex and shows a lot variation depending on pH and chemical properties; in this way no P source can be used as a universal selection factor. Thus, the best way to assess how these microbes are evaluated is in which type of soil (alkaline, acid, or rich in organic matter) the PSMs will be applied. Bashan et al. (2013) suggested the use of specific compounds such as calcium phosphate (alkaline soils), iron/aluminum phosphate (acidic soils), and phytases (organic soils). The PSMs that show greater solubilization *in vitro* are selected for field trials before the production of biofertilizer (Sharma et al. 2013).

Excessive numbers of microorganisms show PSM capacity, and include bacteria, fungi, and actinomycetes. Among the bacterial communities, *Pseudomonas* and *Bacillus* have been described as PSMs. Some of the bacteria with PSM skills are *Rhodococus*, *Arthrobacter*, *Serratia*, *Chryseobacterium*, *Gordonia*, *Phyllobacterium*, *Delftia* sp. (Wani et al. 2005), *Enterobacter* sp., *Pantoea*, *Klebisiella* (Chung et al. 2005), *Vibrio proteolyticus*, *Xanthobacter agilis* (Vazquez et al. 2000) (Sharma et al. 2013), and *Burkholderia* (Inui-Kishi et al. 2012).

There are some nitrogen-fixing microorganisms such as rhizobia that have also shown PSM activity (Ahemad and Kibret 2014; Zaidi et al. 2009). Among fungi *Penicillium* and *Aspergillus* represent the most effective solubilizers (Reyes et al. 2002).

6.4.2 Mechanisms of P-Solubilization by Soil Microbial Communities

Microorganisms are able to increase the ability of plants to obtain P from soil through different mechanisms, such as increased root extension by mycorrhizal associations, or by hormonal stimulation of root hair development and root growth (Hayat et al. 2010; Richardson and Simpson 2011). Another method is through metabolic mechanisms that are efficient in providing unavailable P present in soil as organic and inorganic forms (Fig. 6.2). Moreover, the existence of labile C in these microorganisms appears as a reservoir of P, through immobilization. Thus, the

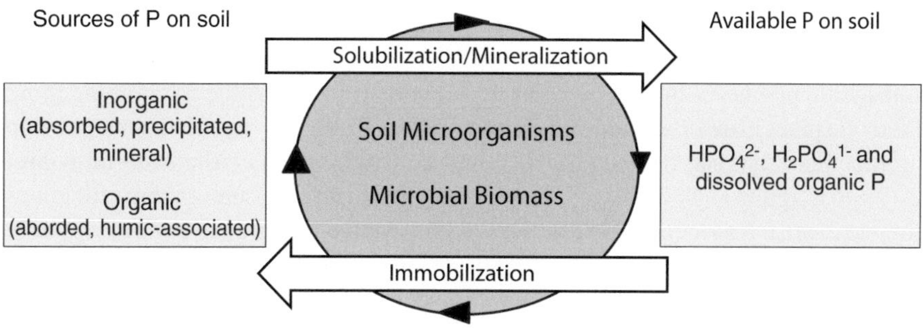

Fig. 6.2 Diagram of phosphorus cycle (Adapted from Richardson and Simpson 2011)

dispensation of P retained by microbes occurs when the cell dies (Richardson and Simpson 2011).

6.4.3 Inorganic P-Solubilization

The majority of available inorganic phosphate is $H_2PO_4^-$, which normally occurs at a lower pH. Nevertheless, with the increase in pH, other phosphate forms such as HPO_4^{2-} and HPO_4^{3-} become more predominant.

The availability of inorganic P through PSMs occurs mainly by organic acid production (Zaidi et al. 2009). Other ways include: lowering pH by H^+ extrusion (Parks et al. 1990), improved capitation of cations ligated with P, or establishment of a soluble compound with metal ions (calcium, aluminum, and iron) that releases P (Maliha and Samina 2004) and the release of protons accompanying respiration or NH4+ assimilation when the solublization occurs without acid production (Illmer and Schimer 1995). The effect on pH is due the liberation of organic acid, originated from microbial metabolism, mostly by oxidative respiration or fermentation processes (Trolove et al. 2003). The predominant acids that are related to PSM activity are gluconic, oxalic, citric, succinic (Khan et al. 2007), lactic, tartaric (Venkateswarlu et al. 1984; Khan et al. 2007), and aspartic (Venkateswarlu et al. 1984) acids. There are also inorganic acids (sulphur and nitric acid) (Siqueira and Franco 1988). One example of higher potential in solubilizing P is caused by gluconic and 2-keto gluconic acids synthetized by direct oxidation in *Erwinia herbicola* and *Pseudonomas cepacea* (Goldstein et al. 1993; Goldstein 1994; Goldstein 1995). The capacity of PSMs to utilize inorganic phosphate is regulated by specific genes. Understanding these genes is very important because it is possible to use them in biotechnology applications. Genes correlated with PSMs are glucose dehydrogenase (GDH), gluconate dehydrogenase (GADH), and pirroloquinoline–quinone (PQQ), which are already identified and cloned in different bacteria (Sashidhar and Podile 2010). However, the recombinant bacteria that express these genes need to be studied in more depth and approved by regulatory laws, because the use of these PSMs can enhance the eutrophication of rivers (Siqueira et al. 2004). Fraga et al. (2001) observed accumulation of extracellular phosphatase when the napA phosphatase gene from *Morganella morganii* was cloned in *Burkholderia cepacia* IS-16 (strain used for inoculant). There may be more advantages in developing genetically modified microorganisms (PSMs) over transgenic plants for improved plant development. It is possible to combine more than one plant growth–promoting characteristic in a unique organism, and develop an inoculant that can be used for different cultivated plants (Ahemad and Kibret 2014; Rodriguez et al. 2006). Molecular genetics yield information that elucidates the mechanisms associated with PSMs. Comparative genomic and transcriptomic sequencing of the microbiome, and differential gene expression analyses have found potential targets such as enzymes, metabolites, and transport proteins that are related to the PSM process that leads to the enhancement of P availability and use by plants (Krishnaraj and Dahale 2014).

6.4.4 Organic P-Solubilization

While inorganic P becomes unavailable by precipitation and chemical adsorption, the organic P is retained in the organic matter of soil (Sharma et al. 2013).

Organic P represents 30–80% of the total soil (Menezes-Blackburn et al. 2013). It is found as inositol phosphate (30–50% of total amount of organic P), nucleic acids, and nucleotides (3–5%), phospholipids, and others (low quantities) (Siqueira et al. 2004). The use of organic P by plants and microbes requires hydrolysis of organic P from soil (Richardson and Simpson 2011; Sharma et al. 2013). Hydrolysis is known as the mineralization of organic P in soil and occurs mainly by the action of phosphatase enzymes (Fig. 6.3).

In general, fungi have more hydrolytic activity on phytate than bacteria; however, bacteria and plants can produce phytate. The nucleases are mainly produced by rhizospheric microorganisms and phospholipases by actinomycetes (Siqueira et al. 2004).

Phytate is the main stock form (60–80%) of organic P in several soils and plant tissues (1–5% of weight) (Singh and Satyanarayana 2011). In plants, phytases cause the liberation of P from phytate degradation. However, the capacity of plants to obtain P immediately from phytate is very limited. *Arabidopisis* plants supplemented with phytate were considerably benefited with P when they were engineered with the phytase gene (phyA) originated from *Aspergillus niger* (Richardson et al. 2005).

Phosphatases are a class of abundant enzymes that are used by PSMs and are widely present in studies. These enzymes can be set into acid and alkaline phosphatase, according to soil characteristic (acid or alkaline) (Jorquera et al. 2008). In plants the production of acid enzymes is more predominant than alkaline phosphatase, suggesting that this is a specific characteristic of PSMs (Criquet et al. 2004).

Microorganisms are the main source of phosphorus mineralization enzymes (Richardson 1994). The activity and synthesis process of these enzymes depends on environmental conditions that are suppressed with high contents of phosphorus or stimulated in limited conditions. In conditions with low availability of P, bacteria have the ability to acquire P in their biomass, which at the final cycle will be mineralized and become available to plants and other organisms (Gyaneshwar et al. 2002).

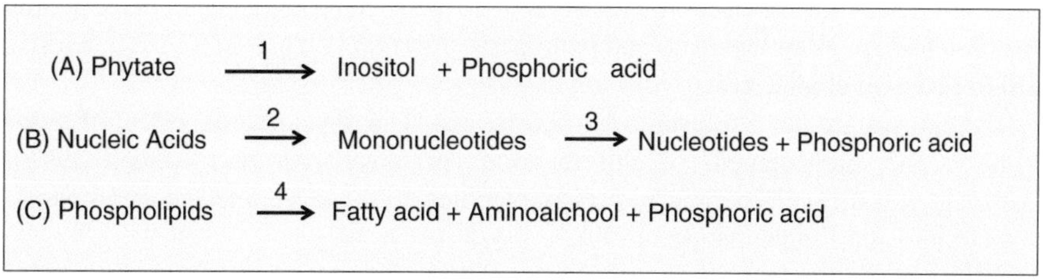

Fig. 6.3 Mineralization reaction of organic phosphorus by (1) phytases, (2) nucleases, (3) nucleotidases, (4) phospholipases (Adapted from Siqueira et al. 2004)

6.5 Phytohormone Production

A phytohormone is defined as an organic signal molecule synthesized in plant organs or tissues that can be translocated to other regions and presents a specific response. However, phytohormones can also be active in tissues where they are created (Baca and Elmerich 2007). Their production has been studied as one of the main instruments by which PGPMs may increase plant growth (Iqbal and Hasnain 2013).

Auxins, cytokinins, gibberellins, ethylene, and abscisic acid are classes of widely recognized phytohormones (Zahir et al. 2004).

In nature, two modes of phytohormones are accessible to plants. They are endogenous production by the plant tissues, and phytohormones that are made available by associated microbes (Patten and Glick 1996).

Fungi inoculants are advantageous compared with bacterial ones because of their efficiency at spreading over the rhizosphere. *Trichoderma* species represents a class of fungi found in the rhizosphere. *Trichoderma* strains can colonize plant roots and improve plant progress in growth and development. These effects are influenced by microbial production of indole-3-acetic acid (IAA) and indolic compounds (Ortíz-Castro et al. 2009).

During plant growth and development, signaling molecules intermediate the contact with microorganisms, performing a relevant communication. Microbes have the means to recognize a plant host and start colonizing the rhizosphere by production of plant growth–regulating substances like phytohormones. Furthermore, these microbial-produced compounds are perceived by plants, which respond and further influence the type of microorganisms found. This represents a molecular conversation that determines the relationship between plants and microbes from pathogenesis to symbiosis (Bais et al. 2004).

Many bacteria and fungi can produce auxins using different pathways, which increases the potential to form associations with plants. Moreover, epiphytic and rhizospheric microfloras of plants are of utmost relevance in the conversion of tryptophan (which is present in plant exudates) into IAA (Tsavkelova et al. 2006).

The main known phytohormone is IAA. At least 80% of the rhizospherical bacteria synthetize IAA (Patten and Glick 1996). In addition to IAA, other indolic compounds that are physiologically active for plants are also produced by rhizospheric microbes (Lebuhn et al. 1997; El-Khawas and Adachi 1999).

On the other hand, IAA and cytokinins derived from bacteria are also related to the virulence of several interactions between microorganisms like genus *Agrobacterium*, *Pseudomonas*, and pathogenic *Erwinia* (Lichter et al. 1995; Morris 1986; Spaepen et al. 2007).

Cytokinins are other relevant phytohormones. The physiological effect of cytokinin is the enhancement of cell division (Frankenberger and Arshad 1995). Although it is difficult to identify these molecules, they can be detected using bioassays (Nieto and Frankenberger 1990).

Microorganisms have the ability to synthetize kinetin, zeatin, isopentenyladenine, and other cytokinin derivatives. Rhizobacteria of the genera *Azotobacter*, *Azospirillum*, *Arthrobacter*, *Bacillus*, *Rhizobium*, *Pseudomonas*, and also streptomycetes are capable of synthetizing cytokinins (Tsavkelova et al. 2006).

PGPMs also produce gibberellins (GAs). There are over 89 known GAs (Dobbelaere et al. 2003), which are numbered GA1 through GA89 in approximate order of their discovery (Frankenberger and Arshad 1995; Arshad and Frankenberger 1998). The most accepted gibberellin is GA3, while the most active is GA1. The main physiological role of gibberellins is stem elongation and increased internode length (Davies 1995).

The fourth phytohormone to be discovered is abscisic acid (ABA). This was detected by radioimmunoassay or thin-layer chromatography in supernatant cultures of plant-associative bacteria *Azospirillium* and *Rhizobium* sp. (Dangar and Basu 1987; Dobbelaere et al. 2003). ABA responses are related to stomatal closure and root proliferation. Therefore, its presence in the rhizosphere is of paramount relevance for plant growth in water-deficient environments (Frankenberger and Arshad 1995).

Ethylene is another important phytohormone. Its role is related to root elongation inhibition, auxin transport, senescence, and abscission promotion of various organs, and fruit ripening (Bleecker and Kende 2000; Glick et al. 2007). Reducing levels of ethylene in plants may be one of the growth-promoting activities of PGPMs. These reductions are due to the activity of enzyme 1-aminocyclopropane-1-carboxylate (ACC) deaminase, which reduces ACC, the precursor of ethylene (Yang and Hoffman 1984).

Cadaverine, a poliamine that enhances root growth and helps to reduce osmotic stress, was reported in rice inoculated with *Azospirillim brasilense* Az39 (Cassán et al. 2009).

6.5.1 Applications

Associative rhizobacteria may provide favorable progress in plant development due to production of phytohormones, nitrogen fixation, biosynthesis of antimicrobial substances, and enhanced water or mineral nutrition uptake (Saharan and Nehra 2011).

There are many reports presenting the beneficial impact of bacteria inoculation on orchids. An early paper was published by Knudson (1922), who inoculated the seeds of *Epidendrum* and *Laeliocattleya* with a diazotrophic strain, *Rhizobium leguminosarum*, to stimulate germination. Wilkinson et al. (1989, 1994) described the propagation of *Pterostylis vittata* seeds co-inoculated with a mycorrhizal fungus and strains of orchid-associated bacteria belonging to *Pseudomonas putida*, *Bacillus sphaericus*, *B. cereus*, and *Arthrobacter* sp.

The *in vitro* germination of mature seeds of the species *Dendrobium moschatum* was enhanced by inoculation of bacterial cultures (Tsavkelova et al. 2007). Seed germination was enhanced by *Mycobacterium* sp. and *Sphingomonas* sp. in *in vitro* inoculations.

Strains of *Bacillus* sp. and *Enterobacter* sp. isolated as endophytic bacteria were used to stimulate *Cattleya walkeriana* seedlings. This was done by increasing such characteristics as fresh weight, dry weight, and plant survival during *ex vitro* acclimatization, which is considered the bottleneck stage for orchid seedling propagation (Galdiano Júnior et al. 2011). One of these isolates, *Enterobacter asburiae*, produces an acid ectophosphatase, which can be a mechanism for the solubilization of mineral phosphates (Sato et al. 2016), configuring a PGPM with two growth-promoting activities (IAA production and solubilization of mineral phosphate).

Growth promotion was also observed when seedlings of *Cattleya loddigesii* generated *in vitro* were inoculated with a bacterial suspension of *Paenibacillus lentimorbus* and *P. macerans* strains (Faria et al. 2013). *Sphingomonas paucimobilis* ZJSH1 significantly promoted the growth of *D. officinale* seedlings, enhancing fresh weight and stem length (Yang et al. 2014).

6.6 The Importance of Siderophores

The siderophore is a biological molecule produced by various bacteria and has wide applications in various fields, such as improving soil fertility in agriculture. Bacterial strains that do not use any other means of biocontrol can act as biocontrol agents by using the siderophores that they produce. Therefore, siderophores from PGPMs prevent phytopathogens from acquiring iron and can be a limiting factor for their proliferation (Kloepper et al. 1980).

Available results indicate PGPM siderophores have a much better affinity for iron than to pathogens (Schippers et al. 1987). As a result, lack of iron in the rhizosphere incapacitates proliferation of fungal pathogens. By reason of biocontrol, PGPMs out-compete fungal pathogens for accessible iron.

However, plant growth is not altered by the iron reduction in the rhizosphere caused by the siderophores, because most plants use less iron than microorganisms (O'Sullivan and O'Gara 1992). Also, some plants can bind and consequently make up the biocontrol PGPM iron-siderophore complex (Bar-Ness et al. 1991; Wang et al. 1993).

6.7 Mycorrhizae

Mycorrhizal symbioses are widespread and common in terrestrial ecosystems around the globe, occurring in nearly all soils (Smith and Read 2008). The symbiotic fungi are often crucial for absorbing water and nutrients from the soil and transferring them to plants (Orwin et al. 2011).

Mycorrhizae are grouped into two types: ectomycorrhizae and endomycorrhizae. Endomycorrhizae are defined by dense mycelial sheaths near the roots with intercellular hyphal invasions of the root cortex. They are limited to forest trees in temperate regions. All other plants represent endomycorrhizae, characterized by fungi forming external hyphal networks in the soil and penetrating the cortical cells of roots (Bolan 1991).

Mycorrhizal fungi are known to secrete phytohormones such as GA3, IAA, ABA, zeatin, and zeatin riboside (Wu et al. 2002; Liu et al. 2010). The mycorrhizal fungus *Trichoderma* sp., isolated from *Pleione bulbocodioides*, increased seed germination up to 84.6%, while the control presented lower germination (77.6%) on OMA medium (Yang et al. 2008).

Pathogenic and growth-promoting fungal species are IAA yielders (Tsavkelova et al. 2006). Four different *Fusarium* species isolated as endophytic fungi from *Dendrobium moschatum* also produced IAA (Tsavkelova et al. 2003, 2012), while mycorrhizal fungus in association with *Dendrobium densiflorum* produced vitamins and GA (Wu et al. 2002).

6.8 Alternative Use of Plant Growth–Promoting Microorganisms

Several bacterial genera are necessary elements in maintaining the balance of soils. They are implicated in distinct biotic activities of soil ecosystems, making them a functional for nutrient turnover and enhancing maintenance for agricultural production (Ahemad et al. 2009; Chandler et al. 2008). The association of plants with bacteria can be considered good, pernicious, or neutral depending on the motif of their action on plant growth (Dobbelaere et al. 2003).

The use of PGPMs has become a habitual practice in various regions of the world. PGPMs are native to the soil and plant rhizosphere, and play an important function in biological control of plant pathogens, suppressing a broad spectrum of bacteria, fungi, and nematodes. PGPMs can also give protection against viral diseases (Sivasakthivelan et al. 2013).

Bacteria in the genera *Agrobacterium*, *Bacillus*, *Burkholderia*, *Pseudomonas*, and *Streptomyces* are frequently studied biocontrol agents. They eliminate plant disease through at least one mechanism: induction of systemic resistance and production of siderophores or antibiotics (Tenuta 2003). Among others, actinobacteria commonly inhabit the rhizosphere making it possible to characterize them as PGPMs (Franco-Correa and Chavarro-Anzola 2016).

These microorganisms are found very close to the roots epidermis, which secretes signal molecules for defense versus invasion of distinct microorganisms in the root area. At this stage, the distinction takes place between symbiotic, associative, pathogenic, or neutralistic association of the microorganisms with the plant (Hayat et al. 2010).

PGPM strains occur in many taxonomic groups and are present in the rhizosphere of plants in any given soil (Kyselková et al. 2009; Almario et al. 2013). This suggests that PGPMs colonize and coexist in the same rhizosphere soil as non-PGPM groups of the bacterial community. Studies have shown the existence of a particular gene (Table 6.1) and of relevant properties of PGPMs, which provide positive effects on plant growth, health, and their ability to inhibit phytopathogens (Bertrand et al. 2001; Barriuso et al. 2005; Upadhyay et al. 2009).

The biological control lineage *Pseudomonas fluorescens* Psd was analyzed for IAA organic synthesis and the logical preparation on it as a PGPM. While the indole pyruvic acid (IPyA) route usually connected with PGPMs was absent, the indole

Table 6.1 Gene functions of plant growth–promoting microorganisms studied in plants

Function	Gene	Phylum	References
Phosphate solubilization	*pqq*B *pqq*C *pqq*D *pqq*E pqqF *pqq*G	Proteobacteria	Bruto et al. (2014)
2,4-*Diacetylphloroglucinol synthesis*	*phl*A *phl*B *phl*C *phl*D		
Hydrogen cyanide synthesis	*hcn*A *hcn*B *hcn*C		
Acetoine/2,3-butanediol synthesis	*budA* bud*B* *bud*C		
Nitric oxide synthesis	*nir*K		
Auxin synthesis	*ipd*C *ppd*C		
ACC deamination	*acd*S		
Nitrogen fixation	*nif*D *nif*H *nif*K		
Biosynthesis of tryptophan	*Trp*		Zahir et al. (2010)
Nitrogenase	*nif*H	Actinobacteria	Valdés et al. (2005) Gauthier et al. (1981)
Nitrogenase	nifDK		Fedorov et al. (2008)

acetamide (IAM) pathway is commonly noticed in plant pathogens and was expressed in the Psd strain. Overexpression of IAM pathway genes iaaM-iaaH, from *P. syringae* subsp. *savastanoi* radically increased IAA levels and demonstrated a prejudicial effect on sorghum root development (Saranraj et al. 2013; Sivasakthivelan and Saranraj 2013).

PGPMs are efficient in secreting molecules as antibiotics into the rhizosphere to control pathogenic microbes, producing iron-chelating molecules (Raaijmakers et al. 2002). They also induce phytoalexin production in association with plants, and have broad acceptance as providing an agricultural advantage (Lifshitz et al. 1986; Halverson and Handelsman 1991). Phytoalexins are antimicrobial compounds with low molecular weights that are both synthesized by and accumulated in plants after exposure to microorganisms (Dakora 1985; Dakora et al. 1993; Van Peer et al. 1990, 1991).

In the rhizosphere of various leguminous and non-leguminous crops, species of *Pseudomonas* and *Bacillus* have been identified (Table 6.2) that help in plant colonization and suppression of phytopathogens (Parmar and Dufresne 2011).

Table 6.2 Plant growth–promoting microorganisms (PGPMs)

PGPM	PGPM and agricultural crop	Plant growth promoting traits	References
Bacillus cereus UW 85	Grain legumes	Lowers the toxicity of chromium to seedlings by reducing Cr (VI) to Cr (III)	Vessey and Buss (2002)
P. fluorescens CHA0	*Arabidopsis* sp.	Increased plant growth	Iavicoli et al. (2003)
Bacillus spp.	Banana	Promoted significantly the root and shoot growth	Jaizme-Vega et al. (2004)
P. putida KD	Tomato and cucumber	Promoted the plant growth, reduced Pb and Cd uptake	Rezzonoco et al. (2005)
P. fluorescens PCL1606	Avocado	Increased plant growth	Cazorla et al. (2006)
Bacillus	Raspberry	Promoted significantly the root and shoot growth	Orhan et al. (2006)
B. pumilus	Wheat variety Orkhon	Increased plant growth	Hafeez et al. (2006)
B. mucilaginosus	Cucumber	Increased plant growth	Han et al. (2006)
B. mucilaginosus	Pepper	Increased plant growth	Supanjani et al. (2006)
P. Marginali	Indian mustard and rape	Increased plant growth	Belimov et al. (2007)
P. oryzihabitans			
P. putida			
Alcaligenes xylosoxidans			
P. brassicacaerum			
Agrobacterium amazonense	Rice	Nitrogen accumulation	Rodrigues et al. (2008)
Pseudomonas BA-8	Strawberry	Increased plant growth	Pirlak and Kose (2009)
Bacillus OSU-142			
Bacillus M-3			
Comamonas acidovorans	Kiwi	Increased plant growth	Erturk et al. (2010)

PGPMs produce beneficial effects on plant health by accelerating nutrient availability, assimilation, and growth by suppressing diseases caused by phytopathogens (Franco-Correa and Chavarro-Anzola 2016). In the quest to improve soil fertility and crop yield while reducing the negative impact of chemical fertilizers on the environment, there is a need to exploit PGPMs for beneficial agriculture. Moreover, PGPMs can also prevent the deleterious effects of stresses from the environment (Paul and Nair 2008). They are also able to increase the capacity of plants to sequester heavy metals and can help plants withstand abiotic stresses (Jing et al. 2007; Saharan and Nehra 2011; Tak et al. 2013).

Bio-inoculants with diverse symbiotic (*Bradyrhizobium*, *Mesorhizobium*, *Rhizobium*) and non-symbiotic (*Azomonas*, *Azospirillum*, *Azotobacter*, *Bacillus*, *Klebsiella*, *Pseudomonas*) microbes are used to promote plant growth and development under various stresses like heavy metals (Ma et al. 2009a, b; Wani and Khan

2010), herbicides (Ahemad and Khan 2010), insecticides (Ahemad and Khan 2009), fungicides (Ahemad and Khan 2012a), and salinization (Mayak et al. 2004). Diverse varieties of rhizobacteria help to improve plant nutrition thus increasing plant health or stress tolerance (Vacheron et al. 2013).

Ferreira et al. (2010) showed similarities between cepacian exopolysaccharide (EPS) produced by members of the *Burkholderia cepacia* complex (BCC complex) and the EPS synthesized by *B. graminis*, *B. phytofirmans*, *B. phymatum*, and *B. xenovorans* (Estrada-De Los Santos et al. 2001). According to Ferreira et al. (2010), EPS may have a role in the tolerance of *Burkholderia* species to ion stress and desiccation.

Bloemberg and Lugtenberg (2001) showed the expression of some genes involved in the defensive response as well as genes expressed under conditions of drought, salt, and stress. The experiments with isolates of *B. graminis*, a species that has been isolated from the rhizosphere of pasture, corn, and wheat, resulted in an improvement in shoot height and neck diameter, as well as inducing a protective response to salt and drought stress in tomato plants (Barriuso et al. 2005, 2008).

PGPMs are able to produce gibberellic acid or ABA, or to control the level of these hormones in plants (Richardson et al. 2009; Dodd et al. 2010). The ABA is well known for its involvement in drought stress. Bauer et al. (2013) showed that during water stress, increases in ABA levels cause closing of stomata, thereby limiting water loss.

Habib et al. (2016), studying the ACC deaminase-containing PGPM isolate *Enterobacter* sp. UPMR18, concluded that it could be an effective bio-resource for enhancing salt tolerance and growth of okra plants under salinity stress. Microorganisms synthesizing the ACC deaminase enzyme can cleave ACC to α-ketobutyrate and ammonia, thus decreasing ethylene stress in plants (Rashid et al. 2012).

Ethylene is a simple gaseous hormone critical for many plant developmental stages (Abeles et al. 1992). The ethylene-mediated stress response can be activated by many environmental factors such as heavy metal contamination, high salinity, flooding, drought, and phytopathogens. Ethylene can also inhibit stimulation of cell proliferation and elongation by repressing auxin response factor synthesis (Dugardeyn and Van Der Straeten 2008). PGPMs, besides promoting plant growth by employing certain mechanisms and protecting the plant from salinity, can also increase plant tolerance against stress conditions (Shrivastava and Kumar 2015).

Conclusion

Understanding the mechanisms used by PGPMs enables the elucidation of their impact on nutrient cycling and on the protection of crops against disease. Moreover, functioning analysis, diversity, and gene expression patterns of PGPM populations in soil will be a precondition to developing a management action plan for sustainable agriculture. Future research and understanding of the mechanisms of PGPMs will pave the way to finding more competent rhizobacterial strains. These yet-to-be-found strains may work under diverse agro-ecological conditions to protect the environment, and produce enough food for an increasing world population.

References

Abeles FB, Morgan PW, Salveit MEJ (1992) Ethylene in plant biology. Academic Press, New York

Ahemad M, Khan MS (2009) Effect of insecticide-tolerant and plant growth promoting *Mesorhizobium* on the performance of chickpea grown in insecticide stressed alluvial soils. J Crop Sci Biotechnol 12:213–222

Ahemad M, Khan MS (2010) Growth promotion and protection of lentil (*Lens esculenta*) against herbicide stress by *Rhizobium* species. Ann Microbiol 60:735–745

Ahemad M, Khan MS (2012a) Alleviation of fungicide-induced phytotoxicity in greengram [*Vigna radiata* (L.) Wilczek] using fungicide-tolerant and plant growth promoting *Pseudomonas* strain. Saudi J Biol Sci 19:451–459

Ahemad M, Khan MS (2012b) Effects of pesticides on plant growth promoting traits of *Mesorhizobium* strain MRC4. J Saudi Soc Agric Sci 11:63–71

Ahemad M, Kibret M (2014) Mechanisms and applications of plant growth promoting rhizobacteria: current perspective. J King Saud Univ Sci 26:1–20

Ahemad M, Khan MS, Zaidi A, Wani PA (2009) Remediation of herbicides contaminated soil using microbes. In: Khan MS, Zaidi A, Musarrat J (eds) Microbes in sustainable agriculture. Nova Science Publishers, New York, pp 261–284

Almario J, Kyselková M, Kopecký J, Ságová-Marecková M, Muller D, Grundmann GL et al (2013) Assessment of the relationship between geologic origin of soil, rhizobacterial communityco mposition and soil receptivity to tobacco black root rot in Savoie region (France). Plant Soil. doi:10.1007/s11104-013-1677-1

Antoun H, Prévost D (2005) Ecology of plant growth promoting rhizobacteria. In: Siddiqui ZA (ed) PGPR: biocontrol and biofertilization. Springer, Dordrecht, pp 1–38

Arjun JK, Harikrishnan K (2011) Metagenomic analysis of bacterial diversity in the rice rhizosphere soil microbiome. Biotechnol Bioinf Bioeng 1(3):361–367

Arshad M, Frankenberger WT Jr (1998) Plant growth regulating substances in the rhizosphere. Microbial production and function. Adv Agron 62:46–51

Baca BE, Elmerich C (2007) Microbial production of plant hormones. In: Elmerich C, Newton WE (eds) Associative and endophytic nitrogen-fixing bacteria and cyanobacterial associations. Springer, Dordrecht, pp 113–143

Bais HP, Park SW, Weir TL, Callaway RM, Vivanco JM (2004) How plants communicate using the underground information superhighway. Trends Plant Sci 9:26–32

Bar-Ness E, Chen Y, Hadar Y et al (1991) Siderophores of *Pseudomonas putida* as an iron source for dicot and monocot plants. In: Chen Y, Hadar Y (eds) Iron nutrition and interactions in plants. Kluwer Academic, Dordrecht, pp 271–281

Barriuso J, Pereyra MT, Lucas Garcia JA, Megias M, Gutierrez Manero FJ, Ramos B (2005) Screening for putative PGPR to improve establishment of the symbiosis *Lactarius deliciosus-Pinus* sp. Microb Ecol 50:82–89

Barriuso J, Ramos Solano B, Fray RG, Camara M, Hartmann A, Gutierrez Manero FJ (2008) Transgenic tomato plants alter quorum sensing in plant growth-promoting rhizobacteria. Plant Biotechnol J 6:442–452

Bashan Y, Kammev AA, de Bashan LE (2013) Tricalcium phosphate is innappopriate as a universal selection factor for isolating and testing phosphate-solubilizing bacteria that enhance plant growth: a proposal for an alternative procedure. Biol Fertil Soils 49:465–479

Bauer H, Ache P, Lautner S, Fromm J, Hartung W, Al-Rasheid KA, Sonnewald S, Sonnewald U, Kneitz S, Lachmann N, Mendel RR, Bittner F, Hetherington AM, Hedrich R (2013) The stomatal response to reduced relative humidity requires guard cell-autonomous ABA synthesis. Curr Biol 1:53–57

Belimov AA, Dodd IC, Safronova VI, Hontzeas N, Davies WJ (2007) *Pseudomonas brassicacearum* strain Am3 containing 1-aminocyclopropane-1-carboxylate deaminase can show both pathogenic and growth-promoting properties in its interaction with tomato. J Exp Bot 58:1485–1495

Berg G, Grube M, Schloter M, Smalla K (2014) Unravelling the plant microbe: looking back and future perspectives. Front Microbiol 5:148
Bertrand H, Nalin R, Bally R, Cleyet-Marel JC (2001) Isolation and identification of the most efficient plant growth-promoting bacteria associated with canola (*Brassica napus*). Biol Fertil Soils 33:152–156
Bhattacharyya PN, Jha DK (2012) Plant growth-promotig rhixobacteria (PGPR): emergence in agriculture. World J Microbiol Biotechnol 28:1327–1250
Bleecker AB, Kende H (2000) Ethylene: a gaseous signal molecule in plants. Annu Rev Cell Dev Biol 16:1–18
Bloemberg GV, Lugtenberg BJ (2001) Molecular basis of plant growth promotion and biocontrol by rhizobacteria. Curr Opin Plant Biol 4:343–350
Bolan NS (1991) A critical review on the role of mycorrhizal fungi in the uptake of phosphorus by plants. Plant Soil 134:189–207
Bruto M, Prigent-Combaret C, Muller D, Moënne-Loccoz Y (2014) Analysis of genes contributing to plant-beneficial functions in plant growth-promoting rhizobacteria and related Proteobacteria. Sci Rep 4:6261
Cassán F, Maiale S, Masciarellia O, Vidal A, Luna V, Ruiz O (2009) Cadaverine production by Azospirillum brasilense and its possible role in plant growth promotion and osmotic stress mitigation. Eur J Soil Biol 45:12–19
Cazorla FM, Duckett SB, Bergstrom FT, Noreen S, Odik R et al (2006) Biocontrol of avocado Dematophora root rot by the antagonistic *Pseudomonas fluorescens* PCL 1606 correlates with the production 2-hexyl-5-propyl resorcinol. Mol Plant-Microbe Interact 19:418–428
Chandler D, Davidson G, Grant WP, Greaves J, Tatchell GM (2008) Microbial biopesticides for integrated crop management: an assessment of environmental and regulatory sustainability. Trends Food Sci Tech 1(9):275–283
Chaparro J, Sheflin A, Manter D, Vivanco J (2012) Manipulating the soil microbiome to increase soil health and plant fertility. Biol Fertil Soils 48:489–499
Chung H, Park M, Madhaiyan M, Seshadri S, Song J, Cho H, Sa T (2005) Isolation and characterization of phosphate solubilizing bacteria from the rhizosphere of crop plants of Korea. Soil Biol Biochem 37(10):1970–1974
Compant S, Clément C, Sessitsch A (2010) Plant growth-promoting bacteria in the rhizo- and endosphere of plants: their role, colonization, mechanisms involved and prospects for utilization. Soil Biol Biochem 42:669–678
Cordell D, Drangert JO, White S (2009) The story of phosphorus: global food security and flood thought. Glob Environ Chang 19:292–305
Criquet S, Ferre E, Farner EM, Le Petit J (2004) Annual dynamics of phosphatase activities in na evergreen oak litter- influence of biotic and abiotic factors. Soil Biol Biochem 36:1111–1118
Dakora FD (1985) Use of intrinsic antibiotic resistance for characterisation and identification of rhizobia from nodules of *Vigna unguiculata* (L) Walp. and *Phaseolus vulgaris* (L). Acta Microbiol Pol 34:187–194
Dakora FD, Joseph CM, Phillips DA (1993) Alfalfa (*Medicago sativa* L.) root exudates contain isoflavonoids in the presence of *Rhizobium meliloti*. Plant Physiol 101:819–824
Dalton DA, Kramer S (2006) Nitrogen-fixing bacteria in non-legumes. Springer, Dordrecht, pp 105–113
Dangar TK, Basu PS (1987) Studies on plant growth substances, IAA metabolism and nitrogenase activity in root nodules of *Phaseolus aureus* Roxb. var. mungo. Biol Plant 29:350–354
Davies PJ (1995) The plant hormones: their nature, occurrence, and functions. In: Davies PJ (ed) Plant hormones: physiology, biochemistry, and molecular biology, 2nd edn. Kluwer, Dordrecht, pp 1–12
Dobbelaere S, Vanderleyden J, Okon Y (2003) Plant growth promoting effects of diazotrophs in the rhizosphere. Crit Rev Plant Sci 22:107–149
Dodd IC, Zinovkina NY, Safronova VI, Belimov AA (2010) Rhizobacterial mediation of plant hormone status. Ann Appl Biol 157:361–379

de Vries FT, Thébault E, Liiri M et al (2013) Soil food web properties explain ecosystem services across European land use systems. Proc Natl Acad Sci U S A 110:14296–14301
Dugardeyn J, Van Der Straeten D (2008) Ethylene: fine-tuning plant growth and development by stimulation and inhibition of elongation. Plant Sci 175:59–70
East R (2013) Soil science comes to life: plants may be getting a little help with their tolerance of drought and heat. Nature 501:18–19
El-Khawas H, Adachi K (1999) Identification and quantification of auxins in culture media of *Azospirillum* and *Klebsiella* and their effect on rice roots. Biol Fertil Soils 28:377–381
Erturk Y, Ercisli S, Haznedar A, Cakmakci R (2010) Effects of plant growth promoting rhizobacteria (PGPR) on rooting and root growth of kiwifruit (*Actinidia deliciosa*) stem cuttings. Biol Res 43:91–98
Estrada-De Los Santos P, Bustillos-Cristales R, Caballero-Mellado J (2001) Burkholderia, a genus rich in plant associated nitrogen fixers with wide environmental and geographic distribution. Appl Environ Microbiol 67:2790–2798
Faria DC, Dias AC, Melo IS, de Carvalho Costa FE (2013) Endophytic bacteria isolated from orchid and their potential to promote plant growth. World J Microbiol Biotechnol 29:217–221
Fedorov D, Ivanova E, Doronina N, Trotsenko Y (2008) A new system of degenerate oligonucleotide primers for detection and amplification of nifHD genes. Microbiologiia 77:247–249
Ferreira AS, Leitao JH, Silva IN, Pinheiro PF, Sousa SA, Ramos CG, Moreira LM (2010) Distribution of cepacian biosynthesis genes among environmental and clinical Burkholderia strains and role of cepacian exopolysaccharide in resistance to stress conditions. Appl Environ Microbiol 76:441–450
Fraga R, Rodriguez H, Gonzalez T (2001) Transfer of the gene encoding the Nap A acid phosphatase from *Morganella morganni* to *Burkholderia cepacia* strain. Acta Biotechnol 21:359–369
Franco-Correa M, Chavarro-Anzola V (2016) Actinobacteria as plant growth-promoting rhizobacteria. doi:10.5772/61291
Frankenberger WTJ, Arshad M (1995) Phitohormones in soil: microbial production and function. Dekker, New York
Galdiano Júnior RF, Pedrinho EAN, Castellane TCL, Lemos EGM (2011) Auxin-producing bacteria isolated from the roots of *Cattleya walkeriana*, an endangered Brazilian orchid, and their role in acclimatization. Rev Bras Ciênc Solo 35:729–737
Galloway JN, Dentener FJ, Capone DG et al (2004) Nitrogen cycles: past, present and future. Biogeochemistry 70:153–226
Gauthier D, Diem HG, Dommergues Y (1981) In Vitro nitrogen fixation by two actinomycete strains isolated from Casuarina nodules. Appl Environ Microbiol 41(1):306–308
Glick BR (2012) Plant growth promoting bacteria: mechanisms and application. Scientifica 2012:963401
Glick BR, Cheng Z, Czarny J, Duan J (2007) Promotion of plant growth by ACC deaminase-producing soil bacteria. Eur J Plant Pathol 119:329–339
Goldstein AH (1995) Recent progress in understanding the molecular genetics and biochemistry of calcium phosphate solubilization by Gram negative bacteria. Biol Agric Hortic 12(2):185–193
Goldstein AH (1994) Involvement of quinoprotein glucose dehydrogenase in the solubilization of exogeneous phosphates by Gram-negative bacteria. In: Torriani-Gorini A, Yagliard E, Silver S (eds) Phosphate in microorganisms: cellular and molecular biology. ASM Press, Washington, DC, pp 197–203
Goldstein AH, Rogers RD, Mead G (1993) Mining by microbe. Bio/Technology 11:125–1254
Gray EJ, Smith DL (2005) Intracellular and extracellular PGPR: commonalities and distinctions in the plant-bacterium signaling processes. Soil Biol Biochem 37:395–412
Gutierrez RA (2012) Systems biology for enhanced plant nitrogen nutrition. Science 336:1673–1675
Gyaneshwar P, Kumar GN, Parekh LJ, Poole PS (2002) Role of soil microorganisms in improving P nutrition of plants. Plant Soil 245:83–93
Haas D, Defago G (2005) Biological control of soil-borne pathogens by fluorescent pseudomonads. Nat Rev Microbiol 3:307–319

Habib SH, Kausar H, Saud HM (2016) Plant growth-promoting rhizobacteria enhance salinity stress tolerance in okra through ROS-scavenging enzymes. BioMed Res Int. doi:10.1155/2016/6284547

Hafeez FY, Yasmin S, Ariani D, Mehboob-ur-Rahman Z, Malik KA (2006) Plant growth-promoting bacteria as biofertilizer. Agron Sustain Dev 26:143–150

Halverson LJ, Handelsman J (1991) Enhancement of soybean nodulation by Bacillus cereus UW85 in the field and in a growth chamber. Appl Environ Microbiol 57:2767–2770

Han H, Supanjani S, Lee KD (2006) Effect of co-inoculation with phosphate and potassium solubilizing bacteria on mineral uptake and growth of pepper and cucumber. Plant Soil Environ 52(3):130–136

Hao X, Cho CM, Racz GJ, Chang C (2002) Chemical retardation of phosphate diffusion in an acid soil as affected by liming. Nutr Cycl Agroecosyst 64:213–224

Hayat R, Safdar Ali S, Amara U, Khalid R, Ahmed I (2010) Soil beneficial bacteria and their role in plant growth promotion: a review. Ann Microbiol 60:579–598

Hungria M, Campo RJ, Mendes IC (2005) Reinoculation increasing soybean grain yield in Brazil. In: Proceedings of the 14th international nitrogen fixation congress, Springer, Dordrecht, p 315

Iavicoli A, Boutet E, Buchala A, Metraux JP (2003) Induced systemic resistance in *Arabidopsis thaliana* in response to root inoculation with *Pseudomonas fluorescens* CHA0. Mol Plant-Microbe Interact 16:851–858

Igual M, Valverde EA, Cervantes E, Velásquez E (2001) Phosphate-solubilizing bactéria as inoculants for agriculture: use of update molecular techniques in their study. Agronomie 21:561–568

Illmer P, Schimer F (1995) Solubilization of inorganic calcium phosphates solubilization mechanisms. Soil Biol Biochem 27:257–263

Inui-Kishi RN, Kishi LT, Picchi SC, Barbosa JC, Lemos MTO, Marcondes J, Lemos EG d M (2012) Phosphorus solubilizing and IAA production activities in plant growth promoting rhizobacteria from Brazilian soils under sugarcane cultivation. ARPN J Eng Appl Sci 7:1446–1454

Iqbal I, Hasnain S (2013) Auxin producing *pseudomonas* strains: biological candidates to modulate the growth of *Triticum aestivum* beneficially. Am J Plant Sci 4:1693–1700

Jaizme-Vega MC, Rodriguez-Romero AS, Guerra MSP (2004) Potential se of rhizobacteria from the Bacillus genus to stimulate the plant growth of micropropagated bananas. Fruits 59(2):83–90

Jing YD, He ZL, Yang XE (2007) Role of soil rhizobacteria in phytoremediation of heavy metal contaminated soils. J Zhejiang Univ Sci B 8:192–207

Jorquera MA, Hernandez MT, Rengel Z, Marschner P, Mora MD (2008) Isolation of culturable phosphor bacteria with both phytate mineralization and phosphate solubilization activity from the rhizosphere of plants grown in a volcanic soil. Biol Fertil Soils 44:1025–1034

Kembel SW, O'Connor TK, Arnold HK, Hubbell SP, Wright SJ, Green JL (2014) Relationships between phyllosphere bacterial communities and plant functional traits in a neotropical forest. Proc Natl Acad Sci U S A 38:13715–13720

Khan MS, Zaidi A, Wani PA (2007) Role of phosphate-solubilizing microorganisms in sustainable agriculture—a review. Agron Sustain Dev 27:29–43

Khan MS, Zaidi A, Ahemad M, Oves M, Wani PA (2010) Plant growth promotion by phosphate solubilizing fungi—current perspective. Arch Agron Soil Sci 56:73–98

Kim J, Rees DC (1994) Nitrogenase and biological nitrogen fixation. Biochemist 33:389–397

Kloepper JW (1994) Plant growth-promoting rhizobacteria (other systems). In: Okon Y (ed) Azospirillum/plant associations. CRC Press, Boca Raton, pp 111–118

Kloepper JW, Leong J, Teintze M, Schroth MN (1980) Enhanced plant growth by siderophores produced by plant growth-promoting rhizobacteria. Nature 286:5776

Knudson L (1922) Nonsymbiotic germination of orchid seeds. Bot Gaz 73:1–25

Köberl M, Schimdt R, Ramadan EM, Bauer R, Berg G (2013) The microbiome of medicinal plants: diversity and importance for plant growth, quality, and health. Front Microbiol 4:1–9

Krishnaraj PU, Dahale SK (2014) Mineral phosphate solubilization: concepts and prospects in sustainable agriculture. Proc Indian Natl Sci Acad 80:389–405

Kyselková M, Kopecký J, Frapolli M et al (2009) Comparison of rhizobacterial community composition in soil suppressive or conducive to tobacco black root rot disease. ISME J 3:1127–1138

Lebuhn M, Heulin T, Hartmann A (1997) Production of auxin and other indolic and phenolic compounds by *Paenibacillus polymyxa* strains isolated from different proximity to plant roots. FEMS Microbiol Ecol 22:325–334

Lichter A, Barash I, Valinsky L, Manulis S (1995) The genes involved in cytokinin biosynthesis in *Erwinia herbicola* pv. gypsophilae, characterization and role in gall formation. J Bacteriol 177:4457–4465

Lifshitz R, Kloepper JW, Scher FM et al (1986) Nitrogen-fixing pseudomonads isolated from roots of plants grown in the Canadian High Arctic. Appl Environ Microbiol 51:251–255

Liu HX, Luo YB, Liu H (2010) Studies of mycorrhizal fungi of Chinese orchids and their role in orchid conservation in China – a review. Bot Rev 76:241–262

Long SR (1996) Rhizobium symbiosis: nod factors in perspective. Plant Cell 8:1885–1898

Lundberg DS, Lebeis SL, Paredes SH et al (2012) Defining the core *Arabdopsis thaliana* root. Nature 488:86–90

Ma Y, Rajkumar M, Freitas H (2009a) Isolation and characterization of Ni mobilizing PGPB from serpentine soils and their potential in promoting plant growth and Ni accumulation by Brassica spp. Chemosphere 75:719–725

Ma Y, Rajkumar M, Freitas H (2009b) Improvement of plant growth and nickel uptake by nickel resistant-plant-growth promoting bacteria. J Hazard Mater 166:1154–1161

Mabood F, Zhou X, Smith DL (2014) Microbial signaling and plant growth promotion. Can J Plant Sci 94:1051–1063

Maeder P, Fliessbach A, Dubois D et al (2002) Soil fertility and biodiversity in organic farming. Science 296:1694–1697

Maliha R, Samina K, Najma A et al (2004) Organic acids production and phosphate solubilization by phosphate solubilizing microorganisms under in vitro consitions. Pak J Biol Sci 7:187–196

Mayak S, Tirosh T, Glick BR (2004) Plant growth-promoting bacteria confer resistance in tomato plants to salt stress. Plant Physiol Biochem 42:565–572

Mendes LW, Tsai SM, Navarrete AA et al (2015) Soil-borne microbiome: linking diversity to function. Microb Ecol 70:255–265

Menezes-Blackburn D, Jorqueira MA, Greiner R et al (2013) Phytases and phytase-labile organic phosphorus in manures and soils. Crit Rev Environ Sci Technol 43:916–954

Morris RO (1986) Genes specifying auxin and cytokinin biosynthesis in pathogens. Annu Rev Plant Physiol 37:509–538

Nieto KF, Frankenberger WT Jr (1990) Microbial production of cytokinins. In: Bollag JM, Stotzky G (eds) Soil biochemestry, vol 6. Dekker, New York, pp 191–248

O'Sullivan DJ, O'Gara F (1992) Traits of fluorescent *Pseudomonas* spp. involved in suppression of plant root pathogens. Microbiol Rev 56:662–676

Oldroyd GED, Downie JA (2004) Calcium, kinases and nodulation signaling in legumes. Nature 5:566–576

Oldroyd GED, Downie JA (2008) Coordinating nodule morphogenesis with rhizobial infection in legumes. Annu Rev Plant Biol 59:519–546

Oldroyd GED, Murray JD, Poole PS, Downie JA (2011) The rules of engagement in the legume-rhizobial symbiosis. Annu Rev Genet 45:119–144

Olivares J, Bedmar EJ, Sanjuán J (2013) Biological nitrogen fixation in the context of global change. MPMI 26:486–494

Oliveira CA, Sa NMH, Gomes EA et al (2009) Assessment of the mycorrhizal community in the rhizosphere of maize (*Zea mays* L.) genotypes contrasting for phosphorus efficiency in the acid savannas of Brazil using denaturing gradient gel electrophoresis (DGGE). Appl Soil Ecol 41:249–258

Orhan E, Esitken A, Ercisli S, et al (2006) Effects of plant growth promoting rhizobacteria (PGPR) on yield, growth and nutrient contents in organically growing raspberry. Sci Hortic 111(1):38–43

Ortíz-Castro R, Contreras-Cornejo HA, Macías-Rodríguez L, López-Bucio J (2009) The role of microbial signals in plant growth and development. Plant Signal Behav 4:701–712

Orwin KH, Kirschbaum MUF, St John MG, Dickie IA (2011) Organic nutrient uptake by mycorrhizal fungi enhances ecosystem carbon storage: a model-based assessment. Ecol Lett 14: 493–502

Parks EJ, Olson GJ, Brinckman FE, Baldi F (1990) Characterization by high performance liquid chromatography (HPLC) of the solubilization of phosphorus in iron ore by fungus. J Ind Microbiol Biotechnol 5:183–189

Parmar N, Dufresne J (2011) Beneficial interactions of plant growth promoting rhizosphere microorganisms. In: Singh A, Parmar N, Kuhad RC (eds) Bioaugmentation, biostimulation and biocontrol. Springer-Verlag, Berlin Heidelberg, pp 27–42

Patten CL, Glick BR (1996) Bacterial biosynthesis of indole-3-acetic acid. Can J Microbiol 42:207–220

Paul D, Nair S (2008) Stress adaptations in a plant growth promoting rhizobacterium (PGPR) with increasing salinity in the coastal agricultural soils. J Basic Microbiol 48:378–384

Peix A, Velazquez E, Martýnez-Molina E (2007) Molecular methods for biodiversity analysis of phosphate solubilizing microorganisms (PSM). In: Velazquez E, Rodriguez-Barrueco C (eds) First international meeting on microbial phosphate solubilization. Springer, Berlin, p 97–100

Penrose DM, Glick BR (2003) Methods for isolating and characterizing ACC deaminase containing plant growth promoting rhizobacteria. Physiol Plant 118:10–15

Peralta H, Mora Y, Salazar E, Encarnacion S, Palacios R, Mora J (2004) Engineering the nifH promoter region and abolishing poly-Â - hydroxybutyrate accumulation in Rhizobium etli enhance nitrogen fixation in symbiosis with Phaseolus vulgaris. Appl Environ Microbiol 70(6):3272–3281

Perret X, Staehelin C, Broughton WJ (2000) Molecular basis of symbiotic promiscuity. Microbiol Mol Biol Rev 64:180–201

Philippot L, Spor A, Hénault C et al (2013) Loss in microbial diversity affects nitrogen cycling in soil. ISME J 7:1609–1619

Pirlak M, Kose M (2009) Effects of plant growth promoting rhizobacteria on yield and some fruit properties of strawberry. J Plant Nutr 32:1173–1184

Qi X, Wang E, Xing M, Zhao W, Chen X (2012) Rhizosphere and nonrhizosphere bacterial community composition of the wild medicinal plant *Rumex patientia*. World J Microbiol Biotechnol 28:2257–2265

Raaijmakers JM, Vlami M, de Souza JT (2002) Antibiotic production by bacterial biocontrol agents. Antonie Van Leeuwenhoek 81:537–547

Rashid S, Charles TC, Glick BR (2012) Isolation and characterization of new plant growth-promoting bacterial endophytes. Appl Soil Ecol 61:217–224

Rastogi G, Sbodio A, Tech JJ, Suslow TV, Coaker GL, Leveau JHJ (2012) Leaf microbiota in an agroecosystem: spatiotemporal variation in bacterial community composition on field-grown lettuce. ISME J 6:1812–1822

Rathore P (2014) A review on approaches to develop plant growth promoting rhizobacteria. Intl J Recent Sci Res 5:403–407

Raymond J, Siefert JL, Cr S, Blankendhip RE (2004) The natural history of nitrogen fixation. Mol Biol Evol 21:541–554

Reyes I, Bernier L, Antoun H (2002) Rock phosphate solubilization and colobization of maize rizosphere by wild and genetically modified strains of *Pennicilium rugulosum*. Microb Ecol 44:39–48

Rezzonoco F, Binder C, Defago G, Moenne-Loccoz Y (2005) The type III secretion system of biocontrol *Pseudomonas fluorescencs* KD targets the phytopathogenic chromista *Pythium ultimum* and promotes cucmuber protection. Mol Plant-Microbe Interact 9:991–1001

Richardson AE (1994) Soil microorganisms and phosphorus availability. In: Pankhurst CE, Doubeand BM, Gupta WSR (eds) Soil biot: management in sustainable farming systems. CSIRO, Victoria, pp 50–62

Richardson AE, Simpson RJ (2011) Soil microorganism mediating phosphorus availability. Plant Physiol 156:989–996

Richardson AE, George TS, Hens M, Simpson RJ (2005) Utilization of soil organic phosphorus by higher plants. In: Turner BL, Frossard E, Baldwin DS (eds) Organic phosphorus in the environment. CABI, Wallingford, pp 165–184

Richardson AE, Baréa JM, McNeill AM, Prigent-Combaret C (2009) Acquisition of phosphorus and nitrogen in the rhizosphere and plant growth promotion by microorganisms. Plant Soil 321:305–339

Rodrigues EP, Rodrigues CS, de Oliveira ALM, Baldani VL, Teixeira da Silva JA (2008) *Azospirillum amazonense* inoculation: effects on growth, yield and N2 fixation of rice (*Oryza sativa* L.) Plant Soil 302:249–261

Rodriguez H, Fraga R (1999) Phosphate solubilizing bacteria and their role in plant growth promotion. Biotechnol Adv 17:319–339

Rodriguez H, Fraga R, Gonzalez T, Bashan Y (2006) Genetics of phosphate solubilization and its potential applications for improving plant growth-promoting bacteria. Plant Soil 287:15–21

Rubio LM, Ludden PW (2008) Biosynthesis of the iron-molybdenum cofactor of nitrogenase. Annu Rev Microbiol 62:93–111

Saharan BS, Nehan V (2011) Plant growth promoting rhizobacteria: a critical review. Life Sci Med Res 21:1–30

Sala OE, Chapin FS, Armesto JJ et al (2000) Biodiversity—global biodiversity scenarios for the year 2100. Science 287:1770–1774

Saranraj P, Sivasakthivelan P, Siva SS (2013) Prevalence and production of plant growth promoting substance by *Pseudomonas fluorescens* isolated from paddy rhizosphere soil of Cuddalore district, Tamil Nadu, India. Afr J Basic Appl Sci 5(2):95–101

Sashidhar B, Podile AR (2010) Mineral phosphate solubilization by rhizosphere bacteria and scope for manipulation of the direct oxidation pathway involving glucose dehydrogenase. J Appl Microbiol 109:1–12

Sato VS, Galdiano RF, Rodrigues GR, Lemos EGM, Pizauro JM (2016) Kinetic characterization of a novel acid ectophosphatase from *Enterobacter asburiae*. J Microbiol 54:106–113

Schippers B, Bakker AW, Bakker AHM (1987) Interactions of deleterious and beneficial rhizosphere microorganisms and the effect of cropping practice. Annu Rev Phytopathol 25:339–358

Sessitsch A, Howieson JG, Perret X, Antoun H, Martínez Romero E (2002) Advances in rhizobium research. Crit Rev Plant Sci 21:323–378

Sharma SK, Johri BN, Ramesh A, Joshi OP, Prasad SVS (2011) Selection of plant growth-promoting *Pseudomonas* spp. that enhanced productivity of soybean-wheat cropping system in central India. J Microbiol Biotechnol 21:1127–1142

Sharma SB, Sayyed RZ, Trivedi MH, Gobi TA (2013) Phosphate solubilizing microbes: sustainable approach for managing phosphorus deficiency in agricultural soils. SpringerPlus 2:587

Shrivastava P, Kumar R (2015) Soil salinity: a serious environmental issue and plant growth promoting bacteria as one of the tools for its alleviation. Saudi J Biol Sci 22(2):123–131

Singh G, Mukerji KG (2006) Root exudates as determinant of rhizospheric microbial biodiversity. In: Mukerji KG, Manoharachary C, Singh J (eds) Microbial activity in the rhizosphere. Spinger, Berlin, pp 39–53

Singh B, Satyanarayana T (2011) Microbial phytases in phosphorus acquisition and plant growth promotion. Physiol Mol Biol Plants 17:93–103

Siqueira JO, Franco AA (1988) Biotecnologia do solo: fundamentos e perspectivas. ESAL/FAEPE, Lavras. (in Portuguese)

Siqueira JO, Andrade AT, Faquin V (2004) O papel dos microrganismos na disponibilização e aquisição de fósforo pelas plantas. In: Yamada T, Stipp e Abdalla SR (eds) Fósforo na agricultura brasileira. Potafos publishers, Piracicaba, pp 117–156

Sivasakthivelan P, Saranraj P (2013) *Azospirillum* and its formulations: a review. Intl J Microbiol Res 4(3):275–287

Smalla K, Wieland G, Buchner A et al (2001) Bulk and rhizosphere soil bacterial communities studied by denaturing gradient gel electrophoresis: plant-dependent enrichment and seasonal shifts revealed. Appl Environ Microbiol 67:4742–4751
Smith SE, Read DJ (2008) Mycorrhizal symbiosis, 3rd edn. Academic Press, Amsterdam
Smith DL, Zhou X (2014) An effective integrated research approach to study climate change in Canada. Can J Plant Sci 94:995–1008
Smith DL, Subramanian S, Lamont JR, Bywater-Ekegärd M (2015) Signalling in the phytomicrobiome: breath and potential. Front Plant Sci 6:1–8
Spaepen S, Vanderleyden J, Remans R (2007) Indole-3-acetic acid in microbial and microorganism-plant signaling. FEMS Microbiol Rev 31:425–448
Supanjani HS, Han JS, Jung KD, Lee KD (2006) Rock phosphate potassium and rock solubilizing bacteria as alternative, sustainable fertilizers. Agron Sustain Dev 26(4):233–240
Tak HI, Ahmad F, Babalola OO (2013) Advances in the application of plant growth-promoting rhizobacteria in phytoremediation of heavy metals. In: Whitacre DM (ed) Reviews of environmental contamination and toxicology, vol 223. Springer Science Bussiness Media, New York, pp 33–52
Tenuta M (2003) Plant growth promoting rhizobacteria: prospects for increasing nutrient acquisition and disease control. http://www.umanitoba.ca/faculties/afs/MAC_proceedings/2003/pdf/tenuta_rhizobacteria.pdf. Accessed 24 Feb 2016
Tilman D, Fargione J, Wolff B, D'Antonio C, Dobson A, Howarth R, Schindler D, Schesinger WH, Simberloff D, Wackhamer D (2001) Forecasting agriculturally driven global environmental change. Science 292:281–284
Trabelsi D, Mhamdi R (2013) Microbial inoculants and their impact on soil microbial communities: a review. Biomed Res 1:13
Trolove SN, Hedley MJ, Kirk GJD, Bolan NS, Loganathan P (2003) Progress in selected areas of rhizosphere research on P acquisition. Aust J Soil Res 41(3):471
Tsavkelova EA, Cherdyntseva TA, Netrusov AI (2003) Phytohormones production by the fungi associated with orchids. Mycol Phytopathol 37:75–83
Tsavkelova EA, Klimova SY, Cherdyntseva TA, Netrusov AI (2006) Microbial producers of plant growth stimulators and their practical use: a review. Appl Biochem Microbiol 42:117–126
Tsavkelova EA, Cherdyntseva TA, Klimova SY, Shestakov AI, Botina SG, Netrusov AI (2007) Orchid-associated bacteria produce indole-3-acetic acid, promote seed germination, and increase their microbial yield in response to exogenous auxin. Arch Microbiol 188:655–664
Tsavkelova E, Oeser B, Oren-Young L, Israeli M, Sasson Y, Tudzynski B, Sharon A (2012) Identification and functional characterization of the genes for indole-3-acetamide-mediated IAA biosynthesis in plant-associated *Fusarium* species. Fungal Genet Biol 49:48–57
Upadhyay SK, Singh DP, Saikia R (2009) Genetic diversity of plant growth promoting rhizobacteria isolated from rhizospheric soil of wheat under saline condition. Curr Microbiol 59:489–496
Vacheron J, Desbrosses G, Bouffaud ML, Touraine B, Moënne-Loccoz Y, Muller D, Legendre L, Wisniewski-Dyé F, Prigent-Combaret C (2013) Plant growth-promoting rhizobacteria and root system functioning. Front Plant Sci 4:356
Valdés M, Pérez N, Estrada P, Caballero J, Peña J, Normand P, Hirsch A (2005) Non-Frankia actinomycetes isolated from surface-sterilized roots of *Casuarina equisetifolia* fix nitrogen. Appl Environ Microbiol 71(1):460–466
Van Peer R, Punte HL, de Weger LA, Schippers B (1990) Characterization of root surface and endorhizosphere pseudomonads in relation to their colonization of roots. Appl Environ Microbiol 56:2462–2470
Van Peer R, Nieman GJ, Schippers B (1991) Induced resistance and phytoalexin accumulation in biological control of Fusarium wilt of carnation by *Pseudomonas* sp. strain WCS417r. Phytopathology 81:728–734
Vazquez P, Holguin G, Puente ME, Lopez-Cortes A, Bashan Y (2000) Phosphate-solubilizing microorganisms associated with the rhizosphere of mangroves in a semiarid coastal lagoon. Biol Fertil Soils 30(5-6):460–468
Venkateswarlu B, Rao AV, Raina P, Ahmad N (1984) Evaluation of phosphorus solubilization by microorganisms isolated from arid soil. J Ind Soc Soil Sci 32:273–277

Vessey JK (2003) Plant growth promoting rhizobacteria as biofertilizers. Plant Soil 255:571–586
Vessey JK, Buss TJ (2002) Bacillus cereus UW85 inoculation effects on growth, nodulation, and N accumulation in grain legumes. Controlled environmental studies. Can J Plant Sci 82:282–290
Wang Y, Brown HN, Crowley DE, Szaniszlo PJ (1993) Evidence for direct utilization of a siderophore, ferrioxamine B, in axenically grown cucumber. Plant Cell Environ 16:579–585
Wang L, Zhang L, Liu Z, Zhao D, Liu X, Zhang B, Xie J, Hong Y, Li P, Chen S, Dixon R, Li J (2013) A minimal nitrogen fixation gene cluster from P*aenibacillus* sp. WLY78 enables expression of active nitrogenase in *Escherichia coli*. PLoS Genet 3:1–11
Wani PA, Khan MS (2010) *Bacillus* species enhance growth parameters of chickpea (*Cicer arietinum* L.) in chromium stressed soils. Food Chem Toxicol 48:3262–3267
Wani PA, Zaidi A, Khan AA, Khan MS (2005) Effect of phorate on phosphate solubilization and indole acetic acid (IAA) releasing potentials of rhizospheric microorganisms. Ann Plant Protect Sci 13:139–144
Wilkinson KG, Dixon KW, Sivasithamparam K (1989) Interaction of soil bacteria, mycorrhizal fungi and orchid seed in relation to germination of Australian orchids. New Phytol 112:429–435
Wilkinson KG, Dixon KW, Sivasithamparam K, Ghisalberti EL (1994) Effect of IAA on symbiotic germination of an Australian orchid and its production by orchid-associated bacteria. Plant Soil 159:291–295
Wu JP, Qian J, Zheng SZ (2002) A preliminary study on ingredient of secretion from fungi of orchid mycorrhizal. Chin J Appl Ecol 13:845–848
Yang SF, Hoffman NE (1984) Ethylene biosynthesis and its regulation in higher plants. Annu Rev Plant Physiol 35:155–189
Yang YL, Liu ZY, Zhu GS (2008) Study on symbiotic seed germination of *Pleione bulbocodioides* (Franch) Rolfe. Microbiology 35:909–912. (in Chinese with English abstract)
Yang S, Zhang X, Cao Z, Zhao K, Wang S, Chen M, Hu X (2014) Growth-promoting *Sphingomonas paucimobilis* ZJSH1 associated with *Dendrobium officinale* through phytohormone production and nitrogen fixation. Microb Biotechnol 7:611–620
Zahir AZ, Arshad M, Frankenberger WT Jr (2004) Plant growth promoting rhizobacteria: applications and perspectives in agriculture. Adv Agron 81:97–168
Zahir ZA, Shah MK, Naveed M, Akhter MJ (2010) Substrate dependent auxin production by *Rhizobium phaseoli* improves the growth and yield of *Vigna radiata* L. under salt stress conditions. J Microbiol Biotechnol 20:1288–1294
Zaidi A, Khan MS, Ahemad M, Oves M (2009) Plant growth promotion by phosphate solubilizing bacteria. Acta Microbiol Immunol Hung 56:263–284
Zhu F, Qu L, Hong X, Sun X (2011) Isolation and characterization of phosphate solubilizing Halophilic Bacterium *Kushneria* sp. YCWA 18 from Daqiao Saltern on the Coast of Yellow Sea of China. Evid Based Complement Alternat Med 2011:1–6

Rhizobacterial Biofilms: Diversity and Role in Plant Health

7

Mohd. Musheer Altaf, Iqbal Ahmad, and Abdullah Safar Al-Thubiani

Abstract

The diverse nature of rhizobacteria and their interaction with plant roots involves complex processes and provides a unique microbial niche in the rhizosphere both beneficial and harmful to plant health depending on nature of bacteria. Biofilms are defined as the bacterial populations which stick to living and nonliving surfaces and encased in a self-produced extracellular polymeric substances (EPS). Both disease-causing and beneficial plant growth-encouraging bacteria may form biofilm on abiotic and biotic surfaces including plant surface and in soil. It is now well known that a microbe under natural condition forms mixed/polymicrobial biofilm. The process of biofilm development and their regulation are well studied among human pathogenic bacteria such as *Pseudomonas aeruginosa*. However, recent investigations indicated an increased interest in the research on biofilm on plant-associated rhizobacteria such as *Azotobacter*, *Acinetobacter*, *Bacillus*, *Burkholderia*, *Klebsiella*, *Pantoea*, *Pseudomonas* and *Rhizobium*. In this chapter we have made an attempt to review recent studies on rhizobacterial biofilms and their possible impact on plant health under natural and stress conditions.

M.M. Altaf (✉) • I. Ahmad
Department of Agricultural Microbiology, Aligarh Muslim University, Aligarh 202002, India
e-mail: mohdmusheer@rediffmail.com

A.S. Al-Thubiani
Department of Biology, College of Applied Sciences, Umm Al-Qura University, Makkah 21955, Saudi Arabia

V. Kumar et al. (eds.), *Probiotics in Agroecosystem*,
DOI 10.1007/978-981-10-4059-7_7

7.1 Introduction

Biofilms are bacterial populations associated with living and nonliving surfaces and encased in self-produced medium of extracellular polymeric substances (EPS). The chemical nature of EPS is heterogeneous largely made up of polysaccharides, extracellular proteins and enzymes, DNA and other substances. A significant benefit of the biofilm lifestyle for rhizobacteria is safety against drought or osmotic stress. The rhizosphere serves as a dynamic microbial habitat for environmental microniches and is continuously studied to understand microbial diversity and their regulation. Both traditional (culture-based) and molecular (culture-independent) approaches (polyphasic) are used to understand such niches (Hinsinger et al. 2009; Bogino et al. 2013). The rhizosphere is the soil niche influenced by plant root and root exudates. The colonization of root by soil microorganisms is largely associated with the movement of bacteria from bulk soil to rhizospheric soil/rhizoplane. Successful establishment of rhizobacteria in association with plant roots depends on the adherence and microcolonies formation. Biofilm formation contributes a necessary step in bacterial survival and physiology. The bacterium has to interact with several microorganisms and should set up themselves as polymicrobial biofilm at the rhizosphere level (Fujishige et al. 2006; Compant et al. 2010). The interactions of biofilm bacteria with plants may be positive or negative; biofilm also results in nutrient turnover and reduction of biotic and abiotic stress factors (Angus and Hirsch 2013).

Recent studies on microbial biofilms associated with plants have gain momentum in the last few years. Excellent review articles have been published on bacterial biofilms highlighting agricultural, environmental and ecological significance in the both positive and negative ways (Morris and Monier 2003; Danhorn and Fuqua 2007; Angus and Hirsch 2013; Vlamakis et al. 2013).

Importance of polymicrobial biofilm and their role in plant health and disease have been recently reviewed (Angus and Hirsch 2013). However, the diversity of biofilm-forming rhizobacteria and their role on plant health is less explored. In this article we address the function of plant growth-promoting bacteria in forming biofilm and the impact of plant root exudates and stress conditions on biofilm and plant health. Further general methods of studying biofilm and its characterization have also been briefly elaborated.

7.2 Common Steps in Biofilm Formation

Biofilms are multi-structured, mixed population of bacterial cells encased in an EPS medium that are permanently adhered to a living and nonliving surface. Basic and fundamental research on biofilm is extensively studied on medical and environmental pathogens. Pioneer work was contributed by John William Costerton group who is regarded as father of biofilm (Geesey et al. 1977; Costerton et al. 1978). Plant-associated biofilm by plant pathogens and other plant-associated bacteria have been investigated in the last few years (Timmusk et al. 2011; Beauregard et al. 2013;

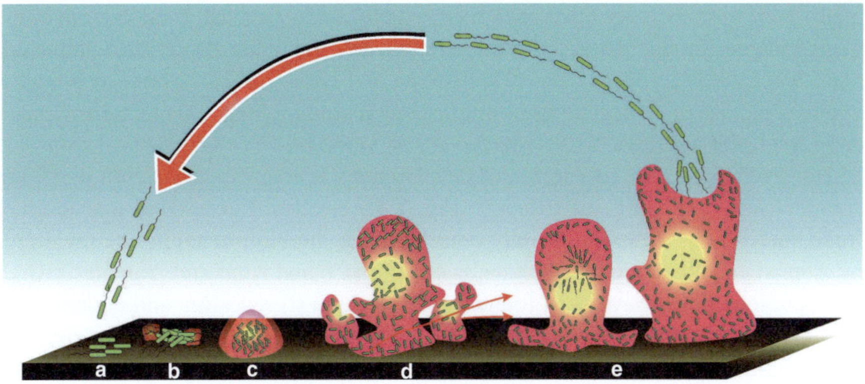

Fig. 7.1 Steps associated with biofilm formation. (**a**) Adherence of bacteria to barren surface. (**b**) Secretion of extracellular polymeric substances (*EPS*) and irreversible attachment of cells. (**c**, **d**) Maturation and development of biofilm architecture. (**e**) Dispersion of cells (*planktonic*) from biofilm (*Red arrows* showing flowing system)

Lugtenberg 2015). The process of biofilm formation involves several steps. An initial step implicated in biofilm development includes the scattering of surface attached cells by surface motility and multiplication. Originally, bacterial cells are small motile sticks, but as the biofilm grows, they change their appearance to lengthy chains of non-motile cells that stick together and to the surface by producing EPS (Branda et al. 2006). The EPS is necessary for the entireness of the biofilm, as it grasps the community together (Flemming and Wingender 2010). Upon maturation the microcolonies increase in size, and the colony is protected and organized by EPS. The maturing biofilm contains EPS producers and motile cells and is socially organized within the maturing biofilm (Lopez et al. 2010). Sequence involved in the biofilm development is depicted in Fig. 7.1.

The occurrence and localization of the different bacterial cell varieties are vibrant, and they give the impression to be a planned progression of segregation such that motile bacterial cells turn into matrix-producing cells. Notably, this method of segregation is not final; as ecological circumstances alter, it is possible for bacterial cells to change their gene expression (in the case of motile or matrix-producing cells) or to germinate (in the case of spores). Under laboratory environment, biofilms have a restricted natural life, and ultimately the biofilm was dispersed in reply to self-developed signals (Kolodkin-Gal et al. 2010).

7.3 Methods of Studying Biofilm

Biofilm is a bacterial lifestyle widespread in microbial world. Nearly in all previous microbiology studies, microorganisms have been studied as free-living cells. Our understanding of biofilms, their physiology and structures is based on microscopic techniques. Research related to biofilm develops in several directions. *In vitro* methods of biofilm study have been well established in medical, food and environmental areas. Similar methods are now applicable for other bacteria of soil- and plant-associated biofilms (Otto 2013; Angus and Hirsch 2013). These techniques involved the use of several types of microscopy including light and electron microscopy (Figs. 7.2 and 7.3).

However, with new research and development in microscopic techniques like confocal laser scanning microscopy (Fig. 7.4), electron microscopy and atomic force microscopy have significantly contributed to the biofilm studies as reviewed by Ahmad and Khan (2012), Pantanella et al. (2013) and Altaf and Ahmad (2016).

These techniques have been found instrumental in (i) revealing and understanding the mechanism and morphology of biofilms embedded in EPS matrix, (ii) assessment of nanoscale arrangement of living bacterial cells in biofilm and (iii) structural and functional characteristics of biofilm. A brief account on various common methods used for studying biofilms applied by different workers is depicted in Table 7.1.

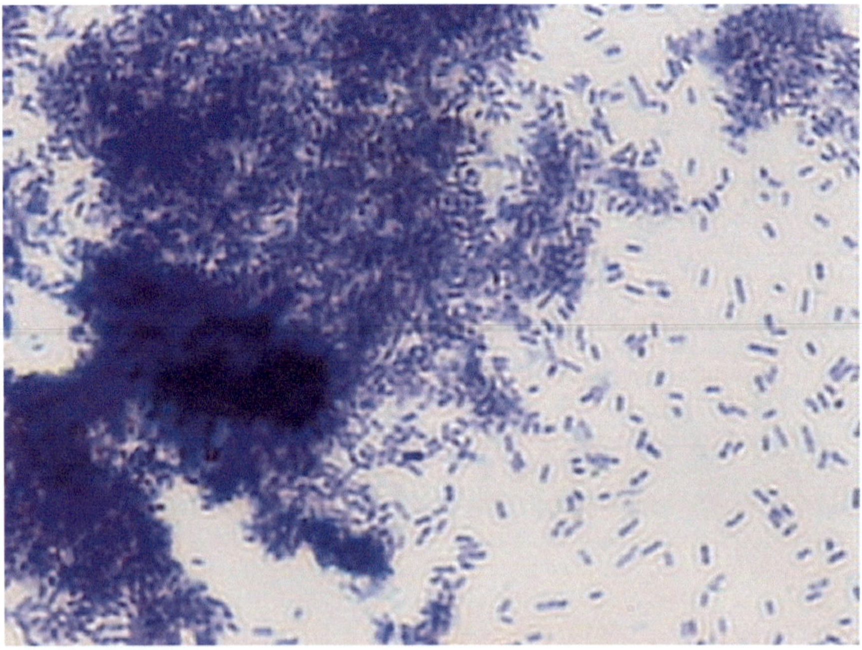

Fig. 7.2 Representative photograph of *Pseudomonas fluorescens* biofilm formed on glass coverslip as viewed by light microscopy (magnifcation 100×)

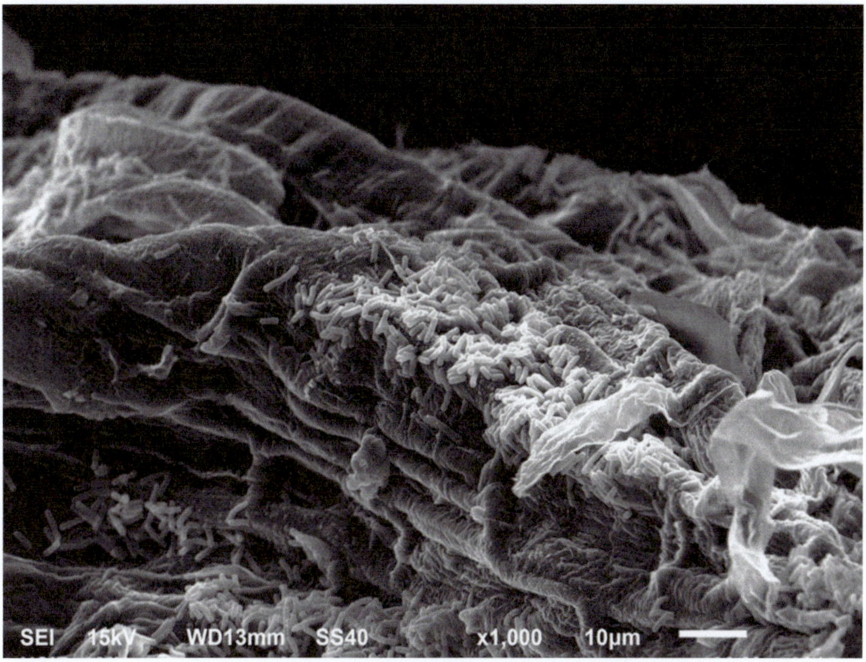

Fig. 7.3 Scanning electron micrograph of *Pseudomonas fluorescens* colonization and biofilm formation on root of *Triticum aestivum*

Fig. 7.4 Confocal laser scanning microscopy image of *Klebsiella pneumoniae* biofilm formed on wheat root. Biofilm stained with acridine orange (magnification 20×. Scale bar-50 μm)

Table 7.1 Methods used for the study of microbial biofilm

Assays	Applications	References
Congo red agar method	Biofilm detection on agar media supplemented with Congo red dye	Freeman et al. (1989), Kaiser et al. (2013)
Crystal violet (CV) assay	Assessment of biofilm formation on abiotic surface like polystyrene plate, catheter disc, glass coverslips/slides, stainless steel	Christensen et al. (1985), O'Toole and Kolter (1998), Nweze et al. (2012), Peters et al. (2013), Craveiro et al. (2015)
BioFilm Ring Test	Evaluation of biofilm formed by clinical isolates	Chavant et al. (2007), Crémet et al. (2013)
1,9-Dimethyl-methylene blue (DMMB) assay	Colorimetric assay for biofilm detection	Toté et al. (2008)
Fluorescein-di-acetate (FDA) assay	Quantification of *Candida albicans* biofilm	Honraet et al. (2005), Tawakoli et al. (2013)
LIVE/DEAD BacLight assay	Detection of live and dead cells in biofilm	Boulos et al. (1999), Leuko et al. (2004)
Resazurin assay	Detection and estimation of living cells in biofilm	Sandberg et al. (2009)
XTT assay	Detection and quantification of living cells in biofilm	Ramage et al. (2001), Adam et al. (2002)
BioTimer assay (BTA)	Colorimetric assay for counting of viable bacteria in biofilm	Pantanella et al. (2008)
qRT-PCR assay	Identification of specific genetic sequences of bacterial species in biofilm	Xie et al. (2011)
Fluorescence in situ hybridization (FISH) assay	Study of multispecies biofilm in natural environment	Almeida et al. (2011)
Light microscopy	For rapid, inexpensive qualitative and quantitative information of biofilm	Ahmad and Khan (2012)
Confocal laser scanning microscopy (CLSM)	*In vitro* and *in vivo* studies of 3D structure of biofilm	Gorman et al. (1994), Ahmad and Khan (2012), Janczarek et al. (2015), Altaf and Ahmad (2016)
Confocal-RAMAN microscopy	For acquiring information on chemical fingerprint of different biofilms	Yuanqing and Tong (2012)
Transmission electron microscopy (TEM)	To study internal structure of biofilms and intracellular features	Ahmad and Khan (2012)
Scanning electron microscopy (SEM)	To visualize the surface of biofilm formed on biotic and abiotic surfaces	Ahmad and Khan (2012), Nongkhlaw and Joshi (2014), Altaf and Ahmad (2016)
Cryo-SEM	To acquire high-magnification photographs of biofilm nearer to the original condition of the sample	Allan-Wojtas et al. (2010), Kumar et al. (2012)

Table 7.1 (continued)

Assays	Applications	References
Environmental SEM (ESEM)	Imaging in gaseous environment of hydrated and nonconductive bacterial biofilms	Karcz et al. (2012), Timmusk et al. (2014)
Scanning transmission X-ray microscopy (STXM)	Used to investigate the composition of bacterial cells and biofilms with nominal resolution of 25 nm	Dynes et al. (2006), Behrens et al. (2012)
Atomic force microscopy (AFM)	Applied to visualize the surface of bacteria and biofilm, quantitative measurement and mapping of biofilm elasticity	Wright et al. (2010), Dong et al. (2011), Ahmad and Khan (2012)

7.4 Bacterial Characteristics Under Planktonic and Biofilm Mode

For several years microorganisms in environment have been studied as free-living cells, which help in the easy characterization of the microorganisms. On the other hand, free-living method of development is not the normal condition for microorganisms, and concern should be needed then to infer these results in their normal state. Throughout the previous decades, in-depth knowledge was generated through the comprehensive research that has been performed in the field of biofilms like medical, industrial and environmental and plant root-associated biofilms (Timmusk and Nevo 2011). Generally, biofilms are defined as intricate microbial populations adhered to the surface or boundary encased in an extracellular medium of microbial and host origin to construct a spatially structured three-dimensional configuration (Costerton et al. 1995). Genotypically bacteria enclosed in similar biofilms are naturally different from the planktonic counterpart. Single cells in biofilm population manage their gene expression to regulate cell differentiation (O'Toole et al. 2000; Lopez et al. 2009). In bacterial ecosystems biofilm is a natural phenomenon. Inside the biofilms bacteria have mutual cooperation, and they may be vulnerable to adverse ecological situations. Biofilm acts as a favoured state of survival because bacterial colony provides defence and various mechanisms of existence and increases its robustness. Bacterial cells also secure their right to use resources and habitats that need significant mass and cannot efficiently be consumed by planktonic cells (Monds and O'Toole 2009). Attainment of novel genetic traits, nutrient accessibility, metabolic collaboration and tolerance to high levels of toxicants has also been recommended as ways to enhance community existence in biofilms (Lopez et al. 2010; Buchholz et al. 2010).

Among the plant root-connected bacteria, the aerobic endospore-developing bacteria, mostly those associated to *Bacillus* and related genera, which are universally present in agronomical structure due to the presence of multiple-layer cell wall structure, are potential to develop resistant endospores and to generate an extensive diversity of antibiotic compounds. Utilizing these capabilities, the microorganisms can colonize different habitats in agroecology and can displace other microbes in rhizosphere/rhizoplane. Consequently, the colonization habitats for the microbes are additionally reproducibly constant, and these bacteria are likely to be used in accuracy organization of agroecosystems. For instance, it was revealed that an endospore-developing species *Paenibacillus polymyxa* inhabits as biofilms in rhizoplane (Timmusk et al. 2005). The microbial biofilms can shield plants not only against pathogens as well as against abiotic stress environments (Milošević et al. 2012; Timmusk et al. 2013).

7.5 Functional Diversity of Rhizobacteria and Their Biofilm

In the rhizosphere, ecological limits are common, and among the concentrations of different microbial species which exists in the soil, it is significant to comprehend how polymicrobial communications are implicated in plant growth and development during the creation of biofilms. On the other hand, biofilm formation by plant pathogens has been investigated *in vitro* and on plant surfaces as described by Guimaraes et al. (2011), Fuente et al. (2013) and Zimaro et al. (2014) and is considered as a significant trait of pathogens.

The ability to form biofilm by a large number of nitrogen-fixing rhizobacteria both symbiotic (*Rhizobium alamii*, *R. leguminosarum bv. viciae* 3841, *R. leguminosarum*, *Rhizobium* sp. NGR234, *Rhizobium*, *Sinorhizobium*) and non-symbiotic nitrogen fixers (*Azospirillum brasilense*, *Azorhizobium caulinodans*, *Azotobacter chroococcum*) have been documented by several scientists (Kumar et al. 2007; Shelud'ko et al. 2010; Krysciak et al. 2011; Robledo et al. 2012). Similarly biocontrol microbes such as the species of *Bacillus* and *Pseudomonas* were also documented to form biofilm by a number of workers (Beauregard et al. 2013; Yasmin et al. 2014) that protect the plant from pathogens by biocontrol mechanism in a biofilm mode of growth. Moreover, several other plant growth-promoting rhizobacteria like *Burkholderia cepacia*, *Enterobacter agglomerans*, *Klebsiella pneumoniae*, *Micrococcus* sp. and *Pantoea agglomerans* were also investigated for biofilm formation and their role in plant growth and development (Ji et al. 2010; Liu et al. 2011). Therefore, the recent trends on plant-associated biofilm indicated a rising trend of research in this area and resulted in their understanding.

7.6 Cell-to-Cell Communications and Root Exudates in PGPR Biofilm

Plant-associated bacteria use small signal molecules for cell-to-cell communications called as quorum sensing (QS). The term quorum sensing explains the circumstances where microbes are able to detect and respond to auto-produced signal compounds to accommodate their behaviour in reply to their community size (Fuqua et al. 1994). The capability of microbes to talk and act as a set for community communications like a multi-cellular creature has contributed important support to bacteria in the form of protection against contenders, colonization of host, formation of biofilms and adjustment to altering environmental conditions (Li and Tian 2012). Generally, bacteria elicit quorum sensing only when they acquire a certain level of colony size, after which the target gene is either activated or suppressed. The requirement of a minimum concentration of bacterial cells to enhance plant growth robustly supports the notion that AHL-mediated QS by microbes has a significant role in plant–rhizobacteria communications (Hartmann et al. 2014).

Since the detection of a diversity of QS signal molecules, *N*-acyl-homoserine lactones (AHL) have been explored extensively and have been demonstrated in regulation of different characters, for instance, biofilm development, bioluminescence, conjugation, motility, production of antibiotics, toxins, symbiosis, siderophore production and virulence (Williams 2007; Barriuso et al. 2008). Within biofilms microbes can reply to QS-like compounds generated by other rhizospheric bacteria and through plants and can obliterate the QS molecules formed by other bacterial species (Dong et al. 2002). Elasri et al. (2001) accounted that majority of plant-associated bacteria produce AHL molecules. Role of QS in plant health protection by PGPR has also been reviewed by other worker (Ahmad et al. 2008). Majority of the strains of genera *Agrobacterium*, *Bacillus*, *Erwinia*, *Pantoea*, *Pseudomonas* and *Rhizobium* produce detectable limits of AHL molecules compared to *Xanthomonas* where only few isolates were found to produce few AHL molecules as reported by Suppiger et al. (2013). Direct task of AHL-mediated QS in biofilm development by plant growth-promoting rhizobacteria like *Burkholderia*, *Pseudomonas*, *Serratia*, *Sinorhizobium*, *Pantoea* and *Bacillus* has been reported by Morohoshi et al. (2007) and Beauregard et al. (2013).

It is now commonly believed that root exudates participate in important functions in plant–microbe communications (Bais et al. 2004; Sun et al. 2012). Root exudates not only provide nutrition to rhizospheric microorganisms but also work as signal in attracting and repelling soil microorganisms (Badri and Vivanco 2009). The content of plant root discharge depends on plant type, species, age and environmental factors. Moreover, the composition of root discharge participates directly in selection of microbial populations from the pool of soil microbiota (Hartmann et al. 2009; Schnitzer et al. 2011).

Rhizospheric microenvironments are largely influenced by the presence of plant root exudates (RE). For example, carbon-rich photosynthates in RE help in survival of large microbial populations and activity in rhizosphere, compared to bulk soil. The destiny of rhizosphere–microbe interactions are largely determined by plant root exudates and their composition (Fan et al. 2012). Generally, most of the root exudates contain free sugars (e.g. glucose, sucrose), amino acids (e.g. glycine, glutamate), organic acids (e.g. citrate, malate and oxalate), fatty acids (e.g. linoleic, linolenic), sterols (e.g. campesterol, cholesterol), vitamins (e.g. p-aminobenzoic acid, biotin), enzymes (e.g. amylase, invertase), flavonones and nucleotides (e.g. adenine, flavonone) and miscellaneous and inorganic compounds (e.g. auxins, scopoletin, hydrocyanic acid) (Curl and Truelove 1986; Uren 2001; Dakora and Phillips 2002; Jones et al. 2009). Chen et al. (2012) highlighted the deciding task that the organic acids showed in plant–microbe interactions. Rudrappa et al. (2008) confirmed that L-malic acid as root exudates from plant roots exclusively invites

Bacillus subtilis FB17. Similarly Ling et al. (2011) also establish the function of watermelon root discharge in recruitment of *Paenibacillus polymyxa* SQR-21.

PGPR are capable of establishing thick biofilm on the surface of root and can respond to root exudates that help in aggregation and formation of stable biofilm (Walker et al. 2004). Several authors documented the direct role of root exudates in biofilm formation by PGPR. Fan et al. (2012) reported that two genes *ycmA* and *luxS* of *Bacillus amyloliquefaciens* responsible for biofilm formation were enhanced by maize root exudates. Comparable incident was described by Espinosa-Urge et al. (2002) and Zhang et al. (2014) for *Pseudomonas putida* and *Bacillus amyloliquefaciens* SQR9.

Root exudates/seedling extracts have been documented in inhibiting quorum sensing of bacteria (Teplitski et al. 2000; Fatima et al. 2010). QS in *Pseudomonas* and other bacteria fully or partially regulate biofilm formation (Li and Tian 2012).

7.7 Understanding Polymicrobial Biofilm

Majority of biofilm-related studies have taken radical approach, where only single-species biofilm were widely studied. However, biofilms in nature exist mainly as multispecies biofilm, where interspecies communications are capable to support the growth, configuration and behaviour of these populations compared to single-species biofilms (Lee et al. 2014). Solitary benefit of population supported biofilms is the conservation of the rhizosphere community itself. Whether sustaining specific plant growth-encouraging bacteria or thwarting pathogens from colonizing plant roots, polymicrobial biofilms offer benefits to the roots of plants that they inhabit. How distant microbial species talk to start and arrange biofilms or manipulate gene expression in other species is hardly inferred (Angus and Hirsch 2013). Lee et al. (2014) developed a model of polymicrobial biofilm containing *Pseudomonas aeruginosa*, *Pseudomonas protegens* and *Klebsiella pneumoniae* and investigated how interspecies communication influences biofilm growth, composition and shock reactions. Elias and Banin (2012) documented that multispecies biofilm cooperation involved cell–cell communications via quorum sensing, metabolic collaboration and competition. He also reported that communications within biofilm species can be antagonistic or synergistic. The synergistic interactions encourage biofilm formation, metabolic cooperation and increased resistance to antibiotics.

7.8 PGPR Biofilm in Plant Disease Suppression

Efficient colonization and biofilm development on plant root surfaces determine the biocontrol potential of PGPR. Efficiency of biocontrol agents depends upon successful colonization of plant surface that plays a crucial role in increasing plant growth. Plant root-associated biofilm can secure the habitation site and work as a reservoir for supply of food in the rhizosphere, therefore minimizing the accessibility of root exudates and other nourishing ingredients for pathogens and their subsequent adherence to root surface (Weller and Thomashow 1994, Bais et al. 2004). Plant growth-promoting rhizobacteria can enhance the growth and yield by several mechanisms such as better mineral nutrient absorption, phytohormone manufacturing and biocontrol mechanism (Trivedi et al. 2011). *Bacillus subtilis*, universally present in soil, is able to increase plant growth, provide protection against fungal pathogen assault and participate in the deterioration of organic polymers in the soil (Vlamakis et al. 2013). Lugtenberg and Kamilova (2009) also demonstrated that *B. subtilis* is used as biocontrol agent against various phytopathogens. Beauregard et al. (2013) analysed *Arabidopsis* root surfaces treated with *B. subtilis* using confocal laser scanning microscopy to disclose a three-sided configuration of *B. subtilis* biofilm. Similarly root-associated pseudomonads have been evaluated comprehensively, and several of them help in the development of host plants or are exploited as biocontrol agents. *Pseudomonas fluorescens* is capable to react quickly to the existence of root discharge in soils, congregating at root inhabitation area, and can set up consistent biofilm system (Couillerot et al. 2009). Haggag and Timmusk (2008) and Chen et al. (2012) explored the task of biofilm-forming *Paenibacillus polymyxa* and *Bacillus subtilis* strains in restricting *Aspergillus niger* and *Ralstonia solanacearum*, respectively, and displayed the significance of biofilms in biocontrol. Similar result was also reported by Dietel et al. (2013) for *Bacillus amyloliquefaciens* FZB42. To make biocontrol efficient and reproducible, competent inhabitation and biofilm formation by means of the biocontrol agent should be established.

7.9 Role of Biofilm in Reducing Toxicity and Abiotic Stress Conditions to Plants and Microbes

Bioremediation employs microorganisms to eliminate, detoxify or immobilize pollutants and eliminates the needs that do not involve adding of dangerous substances. Bioremediation is mainly appropriate for big area wherever pollutant concentration is relatively small and the hydrology of the soil not able to sustain chemical remediation (Pastorella et al. 2012). Biofilm-associated bioremediation confers a skilful and secure choice to bioremediation with free-living microbes since cells in a biofilm can survive and adjust easily to harsh environment as they got protection from EPS (Decho 2000; Hall-Stoodley et al. 2004; Flemming and Wingender 2010). Buchholz et al. (2010) reported that biofilm can tolerate 1,000 times more toxicants compared to their planktonic counterparts. Moreover, the EPS matrix produced by biofilm can help in biodegradation of organic pollutants and heavy metals in less toxic form (Flemming and Wingender 2010). Pal and Paul (2008) and Morel et al. (2009) found that *Stenotrophomonas* biofilms grown with chromium (VI) produce more EPS compared to control biofilms. Recently, Ivanova et al. (2015) found that a consortium of *Rhodococcus erythropolis* S26, *Acinetobacter baumannii* 1B, *Acinetobacter baumannii* 7 and *Pseudomonas putida* F701 was successful in the degradation of oil and efficiently colonizes the roots of barley.

Timmusk et al. (2013) reported that biofilm improved soil collection, developed water stability and increased microbial biomass which consecutively encourage root discharge under stressful conditions. Therefore, there is a tough selective benefit for the secretion of a mucilaginous deposit of extracellular matrix in the rhizosphere, especially under demanding situations. The medium might support to involuntary strength of the biofilm and cooperate among other macromolecules and low molecular mass solutes, providing a large number of microenvironments inside the biofilm.

Survival of agriculturally important microorganisms in rhizosphere under various stress full conditions is an interesting part of investigation, straightforwardly affecting our food safety (Angus and Hirsch 2013). Plant growth-promoting rhizobacteria (PGPR) alleviate most efficiently the influence of abiotic pressure (famine, frostiness, salinity, metal toxicity and intense heat) on plants throughout biofilm formation, which under normal conditions enhance plant development and under traumatic situations help in better survival (Milošević et al. 2012; Bogino et al. 2013). Vanderlinde et al. (2009) reported that the LPS mutant of *Rhizobium leguminosarum* with no biofilm-forming capacity was unable to tolerate the draught conditions. Sandhya et al. (2009) also found that colonization and biofilm formation ability of *Pseudomonas putida* strain GAP-P45 help in mitigation of drought stress effects. Similar results were reported by Timmusk and Nevo (2011). Qurashi and Sabri (2012) reported that two salt tolerant isolates of *Halomonas variabilis* (HT1) and *Planococcus rifietoensis* (RT4) can stimulate chickpea growth through biofilm formation and EPS production. Timmusk et al. (2014) found that application of rhizobacterial isolates can mitigate drought stress in *Triticum aestivum* through biofilm formation and alginate production.

7.10 Conclusion and Future Prospects

In the past decade, considerable amount of research has been carried out on plant–microbe interaction highlighting role of biofilm. Microbes receive various benefits by forming biofilms, along with defence from predation, drought, exposure to antibiotics and better acquirement of nutrients present in the rhizosphere. Biofilms offer better survival equally to beneficial plant growth-promoting rhizobacteria and opportunistic phytopathogens. The biofertilizer industry also gains by developing biofilmed-based PGPR. Interdisciplinary investigations by means of novel advancement will elucidate the approaches in which bacteria move and intermingle in a diversity of surface microenvironments throughout biofilm growth. These kinds of studies are necessary to exploit the potential of biofilm in enhancing plant growth and to fight crop diseases caused by bacteria. Further molecular plant–microbe interactions are required that focus on multispecies biofilm because single-species communications do not reveal the real picture of natural phenomenon. It is expected that better molecular understanding of complex interaction of bacteria with plant root will help in ecological engineering in favour of plant growth promotion and protection in the future.

References

Adam B, Baillie GS, Douglas LJ (2002) Mixed species biofilms of *Candida albicans* and *Staphylococcus epidermidis*. J Med Microbiol 51:344–349

Ahmad I, Khan MSA (2012) Microscopy in mycological research with especial reference to ultra-structures and biofilm studies. In: Méndez-Vilas A (ed) Current microscopy contributions to advances in science and technology. Formatex Research Center, Badajoz, pp 646–659

Ahmad I, Aqil F, Ahmad F, Zahin M, Musarrat J (2008) Quorum sensing in bacteria: potential in plant health protection. In: Ahmad I, Hayat S, Pichtel J (eds) Plant-bacteria interactions. Wiley, Weinheim, pp 129–153

Allan-Wojtas P, Hildebrand PD, Braun PG, Smith-King HL, Carbyn S, Renderos WE (2010) Low temperature and anhydrous electron microscopy techniques to observe the infection process of the bacterial pathogen *Xanthomonas fragariae* on strawberry leaves. J Microsc 239:249–258

Almeida C, Azevedo NF, Santos S, Keevil CW, Vieira MJ (2011) Discriminating multi-species populations in biofilms with peptide nucleic acid fluorescence in situ hybridization (PNA FISH). PLoS One 6(3):e14786

Altaf MM, Ahmad I (2016) Plant growth promoting activities, biofilm formation and root colonization by Bacillus sp. isolated from rhizospheric soils. J Pure Appl Microbio10:109–120

Angus AA, Hirsch AM (2013) Biofilm formation in the rhizosphere: multispecies interactions and implications for plant growth. In: de Bruijn FJ (ed) Molecular microbial ecology of the rhizosphere. Wiley, Hoboken, pp 703–712

Badri DV, Vivanco JM (2009) Regulation and function of root exudates. Plant Cell Environ 32(6):666–681

Bais HP, Park SW, Weir TL, Callaway RM, Vivanco JM (2004) How plants communicate using the underground information superhighway. Trends Plant Sci 9:26–32

Barriuso J, Solano BR, Lucas JA, Lobo AP, Garica-Villaraco A, Gutierrez Manero FJ (2008) Ecology, genetic diversity and screening strategies of plant growth promoting rhizobacteria (PGPR). In: Ahmad I, Pichtel J, Hayat S (eds) Plant-bacteria interaction, strategies and techniques to promote plant growth. Wiley, Weinheim, pp 1–13

Beauregard PB, Chai Y, Vlamakis H, Losick R, Kolter R (2013) *Bacillus subtilis* biofilm induction by plant polysaccharides. Proc Natl Acad Sci U S A 110:E1621–E1630

Behrens S, Kappler A, Obst M (2012) Linking environmental processes to the *in situ* functioning of microorganisms by high-resolution secondary ion mass spectrometry (NanoSIMS) and scanning transmission X-ray microscopy (STXM). Environ Microbiol 14:2851–2869

Bogino P, Abod A, Nievas F, Giordano W (2013) Water-limiting conditions alter the structure and biofilm-forming ability of bacterial multispecies communities in the alfalfa rhizosphere. PLoS One 8(11):e79614

Boulos L, Prévost M, Barbeau B, Coallier J, Desjardins R (1999) LIVE/DEAD BacLight: application of a new rapid staining method for direct enumeration of viable and total bacteria in drinking water. J Microbiol Methods 37:77–86

Branda SS, Chu F, Kearns DB, Losick R, Kolter R (2006) A major protein component of the *Bacillus subtilis* biofilm matrix. Mol Microbiol 59:1229–1238

Buchholz F, Wolf A, Lerchner J, Mertens F, Harms H, Maskow T (2010) Fast and reliable evaluation of bactericidal and bacteriostatic treatment of biofilms using chip calorimetry. Antimicrob Agents Chemother 54(1):312–319

Chavant P, Gaillard-Martinie B, Talon R, Hébraud M, Bernardi T (2007) A new device for rapid evaluation of biofilm formation potential by bacteria. J Microbiol Methods 68(3):605–612

Chen Y, Cao S, Chai Y, Clardy J, Kolter R, Guo J, Losick R (2012) A *Bacillus subtilis* sensor kinase involved in triggering biofilm formation on the roots of tomato plants. Mol Microbiol 85:418–430

Christensen GD, Simpson WA, Younger JJ, Baddour LM, Barrett FF, Melton DM, Beachey EH (1985) Adherence of coagulase-negative staphylococci to plastic tissue culture plates: a quantitative model for adherence of staphylococci to medical devices. J Clin Microbiol 22(6):996–1006

Compant S, Clément C, Sessitsch A (2010) Plant growth-promoting bacteria in the rhizo- and endosphere of plants: their role, colonization, mechanisms involved and prospects for utilization. Soil Biol Biochem 42:669–678

Costerton JW, Geesey GG, Cheng KJ (1978) How bacteria stick. Sci Am 238:86–95

Costerton JW, Lewandowski Z, Caldwell D, Korber DR, Lappinscott HM (1995) Microbial biofilms. Annu Rev Microbiol 49:711–745

Couillerot O, Prigent-Combaret C, Caballero-Mellado J, Moenne-Loccoz Y (2009) *Pseudomonas fluorescens* and closely-related fluorescent pseudomonads as biocontrol gents of soil-borne phytopathogens. Lett Appl Microbiol 48:505–512

Craveiro S, Alves-Barroco C, Barreto Crespo MT, Salvador Barreto A, Semedo-Lemsaddek T (2015) *Aeromonas* biofilm on stainless steel: efficiency of commonly used disinfectants. Int Food Sci Technol 50:851–856

Crémet L, Corvec S, Batard E, Auger M, Lopez I, Pagniez F, Dauvergne S, Caroff N (2013) Comparison of three methods to study biofilm formation by clinical strains of *Escherichia coli*. Diagn Microbiol Infect Dis 75:252–255

Curl EA, Truelove B (1986) The Rhizosphere, Advanced series in agricultural sciences, vol 15. Springer, New York

Dakora FD, Phillips DA (2002) Root exudates as mediators of mineral acquisition in low-nutrient environments. Plant Soil 245:35–47

Danhorn T, Fuqua C (2007) Biofilm formation by plant associated bacteria. Annu Rev Microbiol 61:401–422

Decho AW (2000) Microbial biofilms in intertidal systems: an overview. Cont Shelf Res 20:1257–1273

Dietel K, Beator B, Budiharjo A, Fan B, Borriss R (2013) Bacterial traits involved in colonization of *Arabidopsis thaliana* roots by *Bacillus amyloliquefaciens* FZB42. Plant Pathol J 29:59–66

Dong YH, Gusti AR, Zhang Q, Xu JL, Zhang LH (2002) Identification of quorum quenching N-acyl homoserine lactonases from *Bacillus* species. Appl Environ Microbiol 68:1754–1759

Dong J, Signo KSL, Vanderlinde EM, Yost CK, Dahms TES (2011) Atomic force microscopy of a ctpA mutant in *Rhizobium leguminosarum* reveals surface defects linking CtpA function to biofilm formation. Microbiology 157:3049–3058

Dynes JJ, Tyliszczak T, Araki T, Lawrence JR, Swerhone GDW (2006) Speciation and quantitative mapping of metal species in microbial biofilms using scanning transmission X-ray microscopy. Environ Sci Technol 40:1556–1565

Elasri M, Delorme S, Lemanceau P, Stewart G, Laue B, Glickmann E, Oger PM, Dessaux Y (2001) Acylhomoserine lactone production is more common among plant associated *Pseudomonas* spp. than among soil borne *Pseudomonas* spp. Appl Environ Microbiol 67:1198–1209

Elias S, Banin E (2012) Multi-species biofilms: living with friendly neighbors. FEMS Microbiol Rev 36(5):990–1004

Espinosa-Urgel M, Kolter R, Ramos JL (2002) Root colonization by *Pseudomonas putida*: love at first sight. Microbiology 148(2):341–343

Fan B, Cravalhais LC, Becker A, Fedoseyenko D, Von Wiren N, Borriss R (2012) Transcriptomic profiling of *Bacillus amyloliquefaciens* FZB42 in response to maize root exudates. BMC Microbiol 12:116

Fatima Q, Zahin M, Khan MSA, Ahmad I (2010) Modulation of quorum sensing controlled behavior of bacteria by growing seedling, seed and seedling extracts of leguminous plants. Indian J Microbiol 50:238–242

Flemming HC, Wingender J (2010) The biofilm matrix. Nat Rev Microbiol 8:623–633

Freeman J, Falkiner FR, Keane CT (1989) New method for detecting slime production by coagulase negative *staphylococci*. J Clin Pathol 42:872–874

Fuente LDL, Parker JK, Oliver JE, Granger S, Brannen PM, Santen EV, Cobine PA (2013) The bacterial pathogen *Xylella fastidiosa* affects the leaf ionome of plant hosts during infection. PLoS One 8(5):e62945

Fujishige NA, Kapadia NN, Hirsch AM (2006) A feeling for the microorganism: structure on a small scale. Biofilms on plant roots. Bot J Linn Soc 150:79–88

Fuqua WC, Winans SC, Greenberg EP (1994) Quorum sensing in bacteria: the LuxR–LuxI family of cell density-responsive transcriptional regulators. J Bacteriol 176:269–275

Geesey GG, Richardson WT, Yeomans HG, Irvin RT, Costerton JW (1977) Microscopic examination of natural sessile bacterial populations from an alpine stream. Can J Microbiol 23:1733–1736

Gorman SP, Adair CG, Mawhinney WM (1994) Incidence and nature of peritoneal catheter biofilm determined by electron and confocal laser scanning microscopy. Epidemiol Infect 112:551–559

Guimaraes BG, Barbosa RL, Soprano AS, Campos BM, De Souza TA, Tonoli CC, Leme AF, Murakami MT, Benedetti CE (2011) Plant pathogenic bacteria utilize biofilm growth-associated repressor (BigR), a novel winged-helix redox switch, to control hydrogen sulfide detoxification under hypoxia. J Biol Chem 286:26148–26157

Haggag WM, Timmusk S (2008) Colonization of peanut roots by biofilm forming *Paenibacillus polymyxa* initiates biocontrol against crown rot disease. J Appl Microbiol 104(4):961–969

Hall-Stoodley L, Costerton JW, Stoodley P (2004) Bacterial biofilms: from the natural environment to infectious diseases. Nat Rev Microbiol 2:95–108

Hartmann A, Schmid M, Van Tuinen D, Berg G (2009) Plant-driven selection of microbes. Plant Soil 321:235–257

Hartmann A, Rothballer M, Hense BA, Schröder P (2014) Bacterial quorum sensing compounds are important modulators of microbe-plant interactions. Front Plant Sci 5:131

Hinsinger P, Bengough AG, Vetterlein D, Young IM (2009) Rhizosphere: biophysics, biogeochemistry and ecological relevance. Plant Soil 321:117–152

Honraet K, Goetghebeur E, Nelis HJ (2005) Comparison of three assays for the quantification of *Candida* biomass in suspension and CDC reactor grown biofilms. J Microbiol Methods 63:287–295

Ivanova AA, Vetrova AA, Filonov AE, Boronin AM (2015) Oil biodegradation by microbial-plant associations. Appl Biochem Microbiol 51(2):196–201

Janczarek M, Rachwał K, Cieśla J, Ginalska G, Bieganowski A (2015) Production of exopolysaccharide by *Rhizobium leguminosarum* bv. *trifolii* and its role in bacterial attachment and surface properties. Plant Soil 388:211–227

Ji X, Lu G, Gai Y, Gao H, Lu B, Kong B, Mu Z (2010) Colonization of *Morus alba* L. by the plant-growth-promoting and antagonist bacterium *Burkholderia cepacia* strain Lu10-1. BMC Microbiol 10:243–254

Jones DL, Nguyen C, Finlay RD (2009) Carbon flow in the rhizosphere: carbon trading at the soil-root interface. Plant Soil 321:5–33

Kaiser TDL, Pereira EM, Dos Santos KRN, Maciel ELN, Schuenck RP, Nunes APF (2013) Modification of the Congo red agar method to detect biofilm production by *Staphylococcus epidermidis*. Diagn Microbiol Infect Dis 75:235–239

Karcz J, Bernas T, Nowak A, Talik E, Woznica A (2012) Application of lyophilization to prepare the nitrifying bacterial biofilm for imaging with scanning electron microscopy. Scanning 34:26–36

Kolodkin-Gal I, Romero D, Cao S, Clardy J, Kolter R, Losick R (2010) D-amino acids trigger biofilm disassembly. Science 328:627–629

Krysciak D, Schmeisser C, Preuß S, Riethausen J, Quitschau M, Grond S, Streit WR (2011) Involvement of multiple loci in quorum quenching of autoinducer I molecules in the nitrogen-fixing symbiont *Rhizobium* (*Sinorhizobium*) sp. strain NGR234. Appl Environ Microbiol 77:5089–5099

Kumar R, Bhatia R, Kukreja K, Behl RK, Dudeja SS, Narula N (2007) Establishment of *Azotobacter* on plant roots: chemotactic response, development and analysis of root exudates of cotton (*Gossypium hirsutum* L.) and wheat (*Triticum aestivum* L.) J Basic Microbiol 47:436–439

Kumar AS, Lakshmanan V, Caplan JL, Powell D, Czymmek KJ, Levia DF, Bais HP (2012) Rhizobacteria *Bacillus subtilis* restricts foliar pathogen entry through stomata. Plant J 72:694–706

Lee KWK, Periasamy S, Mukherjee M, Xie C, Kjelleberg S, Rice SA (2014) Biofilm development and enhanced stress resistance of a model, mixed-species community biofilm. ISME J 8:894–907

Leuko S, Legat A, Fendrihan S, Stan-Lotter H (2004) Evaluation of the LIVE/DEAD BacLight kit for extremophilic archaea and environmental hypersaline samples. Appl Environ Microbiol 70:6884–6886

Li YH, Tian X (2012) Quorum sensing and bacterial social interactions in biofilms. Sensors 12:2519–2538

Ling N, Raza W, Ma J, Huang Q, Shen Q (2011) Identification and role of organic acids in watermelon root exudates for recruiting *Paenibacillus polymyxa* SQR-21 in the rhizosphere. Eur J Soil Biol 47:374–379

Liu Y, Wang H, Sun X, Yang H, Wang Y, Song W (2011) Study on mechanisms of colonization of nitrogen-fixing PGPB, *Klebsiella pneumoniae* NG14 on the root surface of rice and the formation of biofilm. Curr Microbiol 62(4):1113–1122

López D, Vlamakis H, Losick R, Kolter R (2009) Cannibalism enhances biofilm development in *Bacillus subtilis*. Mol Microbiol 74(3):609–618

Lopez D, Vlamakis H, Kolter R (2010) Biofilms. Cold Spring Harb Perspect Biol 2:a000398

Lugtenberg B (2015) Principles of plant-microbe interactions: microbes for sustainable agriculture. Springer, Gewerbestrasse

Lugtenberg B, Kamilova F (2009) Plant-growth-promoting rhizobacteria. Annu Rev Microbiol 63:541–556

Milošević NA, Marinković JB, Tintor BB (2012) Mitigating abiotic stress in crop plants by microorganisms. Proc Natl Sci Matica Srpska Novi Sad 123:17–26

Monds RD, O'Toole GA (2009) The developmental model of microbial biofilms: ten years of a paradigm up for review. Trends Microbiol 17:73–87

Morel MA, Ubalde MC, Olivera-Bravo S, Callejas C, Gill PR, Castro-Sowinski S (2009) Cellular and biochemical response to Cr(VI) in *Stenotrophomonas* sp. FEMS Microbiol Lett 291(2):162–168

Morohoshi T, Nakamura Y, Yamazaki G, Ishida A, Kato N, Ikeda T (2007) The plant pathogen *Pantoea ananatis* produces *N*-Acylhomoserinelactone and causes center rot disease of onion by quorum sensing. J Bacteriol 189(22):8333–8338

Morris CE, Monier JM (2003) The ecological significance of biofilm formation by plant-associated bacteria. Annu Rev Phytopathol 41:429–453

Nongkhlaw FMW, Joshi SR (2014) Distribution pattern analysis of epiphytic bacteria on ethnomedicinal plant surfaces: a micrographical and molecular approach. J Microsc Ultrastruct 2:34–40

Nweze EI, Ghannoum A, Chandra J, Ghannoum MA, Mukherjee PK (2012) Development of a 96-well catheter-based microdilution method to test antifungal susceptibility of *Candida* biofilms. J Antimicrob Chemother 67:149–153

O'Toole GA, Kolter R (1998) Initiation of biofilm formation in *Pseudomonas fluorescens* WCS365 proceeds via multiple, convergent signaling pathways: a genetic analysis. Mol Microbiol 28:449–461

O'Toole G, Kaplan HB, Kolter R (2000) Biofilm formation as microbial development. Annu Rev Microbiol 54:49–79

Otto M (2013) Staphylococcal infections: mechanisms of biofilm maturation and detachment as critical determinants of pathogenicity. Annu Rev Med 64:175–188

Pal A, Paul AK (2008) Microbial extracellular polymeric substances: central elements in heavy metal bioremediation. Indian J Microbiol 48(1):49–64

Pantanella F, Valenti P, Frioni A, Natalizi T, Coltella L, Berlutti F (2008) BioTimer assay, a new method for counting *Staphylococcus* spp. in biofilm without sample manipulation applied to evaluate antibiotic susceptibility of biofilm. J Microbiol Methods 75(3):478–484

Pantanella F, Valenti P, Natalizi T, Passeri D, Berlutti F (2013) Analytical techniques to study microbial biofilm on abiotic surfaces: pros and cons of the main techniques currently in use. Ann Ig 25:31–42

Pastorella G, Gazzola G, Guadarrama S, Marsili E (2012) Biofilms: applications in bioremediation. In: Lear G, Lewis GD (eds) Microbial biofilms: current research and applications. Caister Academic Press, Norfolk, pp 73–98

Peters BM, Ward RM, Rane HS, Lee SA, Noverr MC (2013) Efficacy of ethanol against *Candida albicans* and *Staphylococcus aureus* polymicrobial biofilms. Antimicrob Agents Chemother 57:74–82

Qurashi AW, Sabri AN (2012) Bacterial exopolysaccharide and biofilm formation stimulate chickpea growth and soil aggregation under salt stress. Braz J Microbiol 43(3):1183–1191

Ramage G, Vande Walle K, Wickes BL, Lopez-Ribot JL (2001) Standardized method for *in vitro* antifungal susceptibility testing of *Candida albicans* biofilms. Antimicrob Agents Chemother 45:2475–2479

Robledo M, Rivera L, Jiménez-Zurdo JI, Rivas R, Dazzo F, Velázquez E, Molina M, Hirsch AM, Mateos PF (2012) Role of *Rhizobium* endoglucanase CelC2 in cellulose biosynthesis and biofilm formation on plant roots and abiotic surfaces. Microb Cell Factories 11:125

Rudrappa T, Czymmek KJ, Paré PW, Bais HP (2008) Root-secreted malic acid recruits beneficial soil bacteria. Plant Physiol 148:1547–1556

Sandberg ME, Schellmann D, Brunhofer G, Erker T, Busygin I, Leino R, Vuorela PM, Fallarero A (2009) Pros and cons of using resazurin staining for quantification of viable *Staphylococcus aureus* biofilms in a screening assay. J Microbiol Methods 78:104–106

Sandhya V, Ali SKZ, Grover M, Reddy G, Venkateswarlu B (2009) Alleviation of drought stress effects in sunflower seedlings by the exopolysaccharides producing *Pseudomonas putida* strain GAP-P45. Biol Fert Soils 46:17–26

Schnitzer SA, Klironomos JN, Hillerislambers J, Kinkel LL, Reich PB, Xiao K, Rillig MC, Sikes BA, Callaway RM, Mangan SA, van Nes EH, Scheffer M (2011) Soil microbes drive the classic plant diversity-productivity pattern. Ecology 92:296–303

Shelud'ko AV, Shirokov AA, Sokolova MK, Sokolov OI, Petrova LP, Matora LY, Katsy EI (2010) Wheat root colonization by *Azospirillum brasilense* strains with different motility. Microbiology 9(5):688–695

Sun S, Wang J, Zhu L, Liao D, Gu M, Ren L, Kapulnik Y, Xu G (2012) An active factor from tomato root exudates plays an important role in efficient establishment of mycorrhizal symbiosis. PLoS One 7(8):e43385

Suppiger A, Schmid N, Aguilar C, Pessi G, Eberl L (2013) Two quorum sensing systems control biofilm formation and virulence in members of the *Burkholderia cepacia* complex. Virulence 4(5):400–409

Tawakoli PN, Al-Ahmad A, Hoth-Hannig W, Hannig M, Hannig C (2013) Comparison of different live/dead stainings for detection and quantification of adherent microorganisms in the initial oral biofilm. Clin Oral Investig 17(3):841–850

Teplitski M, Robinson JB, Bauer WD (2000) Plants secrete substances that mimic bacterial *N*-acyl homoserine lactone signal activities and affect population density-dependent behaviors in associated bacteria. Mol Plant-Microbe Interact 13(6):637–648

Timmusk S, Nevo E (2011) Plant root associated biofilms. In: Maheshwari DK (ed) Bacteria in agrobiology: plant nutrient management. Springer, Berlin, pp 285–300

Timmusk S, Grantcharova N, Wagner EGH (2005) *Paenibacillus polymyxa* invades plant roots and forms biofilms. Appl Environ Microbiol 71:7292–7300

Timmusk S, Paalme V, Pavlicek T, Bergquist J, Vangala A, Danilas T, Nevo E (2011) Bacterial distribution in the rhizosphere of wild barley under contrasting microclimates. PLoS One 6(3):e17968

Timmusk S, Timmusk K, Behers L (2013) Rhizobacterial plant drought stress tolerance enhancement. J Food Security 1:10–16

Timmusk S, Abd El-Daim IA, Copolovici L, Tanilas T, Kännaste A, Behers L, Nevo E, Seisenbaeva G, Stenström E, Niinemets U (2014) Drought-tolerance of wheat improved by rhizosphere bacteria from harsh environments: enhanced biomass production and reduced emissions of stress volatiles. PLoS One 9(5):e96086

Toté K, Vanden Berghe D, Maes L, Cos P (2008) A new colorimetric microtitre model for the detection of *Staphylococcus aureus* biofilms. Lett Appl Microbiol 46:249–254
Trivedi P, Spann TM, Wang N (2011) Isolation and characterization of beneficial bacteria associated with citrus roots in Florida. Microb Ecol 62(2):324–336
Uren NC (2001) Types, amounts, and possible functions of compounds released into the rhizosphere by soil-grown plants. In: Pinton ZVR, Nannipieri P (eds) The rhizosphere: biochemistry and organic substances at the soil-plant interface. Marcel Dekker, New York, pp 19–40
Vanderlinde EM, Muszynski A, Harrison JJ, Koval SF, Foreman DL, Ceri H, Kannenberg EL, Russell W, Carlson RW, Yost CK (2009) *Rhizobium leguminosarum* biovar *viciae* 3841,deficient in 27-hydroxyoctacosanoate-modified lipopolysaccharide, is impaired in desiccation tolerance, biofilm formation and motility. Microbiology 155:3055–3069
Vlamakis H, Chai Y, Beauregard P, Losick R, Kolter R (2013) Sticking together: building a biofilm the *Bacillus subtilis* way. Nat Rev Microbiol 3:157–168
Walker TS, Bais HP, Déziel E, Schweizer HP, Rahme LG, Fall R, Vivanco JM (2004) *Pseudomonas aeruginosa*-plant root interactions. Pathogenicity, biofilm formation, and root exudation. Plant Physiol 134:320–331
Weller DM, Thomashow LS (1994) Current challenges in introducing beneficial microorganisms into the rhizosphere. In: O'Gara F, Dowling DN, Boesten B (eds) Molecular ecology of rhizosphere microorganisms: biotechnology and release of GMOs. VCH, New York, pp 1–18
Williams P (2007) Quorum sensing, communication and cross-kingdom signalling in the bacterial world. Microbiology 153:3923–3938
Wright CJ, Shah MK, Powell LC, Armstrong I (2010) Application of AFM from microbial cell to biofilm. Scanning 32:134–149
Xie Z, Thompson A, Kashleva H, Dongari-Bagtzoglou A (2011) A quantitative real-time RT-PCR assay for mature *C. albicans* biofilms. BMC Microbiol 11:93
Yasmin S, Hafeez FY, Rasul G (2014) Evaluation of *Pseudomonas aeruginosa* Z5 for biocontrol of cotton seedling disease caused by *Fusarium oxysporum*. Biocontrol Sci Tech 24(11):1227–1242
Yuanqing C, Tong Z (2012) Surface-enhanced Raman scattering (SERS) revealing chemical variation during biofilm formation: from initial attachment to mature biofilm. Anal Bioanal Chem 404:1465–1475
Zhang N, Wang D, Liu Y, Li S, Shen Q, Zhang R (2014) Effects of different plant root exudates and their organic acid components on chemotaxis, biofilm formation and colonization by beneficial rhizosphere-associated bacterial strains. Plant Soil 374:689–700
Zimaro T, Thomas L, Marondedze C, Sgro GG, Garofalo CG, Ficarra FA, Gehring C, Ottado J, Gottig N (2014) The type III protein secretion system contributes to *Xanthomonas citri* subsp. *citri* biofilm formation. BMC Microbiol 14:96

8 How Can Bacteria, as an Eco-Friendly Tool, Contribute to Sustainable Tomato Cultivation?

Vivian Jaskiw Szilagyi Zecchin and Átila Francisco Mógor

Abstract

To contribute to the sustainability of tomato production, the use of bacteria capable of promoting plant growth are discussed in this chapter, with a focus on the bacterial modes of action regarding their biofertilizer and phytostimulation abilities. The bacterial effects on tomato plant development stages, from seed to fruit and finally on yield, are also covered. The bacterial abilities for phosphate solubilization, release of phytase, siderophores, ACC deaminase, and plant hormones, are reported. Their effects on seed germination, seedling growth, plant growth in the field, fruit quality, and yield are also characterized. It is concluded that the use of plant growth–promoting bacteria should be an eco-friendly tool that contributes to sustainable tomato cultivation.

8.1 Introduction

Through its use *in natura* or in processed forms as a part of many food cultures, the tomato (*Solanun lycopersicum*) is one of the most consumed products worldwide and, because of this, is intensively cultivated with large amounts of synthetic input. Valenciano and Uribe (2015) reported the impact of fertilizers and agrochemicals (pesticides and plant growth regulators) on tomato production cost, which besides cost implications, pose risks for the environment, consumers, and growers when used excessively (Aktar et al. 2009).

Looking forward, the use of bacteria capable of promoting plant growth arises as an environmental friendly and sustainable tool to get good tomato yields.

V.J.S. Zecchin (✉) • Á.F. Mógor
Federal University of Paraná, Department of Plant Science and Crop Protection, Curitiba, Paraná, Brazil
e-mail: vivian.szilagyi@gmail.com

V. Kumar et al. (eds.), *Probiotics in Agroecosystem*,
DOI 10.1007/978-981-10-4059-7_8

Among the soil living microorganisms, bacteria are the most common. The highest concentrations are found around the roots of plants (rhizosphere) because of the presence of nutrients including sugars, amino acids, organic acids, and other small molecules from plant root exudates. The bacteria that can promote plant growth include those that are free-living and those that form symbiotic relationships with plants either in the rhizosphere or by endophytic colonization (Glick 2012). The difference between these bacteria is they can promote plant growth directly by facilitating nutrient uptake by plants (i.e., siderophores, phosphate solubilization) or by modulating plant hormone levels (i.e., release of indolic compounds) thus improving plant growth, development, and yield (Szilagyi-Zecchin et al. 2016).

Recent reports have indicated the positive effect of various bacterial strains on promoting tomato plant growth and yield. According to Ahirwar et al. (2015), *Pseudomonas fluorescence* (SS5) enhanced the growth of tomato plants with significant increases in root and shoot length, and fruit yield. Almaghrabi et al. (2013) found beneficial effects of *Bacillus* strains on the yield of tomatoes grown in greenhouses. Szilagyi-Zecchin et al. (2015) reported improvement of tomato seedling growth by inoculation with *Bacillus amyloliquefaciens* subsp. *plantarum* FZB42. Abbamondi et al. (2016) investigating the plant growth–promoting abilities of bacterial strains isolated from different tomato cultivars, detected that inoculation increased the surface area of the plant's root system, improving the capacity to acquire nutrients and water.

Among many others, those results indicated that by using such plant growth–promoting bacteria (PGPB) on tomato production systems, it would be possible to lower the fertilizer and pesticide inputs and grow the plants in a more sustainable way. Below we discuss the plant growth–promoting mechanisms related to tomato plants and PGPB interaction, and those effects on growth and tomato yield.

8.2 Action Mode of Plant Growth-Promoting Bacteria

Beneficial bacteria are a powerful tool for promoting plant growth because they have various action modes to accomplish this. The improvement of plant growth follows two main methods, biofertilizers and phytostimulation. The bacterial biofertilizer improvement in growth is achieved by providing plants with better uptake of nutrients such as phosphorus and iron (Neilands 1957), while phytostimulation is related to plant phytohormone modulation by bacteria.

8.2.1 Phosphate Solubilization and Production of Phytase

Phosphorus (P) is a component of many important molecules such as nucleic acids, phospholipids, and ATP (Taiz and Zeiger 2013). The quantity of phosphorus in soil is generally high, but most of this is insoluble and thus plants cannot utilize it (López-Bucio et al. 2002). Phosphorus limitation affects photosynthesis through changes in the activity of Calvin-cycle enzymes, RuBP regeneration, and/or rubisco activity as long as P plays an important regulatory role in starch and sucrose biosynthesis, and is also part of ATP and NADPH/NADP+ (Marschner 2011).

When chemical fertilization is carried out, it typically uses soluble inorganic phosphorus, the major part of which is immobilized after application and thus wasted because it becomes unavailable to plants (Feng et al. 2004). On the other hand, the phytates, an organic P form that is present in great quantity in soils, might be mineralized by phytases (i.e., a group of enzymes capable of releasing phosphates from phytates) and in this way may be ready to use by the plant (Richardson et al. 2001a; Yao et al. 2012).

A viable way to convert the insoluble forms of inorganic P to a form accessible by plants is the use of phosphate-solubilizing bacteria, which release organic acids such as gluconic and citric acid in the rhizosphere (Richardson et al. 2001b). Therefore bacteria that can produce both phytase and organic acids are very useful in providing P to plants with more efficiency. Many bacterial strains have some of these abilities: *Burkholderia cepacia* DA23 isolated from cultivated soil showed solubilization of insoluble inorganic phosphate (Song et al. 2008), and *Advenella* spp., PB-05, PB-06, and PB-10 and *Cellulosimicrobium* sp. PB-09 were positive for phytase and phosphorus solubilization (Singh et al. 2014).

Turan et al. (2007) used five different phosphorus fertilizers (superphosphate, triple superphosphate, di-ammonium phosphate, phosphoric acid, and rock phosphate) in combination with or without the phosphate-solubilizing bacteria *Bacillus* FS-3. They found that in all fertilizer treatments, FS-3 converted approximately 20% of less available phosphorus into labile forms.

8.2.2 Siderophores

According to the classic definition, siderophores (also called chelating agents) are low molecular weight compounds produced by microorganisms, which allow ions to be transported into the cell (Neilands 1957). Many of them can form complexes with elements such as copper, aluminum, and manganese, but mainly with iron (Fe^{3+}) (Benite et al. 2002).

Iron is an essential nutrient for plants, the deficiency of which can affect plant metabolism, because iron is a co-factor in many important enzymes that participate in processes like respiration, photosynthesis, and nitrogen fixation (Marschner 2011).

In several types of soils, plants do not easily assimilate iron (Fe^{+3}). The Fe3+ form is not available for plants because it is insoluble, forming oxides or hydroxides and limiting their bioavailability (Zuo and Zhang 2011).

The benefits of bacterial siderophores for plants are notorious, especially using *Pseudomonas*. Jurkevitch et al. (1992) while studying why plants inoculated with *Pseudomonas putida* had the chlorotic aspect reduced when iron deficiency was induced, described the action of *P. putida* on iron uptake by plant. This was similar to studies with *Pseudomonas fluorescens*, in which the bacteria improved the performance of Arabdopsis plants by increasing their iron content (Vansuyt et al. 2007).

8.2.3 Auxin

Auxins are a class of plant hormones in which indoleacetic acid (IAA) is the main representative. IAA acts in physiological processes like photosynthesis and pigment formation, responses to light and florescence, resistance to stressful conditions, controlling the processes of vegetative growth specifically in cell elongation, and stimulateing root development, initiating lateral and adventitious root formation (Taiz and Zeiger 2013).

The IAA secreted by bacteria acts in conjunction with the plant's endogenous IAA, with plant response varying according to plant species, IAA concentration, tissue sensitivity, and developmental stage of the plant. For example, the optimal level of IAA for supporting root growth is around five orders of magnitude lower than for shoot growth (Taiz and Zeiger 2013).

Besides that, the bacteria can act on signaling to improve or reduce plant endogenous auxin synthesis. In the case of molecular signals involved in communication with host plants, the microbial auxin has a role in interfering with developmental pathways and altering auxin biosynthesis (Contesto et al. 2010).

Studies indicated that the bacteria were acting on the growth of plants through the possible action of auxins. For example: *Bacillus amyloliquefaciens* FZB42 increased the shoot length of tomato seedlings (Szilagyi-Zecchin et al. 2015), and plants inoculated with wild-type *Pseudomonas putida* increased root length compared with both IAA-deficient bacteria mutant and the uninoculated control, thus relating growth promotion to the IAA released by the wild strain (Xie et al. 1996).

8.2.4 ACC Deaminase and Ethylene

The hormone ethylene has a wide range of biological activities in plants: it affects seed germination, promotes root initiation or inhibits root elongation, has a role in leaf abscission, flower wilting, fruit ripening, and activating the synthesis of other plant hormones and enzymes (Taiz and Zeiger 2013).

There is a connection between 1-aminocyclopropane-1-carboxylate (ACC) deaminase and ethylene, by the pathway in the biosynthesis of ethylene. In the first step a reaction occurs when S-adenosyl-methionine is converted to ACC by the ACC synthetase enzyme (ACCS) reaction. Then, in the second step, the ACC is metabolized by ACC oxidase (ACCO), which needs oxygen and iron, wherein it is activated through CO_2 to produce ethylene (Yang and Hoffman 1984).

Bacteria can produce the enzyme ACC deaminase, which degrades ACC to αketobutyric acid (Honma and Shimomura 1978), which in turn has effects on plants linked to the decrease in ethylene rates. The main visible effect of inoculation with bacteria that produce ACC deaminase is the enhancement of plant root elongation, as seen by Xu et al. (2008) in tomato plants.

The capacity of the bacteria *Pseudomonas putida* to reduce the deleterious effect of exogenous IAA was investigated using seedlings, indicating that the roots grown in the presence of increasing concentrations of IAA (0–10 mg/ml^{-1}) were longer when seeds were previously treated with *P. putida*. This reduction in the detrimental effect of IAA on root elongation could be associated with reduced ethylene production resulting from a decrease of its precursorACC by bacterial degradation of IAA in the rhizosphere, and/or by ACC deaminase activity present in bacteria (Gravel et al. 2007).

8.2.5 Gibberellins

Gibberellins (GAs) consist of a group of hormones involved in cell division and elongation; they have a key role in internode elongation, seed germination, pollen tube growth, and flowering (Taiz and Zeiger 2013).

The bacterial biosynthesis of gibberellins can modify the hormonal balance in plants causing structural changes (Morrone et al. 2009). Plants inoculated with *Bradyrhizobium* sp., a GA producing bacteria, increased the internode elongation (Dobert et al. 1992). Plant dwarf phenotypes induced by paclobutrazol (an inhibitor of gibberellin biosynthesis) were reversed by applications of extracts of *Bacillus pumilus* and *Bacillus licheniformis,* and also by exogenous GA, indicating the effect of bacterial GAs (Gutiérrez-Manero et al. 2001). *Bacillus cereus*, *B. macroides*, and *B. pumilus* promote the growth of plants because GAs were detected in the culture broth of those bacteria (Joo et al. 2004).

8.3 Tomato Growth Promotion in Different Stages of Development

8.3.1 Early Stage of Development

The first stage of plant development is seed germination, where many physiological processes are initiated to generate a new plant. In this stage, the bacteria can act on germination and plantlet growth.

On a paper towel germination test, *Pseudomonas oleovorans* strains improved tomato seed germination, while *Agrobacterium tumefaciens* strains displayed higher seedling fresh weight, improving the tomato seedling's vigor index (Thomas and Upreti 2015). Also, *Pseudomonas aeruginosa* and *Bacillus subtilis* showed improvement in tomato germination rates (Adesemoye et al. 2008).

In the same way, the germination percentage of seed, total length, and dry mass of germinated tomato seedlings were increased by 30.2%, 71.1%, and 270.8%, respectively, compared with those of the uninoculated control 7 days after inoculation by *Rhodopseudomonas* sp. (Koh and Song 2007). The strains *Bacillus subtilis* GBO3, *Bacillus amyloliquefaciens* IN937a, and *Brevibacillus brevis* inoculated on tomato seeds showed enhancement in the seed quality parameters like seed germination and seedling vigor (Girish and Umesha 2005). Many of these results on germination and plantlet growth could be related to the capacity of bacteria to release auxin and gibberellins, as discussed above.

In addition, studies of these first stages of plant development focusing on how the physiology of the plants react to inoculation, show that plantlets grown from tomato seeds inoculated with *Azospirillum brasilense* FT326, 15 days after inoculation, had FT362 localized on roots and in xylematic tissue. This promoted increases in shoot and root fresh weight, and in root hair length in the plantlets. The levels of indole-3-acetic acid (IAA) and ethylene were higher in inoculated plants, but exogenously supplied ethylene mimicked the bacteria effect and the addition of an inhibitor of its synthesis or activity, completely blocked FT326 growth promotion. Therefore, the process of growth promotion triggered by *A. brasilense* involves a signaling pathway in which ethylene is a central regulator (Ribaudo et al. 2006).

8.3.2 Seedling Production

The tomato seedling production period is generally 30–40 days after germination, and is well studied because of its commercial importance. Accelerated seedling growth in the nursery contributes to healthier and vigorous seedlings, which in turn facilitates better establishment of plants in the field, that can reflect a better yield at the end of cycle (Thomas and Upreti 2015). The effects of bacteria at this stage can be seen, because bacteria can change many aspects of the primary and secondary metabolism of the plant (Szilagyi-Zecchin et al. 2016).

In this regard, the effects of *Bacillus amyloliquefaciens* subsp. *plantarum* in two concentrations was evaluated on the production of organic seedlings of "Santa Clara I-5300" and "Cereja 261" tomato cultivars through seed inoculations. This strain, isolated from soil (Krebs et al. 1998), was positive for production of indole compounds and siderophores, enhanced the contents of chlorophyll in the seedling leaves, and promoted shoot growth with additions of 47.7% for Santa Clara and 15.5% for Cereja when compared to the control (Szilagyi-Zecchin et al. 2015).

Besides indolic compounds, some bacteria, such as *Promicromonospora* sp. SE188, isolated from soil, can exhibit a phosphate solubilization potential and release physiologically active (GA1 and GA4) and inactive (GA9, GA12, GA19, GA20, GA24, GA34, and GA53) gibberellins (Kang et al. 2012). Hormone regulation by bacteria was also reported when abscisic acid, a plant stress hormone, was significantly down-regulated in the presence of *Promicromonospora* sp. SE188, and contrarily, salicylic acid was significantly higher compared to the controls. In consequence, 4-week-old tomato seedlings inoculated with SE188 showed a significantly higher shoot length, number of leaves and biomass, with roots significantly longer and the secondary and tertiary root formation more prominent in treated plants (Kang et al. 2012).

In addition, some endophytic bacteria isolated from tomato and chili (*Capsicum annuum*) plants (*Bacillus* sp. BETL9, *Serratia marcescens* BECL8, both phosphate solubilizer and *Bacillus pumilus* BETL13, *Bacillus licheniformis* BECS1 and *Bacillus megaterium* BECS7, all siderophores and IAA releasers) were also tested on tomatoes and chilis, in which 3 weeks after their emergence increases were observed in root and shoot length, and in the number of secondary roots (Amaresan et al. 2012).

The bacteria *Sphingomonas* sp. LK11, isolated from the leaves of *Tephrosia apollinea,* release physiologically active gibberellins GA4 and inactive GA9 and GA20, and also produce IAA. When inoculated on tomato seeds 5-week-old plants showed significant increases in shoot length, chlorophyll content, and shoot and root dry weights (Khan et al. 2014).

In the same hormone-regulation way, Onofre-Lemus et al. (2009) observed that *Burkholderia unamae* was able to endophytically colonize tomato and had ACC deaminase activity. Tomato plants inoculated with the wild-type *B. unamae* strain presented better growth that was reflected in higher shoot and root dry weight than in those plants inoculated with a mutant strain deficient for ACC deaminase activity.

8.3.3 Plant Growth and Development

Researchers focusing on plant development after the seedling stage have in part elucidated how bacterial strains act near to the flowering stage and in the beginning of fruiting.

Adesemoye et al. (2008) comparing seed inoculations of *Pseudomonas aeruginosa* and *Bacillus subtilis*, found that the dry biomass of the tomato plants 65 days after sowing was increased around 31 % by both bacteria. According to Koh and Song (2007), *Rhodopseudomonas* sp., a purple nonsulfur bacteria, produced 5.56 mM/min/mg protein and 67.2 μM/min/mg protein of IAA and 5-aminolevulinic acid (ALA), respectively, which may be one of the mechanisms of tomato plant growth enhancement. Tomato plants cv. Río Fuego cultivated in greenhouses and inoculated with *Bacillus subtilis* BEB-lSbs improved the radical system 50 days after transplant, and showed improved root dry weight, and increased root length by 26 % and 15 %, respectively (Mena-Violante & Olalde-Portugal, 2007).

García et al. (2004), inoculating *Bacillus licheniformis* CECT 5106 isolated from alder tree (*Alnus glutinosa*), and applied on two tomato varieties ('Daniela' and 'Brillante'), found that the bacteria increased the height and the leaf area in both cultivars. Also in greenhouse assays, those authors found that the number and diameter of tomatoes produced in sand and in hydroponic medium were increased significantly by inoculation with CECT 5106.

The *Burkholderia tropica* strain MTo-293, isolated from maize stems, able to colonize the root hairs, root tips, lateral root emergence sites, and stomata of tomato leaves, was used in two tomato crop seasons in greenhouse experiments, and showed a consistent increase in both number and weight of fruit. The number of fruit increased at an average of five fruit per m^2 in the first season and 7.6 fruit per m^2 in the second season (Bernabeu et al. 2015).

8.3.4 Yield

The main aim of the use of bacteria is to improve the yield by inoculation, and also to reduce the use of synthetic fertilizers. As was reported above, this aim is possible, and has also been shown to be feasible according to the following studies.

Gram-positive bacteria strains could be used to improve nutrient uptake as they have abilities like endospore formation that can provide more viability to formulate and survive in inhospitable environments such as sandy, saline soils. Inasmuch as Adesemoye et al. (2008), using 75% of the recommended fertilizer rate in association with the Gram-positive bacteria *Bacillus amyloliquefaciens* IN937a and *Bacillus pumilus* T4 inoculation, had tomato yield and nutrient (nitrogen and phosphorus) uptake equivalent to the full fertilizer rate without inoculation. The cultivar Río Fuego had a higher yield per plant (around 23% more) and higher marketable yield when inoculated with *Bacillus subtilis* BEB-lSbs (Mena-Violante and Olalde-Portugal 2007). Also, the application of *B. amyloliquefaciens* increased the yield of the tomato plants by 8–9% (Gül et al. 2008).

It is not only Gram-positive bacteria that have the agronomic abilities to improve yield, some genus Gram-negative bacteria have also been reported. *Pseudomonas putida* B strain 1, IAA producer, was evaluated to determine the promoting effect on the growth of mature healthy tomato plants, cv. Trust F1, under hydroponic conditions. It has also been shown to improve fruit yields in rockwool (688 g per plant) and in organic medium (630 g per plant) (Gravel et al. 2007). Siderophores from *Chryseobacterium* sp. C138 isolated from the rhizosphere of rice (*Oryza sativa*) are effective in supplying Fe to iron-deficient tomato plants of var. Marglobe by the roots on experiments conducted in greenhouses under iron hydroponic conditions. The media free of bacteria and with bacterial cells applied both significantly increased plant yield and chlorophyll and iron content compared with the positive controls with full Hoagland solution (Guerinot and Ying 1994; Radzki et al. 2013).

Conclusion

Plant growth–promoting bacteria are an eco-friendly tool that should contribute to sustainable tomato cultivation through many mechanisms. These include improved germination, increased plantlet and seedling growth, increased plant growth and development in the field, and consequent yield gains. These beneficial effects happen through many means of action that can act together depending on the bacteria strains present. These effects are related to the biofertilizer bacteria releasing siderophores and phosphate solubilization compounds and thus improving nutrient uptake by plants, and also to the phytostimulaton or biostimulation abilities of bacteria, modulating plant growth by release of hormones and enzymes.

References

Abbamondi GR, Tommonaro G, Weyens N, Thijs S, Sillen W, Gkorezis P, Iodice C, de Melo RW, Nicolaus B, Vangronsveld J (2016) Plant growth-promoting effects of rhizospheric and endophytic bacteria associated with different tomato cultivars and new tomato hybrids. Chem Biol Technol Agric 3(1):1

Adesemoye AO, Obini M, Ugoji EO (2008) Comparison of plant growth-promotion with *Pseudomonas aeruginosa* and *Bacillus subtilis* in three vegetables. Braz J Microbiol 39(3):423–426

Ahirwar NK, Gupta G, Singh V, Rawlley RK, Ramana S (2015) Influence on growth and fruit yield of tomato (*Lycopersicon esculentum* Mill.) plants by inoculation with *Pseudomonas fluorescence* (SS5): possible role of plant growth promotion. Int J Curr Microbiol App Sci 4(2):720–730

Aktar W, Sengupta D, Chowdhury A (2009) Impact of pesticides use in agriculture: their benefits and hazards. Interdiscip Toxicol 2(1):1–12

Almaghrabi OA, Massoud SI, Abdelmoneim TS (2013) Influence of inoculation with plant growth promoting rhizobacteria (PGPR) on tomato plant growth and nematode reproduction under greenhouse conditions. Saudi J Biol Sci 20:57–61

Amaresan N, Jayakumar V, Kumar K, Thajuddin N (2012) Isolation and characterization of plant growth promoting endophytic bacteria and their effect on tomato (*Lycopersicon esculentum*) and chilli (*Capsicum annuum*) seedling growth. Ann Microbiol 62(2):805–810

Benite AMC, de Machado SP, da Machado BC (2002) Sideroforos: "uma Resposta dos Microorganismos". Quím nova 25(6/B):1155–1164

Bernabeu PR, Pistorio M, Torres-Tejerizo G, Luna MF (2015) Colonization and plant growth-promotion of tomato by *Burkholderia tropica*. Sci Hortic 191:113–120

Contesto C, Milesi S, Mantelin S, Zancarini A, Desbrosses G, Varoquaux F, Bellini C, Kowalczyk M, Touraine B (2010) The auxin-signaling pathway is required for the lateral root response of Arabidopsis to the rhizobacterium *Phyllobacterium brassicacearum*. Planta 232(6):1455–1470

Dobert RC, Rood SB, Blevins DG (1992) Gibberellins and the legume-*Rhizobium* symbiosis: I. Endogenous gibberellins of lima bean (*Phaseolus lunatus* L.) stems and nodules. Plant Physiol 98:221–224

Feng K, Lu HM, Sheng HJ, Wang XL, Mao J (2004) Effect of organic ligands on biological availability of inorganic phosphorus in soils. Pedosphere 14(1):85–92

García JAL, Probanza A, Ramos B, Palomino M, Maŋero FJG (2004) Effect of inoculation of *Bacillus licheniformis* on tomato and pepper. Agronomie 24(4):169–176

Girish N, Umesha S (2005) Effect of plant growth promoting rhizobacteria on bacterial canker of tomato. Arch Phytopathol Plant Protect 38(3):235–243

Glick BR (2012) Plant growth-promoting bacteria: mechanisms and applications. Scientifica. doi:10.6064/2012/963401

Gravel V, Antoun H, Tweddell RJ (2007) Growth stimulation and fruit yield improvement of greenhouse tomato plants by inoculation with *Pseudomonas putida* or *Trichoderma atroviride*: possible role of indole acetic acid (IAA). Soil Biol Biochem 39(8):1968–1977

Guerinot ML, Ying Y (1994) Iron: nutritious, noxious, and not readily available. Plant Physiol 104(3):815–820

Gül A, Kidoglu F, Tüzel Y (2008) Effects of nutrition and *Bacillus amyloliquefaciens* on tomato (*Solanum lycopersicum*, L.) growing in perlite. Span J Agric Res 6(3):422–429

Gutiérrez-Manero FJ, Ramos-Solano B, Probanza A, Mehouachi J, Tadeo R, Talon FM (2001) The plant growth promoting rhizobacteria *Bacillus pumilus* and *Bacillus licheniformis* produce high amounts of physiologically active gibberellins. Physiol Plant 111:206–211

Honma M, Shimomura T (1978) Metabolism of 1-aminocyclopropane-1-carboxylic acid. Agric Biol Chem 42:1825–1831

Joo GJ, Kim YM, Lee IJ, Song KS, Rhee IK (2004) Growth promotion of red pepper plug seedlings and the production of gibberellins by *Bacillus cereus*, *Bacillus macroides* and *Bacillus pumilus*. Biotechnol Lett 26(6):487–491

Jurkevitch E, Hadar Y, Chen Y (1992) Differential siderophore utilization and iron uptake by soil and rhizosphere bacteria. Appl Environ Microbiol 58(1):119–124

Kang SM, Khan AL, Hamayun M, Hussain J, Joo GJ, You YH, Kim JG, Lee IJ (2012) Gibberellin-producing *Promicromonospora* sp. SE188 improves *Solanum lycopersicum* plant growth and influences endogenous plant hormones. J Microbiol 50(6):902–909

Khan AL, Waqas M, Kang SM, Al-Harrasi A, Hussain J, Al-Rawahi A, Al-Khiziri S, Ullah I, Ali L, Jung HY, Lee IJ (2014) Bacterial endophyte *Sphingomonas* sp. LK11 produces gibberellins and IAA and promotes tomato plant growth. J Microbiol 52(8):689–695

Koh RH, Song HG (2007) Effects of application of *Rhodopseudomonas* sp. on seed germination and growth of tomato under axenic conditions. J Microbiol Biotechnol 17(11):1805–1810

Krebs B, Höding B, Kübart SM, Workie A, Junge H, Schmiedeknecht G, Grosch R, Bochow H, Hevesi M (1998) Use of *Bacillus subtlis* as biocontrol agent activities and characterization of *Bacillus subtilis* strains. J Plant Dis Protect 105:181–197

López-Bucio J, Hernández-Abreu E, Sánchez-Calderón L, Nieto-Jacobo MF, Simpson J, Herrera-Estrella L (2002) Phosphate availability alters architecture and causes changes in hormone sansitivity in the *Arabidopsis* root systems. Plant Physiol 129:244–256

Marschner H (ed) (2011) Mineral nutrition of higher plants, 3rd edn. Academic press Elsevier, London

Mena-Violante HG, Olalde-Portugal V (2007) Alteration of tomato fruit quality by root inoculation with plant growth-promoting rhizobacteria (PGPR): *Bacillus subtilis* BEB-13bs. Sci Hortic 113(1):103–106

Morrone D, Chambers J, Lowry L, Kim G, Anterola A, Bender K, Peters RJ (2009) Gibberellin biosynthesis in bacteria: separate *ent*-copalyl diphosphate and *ent*-kaurene synthases in *Bradyrhizobium japonicum*. FEBS Lett 583(2):475–480

Neilands JB (1957) Some aspects of microbial iron metabolism. Bacteriol Rev 21:101–111

Onofre-Lemus J, Hernández-Lucas I, Girard L, Caballero-Mellado J (2009) ACC (1-aminocyclo propane-1-carboxylate) deaminase activity, a widespread trait in *Burkholderia* species, and its growth-promoting effect on tomato plants. Appl Environ Microbiol 75(20):6581–6590

Radzki W, Manero FG, Algar E, Lucas García JA, García-Villaraco A, Ramos Solano B (2013) Bacterial siderophores efficiently provide iron to iron-starved tomato plants in hydroponics culture. Antonie Van Leeuwenhoek 104(3):321–330

Ribaudo CM, Krumpholz EM, Cassán FD et al (2006) *Azospirillum* sp. promotes root hair development in tomato plants through a mechanism that involves ethylene. J Plant Growth Regul 25(2):175–185

Richardson AE, Hadobas PA, Hayes JE (2001a) Extracellular secretion of *Aspergillus* phytase from *Arabidopsis* roots enables plants to obtain phosphorus from phytate. Plant J 25(6):641–649

Richardson AE, Hadobas PA, Hayes JE, O'hara CP, Simpson RJ (2001b) Utilization of phosphorus by pasture plants supplied with myoinositol hexaphosphate is enhanced by the presence of soil microorganisms. Plant Soil 229(1):47–56

Singh P, Kumar P, Agrawal S (2014) Evaluation of phytase producing bacteria for their plant growth promoting activities. Int J Microbiol. doi:10.1155/2014/426483

Song OR, Lee SJ, Lee SC, Kim KK, Choi YL (2008) Solubilization of insoluble inorganic phosphate by *Burkholderia cepacia* DA23 isolated from cultivated soil. Braz J Microbiol 39:151–156

Szilagyi-Zecchin VJ, Mógor ÁF, Ruaro L, Röder C (2015) Crescimento de mudas de tomateiro (*Solanum lycopersicum*) estimulado pela bactéria *Bacillus amyloliquefaciens* subsp. *plantarum* FZB42 em cultura orgânica. Rev Ciênc Agrar 38(1):26–33

Szilagyi-Zecchin VJ, Mógor ÁF, Figueiredo GGO (2016) Strategies for characterization of agriculturally important bacteria. In: Singh DP, Singh HB, Prabha R (eds) Microbial inoculants in sustainable agricultural productivity. Springer, New Delhi, pp 1–21

Taiz L, Zeiger E (eds) (2013) Plant physiology, 5th edn. Sinauer Associates, Sunderland

Thomas P, Upreti R (2015) Evaluation of tomato seedling root-associated bacterial endophytes towards organic seedling production. Org Agric 6(2):89–98

Turan M, Ataoglu N, Sahin F (2007) Effects of *Bacillus* FS-3 on growth of tomato (*Lycopersicon esculentum* L.) plants and availability of phosphorus in soil. Plant Soil Environ 53(2):58

Valenciano J. De Pablo, Uribe TJ (2015) Control system of management for intensive cultivation activity in tomato production: Spanish case. J Agric Sci Technol 17(1): 11021

Vansuyt G, Robin A, Briat JF, Curie C, Lemanceau P (2007) Iron acquisition from Fepyoverdine by *Arabidopsis thaliana*. Mol Plant-Microbe Interact 20(4):441–447

Xie H, Pasternak JJ, Glick BR (1996) Isolation and characterization of mutants of the plant growthpromoting rhizobacterium *Pseudomonas putida* GR122 that overproduce indoleacetic acid. Curr Microbiol 32(2):67–71

Xu SL, Rahman A, Baskin TI, Kieber JJ (2008) Two leucine-rich repeat receptor kinases mediate signaling, linking cell wall biosynthesis and ACC synthase in Arabidopsis. Plant Cell 20:3065–3079

Yang S, Hoffman NE (1984) Ethylene biosynthesis and its regulation in higher plants. Annu Rev Plant Physiol 35:155–189

Yao MZ, Zhang YH, Lu WL, Hu MQ, Wang W, Liang AH (2012) Phytases: crystal structures, protein engineering and potential biotechnological applications. J Appl Microbiol 112(1):1–14

Zuo Y, Zhang F (2011) Soil and crop management strategies to prevent iron deficiency in crops. Plant Soil 339:83–95

Development of Future Bio-formulations for Sustainable Agriculture

9

Veluswamy Karthikeyan, Kulliyan Sathiyadash, and Kuppu Rajendran

Abstract

Integrated agriculture based on the basis of the use of chemical fungicides, pesticides, herbicides, and fertilizers. From the point of environmental and human health concern, the prolonged usage of these chemicals is causing drastic damage. The development of biological control methods of plant disease control is cost effecting and eco-friendly approach integrated and sustainable agriculture. Compared to chemical control, the biological control of plant disease is a presently prominent strategy for disease control and growth promotion by means of increasing shoot and root length. Biocontrol agents are proved as potential agents for disease management. The other parts of the beneficial microbes are efficient plant growth promoters (PGPR). Many microorganisms from the rhizosphere can positively influence plant growth and plant health and are referred to as PGPR.

9.1 Introduction

Hary Smith of the University of California was the pioneer of the term biological control; the definition appears as "the suppression of insect population by the actions of their native or induced enemies," and the term biological control was defined by Garret as "the reduction in disease through the agency of one or more, living organisms other than the host or man" (Baker 1968). Later, Cook and Baker (1983) described it as "the reduction of inoculums density or disease-producing activities of a pathogen or a parasite in its active or dormant state, by one or more organisms other than man." The modified definition for biological control by the

V. Karthikeyan (✉) • K. Sathiyadash • K. Rajendran
Centre for Research, Department of Botany and Biotechnology, Thiagarajar College, Madurai, Tamil Nadu, India
e-mail: karthickeyanv@yahoo.com

V. Kumar et al. (eds.), *Probiotics in Agroecosystem*,
DOI 10.1007/978-981-10-4059-7_9

U.S. National Academy of Sciences is " the use of natural or modified organism, gene or gene products to reduce the effects of undesirable organisms, and to favor desirable organisms such as crops, trees, animals, beneficial insects and microorganisms." So far the bacterial antagonists agents like *Pseudomonas* and *Bacillus* are good candidates for biological control.

9.2 *Bacillus*

The gram-positive, endospore-forming *Bacillus* strains are the bacteria that are tolerant to heat and dehydration, which are highly suitable for field application for crop production. The *Bacillus* spp. promote plant biological control (Emmert and Handelsman 1999). Recently this group of bacteria has been employed widely as introduced antagonists for the biocontrol of various pathogens (Hervas et al. 1988; Landa et al. 2001).

B. subtilis have been reported as good biological control agents against fungal pathogens such as *Botrytis*, *Phytophora*, *Pythium* (Li et al. 1998), and *Aspergillus niger* (Sailaja et al. 1998; Manjula and Podile 2001). *Bacillus cereus* UW 85 is a prominent biocontrol agent against damping off alfalfa (Kazmar et al. 2000), tomato (Smith et al. 1999), cucumber root rot (Smith et al. 1993), and sclerotina blight of peanut (Phipps 1992). The *Bacillus* is also against bacterial leaf blight (Vasudevan 2002), blast and sheath blight (Kavitha 2002) of rice. Brindha Priyadarisini and Gnanamanickam (1999). Along with the disease suppression, the bacillus strains enhance the growth and yields of rice (Vasudevan and Gnanamanicakam 2002; Veluswamy 2003) (Fig. 9.1).

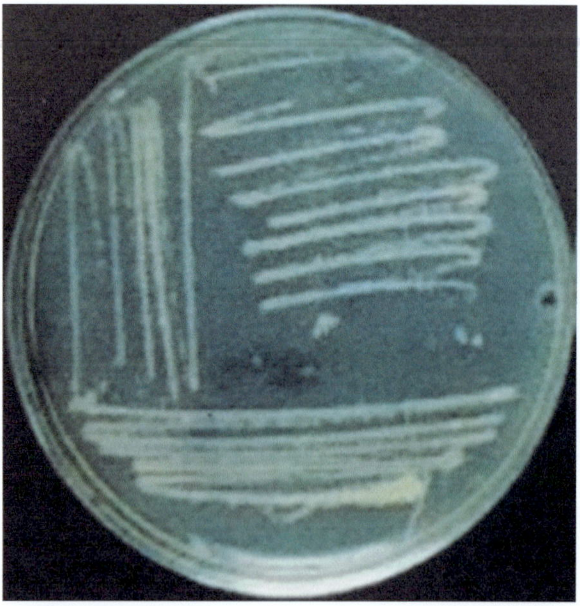

Fig. 9.1 *Bacillus* bacterium morphology on Petri plate

9.3 *Pseudomonas*

The pseudomonades are gram-negative, motile rods, chemoheterotrophic with polar flagella, and are grouped in rRNA homology group (Palleroni et al. 1973), and they have minimum nutritional requirements; they are efficient colonizers and are widely available in rice rhizosphere. These rhizobacteria associate the plant growth by secreting plant hormones (Lifshitz et al. 1987; Yoshikawa et al. 1993; Jacobson et al. 1994; Hong et al. 1991; Glick et al. 1997). The increased root length of certain crops has been definitely associated with the activity of these plant growth-promoting rhizobacteria (PGPR) (Schroth and Hancock 1981; Barret et al. 1986).

The fluorescent and non-fluorescent strains of antagonistic bacteria coupled with upland and lowland rice rhizosphere soils have been found effectual in vitro, greenhouse, and the field against *R. solani* (sheath blight). Around 23 bacterial antagonists strains belongs to the genera, *Bacillus*, *Pseudomonas*, *Serratia*, *Erwinia*, etc. institute to inhibit the mycelia growth of *R. solani*, while some of them also inhibited the growth of other fungal pathogens like *Sclerotium oryzae* (stem rot), *B. oryzae* (brown spot), *M. grisea* (blast), *Sarocladium oryzae* (sheath rot), and *Fusarium fujikuroi* (bakanae). Laboratory studies also exposed that a great number of bacterial strains acquire the ability to defend rice plants from disease such as blast, sheath blight, sheath rot, and stem rot. Forty bacterial isolates antagonistic to the rice sheath blight pathogen have been recognized so far (Vasudevan 2002).

Significant fungal antagonists comprise *Trichoderma* spp., *Pencillium*, *Myrothecium verrucaria*, *Chaetomium globosum*, and *Laetisaria arvalis*. Hashioka and Fukita (1969) reported that the hyperparasitic potential of *Gliocladium deliquescens* on *M. grisea* and the mechanism of parasitism were well characterized. Chung (1988) found that seed bacterization and seed treatment with *Trichoderma* decreased the occurrence of seedling blast. Gnanamanickam and Mew (1992) studied the possibilities of using antagonistic microorganisms such as *Pseudomonas fluorescens* and *Bacillus* spp. for controlling blast disease. There were some peculiar mechanisms which were observed for disease suppression as follows.

Plants are uncovered to quite a lot of phytopathogenic microorganisms; they exhibit forbearance to these pathogens, due to their diverse structures and biochemical mechanisms (such as phytotoxins, phenols, protease inhibitors, cyanogenic glycosides, and hydrolyses) (Caramori et al. 2004; Oliveira et al. 2003; Pascholati et al. 1992). The preceding study has demonstrated that pathogen-related proteins like PAL and β-1,3-glucanases implicated in the defense of the plant against bacterial, viral, and fungal pathogens.

The generally documented mechanism of biocontrol mediated by PGPR is competition for an ecological niche or production of inhibitory allele chemicals and stimulation of induced systemic resistance (ISR) in host. ISR induced by PGPR has not yet been reported as a biological control of disease but is being paid attention to because it has led to disease incidence reduction and supports for plant growth in

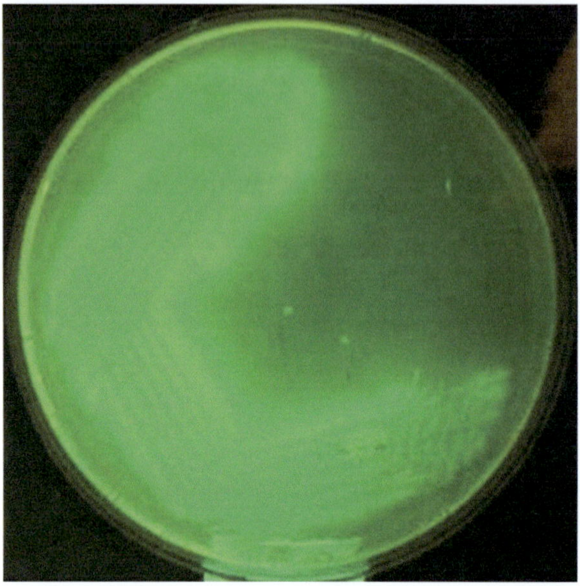

Fig. 9.2 *Pseudomonas* strain under UV light illumination

field condition (Chanway et al. 2000). ISR is brought about by PGPRs through reinforcement of physical and mechanical potency of the cell wall as well as altering the physiological and biochemical reaction of host principal to the synthesis of defense chemicals against the pathogen. PGPRs could act as strong elicitors of plant defense reaction (M'Piga et al. 1997) (Fig. 9.2).

9.4 Production of Siderophores

Under iron-inadequate conditions, the subsequent competition with DRMO (deleterious rhizosphere microorganisms) for ferric iron content in the rhizosphere, the root-colonizing fluorescent pseudomonads produce siderophores. The action of these Fe-binding molecules enhances the plant growth. The siderophores of PGPR have superior attraction to iron than those of DRMO, which lack the iron absorption (or siderophore receptor) systems of these PGPR siderophores. Meanwhile, siderophore production can control the propagation of phytopathogens (Dowling et al. 1996) by depriving the pathogen of iron available in the rhizosphere (Buysons et al. 1996; O'sullivan and O'Gara 1992; Kloepper et al. 1980).

In adding together, the PGPR strains of bacteria secrete some growth hormone-like compounds for the plant growth or they make possible the uptake of nutrients from the environment (Glick et al. 1999).

9.5 Production of 2,4-Diacetylphloroglucinol

The function of Phl in biocontrol of plant disease was reported some 25 years before (Garagulya et al. 1974; Pidoplichko and Garagulya 1974). Recently the Phl was reported to be an effective principle contributing to the "take-all decline" in the USA. Take-all disease is caused by *Gaeumannomyces graminis*. Phl is also effective in the suppression of tobacco black root rot disease caused by *Thielaviopsis basicola* (Keel et al. 1990, 1991, 1992; Vincent et al. 1991). Two parts of chromosomal DNA necessary for Phl production were cloned and DNA probes are obtainable (Vincent et al. 1991; Keel et al. 1992).

Furthermore, a gene, gacA, regulating the production of several metabolites including Phl during the stationary growth phase, has been identified and sequenced (Laville et al. 1992). Phl-producing *Pseudomonas* strains were found to be more consistent biocontrol strains. The derivatives of Phloroglucinol with antifungal activity have been found in some plant species, and many play a vital role as a biochemical defense mechanism against fungi (Tomas Loronte et al. 1989). In the year 2001–2005, a number of Phl-producing strains of *P. fluorescens* were identified in the rhizosphere of rice in southern India, and these were also implicated in the suppression of rice bacterial blight up to 63% (Velusamy and Gnanamanickam 2003). The application of these to rice led also to enhanced growth. *P. fluorescens* strains produce also other fungi toxic compounds; pyoluteorin is more effective on cucumber disease caused by *Pythium ultimum*, and this strain shows an adverse toxic effect on cress and sweet corn (Maurhofer et al. 1992).

9.6 Hydrogen Cyanide (HCN)

The hydrogen cyanide is a secondary metabolite generally produced by rhizosphere pseudomonads (Bakker et al. 1987; Lambers 1980; Schippers et al. 1990). The level of HCN production is quite linked to available nutritional precursors such as methionine, glycine, and root exudates containing cyanogenic glycosides (Alstrom and Burns 1989; Schippers et al. 1990; Owen and Zdor 2001). From *P. fluorescens* CHAO, a negative mutant for HCN, was produced by gene replacement technique, and it afforded a greater level of protection of tobacco plants than the wild-type CHAO strain. The capability of the mutant strain to restrain black root disease of tobacco was studied by complementation analysis.

Literature has clearly demonstrated the potential of antibiotic-producing bacterial agents (Gnanamanickam et al. 1994; Viji and Gnanamanickam 1996) for blast control. Also the possible mechanism of induced systemic resistance (ISR) by using biocontrol agents (Krishnamurthy and Gnanamanickam 1998) to control blast appears to hold much promise.

9.7 Rice Blast

Half of the world population depends on rice as their staple food. The annual yield losses of rice production are majorly due to diseases. Biological control is cost effective, eco-friendly and an alternative strategy of disease management (Roy Manidipa et al. 2013). Rice blast disease caused by *Magnaporthe grisea* Barr (anamorph *Pyricularia grisea* Sacc. synonym *Magnaporthe grisea* Cav.) is disseminated in about 85 countries. When the blast disease occurs in epidemic magnitude, the yield loss can be as high as 50% (Babujee and Gnanamanickam 2000).

Magnaporthe grisea is one of the major production constraints particularly in tropical Asia where rice farmers generally have less access to chemical fungicides and often cannot afford the same economically (Gnanamanickam and Mew 1992). Extensive studies were made about the interaction between biocontrol agents and pathogens in rice (Zarandi et al. 2009), and application of biocontrol agents to control certain commercially valuable crops also has been reported (de Vasconcellos and Cardoso 2009; Anand et al. 2010).

Magnaporthe oryzae is an ascomycete fungus which infects the plants by developing specialized infection structures known as appressoria. The dome-shaped aspersorium generates massive physical force and turgor pressure, allowing the fungus to penetrate into host tissue by breaking host cuticle to invade plant tissue (Talbot 2003). The aptitude of rice blast fungus to develop new races is responsible for the limited success in control of blast through breeding resistance (Gnanamanikam and Mew 1992). The conventional method for the fungal disease control that relies mainly on synthetic fungicides has become less acceptable nowadays due to their increase incidence of development of resistance up on prolonged usage (Fernandez-Acero et al. 2006; MgGrath 2001).

Due to the lack of specificity toward the target pathogen leads to the adverse effect on environment, beneficial microbes, humans and animals (Leroux 2003). Thus, there is a necessity for alternative disease management methodology that provides effective control without their side effects on the environment. Use of beneficial bacteria as a biological control to inhibit plant disease offers a potential eco-friendly alternative to the extensively used chemical fungicides (Kiss 2003; Nagorska et al. 2006; Romero et al. 2007). Some bacterial species, viz., *Pseudomonas* (Lee et al. 2003; Vleesschauwer et al. 2008), *Streptomyces* (Gomes et al. 2001), and *Bacillus* (Cottyn et al. 2001; Leelasuphakul et al. 2006; Tendulkar et al. 2007), have been reported for their antagonistic action against *Magnaporthe grisea*, a causative agent of rice blast. Among them, potentiality of biocontrol agents of *Pseudomonas fluorescence*, *Bacillus subtilis* strain IK-1080, and *Streptomyces sindeneusis* have been demonstrated for the rice blast disease suppression (Gnanamanikam and Mew 1992; Yoshihiro et al. 2003; Zarandi et al. 2009). *B. subtilis* strains have been found to suppress the growth of 23 types of plant pathogens in vitro by virtue of their capability to produce a wide variety of biologically active compounds (Stein 2005; Nagorska et al. 2007).

Numerous research results have described the potential of *B. subtilis* strains as biocontrol agents for diversity of plant disease (Bais et al. 2004; Mizumoto and Shoda 2007; Cavglieri et al. 2004; Nagorska et al. 2007). The Pseudomonas belongs to the family Pseudomonadaceae, rod shaped gram negative bacteria having one or more flagella, non-spore-producing aerobic some bacteria have anaerobic respiration with nitrate as their terminal electron acceptor (Palleroni 2008).

In order to protect themselves from pathogen attack, the plant possesses different inducible defense mechanisms. Some nonpathogenic, root-colonizing rhizobacteria may also activate disease resistance in the host; this phenomenon is termed as induced systemic resistance (ISR) (Van Loon et al. 1998). The triggering of disease resistance by nonpathogenic bacteria depends on plant species to a different level. Biocontrol is therefore being measured as a substitute or supplemental way of decreasing the use of chemicals in agriculture. The potential use of plant-associated rhizobacteria as agents stimulating plant growth, soil administration, and plant fitness is well documented (Glick 1995; Sturz et al. 2000). The widest groups of such bacteria are PGPR (Klopper et al. 1997) which colonize the root plane and closely adhering soil boundary of the rhizosphere.

Rhizobacteria have a vast potential in agriculture to employ as biofertilizer, biocontrol agent, and bioremediation due to their plant growth-promoting capacity, antagonistic activity, and deprivation of pollutants (Ahmad et al. 2008).The bacterial strains, either singly or as a mixture, were assessed to their competence in suppressing rice blast under greenhouse conditions. Spore suspension of *Magnaporthe grisea* with a spore load of 10^4 conidia ml^{-1} was sprayed on the plants, which caused more than 75% infection under glass house condition which was recorded. In addition, growth parameters like plant height, root length, tiller number, and biomass were also measured. The observation on the present disease occurrence of rice blast production was observed on 45 days after planting (maximum tillering) as grades 0–5. (Sriram et al. 1999). Microbes play a vital role in seed germination and seedling establishment. An outsized number of confirmations suggest that PGPRs enhance the growth, crop yield, and seed emergence and assist to the protection of plants against definite pathogens and pests (Dey et al. 2004; Herman et al. 2008; Kloepper et al. 2004; Minorsky 2008). This relationship between bacteria and plants has been well recognized (Holland et al. 2002).The recent study robustly supports the improvement of biocontrol strategies using endophytic bacteria strains to lessen the damage caused by plant pathogens in cost-effectively significant crops like paddy (Lucas et al. 2009).

The in vitro growth inhibition of *Magnaporthe grisea* is associated with in vivo pathogenicity disease suppression between the two isolates of bacteria. *P. pseudoalcaligenes* showed a large zone of inhibition in its vicinity than *B. pumilus*. Marjan et al. demonstrated that there was a direct association between in vitro antagonism and in vivo disease suppression by *Pseudomonas* in radish (de Bore et al. 1999).

Brevibacillus was established as novel genera arising from the reclassification of the *Bacillus brevis* assembly of species (Shida et al. 1994). The pathogenicity prospective of *B. laterosporus* against insects has been demonstrated earlier (Ruiu et al. 2008). *B. laterosporus* has the possibility to be used as a biological control agent in comparison with strains of *Bacillus thuringiensis* and *Bacillus sphaericus*, which demonstrated a very broad range of biological activities (de Oliveira et al. 2004). Plants respond to a variety of chemical stimuli produced by various soil- and plant-related microbes, and these stimulus can either provoke or condition plant defenses through biochemical changes that amplify resistance against subsequent infection by pathogen (Zhang and Reddy 2001; Baysal et al. 2008).

9.8 Mechanism of Blast Suppression

Strains of *Pseudomonas fluorescens* produce many antifungal antibiotics that inhibit germination of conidia of the blast pathogen. The exact chemical scenery of the antibiotics was not recognized. However, later the metabolite appeared to be a phenazine. Transposon (Tn5)-derived mutants that lacked the production of this antifungal antibiotic (afa minus mutants) were less effective in protecting rice from blast (leaf and neck blasts) and sheath blight, while the afa + wild-type strains of Pfl-14 suppressed blast and sheath blight up to 80–82% in the field (Chatterjee et al. 1996; Valasubramanian 1994) (Fig. 9.3).

Many fluorescent pseudomonads and some other plant growth-facilitating rhizobacteria cause induced systemic resistance (ISR) in rice in reaction to treatments with *P. fluorescens* strains Pf7–14 and Ppvi4i. This is an imperative mechanism of biological suppression of blast. Treatment increases the level of salicylic acid that increases ISR, which in turn suppresses rice blast by up to 25% (Krishnamurthy and Gnanamanickam 1998). The initiation of systemic resistance in rice to bacterial blight by 1,2,3-benzothiadiazole 7-carbothioic acid-S-methyl ester (BTH) treatments was done by Karthikeyan and Gnanamanickam in 2011. In 2008 the fertility status of *Setaria* infecting *Magnaporthe grisea* isolates with standard testers were identified by Karthikeyan and Gnanamanickam (2008a), and its virulence characteristic analysis and identification of new pathotypes of rice blast fungus from India were also reported by Karthikeyan et al. in 2013.

Fig. 9.3 Leaf blast symptoms

9.9 Setaria Blast

Biocontrol mechanism to suppress bacterial and fungal pathogens of rice crop by *Pseudomonas* sp. generally participates in the production of antibiotics, volatile compounds, siderophoers, hydrocyanic acid (HCN), enzymes, and phytohormones. There were many fungicides reported as follows. Among five fungicides, viz., Carbendazim, Thiophanate-methyl, Mancozeb, Fosetyl-aluminum, and copper oxychloride, employed against the *Magnaporthe oryzae*, only Mancozeb appeared as the extremely efficient fungicide that completely inhibited the mycelial growth of the fungus. All other fungicides showed modest effect at superior concentrations. Among the extracts of garlic (*Allium sativum* L.), neem (*Azadirachta indica* L.), and calotropis (*Calotropis procera* L.) when used against *M. oryzae* by food poisoning method, only higher dose of garlic completely inhibited the mycelial growth of the test fungus. Six biocontrol agents, viz., *Trichoderma harzianum*, *Trichoderma polysporum*, *Trichoderma pseudokoningii*, *Gliocladium virens*, *Paecilomyces variotii*, and *Paecilomyces lilacinus*, were used. Maximum mycelial inhibition of *M. oryzae* was induced by *P. lilacinus* followed by *Trichoderma* spp. (Hajano et al. 2012).

A combined biological and chemical control of millet blast disease with antagonistic rhizobacteria *Pseudomonas fluorescens* and resistance inducing chemical salicylic acid was evaluated. In pot trial, *Pseudomonas fluorescens* tested in combination with salicylic acid was highly efficient in management of rice blast diseases. Biological control of *Setaria* blast (*Magnaporthe grisea*) with chosen bacterial strains of *Bacillus* and *Pseudomonas* species were reported by Karthikeyan and Gnanamanickam in 2008a, b. Biological control of *Pyricularia oryzae* with *P. fluorescens* was effective but less so than chemical alone at the standard dose. However, combination of the antagonistic rhizobacteria with chemical dose was as effective as the standard chemical alone. Application of *P. fluorescens* along with salicylic acid significantly increased the disease resistance. Further, there were increases in activities of polyphenol oxidase and showed least activities of peroxidase and ascorbic acid oxidase treated with *P. fluorescens* plus salicylic acid. From south India biological control of blast disease (*Magnaporthe grisea*) on minor millets using potential antagonistic bacteria prevails in selected areas of Madurai District by Pal Pandi et al. (2016). The results indicate that the combined biological and chemical inoculation showed a better response to fight against rice blast pathogen *P. oryzae* than the treatment alone.

Application of valuable bacteria as seedling root dip and spraying method to protect against the disease may be a substitute approach to chemical control. The subsequent bacteria consortium that may manage blast disease on millet plants includes Bacillus firmus E65, Serratia marcescens E31, Pseudomonas aeruginosa C32b, Bacillus cereus II.14, and its combination for their suppression capacity against P. oryzae under in vitro conditions. The outcome showed that A2 (*Bacillus firmus* E65) and A6 consortium (*Bacillus firmus* E65, *Bacillus cereus II*.14, and *Pseudomonas aeruginosa* C32b) considerably abridged the mycelial growth of *P. oryzae* with the percentage inhibition of 73–85% and 66–83%, correspondingly. Further greenhouse testing conducted with use of formulated research of the two selected best treatments using talc, bentonite, palm oil, and suspension-based carriers showed that spraying with suspension formulation had high-quality effect in suppressing blast disease compared with that of other carriers evaluated (Suryadi et al. 2013).

Millet-associated microorganisms, *Bacillus* spp. and *Pseudomonas* spp., were isolated and tested for antagonism against the rice blast fungus *Pyricularia oryzae*. Field test of four selected strains in an upland rice farm in the Philippines afforded significant leaf blast reduction with rice var. UPLRi-5. Strain 7-14, identified as *P. fluorescens*, was the most effective. The crude extract prepared from this strain provided 70–100% inhibition of conidial germination at 1.0 ppm. The anti-blast extract also protected IR 50 rice seedlings from infection of *P. oryzae*. Evidence suggested that siderophore was unlikely to be involved in the mechanism of strain 7-14 in its antagonism against the rice blast fungus. In vitro test showed that inhibition of *P. oryzae* by strain 7-14 was not reversed by Fe amendments. Thus, the anti-blast antibiotic, instead of siderophore production from *Pseudomonas* strain 7-14, protected rice seedlings from infection by *P. oryzae* (Gnanamanickam and Mew 1992).

Besides the bacterium some of the soil actinomycetes predominantly *Streptomyces* spp. have antagonistic action against s wide range of plant pathogens. In the current decades, high benefit as biocontrol agents is being paid attention to. In search for finding such principles, in vitro suppression of *Magnaporthe oryzae*, the causal agent of millet blast disease, was studied by use of *Streptomyces sindeneusis* isolate 263 in greenhouse. Spray of rice seedling leaves with mixed spore suspension of the pathogen and *S. sindeneusis* isolate 263 resulted in strong inhibition of the pathogen and inhibition of leaf symptoms. Proliferation of the antagonist crude sap was performed in aqueous cultures, and bioactivity was monitored in shacked cultures. Using PCR method, the blast disease caused by *Magnaporthe grisea* in *Setaria italica* was identified rapidly by Karthikeyan and Gnanamanickam (2005) (Fig. 9.4).

Fig. 9.4 Setaria blast

9.10 Bacterial Blight and Sheath Blight

Significant yield losses from diseases still occur in rice in spite of continuous improvements in rice breeding bacterial blight (BB) caused by *Xanthomonas oryzae pv. oryzae* and sheath blight (shb) caused by *Rhizoctonia solani* which are some of the most devastating rice diseases of global occurrence and are particularly destructive causing annual crop losses that range from 10% to 50% in tropical Asia (Mew 1987). In India, planting of resistant rice cultivars has been the most successful disease management strategy. However, breakdown of varieties carrying a single R-gene is more frequent in the field because of the rapid evolution of subpopulations of the pathogen the overcome these resistances (Venkatesan and Gnanamanickam 1999; Brindha Priyadarisini et al. 2003). In recent years, we had constructed, through molecular marker-assisted backcross breeding and transformation, transgenic elite indica rices (cv.CO39 and IR50) that carry a pyramid of genes for blast and resistance (Pi-1 + Pi-z + Xa21) rice cultivars BB (Narayanan et al. 2002, 2004). There were very limited studies done on the biological suppression of BB and ShB in indica rices with specific antibiotic (2,4-DAPG) producing bacterial agents.

The polyketide metabolite 2,4 is one of the majority efficient antimicrobial metabolite produced by fluorescent pseudomonads and is successful against bacteria, fungi, and helminthes. Recently, we identified using the PCR-based screening method (Raaijmakers et al. 1997) fluorescent pseudomonads which produced 2,4-diacetylphloroglucinol (DAPG) in our tropical rice rhizospheres (Velusamy and Gnanamanickam 2003; Velusamy et al. 2004, 2006).

There has been no preceding account from India on the production of DAPG by plant-associated bacteria or on its inhibition of rice bacterial blight and sheath blight DAPG production which has been an extremely well-known mechanism in the biological control of some of the major fungal pathogens of the temperate regions and has assumed much importance as the factor that contributed to the "take-all" decline in wheat (Raaijmakers et al. 1999; Raaijmakers and Weller 1998; Raaijmakers et al. 1997). Further, its antibacterial activity against soft-rot *Erwinia* was also formerly known. Yet, in the present study, DAPG production has been implicated as an antibacterial and antifungal compound implicated in the suppression of the most important and devastating bacterial and fungal crop diseases of the tropics.

9.11 Fusarium

"Bakanae" disease of rice is caused by *Gibberella fujikuroi* (Sawada) Wollenworth (teleomorph), a disease of rice first reported in Japan, and now extensively distributed in Asia. On rice, *G. fujikuroi* induces foot rot, seedling elongation, seedling rot, grain sterility, and grain discoloration (Ou 1985). Apart from that the crop damage up to 20% to 50% because of reduced tillering, partially filled grains, drying of leaves. During later stage of infection, dry seedlings, sterile, or empty grains appears.

The fungus not only causes substantial damage on many plants but also is parasitic on plants without producing noticeable symptoms (Hsieh et al. 1977). It can be isolated even from kernels that are healthy in appearance. Rice seedlings that grow from these infected seeds tended to display bakanae symptoms (Padwick 1950). Currently, the most common management practice for bakanae is seed treatment with fungicides like thiram, thiophanate-methyl, or benomyl which is effective before planting. The rapid development of resistance against benomyl, carbendazim and resistance of the fungal pathogen to the fungicides has also been reported (Ogawa 1988). Worldwide occurring major groups of *Fusarium* associated with maize and rice with the ability to produce fumonisins were confirmed (Moretti et al. 2004, 2007).

There are some reports available about contamination of grains with fumonisins being associated with human esophageal cancer in South Africa (Nelson et al. 1993). Real-time RT-PCR assay of gene fumonisin *fum5* responsible for mycotoxin production on rice was done by Karthikeyan and Rajendran (2016). Fumonisins also cause leukoencephalomalacia in horses, cancer in experimental animals, and lung edema in swine. PCR-based detection of fumonisin producing strains of *Fusarium verticillioides* and gene related to toxin production was reported by Karthikeyan et al. (2011). The potentiality of *Fusarium verticillioides* to produce fumonisin and its responsible gene detection assay was reported by Karthikeyan et al. (2008).

There are some reports available about contamination of grains with fumonisins which has been associated with human esophageal cancer in South Africa (Nelson et al. 1993). Unknowingly the infected rice grains and rice straw serve as disease inoculums and enter our food chain. So the present urge is to eliminate the mycotoxin contamination in rice.

Conclusion

The soil dweller microbiome especially plant growth-promoting microbes (PGPM) has been reported to possess many beneficial properties and potential to augment the plant growth and production using direct and indirect approaches. These are provision of providing insoluble phosphate and potassium and making available the unavailable atmospheric nitrogen. Other mechanisms such as by providing vital micronutrients to plants and theryby indirectly suppressing the plant diseases in different cropping systems using simple techniques of excreting antibiotics. The potential ability of PGPM to produce proteinaceous compounds, siderophores, cyanogenic compounds, and potential chemicals in the form of antibiotics in order to control phytopathogens could be of applied significance in balanced crop production. The PGPM gifted with growth-promoting and pathogen inhibitory properties might be useful in formulating novel bioinoculants which will offer an inexpensive, low-cost, reasonable, and appealing substitute to the costly agrochemicals, and consequently such PGPM are likely to enhance crop production without any adverse impact on the ecosystem.

At the plant population level, the adaptive capacity of plant and pathogen populations may prove to be one of the most significant forecasters of the significance of climate change effects. The ecologists are now trying to figure out the role of plant disease in ecosystem processes and the challenge of scaling up from individual infection probabilities to epidemics and broader impacts (Garrett et al. 2006). Therefore, the following implications have been suggested for prospective balanced cropping system:

1. The biological control is a potential strategy.
2. The biological control is slow but a persistent method to control pathogens for a long time.
3. Therefore, the biological controls are cost effective in the long-term application, although the biocontrol agents do act as potential bioinoculants.

There were many success stories to control the many plant pathogens and enhance the plant growth by using potential benefits of microbes. Integrated sustainable development of future bio-formulations leads to effective management of plant pathogens and also by enhancing the plant growth. The future bio-formulation will rely upon the efficacy of microbes related to their beneficial aspects and their long time persistence at storage and field.

References

Ahmad F, Ahmad I, Khan MS (2008) Screening of free- living rhizospheric bacteria for their multiple plant growth promoting activities. Microbiol Res 168:173–181

Alstrom S, Burns RG (1989) Cyanide production by rhizobacteria as a possible mechanism of plant growth inhibition. Biol Fertil Soils 7:232–238

Anand T, Chandrasekaran A, Kuttalam S, Senthilraja G, Samiyappan R (2010) Integrated control of fruit rot and powdery mildew of chilli using the biocontrol agent Pseudomonas fluorescens and a chemical fungicide. Biol Control 52:1–7

Bais HP, Fall R, Vivanco JM (2004) Biocontrol of *Bacillus subtilis* against infection of Arabidopsis roots by *Pseudomonas syringae* is facilitated by Biofilm formation and surfactin production. Plant Physiol 134:307–319

Bakker PAHM, Bakker AW, Marugg JD, Weisbeek PJ, Schippers B (1987) Bioassay for studying the role of siderophores in potato growth stimulation by pseudomonas spp. in short potato rotations. Soil Boil Biochem 19:443–449

Baysal O, Caliskan M, Yesilova O (2008) An inhibitory effect of a new *Bacillus subtilis* strain (EU07) against *Fusarium oxysporum* f. sp. *radicis-lycopersici*. Physiol Mol Plant Pathol 73:25–32

Baker R (1968) Mechanisms of biological control of soil-borne pathogens. Annu Rev Phytopathol 6:263–294

Barret EL, Solanes RE, Tang JS, Polleroni NJ (1986) *Pseudomonas fluorescens* biovar V: its resolution into distinct component groups and the relationship of these groups to other *P. fluorescens* biovars, to *P. putida* and to *Psychrotrophic pseudomonads* associated with food spoilage. J Gen Microbiol 132:2707–2721

Buysens S, Heugens K, Poppe J, Hofte M (1996) Involvement of pyochelin and pyovirdin in suppression of Pythium - induced damping off of tomato by *Pseudomonas aeruginosia* 7NSK2. Appl Environ Microbiol 62:865–871

Brindha Priyadarisini V, Gnanamanickam SS (1999) An over view of bacterial blight disease of rice and strategies for its management. Current Science 77(10)
Caramori SS, Lima CSL, Fernandes KF (2004) Biochemical characterization of selected plant species from Brazilian savannas. Braz Arch Bio Techn 47:253–259
Cavaglieri L, Passone A, Etcheverry M (2004) Screening procedures for selecting rhizobacteria with biocontrol effects upon *Fusarium verticillioides* growth and fumonisin B1 production. Res Microbiol 155:747–754
Chanway CP et al (2000) Endophytic colonization and field responses of hybrid spruce seedlings after inoculation with plant growth-promoting rhizobacteria. For Ecol Manag 133(1):81–88
Chatterjee A, Valasubramanian R, Ma W-L Vachhani AK, Gnanamanickam SS, Chatterjee AK (1996) Isolation of ant mutants of *Pseudomonas fluorescens* strain Pf7-14 altered in antibiotic production, cloning of ant+ DNA, and evaluation of the role of antibiotic production in the control of blast and sheath blight of rice. Bio Control 7: 185–195. Cont 54: 807–816
Chung HS (1988) Recent advances of studies on rice fungal disease. World situation of rice disease and recent advances in their studies. (Abstr.) 5th Int Phytopathol Cong Kyotopp: 6
Cook RJ, Baker KF (1983) The nature and practice of biological control of plant pathogens. Amer Phytopathol Soc, St Paul, p 539
Cottyn B, Regalado E, Lanoot B, De Cleene M, Mew TW, Swings J (2001) Bacterial populations associated with rice seed in the tropical environment. Phytopathology 91:282–292
de Boer M, van Der Sliuis I, Van Loon LC, Bakker AHM (1999) Combining florescent pseudomonas spp. strains to enhance suppression of fusarium wilt of radish. Eur J Plant Pathol 105:201–210
de Vasconcellos RLF, Cardoso EJBN (2009) Rhizospheric *Streptomycetes* as potential biocontrol agents of *Fusarium* and *Armillaria* pine rot and as PGPR for *Pinus taeda*. Bio Control 54:807
Glick BR, Patten CL, Holgin G, Penrose DM (1999) Biochemical and genetic mechanisms used byplant growth promoting bacteria. Imperial College Press, London, 267 p
Dey R, Pal KK, Bhatt DM, Chauhan SM (2004) Growth promotion and yield enhancement of peanut (*Arachis hypogaea* L.) by application of plant growth promoting rhizobacteria. Microbiol Res 159:317–394
Dowling DN, Sexton R, Fenton A, Delany I, Fedi S, McHugh B, Callanan M, Moenne-Loccoz Y, O'Gara F (1996) Iron regulation in plant associated *Pseudomonas fluorescens* M114: implications for biological control. In: Nakazawa T, Furukawa K, Haas D, Silver S (eds) Molecular biology of pseudomonads. American Society for Microbiology Press, Washington, DC, pp 502–511
Emmert EAB, Handlesman J (1999) Biological control of plant disease. A (gram) positive perspective. FEMS Microbiol Lett 171:1–9
Fernández-Acero FJ, Jorge I, Calvo E, Vallejo I, Carbú M, Camafeita LE, López JA, Cantoral JM, Jorrin J (2006) Two-dimensional electrophoresis protein profile of the phytopathogenic fungus *Botrytis cinerea*. Proteomics 6:S88–S96
Galagulya AD, Kiprianova EA, Bioko OI (1974) Antibiotic effect of bacteria from the genus pseudomonas on phytopathogenic fungi. Mikrobiologichnii Zhurnal (Kiev) 36:197–202
Garrett KA, Dendy SP, Frank EE, Rouse MN, Travers SE (2006) Climate change effects on plant disease:genomes to ecosystems. Annu Rev Phytopathol 144:489–509. doi:10.1146/annurev.phyto.44.070505.143420
Glick BR (1995) The enhancement of plant growth by free-living bacteria. Can J Microbiol 41:109–117
Glick BR, Liu C, Ghosh S, Dumbroff EB (1997) The effect of plant growth-promoting rhizobacterium pseudomonas putida GR12-2 on the development of canola seedling subjected to various stresses. Soil Biol Biochem 29:1233–1239
Gnanamanickam SS, Mew TW (1992) Biological control of blast disease of rice (Oryza Sativa, L.) with antagonistic bacteria and its mediation by a pseudomonas antibiotic. Ann Phytopathol Soc Jpn 58:380–385
Gnanamanickam SS, Valasubramanian R, Chatterjee A, Chatterjee AK, Mew TW (1994) Antibiotic production mediates the biological control of rice blast by *Pseudomonas fluorescens*. In: Zeigler RS, Leong SA, Tang PS (eds) Rice blast disease. CAB International, Wallingford, pp 601–602

Gomes RC, Semedo LTAS, Soares RMA, Linhares LF, Ulhoa CJ, Alviano CS, Coelho RRR (2001) Purification of a thermostable endochitinase from Streptomyces RC1071 isolated from a cerrado soil and its antagonism against phytopathogenic fungi. J Appl Microbiol 90:653–661

Hashioka Y, Fukita T (1969) Ultrastructural observations on mycoparasitism of Trichoderma, Gliocladium and Acremonium on phytopathogenic fungi. Rep Tottori Myco Inst 7:8–18

Hsieh WH, Smith SN, Snyder WC (1977) Mating groups in *Fusarium moniliforme*. Phytopathology 67:1041–1043

Herman MAB, Nault BA, Smart CD (2008) Effects of plant growth promoting rhizobacteria on bell pepper production and green reach aphid infestations in New York. Crop Prot 27:996–1002

Hervas A, Landa B, DatnoffLE Jimmenez-Diaz RM (1988) Effects of commercial and indigenous microorganisms on Fusarium development in chick pea. Bio Control 13:166–176

Holland MA, Long RLG, Polacco JC (2002) Methylobacterium Spp. Phylloplane bacteria involved in cross – talk with the plant host? In: Lindow SE, Hecht-Poinar EL, Elliot VJ (eds) Phyllosphere microbiology. APS Press, St. Paul, pp 125–135

Hong Y, Pasternak JJ, Glick BR (1991) Biological consequences of plasmid transformation of plant growth-promoting rhizabacterium *Pseudomonas putida* GR12-2. Can J Microbiol 37:796–799

Hajano JUD, Lodhi M, Pathan MA, Khanzada A, Shah GS (2012) In-vitro evaluation of fungicides, plant extracts and bio-controlagents against rice blast pathogen *Magnaportheoryzae* couch. Pak J Bot 44(5):1775–1778

Jacobson CB, Pasternak JJ, Glick BR (1994) Partial purification and characterization of ACC deaminase from the plant growth-promoting rhizobacterium *Pseudomonas putida* GR 12-2. Can J Microbiol 40:1019–1025

Karthikeyan V, Gnanamanickam SS (2011) Induction of systemic resistance in rice to bacterial blight by 1,2,3-Benzothiadiazole 7-carbothioic acid-S-methyl ester (BTH) treatments. Arch Phytopathol Plant Protec 44(3):269–281

Karthikeyan V, Gnanamanickam SS (2008a) Biological control of *Setaria* blast (*Magnaporthe grisea*) with bacterial strains. Crop Protect 27(2):263–267

Karthikeyan V, Gnanamanickam SS (2008b) Determining the fertility status of *Setaria* infecting *Magnaporthe grisea* isolates with standard testers and identification of tolerant cultivar of *Setaria italica*. Mycopathologia 166(4):227–233

Karthikeyan V, Gnanamanickam SS (2005) Rapid and specific disease detection of blast fungus in *Setaria italica* through PCR method. Curr Sci 8:349–350

Karthikeyan V, Titone P, Spadaro D, Gullino ML, Garibaldi A (2008) Detection of fumonisin producing strains of *Fusarium verticillioides* and gene related to toxin production. J Plant Pathol 90:S2.32

Karthikeyan V, Rajarajan R, Patharjan S, Karthikeyan P, Saravanakumar P, Siva M, Aruna Bhavani PS, Palani P (2011) PCR based detection of fumonisin producing strains of Fusarium verticillioides and gene related to toxin production. Curr Bot 2(3):34–37

Karthikeyan V, Rajendran K (2016) Real-time RT-PCR assay of gene fumonisin *fum5* responsible for mycotoxin production on rice. Inter Res J Nat App Sci 3:35–45

Karthikeyan V, Rajarajan R, Gnanamanickam SS (2013) Virulence characteristic analysis and identification of new pathotypes of rice blast fungus from India. Life Sci Feed 2:13–17

Kavitha S (2002) Strategies for management of rice blast and sheath blight with bacterial biocontrol agents in combination with major genes for disease resistance. PhD Thesis, University of Madras

Kazmar ER, Goodman R, Grau CR, Johnson DW, Nordhiem EY, Undesander DJ, Handelsaman J (2000) Regression analysis for evaluating the influence of *Bacillus cereus* on alfalfa yield under variable disease intensity. Phytopathology 90:807–816

Keel C, Schnider U, Maurhoffer M, Voisard C, Laville J, Burger U, Writhner P, Hass D, Defago G (1992) Suppression of root diseases by Pseudomonas Fluorescens CHAO: importance of the bacterial secondary metabolite 2,4-diacetylphoroglucinol. Mol Plant Mic Interact 5:4–13

Keel C, Maurhofer M, Oberh-Nsli TH, Voisard C, Haas D, Defago G (1991) Role of 2,4-diacetylphloroglucinol in the suppression of take-all of wheat by a strain of Pseudomonas lluorescens. In: Beemster ABR, Bollen GJ, Gerlach M, Ruissen MA, Schippers B, Tempel A (eds) Developments in Agricultural and Managed-Forest Ecology 23, Biotic Interactions and Soil-Borne Diseases. Elsevier, Amsterdam, pp 335–338

Keel C, Wirthner P, Oberhansli T, Voisard C, Burger U, Hass D, Defage G (1990) Pseudomonads as antagonists of plant pathogens in the rhizosphere: role of the antibiotic 2,4-diacetylphloroglucinol on the suppression of black root rot of tobacco. Symbiosis 9:327–341

Kiss L (2003) A review of fungal antagonists of powdery mildews and their potential as biocontrol agents. Pest Manag Sci 59:475–483

Kloepper JW, Leong J, Teintze M, Schroth MN (1980) Enhanced plant growth by siderophores produced by plant growth promoting rhizobacteria. Nature 286:885–886

Klopper JW, Tuhnder GW, Wei G (1997) Multiple disease protection by rhizobacteria that induce systemic resistance – historical precedence. Phytopathology 87:136–137

Kloepper JW, Ryu CM, Zhang SA (2004) Induced systemic resistance and promotion of plant growth by *Bacillus* spp. Phytopathology 94:1259–1266

Krishnamurthy K, Gnanmanickam SS (1998) Biological control of rice blast by Pseudomonas *fluorescens* strain Pf7-14: evaluation of a marker gene and formulations. Bio Control 13:158–165

Lambers H (1980) The physiological significance of cyanide-resistant respiration in higher plants. Plant Cell Environ 3:293–302

Landa B, Navas-Cortes JA, Hervas A, Jimmenez-Diaz RM (2001) Influences on temperature and inoculums density of *Fusarium oxysporum* f.sp. *ciceris* on suppression of Fusarium wilt of chick pea by rhizosphere bacteria. Phytopathology 91:807–816

Laville J, Voisard C, Keel C, Maurhofer M, Defage G, Hass D (1992) Global, stationary-phase control in Pseudomonas Fluorescens mediating antibiotic synthesis and suppression of black root rot of tobacco. Proc Natl Acad Sci U S A 89:1562–1566

Babujee L, Gnanamanickam SS (2000) Molecular tools for characterization of rice blast pathogen (*Magnaporthe grisea*) population and molecular marker-assisted breeding for disease resistance. Curr Sci 78(3):248–257

Lee JY, Moon SS, Hwang BK (2003) Isolation and antifungal and Antioomycete activities of aerugine produced by *Pseudomonas fluorescens* strain MM-B16. Appl Environ Microbiol 69:2023–2031

Leelasuphakul W, Sivanunsakul P, Phongpaichit S (2006) Purification, characterization and synergistic activity of ß 1-3-glucanase and antibiotic extract from an antagonistic *Bacillus subtilis* NSRS 89-24 against rice blast and sheath blight. Enzym Microb Technol 38:990–997

Leroux P (2003) Modes of action of agrochemicals against plant pathogenic organisms. Critical Rev Biol 326:9–21

Li H, White D, Lamza KA, Berger F, Liefert C (1998) Biological control of *Botrytis, Phytophora* and *Bacillius subtilis* COT1 and OL27 of micropropogated plants in high-humidity fogging glass house. Plant Cell 52:109–112

Lifshitz R, Kloepper JW, Kozlowski M, Simonson C, Carison J, Tipping EM, Zeleska I (1987) Growth promotion of canola (rapeseed) seeding by a strain of *Pseudomonas putida* under genobiotic conditions. Can J Microbiol 33:390–395

Lucas JA, Ramos Solano B, Montes F, Ojeda J, Megias M, Gutierrez Manero FJ (2009) Use of two PGPR strains in the integrated management of blast disease in rice (*Oryza sativa*) in southern Spain. Field Crop Res 114:404–410

Manjula K, Podile AR (2001) Chitin-supplemented formulations improve biocontrol and plant growth-promoting efficacy of *Bacillus subtilis* AF1. Can J Microbiol 47:618–625

Maurhofer M, Keel C, Schnider U, Voisard C, Hass D, Defago G (1992) Influence of enhanced antibiotic production in *Pseudomonas fluorescens* strain CHAO on diseases suppressive capacity. Phytopathology 82:190–195

Mew TW (1987) Current status and future prospects of research on bacterial blight of rice. Annu Rev Phytopathol 25:359–382
MgGrath MT (2001) Fungicide resistance in cucurbit powdery mildew, experiences and challenges. Plant Dis 85:236–245
Minorsky PV (2008) On the inside plant physiology. Plant Physiol 146:323–324
M'Piga P, Belanger RR, Paulitz TZ, Benhanau N (1997) Increased resistance to *Fusarium oxysporum* f. sp. radicis-lycopersici in tomato plants treated with the endophytic bacterium *Pseudomonas fluorescence,* strain 63-28. Physiol Mol Plant Pathol 50:301–320
Mizumoto S, Shoda M (2007) Medium optimization of Antifugal lipopeptide, iturin A, production by Bacillus subtilis in solid state fermentation by response surface methodology. Appl Microbiol Biotechnol 76:101–108
Moretti A, Mule G, Susca A, Gonzalez-Jaen MT, Logrieco A (2004) Toxin profile, fertility and AFLP analysis of *Fusarium verticillioides* from banana. Eur J Plant Pathol 110:601–609
Moretti A, Somma S, Picco AM, Rodolfi M, Ritieni A (2007) Biological, molecular and toxin characterization of Fusarium species isolated from bakanae diseased rice plants. Proceedings of 4th 259 International Rice conference, pp 164–165
Nagorska K, Bikowski M, Obuchowki M (2006) Multicellular behaviour and production of a wide variety of toxic substances support usage of bacillus subtilis as a powerful biocontrol agent. (review). Acta Biochim Pol 54(3):495–508
Nagorska K, Bikowski M, Obuchowki M (2007) Multicellular behavior and production of a wide variety of toxic substances support usage of *Bacillus subtilis* as a powerful biocontrol agent. (review). Acta Biochim Pol 54(3):495–508
Narayanan NN, Baisakh N, Oliva NP, VeraCruz CM, Gnanamanickam SS, Datta K, Datta SK (2002) Molecular breeding for the development of blast and bacterial blight resistance in rice cv. IR50. Crop Sci 42:2072–2079
Narayanan NN, Baisakh N, Oliva NP, VeraCruz CM, Gnanamanickam SS, Datta K, Datta SK (2004) Molecular breeding: marker-assisted selection combined with biolistic transformation for blast and bacterial blight resistance in indica rice (cv. CO39). Mol Breed 14:61–71
Oliveira AS, Xavier-Filho J, Sales MP (2003) Cysteine proteinases and cystatins. Braz Arch Bio Technol 46:91–104
Nelson PE, Toussoun TA, Marasas WFO (1993) Fusarium species. An illustrated annual for identification. The Pennsylvania State University Press, University Park and London, p 193
Ou SH (1985) Rice disease, 2nd edn. Commonwealth Mycol. Inst, Kew
O'Sullivan DJ, O'Gara F (1992) Traits of fluorescent pseudomonas spp. involved in suppression of plant root pathogens. Microbiol Rev 56:662–676
de Oliveira EJ, Rabinovitch L, Monnerat RG, Passos LKJ, Zahner V (2004) Molecular characterization of *Brevibacillus laterosporus* and its potential use in biological control. Appl Environ Microbiol 70:6657–6664
Ogawa K (1988) Damage by "bakanae" disease and its chemical control. Jpn Pestic Inf 52:13–15
Owan A, Zdor R (2001) Effect of cyanogenic rhizobacteria on the growth of velvet leaf (Abutilon Theophrasti) and corn (Zea Mays) in autoclaved soil and the influence of supplemented glycine. Soil Biol Biochem 33:801–809
Padwick GW (1950) Manual of rice diseases. Commonwealth Mycol. Inst, Kew, pp 74–84
Pal Pandi G, Karthikeyan V, Rajendran K (2016) Biological control of blast disease (*Magnaporthe grisea*) on minor millets using potential antagonistic bacteria prevailing in selected areas of Madurai District. Inter Res J Nat App Sci 3:3-13-34
Palleroni NJ, Kunisawa R, Contopoulou R, Doudoroff M (1973) Nucleic-acid homologies in genus Pseudomonas. Int J Syst Bacteriol 23:333–339
Palleroni NJ (2008) The road to the taxonomy of pseudomonas. In: Cornelis P (ed) Pseudomonas: genomics and molecular biology. Caister Academic Press, Norfolk, pp 1–18
Pascholati SF, Yoshioka H, Kunoh H, Nicholson RL (1992) Preparation of the infection court by *Erysiphe graminis* f. sp. *hordei*—cutinase is a component of the conidial exudates. Physiol Mol Plant Pathol 41:53–59

Phipps PM (1992) Evaluation of biological control agents for control of sclerotina blight of peanut. Biol Cult Tests Con Plant Dis 7:60

Pidoplichko VN, Garagulya AD (1974) Effect of antagonistic bacteria on development of wheat root rot. Mikrobiol Zh (Kiev) 36:599–602

Raaijmakers JM, Weller DM, Thomashow LS (1997) Frequency of antibiotic producing Pseudomonas spp. in natural environments. Appl Environ Microbiol 63:881–887

Raaijmakers JM, Weller DM (1998) Natural plant protection by 2, 4- Diacetylphloroglucinol producing Pseudomonas spp. in take-all decline soils. Mol Plant-Microbe Interact 11:144–152

Raaijmakers JM, Bonsall RF, Weller DM (1999) Effect of population density of *Pseudomonasfluorescens* on production of 2, 4-diacetylphloroglucinol in the rhizosphere of wheat. Phytopathology 89:470–475

Romero D, Vicente A, Rakotoaly RH, Dufour SE, Veening JW, Arrebola E, Cazorla FM, Kuipers O, Paquot M, Garcia AP (2007) The iturin and fengycin families of lipopeptides are key factors in antagonism of *Bacillus subtilis* towards *Podosphaera fusca*. Mol Plant-Microbe Interact 118:323–327

Manidipa R, Dutta SG, Venkata RC (2013) Pseudomonads: potential biocontrol agents of rice diseases. Res J Agric For Sci 1(9):19–25

Ruiu L, Satta A, Floris I (2008) Immature house fly (*Musca domestica*) control in breeding sites with a new *Brevibacillus laterosporus* formulation. Environ Entomol 37:505–509. doi:10.1603/0046-225X(2008)37[505:IHFMDC]2.0.CO;2

Sailaja PR, Podile AR, Reddanna P (1998) Biocontrol strains of *Bacillus subtilis* AF1. Rapidly induces lipoxygenase in groundnut (*Arachis hypoge* L.) compares to crown rot pathogen *Aspergillus niger*. Eur J Plant Pathol 104:125–132

Schippers B, Bakker A, Bakker P, Van Peer R (1990) Beneficial and deleterious effects of HCN-producing pseudomonads on rhizosphere interactions. Plant Soil 129:75–83

Schroth MN, Hancock JG (1981) Disease suppressive soil and root colonizing bacteria. Science 216:1376–1381

Shida O, Takagi H, Kadowaki K, Yano H, Abe M, Udaka S, Komagata K (1994) *Bacillus aneurinolyticus*sp. nov., nom. rev. Int J Syst Bacteriol 144:143–150

Smith JR, Handelsaman J, Goodman R (1999) Genetics basis in plants for interaction with disease suppressive bacteria. Proc Natl Acad Sci U S A 96:4786–4790

Smith KP, Harvey MJ, Handelsman J (1993) Suppression of cottony leak of cucumber with Bacillus cereus UW85. Plant Dis 77:139–142

Sriram PP, Shine YC, Park CS, Chung YR (1999) Biological control of Fusarium wilts of Cucumber by chitinolytic bacteria. Phytopathology 89:92–99

Stein T (2005) *Bacillus subtilis* antibiotics: structures, syntheses and specific functions. Mol Microbiol 56:845–857

Sturz AV, Christie BR, Nowak J (2000) Bacterial endophytes: potential role in developing sustainable systems of crop production. Critical Rev Plant Sci 19:1–30

Suryadi Y, Susilowati DN, Riana E, Mubarik NR (2013) Management of rice blast disease (*Pyriculariaoryzae*) using formulated bacterial consortium plant science. Emir J Food Agric 25(5):349–357

Talbot NJ (2003) On the trail of a cereal killer: exploring the biology of *Magnaporthe grisea*. Annu Rev Microbiol 57:177–202

Tendulkar SR, Saikumari YK, Patel V, Raghotama S, Munshi TK, Balaram P, Chatoo BB (2007) Isolation, purification and characterization of an antifungal molecule produced by *Bacillus licheniformis*BC98, and its effect on phytopathogen *Magnaporthe grisea*. J Appl Microbiol 103:2331–2339

Tomas-Lorente F, Iniesta- Sanmartin E, Tomas-barberan FA, Trowitzsch-Kienast W, Wray V (1989) Antifungal phloroglucinol derivatives and lipophilic flavonoids from Helichrysum decumbens. Phytochemistry 28:1613–1615

Valasubramanian R (1994) Biological control of rice blast with Pseudomonas fluorescens Migula: role of antifungal antibiotic in disease suppression. PhD Thesis, University of Madras

Vasudevan P, Gnanmanickam SS (2002) Suppression of rice bacterial blight and enhancement of plant growth by Bacillus spp. Proc. Int'I Rice Cong. Sep, 16–20, Beijing China, p 449

Vasudevan P (2002) Isolation and characterization of Bacillus spp from the rice rhizosphere and their role in biological control of bacterial blight of rice caused by *Xanthomonas oryzae pv. Oryzae*. PhD Thesis, University of Madras

Van Loon LC, Bakker PAHM, Pieterse CMJ (1998) Systemic resistance induced by rhizosphere bacteria. Annu Rev Phytopathol 36:453–483

Velusamy P (2003) Biological control of bacterial blight of rice by plant-associated bacteria: molecular and genetic analyses of metabolites for their role in disease control. PhD Thesis, University of Madras

Velusamy P, Defago G, Thomashow LS, Gnanamanickam SS (2004) Role of 2,4-diacetylphloroglucinol (DAPG) for plant disease control: its importance to rice bacterial blight suppression in India. In: Biotechnological approaches for the integrated management of crop diseases. Daya Publishing House, New Delhi, pp 182–191

Velusamy P, Immanuel JE, Gnanamanickam SS, Thomashow L (2006) Biological control of rice bacterial blight by plant-associated bacteria producing 2,4-diacetylphloroglucinol. Can J Microbiol 52:56–65

Velusamy P, Gnanamanickam SS (2003) Plant-associated bacteria 2,4-diacetylphloroglucinol (DAPG) production and suppression of rice bacterial blight in India. Curr Sci 85:1270–1273

Venkatesan BP, Gnanamanickam SS (1999) Occurrence of sub population of Xanthomonos oryzae with virulence to rice cultivar IRBB21 (Xa-21) in Southern India. Plant Dis 83:781

Viji S, Gnanamanickam SS (1996) Biological control of blast disease of finger millet (*Eleusine coracana* L.) and an analysis of the fertility status of *Magnaporthe grisea*. Curr Sci 71:144–147

Vincent MN, Harrison LA, Bracki JM, Kovacevich PA, Mukerji P, Weller DM, Pierson EA (1991) Genetic analysis of the antifungal activity of a soilborne *Pseudomonas aureofaciens* strain. Appl Environ Microbiol 57:2928–2934

Vleesschauwer D, Djavaheri M, Bakker P, Hofte M (2008) *Pseudomonas fluorescens* WCS374r-induced systemic resistance in rice against *Magnaporthe oryzae* is based on Pseudobactin-mediated priming for a salicylic acid-repressible multifaceted defense response. Plant Physiol 148(4):1996–2012

Yoshikawa M, Wakabayashi K, Suzuki H, Iwamura H (1993) Succinic and lactic acid as plant growth promoting compounds produced by rhizospheric *Pseudomonas putida*. Can J Microbiol 39:1150–1154

Yoshihoro T, Hayato H, Futoshi K (2003) Biological control of rice blast disease by Bacillus subtilis IK-1080. Ann Phytopathol Soc Japan 69(2):85–93

Zarandi ME, Shahidi Bonjar GH, Padash Dehkaei F, Ayatollahi Moosavi SA, Rashid Farokhi P, Aghighi S (2009) Biological control of rice blast (*Magnaporthe oryzae*) by use of *Streptomyces sindeneusis* isolate 263 in greenhouse. AMJ Appl Sci 6:194–199

Zhang S, Reddy MS (2001) Lack of induced systemic resistance in peanut to late leaf spot disease by plant growth-promoting Rhizobacteria and chemical elicitors. Plant Dis 85:879–884

Plant Growth-Promoting Rhizobacteria and Its Role in Sustainable Agriculture

10

Sunita J. Varjani and Khushboo V. Singh

Abstract

Soil is a dynamic living matrix which serves as a potential resource of food and agricultural applications. It is also a matrix for maintenance of life processes. Sustainable agriculture today has become an indispensable move for plant health which is continuously and invariably affected by pathogenic organisms. These pathogenic organisms have become a major and chronic threat to sustainable agriculture and ecosystem stability worldwide. Increase in yield, growth of plants, and protection from weeds and pathogens contribute to plant health by enormous use of chemical fertilizers. However the use of such chemical fertilizers has also lead to the side effects of agrochemicals. With respect to these considerations and public concerns about side effects of agrochemicals, today there is an urgent need for the biological agents which could be accepted worldwide. The use of plant growth-promoting rhizobacteria (PGPR) is a better alternative to resolve this problem. PGPR exerts potential health benefits to plant health thereby playing important role in increasing soil fertility, plant growth promotion, and suppression of phytopathogens in order to build an eco-friendly sustainable agriculture. This review provides an eco-friendly approach to increase crop production and plant health by growth-promoting rhizobacteria with global applicability.

S.J. Varjani (✉)
School of Biological Sciences and Biotechnology, University and Institute of Advanced Research, Gandhinagar, Gujarat, India
e-mail: drsvs18@gmail.com

K.V. Singh
Department of Microbiology, Gujarat University, Ahmedabad, Gujarat, India

V. Kumar et al. (eds.), *Probiotics in Agroecosystem*,
DOI 10.1007/978-981-10-4059-7_10

10.1 Introduction

In nature plants are always affected by biotic and abiotic factors corresponding to their environment which directly or indirectly affect the plant growth and health (Vessey 2003). A profound diversity of microbial community mostly bacteria thrive in the soil. It has been reported that soil bacteria in communication with rhizospheric organisms exert beneficial effects to plants. These rhizospheric microorganisms are the organisms that colonize the plant roots (especially legumes) with high density. Kloepper and Schroth (1981) termed these beneficial rhizobacteria as plant growth-promoting rhizobacteria (PGPR). PGPR can be defined as the indispensable part of rhizosphere biota that when grown in association with the host plants can stimulate the growth of the host (Vejan et al. 2016).

The rhizospheric zone is the area surrounding the roots of plants which is highly colonized by the rhizobacteria. This particular region surrounding the live creatures is affected qualitatively and quantitatively by vital activities of root like breath and root secretion (Alizadeh and Ordookhani 2011). Rhizosphere constitutes a mixture of solid particles, an active community of microbes, and nutritional constituents which enhance plant growth by a wide variety of mechanisms. Various factors responsible for plant growth promotion by PGPR are root hair proliferation, hair deformation and branching, increase in seedling emergence, early nodulation and nodule functioning, enhanced leaf surface area, biomass, vigor, increasing plant hormonal level, mineral and water uptake, promoted accumulation of carbohydrates, and yield in various plant species (Podile and Kishore 2006). Between the rhizobacteria and growing plants, three types of relationships, viz., positive, negative, and neutral can be observed (Whipps 2001).

An incredible demand for the use of PGPR biofertilizers is observed day by day with an increase in the importance of organic agriculture with minimum input of chemicals. The rhizospheric population in and around the soil roots is highly variable and is greatly dependent on the crop species cultivated and plant health (Tilak et al. 2005). Sustainable agriculture production can be achieved by emphasizing the use of PGPR as biofertilizer inoculants (Schippers et al. 1995). These biofertilizer inoculants could be a particular isolate or group of organisms that confer plant health. The potential role of PGPR to confer plant health could be imparted through direct or indirect mechanisms. Direct mechanisms facilitate nutrient uptake or increase nutrient availability by nitrogen fixation, solubilization of mineral nutrient, mineralization organic compounds, and production of phytohormones (Arora et al. 2012; Bhardwaj et al. 2014). On the contrary indirect mechanisms include antibiotic production, hydrolytic enzyme production, siderophore production, induced systemic resistance, exopolysaccharide production (Lugtenberg and Kamilova 2009; Tariq et al. 2014).

The present review focuses on microorganisms and PGPR, various mechanisms for PGPR action, role of PGPR in crop production, bioformulation as well as the future research, and development strategies to develop sustainable agriculture.

10.2 Plant Growth-Promoting Rhizobacteria and Microorganisms

Cook (2002) considered PGPR as a significant component in the management of agricultural practices with innate genetic potential. The concept of PGPR has now been confined to the bacterial strains that can fulfill at least two of the three criteria such as aggressive colonization, plant growth stimulation, and biocontrol (Weller et al. 2002; Vessey 2003).

According to Whipps (2001), there are three basic categories of interactions (positive, negative, and neutral) generally exists between the rhizobacteria and growing plants. Rhizobacterial colonizations with plant roots sometimes are commensals in which the bacteria form an innocuous interaction with host plants thereby exerting no growth and overall physiology of the host (Beattie 2006). With respect to negative interactions, phytotoxic substances are produced by phytopathogenic rhizobacteria such as hydrogen cyanide (HCN) or ethylene, thus exerting negative influence on the growth and physiology of the plants. On the contrary to these harmful bacteria, there are different PGPR criteria that can exhibit a positive plant health through direct mechanisms, such as solubilization of nutrients, nitrogen fixation, production of growth regulators, etc., or by stimulation of mycorrhizae development, competitive exclusion of pathogens, or removal of phytotoxic substances through indirect mechanisms (Bashan and de Bashan 2010).

According to the type and degree of association with the plant root cells, PGPRs can be bifurcated into extracellular plant growth-promoting rhizobacteria (ePGPR) and intracellular plant growth-promoting rhizobacteria (iPGPR) (Martinez-Viveros et al. 2010). The bacterial genera including the *Agrobacterium*, *Arthrobacter*, *Azotobacter*, *Azospirillum*, *Bacillus*, *Burkholderia*, *Caulobacter*, *Chromobacterium*, *Erwinia*, *Flavobacterium*, *Micrococcous*, *Pseudomonas*, *and Serratia* belong to ePGPR (Gray and Smith 2005). Endophyte comprises a wide range of soil bacterial genera such as *Allorhizobium*, *Azorhizobium*, *Bradyrhizobium*, *Mesorhizobium*, *and Rhizobium* of the family Rhizobiaceae that generally invades the root systems in crop plants to form nodules (Wang and Martinez-Romero 2000) and stimulates growth either through direct or indirect processes. This group of rhizobacteria is mostly Gram-negative and rod shaped with a lower population being Gram-positive rods, cocci, and pleomorphic. In addition to the above bacterial species, several actinomycetes also occur which serve as one of the major constituents of rhizosphere microbial populations and are also useful as they play major significant ecological roles in soil nutrient cycling (Elliot and Lynch 1995) as well as in plant growth-promoting activities (Merzaeva and Shirokikh 2006). Up till now a number of reports are available on the potential activities of actinomycetes as plant growth-promoting agent (Gomes et al. 2000; Sousa et al. 2008). Actinomycetes strains like *Micromonospora sp.*, *Streptomyces* spp., *Streptosporangium sp.*, and *Thermobifida sp.* are recorded as best species which colonize the plant rhizosphere, showing enormous potential as biocontrol agent against a range of root pathogenic fungi (Franco-Correa et al. 2010).

10.3 Mechanism of Action for Plant Growth Promotion

Plant growth promotion is a well-known process in which organic agriculture is often aided by certain traits of plant growth-promoting rhizobacteria. A number of mechanisms are followed by such potential bacteria to enhance the plant growth and health under diverse drastic environments. Plant growth promotion mediated by rhizobacteria occurs through modification of the whole microbial community inhabited near the rhizospheric niche by the production of certain compounds. Generally plant growth promotion by rhizobacteria is often mediated directly through availability of nutrients to plants through their solubilization or through the production of phytohormone and indirectly through the inhibition of pathogens, thereby increasing crop production, and development of plant health as biocontrol agents or root colonizers (Kloepper and Schroth 1981).

10.3.1 Direct Mechanisms

Plant growth promotion by rhizobacteria exhibits direct mechanisms that facilitate nutrient uptake by nitrogen fixation, solubilization of mineral nutrients (phosphate, iron, potassium), siderophore production, and phytohormone production (ethylene, indole acetic acid, jasmonic acid) (Arora et al. 2012; Bhardwaj et al. 2014).

10.3.1.1 Nitrogen Fixation

Nitrogen is an essential element for all life forms inhabiting on earth and is the most vital nutrient for the plant growth and health. Nearly 78% of nitrogen constitutes the environment but is still unavailable to the plants. Surprisingly today no such plant species has been found which could fix atmospheric dinitrogen into ammonia and expends it directly for its growth. A solution to the above problem is the biological nitrogen fixation (BNF), a process which converts atmospheric nitrogen to plant utilizable forms. This nitrogen is further converted to ammonia by nitrogen-fixing organisms using complex enzyme system, i.e., nitrogenase (Gaby and Buckley 2012).

BNF is a major source of nitrogen for plants, which not only complements and substitutes the mineral fertilizers but can also be an economically beneficial and ecologically sound alternative (Glick et al. 1999). In today's era for sustainable crop production, the use of biological inoculants is gaining popularity in various parts of the world, and biological nitrogen fixation serves as a major source of nitrogen in the agricultural fields. Bacterial genera contributing to symbiotic nitrogen fixation involve *Cyanobacteria*, *Rhizobium*, *Bradyrhizobium*, *Sinorhizobium*, *Azorhizobium*, *Allorhizobium*, *Mesorhizobium*, *and Frankia* (Brock et al. 2000). Paynel et al. (2001) have studied the nitrogen fixation mechanism extensively between rhizobium and leguminous plants. Apart from symbiotic nitrogen fixation in legumes, rhizobia as PGPR are also able to contribute growth promotion in non-legume species (Hoflich 2000). Certain substances produced by rhizobacteria as metabolites include cytokinins, riboflavin, lumichrome, auxin, abscisic acid, and vitamins which promote crop growth (Matiru and Dakora 2004).

10.3.1.2 Phosphate Solubilization

Next to nitrogen, phosphorus remains the most important key element in the plant nutrition. It virtually plays an important role in all the metabolic processes occurring in plant including photosynthesis, energy transfer, signal transduction, macromolecular biosynthesis, and respiration (Khan et al. 2010). Both the organic and inorganic forms are abundantly available in the soil. But plants are not able to utilize phosphate as 95–98% of its proportion is immobilized, insoluble and in the precipitated form. Hence plants obtain phosphorus only in the form of monobasic (H_2PO_4) and dibasic (HPO_4^{2-}) ions (Pandey and Maheshwari 2007). Basic mechanisms employed by the plant for PO_4 solubilization include (1) expulsion of complex and mineral-dissolving compounds like organic acid anions, protons, OH^- ions as well as CO_2; (2) production of extracellular enzymes; and (3) biological phosphate mineralization (Sharma et al. 2013). PGPR genera-contributing phosphate solubilization involves *Arthrobacter, Flavobacterium, Microbacterium, Pseudomonas, Rhizobacterium, Pseudomonas, Rhodococcus, Erwinia, Enterobacter, Bacillus, Burkholderia, and Serratia* (Pandey and Maheshwari 2007).

10.3.1.3 Siderophore Production

Almost all organisms constituting the biosphere need iron as an essential micronutrient. Though iron is the fourth most abundant element on earth, still it is not readily assimilated by any bacterial or plant species, because ferric ion which is predominantly present in nature is sparingly soluble, and hence for assimilation, the amount of iron available by living organisms is extremely low (Ma 2005). Siderophores being low molecular weight are the iron-chelating molecules which are released under iron-deficient conditions. In order to facilitate bioavailability of iron in the biological cells, siderophores possess high binding affinity and specificity of iron (III) (Schalk et al. 2001). Specialized mechanisms are evolved in microorganisms for assimilation of iron and thereby production of low molecular weight compounds, i.e., siderophores (Arora et al. 2013). Presently siderophores are classified into three main functional groups hydroxymates, catecholates, and carboxylates. Both direct and indirect enhancements are achieved by siderophores to accomplish plant growth by PGPR. A large number of PGPR that are capable to absorb the labeled iron include *Aeromonas, Azotobacter, Rhizobium, Serratia, Bacillus, Pseudomonas, and Strepotomyces* (Sujatha and Ammani 2013).

10.3.1.4 Phytohormone Production

A wide range of PGPR traits are found in the rhizospheric zone that produce certain substances that regulate plant growth and improvise plant health. Plant growth-promoting rhizobacteria produce phytohormones such as auxins, indole acetic acid (IAA), gibberellins, and abscisic acid ethylene and are well documented. Phytohormones exert profound impacts on root and shoot elongations and also affect cell proliferation in the root architecture. Such effects were achieved by overproduction of lateral roots and root hairs and subsequently increase in the nutrient and water uptake (Arora et al. 2013). The most common auxin found in plants occurs as indole-3-acetic acid that regulates many aspects of plant growth and

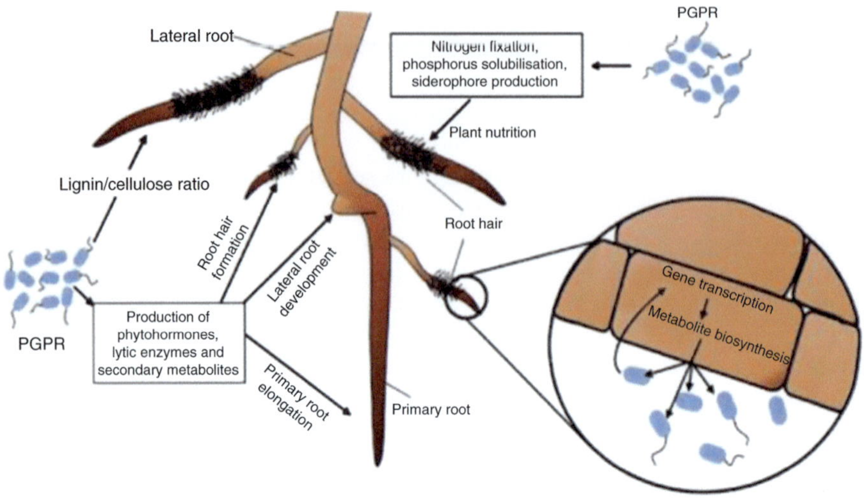

Fig. 10.1 Mechanism of action of PGPR toward growth promotion (Source: Vejan et al. 2016)

development. PGPRs may be synthesized in various organs and can be translocated to other sites where they trigger their morphological, physiological, and biochemical effects which enhance plant growth and development (Hayat et al. 2011). Figure 10.1 represents the possible mode of action used by plant growth-promoting rhizobacteria (PGPR) toward growth promotion in plants. The flow and location of nitrogen fixation, phosphorus solubilization, and siderophore production are shown in this figure (Vejan et al. 2016).

10.3.2 Indirect Mechanisms

Phytopathogenic microorganisms are the prime and the foremost threat to agriculture. This chronic threat to ecosystem stability worldwide subverts the soil ecology, disrupt environment, degrade soil fertility, and consequently show harmful effects on human health, along with contaminating ground water. Today PGPR remains a positive approach toward sustainable agriculture which inspires a wide range of exploitation of PGPR thus leading to reduction in the use of agrochemicals in order to improvise soil fertility by a variety of mechanisms via production of antibiotics, siderophores, HCN, and hydrolytic enzymes (Lugtenberg and Kamilova 2009).

10.3.2.1 Antibiosis

One of the most powerful and studied biocontrol mechanisms of PGPR against phytopathogens is the production of antibiotics. Amphisin, 2,4-diacetylpholoroglucinol (DAPG), oomycin A, phenazine, tropolone, tensin, cyclic lipopeptides, oligomycin A, kanosamine, zwittermicin A, and xanthomycin are some of the antibiotics studied up till now for antibiotics production. These antibiotics prevent the proliferation of plant pathogens (Compant et al. 2005; Looper and Gross 2007). One disadvantage of using such PGPR traits is that they may develop an antibiotic resistance toward specific antibiotics. Glick (2012) have utilized biocontrol agents that synthesize one or more antibiotics as a solution of this problem. Excluding the production of antibiotics, some PGPR also produce volatile compounds like hydrogen cyanide (HCN) for biocontrol of black root rot of tobacco which is caused by *Thielaviopsis basicola* (Sacherer et al. 1994).

10.3.2.2 Induced Systemic Resistance

Induced systemic resistance may be defined as a physiological state of enhanced defensive capacity elicited in response to specific environmental stimuli, and consequently the plants innate defenses are potentiated against subsequent biotic challenges (Avis et al. 2008). Systemic resistance against a broad spectrum of plant pathogens can also be provided by bio-priming plants with some plant growth-promoting rhizobacteria. The application of PGPR can also cause reduction in diseases caused by bacteria, fungi, virus, and in some instances even damage caused by insects and nematodes (Naznin et al. 2012). Moreover jasmonate and ethylene signaling within the plant is also facilitated by induced systemic resistance, and these hormones also stimulate host plant's defense responses against a variety of phytopathogens (Glick 2012). Many individual bacterial components can induce induced systemic resistance such as lipopolysaccharide (LPS), flagella, siderophores, cyclic lipopeptides, homoserine lactones, 2,4-diacetylphloroglucinol, and volatiles like acetoin and 2,3-butanediol (Doornbos et al. 2012).

10.3.2.3 Exopolysaccharide (EPS) Production/Biofilm Formation

A wide spectrum of multifunctional polysaccharides like intracellular, extracellular, and structural polysaccharides is produced by bacteria. The production of EPS is very important for the formation of biofilm. Root colonization can affect the interaction of microbes with roots appendages. The production of EPS by microbes helps in effective colonization of plant roots that helps to hold the free phosphorus from its insoluble form in soil and circulating essential nutrient to the plant for proper growth and development and protect it from attack by foreign pathogens. Other functions of EPS-producing microbes include shielding from desiccation, protection from stress (Qurashi and Sabri 2012), attachment to plant surfaces invasion, and plant defense response in plant-microbe interaction (Tewari and Arora 2014). Some EPS-producing PGPR can also bind cations, including Na^+, suggesting a role in mitigation of salinity stress by reducing the content of Na^+ available for plant uptake (Arora et al. 2013).

10.4 Role of PGPR in Crop Production

The three major staple food crops around the world are rice, wheat, and maize. A variety of PGPR isolates efficiently interact with C3 and C4 plants and can profoundly increase the yield of crops (Kennedy et al. 2004). Around 16–17 kg N is removed by rice crops to produce 1 t dry weight of rice crop with straw. Similarly wheat crop requires about 16–28 kg N which is removed by wheat crops to produce 1 t dry weight of grain with straw (Angus 2001). In case of maize plants, they require 9–11 kg N to produce 1 t biomass. Some of the obligatory anaerobic heterotrophs like clostridia can fix nitrogen in the complete absence of oxygen and are generally isolated from rice fields (Kennedy et al. 2004). Addition of organic source like straw presumably as a result of microbial breakdown of cellulose into cellobiose and glucose can increase the activity of isolates in rice production. A number of microorganisms can increase yield of rice by the application of *Azotobacter*, *Azospirillum lipoferum*, and *Azospirillum brasilense*, *Klebsiella pneumoniae*, *Enterobacter cloaca*, *Citrobacter freundii*, *Pseudomonas putida*, *and P. fluorescence* (Reis et al. 2000; Kennedy et al. 2004).

Some strains can supplement the use of urea-N in wheat production either by BNF or growth promotion. These strains include *Azospirillum*, *Azotobacter*, *Bacillus*, *Herbaspirillum*, *and Klebsiella*. Because of higher grain protein content, the N requirement of wheat is higher than rice depending on inherent soil fertility, amount of applied fertilizer, wheat variety, diseases and other management practices, and environmental conditions; wheat yields can vary widely from 1–7 t ha^{-1} (Angus 2001). The positive effects of *Azospirillum* on maize growth are mainly derived from physiological changes of the inoculated plant roots which enhance water and mineral uptake (Okan and Kapulnik 1986).

10.5 Commercialization of PGPR

The linkages and interaction between scientific organizations and industries decide the success and commercialization of plant growth-promoting rhizobacterial strains. The isolation of antagonistic strains, screening, fermentation methods, mass production, formulation viability, toxicology, industrial linkages, quality control, and field efficacy are the different stages in the process of commercialization under which abundance of work has been performed. Moreover factors which affect the commercial success of PGPR strains involve economical and viable market demand, consistent and broad spectrum market, longer shelf life, safety and stability, low capital costs, and easy availability of career materials (Nandakumar et al. 2001).

10.6 Plant Growth Promontory Bioformulations

Bioformulations may be defined as biologically active products in which one or more effective and beneficial microbial strains are present in an easy to use and economical carrier material. Most of the applications of bioformulations are meant under field applications for which the use of suitable carrier materials is very important in order to maintain cell viability under adverse environmental conditions.

A good quality and efficient formulation promotes survival of bacteria maintaining available population sufficient to exude growth-promoting effects on plants. Therefore plant growth-promoting rhizobacterial bioformulations refer to microbial preparation that may be considered as partial or complete substitute for chemical fertilizers and pesticides and most importantly offer an environmentally sustainable approach to increase crop production and health (Singh et al. 2014).

10.7 Future Research and Development Strategies for Sustainable Agriculture

The demand of high output yield and enhanced crop production with soil fertility in an eco-friendly manner is the necessary requirement of today's world. Hence further research must be focused on the new concept of rhizo-engineering based on favorable partitioning on exotic molecules, which creates a unique setting for interaction between plant and microbes (Tewari and Arora 2014). Future research in rhizosphere biology depends upon the development of molecular and biotechnological fields to increase our knowledge on rhizosphere biology and to achieve an integrated management of soil microbial populations. Based on the use of bio-inoculants, fresh inoculants must be used in order to increase other high-value crops such as fruits, vegetables, and flowers. With respect to reduce harmful impacts of stress on plant growth, an application of multi-strain consortium is always an effective approach as compared to a single bacterial inoculum. Another effective technology for enhancing plant growth at low temperature could be the addition of ice-nucleating PGPR (Nadeem et al. 2013).

Though research on nitrogen fixation and phosphate solubilization by PGPR is at progressed path, still research on other criteria for organic sustainable agriculture must also be considered. In addition, future marketing of such bio-inoculant active products when utilized in environment as eco-friendly alternatives against the exogenous pathogens or synthetic agrochemicals depends upon efficient protective and biosafety data required for the registration of plant growth-promoting rhizobacterial agents.

Conclusions

The present review indicates the development and formulations of PGPRs in biological promotion of different characteristics of plant growth. Most of the PGPR isolates significantly increased plant height, root length, and dry matter production in various agricultural crops like rice, maize, wheat, etc. To increase the crop yield on sustainable basis, the knowledge of microbial flora in the rhizospheric region of crops is very important. PGPR bio-inoculants not only enhance the plant yield directly but also protect it from the adverse environmental conditions. Therefore consistent use of plant growth-promoting rhizobacteria is an effective approach towards the development of organic sustainable agriculture.

Acknowledgment Authors are grateful to academic and administrative staff of University and Institute of Advanced Research, Gandhinagar, Gujarat, India for their kind support.

References

Alizadeh O, Ordookhani K (2011) Use of N_2 fixing bacteria *Azotobacter, Azospirillum* in optimizing of using nitrogen in sustainable wheat cropping. Adv Environ Biol 5(7):1572–1574

Angus JF (2001) Nitrogen supply and demand in Australian agriculture. Aust J Exp Agric 41:277–288

Arora NK, Tewari S, Singh R (2013) Multifaceted plant-associated microbes and their mechanisms diminish the concept of direct and indirect PGPR's. In: Arora NK (ed) Plant microbe symbiosis: fundamentals and advances. Springer, New Delhi, pp 411–449

Arora NK, Tewari S, Singh R, Lal N, Maheshwari DK (2012) PGPR for protection of plant health under saline conditions. In: Maheshwari DK (ed) Bacteria in agrobiology: stress management. Springer, Berlin, Heidelberg, pp 239–258

Avis TJ, Gravel V, Antoun H, Tweddel RJ (2008) Multifaceted beneficial effects of rhizosphere microorganisms on plant health and productivity. Soil Boil Biochem 40:1733–1740

Bashan Y, de Bashan LE (2010) Chapter two—how the plant growth-promoting bacterium *Azospirillum* promotes plant growth–a critical assessment. Adv Agron 108:77–136

Beattie GA (2006) Plant-associated bacteria: survey, molecular phylogeny, genomics and recent advances. In: Gnanamanickam SS (ed) Plant-associated bacteria. Springer, Dordrecht, pp 1–56. doi:10.1007/978-1-4020-4538-7_1

Bhardwaj D, Ansari MW, Sahoo RK, Tuteja N (2014) Biofertilizers function as key players in sustainable agriculture by improving soil fertility, plant tolerance and crop productivity. Microb Cell Factories 13:66

Brock JL, Albrecht KA, Tilbrook JC, Hay MJM (2000) Morphology of white clover during development of seed to clonal populations in grazed pastures. J Agric Sci 135:103–111

Compant S, Reiter B, Sessitch A, Nowak J, Clement C, Barka EA (2005) Endophytic colonization of *Vitis vinifera* L. by plant growth promoting bacterium *Burkholderia* sp. Strain 45. PsJN. Appl Environ Microbiol 71:1685–1693

Cook RJ (2002) Advances in plant health management in the twentieth century. Annu Rev Phytopathol 38:95–116

Doornbos RF, van Loon LC, Peter AHM, Bakker A (2012) Impact of root exudates and plant defense signaling on bacterial communities in the rhizosphere. Rev Sustain Dev 32: 227–243

Elliot LF, Lynch JM (1995) The international workshop on establishment of microbial inocula in soils: cooperative research project on biological resource management of the organization for economic cooperation and development (OECD). Am J Altern Agric 10:50–73

Franco-Correa M, Quintana A, Duque C et al (2010) Evaluation of actinomycete strains for key traits related with plant growth promotion and mycorrhiza helping activities. Appl Soil Ecol 45:209–217

Gaby JC, Buckley DH (2012) A comprehensive evaluation of PCR primers to amplify the nifH gene of nitrogenase. PLoS One 7:e42149

Glick BR (2012) Plant growth promoting bacteria: mechanisms and applications. Scintifica (Cairo) 2012:963401

Glick BR, Patten CL, Holguin G, Penrose DM (1999) Biochemical and genetic mechanisms used by plant growth promoting bacteria. Imperial College Press, London, p 267

Gomes RC, Semedo LTAS, Soares RMA, Alviano CS, Linhares LF, Coelho RRR (2000) Chitinolytic activity of actinomycetes from a cerrado soil and their potential in biocontrol. Lett Appl Microbiol 30:146–150

Gray EJ, Smith DL (2005) Intracellular and extracellular PGPR: commonalities and distinctions in the plant-bacterium signaling processes. Soil Biol Biochem 37:395–412

Hayat R, Ali S, Amara U, Khalid R, Ahmed I (2011) Soil beneficial bacteria and their role in plant growth promotion. A review. Ann Microbiol 60:579–598
Hoflich G (2000) Colonization and growth promotion of non-legumes by Rhizobium bacteria. Microbial bioecosystems: new frontiers. In: Bell CR, Brylinsky M, Jhonson Green P (eds) Proceedings of the 8th international symposium on microbial ecology. Atlantic Canada Society for microbial ecology, Halifax, pp 827–830
Kennedy IR, Choudhary AIMA, Kecskes ML (2004) Non-symbiotic bacterial diazotrophs in crop farming systems: can their potential for plant growth promotion be better exploited? Soil Boil Biochem 36(8):1229–1244
Khan MS, Zaidi A, Ahemad M, Oves M, Wani PA (2010) Plant growth promotion by phosphate solubilizing fungi – current perspective. Arch Argon Soil Sci 56:73–98
Kloepper JW, Schroth MN (1981) Relationship of in vitro antibiosis of plant growth promoting rhizobacteria to plant growth and the displacement of root microflora. Phytopathology 71:1020–1024
Looper JE, Gross H (2007) Genomic analysis of antifungal metabolite production by *Pseudomonas fluorescence* Pf-5. Eur J Plant Pathol 119:265–278
Lugtenberg B, Kamilova F (2009) Plant growth promoting rhizobacteria. Annu Rev Microbiol 63:541–556
Ma JF (2005) Plant root responses to three abundant soil minerals: silicon, aluminium and iron. Crit Rev Plant Sci 24:267–281
Martinez-Viveros O, Jorquera MA, Crowley DE, Gajardo G, Mora ML (2010) Mechanisms and practical considerations involved in plant growth promotion by rhizobacteria. J Soil Sci Plant Nutr 10:293–319
Matiru VN, Dakora FD (2004) Potential use of rhizobial bacteria as promoters of plant growth for increased yield in landraces of African cereal crops. Afr J Biotechnol 3(1):1–7
Merzaeva OV, Shirokikh IG (2006) Colonization of plant rhizosphere by actinomycetes of different genera. Microbiology 75:226–230. doi:10.1134/S0026261706020184
Nadeem SM, Naveed M, Zahir WA, Asghar HN (2013) Plant-microbe interactions for sustainable agriculture: fundamentals and recent advances. In: Arora NK (ed) Plant microbe symbiosis: fundamentals and advances. Springer, New Delhi, pp 51–103
Nandakumar R, Babu S, Viswanathan R, Sheela J, Raguchander T et al (2001) A new bio-formulation containing plant growth promoting rhizobacterial mixture for the management of sheath blight and enhanced grain yield in rice. BioControl 46:493–510
Naznin HA, Kimura M, Miyazawa M, Hyakumachi M (2012) Analysis of volatile organic compounds emitted by plant growth promoting fungus *phoma* sp. GS83 for growth promotion effects on tobacco. Microbes Environ 28:42–49
Okan Y, Kapulnik Y (1986) Development and function of *Azospirillum* inoculated roots. Plant Soil 90:3–16
Pandey P, Maheshwari DK (2007) Two sp. microbial consortium for growth promotion of *Cajanus cajan*. Curr Sci 92:1137–1142
Paynel F, Murray PJ, Cliquet B (2001) Root exudates: a pathway for short-term N transfer from clover rye grass. Plant Soil 229:235–243
Podile AR, Kishore GK (2006) Plant growth promoting rhizobacteria. In: Gnanamanickam SS (ed) Plant associated bacteria. Springer, Amsterdam, pp 195–230
Qurashi AW, Sabri AN (2012) Bacterial exopolysaccharide and biofilm formation chick pea growth and soil aggregation under salt stress. Braz J Microbiol 43:1183–1191
Reis VM, Baldani JIVLD, Doberener J (2000) Biological dinitrogen fixation in the graminae and palm trees. Crit Rev Plant Sci 19:227–247
Sacherer P, Defago G, Haas D (1994) Extracellular protease and Phospholipase C are controlled by the global regulatory gene gacA in the biocontrol strain *Pseudomonas fluorescence* CHAO. FEMS Microbiol Lett 116:155–160
Schalk IJ, Hennard C, Durgave C, Poole K, Absallah MA, Pattus F (2001) Iron free pyoverdine binds to its outer membrane receptor FpvA in *Pseudomonas aeruginosa*: a new mechanism for membrane iron transport. Mol Microbiol 39:351–360

Schippers B, Scheffer RJ, Lugtenberg JJ, Weisbek PJ (1995) Biocoating of seed with plant growth promoting rhizobacteria to improve plant establishment. Outlook Agric 24:179–185

Sharma SB, Sayyed RZ, Trivedi MH, Gobi TA (2013) Phosphate solubilizing microbes: sustainable approach for managing phosphorus deficiency in agricultural soils. Springer Plus 2:587

Singh S, Gupta G, Khare E, Behal KK, Arora NK (2014) Effect of enrichment material on the shelf life and field efficiency of bioformulation of *Rhizobium* sp. and P-solubilizing *Pseudomonas fluorescence*. Sci Res Report 4:44–50

Sousa CS, Soares ACF, Garrido MS (2008) Characterization of streptomycetes with potential to promote plant growth and biocontrol. Sci Agric 65:50–55. http://dx.doi.org/10.1590/S0103-9016 2008000100007

Sujatha N, Ammani K (2013) Siderophore production by the isolates of fluorescent Pseudomonads. Int J Cur Res Rev 5:1–7

Tariq M, Hameed S, Yasmeen T, Zahid M et al (2014) Molecular characterization and identification of plant growth promoting endophytic bacteria isolated from the root nodules of pea (*Pisum sativum L.*) World J Microbiol Biotechnol 30:719–725

Tewari S, Arora NK (2014) Multifunctional exopolysaccharides from *Pseudomonas aeruginosa* Pf23 involved in amelioration in sunflower under saline conditions. Curr Microbiol 69:484–494

Tilak KVBR, Ranganayaki N, Pal KK, De R, Saxena AK, Nautiyal CS, Mittal S, Tripathi AK, Johri BN (2005) Diversity of plant growth and soil health supporting bacteria. Curr Sci 89:136–150

Vejan P, Rosazlin A, Tumirah K, Salmah I, Amru NB (2016) Role of plant growth promoting rhizobacteria in agricultural sustainability – a review. Molecules 21(573):1–17. doi:10.3390/molecules21050573

Vessey JK (2003) Plant growth promoting rhizobacteria as biofertilizers. Plant Soil 255:571–586

Wang ET, Martinez-Romero E (2000) Sesbania Herbacea-rhizobium huautlense nodulation in flooded soils and comparative characterization of S. Herbacea-nodulating rhizobia in different environments. Microb Ecol 40:25–32

Weller DM, Raaijmakers JM, Gardener BB, Thomashow LS (2002) Microbial populations responsible for specific soil suppressiveness to plant pathogens. Annu Rev Phytopathol 40:309–348

Whipps JM (2001) Microbial interactions and biocontrol in the rhizosphere. J Exp Bot 52:487–511

Simultaneous P-Solubilizing and Biocontrol Activity of Rhizobacteria Isolated from Rice Rhizosphere Soil

11

Saida Aarab, Francisco Javier Ollero, Manuel Megias, Amin Laglaoui, Mohammed Bakkali, and Abdelhay Arakrak

Abstract

Application of beneficial rhizobacteria as plant biofertilizers and biocontrol agents may be a promising alternative to chemical control. To perform this aim, the present work is an evaluation of three phosphate-solubilizing *Aeromonas* strains isolated from the rhizosphere of rice. These selected rhizobacteria were checked for quantitative assay of tricalcium phosphate (TCP) solubilization and P concentrations were between 119.56 and 165.85 mg l^{-1}. Then, they were evaluated for extracellular hydrolytic enzymes production (chitinase, cellulase, amylase, lipase, and protease). The results showed that they were all able to hydrolyze different substrates apart from carboxymethyl cellulose. These test bacteria were checked in vitro as well for antagonism ability against six fungal phytopathogens, *Colletotrichum acutatum*, *Verticillium dahliae*, *Phytophthora cinnamomi*, *Phytophthora cactorum*, *Botryotinia fuckheliana*, and *Fusarium oxysporum*, and also against five phytopathogenic bacteria, *Pseudomonas savastanoi*, *Clavibacter michiganensis*, *Ralstonia solanacearum*, *Erwinia amylovora*, and *Pseudomonas*

S. Aarab
Département de Biologie, Faculté des Sciences et Techniques d'Al Hoceima, B.P.34, Ajdir Al Hoceima, Morocco

Equipe de Recherche de Biotechnologies et Génie des Biomolécules (ERBGB), Faculté des Sciences et Techniques de Tangier, B.P.416, Tangier, Morocco

F.J. Ollero
Departamento de Microbiología, Facultad de Biología, Universidad de Sevilla, 41012 Sevilla, Spain

M. Megias
Departamento de Microbiología y Parasitología, Facultad de Farmacia, Universidad de Sevilla, Apdo Postal 874, 41080 Sevilla, Spain

A. Laglaoui • M. Bakkali • A. Arakrak (✉)
Equipe de Recherche de Biotechnologies et Génie des Biomolécules (ERBGB), Faculté des Sciences et Techniques de Tangier, B.P.416, Tangier, Morocco
e-mail: arakrak_abdelhay@yahoo.fr

V. Kumar et al. (eds.), *Probiotics in Agroecosystem*,
DOI 10.1007/978-981-10-4059-7_11

syringae. GT70 was the only isolate that showed antifungal effect against two fungi, *V. dahliae* and *P. cactorum*. For antibacterial activities, the best inhibition was obtained in the presence of GT70 and PT29 against *R. Solanacearum*. Finally, these bacteria were used for rice inoculation substituting soluble P by TCP. The results demonstrated significant increases in plant length and dry matter, especially in the presence of both strains GT70 and PT29 after 30 days under controlled conditions. However, more evaluation of these isolates under field conditions is needed before recommending them as biofertilizers.

11.1 Introduction

The beneficial bacterial group referred to as plant growth-promoting rhizobacteria (PGPR) can affect positively the plant in three different ways: synthesizing particular compounds for the plants, facilitating the uptake of certain nutrients, and lessening or preventing the plants from diseases (Hayat et al. 2010). The solubilization of mineral phosphate is one of the most important PGP activities. Mineral P solubilization is a common phenotype in several rhizobacteria called "phosphate-solubilizing bacteria" (PSB) (Pérez et al. 2007), and *Aeromonas* is one of the important bacterial genera able to dissolve mineral P. In the context of increasing concern for biofertilization and biocontrol instead of chemicals to assure a sustainable agriculture, the use of PGPR as bioinoculants is of great importance. Besides, many authors have shown that inoculation of rice with PGPR could significantly increase different parameters such as plant height, root shoot biomass, and grain yield and improve plant health (Beneduzi et al. 2008; Lucas et al. 2009; Ng et al. 2012). Conclusively, this present work was conducted to evaluate phosphate-solubilizing *Aeromonas* obtained from rice rhizosphere cultivated in the northwest of Morocco for biocontrol activities.

11.2 Materials and Methods

11.2.1 *Aeromonas* Strains

Three strains of *Aeromonas*, namely, GT70, PT66, and PT29 isolated from rice rhizosphere, were obtained from the collection of the Laboratory of ERBGB, Faculty of Sciences and Technologies of Tangier. These strains have been evaluated in vitro for many PGP activities and their ability to promote the growth of rice and peanuts under greenhouse conditions.

11.2.2 Quantitative Assay of Phosphate Solubilization

The test isolates were inoculated in 50 ml PVK's (Pikovskaya 1948) broth (500 μl of 10^8 CFU ml^{-1}) and negative control consisted of uninoculated broth. All flasks were incubated at 28 ± 2 °C with shaking for 7 days. The cultures were centrifuged at 13,000 rpm for 20 min and the P of supernatant was determined by the colorimetric method as described by Ames (1966). The amount of soluble P was detected from the standard curve of KH_2PO_4. Dissolved P concentration was determined by subtracting the P concentration of control from the final concentration of soluble P obtained in the inoculated broths. The pH was determined using a pH meter.

11.2.3 Evaluation of Extracellular Hydrolytic Enzymes Production

The activities of extracellular hydrolytic enzymes were detected according to a qualitative assay in plates by streaking test bacteria on the medium containing enzyme substrate and detecting the zone of degraded substrate formed around the colony after an incubation period of 5–7 days at 28 °C. Chitinase activity was studied using colloidal chitin, amylase using soluble starch as substrate, cellulase using carboxymethyl cellulose (CMC), lipase using tributyrin, and protease using milk agar.

11.2.4 Evaluation of Antifungal Activities

To screen the PSB for in vitro antifungal activity, they were streaked on potato dextrose agar (PDA) plates at 3 cm in distance opposite to fungal phytopathogens (*Colletotrichum acutatum*, *Verticillium dahliae*, *Phytophthora cinnamomi*, *Phytophthora cactorum*, *Botryotinia fuckheliana*, and *Fusarium oxysporum*) inoculated at the center of the plate. The dishes were incubated at 22 °C for 5 days. The antifungal effect of antagonistic bacteria was characterized as a contact inhibition (C) or a distance inhibition (D).

11.2.5 Evaluation of Antibacterial Activities

To test selected PSB against phytopathogenic bacteria, plates containing Mueller-Hilton medium were inoculated with 100 μl of overnight grown culture of bacterial phytopathogens (*Pseudomonas savastanoi*, *Clavibacter michiganensis*, *Ralstonia solanacearum*, *Erwinia amylovora*, and *Pseudomonas syringae*). When the plates became dry, they were inoculated with the test bacteria and incubated at 28 °C. The inhibition halos around PSB were measured.

11.2.6 Inoculation of Rice

Rice (*Oryza sativa*) was used to evaluate the performance of strains under culture chamber conditions. The seeds were surface and germinated in 1% agar water (w/v) plates for 48–72 h at 25 °C. Each pot (12 cm diameter, 18 cm height) filled with vermiculite mixed with perlite (4:1) and 200 ml of nutrient solution (Rigaud and Puppo 1975) received 230 μl of 10% TCP as the sole source of P and then autoclaved. Every pot was sown by five germinated seeds, and each seed was inoculated directly with 1 ml of bacterial culture (10^8 CFU ml^{-1}) grown in TSB. Uninoculated pots and uninoculated pots containing soluble P in the form PO_4H_2K were used as controls (negative (C–) and positive (C+), respectively). All pots were maintained at 26 ± 2 °C under a 16 h photoperiod. Three replications were maintained for each treatment. Plants were harvested after 30 days and washed and dehydrated at 80 °C for 24 h. The dry weight biomass and shoot size were measured.

11.2.7 Statistical Analysis

The data are reported as means ± standard deviation (SD) for three replications or more. The results were subjected to analysis of variance (ANOVA) according to Fisher protected LSD test ($p < 0.05$) using the Statgraphics Plus version 4.0.

11.3 Results and Discussion

11.3.1 Quantitative Assay of TCP Solubilization

In the present work, three phosphate-solubilizing *Aeromonas* strains isolated from rhizosphere of rice were evaluated for TCP solubilization, and results are showed in the Table 11.1. The concentration of dissolved P was between 119.56 and 165.85 mg l^{-1} (Table 11.1).

Phosphorus is an essential macronutrient for plants and is often a limiting mineral nutrient, and the need of frequent application of P fertilizers has become a costly affair and also environmentally undesirable (Vassilev and Vassileva 2003). The solubilization of inorganic phosphate precipitated in the soil and make it available to plants is one of the most important activities exhibiting by some rhizobacteria termed PSB or phosphobacteria (Chen et al. 2008; Muleta et al. 2013).

This biosolubilization was accompanied by a significant decrease in pH (6.38–5.25) compared to uninoculated control (pH 6.8 ± 2). There is a statistically

Table 11.1 Quantitative test of TCP solubilization by the PSB isolates and pH values of media

	P (mg l^{-1})	pH
PT66	$165.85^a \pm 8.11$	$5.25^a \pm 0.03$
GT70	$119.56^b \pm 3.55$	$6.18^b \pm 0.08$
PT29	$158.33^a \pm 7.15$	$5.25^a \pm 0.01$

Values in lines followed by letter a and b differ significantly according to Fisher-protected LSD test ($p<0.05$)

significant relationship between final pH of the culture media and P concentrations ($r = -0.96, p < 0.01$). According to several studies, this acidification is caused by the production of organic acids by bacteria, and the negative relationship between pH and P indicates the significant role of these organic acids in mineral P solubilization (Chen et al. 2006; Pérez et al. 2007; Keneni et al. 2010).

11.3.2 Extracellular Hydrolytic Enzymes Production

PSB can also stimulate plant growth by other mechanisms such as the inhibition of phytopathogenic microorganisms by the production of biocontrol compounds (Bhattacharyya and Jha 2012). So, these test strains were checked for extracellular hydrolytic enzymes production. Chitinase, amylase, lipase, and protease activities were detected for all bacteria, while no strain was able to degrade CMC (Table 11.2).

11.3.3 Antagonist Effect Against Phytopathogenic Fungi and Bacteria

GT70 strain was the only isolate that showed an antifungal effect. It inhibited the growth of *V. dahliae* and *P. cactorum* (Table 11.3). On the contrary, PT66 and PT29 strains had no antifungal effect even if they have proven capable of solubilizing chitin.

Furthermore, GT70 and PT29 strains showed suppression of *R. solanacearum* as compared to PT66 isolate. But no important antibacterial effect was obtained against the rest of the phytopathogenic bacteria (Table 11.4).

Table 11.2 Extracellular hydrolytic enzymes production by test bacteria

Isolates	Chitinase	Amylase	CMCase	Lipase	Protease
PT66	+	+	–	+	+
GT70	+	+	–	+	+
PT29	+	+	–	+	+

Table 11.3 Effect of test bacteria on growth in vitro of phytopathogenic fungi

	C. acutatum	*V. dahliae*	*P. cinnamomi*	*P. cactorum*	*B. fuckheliana*	*F. oxysporum*
PT66	–	–	–	–	–	–
GT70	–	+ (D)	–	+ (C)	–	–
PT29	–	–	–	–	–	–

D distance antagonism, *C* contact antagonism

Table 11.4 Effect of test bacteria on growth in vitro of phytopathogenic bacteria

	P. savastanoi	*C. michiganensis*	*R. solanacearum*	*E. amylovora*	*P. syringae*
	Halo (mm)	Halo (mm)	Halo (mm)	Halo (mm)	Halo (mm)
PT66	–	–	–	–	–
GT70	+/–	–	3[a]	–	–
PT29	–	+/–	3[a]	–	–

Values in lines followed by letter a differ significantly according to Fisher-protected LSD test ($p<0.05$)

These results confirm what was reported earlier in literature about the efficiency of some rhizobacteria in biological control against both fungal and bacterial phytopathogens (Haas and Defago 2005). Besides, it was reported that the both isolates GT70 and PT29 are able to produce siderophores, while PT66 does not (Aarab et al. 2015). Thus, the significant antagonism effect obtained in the presence of GT70 and PT29 strains might be due to a synergistic combination of different metabolites such as siderophores, antibiotics, and various hydrolytic exoenzymes, which has been confirmed also by previous study (Shyamala and Sivakumaar 2012).

11.3.4 Effect of Seed Inoculation on Rice Growth

In the current study, P was the limiting factor for plant inoculation experiment (*Oryza sativa*, Puntal variety), as it is clearly shown by rice response to P fertilization (positive control, C+). The inoculation results showed a significant stimulating effect of strains on rice growth ($p<0.05$) compared to negative control (C–). Shoot length significantly increased in the presence of three test bacteria (Fig. 11.1). For dry matter, the best results were obtained in the presence of PT29 followed by GT70 compared to non-inoculated control (C–). However, PT66 isolate had no effect on dry biomass of rice plants (Fig. 11.1).

This stimulating effect of PSB exhibiting other phytobeneficial traits on rice growth is well documented (Lucas et al. 2009; Ng et al. 2012; Ramyasmruthi et al. 2012). Moreover, it has been shown that the inoculation with PSB stimulates the growth of other cereals such as maize (Hameeda et al. 2008; Frank and Julius 2012) and wheat (Afzal and Bano 2008). Recently, Bouhraoua et al. (2015) reported that inoculation of peanut plants (*Arachis hypogaea* L.) with GT70 has increased root dry biomass.

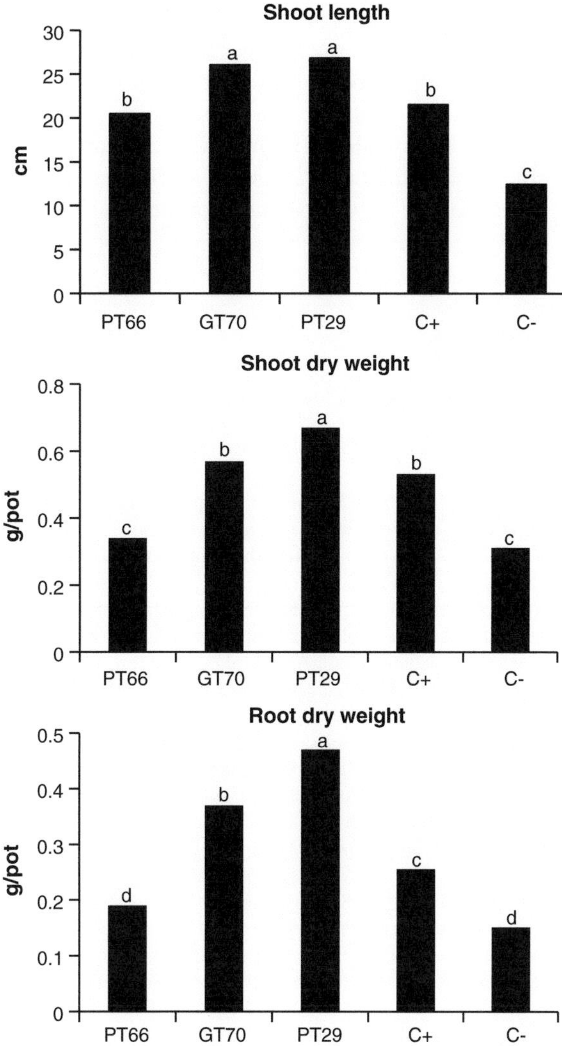

Fig. 11.1 Shoot length, dry matter of shoot, and root of inoculated plants after 30 days of growth in pots. The letters on the bars of the same parameter indicate significant differences according to Fisher protected LSD test ($p < 0.05$)

Conclusion

The results of this study make some isolates of PSB living in rice rhizosphere attractive as biofertilizers with biocontrol aspect, especially PT29 and GT77. These multiple intrinsic characteristics give these isolates a particular interest for their use in biological approaches for agriculture improvement. However, further studies are required to assess their effect under field conditions before they are recommended as bioinoculants for the plant-soil system.

References

Aarab S, Ollero FJ, Megías M, Laglaoui A, Bakkali M, Arakrak A (2015) Isolation and screening of bacteria from rhizospheric soils of rice fields in Northwestern Morocco for different plant growth promotion (PGP) activities: an *in vitro* study. Int J Curr Microbiol App Sci 4(1):260–269

Afzal A, Bano A (2008) *Rhizobium* and phosphate solubilizing bacteria improve the yield and phosphorus uptake in wheat (*Triticum aestivum*). Int J Agri Biol 10:85–88

Ames BN (1966) Assay of inorganic phosphate, total phosphate and phosphatases. Method Enzymol 8:115–118

Beneduzi A, Peres D, Vargas LK, Bodanese-Zanettini MH, Passaglia LMP (2008) Evaluation of genetic diversity and plant growth promoting activities of nitrogen-fixing bacilli isolated from rice fields in South Brazil. Appl Soil Ecol 39:311–320

Bhattacharyya PN, Jha DK (2012) Plant growth-promoting rhizobacteria (PGPR): emergence in agriculture. World J Microbiol Biotechnol 28:1327–1350

Bouhraoua D, Aarab S, Laglaoui A, Bakkali M, Arakrak A (2015) Effect of PGPR on growth and mycorrhization of KT22's peanut variety (*Arachis hypogaea* L.) grown in the northwest of Morocco. Am J Res Commun 3(2):12–24

Chen YP, Rekha PD, Arun AB, Shen FT, Lai WA, Young CC (2006) Phosphate solubilizing bacteria from subtropical soil and their tricalcium phosphate solubilizing abilities. Appl Soil Ecol 34:33–41

Chen Z, Ma S, Liu L (2008) Studies on phosphorus solubilizing activity of a strain of phosphobacteria isolated from chestnut type soil in China. Bioresour Technol 99:6702–6707

Frank O, Julius O (2012) Some characteristics of a plant growth promoting *Enterobacter* sp. isolated from the roots of maize. Adv Microbiol 2:368–374

Haas D, Defago G (2005) Biological control of soil-borne pathogens by fluorescent pseudomonads. Nat Rev Microbiol 3(4):307–319

Hameeda B, Harini G, Rupela OP, Wani SP, Reddy G (2008) Growth promotion of maize by phosphate-solubilizing bacteria isolated from composts and macrofauna. Microbiol Res 163:234–242

Hayat R, Safdar A, Amara U, Khalid R, Ahmed I (2010) Soil beneficial bacteria and their role in plant growth promotion: a review. Ann Microbiol 60(4):579–598

Keneni A, Assefa F, Prabu PC (2010) Isolation of phosphate solubilizing bacteria from the rhizosphere of Faba bean of Ethiopia and their abilities on solubilizing insoluble phosphates. J Agric Sci Tech 12:79–89

Lucas JA, Solano BR, Montes F, Ojeda J, Megias M, Mañero FJG (2009) Use of two PGPR strains in the integrated management of blast disease in rice (*Oryza sativa*) in Southern Spain. Field Crop Res 114:404–410

Muleta D, Assefa F, Börjesson E, Granhall U (2013) Phosphate-solubilising rhizobacteria associated with *Coffea arabica* L. in natural coffee forests of southwestern Ethiopia. J Saudi Soc Agric Sci 12:73–84

Ng LC, Sariah M, Sariam O, Radziah O, Abidin MAZ (2012) Rice seed bacterization for promoting germination and seedling growth under aerobic cultivation system. Aust J Crop Sci 6(1):170–175

Pérez E, Sulbaràn M, Ball MM, Yarzàbal LA (2007) Isolation and characterization of mineral phosphate-solubilizing bacteria naturally colonizing a limonitic crust in the south-eastern Venezuelan region. Soil Biol Biochem 39:2905–2914

Pikovskaya RI (1948) Mobilization of phosphorous in soil in connection with vital activity of some microbial species. Microbiologia 17:362–370

Ramyasmruthi S, Pallavi O, Pallavi S, Tilak K, Srividya S (2012) Chitinolytic and secondary metabolite producing *Pseudomonas fluorescens* isolated from *Solanaceae* rhizosphere effective against broad spectrum fungal phytopathogens. Asian J Plant Sci Res 2(1):16–24

Rigaud J, Puppo A (1975) Indole-3-acetic acid catabolism by soybean bacteroids. J Gen Microbiol 88:223–228

Shyamala L, Sivakumaar PK (2012) Antifungal activity of rhizobacteria isolated from rice rhizosphere soil against rice blast fungus *Pyricularia oryzae*. Int J Pharm Biol Arch 3(3):692–696

Vassilev N, Vassileva M (2003) Biotechnological solubilization of rock phosphate on media containing agro-industrial wastes. Appl Microbiol Biotechnol 61:435–440

Efficient Nutrient Use and Plant Probiotic Microbes Interaction

12

Moses Awodun, Segun Oladele, and Adebayo Adeyemo

Abstract

The use of organic and inorganic fertilizers excessively to enhance soil fertility and crop productivity in the quest towards food security achievement has resulted in detrimental environmental effects, water and soil pollution and ecosystem imbalance. Exploitation of eco-friendly beneficial microbes, emerging microbial bioengineering technologies and the identification of novel microbial gene resources which can be used in transgenic and designer plant technologies for efficient nutrient management have been touted to be the next transformational revolution in food security and mitigation of agro-environmental degradation. Research studies in this direction have demonstrated that these plant–microbe interaction technologies can enhance plant uptake of nutrients, increase the use efficiency of organic and inorganic fertilizers and a decrease in farming cost. The focus of this chapter is on recent findings in areas of plant–microbe interaction for nutrient management, major beneficial microbes, their mechanism of action and plant–microbe interactions using transgenic technologies for efficient nutrient use.

12.1 Introduction

Soil fertility depletion and crop productivity are a global challenge which requires concerted effort towards the realization of food security for an escalating world population. Soil quality is a critical component for food security and sustenance of

M. Awodun (✉) • S. Oladele • A. Adeyemo
Department of Crop, Soil and Pest Management, Federal University of Technology, Akure, Ondo State, Nigeria

Department of Agronomy, Adekunle Ajasin University, Akungba-Akoko, Ondo State, Nigeria
e-mail: m_awodun@yahoo.com; maawodun@futa.edu.ng

V. Kumar et al. (eds.), *Probiotics in Agroecosystem*,
DOI 10.1007/978-981-10-4059-7_12

the living terrestrial ecosystem and soil biota. However, soil quality has been eroded globally due to intensive crop production, soil erosion, excessive use of agrochemicals and heavy machinery usage coupled with climate change, fluctuating precipitation patterns and incidence of extreme weather events. It is estimated that by 2050, there is a need for food production to increase by over 50% (Godfray et al., 2010). Major challenge in meeting this anticipated increase in food demand will be the development of a sustainable enhanced crop production framework under an environmentally sound approach (Gruhn et al., 2000). One of the most realistic approaches in increasing crop productivity is conventional crop production practice (Singh and Macdonald 2014). However, it comes with an environmental cost and associated health problems. The use of synthetic fertilizers and organic materials to improve soil fertility and farm productivity has often negatively affected the biogeochemical cycles in a complex way (Perrott et al., 1992; Steinshamn et al., 2004 cited by Adesemoye et al., 2009a). Despite the deleterious environmental effects, an upsurge in amount of fertilizers used for intensive crop production worldwide is being projected to increase in other to meet increased food demands (Vitousek et al., 1997; Frink et al., 1999). However, recent statistics suggest declining or stagnating crop yields in developed and developing areas of the world despite heavy chemical fertilization.

This has been attributed to the deterioration in soil structure and fertility level of arable land in developed countries (Cassman et al., 2010), which resulted in stagnant and reduced crop yield. In the developing countries, scarce and expensive nutrient input affects farm productivity with associated decline in crop yield quantity and quality. Multiple nutrient limitations have been identified as a factor limiting nutrient-use efficiency required for efficient crop productivity (Singh and Macdonald 2014). For example, phosphorus (P) deficiency or unavailability as a result of chemical fixation into the A horizon of the soil layer could reduce crop performance due to its deleterious impact on nitrogen (N)-use efficiency (Bell and Lester 2011). The relationships that occur between soil characteristics and biological environments are important factors essential for crop productivity (NRC 2008). Improving nutrient-use efficiency is a necessity for increased crop productivity as current nutrient-use efficiency is very low in various cropping systems across the world. The phosphorus (P) and nitrogen (N)-use efficiency is considered below 30% and 50% of total inputs, respectively (Holford 1997). The major constraints in nutrient-use efficiency of plants arise from unavailability of nutrients for plant as nutrients applied to crops undergo biochemical transformation and become less available for plant uptake (Singh and Macdonald 2014). Nitrogenous fertilizers are utilized by microbes as a substrate for nitrification and denitrification (Inselsbacher et al. 2010). This can lead to a significant amount of applied-N being immobilized, leached as nitrate into water bodies causing eutrophication or released into the atmosphere as N_2O, a potent GHG. Furthermore, applied P fertilizers are distributed into different pools, and the majority of this becomes unavailable for plant uptake (Richardson and Simpson 2011). It is obvious that increasing nutrient-use efficiency will reduce cost and improve environmental benefits by reducing GHG emissions, leaching and losses, thus improving soil quality (Singh and Macdonald

2014). Regrettably, beneficial plant–microbe interaction has often been overlooked in breeding programmes for efficient nutrient use by plants despite its promising environmentally friendly potentials and important ecosystem functions for plant and soils. Due to these concerns, researchers and crop breeding scientists have directed their attention towards breeding nutrient efficient plants with dense roots attracting beneficial rhizospheric microbial communities for efficient nutrient and water uptake, thereby increasing crop yields, farm productivity and reducing environmental degradation. Research has repeatedly demonstrated the important roles beneficial plant growth-promoting microorganisms play in exerting various mechanisms such as biological nitrogen fixation, growth hormone production, phosphate solubilisation siderophore production, hydrolytic enzymes production and plant protection activity when activated, solely applied or synergised with starter nutrients (Bunemann et al., 2006; Huang et al., 2014; Tailor and Joshi 2014). In this regard, an integrated nutrient management (INM) plan is being promoted worldwide to reduce the negative impacts of nutrient loss (Adesemoye et al., 2008). Integrated nutrient management plan promotes Oladele and Awodun 2014b; Zou et al., 2014; Srivastava et al. 2015; Zhang et al., 2015); however AMF colonization performance could be determined by soil phosphorus pools (Stewart et al., 2005; Liu et al., 2014). Research has demonstrated that PGPF, PGPR or co-inoculants of PGPR and AMF can enhance nutrient-use efficiency of fertilizers and reduce chemical fertilizer rates (Bhardwaj et al., 2014). Studies have also demonstrated that inoculating banana with *Azotobacter* could substitute for up to 50% nitrogen requirement of banana (Tiwari et al., 1999) and 25% phosphorus requirement of papaya (Padma and Kandasamy 1999). Adesemoye et al., 2009a reported a synergistic interaction between PGPR and AMF with a better performance when compared to 70% synthetic fertilizer for P uptake. Similar trends was also observed in N uptake on a plant tissue analysis which presented that between 75% and 90%, fertilizer plus inoculants were significantly comparable to 100% synthetic fertilizer (Adesemoye et al. 2009b). A lot of study carried out worldwide has affirmed that a synergized plant–microbe interaction could spur plants to mobilize and accumulate required nutrients with regard to metabolic nutrient demand (Berg 2009; Shylaja and Rao 2012; Wu and Srivastava 2012). The outstanding diversity of beneficial microorganism which has been identified but yet to be put to use is dwarfed by the untapped potential of the microbial world. Reconnoitring this massive untapped microbial community is expected to bring to light new microbes with potentials for efficient plant nutrient management and crop productivity. The scientific community is now in a position to produce a timeless revolution where food security is achieved in a sustainable and environmentally friendly manner by selecting and associating crop plants with appropriate beneficial microbial communities to reduce or replace reliance on chemical inputs and breeding crop plants for positive microbe associations coupled with conservative soil management strategies. This chapter focuses on recent developments in areas of plant–microbe interaction for nutrient management, major beneficial microbes and their mechanism of action and plant–microbe interactions using transgenic technologies for efficient nutrient use.

12.2 Potentials of Beneficial Microbes in Nutrient Management and Use Efficiency: Recent Findings

In recent time, scientists have placed emphasis on exploiting safe, cost-effective and eco-friendly beneficial microorganisms for efficient nutrient management in sustainable crop production. These diverse naturally occurring beneficial microbes modify and enhance soil physical and chemical properties, plant nutrient-use efficiency, soil microbial biodiversity, soil health and crop productivity when introduced into the soil ecosystem (Sahoo et al., 2013). Beneficial microbial population includes PGPR, PGPF, nitrogen-fixing cyanobacteria, nitrogen-fixing rhizobia and mycorrhizal fungi. Dhanasekar and Dhandapani (2012) reported that efficient strains of *Azotobacter*, *Azospirillum*, Phosphobacter and Rhizobacter provided a vital and significant amount of nitrogen to *Helianthus annuus* while increasing plant height, leaf numbers, stem circumference, percentage of seed filling and seed dry weight. Furthermore, inoculation of rice plants with *Azotobacter*, *Azospirillum*, mycorrhizal fungi and *Rhizobium* improved the physiology, root morphology and plant tissue nutrient composition (Mishra and Sinha 2000; Choudhury and Kennedy 2004; Oladele and Awodun 2014a, b; Oladele 2015). *Azotobacter* with its diverse species such as *A. chroococcum*, *A. vinelandii*, *A. beijerinckii*, *A. nigricans*, *A. armeniacus* and *A. paspali* has also been reported earlier to have played a vital role in nitrogen fixation, nitrogen cycle in nature as well as a variety of metabolic functions. It also has the ability to produce vitamins such as thiamine and riboflavin (Revillas et al., 2000; Sahoo et al., 2014a). *Azospirillum* is another motile free-living aerobic bacterium that can survive under waterlogged conditions. Diverse species of the genus *Azospirillum* which includes *A. lipoferum*, *A. brasilense*, *A. amazonense*, *A. halopraeferens* and *A. irakense* has been reported to improve productivity of various inoculated crops. Research has shown that crops inoculated with *Azospirillum* experience changes in root morphology by producing plant growth-regulating substances via siderophore production (Sahoo et al., 2014b), which increases the number of root hairs formation and lateral roots to provide more root surface area to absorb sufficient water and nutrients. Furthermore, co-inoculation of *Azospirillum brasilense*, Rhizobium *meliloti* and 2, 4 D proved complementary as increased effect on grain yield, N, P and K content of *Triticum aestivum* was observed (Askary et al., 2009). *Azospirillum*-inoculated plants have been reported to have enhanced root growth and activities (i.e. acidification of the root surroundings) that increase phosphorous and other macronutrient and micronutrient uptake (Dobbelaere and Okon 2007). *Azospirillum* spp. has also been reported to enhance plant N uptake and plant growth promotion. Its nitrogen fixation abilities, phytohormone production, water adsorption, mineral uptake, proton and organic acid exudates were first reported by Dobbelaere et al., (2001) and Bashan et al., (2004). A significant increase in grain yields of lentil, pea, alfalfa and sugar beet rhizosphere, berseem, ground nut and soybean across different ecological zones and soil types has been reported due to inoculation with rhizobium inoculants. Rhizobium isolates from wild rice have also been reported to fix nitrogen in rice plants, thereby promoting growth and development (Hussain et al., 2002; Peng et al., 2008; Sharma et al., 2011; Grossman et al., 2011; Ramachandran et al. 2011; Rashid et al., 2012). PGPRs have demonstrated significant roles in N cycling and plant utilization of fertilizer N in plant–soil interactions

(Adesemoye et al., 2009a). Shaharoona et al., (2008) reported that pot and field trials with inoculation of *Pseudomonas fluorescens* (strain ACC50) and *P. fluorescens* biotype F (strain ACC73) showed increased use efficiency of N and P in all applied NPK fertilizer levels in wheat with strain ACC50 causing 115%, 52%, 26% and 27% increase over the non-inoculated control at NPK application rates of 25%, 50%, 75% and 100% of recommended doses, respectively. A field experiment conducted in India assessed the effectiveness of PGPR (*Azotobacter chroococcum* and *A. brasilense*) and AMF (*Glomus mosseae* and *G. fasciculatum*) on the growth, nutrient uptake and biomass production of pomegranate (*Punica granatum* L.). Strains were applied individually or in combinations. Results showed that dual inoculation of PGPR and AMF led to higher biomass production and increase in the uptake of N as well as P, K, Ca and Mg in pomegranate seedling. The suggested results from improved symbiotic N_2 fixation lead to increase in N and P uptake and improved phosphatase activity (Aseri et al., 2008). *Pseudomonas, Bacillus, Micrococcus, Flavobacterium, Fusarium, Sclerotium, Aspergillus, Penicillium, Enterobacter* and *Burkholderia vietnamiensis* have all been reported to be active in the phosphate-solubilising processes in the soil (Park et al., 2010; Pindi and Satyanarayana 2012). PGPR such as *Pseudomonas* and *Acinetobacter* strains have also been reported to enhance uptake of Fe, Zn, Mg, Ca, K and P by crop plants. Kohler et al. (2008) reported the effects of PGPR (*Pseudomonas mendocina Palleroni*) and AMF (G. *intraradices* and G. *mosseae*) on uptake of N, P, Fe, Ca and Mn (manganese) in lettuce (*Lactuca sativa* L. cv. Tafalla) under three different levels of water stress in Spain. Sheng and He (2006) cited by Adesemoye et al. (2009b) reported improved uptake of K through the inoculation of PGPR *B. edaphicus* strains NBT and suggested that the production of organic acids (citric, oxalic, tartaric, succinic and α-ketogluconic) by the strain and its mutants lead to chelation of metals and mobilization of K from K-containing minerals. Beneficial microorganism with such genus as *Aspergillus, Bacillus* and *Clostridium* is also found to be efficient in potassium solubilisation in the soil and mobilization for different crop uptake (Mohammadi and Sohrabi 2012). Nitrogen-fixing cyanobacteria such as *Aulosira, Tolypothrix, Scytonema, Nostoc, Anabaena* and *Plectonema* are also important contributors of nitrogen and growth-promoting substance to associated plants. Plant-associated beneficial microorganisms can supply macronutrients and micronutrients as required by associated plants. PGPRs are known to metabolize root exudates of associated plants and in turn provide nitrogen to the plant for amino acid synthesis in a nonparasitic beneficial relationship termed symbiosis. Free-living bacteria like *Azospirillum, Burkholderia* and *Stenotrophomonas* are known to be active in this regard with intrinsic ability to fix nitrogen and sulphate, which can be provided to the plant via oxidation mechanism (Banerjee and Yesmin 2002; Dobbelare et al., 2003). Apart from fixing atmospheric nitrogen and solubilising phosphate, PGPRs could also help sequester iron for plants by releasing siderophores (Raaijmakers et al., 1997; Bakker et al., 2007); producing plant hormones (Gutierrez-Manero et al., 2001) such as gibberellins, cytokinins and auxins and synthesizing the enzyme 1-amino cyclopropane-1-carboxylate (ACC) deaminase, which lowers plant levels of ethylene, thereby reducing environmental stress on plants (Glick et al., 2007). Most agricultural soils have large amounts of inorganic and organic P, which are immobilized and mostly unavailable, making many soils to be P deficient. PGPR and AMF can play significant roles in the solubilisation of inorganic phosphate and mineralization of organic phosphates (Tawaraya et al., 2006). Organic P usually accounts for 30–65% of total P in soils and must be converted to inorganic or low-molecular-weight organic acid compounds before they can be assimilated by plants (Adesemoye et al. 2009a). Beneficial microbes could also liberate phosphorous from organic compounds such as phytates and thus indirectly promote plant growth. Inorganic nutrients that is poorly soluble can be made available through the solubilisation of bacterial siderophores and the secretion of organic acids (Unno et al., 2005 cited by Berg 2009). AMF's ability to enhance water uptake, nutrient uptake (high-affinity for P uptake) and content and plant growth has been widely reported over the years. AMF's ability to mine the soil for available P through their extra radical hyphae with large surface areas acting as an interface between the soil and plant roots has enhanced P nutrition in plants. Liu et al., (2000) reported an increase in acquisition of Fe, Zn, Cu and Mn in mycorrhized maize; Giri and Mukerji (2004) also reported significant increase in Mg concentrations in seedling tissues of *Sesbania aegyptiaca* and *S. grandiflora* after application of AMF *Glomus macrocarpum*, compared with non-mycorrhized seedlings in saline soil. Bearing in mind the capacity of both PGPRs and AMF or PGPRs and PGPFs to help plants in uptake of nutrients, a quadruple interaction of PGPR–plant–AMF/PGPF has also been explored especially with the suggestion that AMF could possibly act as a vehicle to spread PGPR throughout the rhizosphere (Morrissey et al., 2004; Adesemoye et al., 2009b). In exploring the interactions between PGPR/PGPF and AMF for better plant-use efficiency of inorganic fertilizers or organic fertilizers, synergism is likely, but one must be aware that un-synergetic interactions between these microbes could also be a possibility (Fig. 12.1).

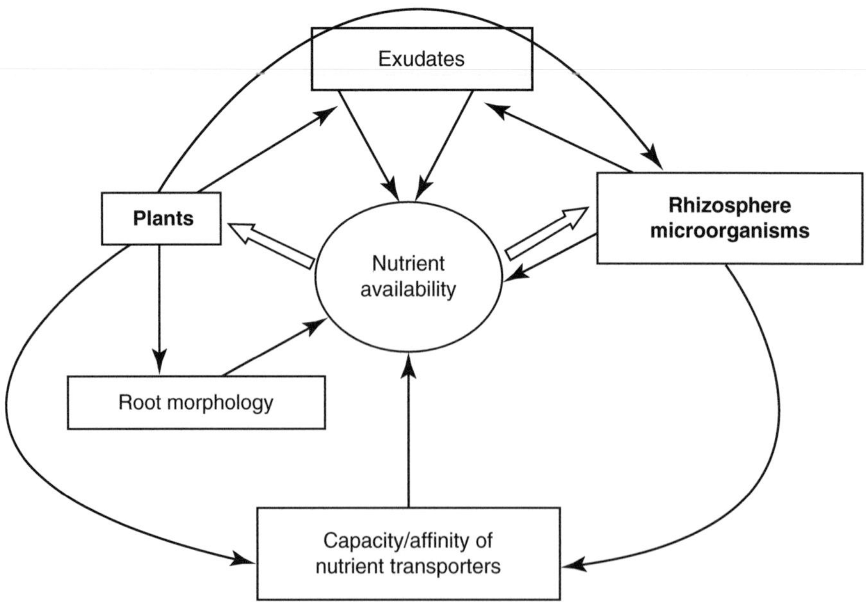

Fig. 12.1 Mechanisms involved in enhancing nutrient availability and uptake (Diagram culled from Rengel and Marschner 2005)

12.3 Integrated Nutrient Management: Microbe–Plant–Fertilizer Tripartite Interaction

Researchers and policymakers have suggested integrated nutrient management (INM) as the key towards mitigating agro-environmental problems. It is pertinent to note that this plan (INM) does not intend to eradicate inorganic fertilizer use permanently but to reduce its use and diminish the negative impacts of excessive use of inorganic fertilizers. The INM plan only encourages the combined use of low chemical input, organic input and beneficial microbes (biofertilizers) for efficient nutrient management and nutrient-use efficiency not compromising crop productivity while safeguarding the environment significantly on a long term and sustainable basis. A long-term field study conducted in India using the INM plan and *Citrus reticulata* as test crop on a Vertic Ustochrept showed much better effectiveness of beneficial microbes when used with inorganic fertilizers, organic manure and farmyard manure which resulted in increased fruit yield, fruit quality and improved soil quality (Srivastava et al. 2002, 2015). A study conducted on banana (*Musa acuminata* L.) using INM plan in a tripartite combination of farmyard manure (FYM) 12 kg/plant – *Azospirillum* sp. 50 g/plant – phosphate-solubilising bacteria 50 g/plant *T. harzianum* 50 g/plant resulted in fruit yield increase and quality (Hazarika and Ansari 2010). Patil and Shinde (2013), using the INM plan, validated Hazarika and Ansari (2010) findings on *Musa acuminata* L. by reporting significant increase in fruit yield and quality when 50% recommended dose of chemical fertilizer (RDM), FYM 20 kg/plant, *Azotobacter sp.* 50 g/plant, phosphate-solubilising bacteria 50 g/plant and VAM 250 g/plant was applied. Furthermore, a study on *Mangifera indica* L. conducted by Singh and Banik (2011) using the IPN plan which consists of 500 g N, 250 g P_2O_5, 250 K_2O g/plant 50 kg FYM and *Azospirillum* sp. 250 g/plant recorded significant increase in flowering, fruit setting, fruit and yield quality. This findings were also validated by Hasan et al. (2012) who conducted similar research on *Mangifera indica* L. using the IPN plan consisting of 250 N, 425 P2O5, 1000 K_2O, *Azospirillum* sp. 250 g/plant, PSB 250 g/plant, $ZnSO_4$ 100 g/plant and Borax 100 g/plant, which observed significant increase in fruit setting and fruit quality. Similar trend was reported by Yadav et al. (2007) on *Emblica officinalis* Gaertn 50% of N-P-K-S 105 kg N – 7.20 kg P_2O_5–125.25 kg K_2O/ha, beneficial *Azotobacter* sp., *Azospirillum* sp., phosphate-solubilising bacteria and FYM (2 tonnes/ha) and validated by Mandal et al. (2013) 100 g N, 25 g P_2O_5, 150 g K_2O/plant, FYM 10 kg/plant and phosphate-solubilising bacteria 50 g/plant with significant increase observed in fruit setting, yield and quality. In a field study conducted by Adesemoye et al. (2008) using an INM plan, it was reported that PGPR enhanced maize plant N uptake which impacted plant growth. The increase in maize plant N content is believed to have resulted from increased fertilizer N utilization efficiency. Results from these findings have revealed significant results which affirm that organic/inorganic/beneficial microbe synergy could be the future key for effective nutrient management and sustainable crop productivity.

12.4 Mechanism of Action of Beneficial Microbes

The association of AMF with roots of colonized plants serves as a prototypical system in studying and understanding the mechanism behind colonization pattern and growth stimulation in the root cells of infected plants (Wu et al., 2013; Bhardwaj et al., 2014). Genome sequencing of ectomycorrhizae fungi assisted in the identification of conditions that aided the development of mycorrhiza and its function in the plant cell (Bonfante and Genre 2010; Bhardwaj et al., 2014). About 15 genes were identified as putative hexose transporters as upregulated during symbiosis (Bonfante and Genre 2010). The movement and regulation of transporter genes during symbiosis specified the action of transportation of useful compounds like oligopeptides, amino acids and polyamines through the symbiotic relationship from one organism to other. Free-living mycelium can take nitrate and ammonium from soil, and subsequently, these compounds reach the mantle and hartig net and are transferred to the plants (Bhardwaj et al., 2014). The ability of *G. versiforme* to absorb phosphate from the soil for onward transmission to colonized plant has been linked to inorganic phosphate (Pi) transporters inherent on its hyphae, while the presence of glutamine synthase gene found in *G. intraradices* reinforces the prospect of nitrogen absorption in fungal hyphae that can be transported later to the plant (Salvioli et al., 2012). The pathways that prepare plant for both AM and Rhizobium colonization have some common points; a Nod factors of Rhizobium similar to bioactive compounds called Myc factors are suggested to be secreted by Rhizobium and Mycorrhiza and perceived by host roots for the activation of signal transduction pathway or common symbiosis (SYM) pathway (Kosuta 2003; Roberts et al., 2013 cited by Bhardwaj et al., 2014). In Fig. 12.2, bioactive ligands called Myc factors

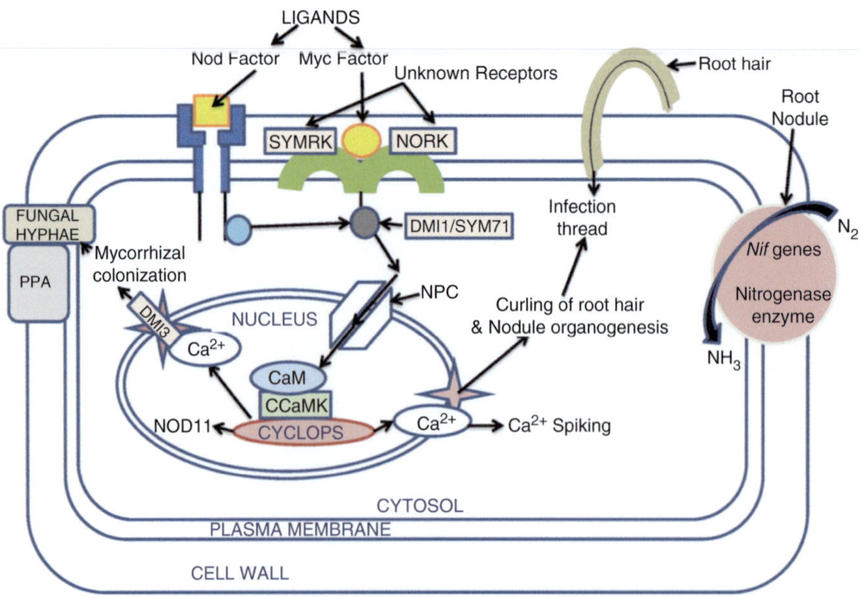

Fig. 12.2 Diagrammatic representation of mechanism of some beneficial microbes in the root cell (Diagram culled from Bhardwaj et al. 2014)

and Nod factors were shown to be secreted by mycorrhiza and Rhizobium perceived by host roots in triggering the signal transduction pathway which initiates further signal transduction pathway through unknown receptors (SYMRK and NORK) which trigger the release of Ca^{2+} in the cytosol, receptors like kinases or other kinase-related proteins like DMI and SYM71 in phosphorylating their substrates. Nuclear pore complex (NPC) and some of its proteins (NUP) play a role in calcium spiking. The periodic oscillation of calcium ions inside and outside the nucleus is being maintained by the role of DMI proteins (Bhardwaj et al. 2014). The production of nitric oxide (NO) is being induced by the production of PGPR, which acts as a second messenger to trigger a complex signalling network leading to improved root growth for nutrient interception and metabolic processes (Molina-Favero et al. 2007). The ability and potential of PGPRs have been enhanced by the introduction of genes involved in the direct oxidation (DO) pathway and mineral phosphate solubilisation (MPS) into some useful strains of PGPRs. *Acinetobacter calcoaceticus* and *E. coli*, *G. oxydans* and *Enterobacter asburiae* are some of the microorganisms where the gene encoding glucose dehydrogenase (gcd) involved in the DO pathway was cloned and characterized from (Tripura et al., 2007). *Rhizobium* and *Bacillus* were also found to synthesize IAA at different conditions such as pH, temperature and in the presence of agro-waste as substrate (Sudha et al., 2012). Nitrogen fixation genes are universally used by scientists through induction of nif genes to create transgenic plants that are able to fix atmospheric nitrogen independently. The induction of these nif genes in nitrogen fixing bacteria can only take place under low availability of nitrogen and oxygen in the rhizosphere (Santos et al., 2012). However, the effectiveness of nitrogen fixation is reliant on the utilization of carbon (Sevilla et al., 2001).

12.5 Transgenic Approach: Plant–Microbe Interactions for Nutrient-Use Efficiency

It is well established that plant–microbe interactions are the main key for primary productivity. Previous researches have highlighted different methodologies in harnessing microbe–plant interactions for increased crop productivity (Altomare and Tringovska 2011; Shen et al., 2013); however these biotechnological approach has not been fully utilized. Rhizospheric microbial interactions are moderated by a number of signals released by plant roots to communicate with soil microbes. Identifying these signals and harnessing them to improve interaction between beneficial microbes and plant roots can maximize available nutrient and use efficiency. A way to go is to genetically modify plants to enhance plant microbial signalling (Singh and Macdonald 2014). A transgenic plant encourages colonization and importunity of beneficial microbes through altered root exudation (Abhilash et al., 2012). Transgenic technology approach can be used to reduce nitrogen inputs by genetic engineering of nonleguminous crops to form N_2-fixing nodules with Rhizobia, resulting in N acquisition through N_2 fixation as each pathway is mediated by multiple genes. A successful transfer of nodule-forming capability in nonlegume crops requires a complete understanding of interactions between multiple plant and bacterial genes (Jones et al., 2007; NRC 2008). Another potential approach for N availability increase in plants includes engineering crops with N-fixing (*nif*) genes. *Nif* genes encode nitrogenase enzyme, an important enzyme in the fixation of N_2 present in a number of free-living and symbiotic bacteria. For this approach to be a reality engineering of *nif* genes in plants will require proper understanding of the interactions between different genes and N_2-fixing chemistry as quite a number of genes need to be transferred into plants in order to achieve N_2 fixation. Crop plants can also be engineered with some P mineralizing and solubilising genes from soil bacteria as several contains acid phosphatase enzymes for P solubilisation. It is envisioned that alkaline phosphatase and phytase genes can be harnessed through transgenic technics so that crops can directly access organic and/or fixed phosphorus and thus improve phosphorus uptake (Singh and Macdonald 2014). Recently some success story has been recorded in transferring phytase and phosphatase genes in transgenic plants (Tian et al., 2012 and Wang et al., 2013); however effectiveness of such an approach on a commercial scale remains elusive. However, it is not all bleak as scientist can be inspired by the success recorded from transgenic plants created worldwide for pest management through herbicide resistance genes, engineered from soil bacteria genes such as *Bacillus thuringiensis* and *Agrobacterium* strain CP4, respectively (Romeis et al., 2006). Using metagenomics, proteomics and metabolomics, it is now possible to identify and isolate genetic resources for maximizing nutrient cycling and nutrient-use efficiency without nurturing soil microbes which is a limitation of current technologies; metagenomics can offer key information on new genetic resources for novel traits in soils. Genes then can be either isolated or synthesized and be used for transgenic application. Additionally, combination of metagenomics along with conventional measures of soil properties could also be used to regulate the soil ability to provide nutrients for crops under low-input farming and fertilizer-use efficiency under conventional faming (Abhilash et al., 2012; Singh and Macdonald 2014). Another approach is development of "designer plants" currently in use for bioremediation (Abhilash et al., 2012), with better root traits which can access different zones in soils for nutrient and attract beneficial microbes, PGPRs and PGPFs with direct positive impacts on nutrient-use efficiency and crop productivity (Maity et al., 2012; Rayu et al., 2012). The designer plant exploits the strength and interactions between plants, rhizosphere bacteria, fungi and endophytical bacteria either through conventional or transgenic breeding. This preferentially attract N_2-fixing bacteria, mycorrhizal fungi and P-solubilising bacteria on roots for accessing organic N and P which can be manipulated with endosymbionts that fix N_2 inside plant tissues. It is believed that if all the above approaches are combined, this technology could have momentous impact on nutrient management and use, thereby increasing crop productivity (Singh and Macdonald 2014). However, there are constraints in developing these technologies appropriately on large scale.

Conclusion

Poor soil quality and environmental degradation are some of the challenges affecting crop productivity and plant nutrient management. Improving crop efficiency through better nutrient management and use efficiency in arable soils is one of those key challenges. Solving food security issues for the escalating world population requires a combination of conventional and biotechnological approach. Harnessing the potentials of beneficial microbes can help solve food security challenge of an increasing global population at a period when crop productivity is bedevilled with nutrient management and use efficiency inadequacies. Understanding the tripartite interactions between beneficial microbe, fertilizers and plants becomes a necessity. This interaction would help increase nutrient uptake into plant tissues, thereby reducing GHG emission, nutrient run-off or leaching and increased nutrient-use efficiency as lower rates of fertilizers will be required for increased crop productivity. Research efforts are being directed towards the development of transgenic, transfer of N_2-fixing genes and designer plant technology for improved nutrient-use efficiency and nutrient management; however its development on a commercial scale for field application could be challenging. A breakthrough in this direction will serve well to ameliorate nutrient-use efficiency, crop productivity, farm profitability, soil quality, reduction in environmental burden of fertilizer inputs and less stress on soil and ecosystems.

References

Abhilash PC, Powell JR, Singh HB, Singh BK (2012) Plant microbe interactions: novel applications for exploitation in multipurpose remediation technologies. Trends Biotechnol 30:416–420

Adesemoye AO, Torbert HA, Kloepper JW (2008) Enhanced plant nutrient use efficiency with PGPR and AMF in an integrated nutrient management system. Can J Microbiol 54:876–886

Adesemoye O, Anthony JW, Kloepper JW (2009a) Plant–microbes interactions in enhanced fertilizer-use efficiency. Appl Microbiol Biotechnol 85:1–12. doi:10.1007/s00253-009-2196-0

Adesemoye AO, Torbert HA, Kloepper JW (2009b) Plant growth promoting rhizobacteria allow reduced application rates of chemical fertilizers. Microb Ecol 58:921–929. doi:10.1007/s00248-009-9531-y

Altomare C, Tringovska I (2011) Beneficial soil microorganisms, an ecological alternative for soil fertility management. In: Lichtfouse E (ed) Genetics, biofuels and local farming system. Springer, Netherlands, pp 161–214

Aseri GK, Jain N, Panwar J, Rao AV, Meghwal PR (2008) Biofertilizers improve plant growth, fruit yield, nutrition, metabolism and rhizosphere enzyme activities of pomegranate (*Punica granatum* L.) in Indian Thar desert. Sci Hortic 117:130–135

Askary M, Mostajeran A, Amooaghaei R, Mostajeran M (2009) Influence of the co-inoculation *Azospirillum brasilense* and *Rhizobium meliloti* plus 2,4-D on grain yield and N, P, K content of *Triticum aestivum* (cv. Baccros and Mahdavi). Am Eurasian J Agric Environ Sci 5:296–307

Bakker PAHM, Pieterse CMJ, van Loon LC (2007) Induced systemic resistance by fluorescent *Pseudomonas* spp. Phytopathology 97:239–243

Banerjee M, Yesmin L (2002) Sulfur-oxidizing plant growth promoting Rhizobacteria for enhanced canola performance US Patent

Bashan Y, Holguin G, de-Bashan LE (2004) *Azospirillum* plant relationships: physiological, molecular, agricultural, and environmental advances (1997–2003). Can J Microbiol 50:521–577

Bell M, Lester D (2011) Observations on a general decline in fertility of multiple nutrients: Grain Research and Development Corporation update paper, 2011

Berg G (2009) Plant-microbe interactions promoting plant growth and health: perspectives for controlled use of microorganisms in agriculture. Appl Microbiol Biotechnol 84:11–18

Bhardwaj D, Ansari MA, Sahoo RK, Tuteja N (2014) Biofertilizers function as key player in sustainable agriculture by improving soil fertility, plant tolerance and crop productivity: a review. Microb Cell Factories 13:66–74

Bonfante P, Genre A (2010) Mechanisms underlying beneficial plant-fungus interactions in mycorrhizal symbiosis. Nat Commun 27:1–48

Bunemann EK, Schwenka GD, Van Zwiten L (2006) Impact of agricultural inputs on soil organisms – a review. Australian J Soil Res 44:379–408

Cassman KG, Grassini G, van Wart J (2010) Handbook of climate change and agroecosystems: impacts, adaptation and mitigation: ICP series on climate change impacts, adaptation and mitigation. In: Hillel D, Rosenzweig C (eds) Crop yield potential, yield trends, and global food security in a changing climate. Imperial College Press, London, pp 37–52

Choudhury MA, Kennedy IR (2004) Prospects and potentials for system of biological nitrogen fixation in sustainable rice production. Biol Fertil Soils 39:219–227

Dhanasekar R, Dhandapani R (2012) Effect of biofertilizers on the growth of *Helianthus annuus*. Int J plant Ani Environ Sci 2:143–147

Dobbelaere S, Okon Y (2007) The plant growth promoting effects and plant responses. In: Elmerich C, Newton WE (eds) Nitrogen fixation: origins, applications and research progress, Associative and endophytic nitrogen-fixing bacteria and cyanobacterial associations, vol V. Springer, Heidelberg, pp 145–170

Dobbelaere S, Croonenborghs A, Thys A, Ptacek D, Vanderleyden J, Dutto P, Labandera-Gonzalez C, Cabellero-Mellado J, Aguirrre JF, Kapulnik Y, Berner S, Burdman S, Kadour D, Sarig S, Okon Y (2001) Responses of agronomically important crops to inoculation with *Azospirillum*. Australian J Plant Physiol 28:871–879

Dobbelare S, Vanderleydern J, Okon Y (2003) Plant-growth promoting effects of diazotrophs in the rhizosphere. Crit Rev Plant Sci 22:107–149

Frink CR, Waggoner PE, Ausubel JH (1999) Nitrogen fertilizer: retrospect and prospect. Proc Natl Acad Sci 96:1175–1180

Giri B, Mukerji KG (2004) Mycorrhizal inoculant alleviates salt stress in *Sesbania aegyptiaca* and *Sesbania grandiflora* under field conditions: evidence for reduced sodium and improved magnesium uptake. Mycorrhiza 14:307–312

Glick BR, Todorovic B, Czarny J, Cheng Z, Duan J, McConkey B (2007) Promotion of plant growth by bacterial ACC deaminase. Crit Rev Plant Sci 26:227–242

Godfray HCJ, Crute IR, Haddad L, Lawrence D, Muir JF, Nisbett N (2010) The future of the global food system. Philos Trans R Soc Lond Ser B Biol Sci 365:2769–2777

Grossman JM, Schipanski ME, Sooksanguan T, Drinkwater LE (2011) Diversity of rhizobia nodulating soybean *Glycine max* (Vinton) varies under organic and conventional management. Appl Soil Ecol 50:14–20

Gruhn P, Goletti F, Yudelman M (2000) Integrated nutrient management, soil fertility, and sustainable agriculture: current issues and future challenges. Food, agriculture, and the environment- discussion paper 32. International Food Policy Research Institute, Washington, DC, pp 15–16

Gutierrez-Manero FJ, Ramos-Solano B, Probanza A, Mehouachi J, Tadeo FR, Talon M (2001) The plant-growth promoting rhizobacteria *Bacillus pumilus* and *Bacillus licheniformis* produce high amounts of physiologically active gibberellins. Physiol Plant 111:206–211

Hasan MA, Manna M, Dutta P, Bhattacharya K, Mandal S, Banerjee H, Ray SK, Jha S (2012) Integrated nutrient management in improving fruit quality of Mango 'Himsagar'. IX International Mango Symposium, ISHS *Acta Horticulturae* 992

Hazarika BN, Ansari S (2010) Effect of integrated nutrient management on growth and yield of banana cv. Jahaji. Indian J Hort 67:270–273

Holford ICR (1997) Soil phosphorus: its measurement, and its uptake by plants. Aust J Soil Res 35:227–239

Huang YM, Srivastava AK, Ying-Ning Z, Qiu-Dan N, Yu H, Qiang-Sheng W (2014) Mycorrhizal induced calmodulin mediated changes in antioxidant enzymes and growth response of drought stressed trifoliate orange. Front Microbiol 5:682–688

Hussain N, Mujeeb F, Tahir M, Khan GD, Hassan NM, Bari A (2002) Effectiveness of *Rhizobium* under salinity stress. Asian J Plant Sci 1:12–14.42

Inselsbacher E, Hinko-Najera Umana N, Stange FC, Gorfer M, Schueller E, Ripka K et al (2010) Short term competition between crop plants and soil microbes for inorganic N fertilizer. Soil Biol Biochem 42:360–372

Jones KM, Kobayashi H, Davies BW, Taga ME, Walker GC (2007) How rhizobial symbionts invade plants: the Sinorhizobium-Medicago model. Nat Rev Microbiol 5:619–633

Kohler J, Hernandez JA, Caravaca F, Roldan A (2008) Plant growth promoting rhizobacteria and arbuscular mycorrhizal fungi modify alleviation biochemical mechanisms in water stressed plants. Funct Plant Biol 35:141–151

Kosuta S (2003) Diffusible factor from arbuscular mycorrhizal fungi induces symbiosis specific expression in roots of *Medicago truncatula*. Plant Physiol 131:952–962

Liu A, Hamel C, Hamilton RI, Ma BL, Smith DL (2000) Acquisition of Cu, Zn, Mn, and Fe by mycorrhizal maize (*Zea Mays* L.) grown in soil at different P and micronutrient levels. Mycorrhiza 9:331–336

Liu C-Y, Srivastava AK, Qiang-Sheng W (2014) Effect of auxin inhibitor and AMF inoculation on growth and root morphology of trifoliate orange (*Poncirus trifoliata*) seedling. Indian J Agric Sci 84(11):1342–1346

Maity A, Jadhav VT, Chandra R (2012) Exploration of microbial wealth for sustainable horticultural production. Int J Bio Resourc Stress Manage 3:489–500

Mandal KK, Rajak A, Debnath S, Hasan MA (2013) Integrated nutrient management in aonla cv A-7 in the red lateritic region of West Bengal. J Crop Weed 9:121–123

Mishra DS, Sinha AP (2000) Plant growth promoting activity of some fungal and bacterial agents on rice seed germination and seedling growth. J Trop Agric 77:188–191

Mohammadi K, Sohrabi Y (2012) Bacterial Biofertilizers for sustainable crop production: a review. J Agric Biol Sci 7:307–316

Molina-Favero C, Mónica Creus C, Luciana Lanteri M, Correa-Aragunde N, Lombardo MC, Barassi AC, Lamattina L (2007) Nitric oxide and plant growth promoting rhizobacteria: common features influencing root growth and development. Adv Bot Res 46:1–33

Morrissey JP, Dow M, Mark GL, O'Gara F (2004) Are microbes at the root of a solution to world food production? Rational exploitation of interactions between microbes and plants can help to transform agriculture. EMBO Rep 5:922–926

NRC (2008) Emerging technologies to benefit farmers in sub-Saharan Africa and South Asia. The National Academic Press, Washington, DC

Oladele S (2015) Mycorrhizal fungus (*Glomus mosseae*) inoculation effects on performance and root biomass development of cacao seedlings in the nursery. Agric Forest 61(3):69–76

Oladele S, Awodun M (2014a) Influence of mycorrhizae and *Rhizobium* inoculation on growth, nutrient uptake and proximate composition of upland rice cultivars. J Nat Sci Res 4:24–30

Oladele S, Awodun M (2014b) Response of lowland rice to biofertilizer inoculation and their effects on growth and yield in Southwestern Nigeria. J Agric Environ Sci 3(2):371–390

Padma TMR, Kandasamy D (1999) Effect of interaction between VA-mycorrhizae and graded levels of phosphorous on the growth of papaya (*Carica papaya*). In Proceedings of the National Conference on trends in mycorrhizal research, 14–16 February, Haryana Agricultural University, Hisar, Haryana, pp 133-134

Park J, Bolan N, Megharaj M, Naidu R (2010) Isolation of phosphate-solubilizing bacteria and characterization of their effects on lead immobilization. Pedologist 53:67–75

Patil VK, Shinde BN (2013) Studies on integrated nutrient management on growth and yield of banana cv Ardhapuri (Musa AAA). J Hort Forest 5(9):130–138

Peng G, Yuan Q, Li H, Zhang W, Tan Z (2008) *Rhizobium oryzae* sp. nov., isolated from the wild rice *Oryza alta*. Int J Syst Evol Microbiol 58:2158–2163

Perrott KW, Sarathchandra SU, Dow BW (1992) Seasonal and fertilizer effects on the organic cycle and microbial biomass in a hill country soil under pasture. Austr J Soil Res 30:383–394

Pindi PK, Satyanarayana SDV (2012) Liquid microbial consortium- a potential tool for sustainable soil health. J Biofertil Biopest 3:4–12

Raaijmakers JM, Weller DM, Thomashow LS (1997) Frequency of antibiotic producing *Pseudomonas* spp. in natural environments. Appl Environ Microbiol 63:881–887

Ramachandran VK, East AK, Karunakaran R, Downie JA, Poole SP (2011) Adaptation of *Rhizobium leguminosarum* to pea, alfalfa and sugar beet rhizosphere investigated by comparative transcriptomics. Genome Biol 12:106–109

Rashid MH, Schafer H, Gonzalez J, Wink M (2012) Genetic diversity of rhizobia nodulating lentil (*Lens culinaris*) in Bangladesh. Syst Appl Microbiol 35:98–109

Rayu S, Karpouzas DG, Singh BK (2012) Emerging technologies in bioremediation: constraints and opportunities. Biodegradation 23:917–926

Rengel Z, Marschner P (2005) Nutrient availability and management in the rhizosphere; exploiting genotypic differences; a review. New Phytol 168:305–312

Revillas JJ, Rodelas B, Pozo C, Martinez-Toledo MV, Gonzalez LJ (2000) Production of B-group vitamins by two *Azotobacter* strains with phenolic compounds as sole carbon source under diazotrophic and adiazotrophic conditions. J Appl Microbiol 89:486–493

Richardson AE, Simpson RJ (2011) Soil microorganisms mediating phosphorus availability update on microbial phosphorus. Plant Physiol 156:989–996

Roberts NJ, Morieri G, Kalsi G, Rose A, Stiller J, Edwards A, Xie F, Gresshoff PM, Oldroyd GE, Downie JA, Etzler ME (2013) Rhizobial and mycorrhizal symbioses in *Lotus japonicus* require lectin nucleotide phosphohydrolase, which acts upstream of calcium signaling. Plant Physiol 161:556–567

Romeis J, Meissle M, Bigler F (2006) Transgenic crops expressing *Bacillus thuringiensis* toxins and biological control. Nat Biotechnol 24:63–71

Sahoo RK, Bhardwaj D, Tuteja N (2013) Biofertilizers: a sustainable eco-friendly agricultural approach to crop improvement. In: Tuteja N, Gill SS (eds) Plant acclimation to environmental stress. Springer Science Plus Business Media, New York, pp 403–432

Sahoo RK, Ansari MW, Dangar TK, Mohanty S, Tuteja N (2014a) Phenotypic and molecular characterization of efficient nitrogen fixing *Azotobacter* strains of the rice fields. Protoplasma 251(3):511–523. doi:10.1007/s00709-013-0547-2

Sahoo RK, Ansari MW, Pradhan M, Dangar TK, Mohanty S, Tuteja N (2014b) Phenotypic and molecular characterization of efficient native Azospirillum strains from rice fields for crop improvement. Protoplasma 251(4):943–953. doi:10.1007/s00709-013-0607-7

Salvioli A, Zouari I, Chalot M, Bonfante P (2012) The arbuscular mycorrhizal status has an impact on the transcriptome profile and amino acid composition of tomato fruit. BMC Plant Biol 12:44–48

Santos VB, Araujo SF, Leite LF, Nunes LA, Melo JW (2012) Soil microbial biomass and organic matter fractions during transition from conventional to organic farming systems. Geoderma 170:227–231

Sevilla M, Burris RH, Gunapala N, Kennedy C (2001) Comparison of benefit to sugarcane plant growth and 15n2 incorporation following inoculation of sterile plants with *Acetobacter diazotrophicus* wild-type and Nif–mutant strains. Mol Plant-Microbe Interact 14:358–366
Shaharoona B, Naveed M, Arshad M, Zahir ZA (2008) Fertilizer dependent efficiency of Pseudomonads for improving growth, yield, and nutrient use efficiency of wheat (*Triticum aestivum* L.) Appl Microbiol Biotechnol 79:147–155
Sharma P, Sardana V, Kandola SS (2011) Response of groundnut (*Arachis hypogaea L.)* to *Rhizobium* Inoculation. Libyan Agric Res Centre J Int 2:101–104
Shen J, Li C, Mi G, Li L, Yuan L, Jiang R (2013) Maximizing root/rhizosphere efficiency to improve crop productivity and nutrient use efficiency in intensive agriculture of China. J Exp Bot 64:1181–1192
Sheng XF, He LY (2006) Solubilization of potassium-bearing minerals by a wild type strain of *Bacillus edaphicus* and its mutants and increased potassium uptake by wheat. Can J Microbiol 52:66–72
Shylaja M, Rao MS (2012) *In vitro* compatibility studies of *Trichoderma harzianum* with inorganic fertilizers. Nematol Mediterr 40:51–54
Singh SR, Banik BC (2011) Response of integrated nutrient management on flowering, fruit setting, yield and fruit quality in mango cv. Himsagar (*Mangifera indica* L). Asian J Hort 6:151–154
Singh B, Macdonald C (2014) Harnessing plant-microbe interactions for enhancing farm productivity. Bioeng Landes Bio Sci 5(1):5–9
Srivastava AK, Singh S, Marathe RA (2002) Organic citrus: soil fertility and plant nutrition. J Sustain Agric 19:5–29
Srivastava AK, Malhotra SK, Krishna Kumar NK (2015) Exploiting nutrient microbe synergy in unlocking productivity potential of perennial fruits: a review. Indian J Agric Sci 85(4):459–481
Steinshamn H, Thuen E, Bleken MA, Brenoe UT, Ekerholt G, Yri C (2004) Utilization of nitrogen (N) and phosphorus (P) in an organic dairy farming system in Norway. Agric Ecosyst Environ 104:509–522
Stewart LI, Hamel C, Hogue R, Moutoglis P (2005) Response of strawberry to inoculation with arbuscular myccorrhizal fungi under very high soil phosphorus conditions. Mycorrhiza 15:612–619
Sudha M, Gowri RS, Prabhavati P, Astapriya P, Devi SY, Saranya A (2012) Production and optimization of indole-acetic-acid by indigenous micro flora using agro waste as substrate. Pak J Biol Sci 15:39–43
Tailor AJ, Joshi BH (2014) Harnessing plant growth promoting rhizobacteria beyond nature: a review. J Plant Nutr 37(1):534–537
Tawaraya K, Naito M, Wagatsuma T (2006) Solubilization of insoluble inorganic phosphate by hyphal exudates of arbuscular mycorrhizal fungi. J Plant Nutr 29:657–665
Tian J, Wang X, Tong Y, Chen X, Liao H (2012) Bioengineering and management for efficient phosphorus utilization in crops and pastures. Curr Opin Biotechnol 23:866–871
Tiwari DK, Hasan MA, Chattopadhyay PK (1999) Leaf nutrient and chlorophyll content in banana (Musa AAA) under influence of *Azotobacter* and *Azospirillum* inoculation. Environ Ecol 17:346–350
Tripura CB, Reddy SP, Reddy MK, Sashidhar B, Podile AR (2007) Glucose dehydrogenase of a rhizobacterial strain of *Enterobacter asburiae* involved in mineral phosphate solubilization shares properties and sequence homology with other members of enterobacteriaceae. Indian J Microbiol 47:126–131
Unno Y, Okubo K, Wasaki J, Shinano T, Osaki M (2005) Plant growth promotion abilities and microscale bacterial dynamics in the rhizosphere of lupin analysed by phytate utilization ability. Environ Microbiol 7:396–404

Vitousek PM, Aber JD, Howarth RW, Likens GE, Matson PA, Schindler DW, Schlesinger WH, Tilman DG (1997) Technical report: human alteration of the global nitrogen cycle: sources and consequences. Ecol Appl 7:737–750

Wang Y, Ye X, Ding G, Xu F (2013) Over expression of phyA and appA genes improves soil organic phosphorus utilisation and seed phytase activity in *Brassica napus*. PLoS One 8:e60801

Wu QS, Srivastava AK (2012) Rhizosphere microbial communities: isolation, characterization and value addition for substrate development. In: Srivastava AK (ed) Advances in citrus nutrition. Springler, Netherlands, pp 169–2194

Wu QS, Srivastava AK, Ying-Ning Z (2013) AMF-induced tolerance to drought stress in citrus. A review. Sci Hort 164:77–87

Yadav R, Singh HB, Singh HK, Yadav AL (2007) Effect of integrated nutrient management on productivity and quality of aonla (*Emblica officinalis* Gaetm.) fruits. Plant Arch 7(1&2):881–883

Zhang ZZ, Srivastava AK, Qiang-Sheng W, Guo-Huai L (2015) Growth performance and rhizospheric traits of peach (*Prunus persica*) in response to mycorrhization on replant versus non replant soil. Indian J Agric Sci 85(1):125–130

Zou YN, Srivastava AK, Wu QS, Huang YM (2014) Increased tolerance of trifoliate orange (*Poncirus trifoliata*) seedlings to waterlogging after inoculation with arbuscular mycorrhizal fungi. J Anim Plant Sci 24(5):415–420

Exploring the Plant Microbiome Through Multi-omics Approaches

13

Rubén López-Mondéjar, Martin Kostovčík, Salvador Lladó, Lorena Carro, and Paula García-Fraile

Abstract

Like many other high organisms, plants harbour a microbiome. The plant microbiome can be defined as the communities of microbial symbionts (microbiota) plus their collective genetic material, which determines the properties of the interactions between the microbes themselves and with their host.

The plant microbiome is crucial in plant health and crop yields. The understanding and management of the plant microbiome have the potential to decrease plant diseases and increase agricultural production; this can allow a reduction in the use of chemical fertilisers and pesticides in fields, thereby increasing the production of food to sustain the human population while simultaneously protecting the environment and human health. Consequently, many scientific studies in recent years have focused on unravelling the secrets of the plant microbiome, and the development of several omics techniques has greatly contributed to this aim.

In this chapter, we will review the methodologies of the application of high-throughput sequencing techniques in performing metagenomics studies focused on the microbiota as a part of the plant microbiome, we will provide an overview of the most recent research on this topic, and we will review other omics approaches important in deciphering and understanding the plant microbiome in full, presenting some of the main goals addressed to date.

R. López-Mondéjar • M. Kostovčík • P. García-Fraile (✉)
Institute of Microbiology of the CAS, v. v. i., Průmyslová 595, 25242 Vestec, Czech Republic
e-mail: paulagf81@usal.es

S. Lladó
Institute of Microbiology of the CAS, v. v. i., Vídeňská 1083, 142 20 Prague 4, Czech Republic

L. Carro
School of Biology, Newcastle University, Ridley Building, Newcastle upon Tyne NE1 7RU, UK

V. Kumar et al. (eds.), *Probiotics in Agroecosystem*,
DOI 10.1007/978-981-10-4059-7_13

13.1 Introduction

Currently, eukaryotes are considered as a whole only when their microbiomes are included (Berg et al. 2014; Hirsch and Mauchline 2012; Muller et al. 2016; Rosenberg and Zilber-Rosenberg 2016) because microbial genomes contribute to the anatomy, physiology, immunity, development, behaviour, and genetic variation of their host. Moreover, the inclusion of microbial genes in a host's hologenome or pan-genome is a powerful mechanism for evolution, making it occur much faster because the host is able to more easily adapt to its environment (Rosenberg and Zilber-Rosenberg 2016).

All plant tissues are susceptible to hosting microbial communities: the rhizosphere (region of soil influenced by plant root exudates), the phyllosphere (plant aerial surfaces), and the endosphere (internal tissues) (Fig. 13.1).

Microorganisms in the rhizosphere and phyllosphere are considered to be epiphytes, and those from the endosphere are known as endophytes. These microbes can establish beneficial, neutral, or detrimental associations with their host plants. A *plant microbiome* can be defined as the communities of plant-microbial symbionts (microbiota) plus their collective genetic material, which determines the properties of the interactions between the microbes themselves and with their host.

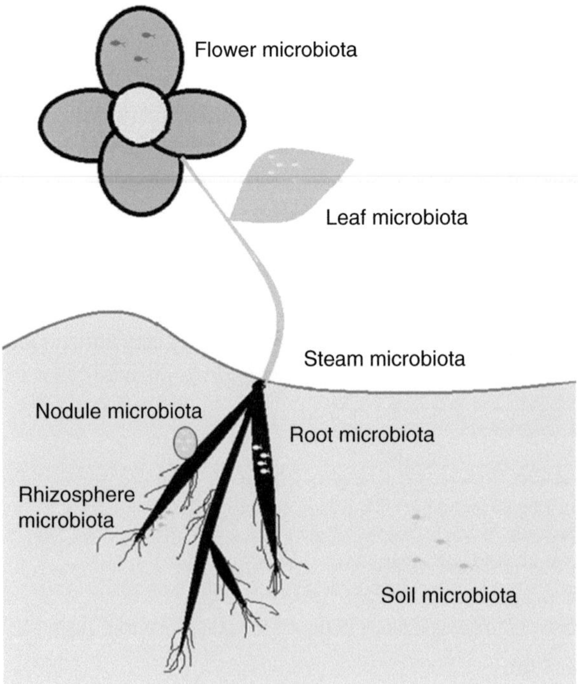

Fig. 13.1 Compartmentalisation of microbiomes in plants

The plant microbiome is crucial to plant health and productivity (Berendsen et al. 2012). The understanding and management of the plant microbiome have the potential to decrease the effects of plant diseases and increase crop yields, therefore reducing the need for agricultural chemicals such as pesticides and fertilisers, resulting in environmental and health benefits while increasing the production of food for sustaining the world's growing population (García-Fraile et al. 2015). Therefore, the study of the plant microbiome has garnered much attention in recent years, and the scientific community has made much progress in unravelling the secrets of the plant microbiome, especially by means of omics approaches.

Unlike traditional techniques, which focus on one or a few molecules at one time, *omics approaches* are *high-throughput technologies* used in the study of different molecules, with genomics for the study of genes, transcriptomics for the study of transcripts, proteomics for the study of proteins, and metabolomics for the study of metabolites being the most commonly applied (Fig. 13.2).

In addition, the prefix 'meta' implies that the omics method attempts to measure all genes, transcripts, proteins or metabolites in a given community sample.

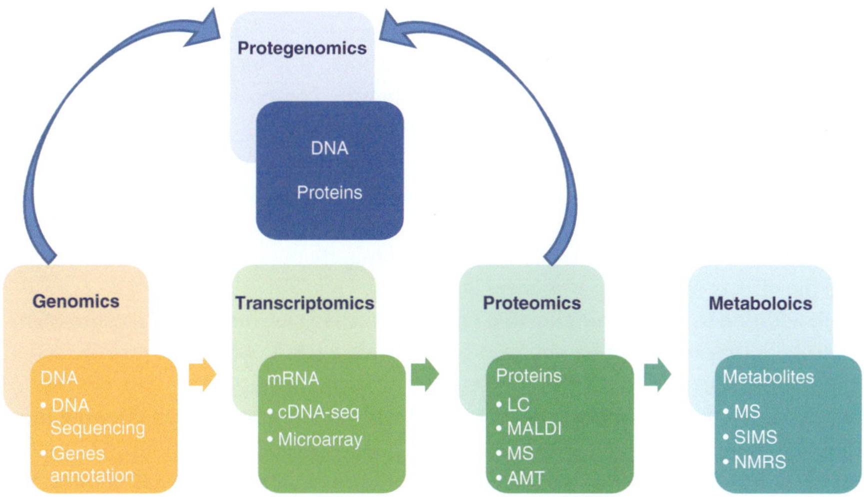

Fig. 13.2 Schematic representation of the omics technologies used in plant microbiome studies, with an indication of the focal molecules and methods used, and symbolised as the DNA information flow in cell metabolism. Genomics: studies of DNA sequences using DNA sequencing and bioinformatics for gene annotation. Transcriptomics: focuses on mRNA for the study of gene expression using cDNA sequencing or microarray hybridisation. Proteomics: unravels protein sequences using liquid chromatography or matrix-assisted laser desorption/ionisation combined with mass spectrometry or accurate mass and time analysis. Metabolomics: studies cell metabolites by means of several analytical techniques such as mass spectrometry, secondary ion mass spectrometry and nuclear magnetic resonance spectroscopy. Proteogenomics: combines the information inferred from genomics and proteomics. *cDNA-Seq* sequencing of complementary DNA (synthesised from a messenger RNA (mRNA)), *LC* liquid chromatography, *MALDI* matrix-assisted laser desorption/ionisation, *MS* mass spectrometry, *AMT* accurate mass and time, *SIMS* secondary ion mass spectrometry, *NMRS* nuclear magnetic resonance spectroscopy

In this chapter, we will review the methodologies of the application of high-throughput sequencing techniques in metagenomic studies that focus on the microbiota as a part of the plant microbiome, we will provide an overview of the most recent research on this topic, and we will provide a review of other omics approaches important in deciphering and understanding the plant microbiome in full, presenting some of the main goals addressed to date.

13.2 High-Throughput Sequencing for Exploring Plant Microbiota

Most plant-microbial communities, especially uncultured microorganisms, are not yet well known. High-throughput sequencing methods, also known as next-generation sequencing (NGS) techniques, are crucial for the rapid advancement of the unravelling of the microbial communities that establish symbioses with plants and are therefore necessary in building up the catalogue of microorganisms that form the plant microbiome.

Advancements in NGS technologies in the last decade have enormously shifted views regarding the complexity of plant-associated microbiomes and are fully uncovering their diversity. Thanks to these methods, it is finally possible to answer some of the fundamental questions concerning the composition of these microbiomes, making it then possible to build further knowledge. So the elementary question 'Who is there?' can now be answered, not only at a full magnitude and for a fraction of the cost compared to previous tedious and time- and effort-demanding cultivation techniques but also on a timescale not conceivable at any previous time. Moreover, fine-scale temporal changes in community composition can be observed and the effects of short-term processes detected.

In addition to the information that DNA can provide us, further insight into plant-microbe interactions can be obtained by evaluating RNA, which mirrors the actual underlying processes and their executors. Complementary data can be further obtained by studying the full metabolomes and proteomes of given communities, thus providing a full picture of the basis of plant-microbe interactions. NGS technologies are thus opening new avenues for high-resolution, low-cost studies and represent a holistic approach to studies of complex biological systems such as plant microbiomes.

As mentioned previously, the first step in the study of any community is generally the identification of its composition. For this purpose, the most frequently used practice is the amplification of a standard barcoding marker from the DNA isolated from the whole community and the subsequent reference database comparison of sequenced amplicon libraries (for a list of phyllosphere community amplicon-based studies, see Knief (2014)). The generally accepted marker for the identification of fungal communities is the internal transcribed spacer of the rRNA cluster (ITS rRNA, Schoch et al. (2012)), whereas the 16S rRNA gene is the selected gene marker for bacterial studies (Pace 1997). Several alternative markers are being developed and tested that could be used separately or in combination with former markers to provide complementary analyses of given communities (Stockinger et al. 2014; Stielow et al. 2015; Větrovský et al. 2015; Links et al. 2012). Most of the ongoing initiatives devoted to barcode enhancement are nevertheless focused solely on redesigning novel primers for former widely used markers to better avoid known taxonomic biases (Toju et al. 2012; Ihrmark et al. 2012; Bokulich and Mills 2013; Walters et al. 2015; Parada et al. 2016; Apprill et al. 2015; Takahashi et al. 2014).

Until recently, the most commonly used platform for amplicon-targeted NGS was the 454 instrument from Roche or its benchtop version with the latest updates, named the 454 GS Junior + system. With the employment of the newest kits for DNA amplification, library preparation and sequencing, these instruments are capable of delivering reads with a mode read length of 700 bp, reaching up to 1000 bp. Length superiority over the competitors at the time was one of the reasons for choosing this platform. The total output of the 454 GS instrument, however, is only 1 million reads per run (100,000 reads for GS Junior+), and thus, the cost per base is on the opposite side of the spectrum from all other NGS platforms available (see Table 13.1 for an overview of the output and cost per base of different NGS platforms, adopted from Loman et al. (2012)); this together with the still-increasing read length provided by other platforms, principally Illumina, represented specifically by its MiSeq instrument (reaching up to 2 × 300 bp as of June 2016), which is best suited for amplicon sequencing, caused the replacement of the 454 platform (Caporaso et al. 2012; Kozich et al. 2013).

Another platform used for amplicon sequencing is the Ion Torrent PGM system which is, in terms of read length and cost per base, somewhere in between the above-mentioned systems. Considering the overall advantages of MiSeq and the Illumina system in general, we will further discuss specificities that are primarily inherent to

Table 13.1 Comparison of the per base cost and run properties of the most frequently used platforms for amplicon-targeted sequencing

Platform	List price	Approximate cost per run	Minimum throughput (read length)	Run time	Cost/Mb	Mb/h
454 GS Junior	$108,000	$1100	35 Mb (400 bases)	8 h	$31	4.4
Ion Torrent PGM						
(314 chip)	$80,490	$225	10 Mb (100 bases)	3 h	$22.5	3.3
(316 chip)		$425	100 Mb (100 bases)	3 h	$4.25	33.3
(318 chip)		$625	1000 Mb (100 bases)	3 h	$0.63	333.3
MiSeq	$125,000	$750	1500 Mb (2 × 150 bases)	27 h	$0.5	55.5

Adapted from Loman et al. (2012)

this specific platform. Moreover, Roche has already announced that support for the 454 platform will be shut down by mid-2016 (http://www.fiercediagnostics.com/story/roche-close-454-life-sciences-it-reduces-gene-sequencing-focus/2013-10-17).

For amplicon-targeted sequencing, the first step in library preparation, as the name of the procedure implies, is the amplification of a targeted region from the community-extracted DNA. Subsequently, DNA adapters necessary for the annealing of amplicons to the glass slide are added either by means of ligation or are already incorporated in the so-called fusion primers used in amplification. Several library preparation kits are available in the marketplace together with a range of protocols to assist with the pre-processing of amplicon libraries (for the overview, see Kucuktas and Liu 2010). Amplicon sequencing is generally regarded as a semiquantitative method (Amend et al. 2010), and to harness its maximal potential in recovering a quantitative picture of the studied community, several issues need to be taken into consideration. In addition to the bioinformatic processing of the data, which will be discussed later, PCR amplification by itself should be adjusted to avoid the magnification of the underlying effects of unbalanced primer efficiency in multi-template PCR as well as the depletion of some combinations in the case of degenerate primers (Polz and Cavanaugh 1998). This can be achieved by limiting the number of PCR cycles so that rare templates are not over-amplified, and the reaction is far from reaching saturation (Kanagawa 2003). Decreasing the number of cycles also prevents the excessive formation and propagation of chimaeras, which can be further restricted by extending the elongation phase of PCR (Haas et al. 2011; Wang and Wang 1996; Meyerhans et al. 1990). In the case of employing fusion primers for library preparation, a two-step PCR amplification procedure is required in which the first amplification is performed with only gene-specific primers and afterwards a second PCR consisting of a few cycles is performed utilising fusion primers with adapters and barcodes, which are added to the PCR products (Kauserud et al. 2012; Berry et al. 2011). Essentially all steps of the library preparation process inevitably introduce biases, which can be accounted for either in bioinformatic processing or during data interpretation (van Dijk et al. 2014). A prerequisite condition for a successful unbiased PCR library preparation is clearly a robust genomic DNA extraction. Several protocols tested in NGS applications are available, and the choice of the most suitable one depends on the specificities of the given project (Venter et al. 2004; Delmont et al. 2011; Knight et al. 2012; Burke et al. 2009; Thomas et al. 2010). Specific procedures need to be performed particularly if the objective of the study is to evaluate the community associated with a host and thus the risk of host DNA dominating the extraction is high (Feehery et al. 2013). The use of selective primers that minimise the impact of host DNA might contribute to the robust amplification of the DNA of the studied communities (Anderson et al. 2003; Smit et al. 1999). Another alternative solution might be the use of blocking primers to prevent the amplification of unwanted DNA (Powell et al. 2012; Wilcox et al. 2014). Additional amplification refinements include repeating the single PCR reaction several times to account for stochastic amplification biases (Polz and Cavanaugh 1998; Ihrmark et al. 2012). The products of these repeated reactions are subsequently pooled. To avoid quantitative biases in library preparation, the enzymes used in the PCR reaction are also assessed for their performance to identify those that are the least biased (Quail et al. 2012).

To fully make use of ever-increasing sequencing output, the multiplexing of samples is necessary in library preparation. This is achieved by adding short, sequence-based sample-specific indices (also called barcodes or tags) to the amplified PCR fragments (Craig et al. 2008; Meyer et al. 2007; Meyer and Kircher 2010). Thus, the sequence reads from multiple libraries can be computationally sorted to the former samples based on a given sample-specific index. At the same time, multiplexing is a source of another type of error connected to the misidentification of the sample of origin. This can happen due to sequencing errors in the index region, errors accumulated during PCR amplification or errors that arise during index synthesis. To minimise such negative impacts, sorting indices are being designed to differentiate among one another using multiple bases so that the chance of random index switching is negligible (Frank 2009; Parameswaran et al. 2007; Degnan and Ochman 2012; Hamady et al. 2008; Faircloth and Glenn 2012). Nevertheless, sequence sorting could be confounded by cross-contamination of indices during processing or by recombination events during so-called 'tag switching' in PCR (Carlsen et al. 2012). Thus, the double-indexing strategy was devised as a means to reduce these types of biases (Kircher et al. 2012). The design of indices should also be done carefully, as an inappropriate base composition might have negative impacts on the even amplification of differently labelled samples in addition to the confounding effects on read sorting.

With the great amount of data acquired by NGS technologies, the limiting factor in their usage and valid interpretation is in bioinformatics analysis (Scholz et al. 2012; Nekrutenko and Taylor 2012; Pop and Salzberg 2008; McPherson 2009). The big data problem is by itself difficult to manage, as the capacities to store, move and transparently share the data are costly and lag behind the pace of innovation of sequencing machines (Dai et al. 2012; Sboner et al. 2011).

There are essentially three options for addressing this issue, including delegating data analysis to commercial companies, conducting the analyses at local computational capacities or employing some of the available cloud computing resources such as Amazon Cloud Web Services (Schatz et al. 2010; Sboner et al. 2011).

General data analysis procedures involve several steps beginning with quality filtering of the raw reads, chimaera detection and purging and eventually OTU (operational taxonomic unit) clustering with subsequent OTU filtering, followed by the taxonomic assignment of representative sequences and an array of diversity analysis and statistical tests. Based on the types of data these general procedures entail, other necessary steps such as paired-end sequence concatenation may be involved. There are a multitude of bioinformatics tools designed to aid with these data processing operations such as individual scripts, fully independent pipelines represented by several types of frequently employed software such as QIIME (Caporaso et al. 2010), mothur (Schloss et al. 2009), and UPARSE (Edgar 2013) and flexible web-based platforms with extendable portfolios that include all types of computational tools, such as that represented by Galaxy (Afgan et al. 2016). The latter platform can be run on one of the many public Galaxy servers or as a local instance.

The first step in the sequencing data analysis workflow, which is subsequently repeated in different forms throughout the analysis, is data filtering. Raw sequence data are quality filtered either by applying a criterion regarding the minimal average Phred score per base or by truncating the reads at a base position where these scores begin to drop under the given threshold (Caporaso et al. 2011; Bokulich et al. 2013). Another frequently applied criterion is a match in the primer part of sequence read with no or minimal difference, and the same rule is applied to the index section of the read as well. Additional quality filters include a minimal and maximal read length and a maximal number of overall and contiguous N characters (Bokulich et al. 2013). The same procedure applies in the case of paired-end datasets wherein the paired reads are first merged, quality scores in the overlapping region are calculated and standard quality filters are subsequently implemented as described. There are a plethora of paired-end mergers available, including PANDAseq (Masella et al. 2012), SHERA (Rodrigue et al. 2010), FLASH (Magoc and Salzberg 2011), COPE (Liu et al. 2012), PEAR (Zhang et al. 2014), UPARSE (Edgar 2013), AdapterRemoval (Lindgreen 2012), IeeHom (Renaud et al. 2014) and the Clip and Merge tool as part of the EAGER pipeline (Peltzer et al. 2016). In some of these tools, the collapsing of paired reads is part of a more complex sequencing read processing procedure that includes autonomous quality filtering, adapter removal and other steps, as in the AdapterRemoval or UPARSE programmes, which also include clustering procedures.

The next step in data processing is OTU picking, or essentially read clustering, which should result in a set of biologically realistic and reliable sequence groups that generally represent the species level but occasionally go as far as strain level by employing a slightly modified procedure of amplicon sequencing called low-error amplicon sequencing (LEA-Seq) (Faith et al. 2013). This modified procedure is based on the redundant sequencing of linear PCR templates, thus exchanging quantity for quality. There are two main approaches to sequence clustering: reference-based clustering and reference-free clustering, or so-called de novo clustering. As the names suggest, the main difference between the two approaches is related to their dependence on the reference database. In reference-based clustering, or close-reference clustering, the building of sequence groups proceeds by searching for the closest hit in a given reference database, and all the reads that pass a minimal similarity filter are grouped around this closest hit. The taxonomic assignment is then provided by the reference sequence. An obvious disadvantage of this protocol is the absolute dependence upon a reference database, resulting in a loss of all the diversity not covered but potentially important in a given database. On the other hand, the major advantage is computing speed and the possibility of comparing datasets based on the different parts of an amplified gene (specifically for 16S rRNA studies, the database is based on the full-length 16S rRNA gene). Reference-free clustering, or so-called de novo clustering, essentially involves comparing reads to one another and then searching for the most optimal configuration of sequence groups that correspond to taxonomical clades or monophyletic groups. There are numerous clustering tools capable of de novo clustering, including UCLUST and USEARCH (Edgar 2010), UPARSE (Edgar 2013), Swarm

(Mahé et al. 2014), SUMACLUST (Mercier et al. 2013), SortMeRNA (Kopylova et al. 2012), CD-HIT (Li et al. 2012), DNACLUST (Ghodsi et al. 2011), SEED (Bao et al. 2011) and many more. Those that are most frequently used (CD-HIT, UCLUST) are built upon the use of a centroid-based greedy clustering algorithm (Rideout et al. 2014). The most frequently applied criterion of sequence similarity within a cluster is >97%, which is generally accepted to represent the variability of intraspecies biological markers (relating to 16S rRNA (Huse et al. 2010) and ITS (Schoch et al. 2012)). In contrast, the UPARSE pipeline (Edgar 2013) works independently of an often misleading and artificial OTU similarity cut-off and simultaneously also filters sequences based on their quality as well as performs chimaera filtering in addition to clustering. This pipeline is thought to be more precise in providing OTU numbers that are more congruent with the true underlying species diversity of a particular system. The recently introduced algorithm update that has been implemented in the USEARCH package named UNOISE uses the expected number of errors as a measure of read quality and an effective means of error rate reduction (Edgar and Flyvbjerg 2015). Afterward, the clustering step filtration of clusters is often performed. Most often, clusters of single reads, or so-called singletons, are discarded, but this threshold is often shifted up to 5 or even 10 reads. This is a somewhat controversial topic, as the minimum cluster size is often selected ad hoc without analytical reasoning, and filtering might cause a reduction in the estimate of real biological diversity instead of filtering artificial variability. Several other procedures have been discussed that might assist in more efficient clustering and error filtering such as the single-linkage pre-clustering procedure described by Huse et al. (2010) and the pre-clustering procedure of Schloss et al. (2011). The final outcome of the clustering procedure is an OTU table representing the presence-absence data for the recovered sequence clusters and includes OTUs and their read abundances.

The ultimate step in the bioinformatic processing of sequence data is OTU taxonomic classification. The standard procedure consists of querying representative sequences of all the identified OTU clusters against a defined database that will depend on the studied subject and marker used. These include curated databases such as the Greengenes database (16S rRNA bacterial data (DeSantis et al. 2006)), the SILVA rRNA database project (16S/18S SSU and 23S/28S LSU data for all three domains of life (Quast et al. 2013)), UNITE (fungal ITS rRNA (Koljalg et al. 2013)) and databases derived from the above-mentioned, such as SilvaMod and many more custom-made databases for the purpose of identifying other marker genes utilised in the study. These are mostly based on the extracted sequence data from INSDC databases employing advanced search queries. To fully exploit the available taxonomic resources, several tools and algorithms are available to proceed with classification procedures. The three main supervised (dependent on the reference database) approaches to classification can be divided between similarity-based methods (homology or alignment dependent), sequence composition methods (k-mers, Markov models) and phylogeny-based methods and their combinations. The most commonly applied programme utilising similarity searches is MEGAN (Huson et al. 2007), which relies on BLAST searches in addition to implementing the lowest common ancestor (LCA) algorithm, the latter of which was subsequently implemented in several other tools. Sequence composition methods are best represented by RDP (Ribosomal Database Project), a naïve Bayesian classification tool (Wang et al. 2007), and phylogeny-employing algorithms are incorporated into software such as pplacer (Matsen et al. 2010), SAP (Munch et al. 2008) or EPA (Berger et al. 2011). Which method to choose depends on several factors including the length of the analysed marker, the proportion of erroneous and missing classifications and the consumption of resources (Porter and Golding 2011). In studies comparing the outcomes of different classifiers, MEGAN was found to be the least erroneous (Porter and Golding 2011, 2012), and the BLAST algorithm was found to have the highest recovery rate in general (Porter and Golding 2011). Phylogeny-utilising methods are best suited for 16S rRNA bacterial studies, as their dependency on alignment building is alleviated by the availability of several publicly available curated alignments (Pruesse et al. 2007; DeSantis et al. 2006). On the other hand, composition-based methods were shown to be the fastest solution and especially suitable for extensive datasets (Porter and Golding 2012). The accuracy of all these methods is dependent, however, on query length (Porter and Golding 2012). The major limitation besides the length of the marker being sequenced still remains the completeness of the reference database.

13.3 Catalogues of Bacterial Communities Studied by Next-Generation Sequencing

Some years after the development of modern high-throughput sequencing techniques, these methods began to be applied to the study of bacterial communities living as plant symbionts. One of the first analyses to examine a bacterial endophyte community based on these new technologies was in potato (*Solanum tuberosum*). In 2010, Manter and collaborators studied the roots of 12 different potato cultivars and found a high diversity of bacterial endophytes that were cultivar dependent. The study also highlights a possible link between plant production and endophyte abundance (Manter et al. 2010). Also using potato as host plant, Kõiv and colleagues analysed the dynamic changes in the endophytic bacterial community in response to infection by *Pectobacterium atrosepticum*, a bacterial pathogen that causes soft rot in numerous economically important crops (Koiv et al. 2015).

The model plant *Arabidopsis thaliana* was chosen in 2012 for the study of its bacterial community using a 454 sequencer (Roche) (Lundberg et al. 2012). Based on the sequencing of the variable regions of the 16 rRNA gene, the bacterial communities of eight different lines were analysed; the study showed *Actinobacteria* as the dominant endophytes followed by *Proteobacteria*, *Firmicutes*, *Bacteroidetes* and *Cyanobacteria* (Fig. 13.3).

The presence of a core microbiome of the plant from the recruitment of soil bacteria that are able to enter the root and survive was shown to be one of the main conclusions of this study, which is in agreement with similar analysis using other hosts, such as humans (Turnbaugh et al. 2009).

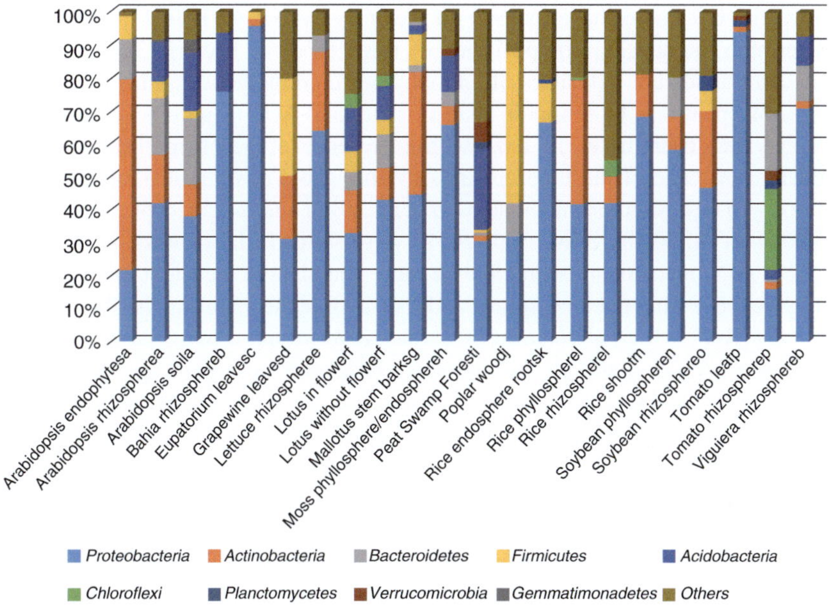

Data from [a]Lundberg *et al.*, 2012; [b]Navarro-Noya *et al.*, 2010; [c]Zhou *et al.*, 2010; [d]Pinto *et al.*, 2014; [e]Krober *et al.*, 2014; [f]Unno and Shinano, 2013; [g]Wang *et al.*, 2008; [h]Bragina *et al.*, 2014; [i]Kanokratana *et al.*, 2011; [j]Van der Lelie *et al.*, 2012; [k]Sessitsech *et al.*, 2011; [l]Knief *et al.*, 2012; [m]Okubo *et al.*, 2014; [n]Delmotte *et al.*, 2009; [o]Mendes *et al.*, 2014; [p]Romero *et al.*, 2014;

Fig. 13.3 Graphical representation of the mayor phyla in plant microbiomes

Recently, the endophytic bacteria of *Aloe vera* were studied by the amplicon sequencing of the V3–V4 regions of 16S rDNA with the Illumina platform (Akinsanya et al. 2015). The analyses showed that the phyla *Proteobacteria*, *Firmicutes*, *Actinobacteria* and *Bacteroidetes* were present in both roots and stems; the study also showed that the most diverse bacterial community is found in the roots, being that 23% of the OTUs detected in root tissue not present in other tissues.

13.3.1 Plant Growth-Promoting Bacteria

Plant growth-promoting bacteria (PGPB) are bacteria that are able to benefit plants by improving their productivity and immunity (García-Fraile et al. 2015). Plant root exudates induce changes in the composition of the rhizosphere and select for specific bacteria of interest (Venturi and Keel 2016). Following this, there are three main factors that have been described that allow these rhizospheric bacteria to become endophytes:

1. Their ability to survive in soil
2. Specific plant factors that determine colonisation and compatibility
3. Microbial factors that enable endophytic survival (Gaiero et al. 2013)

Several NGS studies have focused on the analysis of these PGPB in an attempt to determine the specific compositions of these types of bacterial communities. Some compounds are indicators of endophyte-promoting activity, such as siderophores, phosphatases, IAA (indole-3-acetic acid) and ACC (1-aminocyclopropane-1-carboxylate) deaminase. To analyse the presence of genes implicated in the production of some of these compounds by the whole communities of microbial symbionts in rice, shoot samples (Okubo et al. 2012) and the rhizosphere microbiota (Ikeda et al. 2014; Knief et al. 2012) were studied; the expression of genes related to ACC metabolism seemed to be more abundant in the shoot microbiomes than in the those from the rhizosphere and roots, while the opposite distribution was found for IAA. A high abundance of *Proteobacteria* and a non-friable value for *Actinobacteria* was found in the shoot studies, with no significant differences between them (Okubo et al. 2014); similar profiles, but with a higher percentage of *Actinobacteria*, were found in phyllosphere samples (Knief et al. 2012). However, possible changes in the composition of the microbiome that are induced after inoculation of PGPB must also be taken into account, as was shown in chamomile plants by (Schmidt et al. 2014). Recently, Miyambo and collaborators (2016) studied the bacterial communities associated with fynbos plants using Illumina MiSeq 16S rRNA sequencing, finding several putative plant growth-promoting bacteria among the endophytic bacterial communities, which therefore had the potential to power plant growth and health (Miyambo et al. 2016).

High-throughput DNA sequencing can also be used for the *analysis of temporal and spatial variation* in microbial community composition and structure. Analyses of bacterial populations in several plant tissues (leaves, roots, stem, flowers) and soil samples in the areas of influence of several plants (rhizosphere) have been conducted (Bertani et al. 2016; Junker and Keller 2015; Romero et al. 2014; Trujillo et al. 2015), showing that the composition of the microbiota differs from tissue to tissue (Fonseca-Garcia et al. 2016; Rosenberg and Zilber-Rosenberg 2016). Indeed, the main differences observed in the microbiota of several cacti species were influenced by the plant organ analysed, while plant species, site and season seemed to have less influence on microbial composition (Fonseca-Garcia et al. 2016). In this study, bacteria were shown to be highly present in the rhizosphere and phyllosphere and were less abundant in endophytic tissues, especially in the stem area. However, the main phyla found were similar: *Proteobacteria*, *Firmicutes*, *Actinobacteria*, *Acidobacteria* and *Bacteroidetes*. A recent study conducted by Meaden et al. (2016) that involved analysing the microbiome composition of *Quercus* tree woody tissue at different tree life stages indicates that bacterial community structure varies with life stage, showing a decreased abundance of *Alphaproteobacteria* in older trees (Meaden et al. 2016). Microbial interactions and their importance in the equilibrium of grape plants at several life stages have also been studied to understand their role in grape quality and fermentation for wine production; in general, a high diversity of *Proteobacteria*, *Firmicutes* and *Actinobacteria* was found (Pinto et al. 2014). The communities found in this study were maintained over time in the leaves (at different stages of the vegetative cycle); however, the relative abundances of the taxa varied. Some of the bacteria were found to be related to the wine production process (lactic acid- or acetic acid-producing bacteria), but others do not seem to have a direct relationship with fermentation, or the relationship is at least not known. Shi and collaborators (2014) examined how beetroot (*Beta vulgaris*) endophytic bacteria vary in samples from different locations and during different growth periods using Illumina sequencing of the V3 region of the 16S rRNA gene. Despite the fact that *Alphaproteobacteria*, followed by *Acidobacteria*, *Gemmatimonadetes* and *Actinobacteria*, were dominant in all the samples, the authors describe how endophytic bacterial communities are shaped by both the plant's location and its growth stage; moreover, the authors discovered that the greatest number of OTUs occurs during the tuber growth and rosette formation stages, whereas seedling growth and glucose accumulation were the phases with the lowest number of OTUs (Shi et al. 2014). In contrast, a recent study of the dynamics of bacterial communities in the halophyte plant *Salicornia europaea* using high-throughput sequencing detected that the greatest endophytic bacterial diversity in this plant occurs during the seedling stage, whereas bacterial diversity was decreased during the flowering and fruiting stages (Zhao et al. 2016).

In general, the various studies presented thus far have shown a high abundance of *Proteobacteria* in plant-related ecosystems, with additional significant amounts of *Actinobacteria*, *Bacteroidetes*, *Firmicutes* and *Acidobacteria* (Fig. 13.3). However, the final conclusions regarding the composition of these microbiomes should carefully take into consideration the limitations of the methods related to the following:

1. The variability in the capacity to recover true proportions of DNA due to the different types of membranes and spores of known bacteria and unknown characteristics of undescribed species (Hart et al. 2015), which results in quantities of DNA that are not directly related to bacterial abundances but rather to the ease of DNA extraction
2. The 16S rRNA gene read lengths obtained with the next-generation sequencers, which has been shown to be sufficient for identification to phyla but problematic for genus or species (Franzen et al. 2015)
3. The region of the 16S rRNA gene selected, which can give different percentages at the genus level due to misclassification (Okubo et al. 2012)
4. The sample type, recovery and post-treatment, which can contribute to the generation of biases (Felczykowska et al. 2015; Glassing et al. 2015)

13.4 High-Throughput Sequencing Analyses of Fungal Communities

NGS has also been successfully used for exploring in depth the fungal communities living as a part of the plant microbiome. As in the case of bacteria, fungal communities are important inhabitants of the phyllosphere, rhizosphere and endosphere of plants. The plant-associated fungi are usually categorised according to their functional roles and their relationship with the plant (Peršoh 2015). The most commonly studied are the mycorrhizal fungi from the rhizosphere due to their important role in plant nutrition. Together with mycorrhizal fungi, endophytic (living in plant tissues) and epiphytic (living on the plant surface) fungi are recently gaining a great amount of attention due to their roles in plant protection. In addition, fungal communities associated with plants also include numerous non-mycorrhizal fungi involved in decomposition (saprophytes) and other potentially phytopathogenic fungi (Peršoh 2015). NGS techniques can reveal the high diversity and complexity of such fungal communities that interact with crops. In this sense, numerous studies have been published in recent years that have focused on the description of these communities and have therefore begun to illuminate the soil 'black box'.

The most ubiquitous and important interactions between plants and microorganisms in the rhizosphere are those established between *mycorrhizal fungi* and plant roots (Tkacz and Poole 2015). These fungi provide up to 80% of the N and P the plant needs and are essential for plant growth and survival (van der Heijden et al. 2015). Most plant roots are colonised by multiple mycorrhizal fungi, and most of them can colonise various host plants at the same time, forming mycelial networks in the soil (van der Heijden et al. 2015). Arbuscular mycorrhizae (AM) are the most common type of mycorrhizal fungi inhabiting the microbiome of crop plants (Baum et al. 2015). Due to their important effects on the growth and health of the plant, the exploration of the diversity and structure of AM fungal communities in agricultural ecosystems together with the understanding of how environmental variables affect them is of high relevance (De Beenhouwer et al. 2015b). Because the traditional identification methods based on spore morphology or abundance were problematic and not accurate, the use of NGS techniques has proved to be an ideal approach for studying the AM communities in soil (Higo et al. 2014). High-throughput sequencing has been used extensively for exploring AM communities in crops around the world. Opik et al. (2013) sampled 96 plant species from 25 sites across all continents except Antarctica, showing the considerable diversity of AM among the different continents and climatic zones. Metagenomic studies have been recently applied in olive crops (Montes-Borrego et al. 2014), grapevines (Holland et al. 2013), coffee plants (De Beenhouwer et al. 2015b; De Beenhouwer et al. 2015a), cover crops (Higo et al. 2014), potatoes (Senés-Guerrero and Schüßler 2015), maize (Turrini et al. 2016), wheat (Dai et al. 2014) and apple trees (van Geel et al. 2015), among others. Most of these studies focus on describing the structure and diversity of AM communities, as high AM diversity is expected to be more beneficial to the plant than low diversity. In addition, because AM diversity is known to have declined due to the intensive use of agricultural soils, most of the studies also aim to identify the factors affecting their diversity and composition, such as type of soil or cultivar, and agricultural practices such as level of fertilisation, management intensity gradients or crop rotation (van Geel et al. 2015; Bazghaleh et al. 2015; Higo et al. 2014; Yeoh et al. 2016). These results provide valuable information for understanding the important effects of AM communities inhabiting the rhizosphere and their role in ultimately improving crop yields.

Fungal *endophytes* live inside plant tissues without causing symptoms of disease. As in the case of endophytic bacteria, they are emerging as important microorganisms affecting plant growth (Porras-Alfaro and Bayman 2011). Unlike mycorrhizae, endophytes do not present a clear and specific functional relationship, such as nutrient transfer, with the plant. Many fungal endophytes are known to stimulate plant growth, improve the ability of plants to resist environmental stresses and increase disease resistance, as they are defined as *biological control agents* (BCAs). Therefore, these fungi are highly important as safe and sustainable tools for plant protection. NGS methods have been successfully applied in the study of these fungal communities. For example, the abundance and diversity of indigenous fungal endophytes from different chickpea cultivars were analysed with pyrosequencing (Bazghaleh et al. 2015), identifying genera such as *Trichoderma*, *Mortierella*, *Geomyces* and *Penicillium* and relating the high diversity and richness of these communities with AM fungi that were also found the be key in the reduced levels of *Fusarium* in the roots of some cultivars. Metagenomic studies have allowed the confirmation of the presence of natural antagonists in suppressive soils. Penton et al. (2014) showed that soils that naturally suppressed *Rhizoctonia* were inhabited by numerous genera of endophytic species with the potential for pathogen suppression and mycoparasites such as *Xylaria* sp. In the same way, Nallanchakravarthula et al. (2014) pointed to the dominant endophyte *Leptodontidium orchidicola* as being mainly responsible for the resistance of strawberry to *Verticillium dahliae*. Metagenomic approaches have been successfully used to show that sustainable agricultural management regimes such as crop rotation increase the abundance of natural BCAs that prevent Fusarium wilt in vanilla plants, such as those in the genera *Trichoderma* and *Penicillium* (Xiong et al. 2016). Huang et al. (2015) also showed that other agricultural practices such as reductive soil disinfestation also improve the amount of indigenous biocontrol fungi such as *Podospora* spp. or *Zopfiella* spp., making the soil more disease-suppressive and beneficial to soil nutrient cycling and plant growth.

In addition, new technologies also provide an in-depth analysis of the microbiome of the rhizosphere after treatment with fungi and other BCA (not necessarily endophytes). For example, Hirsch et al. (2013) showed that the application of the entomopathogenic fungus *Beauveria bassiana* had no influence on the fungal communities associated with chili plants. In comparison with previous low-throughput approaches, NGS allows the more precise detection of the changes caused by the application of BCAs to the microbial community (Massart et al. 2015a). The use of NGS approaches for studying BCAs has been extensively reviewed recently (Massart et al. 2015a, b). In summary, these reviews highlight that the new technologies can help in the understanding of the mechanisms of action of the fungal endophytes and biocontrol agents, which will allow us to improve their efficacy in their practical use in agriculture.

Endophytes living in the aerial surfaces of vegetal organs (phyllosphere) also require a special mention. Traditional culture-based methods have revealed that these fungal communities are highly diverse but show less microbial diversity than those in the rhizosphere (Peršoh 2015). However, the application of NGS methods to the study of the phyllosphere is revealing a potential goldmine of undescribed diversity (Porras-Alfaro and Bayman 2011). Together with endophytic fungi, epiphytes also form part of the phyllosphere of plants, inhabiting the aerial surfaces of vegetal organs. In contrast to foliar endophytes, epiphytic fungi can be washed off from these surfaces or removed by sterilisation with chemical agents (Porras-Alfaro and Bayman 2011). However, this distinction has been considered arbitrary because epiphytes can penetrate the leaf tissues after growing on the surface and endophytes in tissues can also be exposed to the surface (Porras-Alfaro and Bayman 2011). Unlike bacterial communities inhabiting the phyllosphere, little is known about the structure and function of foliar fungi (Rastogi et al. 2013). Estimates of the diversity of fungal epiphytes are high, but their population size is thought to be lower than that of epiphytic bacteria (Rastogi et al. 2013). As a result, high-throughput sequencing has been determined to be an accurate approach for exploring these 'rare' inhabitants of foliar parts. In this sense, metagenomic studies are useful for deciphering the factors shaping the structure of these communities. Even though some published studies explore the foliar fungi of forest trees such as *Quercus*, *Fagus* and *Populus*, studies focusing on the fungal communities inhabiting the phyllosphere of crops are still lacking (Rastogi et al. 2013). Peršoh (2015) summarised various studies using NGS approaches, showing, for example, that the composition of the foliar fungal community changes throughout the year, with apparent shifts occurring within a few weeks or a month. These studies also showed that host plant identity, different aboveground organs or environmental conditions such as rainfall and temperature are important factors determining the structure of the endophytic fungal community (Rastogi et al. 2013). In this way, the use of NGS has also been shown to be an accurate approach for the study of succession dynamics occurring in these microbiomes and for understanding the process of plant colonisation by microorganisms, as the composition of the community inhabiting the leaf surface can change rapidly due to high abiotic fluctuations occurring in the phyllosphere (Lebeis 2015; Knief 2014). Omics approaches have also been considered to be essential in the study of the fungal communities inhabiting other microenvironments in plants, such as the anthosphere (flowers), the spermosphere (seeds) and the carposphere (fruits), by contributing to a better understanding of plant growth and health as well as sustainable crop production (Berg et al. 2014; Schiltz et al. 2015). For example, a recent study of the microbiome of wheat seeds and sprouts detected the presence of beneficial fungal taxa (such as *Emericella nidulans*) that were vertically transmitted from seeds to sprouts and to mature wheat plants (Huang et al. 2016).

An important issue regarding endophytic fungi is their potential role as latent pathogens. Several studies have shown that changes in the host or environment can trigger the pathogenicity of a previously asymptomatic endophyte (Porras-Alfaro and Bayman 2011). In this sense, recent studies are shedding light on the ecological function of these fungi. (Busby et al. 2016) showed that foliar fungi are able to alter the severity of *Melampsora* rust disease in *Populus* by identifying pathogen antagonists such as *Trichoderma* as well as pathogen facilitators such as *Alternaria* and *Cladosporium*. These findings illuminate the management of diseases in agricultural plant systems, supporting the idea that modifications of the foliar microbiome can potentially enhance plant growth or suppress disease. However, because a clear distinction between endophytes, epiphytes and pathogens is still lacking, further studies are needed before field application of potentially beneficial fungi can occur.

Finally, fungal communities inhabiting the plant microbiome are normally rich in potential *pathogenic fungi.* As mentioned above, pathogens can be found together with potential antagonists in apparently healthy plants. Miao et al. (2016) combined culture-dependent techniques and Illumina sequencing to show that the rhizosphere of *Panax notoginseng* plants that are apparently free from disease is rich in pathogens (such as *Fusarium* and *Phoma*) as well as antagonistic fungi (such as *Aspergillus versicolor*). These authors suggested that the presence of antagonists is responsible for the suppression of the root rot pathogenic fungi. Similar findings have been shown in the microbiome of grapevine leaves (Pinto et al. 2014). Numerous phyto-pathogens commonly associated with diseases in vineyards (such as *Alternaria, Rhizopus, Ustilago, Phomopsis*, and *Lewia*, among others) were present in the analysed leaves but were potentially suppressed by the presence of fungal (*Aureobasidium*) and bacterial antiphytopathogens. High-throughput sequencing of the fungal communities inhabiting the rhizosphere of apple trees has also identified pathogenic fungi such as *Fusarium, Cylindrocarpon* and *Acremonium* affecting the plant growth (Franke-Whittle et al. 2015). Moreover, this study showed that plant growth was positively correlated with the abundance of potentially antagonistic fungi, identifying some of them as novel rather than previously described BCAs. In the same way, a metagenomic approach using Illumina sequencing was recently successful in identifying up to 17 fungal species as being candidates in affecting strawberry yield decline, in which only four of them were previously confirmed to be pathogens (Xu et al. 2015). Importantly, the authors also exposed the importance of further analyses based on microbial identification at the species level, as a single genus can include beneficial, neutral and pathogenic fungi (Franke-Whittle et al. 2015).

13.5 Other Omics Approaches for Further Analysis of the Plant Microbiome

The use of NGS approaches has been extremely useful and more accurate than low-throughput techniques for characterising the microbial communities inhabiting the plant microbiome. The reduction in cost provided by high-throughput sequencing allows a cheaper and deeper sequencing of the microbial communities. The findings described above highlight the potential of these techniques in identifying the players involved in the health status of plants and in revealing how different environmental factors affect these microbial communities and therefore crop yields. However, most of these studies have focused on generating DNA amplicon sequences and a large amount of data as a means to simply report 'who' is living there (Massart et al. 2015a), which offers information about the catalogue of microorganisms forming part of the microbiome. However, the plant microbiome does not just include the microbiota living in symbiosis with plants; the full set of genes plus the interactions among all the symbionts, as determined by the levels of gene expression and the activities of the gene products, is also part of such microbiomes.

Metagenomic approaches based on the amplicon sequencing of the ribosomal genes of bacteria and fungi or alternative gene markers result in insufficient data for deciphering the complexity of the microbiome associated with plants. As a result, most of the above-referenced works emphasise the need to carry out further studies exploring the role of the microbiome, how it responds to change, how it influences plant health and how its members interact with one another using the whole-genome sequencing of microbial symbionts or even the analysis of whole microbiome genes, transcriptomics and metatranscriptomics, proteomics and metaproteomics, metabolomics and community metabolomics of the entire microbiome. The potential of these other omics approaches to afford a comprehensive analysis of microbial genes and gene products makes them well suited for completing the study of the plant microbiome.

The use of metagenomic techniques based on *shotgun sequencing* is emerging as a new step in the further study of the microbiome (Knief 2014). This approach allows for the understanding of both the microbial community composition and its functional potential by assessing the diversity of functional genes within a given ecosystem (Classen et al. 2015). The presence and persistence of functions in the plant microbiome have been demonstrated to be more valuable information than only taxonomical information (Massart et al. 2015a). The global analysis of the genes present in the microbiome is important for discovering which processes are happening within it by identifying genes potentially involved in plant colonisation, nutrient exchange or the molecular dialogue between the microbes and the plant and among the microbes themselves, among others. Shotgun metagenomics has been already successfully applied in studies of the gut microbiome, but studies focused on the plant microbiome remain scarce. Some recent studies based on the soil microbiome have been published, showing the power of these approaches in reporting changes in the functions of microbial communities (Navarrete et al. 2015; Mendes et al. 2015).

One of the most important limitations of meta-omics approaches is the presence of a high number of sequences coding for genes with unknown function together with sequences from unknown microbes for which no homologous sequences are found in the available public databases (Knief 2014). An ideal solution for overcoming this problem and improving the characterisation of the plant microbiome would be to couple these culture-independent methods with the traditional methods used in the isolation of microorganisms (Lebeis 2014). The *sequencing and analysis of the genomes* of representative pure cultures of the microbes associated with plants are extremely important in improving the databases. For example, one of the first symbiotic bacteria sequenced was the nitrogen-fixing bacterium *Mesorhizobium loti* (Kaneko et al. 2000). After this, the genome sequences of many other rhizobial strains that are able to nodulate legumes have been obtained, and over 30 complete genome sequences are available in the NCBI genome database as well as those of other plant growth-promoting microorganisms.

Regarding root nodule bacteria (RNB), the work performed by Seshadri and collaborators in 2015, in which the authors obtained and compared the genome sequences of 110 RNB from diverse hosts and biogeographical regions to identify the novel genetic determinants of symbiotic associations and the promotion of plant growth, is of outstanding interest (Seshadri et al. 2015). Specifically, the authors performed a subtractive comparative analysis with non-RNB genomes, employed relevant transcriptomic data and leveraged phylogenetic distribution patterns and sequence signatures based on known precepts of symbiotic- and host-microbe interactions. A total of 184 protein families, including known factors for nodulation and nitrogen fixation, and newly discovered proteins with previously unexplored functions, for which a role in host-interaction was predicted, were defined. Their results provide bases for further studies on rhizobial strains that focus on biofertiliser improvement to increase plant productivity and agricultural sustainability. Genome sequences of plant pathogens are also being broadly obtained, with *Xylella fastidiosa* being the first bacterial plant pathogen with a published genome (Simpson et al. 2000) and the rice blast fungus *Magnaporthe grisea* (Dean et al. 2005) being one of the first fungal pathogen genomes sequenced. Advances in NGS techniques have also allowed the complete genome sequencing of some uncultured microbial symbionts, as is the case of the uncultured plant pathogen and insect symbiont '*Candidatus* Liberibacter asiaticus' (Duan et al. 2009). Fortunately, the number of available genomes from endophytes, epiphytes, BCAs, pathogens and AM is rapidly growing (Massart et al. 2015a). As a result, numerous genomes have been published recently, including those of biocontrol fungi (Berger et al. 2016; Baroncelli et al. 2015, 2016; Sun et al. 2015b), bacterial endophytes (Megías et al. 2016; Sun et al. 2015a; Ho and Huang 2015), phytopathogens (Garita-Cambronero et al. 2016; Wibberg et al. 2016; Verma et al. 2016; Lee et al. 2015) and mycorrhizal fungi (Sedzielewska-Toro and Brachmann 2016). Moreover, the development of third-generation sequencing instruments based on nanopore technology and the sequencing of single molecules are producing sequence reads with unprecedented lengths and will help to strongly increase the quality of genome assemblies (Bleidorn 2015). This technology has already been effective in assembling the genome of a strain of *Escherichia coli* de novo using only nanopore sequencing data (Loman et al. 2015). The annotation and characterisation of the genes and proteins contained in these genomes will allow the identification of the pathways and functions expressed by these microbes and the understanding of, for example, the differences between those that are beneficial and those that are pathogenic.

As shown above, the analysis of whole-genome sequences offers insights into the metabolic potential of microbial-plant isolates and allows for the identification of mechanisms associated with symbiosis, pathogenicity and virulence as well as microbe-microbe and microbe-host interactions. Moreover, genetic analysis and comparative genomics will also improve and complement the data and information obtained from meta-omics approaches, allowing a better understanding of the role of plant microbiomes (Pible and Armengaud 2015). However, it is not always easy to identify existing correlations between some functions or taxa and environmental parameters. This is because such studies are based on DNA, which may be obtained from inactive or dormant taxa inhabiting these ecosystems. The presence of these taxa does not normally respond to any changes, but their DNA can persist in the environment for years (Peršoh 2015). Studies based on gene expression and RNA can help to overcome this problem because they focus only on the active players in the community, thereby reflecting responses to influencing factors in a direct way.

Transcriptomics or Expressed Sequence Tag (EST) analysis involves the massive parallel sequencing of cDNA and the generation of information about the expressed genes. Transcriptomic analyses of any of the microbial endosymbionts or the plant host have been very useful in the discernment of molecules and genes involved in microbial-plant symbioses. For example, Reininger and Schlegel (2016) analysed the effect of the presence or absence of *Picea abies* on the transcriptome of its fungal endophytic strain *Phialocephala subalpina* 6_70_1 using Illumina sequencing; the strain was shown to be metabolically very active during the colonisation of its host plant, and its differentially expressed genes were grouped into three functional groups: 'metabolism', 'transport' and 'cell rescue, defence and virulence'. Before NGS appeared on scene, other molecular techniques such as RT-PCR or microarrays were the basis of transcriptomic studies (Reininger and Schlegel 2016). Verhagen and collaborators (2004) obtained the transcriptome of *Arabidopsis* roots colonised by *Pseudomonas fluorescens* WCS417r, an endophyte bacterium capable of inducing systemic resistance in the plant; the authors showed significant changes in the level of expression of 97 genes in the roots that are predicted to be ISR-related genes (Verhagen et al. 2004). In addition, our knowledge regarding the symbiotic processes in legumes has been greatly improved thanks to such transcriptomic analyses. In 2000, the expressed sequence tags from nodules of *Medicago truncatula* that were induced by *Sinorhizobium meliloti* were presented (Gyorgyey et al. 2000). The authors likely did not know it at the time, but this finding would open the door to the study of the molecular dialogue between the members of symbiotic relationships. Another transcriptomic analysis using different *Medicago* plant mutants impaired in the biosynthesis of different nodule cysteine-rich peptides has provided information regarding how this plant has developed various ways to control *Sinorhizobium* bacterial infection at different stages of the symbiotic process. The authors also analysed the differential expression of *Sinorhizobium* genes implicated in cell envelope homeostasis, cell division, stress response, energy metabolism and nitrogen fixation (Lang and Long 2015). The study of the symbiotic relationships of actinorhizal plants, a group of angiosperms with the ability to establish symbioses with nitrogen-fixing bacteria belonging to the genus *Frankia*, has been simplified in a great manner with the development of -omics techniques. Because several genomes of *Frankia* were described and analysed in 2007 (Normand et al. 2007), many subsequent studies have been conducted to elucidate the symbiotic process. Transcriptomic analysis has proven to be a good tool for the identification of specific processes between plants and their symbiotic bacteria, and it has also been used to analyse the interactions between *Frankia* and actinorhizal plants, both from the bacteria side (Alloisio et al. 2010) and from the plant side (Hocher et al. 2011) of the interaction. Free-living bacteria were compared with symbiotic bacterial cells obtained from *Alnus glutinosa* nodules using whole-genome microarrays, showing the limitations of the last strains to assimilate the fixed ammonium (Alloisio et al. 2010). A list of candidate genes possibly implicated in symbiotic processes in relation to transcriptional regulation and signalling processes, drug export, protein secretion and lipopolysaccharide and peptidoglycan biosynthesis was also identified. Hocher et al. (2011) analysed gene expression during the symbiotic process in two actinorhizal plants, *A. glutinosa* and *Casuarina glauca.* These plants presented homologous genes for the nodule-specific signalling pathway. Other genes implicated in the symbiotic process were related to carbon and nitrogen exchange, stress resistance or defence against pathogens. The latter genes were later observed to be *Alnus* symbiotic upregulated peptides with direct implications in symbiosis (Carro et al. 2015, 2016). These types of peptides have also been described in other actinorhizal plants through transcriptomic analysis, such as *Datisca glomerata* (Demina et al. 2013) and *C. glauca* (Carro et al. 2016). Other analyses of these transcriptomic data from *Casuarina* and *Alnus* have allowed the identification of transcriptomic factors implicated in plant-bacteria interactions, with some of them being related to nodule formation and previously also detected in legumes (Diedhiou et al. 2014).

High-throughput sequencing of the whole transcriptome of the microbial community allows for the analysis of shifts at the transcriptional level (Massart et al. 2015a). Recent studies have shown that the functional versatility and function-based diversity of the microbiome are more important factors than mere traditional diversity descriptions (Lakshmanan et al. 2014). Because of this, *metatranscriptomics*, rather than metagenomics, is highly preferred for providing advanced functional insights into microbial communities and has been successfully applied in diverse microbiomes in permafrost and agricultural and forest soils (Hesse et al. 2015; Kim and Liesack 2015; Jiang et al. 2016; Žifčáková et al. 2015). However, metatranscriptomics studies of the full microbial communities closely associated with plants are very limited, with most of them having been reviewed by Knief (2014). Recently, Chapelle et al. (2016) sequenced metagenomic DNA and RNA of the rhizospheric microbiome of sugar beet seedlings grown in a soil suppressive to the fungal pathogen *Rhizoctonia solani.* The authors identified abundant bacterial taxa during the process of fungal invasion, but more importantly, they identified stress-related genes (ppGpp metabolism and oxidative stress) that were upregulated in these bacterial families. The stress responses in the rhizobacterial community caused by the pathogen resulted in not only shifts in microbiome composition but also the activation of antagonistic traits that restrict pathogen infection. Newman et al. (2016) explored the response of the rhizosphere prokaryotic metatranscriptome to glyphosate in corn and soybean. Their results demonstrated that long-term agrichemical use may potentially shift the bacterial community composition to favour more glyphosate-tolerant bacteria and thereby affect gene expression and functions in the rhizospheric microbiome.

In the same way, *metaproteomic* analysis is a powerful approach for identifying microbes and targeting the active functional part of the microbiome. The large-scale characterisation of the entire protein complement of environmental microbiota at a given point in time reveals important metabolic information of the microbial community (Herbst et al. 2016). Advances in next-generation sequencing technologies together with great improvements in the depth and throughput of mass spectrometry have allowed the development of a new research field: *proteogenomics.* Proteogenomics combines the information inferred from proteomic studies, usually based on mass spectrometry data, with genomics and transcriptomics information (Nesvizhskii 2014). The methodology consists of the identification of the peptides in the sample by means of mass spectrometry by searching the six-frame translation of the genome sequence. One of the main applications of this approach is in completing genome annotations. The combination of metagenomics and metaproteomic approaches, also known as community proteogenomics or *metaproteogenomics*, was used by Delmotte and collaborators (2009) to unravel insights into the physiology of the bacteria of soybean, clover and *A. thaliana* phyllospheres; the study allowed for the identification of bacteria present in the phyllosphere together with the identification of abundant proteins in the phyllosphere microbiota, offering insights into the strategies employed by endophytes for their lifestyles in their plant hosts phyllospheres (Delmotte et al. 2009). Despite the fact that proteomic techniques have been successfully applied in the study of some important components of the microbiome, such as endophytes with antagonistic ability (Massart et al. 2015b), the use of metaproteomics for the study of the whole community is still in an early stage of development (Lakshmanan et al. 2014; Armengaud 2016), and few works have been published until now.

Recently, novel technologies have allowed the development of tools for studying the metabolites produced by plants and microbes. The global analysis of these compounds, called *metabolomics*, may help to understand the chemical communication that occurs in the plant microbiome (van Dam and Bouwmeester 2016). Changes in plant foliar and floral metabolomes related to associate microbiomes have been studied by Gargallo-Garriga and associates (2016). The authors analysed the epiphytic and internal metabolomes of *Sambucus nigra* leaves and flower metabolomes with and without associated microbiota (the microbiota were suppressed by the application of antibiotics) by liquid chromatography-mass spectrometry (LC-MS). The results of this study show how the microbiome plays a very important role in the plant metabolome. Metabolomics approaches have special importance in the rhizosphere, where exact mechanisms or signals by which plants shape their microbiome are still unknown. However, these techniques still present limitations due to the difficulty in properly sampling plant exudates, the sensitivity of available platforms and the analysis of complex data (van Dam and Bouwmeester 2016).

The comparison and integration of metagenomics with metatranscriptomics, metaproteomics and metabolomics data may be the key for achieving a complete view of the microbial players and the activities that occur in the plant microbiome and for changing our perception of plant-microbial interactions (Knief 2014).

13.6 Future Perspectives

As we have compiled in this review, plants harbour diverse microbiomes that are integrated by complex microbial communities organised in interconnected networks. Advances in sequencing technologies have expanded our understanding of the structure of the plant microbiome over recent years. However, with almost 300,000 species of plants on earth, there remains much to explore in relation to the diversity of plant-associated microorganisms. In addition, recent studies have primarily focused on the descriptive assessment of the composition of plant-associated microbiota through rRNA gene amplicon sequencing, leading to a greater knowledge of the presence or absence of specific OTUs. Nonetheless, the importance of the structure of such microbiota communities is still largely unknown. For example, minor changes in the abundance of particular OTUs can cause remarkable impacts on plant physiology. Despite the well-known advances in high-throughput sequencing, there remains a need to foster experimental microbiome genetic characterisation to begin answering questions about its functionality. Thus, although metagenomics, metatranscriptomic, metaproteomic and metabolomics approaches are increasingly more available, the lack of high-quality reference databases is still a problem to overcome if this field of research is to be further developed.

Much remains to be understood about plant-associated microbiota regarding their relevant metabolic capabilities related to plant activity and environmental processes. In metagenomics studies, it remains highly challenging for environmental scientists to obtain a comprehensive representation of all the microorganisms present in a particular niche. The remarkable complexity of these microbial communities hampers the de novo assembly of whole genomes from metagenomes, especially for less abundant taxa. Thus, research on environmental isolates of microbial taxa remains the main approach used to describe their ecophysiological traits. Improvement in sampling efforts and culturing methods should produce representative culture collections of the plant microbiota, as many root and leaf microbial communities contain a higher percentage of culturable microbes than soil. After an initial culture-independent survey, the relative abundance of isolated OTUs can be evaluated for the different plant tissues of interest or even in external environments such as the rhizosphere. These collections also have the potential to be developed as microbial inocula to promote plant health in organic applications.

Whole-genome sequencing of isolated plant-associated microbial taxa will allow for a better understanding of their metabolic potential (García-Fraile et al. 2016; Lladó et al. 2016). The combination of genome sequencing with meta-omics technologies offers the opportunity to further address underlying microbiome functions, thereby improving the annotation of metatranscriptomes and metaproteomes and, as a consequence, improving the available databases.

Because most of the currently published studies focus only on bacteria or on fungi, studies including the analysis of both bacterial and fungal communities are needed. The simultaneous analysis of the complete microbial community associated with plants offers a global view of all the taxa together with the potential networks and interactions occurring in these environments (van der Heijden and Hartmann 2016). Moreover, the study of other members such as viruses, archaea and other eukaryotes, as well as the endosymbiotic bacteria living inside hyphae, will provide a more comprehensive picture of the plant microbiome (Lebeis 2015). Unfortunately, such global studies are largely missing. Similarly, global studies including the analysis of the different microbiomes existing within a plant are not yet available. Establishing a 'core' or groups of 'core' plant microbiomes remains an important future task. The exploration of plant microbiomes allows us to define the healthy microbiota for certain crop species and to monitor the changes caused by abiotic and biotic stress. Variations in the 'core' microbiome would allow us to predict the effects of plant diseases or their reduction on crop yields. However, this task would require long-term and intensive studies (Chagnon and Bainard 2015) and the use of different pathogens or different methods of agricultural management together with proper and robust experimental designs (Berg et al. 2016). Even if these approaches may appear unfeasible, similar directions have been taken in the study of the human gut microbiome. For the first time, it is possible to study the microbial communities associated with plants at a high resolution and to obtain a holistic view of these communities. The linking of information obtained from multiple meta-approaches will provide models to explain the interactions occurring in these plant microbiomes as well as their responses to environmental factors such as pathogen attacks and diseases, agricultural practices and climate change (Knief 2014).

The essential next step is to obtain control over plant-beneficial functions encoded in the plant microbiome. The description of how these beneficial functions of the microbiome operate at the molecular level and revealing which plant genes play a role in acquiring the profits of these functionalities will be of paramount importance in the design of future crops. Such genomic design, concomitantly with microbial agriculture, will enhance crop production with a lower input of dangerous chemicals, thereby fostering crop sustainability. Understanding the key factors necessary for healthy plant-microbe associations would contribute to the conscious selection of microorganisms for use as soil inoculum. This could be included in new management practices for sustainable food production and climate change mitigation. Such practices include those with a main objective of fostering quality plant microbiomes to enhance crop yields while also improving soil quality and nutrient cycling functions such as C sequestration. However, many knowledge gaps must be overcome. Microbial inoculants may affect the autochthonous soil microbiota in different ways; thus, we need to assess the survival of inoculated strains in the natural environment and carry out risk assessment analyses, especially in the case of possible genetically modified microorganisms. To establish successful inoculations, we also need to understand the effects of abiotic factors such as pH or soil texture on allochthonous strains. Furthermore, it is also necessary to improve our understanding of microbe-microbe interactions. The individual members of a microbial inoculum might not produce additive effects depending on soil conditions or plant type but could create competitive situations that have no benefit to the plants.

Plant microbiomes are particularly relevant to the production of microbial inocula to enhance crop production in depressed world regions where limited resources are available for irrigation, fertilisation or treating plant diseases. Microbial inocula formed by drought-tolerant, halophytic and N-fixing strains might make it possible to produce crops in arid or saline soils. Another possible benefit of these types of associations may be to enhance crop nutritional value by introducing microbial strains with metabolic capabilities to supplement vitamins, proteins and antioxidants in the host plant. In addition, healthy plant microbiomes may also help plant hosts adapt to climate change due to their ability to change rapidly in response to environmental changes. In sum, optimising the use of plant microbiome functionalities by providing allochthonous microbial inoculants will help to foster agricultural production in less developed countries.

Acknowledgements The Czech Science Foundation (GAČR) under the grant number 16-15293Y and BIOCEV, the Biotechnology and Biomedicine Centre of the Academy of Sciences and Charles University provided funding to Rubén López-Mondejar, Martin Kostovčík and Paula García-Fraile, and Salvador Lladó was funded by the Czech Science Foundation (GAČR) under grant number 14-09040P. Lorena Carro thanks the School of Biology (Newcastle University) for a post-doctoral contract.

References

Afgan E, Baker D, van den Beek M, Blankenberg D, Bouvier D, Cech M, Chilton J, Clements D, Coraor N, Eberhard C, Gruning B, Guerler A, Hillman-Jackson J, Von Kuster G, Rasche E, Soranzo N, Turaga N, Taylor J, Nekrutenko A, Goecks J (2016) The galaxy platform for accessible, reproducible and collaborative biomedical analyses: 2016 update. Nucleic Acids Res 44(W1):W3–W10. doi:10.1093/nar/gkw343

Akinsanya MA, Goh JK, Lim SP, Ting AS (2015) Metagenomics study of endophytic bacteria in Aloe vera using next-generation technology. Genom Data 6:159–163. doi:10.1016/j.gdata.2015.09.004

Alloisio N, Queiroux C, Fournier P, Pujic P, Normand P, Vallenet D, Medigue C, Yamaura M, Kakoi K, Kucho K (2010) The Frankia alni symbiotic transcriptome. Mol Plant Microbe Interact 23(5):593–607. doi:10.1094/MPMI-23-5-0593

Amend AS, Seifert KA, Bruns TD (2010) Quantifying microbial communities with 454 pyrosequencing: does read abundance count? Mol Ecol 19(24):5555–5565

Anderson IC, Campbell CD, Prosser JI (2003) Potential bias of fungal 18S rDNA and internal transcribed spacer polymerase chain reaction primers for estimating fungal biodiversity in soil. Environ Microbiol 5(1):36–47

Apprill A, McNally S, Parsons R, Weber L (2015) Minor revision to V4 region SSU rRNA 806R gene primer greatly increases detection of SAR11 bacterioplankton. Aquat Microb Ecol 75(2):129–137. doi:10.3354/ame01753

Armengaud J (2016) Next-generation proteomics faces new challenges in environmental biotechnology. Curr Opin Biotechnol 38:174–182. doi:10.1016/j.copbio.2016.02.025

Bao E, Jiang T, Kaloshian I, Girke T (2011) SEED: efficient clustering of next-generation sequences. Bioinformatics 27(18):2502–2509. doi:10.1093/bioinformatics/btr447

Baroncelli R, Piaggeschi G, Fiorini L, Bertolini E, Zapparata A, Pè ME, Sarrocco S, Vannacci G (2015) Draft whole-genome sequence of the biocontrol agent Trichoderma harzianum T6776. Genome Announc 3(3):e00647–e00615. doi:10.1128/genomeA.00647-15

Baroncelli R, Zapparata A, Piaggeschi G, Sarrocco S, Vannacci G (2016) Draft whole-genome sequence of Trichoderma gamsii T6085, a promising biocontrol agent of Fusarium head blight on wheat. Genome Announc 4(1):e01747–e01715. doi:10.1128/genomeA.01747-15

Baum C, El-Tohamy W, Gruda N (2015) Increasing the productivity and product quality of vegetable crops using arbuscular mycorrhizal fungi: a review. Sci Hortic 187:131–141. doi:10.1016/j.scienta.2015.03.002

Bazghaleh N, Hamel C, Gan Y, Tar'an B, Knight JD (2015) Genotype-specific variation in the structure of root fungal communities is related to chickpea plant productivity. Appl Environ Microbiol 81(7):2368–2377. doi:10.1128/AEM.03692-14

Berendsen RL, Pieterse CM, Bakker PA (2012) The rhizosphere microbiome and plant health. Trends Plant Sci 17(8):478–486. doi:10.1016/j.tplants.2012.04.001

Berg G, Grube M, Schloter M, Smalla K (2014) Unraveling the plant microbiome: looking back and future perspectives. Front Microbiol 5:148. doi:10.3389/fmicb.2014.00148

Berg G, Rybakova D, Grube M, Koberl M (2016) The plant microbiome explored: implications for experimental botany. J Exp Bot 67(4):995–1002. doi:10.1093/jxb/erv466

Berger H, Yacoub A, Gerbore J, Grizard D, Rey P, Sessitsch A, Compant S (2016) Draft genome sequence of biocontrol agent Pythium Oligandrum strain Po37, an Oomycota. Genome Announc 4(2):e00215–e00216. doi:10.1128/genomeA.00215-16

Berger SA, Krompass D, Stamatakis A (2011) Performance, accuracy, and web server for evolutionary placement of short sequence reads under maximum likelihood. Syst Biol 60(3): 291–302. doi:10.1093/sysbio/syr010

Berry D, Ben Mahfoudh K, Wagner M, Loy A (2011) Barcoded primers used in multiplex amplicon pyrosequencing bias amplification. Appl Environ Microbiol 77(21):7846–7849. doi:10.1128/Aem.05220-11

Bertani I, Abbruscato P, Piffanelli P, Subramoni S, Venturi V (2016) Rice bacterial endophytes: isolation of a collection, identification of beneficial strains and microbiome analysis. Environ Microbiol Rep 8(3):388–398. doi:10.1111/1758-2229.12403

Bleidorn C (2015) Third generation sequencing: technology and its potential impact on evolutionary biodiversity research. Syst Biodivers 14:1–8. doi:10.1080/14772000.2015.1099575

Bokulich NA, Mills DA (2013) Improved selection of internal transcribed spacer-specific primers enables quantitative, ultra-high-throughput profiling of fungal communities. Appl Environ Microbiol 79(8):2519–2526. doi:10.1128/AEM.03870-12

Bokulich NA, Subramanian S, Faith JJ, Gevers D, Gordon JI, Knight R, Mills DA, Caporaso JG (2013) Quality-filtering vastly improves diversity estimates from Illumina amplicon sequencing. Nat Methods 10(1):57–U11. doi:10.1038/Nmeth.2276

Burke C, Kjelleberg S, Thomas T (2009) Selective extraction of bacterial DNA from the surfaces of macroalgae. Appl Environ Microbiol 75(1):252–256. doi:10.1128/AEM.01630-08

Busby PE, Peay KG, Newcombe G (2016) Common foliar fungi of Populus trichocarpa modify Melampsora rust disease severity. New Phytol 209(4):1681–1692. doi:10.1111/nph.13742

Caporaso JG, Kuczynski J, Stombaugh J, Bittinger K, Bushman FD, Costello EK, Fierer N, Pena AG, Goodrich JK, Gordon JI, Huttley GA, Kelley ST, Knights D, Koenig JE, Ley RE, Lozupone CA, McDonald D, Muegge BD, Pirrung M, Reeder J, Sevinsky JR, Turnbaugh PJ, Walters WA, Widmann J, Yatsunenko T, Zaneveld J, Knight R (2010) QIIME allows analysis of high-throughput community sequencing data. Nat Methods 7(5):335–336. doi:10.1038/nmeth.f.303

Caporaso JG, Lauber CL, Walters WA, Berg-Lyons D, Huntley J, Fierer N, Owens SM, Betley J, Fraser L, Bauer M, Gormley N, Gilbert JA, Smith G, Knight R (2012) Ultra-high-throughput microbial community analysis on the Illumina HiSeq and MiSeq platforms. ISME J 6(8): 1621–1624. doi:10.1038/ismej.2012.8

Caporaso JG, Lauber CL, Walters WA, Berg-Lyons D, Lozupone CA, Turnbaugh PJ, Fierer N, Knight R (2011) Global patterns of 16S rRNA diversity at a depth of millions of sequences per sample. Proc Natl Acad Sci U S A 108:4516–4522. doi:10.1073/pnas.1000080107

Carlsen T, Aas AB, Lindner D, Vrålstad T, Schumacher T, Kauserud H (2012) Don't make a mista (g) ke: is tag switching an overlooked source of error in amplicon pyrosequencing studies? Fungal Ecol 5(6):747–749

Carro L, Pujic P, Alloisio N, Fournier P, Boubakri H, Hay AE, Poly F, Francois P, Hocher V, Mergaert P, Balmand S, Rey M, Heddi A, Normand P (2015) Alnus peptides modify membrane porosity and induce the release of nitrogen-rich metabolites from nitrogen-fixing Frankia. ISME J 9(8):1723–1733. doi:10.1038/ismej.2014.257

Carro L, Pujic P, Alloisio N, Fournier P, Boubakri H, Poly F, Rey M, Heddi A, Normand P (2016) Physiological effects of major upregulated Alnus glutinosa peptides on Frankia sp. ACN14a. Microbiology 162(7):1173–1184. doi:10.1099/mic.0.000291

Chagnon PL, Bainard LD (2015) Using molecular biology to study mycorrhizal fungal community ecology: limits and perspectives. Plant Signal Behav 10(7):e1046668. doi:10.1080/15592324.2015.1046668

Chapelle E, Mendes R, Bakker PA, Raaijmakers JM (2016) Fungal invasion of the rhizosphere microbiome. ISME J 10(1):265–268. doi:10.1038/ismej.2015.82

Classen AT, Sundqvist MK, Henning JA, Newman GS, Moore JAM, Cregger MA, Moorhead LC, Patterson CM (2015) Direct and indirect effects of climate change on soil microbial and soil microbial-plant interactions: what lies ahead? Ecosphere 6(8):1–21. doi:10.1890/es15-00217.1

Craig DW, Pearson JV, Szelinger S, Sekar A, Redman M, Corneveaux JJ, Pawlowski TL, Laub T, Nunn G, Stephan DA, Homer N, Huentelman MJ (2008) Identification of genetic variants using bar-coded multiplexed sequencing. Nat Methods 5(10):887–893. doi:10.1038/Nmeth.1251

Dai L, Gao X, Guo Y, Xiao J, Zhang Z (2012) Bioinformatics clouds for big data manipulation. Biol Direct 7(1):43; discussion 43. doi:10.1186/1745-6150-7-43

Dai M, Hamel C, Bainard LD, Arnaud MS, Grant CA, Lupwayi NZ, Malhi SS, Lemke R (2014) Negative and positive contributions of arbuscular mycorrhizal fungal taxa to wheat production and nutrient uptake efficiency in organic and conventional systems in the Canadian prairie. Soil Biol Biochem 74:156–166. doi:10.1016/j.soilbio.2014.03.016

De Beenhouwer M, Muleta D, Peeters B, Van Geel M, Lievens B, Honnay O (2015a) DNA pyrosequencing evidence for large diversity differences between natural and managed coffee mycorrhizal fungal communities. Agron Sustain Dev 35(1):241–249. doi:10.1007/s13593-014-0231-8

De Beenhouwer M, Van Geel M, Ceulemans T, Muleta D, Lievens B, Honnay O (2015b) Changing soil characteristics alter the arbuscular mycorrhizal fungi communities of Arabica coffee (Coffea arabica) in Ethiopia across a management intensity gradient. Soil Biol Biochem 91:133–139. doi:10.1016/j.soilbio.2015.08.037

Dean RA, Talbot NJ, Ebbole DJ, Farman ML, Mitchell TK, Orbach MJ, Thon M, Kulkarni R, Xu JR, Pan H, Read ND, Lee YH, Carbone I, Brown D, Oh YY, Donofrio N, Jeong JS, Soanes DM, Djonovic S, Kolomiets E, Rehmeyer C, Li W, Harding M, Kim S, Lebrun MH, Bohnert H, Coughlan S, Butler J, Calvo S, Ma LJ, Nicol R, Purcell S, Nusbaum C, Galagan JE, Birren BW (2005) The genome sequence of the rice blast fungus Magnaporthe grisea. Nature 434(7036):980–986. doi:10.1038/nature03449

Degnan PH, Ochman H (2012) Illumina-based analysis of microbial community diversity. ISME J 6(1):183–194. doi:10.1038/ismej.2011.74

Delmont TO, Robe P, Clark I, Simonet P, Vogel TM (2011) Metagenomic comparison of direct and indirect soil DNA extraction approaches. J Microbiol Methods 86(3):397–400. doi:10.1016/j.mimet.2011.06.013

Delmotte N, Knief C, Chaffron S, Innerebner G, Roschitzki B, Schlapbach R, von Mering C, Vorholt JA (2009) Community proteogenomics reveals insights into the physiology of phyllosphere bacteria. Proc Natl Acad Sci U S A 106(38):16428–16433. doi:10.1073/pnas.0905240106

Demina IV, Persson T, Santos P, Plaszczyca M, Pawlowski K (2013) Comparison of the nodule vs. root transcriptome of the actinorhizal plant Datisca glomerata: actinorhizal nodules contain a specific class of defensins. PLoS One 8(8):e72442. doi:10.1371/journal.pone.0072442

DeSantis TZ, Hugenholtz P, Larsen N, Rojas M, Brodie EL, Keller K, Huber T, Dalevi D, Hu P, Andersen GL (2006) Greengenes, a chimera-checked 16S rRNA gene database and workbench compatible with ARB. Appl Environ Microbiol 72(7):5069–5072. doi:10.1128/Aem.03006-05

Diedhiou I, Tromas A, Cissoko M, Gray K, Parizot B, Crabos A, Alloisio N, Fournier P, Carro L, Svistoonoff S, Gherbi H, Hocher V, Diouf D, Laplaze L, Champion A (2014) Identification of potential transcriptional regulators of actinorhizal symbioses in Casuarina glauca and Alnus glutinosa. BMC Plant Biol 14:342. doi:10.1186/s12870-014-0342-z

Duan Y, Zhou L, Hall DG, Li W, Doddapaneni H, Lin H, Liu L, Vahling CM, Gabriel DW, Williams KP, Dickerman A, Sun Y, Gottwald T (2009) Complete genome sequence of citrus huanglongbing bacterium, 'Candidatus Liberibacter asiaticus' obtained through metagenomics. Mol Plant Microbe Interact 22(8):1011–1020. doi:10.1094/MPMI-22-8-1011

Edgar RC (2010) Search and clustering orders of magnitude faster than BLAST. Bioinformatics 26(19):2460–2461. doi:10.1093/bioinformatics/btq461

Edgar RC (2013) UPARSE: highly accurate OTU sequences from microbial amplicon reads. Nat Methods 10(10):996–998. doi:10.1038/nmeth.2604

Edgar RC, Flyvbjerg H (2015) Error filtering, pair assembly and error correction for next-generation sequencing reads. Bioinformatics 31(21):3476–3482. doi:10.1093/bioinformatics/btv401

Faircloth BC, Glenn TC (2012) Not all sequence tags are created equal: designing and validating sequence identification tags robust to indels. PLoS One 7(8):e42543

Faith JJ, Guruge JL, Charbonneau M, Subramanian S, Seedorf H, Goodman AL, Clemente JC, Knight R, Heath AC, Leibel RL, Rosenbaum M, Gordon JI (2013) The long-term stability of the human gut microbiota. Science 341(6141):1237439. doi:10.1126/science.1237439

Feehery GR, Yigit E, Oyola SO, Langhorst BW, Schmidt VT, Stewart FJ, Dimalanta ET, Amaral-Zettler LA, Davis T, Quail MA, Pradhan S (2013) A method for selectively enriching microbial DNA from contaminating vertebrate host DNA. PLoS One 8(10):e76096. doi:10.1371/journal.pone.0076096

Felczykowska A, Krajewska A, Zielinska S, Los JM (2015) Sampling, metadata and DNA extraction – important steps in metagenomic studies. Acta Biochim Pol 62(1):151–160

Fonseca-Garcia C, Coleman-Derr D, Garrido E, Visel A, Tringe SG, Partida-Martinez LP (2016) The cacti microbiome: interplay between habitat-filtering and host-specificity. Front Microbiol 7:150. doi:10.3389/fmicb.2016.00150

Frank DN (2009) BARCRAWL and BARTAB: software tools for the design and implementation of barcoded primers for highly multiplexed DNA sequencing. BMC Bioinformatics 10(1):362. doi:10.1186/1471-2105-10-362

Franke-Whittle IH, Manici LM, Insam H, Stres B (2015) Rhizosphere bacteria and fungi associated with plant growth in soils of three replanted apple orchards. Plant Soil 395(1–2):317–333. doi:10.1007/s11104-015-2562-x

Franzen O, Hu J, Bao X, Itzkowitz SH, Peter I, Bashir A (2015) Improved OTU-picking using long-read 16S rRNA gene amplicon sequencing and generic hierarchical clustering. Microbiome 3:43. doi:10.1186/s40168-015-0105-6

Gaiero JR, McCall CA, Thompson KA, Day NJ, Best AS, Dunfield KE (2013) Inside the root microbiome: bacterial root endophytes and plant growth promotion. Am J Bot 100(9):1738–1750. doi:10.3732/ajb.1200572

Garcia-Fraile P, Llado S, Benada O, Cajthamal T, Baldian P (2016) Terracidophilus gabretensis gen nov an abundant and active forest soil bacterium important in organic matter transformation. Appl Environ Microb. 82:560–9.

García-Fraile P, Menéndez E, Rivas R (2015) Role of bacterial biofertilizers in agriculture and forestry. AIMS Bioengineering 2(3):183–205. doi:10.3934/bioeng.2015.3.183

Gargallo-Garriga A, Sardans J, Pérez-Trujillo M, Guenther A, Llusià J, Rico L, Terradas J, Farré-Armengol G, Filella I, Parella T, Peñuelas J (2016) Shifts in plant foliar and floral metabolomes in response to the suppression of the associated microbiota. BMC Plant Biology1600 doi:10.1186/s12870-016-0767-7

Garita-Cambronero J, Palacio-Bielsa A, Lopez MM, Cubero J (2016) Draft genome sequence for virulent and avirulent strains of Xanthomonas arboricola isolated from Prunus spp. in Spain. Stand Genomic Sci 11:12. doi:10.1186/s40793-016-0132-3

Ghodsi M, Liu B, Pop M (2011) DNACLUST: accurate and efficient clustering of phylogenetic marker genes. BMC Bioinformatics 12(1):271. doi:10.1186/1471-2105-12-271

Glassing A, Dowd SE, Galandiuk S, Davis B, Jorden JR, Chiodini RJ (2015) Changes in 16s RNA Gene microbial community profiling by concentration of prokaryotic DNA. J Microbiol Methods 119:239–242. doi:10.1016/j.mimet.2015.11.001

Gyorgyey J, Vaubert D, Jimenez-Zurdo JI, Charon C, Troussard L, Kondorosi A, Kondorosi E (2000) Analysis of Medicago truncatula nodule expressed sequence tags. Mol Plant Microbe Interact 13(1):62–71. doi:10.1094/MPMI.2000.13.1.62

Haas BJ, Gevers D, Earl AM, Feldgarden M, Ward DV, Giannoukos G, Ciulla D, Tabbaa D, Highlander SK, Sodergren E, Methe B, DeSantis TZ, Petrosino JF, Knight R, Birren BW,

Consortium HM (2011) Chimeric 16S rRNA sequence formation and detection in sanger and 454-pyrosequenced PCR amplicons. Genome Res 21(3):494–504. doi:10.1101/gr.112730.110

Hamady M, Walker JJ, Harris JK, Gold NJ, Knight R (2008) Error-correcting barcoded primers allow hundreds of samples to be pyrosequenced in multiplex. Nat Methods 5(3):235

Hart ML, Meyer A, Johnson PJ, Ericsson AC (2015) Comparative evaluation of DNA extraction methods from feces of multiple host species for downstream next-generation sequencing. PLoS One 10(11):e0143334. doi:10.1371/journal.pone.0143334

Herbst FA, Lunsmann V, Kjeldal H, Jehmlich N, Tholey A, von Bergen M, Nielsen JL, Hettich RL, Seifert J, Nielsen PH (2016) Enhancing metaproteomics-the value of models and defined environmental microbial systems. Proteomics 16(5):783–798. doi:10.1002/pmic.201500305

Hesse CN, Mueller RC, Vuyisich M, Gallegos-Graves LV, Gleasner CD, Zak DR, Kuske CR (2015) Forest floor community metatranscriptomes identify fungal and bacterial responses to N deposition in two maple forests. Front Microbiol 6:337. doi:10.3389/fmicb.2015.00337

Higo M, Isobe K, Drijber RA, Kondo T, Yamaguchi M, Takeyama S, Suzuki Y, Niijima D, Matsuda Y, Ishii R, Torigoe Y (2014) Impact of a 5-year winter cover crop rotational system on the molecular diversity of arbuscular mycorrhizal fungi colonizing roots of subsequent soybean. Biol Fertil Soils 50(6):913–926. doi:10.1007/s00374-014-0912-0

Hirsch J, Galidevara S, Strohmeier S, Devi KU, Reineke A (2013) Effects on diversity of soil fungal community and fate of an artificially applied Beauveria bassiana strain assessed through 454 pyrosequencing. Microb Ecol 66(3):608–620. doi:10.1007/s00248-013-0249-5

Hirsch PR, Mauchline TH (2012) Who's who in the plant root microbiome? Nat Biotechnol 30(10):961–962. doi:10.1038/nbt.2387

Ho Y-N, Huang C-C (2015) Draft genome sequence of Burkholderia cenocepacia strain 869T2, a plant-beneficial endophytic bacterium. Genome Announc 3(6):e01327–e01315

Hocher V, Alloisio N, Auguy F, Fournier P, Doumas P, Pujic P, Gherbi H, Queiroux C, Da Silva C, Wincker P, Normand P, Bogusz D (2011) Transcriptomics of actinorhizal symbioses reveals homologs of the whole common symbiotic signaling cascade. Plant Physiol 156(2):700–711. doi:10.1104/pp.111.174151

Holland TC, Bowen P, Bogdanoff C, Hart MM (2013) How distinct are arbuscular mycorrhizal fungal communities associating with grapevines? Biol Fertil Soils 50(4):667–674. doi:10.1007/s00374-013-0887-2

Huang X, Liu L, Wen T, Zhu R, Zhang J, Cai Z (2015) Illumina MiSeq investigations on the changes of microbial community in the Fusarium oxysporum f.Sp. cubense infected soil during and after reductive soil disinfestation. Microbiol Res 181:33–42. doi:10.1016/j.micres.2015.08.004

Huang Y, Kuang Z, Wang W, Cao L (2016) Exploring potential bacterial and fungal biocontrol agents transmitted from seeds to sprouts of wheat. Biol Control 98:27–33. doi:10.1016/j.biocontrol.2016.02.013

Huse SM, Welch DM, Morrison HG, Sogin ML (2010) Ironing out the wrinkles in the rare biosphere through improved OTU clustering. Environ Microbiol 12(7):1889–1898. doi:10.1111/j.1462-2920.2010.02193.x

Huson DH, Auch AF, Qi J, Schuster SC (2007) MEGAN analysis of metagenomic data. Genome Res 17(3):377–386. doi:10.1101/gr.5969107

Ihrmark K, Bodeker IT, Cruz-Martinez K, Friberg H, Kubartova A, Schenck J, Strid Y, Stenlid J, Brandstrom-Durling M, Clemmensen KE, Lindahl BD (2012) New primers to amplify the fungal ITS2 region–evaluation by 454-sequencing of artificial and natural communities. FEMS Microbiol Ecol 82(3):666–677. doi:10.1111/j.1574-6941.2012.01437.x

Ikeda S, Sasaki K, Okubo T, Yamashita A, Terasawa K, Bao Z, Liu D, Watanabe T, Murase J, Asakawa S, Eda S, Mitsui H, Sato T, Minamisawa K (2014) Low nitrogen fertilization adapts rice root microbiome to low nutrient environment by changing biogeochemical functions. Microbes Environ 29(1):50–59

Jiang Y, Xiong X, Danska J, Parkinson J (2016) Metatranscriptomic analysis of diverse microbial communities reveals core metabolic pathways and microbiome-specific functionality. Microbiome 4(1):2. doi:10.1186/s40168-015-0146-x

Junker RR, Keller A (2015) Microhabitat heterogeneity across leaves and flower organs promotes bacterial diversity. FEMS Microbiol Ecol 91(9):fiv097. doi:10.1093/femsec/fiv097

Kanagawa T (2003) Bias and artifacts in multitemplate polymerase chain reactions (PCR). J Biosci Bioeng 96(4):317–323. doi:10.1016/S1389-1723(03)90130-7

Kaneko T, Nakamura Y, Sato S, Asamizu E, Kato T, Sasamoto S, Watanabe A, Idesawa K, Ishikawa A, Kawashima K, Kimura T, Kishida Y, Kiyokawa C, Kohara M, Matsumoto M, Matsuno A, Mochizuki Y, Nakayama S, Nakazaki N, Shimpo S, Sugimoto M, Takeuchi C, Yamada M, Tabata S (2000) Complete genome structure of the nitrogen-fixing symbiotic bacterium Mesorhizobium loti (supplement). DNA Res 7(6):381–406

Kauserud H, Kumar S, Brysting AK, Norden J, Carlsen T (2012) High consistency between replicate 454 pyrosequencing analyses of ectomycorrhizal plant root samples. Mycorrhiza 22(4):309–315. doi:10.1007/s00572-011-0403-1

Kim Y, Liesack W (2015) Differential assemblage of functional units in paddy soil microbiomes. PLoS One 10(4):e0122221. doi:10.1371/journal.pone.0122221

Kircher M, Sawyer S, Meyer M (2012) Double indexing overcomes inaccuracies in multiplex sequencing on the Illumina platform. Nucleic Acids Res 40(1):e3. doi:10.1093/nar/gkr771

Knief C (2014) Analysis of plant microbe interactions in the era of next generation sequencing technologies. Front Plant Sci 5:216. doi:10.3389/fpls.2014.00216

Knief C, Delmotte N, Chaffron S, Stark M, Innerebner G, Wassmann R, von Mering C, Vorholt JA (2012) Metaproteogenomic analysis of microbial communities in the phyllosphere and rhizosphere of rice. ISME J 6(7):1378–1390. doi:10.1038/ismej.2011.192

Knight R, Jansson J, Field D, Fierer N, Desai N, Fuhrman J, Hugenholtz P, Meyer F, Stevens R, Bailey M (2012) Designing better metagenomic surveys: the role of experimental design and metadata capture in making useful metagenomic datasets for ecology and biotechnology. Nat Biotechnol 30(6):513–512

Koiv V, Roosaare M, Vedler E, Kivistik PA, Toppi K, Schryer DW, Remm M, Tenson T, Mae A (2015) Microbial population dynamics in response to Pectobacterium atrosepticum infection in potato tubers. Sci Rep 5:11606. doi:10.1038/srep11606

Koljalg U, Nilsson RH, Abarenkov K, Tedersoo L, Taylor AFS, Bahram M, Bates ST, Bruns TD, Bengtsson-Palme J, Callaghan TM, Douglas B, Drenkhan T, Eberhardt U, Duenas M, Grebenc T, Griffith GW, Hartmann M, Kirk PM, Kohout P, Larsson E, Lindahl BD, Luecking R, Martin MP, Matheny PB, Nguyen NH, Niskanen T, Oja J, Peay KG, Peintner U, Peterson M, Poldmaa K, Saag L, Saar I, Schuessler A, Scott JA, Senes C, Smith ME, Suija A, Taylor DL, Telleria MT, Weiss M, Larsson KH (2013) Towards a unified paradigm for sequence-based identification of fungi. Mol Ecol 22(21):5271–5277. doi:10.1111/mec.12481

Kopylova E, Noe L, Touzet H (2012) SortMeRNA: fast and accurate filtering of ribosomal RNAs in metatranscriptomic data. Bioinformatics 28(24):3211–3217. doi:10.1093/bioinformatics/bts611

Kozich JJ, Westcott SL, Baxter NT, Highlander SK, Schloss PD (2013) Development of a dual-index sequencing strategy and Curation pipeline for analyzing amplicon sequence data on the MiSeq Illumina sequencing platform. Appl Environ Microbiol 79(17):5112–5120. doi:10.1128/Aem.01043-13

Kucuktas H, Liu ZJ (2010) Library construction for next generation sequencing. In: Next generation sequencing and whole genome selection in aquaculture. Wiley, Chichester, pp 57–67

Lakshmanan V, Selvaraj G, Bais HP (2014) Functional soil microbiome: belowground solutions to an aboveground problem. Plant Physiol 166(2):689–700. doi:10.1104/pp.114.245811

Lang C, Long SR (2015) Transcriptomic analysis of Sinorhizobium meliloti and Medicago truncatula Symbiosis using nitrogen fixation-deficient nodules. Mol Plant Microbe Interact 28(8):856–868. doi:10.1094/MPMI-12-14-0407-R

Lebeis SL (2014) The potential for give and take in plant-microbiome relationships. Front Plant Sci 5:287. doi:10.3389/fpls.2014.00287

Lebeis SL (2015) Greater than the sum of their parts: characterizing plant microbiomes at the community-level. Curr Opin Plant Biol 24:82–86. doi:10.1016/j.pbi.2015.02.004

Lee I-M, Shao J, Bottner-Parker KD, Gundersen-Rindal DE, Zhao Y, Davis RE (2015) Draft genome sequence of "Candidatus Phytoplasma pruni" strain CX, a plant-pathogenic bacterium. Genome Announc 3(5):e01117–e01115. doi:10.1128/genomeA.01117-15

Li W, Fu L, Niu B, Wu S, Wooley J (2012) Ultrafast clustering algorithms for metagenomic sequence analysis. Brief Bioinform 13(6):656–668. doi:10.1093/bib/bbs035

Lindgreen S (2012) AdapterRemoval: easy cleaning of next-generation sequencing reads. BMC Res Notes 5(1):337. doi:10.1186/1756-0500-5-337

Links MG, Dumonceaux TJ, Hemmingsen SM, Hill JE (2012) The Chaperonin-60 universal target is a barcode for bacteria that enables de novo assembly of metagenomic sequence data. PLoS One 7(11):e49755. doi:10.1371/journal.pone.0049755

Liu BH, Yuan JY, Yiu SM, Li ZY, Xie YL, Chen YX, Shi YJ, Zhang H, Li YR, Lam TW, Luo RB (2012) COPE: an accurate k-mer-based pair-end reads connection tool to facilitate genome assembly. Bioinformatics 28(22):2870–2874. doi:10.1093/bioinformatics/bts563

Llado S, Benada O, Cajthamal T, Baldian P, Garcia-Fraile P (2016) Silvibacterium bohemicum gen nov sp nov a new acidophilic bacterium isolated from coniferous soil in the Bohemian National Park. Syst Appl Microb. 39:14–9

Loman NJ, Misra RV, Dallman TJ, Constantinidou C, Gharbia SE, Wain J, Pallen MJ (2012) Performance comparison of benchtop high-throughput sequencing platforms. Nat Biotechnol 30(5):434–439. doi:10.1038/nbt.2198

Loman NJ, Quick J, Simpson JT (2015) A complete bacterial genome assembled de novo using only nanopore sequencing data. Nat Methods 12(8):733–735. doi:10.1038/nmeth.3444

Lundberg DS, Lebeis SL, Paredes SH, Yourstone S, Gehring J, Malfatti S, Tremblay J, Engelbrektson A, Kunin V, del Rio TG, Edgar RC, Eickhorst T, Ley RE, Hugenholtz P, Tringe SG, Dangl JL (2012) Defining the core Arabidopsis thaliana root microbiome. Nature 488(7409):86–90. doi:10.1038/nature11237

Magoc T, Salzberg SL (2011) FLASH: fast length adjustment of short reads to improve genome assemblies. Bioinformatics 27(21):2957–2963. doi:10.1093/bioinformatics/btr507

Mahé F, Rognes T, Quince C, de Vargas C, Dunthorn M (2014) Swarm: robust and fast clustering method for amplicon-based studies. PeerJ 2:e593

Manter DK, Delgado JA, Holm DG, Stong RA (2010) Pyrosequencing reveals a highly diverse and cultivar-specific bacterial endophyte community in potato roots. Microb Ecol 60(1):157–166. doi:10.1007/s00248-010-9658-x

Masella AP, Bartram AK, Truszkowski JM, Brown DG, Neufeld JD (2012) PANDAseq: PAired-eND assembler for Illumina sequences. BMC Bioinformatics 13(31). doi:10.1186/1471-2105-13-31

Massart S, Martinez-Medina M, Jijakli MH (2015a) Biological control in the microbiome era: challenges and opportunities. Biol Control 89:98–108. doi:10.1016/j.biocontrol.2015.06.003

Massart S, Perazzolli M, Höfte M, Pertot I, Jijakli MH (2015b) Impact of the omic technologies for understanding the modes of action of biological control agents against plant pathogens. BioControl 60(6):725–746. doi:10.1007/s10526-015-9686-z

Matsen FA, Kodner RB, Armbrust EV (2010) Pplacer: linear time maximum-likelihood and Bayesian phylogenetic placement of sequences onto a fixed reference tree. BMC Bioinformatics 11:538. doi:10.1186/1471-2105-11-538

McPherson JD (2009) Next-generation gap. Nat Methods 6(11 Suppl):S2–S5. doi:10.1038/nmeth.f.268

Meaden S, Metcalf CJ, Koskella B (2016) The effects of host age and spatial location on bacterial community composition in the English Oak tree (Quercus robur). Environ Microbiol Rep. doi:10.1111/1758-2229.12423

Megías E, Megías M, Ollero FJ, Hungria M (2016) Draft genome sequence of Pantoea ananatis strain AMG521, a Rice Plant growth-promoting bacterial endophyte isolated from the Guadalquivir marshes in southern Spain. Genome Announc 4(1):e01681

Mendes LW, Tsai SM, Navarrete AA, de Hollander M, van Veen JA, Kuramae EE (2015) Soil-borne microbiome: linking diversity to function. Microb Ecol 70(1):255–265. doi:10.1007/s00248-014-0559-2

Mercier C, Boyer F, Bonin A, Coissac E (2013) SUMATRA and SUMACLUST: fast and exact comparison and clustering of sequences. In: Programs and Abstracts of the SeqBio 2013 workshop. Abstract, Citeseer, pp 27–29

Meyer M, Kircher M (2010) Illumina sequencing library preparation for highly multiplexed target capture and sequencing. Cold Spring Harb Protoc 2010(6):pdb prot5448. doi:10.1101/pdb.prot5448

Meyer M, Stenzel U, Myles S, Prufer K, Hofreiter M (2007) Targeted high-throughput sequencing of tagged nucleic acid samples. Nucleic Acids Res 35(15):e97. doi:10.1093/nar/gkm566

Meyerhans A, Vartanian JP, Wain-Hobson S (1990) DNA recombination during PCR. Nucleic Acids Res 18(7):1687–1691

Miao CP, Mi QL, Qiao XG, Zheng YK, Chen YW, Xu LH, Guan HL, Zhao LX (2016) Rhizospheric fungi of Panax notoginseng: diversity and antagonism to host phytopathogens. J Ginseng Res 40(2):127–134. doi:10.1016/j.jgr.2015.06.004

Miyambo T, Makhalanyane TP, Cowan DA, Valverde A (2016) Plants of the fynbos biome harbour host species-specific bacterial communities. FEMS Microbiol Lett. doi:10.1093/femsle/fnw122

Montes-Borrego M, Metsis M, Landa BB (2014) Arbuscular mycorrhizal fungi associated with the olive crop across the Andalusian landscape: factors driving community differentiation. PLoS One 9(5):e96397. doi:10.1371/journal.pone.0096397

Muller CA, Obermeier MM, Berg G (2016) Bioprospecting plant-associated microbiomes. J Biotechnol. doi:10.1016/j.jbiotec.2016.03.033

Munch K, Boomsma W, Huelsenbeck JP, Willerslev E, Nielsen R (2008) Statistical assignment of DNA sequences using Bayesian Phylogenetics. Syst Biol 57(5):750–757. doi:10.1080/10635150802422316

Nallanchakravarthula S, Mahmood S, Alstrom S, Finlay RD (2014) Influence of soil type, cultivar and Verticillium dahliae on the structure of the root and rhizosphere soil fungal microbiome of strawberry. PLoS One 9(10):e111455. doi:10.1371/journal.pone.0111455

Navarrete AA, Tsai SM, Mendes LW, Faust K, de Hollander M, Cassman NA, Raes J, van Veen JA, Kuramae EE (2015) Soil microbiome responses to the short-term effects of Amazonian deforestation. Mol Ecol 24(10):2433–2448. doi:10.1111/mec.13172

Nekrutenko A, Taylor J (2012) Next-generation sequencing data interpretation: enhancing reproducibility and accessibility. Nat Rev Genet 13(9):667–672. doi:10.1038/nrg3305

Nesvizhskii AI (2014) Proteogenomics: concepts, applications and computational strategies. Nat Methods 11(11):1114–1125. doi:10.1038/nmeth.3144

Newman MM, Lorenz N, Hoilett N, Lee NR, Dick RP, Liles MR, Ramsier C, Kloepper JW (2016) Changes in rhizosphere bacterial gene expression following glyphosate treatment. Sci Total Environ 553:32–41. doi:10.1016/j.scitotenv.2016.02.078

Normand P, Lapierre P, Tisa LS, Gogarten JP, Alloisio N, Bagnarol E, Bassi CA, Berry AM, Bickhart DM, Choisne N, Couloux A, Cournoyer B, Cruveiller S, Daubin V, Demange N, Francino MP, Goltsman E, Huang Y, Kopp OR, Labarre L, Lapidus A, Lavire C, Marechal J, Martinez M, Mastronunzio JE, Mullin BC, Niemann J, Pujic P, Rawnsley T, Rouy Z, Schenowitz C, Sellstedt A, Tavares F, Tomkins JP, Vallenet D, Valverde C, Wall LG, Wang Y, Medigue C, Benson DR (2007) Genome characteristics of facultatively symbiotic Frankia sp. strains reflect host range and host plant biogeography. Genome Res 17(1):7–15. doi:10.1101/gr.5798407

Okubo T, Ikeda S, Yamashita A, Terasawa K, Minamisawa K (2012) Pyrosequence read length of 16S rRNA gene affects phylogenetic assignment of plant-associated bacteria. Microbes Environ 27(2):204–208

Okubo T, Ikeda S, Sasaki K, Ohshima K, Hattori M, Sato T, Minamisawa K (2014) Phylogeny and functions of bacterial communities associated with field-grown rice shoots. Microbes Environ. 29, 329–332. doi:10.1264/jsme2.ME14077

Opik M, Zobel M, Cantero JJ, Davison J, Facelli JM, Hiiesalu I, Jairus T, Kalwij JM, Koorem K, Leal ME, Liira J, Metsis M, Neshataeva V, Paal J, Phosri C, Polme S, Reier U, Saks U, Schimann H, Thiery O, Vasar M, Moora M (2013) Global sampling of plant roots expands the described molecular diversity of arbuscular mycorrhizal fungi. Mycorrhiza 23(5):411–430. doi:10.1007/s00572-013-0482-2

Pace NR (1997) A molecular view of microbial diversity and the biosphere. Science 276(5313):734–740

Parada AE, Needham DM, Fuhrman JA (2016) Every base matters: assessing small subunit rRNA primers for marine microbiomes with mock communities, time series and global field samples. Environ Microbiol 18(5):1403–1414. doi:10.1111/1462-2920.13023

Parameswaran P, Jalili R, Tao L, Shokralla S, Gharizadeh B, Ronaghi M, Fire AZ (2007) A pyrosequencing-tailored nucleotide barcode design unveils opportunities for large-scale sample multiplexing. Nucleic Acids Res 35(19):e130. doi:10.1093/nar/gkm760

Peltzer A, Jager G, Herbig A, Seitz A, Kniep C, Krause J, Nieselt K (2016) EAGER: efficient ancient genome reconstruction. Genome Biol 17(1):60. doi:10.1186/s13059-016-0918-z

Penton CR, Gupta VV, Tiedje JM, Neate SM, Ophel-Keller K, Gillings M, Harvey P, Pham A, Roget DK (2014) Fungal community structure in disease suppressive soils assessed by 28S LSU gene sequencing. PLoS One 9(4):e93893. doi:10.1371/journal.pone.0093893

Peršoh D (2015) Plant-associated fungal communities in the light of meta'omics. Fungal Divers. doi:10.1007/s13225-015-0334-9

Pible O, Armengaud J (2015) Improving the quality of genome, protein sequence, and taxonomy databases: a prerequisite for microbiome meta-omics 2.0. Proteomics 15(20):3418–3423. doi:10.1002/pmic.201500104

Pinto C, Pinho D, Sousa S, Pinheiro M, Egas C, Gomes AC (2014) Unravelling the diversity of grapevine microbiome. PLoS One 9(1):e85622. doi:10.1371/journal.pone.0085622

Polz MF, Cavanaugh CM (1998) Bias in template-to-product ratios in multitemplate PCR. Appl Environ Microbiol 64(10):3724–3730

Pop M, Salzberg SL (2008) Bioinformatics challenges of new sequencing technology. Trends Genet 24(3):142–149. doi:10.1016/j.tig.2007.12.006

Porras-Alfaro A, Bayman P (2011) Hidden fungi, emergent properties: endophytes and microbiomes. Annu Rev Phytopathol 49:291–315. doi:10.1146/annurev-phyto-080508-081831

Porter TM, Golding GB (2011) Are similarity- or phylogeny-based methods more appropriate for classifying internal transcribed spacer (ITS) metagenomic amplicons? New Phytol 192(3): 775–782. doi:10.1111/j.1469-8137.2011.03838.x

Porter TM, Golding GB (2012) Factors that affect large subunit ribosomal DNA amplicon sequencing studies of fungal communities: classification method, primer choice, and error. PLoS One 7(4):e35749. doi:10.1371/journal.pone.0035749

Powell SM, Chapman CC, Bermudes M, Tamplin ML (2012) Use of a blocking primer allows selective amplification of bacterial DNA from microalgae cultures. J Microbiol Methods 90(3):211–213. doi:10.1016/j.mimet.2012.05.007

Pruesse E, Quast C, Knittel K, Fuchs BM, Ludwig WG, Peplies J, Glockner FO (2007) SILVA: a comprehensive online resource for quality checked and aligned ribosomal RNA sequence data compatible with ARB. Nucleic Acids Res 35(21):7188–7196. doi:10.1093/nar/gkm864

Quail MA, Otto TD, Gu Y, Harris SR, Skelly TF, McQuillan JA, Swerdlow HP, Oyola SO (2012) Optimal enzymes for amplifying sequencing libraries. Nat Methods 9(1):10–11. doi:10.1038/nmeth.1814

Quast C, Pruesse E, Yilmaz P, Gerken J, Schweer T, Yarza P, Peplies J, Glockner FO (2013) The SILVA ribosomal RNA gene database project: improved data processing and web-based tools. Nucleic Acids Res 41(D1):D590–D596. doi:10.1093/nar/gks1219

Rastogi G, Coaker GL, Leveau JH (2013) New insights into the structure and function of phyllosphere microbiota through high-throughput molecular approaches. FEMS Microbiol Lett 348(1):1–10. doi:10.1111/1574-6968.12225

Reininger V, Schlegel M (2016) Analysis of the Phialocephala subalpina transcriptome during colonization of its host plant Picea abies. PLoS One 11(3):e0150591. doi:10.1371/journal.pone.0150591

Renaud G, Stenzel U, Kelso J (2014) leeHom: adaptor trimming and merging for Illumina sequencing reads. Nucleic Acids Res 42(18):e141. doi:10.1093/nar/gku699

Rideout JR, He Y, Navas-Molina JA, Walters WA, Ursell LK, Gibbons SM, Chase J, McDonald D, Gonzalez A, Robbins-Pianka A, Clemente JC, Gilbert JA, Huse SM, Zhou HW, Knight R,

Caporaso JG (2014) Subsampled open-reference clustering creates consistent, comprehensive OTU definitions and scales to billions of sequences. PeerJ 2:e545. doi:10.7717/peerj.545

Rodrigue S, Materna AC, Timberlake SC, Blackburn MC, Malmstrom RR, Alm EJ, Chisholm SW (2010) Unlocking short read sequencing for metagenomics. PLoS One 5(7):e11840. doi:10.1371/journal.pone.0011840

Romero FM, Marina M, Pieckenstain FL (2014) The communities of tomato (Solanum lycopersicum L.) leaf endophytic bacteria, analyzed by 16S-ribosomal RNA gene pyrosequencing. FEMS Microbiol Lett 351(2):187–194. doi:10.1111/1574-6968.12377

Rosenberg E, Zilber-Rosenberg I (2016) Microbes drive evolution of animals and plants: the hologenome concept. MBio 7(2):e01395. doi:10.1128/mBio.01395-15

Sboner A, Mu XJ, Greenbaum D, Auerbach RK, Gerstein MB (2011) The real cost of sequencing: higher than you think! Genome Biol 12(8):125. doi:10.1186/gb-2011-12-8-125

Schatz MC, Langmead B, Salzberg SL (2010) Cloud computing and the DNA data race. Nat Biotechnol 28(7):691–693. doi:10.1038/nbt0710-691

Schiltz S, Gaillard I, Pawlicki-Jullian N, Thiombiano B, Mesnard F, Gontier E (2015) A review: what is the spermosphere and how can it be studied? J Appl Microbiol 119(6):1467–1481. doi:10.1111/jam.12946

Schloss PD, Gevers D, Westcott SL (2011) Reducing the effects of PCR amplification and sequencing artifacts on 16S rRNA-based studies. PLoS One 6(12):e27310. doi:10.1371/journal.pone.0027310

Schloss PD, Westcott SL, Ryabin T, Hall JR, Hartmann M, Hollister EB, Lesniewski RA, Oakley BB, Parks DH, Robinson CJ, Sahl JW, Stres B, Thallinger GG, Van Horn DJ, Weber CF (2009) Introducing mothur: open-source, platform-independent, community-supported software for describing and comparing microbial communities. Appl Environ Microbiol 75(23):7537–7541. doi:10.1128/AEM.01541-09

Schmidt R, Köberl M, Mostafa A, Ramadan EM, Monschein M, Jensen K, Bauer R, Berg G (2014) Effects of bacterial inoculants on the indigenous microbiome and secondary metabolites of chamomile plants. Front Microbiol 5:64

Schoch CL, Seifert KA, Huhndorf S, Robert V, Spouge JL, Levesque CA, Chen W, Fungal Barcoding C, Fungal Barcoding Consortium Author L (2012) Nuclear ribosomal internal transcribed spacer (ITS) region as a universal DNA barcode marker for fungi. Proc Natl Acad Sci U S A 109(16):6241–6246. doi:10.1073/pnas.1117018109

Scholz MB, Lo C-C, Chain PSG (2012) Next generation sequencing and bioinformatic bottlenecks: the current state of metagenomic data analysis. Curr Opin Biotechnol 23(1):9–15. doi:10.1016/j.copbio.2011.11.013

Sedzielewska-Toro K, Brachmann A (2016) The effector candidate repertoire of the arbuscular mycorrhizal fungus Rhizophagus Clarus. BMC Genomics 17(1):101. doi:10.1186/s12864-016-2422-y

Senés-Guerrero C, Schüßler A (2015) A conserved arbuscular mycorrhizal fungal core-species community colonizes potato roots in the Andes. Fungal Diversity 77(1):317–333. doi:10.1007/s13225-015-0328-7

Seshadri R, Reeve WG, Ardley JK, Tennessen K, Woyke T, Kyrpides NC, Ivanova NN (2015) Discovery of novel plant interaction determinants from the genomes of 163 root nodule bacteria. Sci Rep 5:16825. doi:10.1038/srep16825

Shi Y, Yang H, Zhang T, Sun J, Lou K (2014) Illumina-based analysis of endophytic bacterial diversity and space-time dynamics in sugar beet on the north slope of Tianshan mountain. Appl Microbiol Biotechnol 98(14):6375–6385. doi:10.1007/s00253-014-5720-9

Simpson AJ, Reinach FC, Arruda P, Abreu FA, Acencio M, Alvarenga R, Alves LM, Araya JE, Baia GS, Baptista CS, Barros MH, Bonaccorsi ED, Bordin S, Bove JM, Briones MR, Bueno MR, Camargo AA, Camargo LE, Carraro DM, Carrer H, Colauto NB, Colombo C, Costa FF, Costa MC, Costa-Neto CM, Coutinho LL, Cristofani M, Dias-Neto E, Docena C, El-Dorry H, Facincani AP, Ferreira AJ, Ferreira VC, Ferro JA, Fraga JS, Franca SC, Franco MC, Frohme M, Furlan LR, Garnier M, Goldman GH, Goldman MH, Gomes SL, Gruber A, Ho PL, Hoheisel JD, Junqueira ML, Kemper EL, Kitajima JP, Krieger JE, Kuramae EE, Laigret F, Lambais MR, Leite LC, Lemos EG, Lemos MV, Lopes SA, Lopes CR, Machado JA, Machado MA, Madeira

AM, Madeira HM, Marino CL, Marques MV, Martins EA, Martins EM, Matsukuma AY, Menck CF, Miracca EC, Miyaki CY, Monteriro-Vitorello CB, Moon DH, Nagai MA, Nascimento AL, Netto LE, Nhani A Jr, Nobrega FG, Nunes LR, Oliveira MA, de Oliveira MC, de Oliveira RC, Palmieri DA, Paris A, Peixoto BR, Pereira GA, Pereira HA Jr, Pesquero JB, Quaggio RB, Roberto PG, Rodrigues V, de M Rosa AJ, de Rosa VE Jr, de Sa RG, Santelli RV, Sawasaki HE, da Silva AC, da Silva AM, da Silva FR, da Silva WA Jr, da Silveira JF, Silvestri ML, Siqueira WJ, de Souza AA, de Souza AP, Terenzi MF, Truffi D, Tsai SM, Tsuhako MH, Vallada H, Van Sluys MA, Verjovski-Almeida S, Vettore AL, Zago MA, Zatz M, Meidanis J, Setubal JC (2000) The genome sequence of the plant pathogen Xylella fastidiosa. The Xylella fastidiosa consortium of the Organization for Nucleotide Sequencing and Analysis. Nature 406(6792):151–159. doi:10.1038/35018003

Smit E, Leeflang P, Glandorf B, van Elsas JD, Wernars K (1999) Analysis of fungal diversity in the wheat rhizosphere by sequencing of cloned PCR-amplified genes encoding 18S rRNA and temperature gradient gel electrophoresis. Appl Environ Microbiol 65(6):2614–2621

Stielow J, Lévesque C, Seifert K, Meyer W, Irinyi L, Smits D, Renfurm R, Verkley G, Groenewald M, Chaduli D (2015) One fungus, which genes? Development and assessment of universal primers for potential secondary fungal DNA barcodes. Persoonia (Molecular Phylogeny and Evolution of Fungi) 35(1):242–263

Stockinger H, Peyret-Guzzon M, Koegel S, Bouffaud ML, Redecker D (2014) The largest subunit of RNA polymerase II as a new marker Gene to study assemblages of arbuscular mycorrhizal fungi in the field. PLoS One 9(10):e107783. doi:10.1371/journal.pone.0107783

Sun Z, Hsiang T, Zhou Y, Zhou J (2015a) Draft genome sequence of bacillus amyloliquefaciens XK-4-1, a plant growth-promoting endophyte with antifungal activity. Genome Announc 3(6):e01306–e01315

Sun ZB, Sun MH, Li SD (2015b) Draft genome sequence of Mycoparasite Clonostachys rosea strain 67-1. Genome Announc 3(3):e00546–e00515. doi:10.1128/ genomeA.00546-15

Takahashi S, Tomita J, Nishioka K, Hisada T, Nishijima M (2014) Development of a prokaryotic universal primer for simultaneous analysis of bacteria and archaea using next-generation sequencing. PLoS One 9(8):e105592. doi:10.1371/journal.pone.0105592

Thomas T, Rusch D, DeMaere MZ, Yung PY, Lewis M, Halpern A, Heidelberg KB, Egan S, Steinberg PD, Kjelleberg S (2010) Functional genomic signatures of sponge bacteria reveal unique and shared features of symbiosis. ISME J 4(12):1557–1567. doi:10.1038/ismej.2010.74

Tkacz A, Poole P (2015) Role of root microbiota in plant productivity. J Exp Bot 66(8):2167–2175. doi:10.1093/jxb/erv157

Toju H, Tanabe AS, Yamamoto S, Sato H (2012) High-coverage ITS primers for the DNA-based identification of ascomycetes and basidiomycetes in environmental samples. PLoS One 7(7):e40863. doi:10.1371/journal.pone.0040863

Trujillo ME, Riesco R, Benito P, Carro L (2015) Endophytic Actinobacteria and the interaction of Micromonospora and nitrogen fixing plants. Front Microbiol 6:1341. doi:10.3389/fmicb.2015.01341

Turnbaugh PJ, Hamady M, Yatsunenko T, Cantarel BL, Duncan A, Ley RE, Sogin ML, Jones WJ, Roe BA, Affourtit JP, Egholm M, Henrissat B, Heath AC, Knight R, Gordon JI (2009) A core gut microbiome in obese and lean twins. Nature 457(7228):480–484. doi:10.1038/nature07540

Turrini A, Sbrana C, Avio L, Njeru EM, Bocci G, Bàrberi P, Giovannetti M (2016) Changes in the composition of native root arbuscular mycorrhizal fungal communities during a short-term cover crop-maize succession. Biol Fertil Soils. doi:10.1007/s00374-016-1106-8

van Dam NM, Bouwmeester HJ (2016) Metabolomics in the rhizosphere: tapping into below-ground chemical communication. Trends Plant Sci 21(3):256–265. doi:10.1016/j.tplants.2016.01.008

van der Heijden MG, Martin FM, Selosse MA, Sanders IR (2015) Mycorrhizal ecology and evolution: the past, the present, and the future. New Phytol 205(4):1406–1423. doi:10.1111/nph.13288

van der Heijden MGA, Hartmann M (2016) Networking in the plant microbiome. PLoS Biol 14(2):e1002378. doi:10.1371/journal.pbio.1002378.t001 van Dijk EL, Jaszczyszyn Y, Thermes C (2014) Library preparation methods for next-generation sequencing: tone down the bias. Exp Cell Res 322(1):12–20. doi:10.1016/j.yexcr.2014.01.008

van Geel M, Ceustermans A, van Hemelrijck W, Lievens B, Honnay O (2015) Decrease in diversity and changes in community composition of arbuscular mycorrhizal fungi in roots of apple trees with increasing orchard management intensity across a regional scale. Mol Ecol 24(4): 941–952. doi:10.1111/mec.13079

Venter JC, Remington K, Heidelberg JF, Halpern AL, Rusch D, Eisen JA, Wu D, Paulsen I, Nelson KE, Nelson W, Fouts DE, Levy S, Knap AH, Lomas MW, Nealson K, White O, Peterson J, Hoffman J, Parsons R, Baden-Tillson H, Pfannkoch C, Rogers YH, Smith HO (2004) Environmental genome shotgun sequencing of the Sargasso Sea. Science 304(5667):66–74. doi:10.1126/science.1093857

Venturi V, Keel C (2016) Signaling in the rhizosphere. Trends Plant Sci 21(3):187–198. doi:10.1016/j.tplants.2016.01.005

Verhagen BW, Glazebrook J, Zhu T, Chang HS, van Loon LC, Pieterse CM (2004) The transcriptome of rhizobacteria-induced systemic resistance in arabidopsis. Mol Plant Microbe Interact 17(8):895–908. doi:10.1094/MPMI.2004.17.8.895

Verma S, Gazara RK, Nizam S, Parween S, Chattopadhyay D, Verma PK (2016) Draft genome sequencing and secretome analysis of fungal phytopathogen Ascochyta rabiei provides insight into the necrotrophic effector repertoire. Sci Rep 6:24638. doi:10.1038/srep24638

Větrovský T, Kolařík M, Žifčáková L, Zelenka T, Baldrian P (2015) The rpb2 gene represents a viable alternative molecular marker for the analysis of environmental fungal communities. Mol Ecol Resour 16(2):388–401

Walters W, Hyde ER, Berg-Lyons D, Ackermann G, Humphrey G, Parada A, Gilbert JA, Jansson JK, Caporaso JG, Fuhrman JA (2015) Improved Bacterial 16S rRNA Gene (V4 and V4–5) and Fungal Internal Transcribed Spacer Marker Gene Primers for Microbial Community Surveys. mSystems 1(1). pii: e00009–15

Wang GCY, Wang Y (1996) The frequency of chimeric molecules as a consequence of PCR co-amplification of 16S rRNA genes from different bacterial species. Microbiology (UK) 142(5):1107–1114

Wang Q, Garrity GM, Tiedje JM, Cole JR (2007) Naive Bayesian classifier for rapid assignment of rRNA sequences into the new bacterial taxonomy. Appl Environ Microbiol 73(16):5261–5267. doi:10.1128/Aem.00062-07

Wibberg D, Andersson L, Rupp O, Goesmann A, Puhler A, Varrelmann M, Dixelius C, Schluter A (2016) Draft genome sequence of the sugar beet pathogen Rhizoctonia solani AG2-2IIIB strain BBA69670. J Biotechnol 222:11–12. doi:10.1016/j.jbiotec.2016.02.001

Wilcox TM, Schwartz MK, McKelvey KS, Young MK, Lowe WH (2014) A blocking primer increases specificity in environmental DNA detection of bull trout (Salvelinus confluentus). Conserv Genet Resour 6(2):283–284. doi:10.1007/s12686-013-0113-4

Xiong W, Zhao Q, Xue C, Xun W, Zhao J, Wu H, Li R, Shen Q (2016) Comparison of fungal community in black pepper-vanilla and vanilla monoculture systems associated with vanilla Fusarium wilt disease. Front Microbiol 7:117. doi:10.3389/fmicb.2016.00117

Xu X, Passey T, Wei F, Saville R, Harrison RJ (2015) Amplicon-based metagenomics identified candidate organisms in soils that caused yield decline in strawberry. Hortic Res 2:15022. doi:10.1038/hortres.2015.22

Yeoh YK, Paungfoo-Lonhienne C, Dennis PG, Robinson N, Ragan MA, Schmidt S, Hugenholtz P (2016) The core root microbiome of sugarcanes cultivated under varying nitrogen fertilizer application. Environ Microbiol 18(5):1338–1351. doi:10.1111/1462-2920.12925

Zhang JJ, Kobert K, Flouri T, Stamatakis A (2014) PEAR: a fast and accurate Illumina paired-end reAd mergeR. Bioinformatics 30(5):614–620. doi:10.1093/bioinformatics/btt593

Zhao S, Zhou N, Zhao ZY, Zhang K, Tian CY (2016) High-throughput sequencing analysis of the endophytic bacterial diversity and dynamics in roots of the halophyte Salicornia europaea. Curr Microbiol 72(5):557–562. doi:10.1007/s00284-016-0990-3

Žifčáková L, Větrovský T, Howe A, Baldrian P (2015) Microbial activity in forest soil reflects the changes in ecosystem properties between summer and winter. Environ Microbiol 18(1): 288–301. doi:10.1111/1462-2920.13026

Microbial Inoculants: A Novel Approach for Better Plant Microbiome Interactions

14

Satwant Kaur Gosal and Jupinder Kaur

Abstract

The rhizospheric interactions between plant and the microbiome influence fertility of soil, growth, development, and yield of crop plants. The interplay between plant and the microbes provides various services to the plant which are helpful for the production of agricultural crops in a sustainable manner. Among all the factors, the most influential factor that can improve plant microbiome interplay is the soil microbial community of the rhizosphere, where soil microbes, soil, and the plant roots interact with each other. Microbial interventions to improve plant microbiome interactions involve the introduction of microbial inoculants, which consist of naturally occurring diverse microflora of soil that improve health of crop plant and can protect the host plant from stresses and diseases through a diverse range of mechanisms. The use of beneficial microbes as inoculants for production of crops increases the diversity of microorganisms in soil and also ensures the production of sufficient food for the growing human population. The soil microbiome and the plants work together in coordination with each other for the benefit of plant and soil. The number of functional characters of microbes such as fixation of molecular nitrogen, solubilization of inorganic phosphate and production of iron chelating agents, and plant growth promoting hormones are used as plant growth promotion traits for the selection of microbial isolates to be used as bio-inoculants. Microbial inoculants are economic and easy to use and their incorporation reduces the dependence on chemical fertilizers. Thus, their application protects the environment from adverse effects of inorganic fertilizers. But, there are number of natural factors that influence and limit the effectiveness and efficiency of inoculated microbes under field conditions. The use of microbial inoculants for sustainable agriculture will be an environmental benign approach to improve plant microbiome interactions for nutrient management and ecological functions.

S.K. Gosal (✉) • J. Kaur
Department of Microbiology, Punjab Agricultural University, Ludhiana, India
e-mail: skgosal@rediffmail.com

V. Kumar et al. (eds.), *Probiotics in Agroecosystem*,
DOI 10.1007/978-981-10-4059-7_14

14.1 Introduction

One of the major challenges for the twenty-first century will be the continued production of fuel and food for the growing human population. According to a report by the United Nations Population Fund, it is estimated that the human population can reach ten billion by 2050. Due to the increasing human population, the productivity of agricultural crops needs to be increased in an efficient and sustainable manner to ensure food safety. Productivity is not only the growth of plant per ha (hectare) in the field. It is also defined by the fitness, productivity, and healthy growth of crop plants. The conventional agriculture plays a vital role in feeding the increasing human population, but it uses pesticides and inorganic fertilizers in large amounts to maintain higher productivity. Furthermore, improvement in agricultural production is not possible without increasing the area under cultivation. This will threaten the diversity of living forms which are already in danger from human activities. Keeping all these points in view, the recent research is more focused on improvement of soil fertility and productivity of crop plants in a sustainable and effective way. Organic farming is one of such methods that increase the food production without causing any harm to natural microflora of soil (Megali et al. 2013). It is mostly dependent on the native soil biota which includes all types of beneficial bacteria and fungi. Microbial inoculants are essential component of organic farming. Use of microbial inoculants improve the nutrient profile of soil by providing all kinds of macro- and micronutrients via biological nitrogen fixation, increasing the availability of essential elements like phosphorous and potassium and production of phytohormones and antibiotics in the soil. The objective of this chapter is to understand the potential role of inoculated microbes in sustainable agriculture to improve plant microbiome interactions, thus resulting in improved soil fertility and crop productivity.

14.2 The Microbiome

Microbes are necessary and integral component of agricultural production. They are required for improvement of fertility and nutrient status of soil, processing and preservation of agricultural products, and recycling of agro-residues like rice straw. Thus, microbes contribute a diverse array of benefits for sustainable agriculture production, by performing nutrient transformations, improving soil and plant health, and increasing the crop productivity. These applications of microorganism highlight their importance in agricultural production and various interactions with plants. The beneficial microorganisms that are used as bio-inoculants to improve plant microbe interactions can have two origins:

1. Plant microbiome
2. Rhizospheric microbiome

14.2.1 Plant Microbiome

Plants are colonized by a large number of microorganisms. The population of microbes colonizing the various plant parts can even exceed the number of cells in plant. The total genome of microbes associated with plants is known as *plant microbiome* (Fig. 14.1). The interactions of the plant microbiome influences the health and productivity of crops by providing diverse range of benefits to host plant, viz., transformation of nutrients, decomposition of organic matter, suppression of diseases, assistance in uptake of nutrients, and control of weeds. So, plant-associated microorganisms, i.e., plant microbiome has positive influence on germination of seeds, development and productivity of crop, maintenance of soil fertility, and protection from stresses. Plants and the microbiome are interdependent on each other for specific traits and functions (Bulgarelli et al. 2013). Carbon (C) which is fixed photosynthetically by plants is deposited into their surroundings, that is, phyllosphere, rhizosphere, and mycorrhizosphere. This carbon is used as feed by the soil biota leading to influence on the activities, abundance, and composition of microbes (Berendsen et al. 2012). The plant microbiome interactions have been studied extensively for various microbes like symbiotic nitrogen fixers (*Rhizobium*), *Mycorrhizae*, and various pathogenic microbes. But, the knowledge regarding influence of inoculation of beneficial microorganisms on the growth, health, and productivity of crop plants is still not sufficient. Hence, the information of the plant microbiome and the various rhizospheric interactions is necessary to identify microbes that can be used as microbial inoculants for efficient and sustainable increase in productivity of crops.

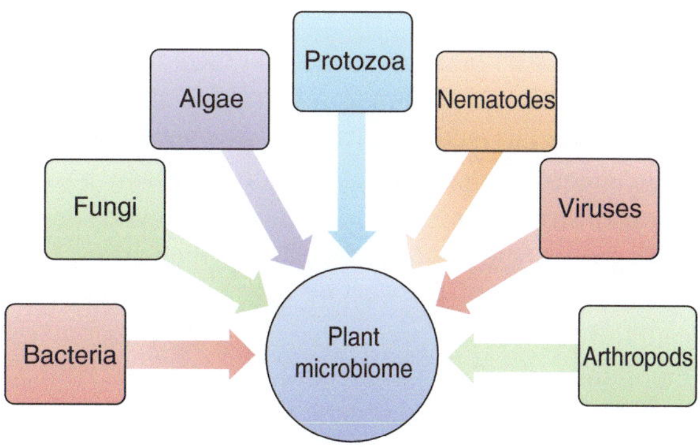

Fig. 14.1 Microbial communities associated to plant

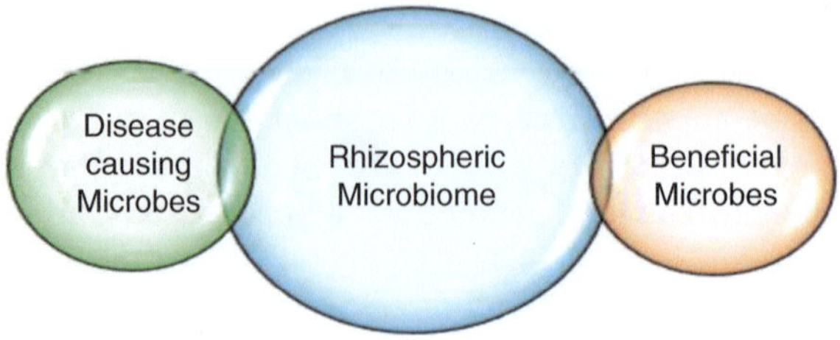

Fig. 14.2 Rhizospheric microbiome

14.2.2 Rhizospheric Microbiome

The total genome of rhizospheric microflora surrounding the root zone of plants is larger as compared to that of plants. This is known as *rhizospheric microbiome* (Fig. 14.2). Rhizosphere is the interface between the soil and roots of plant where the most complex interactions between plants and microorganisms occur. The concept of the rhizosphere was given by Lorenz Hiltner (German biologist). He defined rhizosphere as the region around the plant roots characterized by high activity of microbes (Hartmann et al. 2008). Till date, a lot of research work has been carried out on microbiology of soil by various researchers. But, the knowledge regarding the various associations and interactions occurring in the soil is still not sufficient. Billions of microbes with diverse characters can be present in one gram of rhizospheric soil. These microbes help in improving crop productivity by various plant microbiome interactions. The use of beneficial microbes as inoculants for sustainable agricultural production has become focus of research as their use reduces the dependence on inorganic fertilizers. Incorporation of microbial inoculants can improve crop productivity by exploiting the beneficial plant microbiome interactions.

14.3 Microbes and the Plant Microbiome Interactions

The relationship between plants and their surroundings is very complex. Plants and microbes share an intimate relationship that enables them to coexist. The plant microbiome interactions involve a vast array of microbes and often produce synergistic effects (Mendes et al. 2013). The soil, the host plant, and the microbes present in soil all work together to mediate and influence the various exchanges that contribute to growth and productivity of crop plants (Fig. 14.3). There are large numbers of environmental and soil factors that can improve nutrient profile and soil fertility leading to increased productivity of crop plants. Among all the factors, the most important and influential are soil microbes inhabiting the region surrounding plant roots. Soil factors influence soil biota as well as the root zone of plants which in turn reshape the environment of soil through a dynamic exchange of chemical responses to living and nonliving stimuli. The compounds secreted by plant roots commonly known as root exudates act as substrate for microbes. These root exudates work as signal molecules and create a complex relationship between the microbiome and the

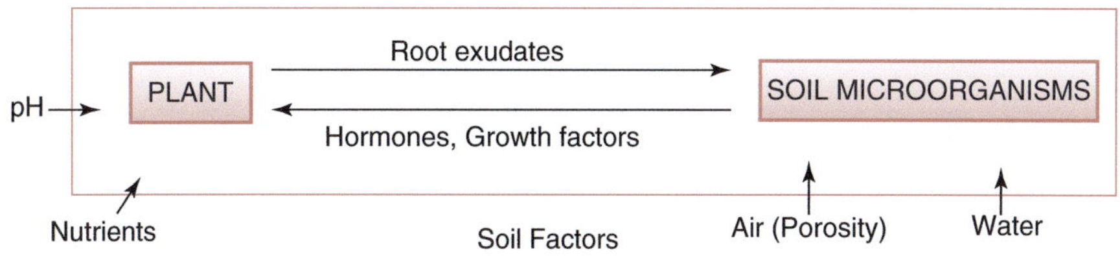

Fig. 14.3 Schematic illustration of the interactions between the soil microbes, the host plant, and the soil

crop plant. The soil microbiome and the plants work together in coordination with each other for the benefit of plant and soil. Various research reports support the fact that the soil microbiome can be reshaped by plant through the secretion of root exudates. The interactions between plants and the soil microbiome fluctuate according to the various factors like development stage of crop, agricultural practices, etc. An understanding of how each component manipulates and influences each other is needed. Research on plant microbiome interactions can help us answer these questions and allow us to see how all these interactions relate and influence one another (Morales and Holben 2011).

14.4 The Effect of Plants on the Plant Microbiome Interplay

The interaction between plants and their surroundings is a dynamic process. In this process, plants monitor their environment and react to changes. Earlier, it was thought that the function of root system is only to provide support to plant and facilitate the water and nutrient uptake. But, this root system plays a vital role in reshaping of plant microbiome interactions. Microorganisms present in the soil emit chemical signals that are received by plants. Plants respond to the signals emitted by microbes through the secretion of root exudates. According to the study conducted by various researchers, the type and composition of root exudates change with change in species of plant and even within the roots of same plant (Chaparro et al. 2012). The diverse compounds released by plants as root exudates include sugars, amino acids, growth factors, aliphatic acids, proteins, and fatty acids (Table 14.1).

In addition to these organic compounds, some other compounds like auxins, scopoletin, glycosides, glucosides, bacterial stimulants and inhibitors, etc. have also been detected in plant root exudates. These root exudates create a unique environment in the rhizosphere. All these different compounds (root exudates) initiate symbiotic and various other interactions within the rhizosphere. The composition and the concentration of root exudates change according to the signals received from the environment and the rhizosphere, age of the plant, soil type, and various environmental factors (Flores et al. 1999). Change in protein composition of the root exudates was observed by De-la-Pena et al. (2010) when the plant grew alone as compared to when the plant interacted with pathogens or applied with microbial

Table 14.1 Various organic compounds secreted by plant roots as exudates

Group of organic compounds	Root exudates
Amino compounds	Glutamine, leucine, tyrosine, asparagine, threonine, aspartic acid, phenylalanine, serine, methionine, arginine, etc.
Fatty acids and sterols	Palmitic acid, stearic acid, oleic acid, cholesterol, campesterol, etc.
Growth factors	Biotin, thiamine, niacin, choline, inositol, pyridoxine, etc.
Organic acids	Tartaric acid, oxalic acid, citric acid, malic acid, acetic acid, etc.
Nucleotides, flavanones, and enzymes	Flavanone, adenine, guanine, invertase, proteinase, etc.
Carbohydrates	Monosaccharides, disaccharides, oligosaccharides, and polysaccharides

inoculants. The soil microorganisms use the compounds secreted by plants as growth substrates, and sometimes, these root exudates can act as antimicrobials also. So, fluctuations in the concentration and composition of the root exudates can have positive, negative, or neutral effect on the microbes residing in rhizosphere.

Rhizodeposition is the main source of organic carbon to enter the soil. Plant uses a large percentage of its energy to produce and release these rhizodeposits. The main purpose of this rhizodeposition is to attract the microbes present in soil and rhizospheric region that service the plant through secreting growth promoting hormones, preventing disease, or acquiring nutrients via the excretions of a biochemically active root system. Hamilton and Frank (2001) demonstrated that a grazing tolerant grass, *Poa pratensis*, is capable of concentrating microbes that facilitate the uptake of a limiting soil resource needed for growth. White lupin, on the other hand, is able to discourage microbial growth by drastically decreasing the soil pH in the rhizosphere via the release of organic acids, lowering the competition for phosphorous (P) acquisition. White lupin also prevents microbial degradation of root exudates important for phosphorous acquisition. An experiment was conducted with *Arabidopsis thaliana* and *Medicago truncatula*. Both the plants were grown in native as well as in nonnative soil. *Arabidopsis* plants or root exudates added alone maintained the native fungus population in its native soil but not in nonnative soil. In nonnative soil, some microbial species increased while others diminished and the same results were observed with Medicago (Chaparro et al. 2012). These results strongly suggest that plant root exudates and the plants themselves are able to affect the composition and total population of soil microflora.

14.5 Influence of Soil Properties on Plant Microbiome Interactions

Soils are highly diverse acting as habitat for diverse communities of microorganisms with as many as 10,000–50,000 species of microbes existing in 1 gm of soil (Schloss and Handelsman 2006). Change in the soil texture, nitrogen (N) content, phosphorous content, and soil pH lead to fluctuations in the abundance and

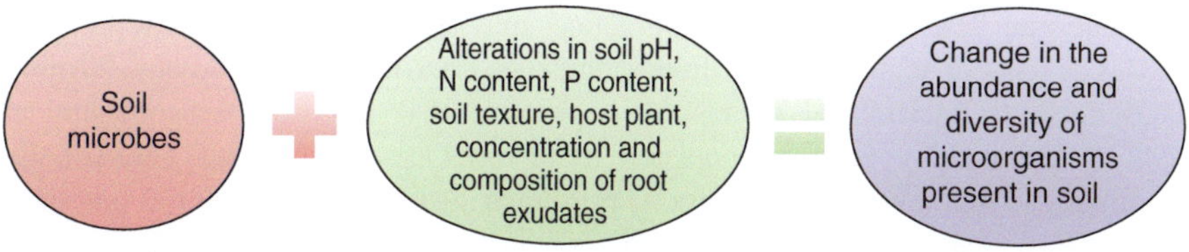

Fig. 14.4 Effect of fluctuation in plant and soil properties on soil microbial communities

composition of bacterial and fungal communities. Soil pH may have the most influence on the soil bacterial population. A strong correlation between hydrogen ion concentration and the abundance and diversity of microbial population was observed by Rousk et al. (2010) in the soil samples collected from an experiment where the pH varied from 4.0 to 8.3, while all other factors and variables that compose soil variability were controlled. A strong positive relation was found between soil pH and bacterial population. The reason for this relation is the sensitivity of bacterial cells to pH, as bacterial taxa exhibit a relatively narrow pH growth tolerance. Other evidence indicates that phosphorous content, altitude, and the ratio of cations in the soil (Ca^{2+}, Mg^{2+}, and Al^{3+}) are more influential (Faoro et al. 2010). So, influence of many soil properties converge to create the ultimate effect on the soil biota and multiple soil factors potentially exhibit synergistic effects leading to reshaping of microbiome (Fig. 14.4).

14.6 Microbiome Reshaping

An intensive agriculture production is necessary to satisfy food requirements for the growing world population. According to information from specialized sources, demand for food products is expected to increase by at least 70% by 2050. At the same time, people are becoming aware that sustainable agricultural practices are fundamental to meet the future world's agricultural demands (Altieri 2004). This is why modern agriculture is being implemented on a global scale and diverse research approaches are being undertaken to meet environmental and economical sustainability issues. A recommended approach is to reshape the soil microbial communities for a sustainable and healthy crop production while preserving the biosphere. Changes in the rhizospheric microflora can take place by:

1. Alterations in the management practices (soil amendments and nutrients application)
2. Artificial inoculation of soil, any plant part or seed with preparations containing live microorganisms

The diversity and composition of microbes present in the soil is affected by agricultural management practices, land use, and degrees of stress and disturbance. Agricultural management practices fall into two general categories, organic and conventional. The choice of farming practices may lend themselves to different processes or steps to achieve a more diverse and even microbiome. The organic farming is defined as agricultural management practice that improves biogeochemical cycling, biodiversity, and biological activity of soil. Organic farming is based on the minimal use of inorganic fertilizers. It promotes sustainable agricultural practices that restore, maintain, and improve the biological integrity of soil (Gold 1995). Organic farming uses no synthetic fertilizers or added inputs to increase productivity, whereas conventional farming does just the opposite. Conventional farming often uses synthetic, chemical fertilizers, and pesticides to benefit crop protection and productivity. The main impact of these different farming systems is on the soil microbiome. For example, conventional agriculture may target plant pathogens through the use of pesticides/fungicides, but it has a potential side effect of reducing soil microbial community diversity and evenness. Whereas, organic agriculture may seek to control plant pathogens through competition and/ or antagonism that promote a more diverse and even microbial community such as the addition of varying types of organic matter. By understanding the influences that combine to create more diverse and even soil microbial communities, fertility and disease resistance can be inherently restored in depleted, disease-stricken soil environments. The other method of microbiome reshaping is artificial inoculation of seed or soil with preparations containing live microorganisms (microbial inoculants).

The problem is no longer to produce more food but also to do so in environmentally and socially sustainable ways. Agriculture practices should emphasize on maximizing the coadaptation between plants and microbes in an effort to promote soil microbial diversity. Although, this may reduce short-term productivity, but it will maximize long-term yields while minimizing resource use. The loss of diversity and evenness of microbiome is detrimental to ecosystem functioning and plant productivity. In a world where the demand for food increases by the second, unhealthy crops with low productivity are unacceptable.

14.7 Need for Microbial Inoculants for Improvement of Plant Microbiome Interactions

The human population is increasing at an alarming rate. This fast increase in human population decreases the area under cultivation. To feed the growing human population, the productivity of agriculture is maintained by continuous and heavy input of inorganic fertilizers. The indiscriminate application of chemical fertilizers to improve soil nutrient status and to increase productivity of crop plants has negatively affected the environment leading to pollution and contamination of the soil. This also results in destruction of indigenous microbes of soil and friendly insects. This decreases the soil fertility and makes the crop plants more susceptible to attack by pathogenic microbes.

Furthermore, due to increasing human population, demand for food is more than its production. According to the various reports, the demand for food products will exceed very much than the productivity (Arun 2007). The use of soil microbes as inoculants for crop production is the most effective and efficient solution to maintain productivity of agricultural products in a sustainable manner. The application of microbial inoculants to enhance crop productivity is eco-friendly and economical technique. Microbial inoculants are cheap and easily accessible. Their use reduces the dependence on chemical fertilizers. The fertilizer use efficiency of crop plants is very less, so microbial inoculants can play key role in integrated nutrient management systems by improving plant microbiome interactions. Integrated application of microbial inoculants along with organic manures improves the nutrient use efficiency of fertilizers leading to sustainable agricultural production.

14.8 Microbial Inoculants

Microbial intervention for increased fertility of soil, improving plant growth, plant microbiome interactions, and biocontrol, involves the introduction of microbial inoculants. Microbial inoculants are solution to numerous problems caused by hiked input of chemical fertilizers. Their use reduces the problems of agriculture and environment because the inoculated microbes possess various functional characters that promote the growth and productivity of plants, enhances the uptake and availability of various macro- and micronutrients and even protect the crop plant from diseases by acting as biocontrol agent. Microbial inoculants popularly known as "*bio-inoculants or biofertilizers*" are economical, pollution-free, environmentally friendly, and safe renewable agricultural amendments, that use microbes which possess the ability of fixing molecular nitrogen, transforming various unavailable nutrients into their available form, production of phytohormones, and iron chelating agents to promote plant health. These preparations contain living cells of beneficial microbes that are applied to seed, soil, any plant part, or composting areas. These beneficial microbes when inoculated, colonize rhizosphere, increase their number, and accelerate various microbial transformations which increase the availability of many essential nutrients. Microbial inoculants are one of the best agricultural amendments for understanding plant microbiome interactions and for improving crop productivity in a sustainable manner. They are becoming center of attraction for various researchers due to the increasing emphasis on maintenance of soil quality and health and to reduce the adverse effects of inorganic fertilizers on agriculture and environment.

Microbial inoculants contain diverse beneficial microbes that occur naturally in soil. Their incorporation to soil, seed, or any plant adds to biodiversity of soil, improves soil quality, availability of nutrients, growth, development, and productivity of plants. The beneficial microbes that can be used as microbial inoculants for agricultural crops includes plant growth promoting rhizobacteria, diazotrophs, phosphorous mobilizing and solubilizing bacteria, bacteria with ability to suppress plant diseases and microbes with ability to degrade complex compounds. Microbial inoculants are now essential and integral part of agriculture. Their integrated application with various organic amendments, agro-residues, and inorganic nutrients is very useful in maintaining the sustainability of various crop productions due to synergistic effects (Sahoo et al. 2013). Microorganisms which are present naturally in soil are not very efficient due to various types of stresses and antagonistic activities. Thus, artificially multiplied cultures of beneficial microbes with plant growth promotion traits play a significant role in accelerating microbial transformations. These organisms increase the fertilizer use efficiency and improve the soil nutrient status and lead to higher productivity in sustainable way (Anonymous 2008).

14.9 Selection of Microorganisms for Inoculation

The success and performance of microbial inoculants under field conditions is dependent on the type and function of microbes used for inoculation. Thus, for better results the choice of microbial inoculants is critical step. For the selection of microbes to be used as inoculant, complete knowledge about the biology and functional activity of microbes is essential. The factors like adaptation to adverse climatic conditions, survival, and persistence in soil after inoculation, and efficiency under field conditions should be considered while selecting microbial inoculants. A microbial strain should possess all the potential features or attributes to be used as a successful inoculant. For example, a bacterium cannot be used as inoculant if it is difficult to multiply or maintain it in laboratory.

Genetic engineering is another tool that can help us to insert desired features to microbial inoculant. By detailed understanding of genetic basis of the physiology of many soil organisms, we can alter the existing organisms by introducing genes from other organisms with desired characteristics. For example, we can insert genes for increased tolerance toward various stresses and diseases. We can alter the soil microbes at gene level. By this, we can reshape the rhizospheric organisms and can make them more efficient for competing with pathogenic microbes and to increase nutrient availability. But, the care should be taken that whether the genetically modified organism is superior to existing microbes, able to survive under field conditions and contributes to sustainable agriculture? Although genetic engineering of microbes is a great technological advance for agricultural production but still the guidelines and regulations are in place in most countries that restrict the use of genetically altered organisms unless scientific evidence for their harmlessness to nontarget organisms is demonstrated.

14.10 Types of Microbial Inoculants

Soil microorganisms play fundamental roles (microbial services) in agriculture mainly by improving plant nutrition and health, as well as soil quality (Lugtenberg 2015). The main microorganisms which are used as inoculants are bacteria, fungi, and blue-green algae. Microbial inoculants help us to get higher productivity of agricultural crops by increasing the availability of essential nutrients in soil. Their use decreases the dependence on chemical fertilizers which is necessary as the sole use of chemical fertilizers adversely affect the health and quality of soil. Generally, agriculturally used microbial inoculants are classified into three categories depending upon microorganisms involved (Fig. 14.5).

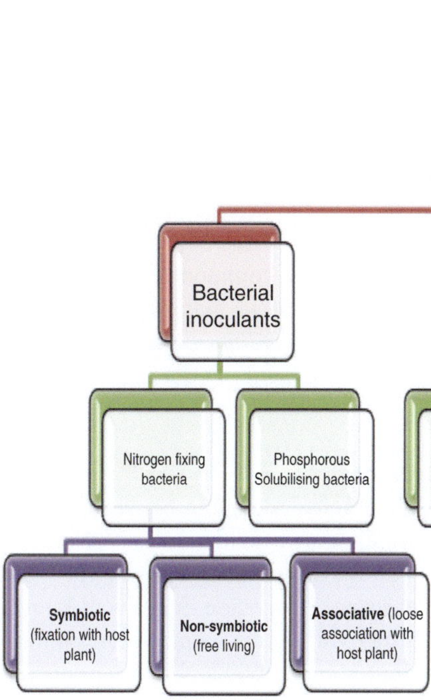

Fig. 14.5 Different types of microbial inoculants

14.10.1 Bacterial Inoculants

Bacteria possess great potential to reshape the plant microbiome for betterment of plant microbiome interactions as they show variety of associations with the plant. Bacterial colonization has been reported in the soil, inside the cell, rhizosphere, phyllosphere, and the rhizoplane. The bacteria commonly used as microbial inoculants are usually inhabitants of rhizosphere having abilities like biological nitrogen fixation, phosphorous solubilization, etc. The presence of these traits in microbial inoculants helps in meeting the nitrogen and phosphorus demands of the host plant. These bacteria are also known as plant growth promoting rhizobacteria (PGPR). Application of PGPRs as microbial inoculants positively influences the plant microbiome interactions, soil nutrient status, growth, and productivity of host plant.

14.10.1.1 Nitrogen-Fixing Microbial Inoculants

Nitrogen is the most important and essential macronutrient required by plant for proper growth and development. The amount of nitrogen in soil gets decreased with time due to loss of this essential nutrient by natural processes such as leaching. The nitrogen cycle has a great impact on soil health, quality, and fertility. The four most important microbial transformations occurring in the nitrogen cycle are biological nitrogen fixation (conversion of atmospheric nitrogen to ammonia), nitrification (oxidation of ammonia to nitrate), denitrification (reduction of oxidized nitrogenous compounds to nitrogen gas), and nitrogen mineralization (conversion of organic nitrogenous compounds to inorganic nitrogen). Microbial inoculants play key role in nitrogen cycle and in utilization of nitrogen present in the fertilizer by host plant. The nitrogen-fixing bacteria or the diazotrophs have the ability to fix molecular nitrogen present in atmosphere (inert) in reduced form, i.e., ammonia in the rhizospheric soil of the nonlegume or the root nodules of the legume plants. This dinitrogen fixation is exclusive prokaryotic phenomenon by virtue of production of the nitrogenase enzyme by the nitrogen-fixing bacteria. The common diazotrophs that exist in the rhizosphere of a variety of plants include in majority *Rhizobium*, *Frankia*, *Azotobacter*, *Beijerinckia*, *Derxia*, *Azospirillum*, *Flavobacterium*, *Gluconacetobacter*, cyanobacterial forms like *Anabaena, Nostoc, Tolypothrix, Cylindrospermum*, etc.

Symbiotic Nitrogen-Fixing Microbial Inoculants

Symbiotic nitrogen fixation (SNF) is specific chemical signaling between legume plant and rhizobial species resulting in the formation of specialized structures called nodules. Nodules act as mini factories to convert the atmosphere nitrogen into plant usable form of ammonia with the help of an enzyme nitrogenase. It is a well-known fact that world's supply of organic nitrogen is met via legume – *Rhizobium* symbiosis. The legume *Rhizobium* association contributes up to 360 kg N/ha/year depending upon the legume species, host *Rhizobium* genotype, agroclimatic conditions, and their interaction (Odame 1997). The success rate of nodulation depends on the compatibility between the inoculated strain and host plant.

Rhizobium Rhizobium (Plate 14.1a) is a symbiotic nitrogen fixer and belongs to family Rhizobiaceae. Inoculation of *Rhizobium* biofertilizer to legume crops enhances nodulation, nitrogen fixation, and yield. In addition to N_2 fixation, most rhizobial strains are also found to exhibit some other plant growth promoting characters such as production of phytohormones which includes release of indoleacetic acid (IAA), phosphate solubilization, and production of iron chelating agents (siderophores). They can also improve the growth and productivity of plant by suppression of diseases and by protecting the plant from variety of stresses. Legume crop is nodulated by specific *Rhizobium*; therefore, only recommended *Rhizobium* biofertilizer for specific legume crop is being advocated to get benefits of inoculation. Nodulation surveys indicate a need for inoculation every season for majority of legume crops cultivated in India. Competition between inefficient native strains appears to be a bottle neck in realizing higher yields from *Rhizobium* inoculation. *Rhizobium* has ability of symbiotic nitrogen fixation in association with some nonlegumes like *Parasponia*. Population of this symbiotic bacterium in the soil depends on the presence of compatible host plant. The decrease in the population of *Rhizobium* has been reported in the absence of legume host. Reinoculation is necessary to increase the colonization of this bacterium for speed up of the nitrogen fixation process. To form nodules, specific strain is required by host plant (Venkatashwarlu 2008).

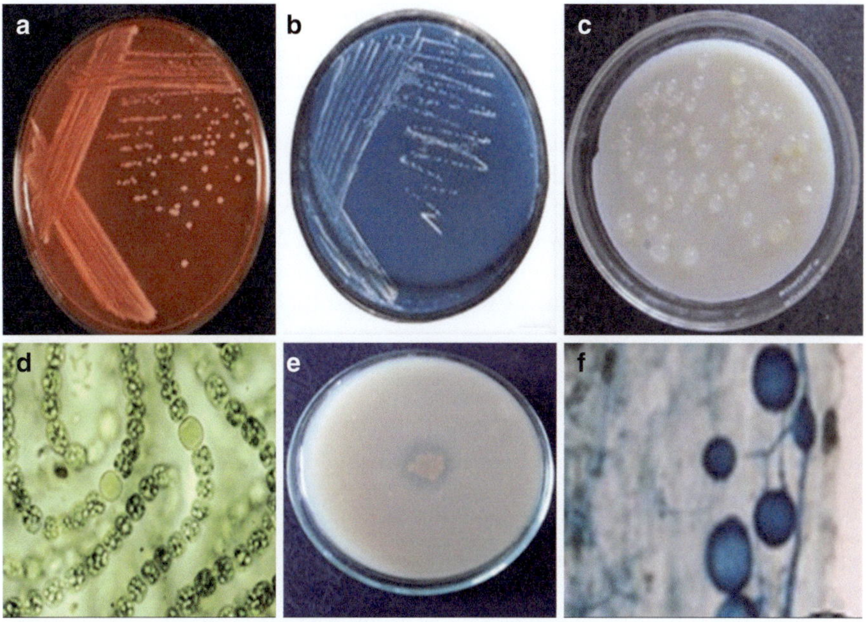

Plate 14.1 (**a**) *Rhizobium* colonies on Congo red medium, (**b**) *Azospirillum* colonies on Dobereiner medium (www.lookfordiagnosis.com), (**c**) *Azotobacter* colonies on Jensen medium, (**d**) Microscopic view of cyanobacteria, (**e**) clear zone of phosphate solubilization by PSB on Pikovskaya's medium, and (**f**) Mycorrhizal spores

Associative Bio-inoculants

The term "*associative symbiosis*" has been used to describe the interaction between *Azospirillum* and other rhizospheric bacteria with host plant. It is the loose association of microsymbiont with the host plant.

Azospirillum Azospirillum (Plate 14.1b, source - www.lookfordiagnosis.com) is motile and aerobic bacterium that can survive in flooded conditions. It is heterotrophic loose associative nitrogen-fixing bacteria that possess various plant growth promoting traits. It is gram-variable bacteria that colonize the root zone of the plant, outside the plant roots, on the surface of the roots, as well as internal cortical and vascular tissues of plant harboring in root zone. It belongs to family Spirilaceae. *Azospirillum* is remarkably versatile; it fixes atmospheric N_2, mineralizes nutrients from soil, sequesters iron (Fe), survives in stressed conditions, and also favors beneficial mycorrhizal plant associations. This associative nitrogen fixer can fix about 10–40 kg N/ha and also produce phytohormones like IAA, gibberellins, etc. which promotes root proliferation, development, and productivity of crop plant (Sahoo et al. 2014). *Azospirillum amazonense*, *Azospirillum halopraeferens*, *Azospirillum brasilense*, etc. are among the species coming under the genera, *Azospirillum*. But, the species most commonly used as microbial inoculants are *Azospirillum lipoferum and Azospirillum brasilense.* This associative nitrogen fixer shows symbiotic relationship mainly with C_4 plants (have C_4-dicarboxylic photosynthetic pathway). Thus, *Azospirillum* biofertilizers are suitable for C_4 crops, viz., maize, sorghum, and other cereals like rice, wheat, barley, and various horticultural crops.

Beneficial effects of *Azospirillum* inoculation on development and productivity of host crop has been reported in glass house as well as field experiments by Saikia et al. (2013). Improvement in the growth and productivity of numerous crops has been reported by inoculation with *Azospirillum* species. Application of *Azospirillum* improves the nutrient level in soil by helping in the absorption of various macro- and micronutrients. This improves status of soil fertility along with the promotion of growth and yield of crops.

Nonsymbiotic Microbial Inoculants

These microorganisms are present freely in the rhizosphere and fix nitrogen in the soil without any host plant. The nonsymbiotic bacteria include genus like *Azotobacter, Clostridium*, etc.

Azotobacter *Azotobacter* (Plate 14.1c) is a key player in the nitrogen cycle. It is from Azotobacteriaceae family. It is aerobic, nonsymbiotic nitrogen-fixing bacteria which is capable of surviving in neutral as well as in soil with pH in alkaline range. Along with the release of phytohormones like IAA, gibberellins, and cytokinins, this heterotrophic bacterium also produce vitamins such as vitamin B_1 (thiamine) and vitamin B_2, i.e., riboflavin (Revillas et al. 2000). The most prevalent species of this genus in soil is *Azotobacter chroococcum.* The other species of this nonsymbiotic genus that can be found in the soil are *Azotobacter vinelandii, Azotobacter macrocytogenes* and *Azotobacter beijerinckii*, etc. Use of *Azotobacter* as microbial inoculants enhances the plant productivity by suppression of diseases and by modifying the root architecture with dense root system. The population of this genus in soil is mostly reported between 10^4 and 10^5 g^{-1} of soil. This might be due to the negative effect of other microbes present in soil as well as due to deficiency of organic matter in soil (Mali and Bodhankar 2009). Application of *Azotobacter* as microbial inoculant in soil improves the seed germination index due to the inhibitory effect of this bacterium on growth of disease causing fungi. The number of *Azotobacter* is more in cultivated soil as compared to soil without any plantation. This nonsymbiotic nitrogen-fixing bacteria can be isolated from rhizosphere of numerous crop plants. This bacterium is used as inoculant for variety of crops such as maize, wheat, rice, etc. (Wani et al. 2013).

Blue-Green Algae (Cyanobacteria) and Azolla The algal microbial inoculants improve soil health and texture by providing valuable nutrients such as N, amino acids, etc. These are photosynthetic and belong to eight different families. Representative genera of blue-green algae (BGA) includes *Nostoc, Anabaena, Tolypothrix, Calothrix, Aulosira*, etc. Cyanobacteria (Plate 14.1d) show symbiotic relationship with fungi, ferns (especially *Azolla*), and flowering plants. These are used as microbial inoculants as they fix molecular nitrogen (20–30 kg/ha) and produce plant growth promoting hormones (Wani and Lee 1995). These nitrogen fixers are also called "paddy organisms" as they are more prevalent in rice fields. Nitrogen is the most essential and important macronutrient required for the production of rice crop. The more than half of the nitrogen demand of lowland rice is met by the process of biological nitrogen fixation carried by microbes and mineralization of soil nitrogen present in the form of organic compounds. For the development of agriculture sector in a sustainable manner and to meet the food demands of growing human population, the enhancement of biological nitrogen fixation process is necessary. *Azolla* is incorporated as green manure to provide nitrogen for paddy production. The benefit of using *Azolla* as microbial inoculant is its fast decomposition. Apart from meeting N demands, these bio-inoculants also increase the availability of micronutrients. In India, the most abundant and prevalent species of *Azolla* is *A. pinnata*. The other species reported in India are *A. caroliniana, A. microphylla, A. filiculoides, and A. mexicana.*

14.10.1.2 Phosphate Solubilizing Bacteria (PSB)

It is well known that phosphate fertilizers are not efficient in agroecosystems due to their fixation in soil with low as well as high pH (acidic and alkaline, respectively). Both acidic as well as alkaline soils are predominant in India (Wani and Lee 2002). Thus, application of crop plant with PSB (Plate 14.1e) and other efficient microbial strains is essential to maintain the productivity of agricultural ecosystem. Phosphate solubilizing as well as P mobilizing microbes are required to maintain the soil biological diversity. These microbes help in transformation of various nutrients into plant utilizable form to get higher productivity. It has been reported by various studies that many bacterial strains can solubilize insoluble inorganic phosphate compounds, such as hydroxyapatite, tricalcium phosphate, dicalcium phosphate, and rock phosphate. The potential P solubilizing bacterial genera found in soil are *Pseudomonas, Bacillus, Rhizobium, Burkholderia, Agrobacterium, Micrococcus, Flavobacterium*, and *Erwinia*. Phosphate solubilizing microbes can be found both in rhizosphere as well as in bulk soil. But, higher population of phosphorous solubilizing bacteria is commonly found in the rhizosphere as compared to bulk soil (non-rhizospheric soil). The bacteria belonging to the genera *Pseudomonas* and *Bacillus* are more commonly present in soil. PSB can be aerobic or anaerobic. The higher population of aerobic PSB is mostly reported in submerged soils.

14.10.2 Fungal Inoculants

Apart from bacteria, fungi can also be used as inoculant for agriculture to improve the plant microbiome interactions. Fungi can form symbiotic relationship with many other organisms and plant, and it can act as parasite also (Behie et al. 2012). Fungal inoculants are of two types:

1. Phosphate mobilizing *Mycorrhizae*
2. Phosphate solubilizing Fungi

14.10.2.1 Phosphate Mobilizing *Mycorrhizae*

The term *Mycorrhiza* referred to "fungus roots." It is a mutualistic symbiosis between fungi and roots of higher plants. In this symbiotic association, host plant provide photosynthetic carbon to fungal partner whereas, fungi helps in growth and development of host plant by providing various essential macronutrients. Mycorrhizal fungi are very efficient in nutrient absorption from the soil system due to its hyphae. *Mycorrhizae* (Plate 14.1f) are of three types based on their presence inside or outside the plant. When they form an external mantle or sheath around the roots then, they are known as ectomycorrhiza, and when they infect the roots of plant, they are known as endomycorrhiza or arbuscular mycorrhizal fungi (AMF). When present both inside and outside the plant roots, they are known as ectendomycorrhiza. The *Mycorrhizae* are well known to help in absorption and accumulation of macronutrients like nitrogen, phosphorus, magnesium, and sulfur from soil to plant apart from enhancing the uptake of essential micronutrients such as zinc, iron, and manganese (Ryan and Angus 2003).

These fungi are member of the family Endogonaceae. At present seven genera are included in the family Endogonaceae, which are *Acaulospora*, *Endogone*, *Gigaspora*, *Glaziella*, *Glomus*, *Modicella*, and *Sclerocystis*. It has been reported that these fungi form symbiotic relationships with many agricultural crops, except with plants belonging to families of Amaranthaceae, Chenopodiaceae, Polygonaceae, Caryophyllaceae, Commelinaceae, Brassicaceae, Juncaceae, and Cyperaceae. The arbuscule-forming *Mycorrhiza* (AMF) is a type of endomycorrhiza which show mutualistic association with many crops. In this symbiotic association, the fungal hyphae move inside the cortical cells of plant root and form branched structures called arbuscules (Behie et al. 2012). *Mycorrhizae* are potential candidates for their use as inoculants to enhance development and productivity of agroecosystem due to their effectiveness and efficiency in uptake of macro- and micronutrients and their role in protection of plants from various stresses (Gianinazzi et al. 2010).

14.10.2.2 Phosphate Solubilizing Fungi

Phosphorus is essential macronutrient required by plant for higher productivity. This macronutrient is present in agricultural soil in the form of both inorganic and organic compounds. But, these forms of phosphorous present in soil ecosystem are unavailable to crop plants. The predominance of unavailable form of P in soil makes the soil phosphorus deficient. Deficiency of P in soil solution limits the agricultural productivity. To overcome the deficiency of phosphorous, transformation of unavailable form of phosphate to its plant utilization form is necessary. This transformation of phosphate is carried out by phosphate solubilizing microbes (PSM). PSM play a key role in increasing the productivity of crop plants by dissolving insoluble phosphate and making it available to crop plants. The transformation of inorganic P to plant available form by fungi has been documented by various researchers. The use of fungi capable of mobilizing P increases the availability of P in soil and helps in management of phosphorus fertilization. Apart from solubilizing P, fungi exhibit some other traits also like suppression of various diseases and production of secondary metabolites, etc. But, their potential role in promoting plant growth and productivity by increasing nutrient absorption has been studied in detail by many researchers.

14.10.2.3 Consortium Biofertilizers

Consortium biofertilizer refers to the use of mixture of two or more microbial cultures for maintenance of soil health and promotion of plant productivity due to variety of growth mechanisms. Co-inoculated microbes work in coordination with each other to produce synergistic effects. They are found to be more effective than application of individual microbe. The integrated application of two or more microbial cultures gives fast and better results. This might be due to the fact that microbes interact with each other and function as coherent groups in rhizosphere. By this, they become capable of colonizing multiple ecological niches. The combined application of more than one plant growth promoting microorganism could be a better strategy for improving productivity of agricultural system in sustainable manner. The common bacterial genera which act as the PGPR include *Pseudomonas* sp., *Paenibacillus polymyxa*, *Achromobacter*, *Arthrobacter*, *Azospirillum*, *Azotobacter*, *Bacillus* sp., *Burkholderia, Ralstonia*, etc. PGPR benefit the host plant by direct or indirect mechanisms. PGPR have promise for maintaining adequate plant nutrition by providing the essential nutrient elements like nitrogen, phosphorous, releasing growth promoting substances like phytohormones, organic acids, siderophores, vitamins, and suppression of phytopathogens by competition and induction of systemic resistance response. Microbial inoculants can increase the growth and productivity of crop plants by increasing the uptake of essential macronutrients. This is due to the modification of root architecture. The modified root system is more efficient in uptake of nutrients leading to higher productivity. Such stimulation of roots not only helps in enhanced exploration but also improve water uptake which results in improvement of plant growth characters followed by enhanced yields.

14.11 Potential Role of Inoculated Microorganisms in Improvement of Plant Microbiome Interactions

The incorporation of microbial inoculants improves the plant microbiome interactions leading to enhanced level of soil fertility and yield. Their incorporation also improves the soil microbial flora and nutrient status. This results in decrease dependence on the use of chemical fertilizers. The improvement of soil microbial flora and soil health was reported by Sharon et al. (2001) with the use of cyanobacteria and associative nitrogen fixer, *Azospirillum.* Application of microbial inoculants also resulted in improved grain yield and harvest index. Significantly higher grain and straw yield of wheat crop has been observed with the inoculation with *Azotobacter*+*Rhizobium*+VAM and phosphate fertilizer. Beneficial effects of *Azolla* have also been reported on the nutrient status of soil. Another study conducted by Raj (2007) report that application of microbial inoculants increases the availability of both macro- and micronutrients. Leguminous plants can meet more than half of their nitrogen demand with the help of symbiotic nitrogen fixation. Another important aspect of microbial inoculants is their ability to suppress plant diseases (biocontrol). Thus, incorporation of bio-inoculants is a step toward sustainable agriculture.

It has been reported that root knot disease of French bean can be controlled using *Trichoderma* (Rahman 2005). The symbiotic bacteria like *Rhizobium* and *Bradyrhizobium* are also useful in suppressing the root knot of mung bean by their antagonistic activity (Khan et al. 2006). Microbial inoculants comprising of diazotrophs, phosphorous, or potassium solubilizing bacteria along with some other PGPRs can improve the growth, development, and productivity of many crop plants (Youssef and Eissa 2014).

14.12 Role of Microbial Inoculants in Improving Soil Health and Productivity of Plants

Microbial inoculants possess potential practical applications for sustainable and integrated approaches to agriculture. These beneficial microbes when inoculated, colonizes rhizosphere, increase their number and help in increasing productivity by protecting the host plant from various stresses and attack of pathogenic microbes. Microbial inoculants can be incorporated to crop plants as an agricultural amendment to increase the population of beneficial microbes in the rhizosphere of plant. Addition of beneficial microorganisms in the form of microbial inoculants to soil, seed, or any plant part can improve availability of nutrients, increase plant growth, confer resistance to abiotic stress, and suppress disease. These inoculated microbes can survive in field soil and compete with pathogenic and antagonistic microorganisms for colonization of vacant niches and nutrients. The potential of microbial inoculation was shown by Adesemoye et al. (2009) through their experiment carried out in glasshouse. They use PGPR and fungi (*Mycorrhizae*) as inoculants for tomato plants and reported that inoculated plants requires 25% less chemical fertilizers as compared to uninoculated plants for same quantity of yield. But, an awareness of the existing soil fertility level is critical to realize the benefits of microbial inoculants, as a diminishing effect is seen when starting N, P, and K levels are high.

Recent discoveries have shown that plants also respond to the application of various microbial inoculants that are able to increase the rate of photosynthesis, provide protection from stresses, increase disease suppression and plant growth, and enhance the efficiency of crop plant. These discoveries offer potential for microbial inoculants applications to improve agricultural production and sustainability. Currently, producers are faced with a need to reduce inputs like water and fertilizer applications while simultaneously increasing production. Combined application of various beneficial microorganisms (consortium microbial inoculants) has been found to more effective and efficient in improving growth, health, and productivity of plant as compared to application of any single microorganism. One example of combined inoculations includes the PGPR *Pseudomonas putida* added in combination with nodule inducing *Sinorhizobium meliloti* in the legume *Medicago sativa*, which resulted in increased nodulation and significantly increased plant biomass (Guiñazú et al. 2009). Use of consortium microbial inoculants also leads to improved nutrient status of plant.

The positive effects of combined application of beneficial microorganisms on the growth of crop plant have been documented by various researchers. These results indicate that the plants and the microbes work in coordination with each other to bring the desired outcome. In some cases, application of a microbial organism that confers benefit may not even be necessary. The same result can also be achieved by the application of a microbial product (elicitor). For example, the use of acetoin (produced by *Bacillus subtilis*) stimulates the induced systemic resistance (ISR) and play key role in disease suppression (Rudrappa et al. 2008). Determining the precise compounds and dosages necessary for application would allow for commercial development of a nonliving application providing the same benefits as the microbial inoculants themselves. Such treatments could avoid some of the potential complications associated with developing commercial PGPR applications such as low survivability due to competition and adverse environmental conditions. Nowadays, researchers emphasize on the studies that how to modify the rhizospheric environment for the better survival of microbial inoculants using rhizospheric engineering.

Conclusion

Intensive agriculture being followed to feed the increasing human population is depleting soil health. The successful management of soil health and plant productivity requires even and balanced microbiome along with environmentally friendly agricultural practices. There is a complex conversation that occurs between soil microbes and plants, mediated by root exudates, but this conversation still needs a lot more translating. Microbial inoculants are economic and environment friendly. Their application to soil or plant improves the soil nutrient status and crop yield in sustainable manner. They also possess the potential to improve the conversations between plants and soil microbes. Future studies should determine what key compounds and root exudates compositions will culture these beneficial microbes that produce healthy and more productive plants.

References

Adesemoye AO, Torbert HA, Kloepper JW (2009) Plant growth-promoting rhizobacteria allow reduced application rates of chemical fertilizers. Microb Ecol 58:921–929

Altieri MA (2004) Linking ecologists and traditional farmers in the search for sustainable agriculture. Front Ecol Environ 2:3542

Anonymous (2008) "Promoting Bio-fertilizers in Indian Agriculture Nilabja Ghosh, Institute of Economic Growth", University Enclave, Delhi 110007, India

Arun KS (2007) Bio-fertilizers for sustainable agriculture. Mechanism of P solubilization, 6th edn. Agribios publishers, Jodhpur, pp 196–197

Behie SW, Zelisko PM, Bidochka MJ (2012) Endophytic insect-parasitic fungi translocate nitrogen directly from insects to plants. Science 336(6088):1576–1577

Berendsen RL, Pieterse CMJ, Bakker P (2012) The rhizosphere microbiome and plant health. Trends Plant Sci 17:478–486

Bulgarelli D, Schlaeppi K, Spaepen S, Ver Loren van Themaat E, Schulze-Lefert P (2013) Structure and functions of the bacterial microbiota of plants. Annu Rev Plant Biol 64:9.1–9.32

Chaparro JM, Sheflin AM, Manter DK, Vivanco JM (2012) Manipulating the soil microbiome to increase soil health and plant fertility. Biol Fertil Soils 48(5):489–499

De-la-Pena C, Badri DV, Lei Z, Watson BS, Brandao MM, Silva-Filho MC, Sumner LW, Vivanco JM (2010) Root secretion of defense- related proteins is development-dependent and correlated with flowering time. J Biol Chem 285:30654–30665

Faoro H, Alves AC, Souza EM, Rigo LU, Cruz LM, Janabi SM, Monteiro RA, Baura VA, Pedrosa FO (2010) Influence of soil characteristics on the diversity of bacteria in the southern Brazilian Atlantic Forest. Appl Environ Microbiol 76:4744–4749

Flores HE, Vivanco JM, Loyola-Vargas VM (1999) 'Radicle' biochemistry: the biology of root-specific metabolism. Trends Plant Sci 4:220–226

Gianinazzi S, Gollotte A, Binet MN, van Tuinen D, Redecker D, Wipf D (2010) Agroecology: the key role of arbuscular mycorrhizas in ecosystem services. Mycorrhiza 20:519–530

Gold MV (1995) Organic production/organic food: information access tools. USDA National Agricultural Library, Alternative Farming Systems Information Centre (http://www.nal.usda.gov/afsic/organic_production)

Guiñazú LB, Andrés JA, Del Papa MF, Pistorio M, Rosas SB (2009) Response of alfalfa (*Medicago sativa* L.) to single and mixed inoculation with phosphate-solubilizing bacteria and *Sinorhizobium meliloti*. Biol Fertil Soils 46:185–190

Hamilton EW, Frank DA (2001) Can plants stimulate soil microbes and their own nutrient supply? Evidence from a grazing tolerant grass. Ecology 82:2397–2402

Hartmann A, Rothballer M, Schmid M (2008) Lorenz Hiltner, a pioneer in rhizosphere microbial ecology and soil bacteriology research. Plant Soil 312(1):7–14

http://www.lookfordiagnosis.com/mesh_info.php?term=Azospirillum&lang=1

Khan A, Zaki MJ, Tariq M (2006) Seed treatment with nematicidal *Rhizobium* species for the suppression of *Meloidogyne javanica* root infection on mung bean. Int J Biol Biotechnol 3(3):575–578

Lugtenberg B (2015) Life of microbes in the rhizosphere. In: Lugtenberg B (ed) Principles of plant microbe interactions. Springer International Publishing Switzerland, Heidelberg, p 715

Mali GV, Bodhankar MG (2009) Antifungal and phytohormone production potential of *Azotobacter chroococcum* isolates from groundnut (Arachis Hypogea L.) rhizosphere. Asian J Exp Sci 23:293–297

Megali L, Glauser G, Rasmann S (2013) Fertilization with beneficial microorganisms decreases tomato defenses against insect pests. Agron Sustain Dev. doi:10.1007/s13593-013-0187-0

Mendes R, Garbeva P, Raaijmakers JM (2013) The rhizosphere microbiome: significance of plant beneficial, plant pathogenic, and human pathogenic microorganisms. FEMS Microbiol Rev 37:634–663

Morales SE, Holben WE (2011) Linking bacterial identities and eco- system processes: can 'omic' analyses be more than the sum of their parts? FEMS Microbiol Ecol 75:2

Odame H (1997) Biofertilizer in Kenya: research, production and extension dilemmas. Biotechnol Dev Monit 30:2023

Rahman M (2005) Effect of BAU-Biofungicide and nematicide Curaterr against root-knot of French bean. M.Sc. Thesis, Department of Plant Pathology, Bangladesh Agricultural University, Mymen singh

Raj SA (2007) Bio-fertilizers for micronutrients. Biofertilizer Newsletter (July). pp. 8–10

Revillas JJ, Rodelas B, Pozo C, Martinez-Toledo MV, Gonzalez LJ (2000) Production of B-group vitamins by two *Azotobacter* strains with phenolic compounds as sole carbon source under diazotrophic and adiazotrophic conditions. J Appl Microbiol 89:486–493

Rousk J, Baath E, Brookes PC, Lauber CL, Lozupone C, Caporaso JG, Knight R, Fierer N (2010) Soil bacterial and fungal communities across a pH gradient in an arable soil. ISME J 4:1340–1351

Rudrappa T, Czymmek KJ, Paré PW, Bais HP (2008) Root-secreted malic acid recruits beneficial soil bacteria. Plant Physiol 148:1547–1556

Ryan MH, Angus JF (2003) Arbuscular mycorrhizae in wheat and field pea crops on a low P soil with increased Zn-uptake but no increase in P-uptake or yield. Plant Soil 250:225–239

Sahoo RK, Ansari MW, Dangar TK, Mohanty S, Tuteja N (2013) Phenotypic and molecular characterization of efficient nitrogen fixing *Azotobacter* strains of the rice fields. Protoplasma. doi:10.1007/s00709-013-0547-2

Sahoo RK, Ansari MW, Pradhan M, Dangar TK, Mohanty S, Tuteja N (2014) Phenotypic and molecular characterization of efficient native *Azospirillum* strains from rice fields for crop improvement. Protoplasma. doi:10.1007/s00709-013-0607-7

Saikia SP, Bora D, Goswami A, Mudoi KD, Gogoi A (2013) A review on the role of *Azospirillum* in the yield improvement of nonleguminous crops. Afr J Microbiol Res 6:1085–1102

Schloss PD, Handelsman J (2006) Toward a census of bacteria in soil. PLoS Comput Biol 2:92

Sharon E, Bar EM, Chet I, Herrera EA, Kleifeld O, Spiegel Y (2001) Biological control of the root-knot nematode *Meloidogyne javanica* by *Trichoderma harzianum*. Phytopathology 91(7):687–963

Venkatashwarlu B (2008) Role of bio-fertilizers in organic farming: Organic farming in rain fed agriculture: central institute for dry land agriculture, Hyderabad, 85–95

Wani SA, Chand S, Ali T (2013) Potential use of *Azotobacter chroococcum* in crop production: an overview. Curr Agric Res J 1:35–38

Wani SP, Lee KK (1995) Microorganisms as biological inputs for sustainable agriculture. In: Thampan PK (ed) Organic agriculture. Peekay Tree Crops Development Foundation, Cochin, pp 39–76

Wani SP, Lee KK (2002) Population dynamics of nitrogen fixing bacteria associated with pearl millet (*P. americanum* L.) In: Biotechnology of nitrogen fixation in the tropics. University of Pertanian, Serdang, pp 21–30

Youssef MMA, Eissa MFM (2014) Biofertilizers and their role in management of plant parasitic nematodes. A review. J Biotechnol Pharm Res 5(1):1–6

Siderophores: Augmentation of Soil Health and Crop Productivity

15

Rizwan Ali Ansari, Irshad Mahmood, Rose Rizvi, Aisha Sumbul, and Safiuddin

Abstract

Microorganisms harbouring in the soil are extremely important in the sustainable agriculture. They play a very crucial role in the sustenance of ecological services/balance. Siderophore-producing microorganisms have enormous range of application for the sustainable crop production. Application of siderophores has recently caught fire of discussion and being used in the plant disease management, maintenance of healthy soil, plant growth promotion, SAR induction, acceleration of phytohormone production, bioaugmentation of heavy metal (HM), etc. Moreover, nearly all living beings shine their cellular reactions such as electron transportation, different metabolic reactions and organic molecule formations with the help of iron. In iron-deprived environment, siderophores are the chief media by which maximum cellular reactions get completed. However, a wide range of variations among the siderophores has been noticed like bacterial siderophores that have extremely high binding affinities than fungi; however, phytosiderophores have less binding affinities than microbial siderophores. A lot of variations among the microbial siderophores such as algae, bacteria, fungi and actinomycetes have been noticed in significant manner. Additionally, beneficial job of siderophore in other sector of agriculture and allied branch of science may not be ignored. However, there are some hurdles such as lack of infrastructure and communication gap among the concerned researcher which has put such important research on hold. Research on siderophore-producing organisms will provide an arena to formulate bioprocess technology which is indeed needed to maximize the production of microbial siderophores because of its wide range of applicability. Overall, siderophores and siderophore-producing organisms are conducive to

R.A. Ansari (✉) • I. Mahmood • R. Rizvi • A. Sumbul • Safiuddin
Section of Plant Pathology and Nematology, Department of Botany,
Aligarh Muslim University, Aligarh, India
e-mail: rizwans.ansari@gmail.com

V. Kumar et al. (eds.), *Probiotics in Agroecosystem*,
DOI 10.1007/978-981-10-4059-7_15

human kind as well as in the sustenance of ecological balance. Thus, this article discusses about the present scenario of research pertaining to siderophore applications in agriculture.

15.1 Introduction

Plants and microbes play an important role in our daily routine. Iron is ranked among the most abundant element in the Earth's crust (Kurth et al. 2016). Besides, many autotrophs face turmoil in acquiring iron due to its insoluble form, which inhibits the bioavailability of iron (Kurth et al. 2016). Generally, living organisms survive by the performance of certain cellular processes during that iron plays a significant role. Iron is the chief constituent for a variety of vital functions such as photosynthesis, enzyme cofactor, redox reagent, respiration, nucleosides and amino acid synthesis. Moreover, microbes and plants flourish themselves under iron-limited conditions by releasing iron chelator called siderophore. Siderophores are low-molecular-weight (<10 kDa) iron-chelating organic molecules, released by microbial communities thriving in the rhizosphere under iron-limited conditions. These iron chelators play a crucial role in the solubilization of iron from inorganic as well as organic molecules. Siderophores help to enhance the plant growth by scavenging iron from the nearby area and make them available to root (Hider and Kong 2010; Maheshwari 2011; Ahmed and Holmstrom 2014; Zhou et al. 2016). Siderophores play a valuable role in plant growth promotion (Yadav et al. 2011; Verma et al. 2011; Trapet et al. 2016), biocontrol agents (Verma et al. 2011; Di Francesco et al. 2016), bioremediation agents (Wang et al. 2011; Ishimaru et al. 2012; Ma et al. 2016) and mineral weathering (Reichard et al. 2005; Buss et al. 2007; Shirvani and Nourbakhsh 2010; Ahmed and Holmstrom 2015). In addition, various plants have been reported to release phytosiderophore that sequester the iron by the roots which assist the Fe complex uptake under iron-deprived state (Kannahi and Senbagam 2014). It has been ascertained that competition for iron in the rhizosphere is governed by the empathy of the siderophores for iron (Bernd and Rehm 2008; Munees and Mulugeta 2014). It has been well known fact that alkaline soils are strong inducers of iron deficiency in plants. Besides, if soil pH exceeds 6.5–7.0, the availability of iron in the soil is considerably reduced; however, calcareous soils, having high pH, diminish the affinity of plants for Fe and hence hinder Fe uptake.

Iron is a vital nutrient necessary for almost all living organism for carrying out various cellular processes (Neilands 1995). Generally, bacteria obtain iron molecule after producing iron chelators, siderophores having high affinity for iron complexing. In cellular context, there are two types of siderophores, and they can be divided into extracellular and intracellular siderophores. Also, there is a large variation in rhizobacteria pertaining to siderophore utilization ability. It has been seen that there were restrictions of siderophore utilization while no such bar were detected in other genus of rhizobacteria (Khan et al. 2009). It is well known that during Fe^{3+} complex, Fe^{3+} is reduced into Fe^{2+} on bacterial membrane which is later on delivered into the cell through a gating mechanism; however, this process leads to sometime loss of siderophores (Rajkumar et al. 2010; Neilands 1995). In this way, siderophores have the capability to solubilize iron from organic compounds or minerals under iron-deprived conditions (Indiragandhi et al. 2008). Besides, siderophores also bind with other HMs which are actively involved in some environmental concerns (Kiss and Farkas 1998; Neubauer et al. 2000). Formation of stable complex with siderophore to HMs enhances the soluble metal concentration (Rajkumar et al. 2010). Therefore, in such way, iron chelators assist in the alleviation of abiotic stress such as HMs imposed on plants. As far as assimilation of iron is concerned, in plants, various possible pathways such as chelation and production of iron, direct accumulation of siderophore–Fe complex and through ligand exchange process have been put forward (Schmidt 1999). Recent studies have generated the information pertaining to enhanced plant growth promotion after inoculation of siderophore-producing PGPR (plant growth-promoting rhizobacteria) (Rajkumar et al. 2010). In addition to PGPR uptake, machinery of the plants also determines the level of significance like application of siderophore-producing bacteria in oat plants under iron-deprived conditions leads to significant plant growth promotion which may be due to plants having the mechanisms for using Fe–siderophore complexes under iron-limited conditions (Crowley and Kraemer 2007). Similar results were also seen in *Arabidopsis thaliana* plants by *Pseudomonas fluorescens* C7 which leads to large accumulation of iron and thereby enhancement in plant growth (Vansuyt et al. 2007).

Siderophore and its derivative have a broad range of significance in sustainable agriculture in the form of enhancement of soil fertility and as potent biocontrol agent for fungal pathogen. Therefore, the present article accounts for the role of siderophores in sustainable agriculture with special emphasis on maintenance of soil health, management of fungal pathogens and crop growth promotion.

15.2 Iron Bioavailability in Iron-Deprived Environment

Generally, iron is found as Fe(III), insoluble under physiological conditions (Powell et al. 1980; Matzanke et al. 1989). Many enzymes and cofactors are responsible for carrying out various cellular processes like respiration, oxygen activation, hydrogen peroxide and hydroxyl radicals' degradation, etc. (Andrews 1998).

Ferrous is more soluble state at neutral pH which is available for living cells for further process. In addition, most of bacterial communities accumulate Fe(II) through divalent metal transporters (Miethke and Marahiel 2007). Moreover, iron is the key element for the life to be processed; however, there are some exceptions such as lactic acid bacteria, and they do not have heme enzymes (Neilands 1995). Additionally, iron may be toxic because high intracellular concentration of ferrous ion starts producing hydroxyl radicals (Crichton and Charloteaux-Wauters 1987). However, such problem no longer exists and can be alleviated with the help of certain antioxidants. The toxicity of the iron may be nullified by the presence of glutathione and endonucleases which repair DNA (Andrews 1998). It has been a well-established fact that iron imports toxicity towards rice plants being grown in lowland. This may be advocated that rice plants accumulate large values of ferrous after reduction of iron oxides and hydroxides which leads to disruption of metabolic process and plants become damaged (Becker and Asch 2005).

Chief iron pool in soil and water ecosystems is comprised by oxides of iron (Kraemer 2004). Production of siderophores is a specialized iron acquisition system which reveals competitive benefit to many microorganisms in biotic and abiotic environments. Plenty of research on biological iron acquisition have stated about significant increase in iron solubility (Kraemer 2004). Availability of iron depends on its properties such as particle size, pH, ionic strength and amount of organic ligands in solution (Kraemer 2004). For instance, Fe(II) quickly oxidizes to Fe(III) at neutral pH and oxic conditions (Stumm and Morgan 1995). In the weak organic ligand, Fe(III) precipitates quickly as a hydrous ferric oxide, and citrate is too weak to bind iron which inhibits Fe(III) precipitation in the culture medium (Konigsberger et al. 2000).

In soil at neutral pH concentrations, the ferric oxide hydrate is around 10^{-17} M (Budzikiewicz 2010). However, living systems require 10^{-6} M, as soon as cells that detect the necessities of iron siderophore production begin (Miethke and Marahiel 2007). Siderophores have a manifold impact on the solubility of iron oxides with a varying range of pH because of extraordinary thermodynamic stability of soluble siderophore–iron complexes.

15.3 Types of Siderophores

A lot of variation have been detected in the structure of siderophores produced by the microbes especially bacteria. They are categorized on co-ordinating atom basis on which they chelate the Fe(III) ion. Hydroxamate, catecholate and carboxylate are important groups of siderophores.

15.3.1 Catecholate

These types of siderophores are produced by not all but only some bacteria. Each catecholate composed of two oxygen atoms with iron forming a hexadentate octahedral complex, a cyclic trimer composed of 2,3-dihydroxy-*N*-benzoylserine is the best example of the catecholate.

For the first time, a tricatechol siderophore, enterobactin, was isolated from *Escherichia coli*, *Aerobacter aerogenes* and *Salmonella typhimurium* (Ward et al. 1999). A bacterium of the family Enterobacteriaceae produces enterobactin; possibly, all strain have the capability to bind with iron. In addition, *S. typhimurium*, *Klebsiella pneumoniae* and *Erwinia herbicola* are well-studied models to produce enterobactin. Enterobactin is blessed with the capacity to trap the iron even from the environment where iron content is far away from its reach (Raymond et al. 2003).

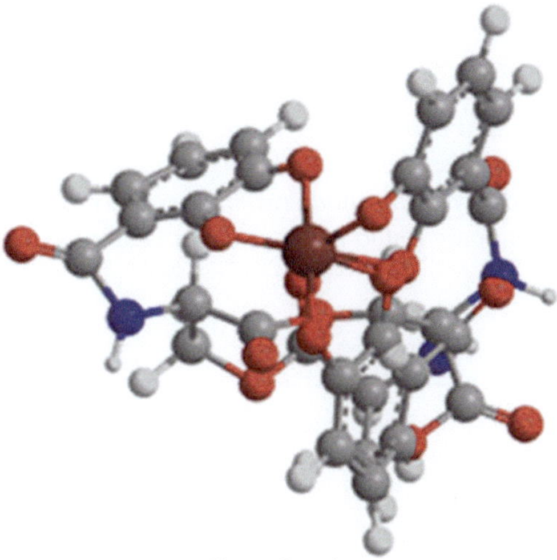

Enterobactin

15.3.2 Hydroxamate

Ferrichrome-type hydroxamate is produced by many soil fungi including some mycorrhiza (Schalk et al. 2011) A plenty of research have provided extensive support that hydroxamate siderophores may provide iron to only certain plant species. Mostly, they are produced by fungi not by bacteria belonging to class Zygomycotina (*Mucorales*), Ascomycotina (*Aspergilli*, *Penicillia*, *Neurospora crassa*) and Deuteromycotina (*Fusarium dimerum*).

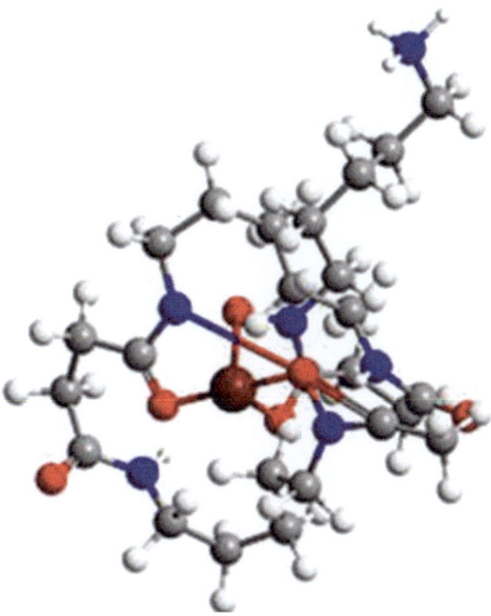

Ferrioxamine B

15.3.3 Carboxylate

This is a special type of siderophore where iron binding is accomplished by hydroxyl carboxylate and carboxylates (Schwyn and Neiland 1987). These siderophores have shown their presence in the group of bacteria as well as fungi. Carboxylates produced by *Rhizobium* and *Staphylococcus* species and members of *Mucorales* are commonly found where iron with carboxyl and hydroxyl groups is coordinated.

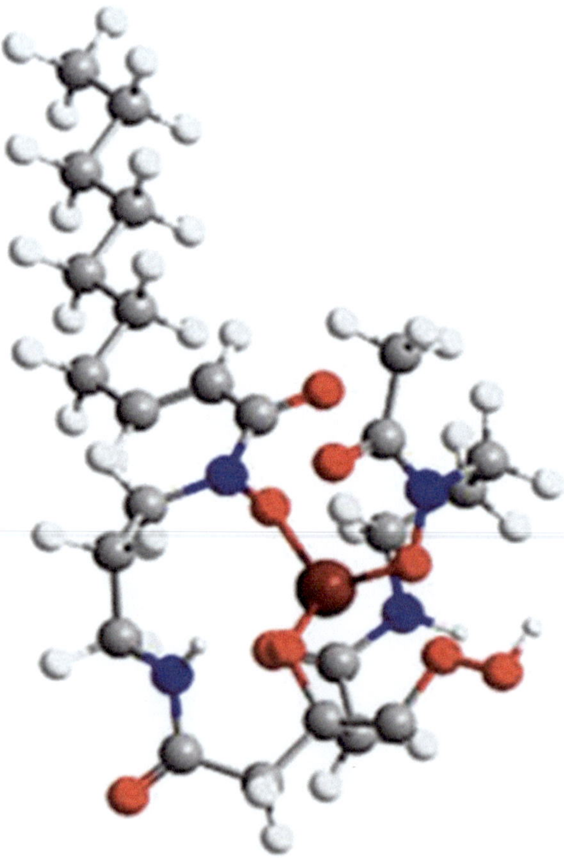

Rhizobactin

15.3.4 Miscellaneous

In addition to the above different siderophores, some have derivatives of mixed ligands of lysine, ornithine and histamine. An array of fluorescent chromopeptide siderophore called as pseudobactin and pyoverdines that contain a dihydroxyquinoline derivative are currently in vogue of research. There are two types of significant siderophore-mediated iron uptake scheme in these bacteria; first it involves the fluorescent siderophore pseudobactin and second it contains the siderophore pyochelin (Meyer 2000; Meneely and Lamb 2007).

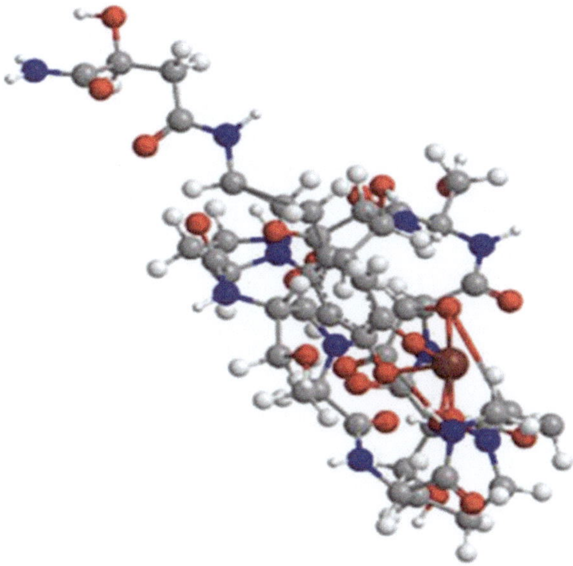

Pyoverdine

15.4 Siderophores from Different Organisms

15.4.1 Plants

Certain plants have acquired-specialized mechanism for iron uptake in plants belonging to the family Poaceae through roots by releasing iron chelators known as phytosiderophores. Plants begin acquisition of iron by different strategies

(Römheld and Marschner 1986). According to one theory, strategy I is used by most non-Poaceae plants having inducible plasma membrane-bound reductase with the significant increase in H^+ release, while in strategy II, a significant increase in phytosiderophores characterized by an enhanced release with highly specific uptake system is reported. Strategy II has several ecological benefits over strategy I such as solubilization of inorganic Fe(III) compounds in the rhizosphere and lowering down of pH. There are lower affinities in phytosiderophores as compared to microbial siderophores which is replenished by high exudation rates by Poaceae plant roots.

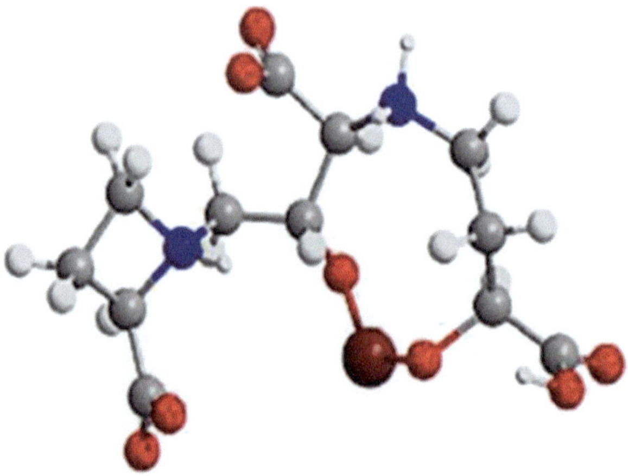

Mugineic acid

15.4.2 Fungi

Fungi are the important source of siderophore-producing microorganisms and ranked after bacteria (Scavino and Pedraza 2013). Common genera of important siderophore-producing fungi are *Aspergillus nidulans*, *A. versicolor*, *Penicillium chrysogenum*, *P. citrinum*, *Mucor*, *Rhizopus* and *Trametes versicolor. Ustilago sphaerogina*, *Saccharomyces cerevisiae*, *Rhodotorula minuta* and *Debaryomyces* species. Majority of the fungi produce a wide range of siderophores covering a large range of physico-chemical properties. These particular characters of siderophores make it capable to overcome the adverse conditions (Winkelmann 2007). A large number of structurally different fungal siderophores are reported having a peptidic ring in common. Generally, all aerobic bacteria and fungi generate siderophores (Neilands and Leong 1986). However, this property reveals a clear picture of benefit for microbes occupying in aerobic environments. For example, many facultative bacteria from paddy field soils are found on siderophore producers (Loaces et al. 2011). However, there are some other microbes having no mechanism to synthesize and produce siderophores such as *Saccharomyces cerevisiae*; however, they utilize the siderophore produced by other species (Eissendle et al. 2003).

15.4.3 Bacteria

Bacteria occupying the metal-contaminated environment are able to accumulate and transport the HMs (Rajkumar and Freitas 2008; Weyens et al. 2009;). Bacterial cell produces polysaccharide sheath that determines metal-binding affinities (Sheng et al. 2008). Normally, four types of siderophores are produced by bacteria, and they are hydroxamate, catecholate, salicylate and carboxylate (Rajkumar et al. 2010). These siderophores play a pivotal role in the accumulation of iron from various organic materials. Certain common siderophore-producing bacteria are *Escherichia coli*, *Salmonella*, *Klebsiella pneumoniae*, *Vibrio cholerae*, *Vibrio anguillarum*, *Aeromonas*, *Aerobacter aerogenes*, *Enterobacter*, *Yersinia* and *Mycobacterium* species (Balagurunathan and Radhakrishnan 2007).

15.4.4 Actinomycetes

Actinomycetes are filamentous bacteria having high quantity of guanine + cytosine (G+C) content which form asexual spores. Generally, they are saprophytic in nature which rely on complex substrate for their development and have the capability to nullify the impact of HMs even their concentration is extremely high. *Actinomadura madurae, Nocardia asteroides and Streptomyces griseus* are important genera suitable for such stressed environment (Khamna et al. 2009; Taj and Rajkumar 2016).

15.4.5 Algae

The production of siderophore has been reported also from some algae. *Anabaena* sp. produces an important siderophore, schizokinen, a dihydroxamate which helps in the facilitation of iron acquisition. In addition, certain siderophores produced by *Anabaena flos-aquae* and *Anabaena cylindrica* have been reported to accumulate copper instead of iron (Balagurunathan and Radhakrishnan 2007).

15.5 Role of Siderophores

The significance of microbial siderophores extends beyond our imagination (Kurth et al. 2016). Applications of iron chelators in sustainable agriculture are enormous especially in certain branches. Siderophores are produced by different bacteria having a wide range of application in different branches of agriculture such as soil science, plant pathology, environmental sciences, etc.

15.5.1 Maintenance of Soil Health

Soil is a dynamic place where trillions of microorganisms such as algae, bacteria, fungi, protozoans, insects, mites and worms complete their life cycle. It has been well studied that 1 gm of soil may carry about 10 billion microorganism (Torsvik and Øvreås 2002; Crecchio et al. 2004).

Soil bioremediation process has been well studied by the use of different types of siderophores just to maintain a healthy environment in soil ecosystem. There are some siderophores which have the ability to bind with metals other than iron. In this context, a wide range of bioreactors have been developed for the solubilization of the HMs (Diels et al. 2009). It has been found that the quantity of some HMs was reduced by 16-folds from its original state in the soil treated with *Cupriavidus metallidurans* which produces citrate siderophores staphyloferrin B (Munzinger et al. 1999; Diels et al. 2009). Similarly, *Pseudomonas azotoformans* have the potential to purify catecholate–hydroxamate siderophore (Nair et al. 2008). A major problem in the selection of the microorganisms is the persistence and metal tolerance limit in the new environment (Thompson et al. 2005; Braud et al. 2015). Conjoint implementation of bioaugmentation with phytoextraction has recently caught a fire of discussion among the researchers. Siderophore-producing microorganisms are well adopted for bioaugmentation because they help in the promotion of biomass as well as accumulation of HMs in various ways. Plant siderophores like mugineic acid and avenic acid are not always be able to fulfil the demand of iron, particularly in HM-polluted soil (Ma et al. 2011). However, some plants have been found to be able to access iron from bacterial siderophores by different mechanisms such as direct accumulation, chelate degradation or ligand exchange process (Schmidt 1999). Many siderophores have been reported to bind with other than iron and help in the accumulation of HMs. The bioaugmenting process of contaminated soil with *Ralstonia metallidurans* and *Pseudomonas aeruginosa* enhanced the capability of accumulation of Cr in *Zea mays* L. by 5.4 times (Takemoto et al. 1978). Similarly, application of *Streptomyces tendae* F4 enhanced the uptake of Cd and Fe in sunflowers and assisted well in plant growth promotions (Dimpka et al. 2009). In this way, it can be apprehended that siderophores may be enough to solubilize the HMs transporting them to the plants which ultimately lower down the HM concentration from the environment. In this way, siderophore helps in normalizing the soil ecosystem which is necessary in the present scenario.

15.5.2 Management of Plant Diseases

Biological control of plant disease has been fascinating and eco-friendly (Lugtenberg and Kamilova 2009). This way illustrates the indirect pathways of plant growth promotion by managing the disease significantly (Glick 2012). The main activity of biocontrols is food competition, colonization, ISR induction and antifungal compound production (Lugtenberg and Kamilova 2009). A large number of rhizobacteria have been found to produce antifungal compounds such as HCN, phenazines, pyrrolnitrin, 2, 4-diacetylphloroglucinol, pyoluteorin, viscosinamide and tensin (Bhattacharyya and Jha 2012). Resistance against some pathogenic bacteria, fungi and viruses is induced due to the interaction between rhizobacteria and plant root, called as induced systemic resistance (ISR). In addition, ISR activates the jasmonate and ethylene signalling pathway (Lugtenberg and Kamilova 2009). ISR involves in the activation of host plant's defence system against a wide range of plant pathogens. There are several other which induce bacterial components, ISR, lipopolysaccharides, flagella, iron-chelating compounds, cyclic lipopeptides, 2,4-diacetylphloroglucinol, homoserine lactones and volatiles like acetoin and 2,3-butanediol (Lugtenberg and Kamilova 2009).

Frequent and haphazard use of pesticides has escorted to the development of pest-resistant strains which facilitate in the transformation of fungicides ineffective. However, microbial metabolites can improve the management strategies of plant pathogens either by augmenting the action of antagonistics or by paving the ways to develop healthier alternatives as compared to synthetic pesticides (Rizvi et al. 2015). Additionally, there is a lot of variation in the production of siderophores. Production of siderophores is correlated with the types of strain and how that specific strain is familiar with target pathogens. The use of mutants that were effective once in siderophore secretion was less effective than the wild-type strains in crop protection (Buysens et al. 1996). Pseudomonads form a line of siderophores pertaining to enhance plant yield through the management of harmful pathogens. It has been found that many rhizobacteria suppress the growth of harmful microorganism by releasing siderophore and other related organic molecules (Husen 2003). In addition, siderophores inhibit the growth of various plant pathogenic fungi, like *Phytophthora parasitica* (Seuk et al. 1988), *Pythium ultimum* (Hamdan et al. 1991), *Fusarium oxysporum* var. *dianthi* (Buysens et al. 1996) and *Sclerotinia sclerotiorum* (Kraemer et al. 2006).

15.5.3 Promotion of Crop Yield

Although most of the soil is blessed with sufficient iron for plant growth, plant iron deficiency is a common problem in some range of soil especially calcareous soil which may be due to low solubility of Fe(III) hydroxide. Calcareous soil harbours around 30% of the world's agricultural land. In such case, some plants (grasses, cereals and rice) secrete phytosiderophores into the soil. Some plant species such as barley and wheat are well efficient to sequester iron by releasing phytosiderophores via their root into the surrounding soil rhizosphere (Hershko et al. 2002). Many studies have advocated that plants are able to incorporate and use Fe^{3+} of siderophores into their biomass. In addition to this, some plants are efficient to assimilate iron through siderophores produced by microorganism harbouring rhizospheric soil. The use of microbial siderophore has been extensively studied and found that this organic molecule has rescued groundnut from iron chlorosis. A significant improvement in some growth attributes and plants health has been extensively observed after the treatment of seeds with siderophorogenic bioinoculants.

A considerable increase in the percentage of germination, and some plant growth attributes including chlorophyll content, has been achieved when seeds were bacterized with siderophore of *Pseudomonas* (Manwar et al. 2001). The effect of bacterial siderophores on plant growth has been seen in various studies. For instance, the use of radiolabelled ferric siderophore as a sole source of iron explained that plants are able to take up the labelled iron; mung bean plants treated with *Pseudomonas* strain GRP3 grown under iron-deprived conditions showed less chlorotic symptoms and a significant chlorophyll level (Sharma et al. 2003). Similarly, considerable enhancements in iron content were recorded in *Arabidopsis thaliana* plant tissues leading to improved plant growth (Vansuyt et al. 2007). Siderophores play a crucial role in the dissolution of iron, making it available for microbial and plant growth.

15.5.3.1 Role of *Pseudomonads*

Siderophores, pseudobactin (pyoverdine), produced from *Pseudomonas* (B10) isolated from suppressive soils when inoculated to soils conducive to Fusarium wilt or take all disease caused by *Gaeumannomyces graminis* transformed them to disease-suppressive soils (Desai and Archana 2011). Moreover, addition of exogenous iron(III) to disease-suppressive soils leads to conversion of them into conductive soils. A large number of bacteria are found effective in biocontrol of plant diseases due to their antagonistic ability to phytopathogenic bacteria or fungi having a higher binding affinity for iron (Raaijmakers et al. 1995; Loper and Henkel 1999). Production of siderophore by *Pseudomonas* spp. has been reported to involve in the control of *G. graminis* var. *tritici* (Kloepper et al. 1980), *F. oxysporum* (Elad and Baker 1985) and *Pythium* spp. (Becker and Cook 1988; Loper 1988). It has been well documented that the antagonistic activity of pseudomonads against phytopathogens leads to a significant enhancement in plant growth and yield (Loper and Henkel 1999) also against detrimental phytopathogens (Becker and Cook 1988; Schippers et al. 1987), thereby increasing plant growth. Siderophores have been also found to be the inducers of defence mechanisms in a wide range of plants. For example, *P. fluorescens* CHA0 was reported to induce systemic acquired resistance (SAR) of tobacco; however, at varying extent, its pvd mutant registered minimum improvement than the wild one (Maurhofer et al. 1994). Some microbial siderophores including pyoverdines have played a pivotal role in the direct improvement of the iron nutrition in many plant species (Crowley et al. 1988; Hordt et al. 2000). A significant enhancement in iron content and uptake has been reported in various horticultural crops (Bar-Ness et al. 1991). Vansuyt et al. (2007) reported that iron chelated to pyoverdine was transported to *A. thaliana* plants in an independent pathway which leads to enhanced plant growth.

15.5.3.2 Role of *Rhizobia*

Rhizobium spp. has impacted a large in cash crop especially on pulses. The information on the advantageous effect of siderophore conferred by a free-living *Rhizobium* strain in the siderophore production and uptake are still meagre. However, available literature have suggested that rhizobial siderophores play a pivotal role in rhizosphere competition possibly in the same manner as pseudomonads do (Joshi et al. 2008). Some rhizobia are efficient enough to produce siderophores leading to plant growth promotion and nodulation (Bai et al. 2002; Dahsti et al. 1998; Rao and Pal 2003). In addition to this, some phytopathogenic bacteria harbouring in the soil have the capability to colonize the rhizosphere, leaving negative effects on plant growth. Besides rhizobial nitrogen fixation, they are also effective as biocontrol agents for the management of certain soilborne phytopathogen enhancements of plant growth by IAA production and accumulation of some minerals and phosphorous (Chakraborty and Purkayastha 1984). A large number of rhizobial strains promote plant growth in one hand, while, on the other hand, inhibit the growth of pathogenic fungi/bacteria. *Rhizobium meliloti* and *B. japonicum* are examples which reduced the detrimental effect of *Macrophomina* disease severity considerably. Reduction of disease severity caused by *Macrophomina phaseolina* was significant over control because of starvation of iron (Arora et al. 2001; Deshwal et al. 2003; Desai and Archana 2011).

15.6 Microbial Interaction

The role of siderophores among organisms' interaction has been well researched and found to be greatly influenced. Production of siderophores modifies the niche area of an organism through various mechanisms such as cooperation, competition, etc. (Scavino and Pedraza 2013). A large number of microbes have the machinery to utilize the Fe(III) siderophore complex synthesized by the siderophore-producing organisms. Several enterobacteria have the receptors for uptaking such siderophores leading to modification of the current environment (Winkelmann 2007). The siderophores produced by bacteria have been reported to get utilized by fungi (Hass 2003; Heymann et al. 2000). Similarly, enterobactin produced by enterobacteria can be used by *Saccharomyces* sp. (Winkelmann 2007).

Microbial interaction is a natural process and necessary for maintaining the ecological balance (Kurth et al. 2016) which may be positive, negative or neutral. There are wide ranges of alteration in interacted microorganism-producing siderophores. For example, bacterial siderophore has higher affinity to bind Fe than the fungi which explain the reason of biocontrol of plant pathogenic fungi (Loper and Henkels 1999). Besides, some siderophore producers are invaded by non-siderophore-producing chelators either from same or different species. Generally, siderophore production is very expensive to a single producer but that enables other cell of the same species present in the vicinity to capture iron siderophore complexes (Harrison et al. 2008). Interestingly, some siderophore-producing microorganisms synthesize some different siderophores just to bypass the cheaters' tactic. *Streptomyces* spp. have distinct type of siderophore production system. They are generally categorized into two types of independent uptake system. For example, ferrioxamine can be used by different organisms, while ferric coelichelin can only be absorbed by *Streptomyces coelicolor* (Challis and Hopwood 2003). Moreover, some microorganisms have the capability to destruct the siderophore leading towards the modification of interaction process. For instance, *Azospirillum* sp. in pure keeps the capability to vandalize the ferrioxamine during iron-free state. In addition to this, it was seen that unculturable bacteria were stimulated and transformed into culturable form in the presence of some siderophore-producing bacteria. Acyl-desferrioxamine, a prominent siderophore, enables the uncultured microorganisms to get flourished themselves and helps in the plant growth promotion (D'Onofrio et al. 2010).

15.7 Environmental Research

Siderophores have the potential ability to settle down a range of ecological issues such as HM accumulation, rust removal, biofouling, dye degradation, sewage treatment and bioleaching, etc. Soil biota promotes mineral weathering by the production of enormous type of siderophores which offer competent Fe acquisition organization due to its high binding affinity for Fe(III) (McGrath et al. 1995; Kraemer 2004). HMs such as Cd, Cr, Cu, Hg, Pb and Ni are commonly found in the soil, but geological and anthropogenic activities have increased the concentration of these HMs to the extents which are beyond the permissible limits. Excessive uptake of HMs is found toxic to living organisms posing significant environmental problem which leads to bad heath of human kinds. Some activities such as mining, smelting of metals, burning of fossil fuels, application of fertilizers and chemicals in agriculture, manufacturing of batteries and other goods produced in industries, sewage sludge and municipal waste disposal are the chief producers of HMs. HMs are deteriorated during phytoremediation; however, it is transformed from one organic molecule composite to another. Thus, changing in their oxidation state, HMs are converted to low carcinogens, easily volatilized and more water soluble (Wang et al. 1989). A large number of microorganisms especially rhizobacteria such as *Bacillus subtilis*, *P. putida* and *Enterobacter cloacae* are being used for the reduction of Cr(VI) to Cr(III) which is less toxic (van der Lelie et al. 1999; Haja et al. 2010). *B. subtilis* has been involved in the reduction of nonmetallic elements such as toxic selenite to less toxic Se (Garbisu et al. 1995). Another instance, *B. cereus* and *B. thuringiensis* enhance the ability of extraction of Cd and Zn from Cd-rich soil and soil polluted with garbage and effluent from metal industry (Ruggiero et al. 2000). It is, therefore, surmised that siderophore production by rhizobacteria has provided the avenues for the extraction of these HMs from the soil ecosystem. This is what siderophore productions are found to play a pivotal role in the accumulation HMs (Von Gunten and Benes 1995). In addition, siderophore production by *A. vinelandii* was markedly enhanced in the presence of Zn(II). Plant growth-promoting rhizobacteria are able to play a significant role in providing the assistance for the phytoremediation of HMs from contaminated soils. Therefore, HMs influence the role of bacteria-producing siderophore which in turn help in the mobilization and extraction of HMs from soil. Siderophores have the ability to resolve these environmental issues such as accumulation of heavy metal from various industries.

Moreover, siderophores are used in the treatment of radioactive waste before long storage (Von Gunten and Benes 1995; Bouby et al. 1998). Some fungi, like *Fusarium* sp., and bacteria, *P. aeruginosa*, are rich in production of siderophores which are able to modify the pH and maximize the chelation of some elements such as uranium (U^{6+}) and thorium (Th^{4+}) (Joshi et al. 2014).

15.8 Mechanisms for Siderophore-Mediated Iron Transport

Microorganisms catch up iron with the help of iron chelator molecules that fulfil the demands of needy plants. To send iron into the cellular machinery, bacteria trap iron-loaded siderophores at the surface of the cell and push them to enter into the cytosol. Siderophore-binding affinities for Fe(III) are extremely high in bacteria which illustrates that these organic molecules can significantly catch up the Fe(III) from a wide range of environment (Stintzi et al. 2000; Bernd and Rehm 2008). To explore the ferric–siderophore complex mechanism and how iron gets trapped by siderophore-producing microorganism, an outline has been presented. Initially, receptors present at the outer membrane specifically bind ferric siderophore and transport them into the periplasm. Thereafter, a system which is basically composed of protein Ton B transduces the energy from proton force into transport-proficient structural changes of the receptor. Lastly, one specific protein present in the periplasm helps in transferring the iron into transporter molecules associated with the cytoplasmic membrane (Fig. 15.1; Sah and Singh 2015). ABC transporter is made up of a protein channel in the membrane of the cytoplasm coupled with a cytoplasmic ATPase which determines ferric siderophore internalization at the expense of cytoplasmic ATP hydrolysis. ABC transporter complex is composed of two distinct proteins, each one has its own function. For instance, the first one separates the membrane which acts as permease and the second one provides energy for transport via hydrolysis reaction. There are certain different transmembrane permeases such as Fhu B (hydroxamate), FepD4 (enterobactins) and Fec CA (ferric dicitrate).

The ferric–siderophore complex is released at the specific site of the cytoplasmic membrane from its vehicle/transport system through reduction reaction. There is then ligand exchange on the cell surface which involves the exchange of iron from ferric pyoverdine to iron-free pyoverdine which is tightly bound with the receptor of FpvA (Schalk et al. 2011).

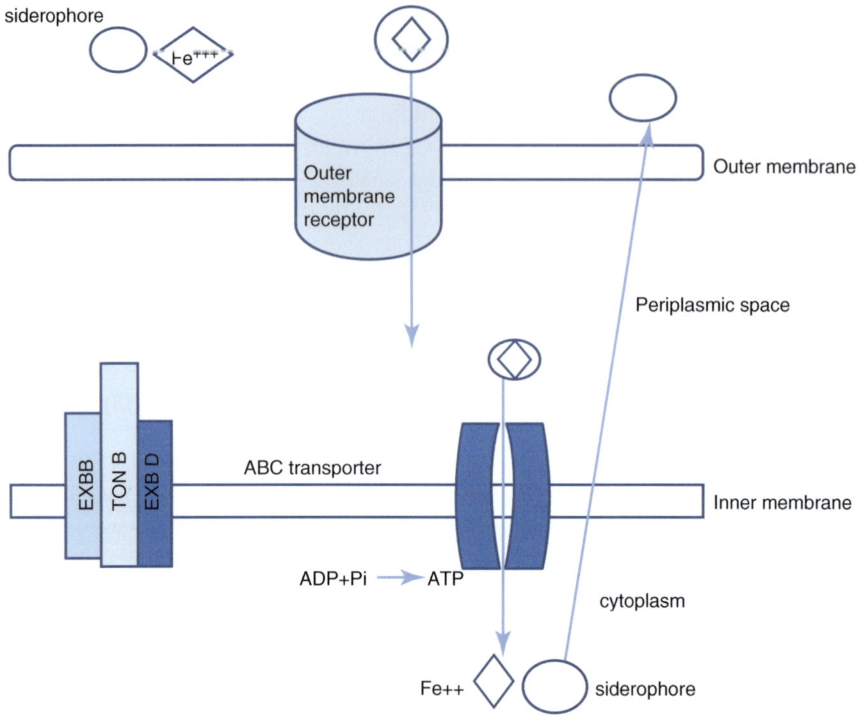

Fig. 15.1 Pathway of iron transport across the outer and inner membrane (Source: Sah and Singh 2015)

15.9 Conclusions and Future Perspective

The information pertaining to siderophore production have suggested that newer avenues related to maximization of siderophore production are needed to be explored. Application of siderophore-producing microorganism has played a pivotal role in maintaining the ecological balance. These microbes have provided a new vista of research towards the utilization of microbes for plant disease protection, plant growth promotion, SAR induction, environmental research and maintenance of soil health. Siderophores have also accelerated the production of many phytohormones such as IAA leading to induction of SAR and growth promotion. It is also summarized that there are a lot of variations in the siderophore-binding affinities which may be due to structural differentiations. However, this variation enables siderophores to quench iron from soil and mobilize them to a specific target. Siderophore-producing microorganisms containing extremely high binding affinities for iron are ecologically sound communities. Therefore, such communities may be determinant of better plant growth. Information pertaining to maintenance of soil health revealed that the contaminants are reduced and less toxic in the siderophore-producing-rich microorganisms. This organic molecule has a significant role in the purification of HM-polluted soil. Environmental research is a separate segment of thrust area of research where it has a wide range of applicability, for example, removal of HMs, purification of oceanic contaminants, elimination of algal bloom, etc. Overall, application of siderophores is conducive to the human welfare as well as in the sustenance of ecological balance. Further emphasis just to promote the siderophore production will open new door for researchers leading to resolve the "yet to be answered" questions.

References

Ahmed E, Holmstrom SJ (2014) Siderophores in environmental research: roles and applications. Microb Biotechnol 7(3):196–208

Ahmed E, Holmström SJ (2015) Microbe–mineral interactions: the impact of surface attachment on mineral weathering and element selectivity by microorganisms. Chem Geol 403:13–23

Andrews SC (1998) Iron storage in bacteria. Adv Microb Physiol 40:281–351

Arora NK, Kang SC, Maheshwari DK (2001) Isolation of siderophore producing strains of *Rhizobium meliloti* and their biocontrol potential against *Macrophomina phaseolina* that causes charcoal rot of groundnut. Curr Sci 81:673–677

Bai YM, D'Aoust F, Smith DL, Driscott BT (2002) Isolation of plant growth-promoting *Bacillus* strains from soybean nodules. Can J Microbiol 48:230–238

Balagurunathan R, Radhakrishnan M (2007) Microbial Siderophores-gateway for iron removal Envis Centre Newsletter. http://www.envismadrasuniv.org/nl20007articles%20sidephore.html

Bar-Ness E, Chen Y, Hadar Y, Marschner H, Rőmheld V (1991) Siderophores of *Pseudomonas putida* as an iron source for dicot and monocot plants. Plant Soil 130:231–241

Becker JO, Cook RJ (1988) Role of siderophores in suppression of Pythium species and production of increased growth response of wheat by fluorescent pseudomonads. Phytopathol 78:778–782

Becker M, Asch F (2005) Iron toxicity in rice-conditions and management concepts. J Plant Nutr Soil Sci 168:558–573

Bernd H, Rehm A (2008) Biotechnological relevance of pseudomonads. In: Bernd H, Rehm A (eds) *Pseudomonas*. Model organism, pathogen, cell factory. Wiley-VCH Verlag GmbH and Co. KGaA, Weinheim, p 377

Bhattacharyya PN, Jha DK (2012) Plant growth-promoting rhizobacteria (PGPR): emergence in agriculture. World J Microbiol Biotechnol 28:1327–1350

Bouby M, Billard I, Maccordick J (1998) Complexation of Th (IV) with the siderophore pyoverdine A. J Alloys Compd 273:206–210

Braud AM, Hubert M, Gaudin P, Lebeau T (2015) A quick rhizobacterial selection tests for the remediation of copper contaminated soils. J Appl Microbiol 119(2):435–445

Budzikiewicz H (2010) Siderophores from bacteria and from fungi. In: Cornelis P, Andrews SC (eds) Iron uptake and homeostasis in microorganisms. Caister Academic, Norfolk, pp 1–16

Buss HL, Luttg A, Brantley SL (2007) Etch pit formation on iron silicate surfaces during siderophore-promoted dissolution. Chem Geol 240:326–342

Buysens S, Heungens K, Poppe J, Hofte M (1996) Involvement of Pyochelin and pioverdin in suppression of *Pseudomonas aeruginosa* 7NSK2. Appl Environ Microbiol 62(3):865–871

Chakraborty U, Purkayastha RP (1984) Role of rhizobiotoxine in protecting soybean roots from *Macrophomina phaseolina* infection. Can J Microbiol 30:285–289

Challis G, Hopwood D (2003) Synergy and contingency as driving forces for the evolution of multiple secondary metabolite production by *Streptomyces* species. Proc Natl Acad Sci 25:14555–14561

Crecchio C, Curci M, Pizzigallo MDR, Ricciuti P, Ruggiero P (2004) Effects of municipal solid waste compost amendments on soil enzyme activities and bacterial genetic diversità. Soil Biol Biochem 36:1595–1605

Crichton RR, Charloteaux-Wauters M (1987) Iron transport and storage. Eur J Biochem 164:485–506

Crowley DE, Kraemer SM (2007) Function of siderophores in the plant rhizosphere. In: Pinton R et al (eds) The rhizosphere, biochemistry and organic substances at the soil-plant interface. CRC Press, Boca Raton, pp 73–109

Crowley DE, Reid CPP, Szaniszlo PJ (1988) Utilization of microbial siderophores in iron acquisition by oat. Plant Physiol 87:685–688

D'Onofrio A, Crawford JM, Stewart EJ, Witt K, Gavrish E, Epstein S, Clardy J, Lewis K (2010) Siderophores from neighboring organisms promote the growth of uncultured bacteria. Chem Biol 17:254–264

Dahsti N, Zhang F, Hynes R, Smith DL (1998) Plant growth promoting rhizobacteria accelerate nodulation and increase nitrogen fixation activity by field grown soybean under short season conditions. Plant Soil 200:205–213

Desai A, Archana G (2011) Role of siderophores in crop improvement. In: Maheshwari DK (ed) Bacteria in agrobiology: plant nutrient management. Springer, Berlin/Heiderlberg, pp 109–137

Deshwal VK, Dubey RC, Maheshwari DK (2003) Isolation of plant growth promoting strains of *Bradyrhizobium* (Arachis) sp. with biocontrol potential against *Macrophomina phaseolina* causing charcoal rot of peanut. Curr Sci 84:443–448

Di Francesco A, Martini C, Mari M (2016) Biological control of postharvest diseases by microbial antagonists: how many mechanisms of action? Eur J Plant Pathol 145:711–718

Diels L, Van Roy S, Taghavi S, Van Houdt R (2009) From industrial sites to environmental applications with *Cupriavidus metallidurans*. Antonie Van Leeuwenhoek 96(2):247–258

Dimkpa CO, Merten D, Svatoš A, Büchel G, Kothe E (2009) Siderophores mediate reduced and increased uptake of cadmium by *Streptomyces tendae* F4 and sunflower (*Helianthus annuus*), respectively. J Appl Microbiol 107(5):1687–1696

Eissendle M, Oberegger H, Zadra I, Hass H (2003) The siderophore system is essential for viability monooxygenase (Sid A) and a non-ribosomal peptide synthesis (Sid C). Mol Microbiol 49:359–375

Elad Y, Baker R (1985) The role of competition for iron and carbon in suppression of chlamydospore germination of *Fusarium* spp. by *Pseudomonas* spp. Phytopathol 75:1053–1059

Garbisu C, González S, Yang WH (1995) Physiological mechanisms regulating the conversion of selenite to elemental selenium by *Bacillus subtilis*. Biofactors 5(1):29–37

Glick BR (2012) Plant growth-promoting bacteria: mechanisms and applications. Scientifica 2012:963401, 15 p

Haas H (2003) Molecular genetics of fungal siderophore biosynthesis and uptake: the role of siderophores in iron uptake and storage. Appl Microbiol Biotechnol 62(4):316–330

Haja AR, Mohideena V, Thirumalai A, Narayananb KR, Zahir Hussaind MI (2010) Bioremediation of heavy metal contaminated soil by the exigobacterium and accumulation of Cd, Ni, Zn and Cu from soil environment. Int J Biol Technol 1(2):94–101

Hamdan H, Weller D, Thomashow L (1991) Relative importance of fluorescens siderophores and other factors in biological control of *Gaeumannomyces graminis* var. Tritici by *Pseudomonas fluorescens* 2-79 and M4-80R. Appl Environ Microbiol 57(11):3270–3277

Harrison F, Paul J, Massey R, Buckling A (2008) Interspecific competition and siderophore- mediated cooperation in *Pseudomonas aeruginosa*. ISME J 2:49–55

Hershko C, Link G, Konijn AM (2002) Cardioprotective effect of iron chelators. In: Hershko C (ed) Iron chelation therapy. Kluwer Academic/Plenum Publishers, New York, pp 77–89

Heymann P, Ernst JF, Winkelmann G (2000) A gene of the major facilitator superfamily encodes a transporter for enterobactin (Enb1p) in *Saccharomyces cerevisiae*. Biometals 13:65–72

Hider RC, Kong X (2010) Chemistry and biology of siderophores. Nat Prod Rep 27(5):637–657

Hordt W, Rőmheld V, Winkelmann G (2000) Fusarinines and dimerum acid, mono- and dihydroxamate siderophores from *Penicillium chrysogenum*, improve iron utilization by strategy I and strategy II plants. Biometals 13:37–46

Husen E (2003) Screening of soil bacteria for plant growth promotion activities in vitro. Indones J Agric Sci 4(1):27–31
Indiragandhi P, Anandham R, Madhaiyan M, Sa TM (2008) Characterization of plant growth-promoting traits of bacteria isolated from larval guts of diamondback moth *Plutella xylostella* (Lepidoptera: Plutellidae). Curr Microbiol 56:327–333
Ishimaru Y, Takahashi R, Bashir K, Shimo H, Senoura T, Sugimoto K, Ono K, Yano M, Ishikawa S, Arao T, Nakanishi H, Nishizawa NK (2012) Characterizing the role of rice in manganese, iron and cadmium transport. Sci Rep 2:286
Joshi F, Chaudhary A, Joglekar P, Archana G, Desai AJ (2008) Effect of expression of *Bradyrhizobium japonicum* 61A152 fegA gene in *Mesorhizobium* sp., on its competitive survival and nodule occupancy on Arachis hypogeal. Appl Soil Ecol 40:338–347
Joshi H, Dave R, Venugopalan VP (2014) Pumping iron to keep fit: modulation of siderophore secretion helps efficient aromatic utilization in *Pseudomonas putida* KT2440. Microbiol 160:1393–1400
Kannahi M, Senbagam N (2014) Studies on siderophore production by microbial isolates obtained from rhizosphere soil and its antibacterial activity. J Chem Pharma Res 6(4):1142–1145
Khamna S, Yokota A, Lumyong S (2009) Actinomycetes isolated from medicinal plant rhizosphere soils: diversity and screening of antifungal compounds, indole-3-acetic acid and siderophore production. World J Microbiol Biotechnol 25:649–655
Khan MS, Zaidi A, Wani PA, Oves M (2009) Role of plant growth promoting rhizobacteria in the remediation of metal contaminated soils. Environ Chem Lett 7:1–19
Kiss T, Farkas E (1998) Metal-binding ability of desferrioxamine B. J Incl Phenom Macrocycl Chem 32:385–403
Kloepper JW, Leong J, Tientze M, Schroth MN (1980) *Pseudomonas* siderophores: a mechanism explaining disease-suppressive soils. Curr Microbiol 4:317–320
Konigsberger LC, Konigsberger E, May PM, Hefter GT (2000) Complexation of iron (III) and iron (II) by citrate. Implications for iron speciation in blood plasma. J Inorg Biochem 78:175–184
Kraemer SM (2004) Iron oxide dissolution and solubility in the presence of siderophores. Aquat Sci 66:3–18
Kraemer SM, Crowley D, Kretzschmar R (2006) Siderophores in plant iron acquisition: geochemical aspects. Adv Agron 91:1–46
Kurth C, Kage H, Nett M (2016) Siderophores as molecular tools in medical and environmental applications. Org Biomol Chem. doi:10.1039/C6OB01400C
Loaces I, Ferrando L, Fernández Scavino A (2011) Dynamics, diversity and function of endophytic siderophore-producing bacteria in rice. Microb Ecol 61:606–618
Loper JE (1988) Role of fluorescent siderophore production in biological *control of Pythium ultimum by a Pseudomonas fluorescens strain.* Phytopathol 78:166–172
Loper JE, Henkel MD (1999) Utilization of heterologous siderophore enhances levels of iron available to *Pseudomonas putida* in rhizosphere. Appl Environ Microbiol 65(12):5357–5363
Lugtenberg B, Kamilova F (2009) Plant-growth-promoting rhizobacteria. Annu Rev Microbiol 63:541–556
Ma Y, Prasad MNV, Rajkumar M, Freitas H (2011) Plant growth promoting rhizobacteria and endophytes accelerate phytoremediation of metalliferous soils. Biotechnol Adv 29(2):248–258
Ma Y, Rajkumar M, Zhang C, Freitas H (2016) Beneficial role of bacterial endophytes in heavy metal phytoremediation. J Environ Manag 174:14–25
Maheshwari DK (2011) Plant growth promoting rhizobacteria: fundamentals and applications. In: Maheshwari DK (ed) Plant growth and health promoting bacteria. Springer, Berlin/Heidelberg, pp 21–42
Manwar AV, Khandelwal SR, Chaudhari BL, Kothari RM, Chincholkar SB (2001) Generic technology for assured biocontrol of groundnut infections leading to its yield improvement. Chem Weekly 46(26):157–158
Matzanke BF, Muller-Matzanke G, Raymond KN (1989) Siderophore-mediated iron transport. In: Loehr TM (ed) Iron carriers and iron proteins, Physical bioinorganic chemistry, vol 5. VCH Publishers, New York, pp 1–121

Maurhofer M, Hase C, Meuwly P, Métraux J-P, Défago G (1994) Induction of systemic resistance of tobacco to tobacco necrosis virus by the root-colonizing *Pseudomonas fluorescens* strain CHA0: influence of the gacA gene and of pyoverdine production. Phytopathol 84:139–146

McGrath SP, Chaudri AM, Giller KE (1995) Long-term effects of metals in sewage sludge on soils, microorganisms and plants. J Ind Microbiol 14(2):94–104

Meneely KM, Lamb AL (2007) Biochemical characterization of an FAD-dependent monooxygenase, the ornithine hydroxylase from *Pseudomonas aeruginosa*, suggests a novel reaction mechanism. Biochemist 46:11930–11937

Meyer JM (2000) Pyoverdines: pigments, siderophores and potential taxonomic markers of fluorescent *Pseudomonas* species. Arch Microbiol 174(3):135–142

Miethke M, Marahiel MA (2007) Siderophore-based iron acquisition and pathogen control. Microbiol Mol Biol Rev 71:413–451

Munees A, Mulugeta K (2014) Mechanisms and applications of plant growth promoting rhizobacteria. Curr Perspec J King Saud Uni Sci 26:1–20

Münzinger M, Taraz K, Budzikiewicz H (1999) Staphyloferrin B, a citrate siderophore of Ralstonia eutropha. Z Naturforsch C 54(11):867–875

Nair A, Juwarkar AA, Devotta S (2008) Study of speciation of metals in an industrial sludge and evaluation of metal chelators for their removal. J Hazard Mater 152(2):545–553

Neilands JB (1995) Siderophores: structure and function of microbial iron transport compounds. J Biol Chem 270:26723–26726

Neilands JB, Leong SA (1986) Siderophores in relation to plant growth and disease. Annu Rev Plant Physiol 37:187–208

Neubauer U, Furrer G, Kayser A, Schulin R (2000) Siderophores, NTA, and citrate: potential soil amendments to enhance heavy metal mobility in phytoremediation. Int J Phytoremediation 2:353–368

Powell PE, Cline GR, Reid CPP, Szaniszlo PJ (1980) Occurrence of hydroxamate siderophore iron chelators in soils. Nature 287:833–834

Raaijmakers JM, Leeman M, Van Oorschot MPM, Van der Sluis I, Schippers B, Bakker PAHM (1995) Dose-response relationships in biological control of fusarium wilt of radish by *Pseudomonas* spp. Phytopathol 85:1075–1081

Rajkumar M, Ae N, Prasad MNV, Freitas H (2010) Potential of siderophore-producing bacteria for improving heavy metal phytoextraction. Trends Biotechnol 28(3):142–149

Rajkumar M, Freitas H (2008) Influence of metal resistant-plant growth promoting bacteria on the growth of *Ricinus communis* in soil contaminated with heavy metals. Chemosphere 71:834–842

Rao DLN, Pal KK (2003) Biofertilizers in oilseeds production: status and future strategies. National Seminar on Stress Management in Oilseeds for attaining self-reliance in vegetable oils. Directorate of Oilseeds Research. Indian Council of Agricultural research, Hyderabad, India. pp 195–220

Raymond KN, Emily AD, Sanggoo SK (2003) Enterobactin: an archetype for microbial iron transport. Proc Natl Acad Sci U S A 100(7):3584–3588

Reichard PU, Kraemer SM, Frazier SW, Kretzschmar R (2005) Goethite dissolution in the presence of phytosiderophores: rates, mechanisms, and the synergistic effect of oxalate. Plant Soil 276:115–132

Rizvi R, Ansari RA, Iqbal A, Ansari S, Sumbul A, Mahmood I, Tiyagi SA (2015) Dynamic role of organic matter and bioagent for the management of *Meloidogyne incognita*–*Rhizoctonia solani* disease complex on tomato in relation to some growth attributes. Cogent Food Agric 1(1):1068523

Römheld V, Marschner H (1986) Mobilization of iron in the rhizosphere of different plant species. In: Tinker B, Laüchi A (eds) Advances in plant nutrition, vol 2. Greenwood publishing, New York, pp 155–204

Ruggiero CE, Neu MP, Matonic JH, Reilly SD (2000) Interactions of Pu with desferrioxamine siderophores can affect bioavailability and mobility. Actinide Research Quarterly, 2nd/3rd Quarter, pp 16–18

Sah S, Singh R (2015) Siderophore: structural and functional characterization–a comprehensive review. Agric (Polnohospodárstvo) 61(3):97–114

Scavino AF, Pedraza RO (2013) The role of siderophores in plant growth-promoting bacteria. In: Maheshwari DK, Saraf M, Aeron A (eds) Bacteria in agrobiology: crop productivity. Springer, Berlin/Heidelberg, pp 265–285

Schalk IJ, Hannauer M, Braud A (2011) Mini review new roles for bacterial siderophores in metal transport and tolerance. Environ Microbiol 13:2844–2854

Schippers B, Bakker AW, Bakker PAHM (1987) Interactions of deleterious and beneficial rhizosphere microorganisms and the effect of cropping practices. Annu Rev Phytopathol 25:339–358

Schmidt W (1999) Mechanisms and regulation of reduction-based iron uptake in plants. New Phytol 141:1–26

Schwyn B, Neilands JB (1987) Universal chemical assay for the detection and determination of siderophores. Anal Biochem 160(1):47–56

Seuk C, Paulita T, Baker R (1988) Attributes associate with increased bio-control activity of fluorescent Pseudomonads. J Plant Pathol 4(3):218–225

Sharma A, Johri BN, Sharma AK, Glick BR (2003) Plant growth-promoting bacterium *Pseudomonas sp.* strain GRP3 influences iron acquisition in mung bean (*Vigna radiata* L. Wilzeck). Soil Biol Biochem 35(7):887–894

Sheng XF, Jiang CY, He LY (2008) Characterization of plant growth-promoting *Bacillus edaphicus* NBT and its effect on lead uptake by Indian mustard in a lead amended soil. Can J Microbiol 54(5):417–422

Shirvani M, Nourbakhsh F (2010) Desferrioxamine-B adsorption to and iron dissolution from paly-gorskite and sepiolite. Appl Clay Sci 48:393

Stintzi A, Barnes C, Xu J, Raymond KN (2000) Microbial iron transport via a siderophore shuttle: a membrane ion transport paradigm. Proc Natl Acad Sci 97(20):10691–10696

Stumm W, Morgan JJ (1995) Aquatic chemistry. Wiley-Interscience, New York

Taj ZZ, Rajkumar M (2016) Perspectives of plant growth-promoting actinomycetes in heavy metal phytoremediation. In: Plant growth promoting actinobacteria. Springer, Singapore, pp 213–231

Takemoto T, Nomoto K, Fushiya S, Ouchi R, Kusano G, Hikino H, Takagi SI, Matsuura Y, Kakudo M (1978) Structure of mugineic acid, a new amino acid possessing an iron-chelating activity from roots washings of water cultured *Hordeum vulgare* L. Proc Japan Acad Ser B Phys Biol Sci 54(8):469–473

Thompson IP, Van Der Gast CJ, Ciric L, Singer AC (2005) Bioaugmentation for bioremediation: the challenge of strain selection. Environ Microbiol 7(7):909–915

Torsvik V, Ovreas L (2002) Microbial diversity and function in soil: from genes to ecosystems. Curr Opin Microbiol 5:240–245

Trapet P, Avoscan L, Klinguer A, Pateyron S, Citerne S, Chervin C, Mazurier S, Lemanceau P, Wendehenne D, Besson-Bard A (2016) The siderophore pyoverdine weakens defense in favour of growth in iron deficient conditions. Plant Physiol 171:675–693

van der Lelie D, Corbisier P, Diels L, Gilis A, Lodewyckx C, Mergeay M, Taghavi S, Spelmans N, Vangronsveld J (1999) The role of bacteria in the phytoremediation of heavy metals. In: Terry N, Banuelos E (eds) Phytoremediation of contaminated soil and water. G Lewis Publishers, Boca Raton, pp 265–281

Vansuyt G, Robin A, Briat JF, Curie C, Lemanceau P (2007) Iron acquisition from Fe-pyoverdine by *Arabidopsis thaliana*. Mol Plant-Microbe Interact 4:441–447

Verma VC, Singh SK, Prakash S (2011) Bio-control and plant growth promotion potential of siderophore producing endophytic *Streptomyces* from *Azadirachta indica A Juss*. J Basic Microbiol 51:550–556

Von Gunten HR, Benes P (1995) Speciation of radionuclides in the environment. Radiochim Acta 69:1–29

Wang P, Mori T, Komori K, Sasatsu M, Toda K, Ohtake H (1989) Isolation and characterization of an *Enterobacter cloacae* strain that reduces hexavalent chromium under anaerobic conditions. Appl Environ Microbiol 55(7):1665–1669

Wang Q, Xiong D, Zhao P, Yu X, Tu B, Wang G (2011) Effect of applying an arsenic-resistant and plant growth promoting rhizobacterium to enhance soil arsenic phytoremediation by *Populus deltoides* LH05-17. J Appl Microbiol 111:1065–1074

Ward TR, Reas L, Serge P, Parel JE, Philipp G, Peter B, Chris O (1999) An iron-based molecular redox switch as a model for iron release from enterobactin via the salicylate binding mode. Inorg Chem 38(22):5007–5017

Weyens N, van der Lelie D, Taghavi S, Newman L, Vangronsveld J (2009) Exploiting plant microbe partnerships to improve biomass production and remediation. Trends Biotechnol 27(10):591–598

Winkelmann G (2007) Ecology of siderophores with special reference to the fungi. Biometals 20:379–392

Yadav S, Kaushik R, Saxena AK, Arora DK (2011) Diversity and phylogeny of plant growth promoting bacilli from moderately acidic soil. J Basic Microbiol 51:98–106

Zhou D, Huang XF, Chaparro JM, Badri DV, Manter DK, Vivanco JM, Guo J (2016) Root and bacterial secretions regulate the interaction between plants and PGPR leading to distinct plant growth promotion effects. Plant Soil 401(1–2):259–272

Growth Stimulation, Nutrient Quality and Management of Vegetable Diseases Using Plant Growth-Promoting Rhizobacteria

16

Almas Zaidi, Mohammad Saghir Khan, Ees Ahmad, Saima Saif, and Asfa Rizvi

Abstract

Vegetables play an important role in human nutrition. And hence, to produce quality vegetables is a major challenge for growers. In order to optimize vegetable production, growers quite often use a heavy dose of agrochemicals without considering the deleterious impact of such chemicals on vegetables. Researchers have tried to minimize the use of agrochemicals in vegetable production vis-a-vis to develop resistant varieties, but all such approaches have been unsuccessful. The excessive use of agrochemicals can be replaced by "biofertilizers" especially plant growth-promoting rhizobacteria (PGPR) for producing safe and healthy vegetables without posing any threat to the environment. Moreover, as a biocontrol agent, PGPR will be useful in the management of vegetable diseases. In this chapter, some successful stories of PGPR applications in growth stimulation of popularly grown vegetables are described. Also, the disease suppressing ability of PGPR is considered and discussed. The strategy of incorporating low cost rhizotechnology in vegetable production system is likely to reduce dependence on chemicals applied by vegetable growers.

16.1 Introduction

Vegetables play an important role in human health, and due to increasing health awareness, there is greater demand of quality vegetables. To fulfill the rising demands of consumers, growers have substantially increased the use of fertilizers to harness optimum vegetable yields (Abayomi and Adebayo 2014). Regular and unbalanced dose of fertilizers in vegetable productions (Guo et al. 2011) have,

A. Zaidi (✉) • M.S. Khan • E. Ahmad • S. Saif • A. Rizvi
Department of Agricultural Microbiology, Faculty of Agricultural Sciences, Aligarh Muslim University, Aligarh, Uttar Pradesh, India
e-mail: alma29@rediffmail.com

V. Kumar et al. (eds.), *Probiotics in Agroecosystem*,
DOI 10.1007/978-981-10-4059-7_16

however, led to decrease in soil fertility, human health (via food chain) problems, ecological risks, and poor quality and lesser vegetable yields (Olowoake and Adeoye 2010). So, to counter such destructive challenges, there is an urgent need to find an alternative option to boost the production without any threat to vegetables. In this regard, PGPR along with (Bhadoria et al. 2005)/without (Sharafzadeh 2012) fertilizers have been used against vegetables such as tomato (Ramakrishnan and Selvakumar 2012), potato (Naderi et al. 2012), cabbage (Turan et al. 2014), cucumber (Isfahani and Besharati 2012), brinjal (Fu et al. 2010), okra (Kumar et al. 2014a), onion (Reetha et al. 2014), and mint (Kaymak et al. 2008). Another important problem associated with vegetable production is the incidence of diseases such as damping-off disease of cucumber caused by *Pythium aphanidermatum* (Elazzazy et al. 2012), wilt of brinjal caused by *Ralstonia solanacearum* (Chakravarty and Kalita 2012), and tomato wilt caused by *Fusarium oxysporum* fsp. *lycopersici* (Loganathan et al. 2014), etc. Conventionally, the disease management strategies involve sanitary/cultural practices and development of resistant varieties and fungicide applications (Sharma and Saikia 2013; Sahar et al. 2013). These disease control measures have, however, not been successful. Management of plant disease using PGPR (Sang et al. 2011) is another most striking strategy by which the ill effects such as environmental pollution, residual toxicity, and fungicidal resistance (Fry and Goodwin 1997) of agrochemicals could inexpensively be avoided. Broadly, PGPR controls plant diseases by producing siderophores (Panhwar et al. 2014), antibiotics (Keel et al. 1992), HCN (Ruangsanka 2014), and lytic enzymes (Nabti et al. 2014). As an example, the bacterial wilt caused by *Ralstonia solanacearum* (Talat and Sijam 2010) has been found to deleteriously affect the brinjal production worldwide. The antagonistic *Pseudomonas fluorescens* applied as suspension dramatically decreased the severity of disease and consequently enhanced the yield attributes and other physiological and biochemical parameters of brinjal plants (Chakravarty and Kalita 2012). Even though PGPR have been found useful against a range of crops, information on production of vegetables is very limited. Considering such gaps, an attempt is made to identify PGPR with multiple growth-promoting activities for ultimate use in vegetable production.

16.2 Vegetables and Human Health

Broadly, vegetables (Fig. 16.1) are fresh and edible portions (such as roots, stems, leaves, fruits, or seeds) of plants which play a significant role in human health (Alertor et al. 2002). Vegetables contain many important ingredients such as vitamins, proteins, carbohydrates, minerals, phenolics, flavonoids, riboflavin, carotenoids, antioxidant enzymes, and dietary fiber (Table 16.1).

Vegetables also provide micronutrients (Table 16.2) and appreciable amounts of phytochemicals. The phytochemicals of *Brassica* vegetables, for example, have been reported to play major part in (i) prevention of oxidative stress, (ii) detoxification of enzymes, (iii) stimulation of immune system, (iv) decrease in the risk of cancers, (v) inhibition of malignant transformation and carcinogenic mutations, and

Fig. 16.1 Commonly used vegetables

Table 16.1 Nutrient composition of some selected vegetables

Vegetables		Nutrient composition (g/100 g)							
Common name	Botanical name	Carbohydrate	Protein	Fats	Moisture	Dietary fiber	Sugars	Ash	Energy (kcal)
Potato	*Solanum tuberosum*	17.47	2.0	0.10	77	2.2	15.44	0.9	81
Tomato	*Lycopersicum esculentum*	3.9	0.9	0.2	94	1.2	2.6	0.9	23
Cabbage	*Brassica Oleracea*	5.80	1.28	0.10	92	2.5	3.20	0.6	25
Broccoli	*Brassica Oleracea*	6.64	2.82	0.37	89.3	2.6	1.7	NA	34
Spinach	*Spinacia oleracea*	4.0	2.1	0.38	91	0.6	0.4	1.1	27
Cauliflower	*Brassica Oleracea* cv Botrytis	5.0	1.9	0.3	92	2.0	1.9	–	25
Okra	*Abelmoschus esculentus* Moench	7.45	2.0	0.19	90	3.1	1.48	NA	33
Onion	*Allium cepa*	9.34	1.1	0.1	89.1	1.7	4.24	NA	40
Brinjal	*Solanum melongena*	5.88	0.98	0.18	NA	3.0	3.53	NA	25
Lettuce	*Lactuca sativum*	3.0	1.2	0.25	93.8	0.7	0.94	0.8	17

Source: Modified from Rizvi et al. (2014), Hanif et al. (2006), and USDA Nutrient Database

(vi) reduction in proliferation of cancer cells (Hennekens 1986). Vegetables inhibit DNA methylation and limit cancer development (Kapusta-Duch et al. 2012). The regular use of vegetables helps to protect humans from esophageal, stomach, pancreatic, bladder, and cervical cancers (Crawford et al. 1994).

Table 16.2 Major and micronutrients found in some commonly used vegetables

	Macro and Micronutrients (mg/100 g)						
Vegetables	Ca	P	Na	K	Cr	Fe	Zn
Potato	12	57	6.0	421	0.007	0.78	0.29
Tomato	13	24	44	237	0.005	0.07	NA
Broccoli	47	66	33	316	NA	0.73	0.41
Cauliflower	22	44	30	299	NA	0.42	0.27
Spinach	99	49	79	558	0.005	2.71	0.53
Okra	82	NA	NA	299	NA	0.61	0.58
Lettuce	35	33	5.0	238	0.005	1.24	0.2
Onion	23	29	NA	146	NA	0.21	0.17
Reddish	25	20	63.9	233	0.008	0.34	0.28
Brinjal	9.0	24	NA	229	NA	0.23	0.16

Source: Modified from Rizvi et al. (2014), Hanif et al. (2006) and USDA Nutrient Database

16.3 Importance of Plant Growth Promoting Rhizobacteria in Sustainable Production of Vegetables

Plant nutrients involving both major (NPK) and micronutrients play important roles in the formation of protein, nucleic acid, and chlorophyll besides affecting cell regulation, flowering and fruiting, energy transfer, maintenance of internal pressure, water potential, respiration, and enzyme action (Ahemad et al. 2009). Deficiency of multiple/even any one plant nutrient may lead to reduction in growth and yields, and in some extreme cases seizure of growth beyond seedling stage (Abd El-Salam et al. 2005). Garden and commercial growers, therefore, apply heavy amounts of fertilizers to fulfill nutrient demands of vegetables grown especially in nutrient deficient soils. In contrast, growers (via food consumption) and soil fertility suffer heavily by excessive use of fertilizers. The lethal effects resulting from higher and injudicious rates of fertilizers in vegetable cultivation thus include (i) altered compositional changes such as (a) reduced ascorbic acid (vitamin C) content, (b) lower sugar content, (c) lower acidity, and (d) variable ratios of essential amino acids; (ii) accumulation of higher level of nitrates especially in leafy vegetables; (iii) reduced volatile production; (iv) altered flavor; (v) increased glutamine levels; (vi) delayed maturity; and (vii) increased weight loss. Considering the deleterious effects of fertilizers, and challenge to produce fresh and healthy vegetables, there is an urgent need to find alternatives that could help to implement need-based nutrient management (NBNM) practices in order to achieve optimum quality vegetables without any dangerous impact of such chemicals on natural microbiota, vegetables, and environment. In this context, the application of PGPR in vegetables production strategies has become a feasible and sound option which facilitates growth by various direct (e.g., N_2 fixation, P solubilization, and phytohormones production) and indirect (producing antifungal metabolites: siderophores, HCN, etc.) mechanisms (Ahemad and Khan 2011). The growth-promoting substances involved in vegetable production synthesized by various PGPR are summarized in Table 16.3.

Table 16.3 Important bioactive compounds released by plant growth-promoting rhizobacteria

Plant growth-promoting rhizobacteria	Bioactive compounds involved in plant growth promotion	References
Pseudomonas fluorescens, P. putida, P. aeruginosa	P solubilization, IAA, gibberellic acid, siderophores, ACC deaminase, HCN, NH_3, cell wall degrading enzymes	Lukkani and Reddy (2014); Ali et al. (2014); Deshwal and Kumar (2013); Bholay et al. (2012)
Bacillus, Paenibacillus	P solubilization, IAA, gibberellic acid (GA3), siderophores, phosphatase, phytase, antifungal antibiotics, catalase	Susilowati and Syekhfani (2014); Sivasakthi et al. (2013); Chen et al. (2012); Sarvani and Reddy (2013)
Burkholderia spp.	N_2 fixation, P solubilization, IAA, siderophores, ACC deaminase, pyrrolnitrin	Nailwal et al. (2014); Martínez-Aguilar et al. (2013)
Azotobacter spp., *A. vinelandii, A. chroococcum*	N_2 fixation, P solubilization, IAA, siderophores	Kumar et al. (2014b); Nosrati et al. (2014); Kanchana et al. (2013); Farajzadeh et al. (2012)
Aeromonas, Sphingomonas, Stenotrophomonas, Achromobacter, Ewingella	P solubilization, IAA, siderophores, ACC deaminase, HCN, NH_3, catalase	Bumunang and Babalola (2014)
Micrococcus	IAA, siderophores, ACC deaminase	Dastager et al. (2010)

16.4 Examples of PGPR Effects on Some Important Vegetable Crops

Production of inexpensive and nutritionally healthy vegetables is of course a major challenge for growers around the world to fulfill the increasing demands of consumers for fresh and pollutant-free vegetables. In terms of production cost and nutritive value, consumer's preference, general adaptability, and extent of cultivation, the widely grown vegetables in different production systems across the globe are tomato, potato, broccoli, spinach, cucumber, pepper, eggplant, cabbage, onion, salad vegetables, etc. Plant growth-promoting rhizobacteria either alone or in combination with other compatible microorganisms have indeed been reported to enhance vegetable production. Realizing the importance of PGPR as an alternative to synthetic chemical fertilizers in vegetables production and experimental reports available so far on the growth stimulation by PGPR, an attempt is made here in the following section to highlight the impact of PGPR on some vegetables grown in different production systems. However, since most of the studies on the role of PGPR in vegetable production have been conducted on tomato and potato, these crops are not included in the present review. Instead, some of the other most commonly consumed vegetables worldwide are considered and discussed.

16.4.1 Broccoli (*Brassica oleracea*)

Broccoli is an edible green plant in the cabbage family (*Brassicaceae*) whose large flowering head is eaten as a vegetable. It is an important winter season vegetable cultivated widely in many European and American countries. For enhancing the growth, yield, and head quality of broccoli, higher rates of P are applied (Brahma and Phookan 2006). In order to reduce the usage of fertilizers, combinations of three levels (50, 75, and 100 kg P_2O_5/fed) of P fertilizers and two biophosphorus fertilizer have been used to inoculate broccoli plants (Abou El-Magd et al. 2013). Gradual increase in P from 50 to 100 kg P_2O_5/fed consistently increased plant height, number of leaves per plant, and fresh weight of leaves, stem, and main head of broccoli plants. Generally, the yield and quality were enhanced by increasing the level of P. Moreover, the phosphorin in the presence of mineral P showed maximum positive synergistic impact on vegetative growth, head yield, and quality of broccoli heads compared to sole application of any fertilizer level or single application of phosphorin. In a study, Yildirim et al. (2011) investigated the effects of root inoculations with *B. cereus* (N_2-fixing), *Brevibacillus reuszeri* (P-solubilizing), and *Rhizobium rubi* (both N_2-fixing and P-solubilizing) on growth, nutrient uptake, and yield of broccoli, grown in field soils and treated with manure and some fertilizers. Bacterial inoculations with manure significantly increased the yield, plant weight, head diameter, chlorophyll content, N, K, Ca, S, P, Mg, Fe, Mn, Zn, and Cu contents of broccoli over control. Among different treatments, manure with *B. cereus*, *R. rubi*, and *B. reuszeri* increased the yield by 17%, 20.2%, and 24.3% and chlorophyll content by 14.7%, 14%, and 13.7%, respectively, over control. It was suggested from this study that seedling inoculation with P-solubilizing (*B. reuszeri*) and both N_2-fixing and P-solubilizing (*R. rubi*) could be employed as an alternative to partially reduce the use of costly fertilizers in broccoli production.

16.4.2 Spinach (*Spinacia oleracea*)

Spinach grown mainly for its foliage is an annual member of Chenopodiaceae family and a valuable vegetable with low calories but serves as a good source of vitamin C, vitamin A, and minerals like iron. The use of biofertilizers has been found as a cheap source for supplying spinach plants with N and P during growth. For example, *P. putid*, *P. fluorescence*, *Vibrio fluvialis*, and *Ewingella americana* with varying PGP activities increased spinach heights ranging between 17.14% and 21.43% over control plants (Hou and Oluranti 2013). Similarly, the single/combined effect of *A. chroccocum* and phosphorein, with varying rates of N and P fertilizers on growth, yield, sex ratio, seeds (yield and quality) of spinach plants cv. *Dokki*, was evaluated (El-Assiouty and Abo-Sedera 2005). Seed inoculation with 300 g phosphorein inoculum/fed along with 40 kg N/fed (100% of the recommended N dose) + 15 or 7.5 kg/fed (66.7 or 33% of the recommended dose of P_2O_5) and seed inoculation with 300 g *Azotobacter* inoculum in the presence of the full dose of P_2O_5 (22.5 kg P_2O_5/fed.) + 50% of the full dose of N (20 kg/fed) gave the highest favorable effect on growth, yield, sex ratio, and higher seed yield with the best quality compared with control. Nitrogen at 40 kg/fed along with 15 kg P_2O_5 and 300 g phosphorein increased plant fresh yield by 27.2% and 42.3%, while seed yield was enhanced by 16.3% and 10.4% in the first and second seasons, respectively, over control. In other experiment, Çakmakç et al. (2007) investigated the effect of N_2-fixing, phytohormone-producing, and P-solubilizing bacteria on growth and enzymes of spinach. The bacterial cultures included in this study were *B. cereus*, *B. licheniformis*, *B. megaterium*, *B. subtilis*, *Bacillus* (OSU-142), *Bacillus* (M-13), *P. putida*, and *Paenibacillus polymyxa*. Inoculation with PGPR increased fresh weight of shoots by 2.2–53.4%, leaf area by 5.3–49.3%, and plant height by 1.9–36.8% over control. Furthermore, a close relationship existed between plant growth and enzyme activities. Sreedevi et al. (2014) also evaluated the impact of PSB alone or in combination with FYM and citrate, on physico-chemical characteristics of spinach. Among all treatments, the sole application of (i) FYM and its combination with (ii) PSB and (iii) PSB + citrate performed better and significantly enhanced the yield weight, vitamin C, β-carotene, minerals, and antinutritional factors relative to 100% recommended dose of fertilizer.

16.4.3 Cauliflower (*Brassica oleracea* L. var. Botrytis)

Cauliflower is one of the most important commercial vegetables belonging to the family *Brassicaceae* and is grown worldwide. Cauliflower is grown round the year for its white and tender curd. It is widely cultivated for its nutritive values, high productivity, and wider adaptability under different ecological conditions. In India, it is grown in area of 248.3 × 10^3 ha with production of 4714.8 × 10^3 t. Cauliflower production demands constant supply of large amounts of major nutrients (NPK) and water for its growth. Currently, the soil fertility is, however, deteriorating rapidly due to frequent and heavy use of chemical fertilizers (Savci 2012). Cauliflower on the other hand requires the use of balanced fertilizer and if not adequately fertilized, considerable yield losses are reported (Prabhakar and Srinivas 1995). It is, therefore, required to reduce the usage of heavy dose of chemical fertilizers in cauliflower production. This can be achieved by applying biofertilizers (Manisha and Korla 2009) and some other organic manures. Although there are many studies that highlight the effect of PGPR on plant growth and yield of some vegetables (Kalita et al. 2015), there is little research on the effect of PGPR on transplant growth and quality of cauliflower. However, in some studies the increase in curd size and yield of cauliflower due to application of biofertilizers along with inorganic fertilizers is reported (Kaushal et al. 2011; Shree et al. 2014). In order to make cauliflower cultivation sustainable and less dependent on chemical fertilizers, there is a need to identify PGPR with multiple plant growth-promoting activities such as they should be able to fix N, solubilize P, and secrete phytohormones that could contribute to the higher production of cauliflower. Considering these, an attempt is made in the following section to highlight the role of PGPR in sustainable production of cauliflower.

The effect of different PGPR strains, for example, *Bacillus megaterium* (TV-3D), *B. megaterium* (TV-91C), *Pantoea agglomerans* (RK-92), *B. subtilis* (TV-17C),

B. megaterium (TV-87A), *B. megaterium* (KBA-10), on growth and quality of cauliflower transplants has been found variable. Such bacterial inoculations when tested under greenhouse conditions increased plant growth parameters such as fresh shoot weight, dry shoot weight, root diameter, root length, fresh root weight, dry root weight, plant height, stem diameter, leaf area, and chlorophyll contents of cauliflower transplant, respectively (Ekinci et al. 2014). Moreover, the concentrations of gibberellic acid (GA), salicylic acid (SA), and IAA were increased by 24%, 90%, and 26%, respectively, compared to control following application of strains KBA-10 and RK-92. The PGPR inoculations also increased the macro- and micronutrient content of cauliflower transplants. Similarly, biochemical estimation of plants treated with consortium prepared from *B. cereus* (MTCC 8297), *Pseudomonas rhodesiae* (MTCC 8299), and *P. rhodesiae* (MTCC 8300) showed high nutrient content such as carbohydrate, protein, lipid, and total amino acid in treated cauliflower (Kalita et al. 2015). In other study, the PGPR isolated from cauliflower rhizosphere induced the production of IAA and could solubilize P. The mixture of PGPR and fertilizers (N and P) significantly increased the number of nonwrapper leaves, curd diameter, curd depth and weight, and yield of cauliflower (Kaushal et al. 2011). Furthermore, four biofertilizers (*Azospirillum*, *Azotobacter*, PSB, and VAM) and two levels (75% and 100%) of N and P of recommended dose of NPK (120 : 60 : 60 kg/ha) in a study were applied to evaluate the growth, yield, and quality parameters of cauliflower. The application of *Azospirillum* with recommended dose of NPK significantly increased plant height, number of leaves/plant, gross weight of plants (without root), average weight of curd, and yield of cauliflower (Singh and Singh 2005). In a field experiment, Bashyal (2011) assessed the response of cauliflower to biofertilizer containing free living nitrogen fixing bacteria *Azospirillum* and *Azotobacter* and different levels of N. Application of N along with the biofertilizer significantly increased morphological, yield, and quality characters compared to application of N alone. The maximum stem height, stem diameter, highest curd height, curd diameter, fresh curd weight, and curd yield were recorded when cauliflower was grown with 120 kg N and 2 kg biofertilizer ha^{-1}. However, cauliflower curd yield recorded at 120 kg N ha^{-1} did not differ significantly with the curd yield obtained at 60 kg N and 2 kg biofertilizer ha^{-1}. The curd initiation and maturity occurred earlier when cauliflower was grown with 30 kg N and 2 kg biofertilizer ha^{-1}. The highest vitamin C content of curds and the most attractive curd color were recorded at 60 kg N and 2 kg biofertilizer ha^{-1}, while the appearance and over all acceptability were recorded at 120 kg N and 2 kg biofertilizer ha^{-1}. Conclusively, the present finding resulted in saving of 60 kg N ha^{-1} without significantly affecting the yield of cauliflower.

16.4.4 Okra (*Abelmoschus esculentus* Moench.)

Okra is yet another important vegetable grown worldwide during summer/rainy season and provides higher amounts of carbohydrates, fats, protein, minerals, and vitamins. An integrated approach involving bioinoculants/bioagents and fertilizers has been employed for okra production (Singh et al. 2010b). For example, Bhushan et al. (2013) evaluated the effect of asymbiotic N_2 fixing *Azotobacter* and fertilizers on growth, fruit, and seed yield of okra cv. Hisar Unnat. Among all treatments, *Azotobacter* in the presence of 50% NPK produced maximum green fruit yield (18,300 kg/ha) and seed yield (3490 kg/ha). Also, a significant increase in plant height (157.2 cm), number of branches/plant (2.2), number of nodes/plant (19.6), number of fruits/plant (14.3), number of pods/plant (14.2), pod weight (13.9 g), number of seeds/pod (57.5), and seed weight/pod (6.1 g) were recorded due to combined application of *Azotobacter* and 50% NPK. The combined effects of biofertilizers, for instance, *Azotobactor*, *Azospirillum*, PSB, and fertilizers such as full dose of N, potash, and half dose of P (Sahu et al. 2014) along with vermicompost, resulted in significantly vigorous growth and also increased yield of okra (Mal et al. 2013). The tested biofertilizers in the presence of FYM (10 t ha^{-1}), NPK (100%) and vermicompost (5 t ha^{-1}) showed maximum increase in plant height (148.97 cm), leaf area (434.99 cm^2), number of nodes (30.16), fruit length (16.45 cm), fruit girth (1.62 cm), single fruit weight (18.7 g), and plant biomass-fresh weight (548.74 q ha^{-1}). The maximum number of fruits per plant on the contrary was recorded for treatment containing biofertilizers + FYM (10 t ha^{-1}) + NPK (75%) + vermicompost (5 t ha^{-1}). Conclusively, the integrated strategy involving the use of diazotrophs, vermicompost, and fertilizers could be a safe option to enhance the overall performance of okra in different production systems.

16.4.5 Onion (*Allium cepa*)

The onion also known as the bulb onion or common onion is the most widely cultivated vegetable species of the genus *Allium*. Considering the chemical threat to the growth and nutritive value of onion, Reetha et al. (2014) reported that the PGPR *P. fluorescens* and *B. subtilis* increased the length of roots and shoots and biomass accumulation in roots and shoots of onion plants relative to control. Similarly, El-Batanomy (2009) evaluated the effect of single and composite culture of *B. circulans*, *Azospirillum lipoferum*, *A. chrococcoum*, *B. polymyxa*, *Rhizobium* sp., and AM-fungi on growth and quality of onion bulbs. Vegetative growth and total bacterial populations in onion rhizosphere were positively affected following PGPR inoculations. Also, highest increase in dry matter and bulb diameter was recorded when all cultures were used together. Mixture of microbial cultures showed highest nitrogenase activity (41.98 μmole C_2H_4/h/g RDW) and mycorrhizal infection (95%) in onion roots. The total NPK (4 :1.97 : 2.91%) in onion dry shoots was found considerably higher in mixed inocula of *B. circulans*, *A. lipoferum*, *A. chrococcoum*, *B. polymyxa*, *Rhizobium* sp., and AM-fungi compared to fertilized control. The co-inoculation of all six cultures together showed highest total carbohydrate (29.23 mg/g) which was followed by co-culture of *Rhizobium* sp. with AM-fungi (28.77 mg/g) and *B. circulans* alone (24.9 mg/g). In other study, significant increase in growth and yield of onion plants due to the synthesis of IAA, siderophores, and P solubilizing activity of *B. subtilis* and *A. chroococcum* is reported (Colo et al.

2014). The longest seedling was observed due to inoculation with *A. chroococcum*, while all inoculated plants had maximum height recorded 60 days after sowing. The onion yield was highest when plants were bacterized with *B. subtilis* and *A. chroococcum*.

16.4.6 Mint (*Mentha piperita* L.)

Mint belonging to family labiatae is accepted as a kind of vegetable which is produced economically both in greenhouse and in field soils. Mint is consumed both fresh and dried and it is mixed as aroma source with different salads (Vural et al. 2000). Mint growers apply chemicals and other growth regulators to stimulate mint rooting. On the contrary, the reports on PGPR use against aromatic plants are scanty. Del Rosario et al. (2015), in a study, inoculated peppermint seedlings with PGPR strains such as *B. subtilis* (GB03), *P. fluorescens* (WCS417r), and *P. putida* (SJ04) using them singly and/or as mixture and measured the growth, chlorophyll content, trichome density, stomatal density, and levels of secondary metabolites in peppermint. The PGPR-inoculated plants had greater shoot and root biomass, leaf area, node number, trichome, and stomatal density. Also, monoterpene content was increased significantly following PGPR inoculation. In other study, Kaymak et al. (2008) determined the effect of *Agrobacterium rubi*, *Burkholderia gladii*, *P. putida*, *B. subtilis*, and *B. megaterium* on root formation, root length, and dry matter content of roots of mint. Generally, root length and dry matter content of roots were greater in PGPR-inoculated cuttings of mint compared to control cuttings under both greenhouse and field environment. Among PGPR, *A. rubi* showed the highest rooting percentage which was followed by *B. megaterium* and *P. putida* relative to control plants, while mixture of bacterial cultures resulted in maximum root length.

16.5 Disease Management of Vegetable Crops Using PGPR

Losses to yield and quality of vegetables due to insects-pests are reported (McCollum 1980). Some of the major diseases of vegetables include bacterial wilt (*R. solanacearu*), bacterial canker (*X. vasicaforia*), leaf curl virus disease, *Fusarium* wilt (*F. oxysporum*), angular leaf spot (*P. syringae* pv. lachrymans), black rot (*X. campestris* pv. campestris), bacterial speck (*P. syringae* pv. tomato), early blight (*A. solani*), and damping-off disease (*Pythium* spp., *Phytophthora* spp. and *Botrytis* spp.) (McCollum 1980). In agronomic practices, several strategies, for example, cultural practices, good sanitation, crop rotation, use of pathogen-free seeds, development of resistant varieties (Saravanakumar et al. 2007), and use of pesticides, are adopted to eradicate/minimize disease infestation. None of these approaches have, however, completely been successful. Focus has, therefore, been shifted toward PGPR to control vegetable diseases. Management of diseases through PGPR compared to other methods is likely to abolish the negative effects such as (i) environmental hazards, (ii) residual toxicity, and (iii) problems of fungicidal resistance in vegetable production. In general, PGPR controls phytopathogens through several mechanisms including competition, secretion of antagonistic substances, for example, siderophores (Ali and Vidhale 2013), HCN (Deshwal and Kumar 2013), antibiotics (Chen et al. 2012), fungal cell wall lysing enzymes, (Saravanakumar et al. 2007) or by ISR (Sangeetha et al. 2010). The mechanisms/substances involved in diseases suppression may act independently or simultaneously. Even though the disease suppressing ability of PGPR have been well documented for a wide range of crops; information on the role of PGPR in the management of vegetable diseases is limited.

Black rot of cauliflower caused by *Xanthomonas campestris* pv. campestris is very destructive disease worldwide and causes great losses to the crop. The biocontrol activity of the two bioagents *P. fluorescens* strain PF-1 and *B. subtilis* strain BS-7 were tested against *X. campestris* pv. campestris. Both the bioagents inhibited the mycelia growth of *X. campestris* pv. campestris and it was suggested from this study that combination of both the bioagents may be applied for better management of black rot diseases (Singh et al. 2010a). In a similar experiment, 10 strains of PGPR (*P. fluorescens* and *Bacillus* spp.) were evaluated for biological control of leaf spot (black spot) of cauliflower caused by the fungus *Alternaria brassicae*. The test PGPR inhibited the growth of *A. brassicae* under in vitro conditions and promoted seed germination and enhanced plant vigor index compared to fungicide and control treatments (Didwania et al. 2013). Girish and Umesha (2005) in other study treated tomato seeds with *B. pumilus* INR7, *B. pumilus* SE34, *B. pumilus* T4, *B. subtilis* GBO3, *B. amyloliquefaciens* IN937a, and *B. brevis* IPC11 to manage bacterial canker disease of tomato. Among PGPR tested, only three strains (IN937a, GBO3, and IPC11) enhanced seed germination and seedling vigor. Sundaramoorthy and Balabaskar (2012) used PGPR strains of *B. subtilis* (EPCO16 and EPC5) and *P. fluorescens* (Pf1, Py15, and Fp7) both in isolation and in combination to assess their biocontrol potential against early blight of tomato incited by *A. solani*. Both strains of *B. subtilis* and *P. fluorescens* were found compatible and the combined application of EPCO16 + Pf1 under in vitro conditions very effectively inhibited the mycelial growth of the pathogen and consequently improved the growth of tomato seedlings relative to the sole application of bacterial strains. Furthermore, the dual application of EPCO16 + Pf1 significantly reduced the early blight of tomato under greenhouse, suggesting that the consortia of biocontrol agents may synergistically enhance the yield of tomato. And hence the dual culture may be recommended in the management of early blight of tomato. In a follow-up study, Loganathan et al. (2014) reported a positive response of *B. subtilis* (BS2) against tomato wilt caused by *F. oxysporum* fsp. *lycopersici* under field conditions. Pretreatment of tomato plants with *B. subtilis* (BS2) significantly induced the activities of defense-related enzymes, for example, peroxidase, polyphenol oxidase, chitinase, and phenylalanine ammonialyase and phenolics when challenged with the pathogen. Also, BS2 improved the fruit quality with lycopene content from 40.34 mg/kg (control) to 76.3 mg/kg and texture from 56.35 *F*max (control) to 90.5 *F*max during harvest and even 15 days after harvest, indicating that PGPR may control both the diseases of plants and can improve the nutritional quality and shelf life of fruits. Maji and Chakrabartty (2014) identified five strains of *Pseudomonas*, namely, *P. aeruginosa*

(T1), *Pseudomonas* sp. (BH25), *Pseudomonas* sp. (AM12), *Pseudomonas* sp. (AM13), and *P. putida* (RG), antagonistic to the pathogen *Ralstonia solanacearum*-Tom5. Bacterial strains were tested for their biocontrol activity against tomato pathogens and the seedling emergence, vigor of the germinated seedlings, and survivability of the seedlings were measured. Among PGPR strains, *Pseudomonas* sp. BH25 was found as a most promising biocontrol agent against *R. solanacearum*-Tom5 in bioassays. Here, the pathogen *R. solanacearum*Tom5 caused only 40% seedling emergence relative to control (76%). Co-culture of antagonist BH25 and pathogen Tom5 (Tom5 : BH25 at 1 : 10), however, further enhanced the percentage of the seedling emergence (75%) which was at par with the control. Also, the combination of BH25 with pathogen improved fresh and dry weights of plants and vigor index of the seedlings over sole application of pathogen and their values were almost similar to those of the control. Vigor index of the seedlings reduced from 935 (control) to 237 (Tom5 treated plants) which, however, reached to 878 when a tenfold high concentration of BH25 was incubated with the pathogen Tom5.

16.6 Conclusion and Future Prospects

Soil dwellers especially PGPR possessed with immense yet variable plant growth-promoting potentials have been found greatly effective in enhancing vegetable production both directly by supplying important nutrients to plants and indirectly by suppressing plant diseases in different production systems. The capability of PGPR to secrete siderophores, cyanogenic compounds, and antibiotics in order to contain phytopathogens could be of practical importance in sustainable production of vegetables. The PGPR endowed with growth promontory and phytopathogens inhibitory properties might be useful in formulating new inoculants, which could offer both an inexpensive and attractive alternative to agrochemicals and consequently such PGPR are likely to improve vegetable production without any adverse impact on environment.

References

Abayomi OA, Adebayo OJ (2014) Effect of fertilizer types on the growth and yield of *Amaranthus caudatus* in Ilorin, Southern Guinea, Savanna Zone of Nigeria. Adv Agric:Article ID 947062, 5 pages

Abd El-Salam IZ, Arafa MM, Shalaby OE (2005) Effect of rock phosphate and rare earth minerals on growth, yield, chemical constituents and active ingredients of hot pepper (*Capsicum annuum, L.*) under new reclaimed soils conditions. Egypt J Appl Sci 20:285–310

Abou El-Magd MM, Asmaa RM, Magda MH, Aisha HA (2013) Effect of different levels of mineral phosphorus fertilizer and bio-phosphorus on vegetative growth, head yield and quality of broccoli. Res J Agric Biol Sci 9:164–169

Ahemad M, Khan MS (2011) Functional aspects of plant growth promoting rhizobacteria: recent advancements. Insight Microbiol 1:39–54

Ahemad M, Zaidi A, Khan MS, Oves M (2009) Biological importance of phosphorus and phosphate solubilizing microorganisms: an overview. In: Khan MS, Zaidi A (eds) Phosphate solubilizing microbes for crop improvement. Nova Science Publishers, New York, pp 1–4

Alertor O, Oshodi AA, Ipinmoroti K (2002) Chemical composition of common leafy vegetables and functional properties of their leaf protein concentrates. Food Chem 78:63–68

Ali SS, Vidhale NN (2013) Bacterial Siderophore and their application: a review. Int J Curr Microbiol App Sci 2(12):303–312

Ali SZ, Sandhya V, Rao LV (2014) Isolation and characterization of drought-tolerant ACC deaminase and exopolysaccharide-producing fluorescent Pseudomonas sp. Ann Microbiol 64:493–502

Bashyal LN (2011) Response of cauliflower to nitrogen fixing biofertilizer and graded levels of nitrogen. J Agric Environ 12:41–50

Bhadoria SKS, Dwivedi YC, Kushwah SS (2005) Effect of *Azotobacter* inoculation with nitrogen levels on quality characters of tomato. J Veg Sci 32:94–95

Bholay AD, Priyanka JU, Borkhataria BV, Dhalkari MV (2012) Fluorescent pseudomonads as plant growth promoting rhizobacteria and their siderophoregenesis. IOSR J Pharm Biol Sci 3:27–32

Bhushan A, Bhat KL, Sharma JP (2013) Effect of *Azotobacter* and inorganic fertilizers on fruit and seed yield of okra cv. Hisar Unnat. Agric Sci Dig 33:135–138

Brahma S, Phookan DB (2006) Effect of nitrogen, phosphorus and potassium on yield and economics of broccoli [*Brassica oleracea* (L.) var. italica] cv. Pusa. Res Crops 7:261–262

Bumunang EW, Babalola OO (2014) Characterization of rhizobacteria from field grown genetically modified (GM) and non-GM maizes. Braz Arch Biol Technol 57:1–8

Çakmakç R, Erat M, Erdoğan U, Dönmez MF (2007) The influence of plant growth–promoting rhizobacteria on growth and enzyme activities in wheat and spinach plants. J Plant Nutr Soil Sci 170:288–295

Chakravarty G, Kalita MC (2012) Biocontrol potential of *Pseudomonas fluorescens* against bacterial wilt of Brinjal and its possible plant growth promoting effects. Ann Biol Res 3:5083–5094

Chen N, Jin M, Qu HM, Chen ZQ, Chen ZL, Qiu ZG, Wang XW, Li JW (2012) Isolation and characterization of *Bacillus* sp. producing broad-spectrum antibiotics against human and plant pathogenic fungi. J Microbiol Biotechnol 22:256–563

Colo J, Hajnal-Jafari TI, Durić S, Stamenov D, Hamidović S (2014) Plant growth promotion rhizobacteria in onion production. Pol J Microbiol 63:83–88

Crawford PB, Obarzanek E, Morrison J, Sabry ZI (1994) Comparative advantage of 3-day food records over 24 recall and 5-day food frequency validated by observation of 9-and 10-year girls. J Am Diet Assoc 94:626–630

Dastager SG, Deepa CK, Pandey A (2010) Isolation and characterization of novel plant growth promoting *Micrococcus* sp NII-0909 and its interaction with cowpea. Plant Physiol Biochem 48:987–992

Del Rosario CL, Santoro MV, Reinoso H, Travaglia C, Giordano W, Banchio E (2015) Anatomical, morphological, and phytochemical effects of inoculation with plant growth- promoting rhizobacteria on peppermint (*Mentha piperita*). J Chem Ecol 41:149–158

Deshwal VK, Kumar P (2013) Production of plant growth promoting substance by Pseudomonads. J Acad Ind Res (JAIR) 2:221–225

Didwania N, Solanki M, Trivedi PC (2013) Biocontrol of alternaria blight of cauliflower by plant-growth promoting rhizobacteria. Asian J Microbiol Biotechnol Environ Exp Sci 15:567–572

Ekinci M, Turan M, Yildirim E, Güneş A, Kotan R, Dursun A (2014) Effect of plant growth promoting rhizobacteria on growth, nutrient, organic acid, amino acid and hormone content of cauliflower (*Brassica oleracea* l. var. *botrytis*) transplants. Acta Sci Pol Hortorum Cultus 13:71–85

El-Assiouty FMM, Abo-Sedera SA (2005) Effect of bio and chemical fertilizers on seed production and quality of spinach (*Spinacia oleracea* L.) Int J Agric Biol 7:947–952

Elazzazy AM, Omar AA, Tarek AAM, Tamer SA (2012) Evaluation of some plant growth promoting rhizobacteria (PGPR) to control *Pythium aphanidermatum* in cucumber plants. Life Sci 9:3147–3153

El-Batanomy N (2009) Synergistic effect of plant-growth promoting rhizobacteria and arbuscular mycorrhiza fungi on onion (*Allium cepa*) growth and its bulbs quality after storage. New Egypt J Microbiol 23:163–182

Farajzadeh D, Yakhchali B, Aliasgharzad N, Sokhandan-Bashir N, Farajzadeh M (2012) Plant growth promoting characterization of indigenous *Azotobacter* isolated from soils in Iran. Curr Microbiol 64:397–403

Fry W, Goodwin S (1997) Re-emergence of potato and tomato late blight in the United States. Plant Dis 81:1349–1357

Fu Q, Liu C, Ding N, Lin Y, Guo B (2010) Ameliorative effects of inoculation with the plant growth-promoting rhizobacterium *Pseudomonas* sp. DW1 on growth of eggplant (*Solanum melongena* L.) seedlings under salt stress. Agric Water Manag 97:1994–2000

Girish N, Umesha S (2005) Effect of plant growth promoting rhizobacteria on bacterial canker of tomato. Arch Phytopathol Plant Protect 38:235–243

Guo Z, He CL, Ma Y, Zhu H, Liu F, Wang D, Sun L (2011) Effect of different fertilization on spring cabbage (*Brassica oleracea* L. var. capitata) production and fertilizer use efficiencies. Agric Sci 2:208–212

Hanif R, Iqbal Z, Iqbal M, Hanif S, Rasheed M (2006) Use of vegetables as nutritional food: role in human health. Am J Agric Biol Sci 1:18–22

Hennekens CH (1986) Micronutrients and cancer prevention. N Engl J Med 315:1288–1289

Hou MP, Oluranti BO (2013) Evaluation of plant growth promoting potential of four rhizobacterial species for indigenous system. J Cent South Univ 20:164–171

Isfahani FM, Besharati H (2012) Effect of biofertilizers on yield and yield components of cucumber. J Biol Earth Sci 2:B83–B92

Kalita M, Bharadwaz M, Dey T, Gogoi K, Dowarah P, Unni BG, Ozah D, Saikia I (2015) Developing novel bacterial based bioformulation having PGPR properties for enhanced production of agricultural crops. Indian J Exp Biol 53(1):56–60

Kanchana D, Jayanthi M, Saranraj P, Sujitha D (2013) Prevelence of *Azotobacter* sp. in chilli (*Capsicum annuum* L.) rhizosphere soil of Cuddalore district, Tamil Nadu, India. Int J Microbiol Res 4:296–299

Kapusta-Duch J, Kopeć A, Piatkowska E, Borczak B, Leszczyńska T (2012) The beneficial effects of *Brassica* vegetables on human health. Rocz Panstw Zakl Hig 63:389–395

Kaushal M, Kaushal R, Thakur BS, Spehia RS (2011) Effect of plant growth-promoting rhizobacteria at varying levels of N and P fertilizers on growth and yield of cauliflower in mid hills of Himachal Pradesh. Int J Farm Sci 1:19–26

Kaymak HC, Yarali F, Guvenc I, Figen Donmez M (2008) The effect of inoculation with plant growth rhizobacteria (PGPR) on root formation of mint (*Mentha piperita* L.) cuttings. Afr J Biotechnol 7:4479–4483

Keel C, Schnider U, Maurhofer M, Voisard C, Laville J, Burger U, Wirthner P, Haas D, Defago G (1992) Suppression of root diseases by *Pseudomonas fluorescens* CHAO: importance of bacterial secondary metabolite, 2,4-diacetylphoroglucinol. Mol Plant-Microbe Interact 5:4–13

Kumar K, Madhuri K, Murugan V, Sakthivel K, Anantharaj A, Singh AK, Gautam RK, Roy SD (2014a) Growth enhancement in vegetable crop by multifunctional resident plant growth promoting rhizobacteria under tropical island ecosystem. Afr J Microbiol Res 8:2436–2448

Kumar A, Kumar K, Kumar P, Maurya R, Prasad S, Singh SK (2014b) Production of indole acetic acid by *Azotobacter* strains associated with mungbean. Plant Arch 14:41–42

Loganathan M, Garg R, Venkataravanappa V, Saha S, Rai AB (2014) Plant growth promoting rhizobacteria (PGPR) induces resistance against *Fusarium* wilt and improves lycopene content and texture in tomato. Afr J Microbiol Res 8:1105–1111

Lukkani NJ, Reddy ECS (2014) Evaluation of plant growth promoting attributes and biocontrol potential of native *fluorescent Pseudomonas* spp. against *Aspergillus niger* causing collar rot of ground nut. Int J Plant Anim Environ Sci 4:267–262

Maji S, Chakrabartty PK (2014) Biocontrol of bacterial wilt of tomato caused by *Ralstonia solanacearum* by isolates of plant growth promoting rhizobacteria. Aust J Crop Sci 8:208–214

Mal B, Mahapatra P, Mohanty S, Mishra HN (2013) Growth and yield parameters of okra (*Abelmoschus esculentus*) influenced by diazotrophs and chemical fertilizers. J Crop Weed 9:109–112

Manisha K, Korla BN (2009) Effect of biofertilizers on growth and yield of cauliflower cv. PSB K-1. Indian J Hort 66:496–501

Martínez-Aguilar L, Salazar-Salazar C, Méndez RD, Caballero-Mellado J, Hirsch AM, Vásquez-Murrieta MS, Estrada-de los Santos P (2013) *Burkholderia caballeronis* sp. nov., a nitrogen fixing species isolated from tomato (*Lycopersicon esculentum*) with the ability to effectively nodulate *Phaseolus vulgaris*. Antonie Van Leeuwenhoek 104:1063–1071

McCollum JP (1980) Producing vegetable crops, vol 34, 3rd edn. Interstate Printers & Publishers, Danville, p 599

Nabti E, Bensidhoum L, Tabli N, Dahel D, Weiss A, Rothballer M, Schmid M, Hartman A (2014) Growth inhibition of barley and biocontrol effect on plant pathogenic fungi by a *Cellulosimicrobium* isolated from salt affected rhizosphere soil in northwestern Algeria. Eur J Soil Biol 61:20–26

Naderi DMR, Ahmadi NH, Bahari B (2012) Assessment of applications of biological fertilizer for potato cultivation. Int J Agric Sci 2:102–107

Nailwal S, Anwar MS, Budhani KK, Verma A, Nailwal TK (2014) *Burkholderia* sp. from rhizosphere of *Rhododendron arboretum*: isolation, identification and plant growth promotory (PGP) activities. J Appl Nat Sci 6:473–479

Nosrati R, Owlia P, Saderi H, Rasooli I, Malboobi MA (2014) Phosphate solubilization characteristics of efficient nitrogen fixing soil *Azotobacter* strains. Iran J Microbiol 6:285–295

Olowoake AA, Adeoye GO (2010) Comparative efficacy of NPK fertilizer and composted organic residues on growth, nutrient absorption and dry matter accumulation in maize. Int J Org Agric Res Dev 2:43–53

Panhwar QA, Naher UA, Jusop S, Othman R, Latif MA, Ismail MR (2014) Biochemical and molecular characterization of potential phosphate solubilizing bacteria in acid sulphate soils and their beneficial effects on rice growth. PLoS One 9(10):PMC4186749

Prabhakar M, Srinivas K (1995) Growth, dry-matter production, yield and water use of cauliflower (*Brassica oleracea* var botrytis subvar cauliflora) in relation to irrigation and nitrogen fertilization. Indian J Agric Sci 65(8):83–91

Ramakrishnan K, Selvakumar G (2012) Effect of biofertilizers on enhancement of growth and yield on tomato (*Lycopersicum esculentum* Mill.) Int J Res Bot 2:20–23

Reetha S, Bhuvaneswari G, Thamizhiniyan P, Ravi TM (2014) Isolation of indole acetic acid (IAA) producing rhizobacteria of *Pseudomonas fluorescens* and *Bacillus subtilis* and enhance growth of onion (*Allium cepa*. L). Int J Curr Microbiol App Sci 3:568–574

Rizvi A, Khan MS, Ahmad E (2014) Inoculation impact of phosphate solubilizing microorganisms on growth and development of vegetable crops. In: Khan MS, Zaidi A, Musarrar J (eds) Phosphate solubilizing microorganisms: principle and application of microphos technology. Springer, Basel, pp 287–297

Ruangsanka S (2014) Identification of phosphate-solubilizing fungi from the asparagus rhizosphere as antagonists of the root and crown rot pathogen *Fusarium oxysporum*. ScienceAsia 40:16–20

Sahar P, Sahi ST, Jabbar A, Rehman A, Riaz K, Hannan A (2013) Chemical and biological management of *Fusarium oxysporum* f.sp. melongenae. Pak J Phytopathol 25:155–159

Sahu AK, Kumar S, Maji S (2014) Effect of biofertilizers and inorganic fertilizers on vegetative growth and yield of okra [*Abelmoschus esculentus* (L.) Moench]. Int J Agric Sci 10:558–561

Sang MK, Kim JD, Kim BS, Kim KD (2011) Root treatment with rhizobacteria antagonistic to *Phytophthora* blight affects anthracnose occurrence, ripening, and yield of pepper fruit in the plastic house and field. Phytopathology 101:666–678

Sangeetha G, Thangavelu R, Usha Rani S, Muthukumar A, Udayakumar R (2010) Induction of systemic resistance by mixtures of antagonist bacteria for the management of crown rot complex on banana. Acta Physiol Plant 32:1177–1118

Saravanakumar D, Kumar CV, Kumar N, Samiyappan R (2007) PGPR-induced defense responses in the tea plant against blister blight disease. Crop Prot 26:556–565

Sarvani B, Reddy RS (2013) *In vitro* screening of native bacillus isolates for plant growth promoting attributes. Int J Bio-Resour Stress Manag 4:298–303

Savci S (2012) An agricultural pollutant: chemical fertilizer. Int J Environ Sci Dev 3:77–80

Sharafzadeh S (2012) Effects of PGPR on growth and nutrients uptake of tomato. Int J Adv Eng Technol 2:27–31

Sharma P, Saikia MK (2013) Management of late blight of potato through chemicals. IOSR J Agric Vet Sci 2:23–36

Shree S, Singh VK, Kumar R (2014) Effect of integrated nutrient management on yield and quality of cauliflower (*Brassica oleracea var. Botrytis* L.) Bioscan 9:1053–1058

Singh V, Singh SS (2005) Effect of inorganic and biofertilizers on production of cauliflower (*Brassica oleracea* L. var. Botrytis). Veg Sci 32:146–149

Singh D, Dhar S, Yadav DK (2010a) Effect of endophytic bacterial antagonists against black rot disease of cauliflower caused by *Xanthomonas campestris* pv. campestris. Indian Phytopathol 63:122–126

Singh JK, Bahadur A, Singh NK, Singh TB (2010b) Effect of using varying level of npk and biofertilizers on vegetative growth and yield of okra (*abelmoschus esculentus* (l.) moench). Veg Sci 37:100–101

Sivasakthi S, Kanchana D, Usharani G, Saranraj P (2013) Production of plant growth promoting substance by *Pseudomonas fluorescens* and *Bacillus subtilis* isolates from paddy rhizosphere soil of Cuddalore District, Tamil Nadu, India. Intern J Microbiol Res 4:227–233

Sreedevi SK, Suma K, Bhuvana CK, Prem KA, Minakshi G, Shankar M, Maruthi S, Vanaja M, Sharma KL (2014) Improving phytochemical and nutritional quality of spinach (*Spinacia oleracea*) through phosphate solubilizing bacteria. Indian J Dryland Agric Res Dev 29:104–107

Sundaramoorthy S, Balabaskar P (2012) Consortial effect of endophytic and plant growth promoting rhizobacteria for the management of early blight of tomato incited by *Alternaria solani*. J Plant Pathol Microbiol 3:145

Susilowati LE, Syekhfani S (2014) Characterization of phosphate solubilizing bacteria isolated from Pb contaminated soils and their potential for dissolving tricalcium phosphate. J Degrad Mining Lands Manag 1:57–62

Talat MM, Sijam K (2010) *Ralstonia solanacearum*: the bacterial wilt causal agent. Asian J Plant Sci 9:385–393

Turan M, Ekinci M, Yildirim E, Güneş A, Karagöz K, Kotan R, Dursun A (2014) Plant growth-promoting rhizobacteria improved growth, nutrient, and hormone content of cabbage (*Brassica oleracea*) seedlings. Turk J Agric For 38:327–333

Vural H, Eiyok D, Duman I (2000) Kültür Sebzeleri (Sebze Yetitirme). Ege Üniversitesi Ziraat Fakültesi Bahçe Bitkileri Bölümü, Bornovazmir

Yildirim E, Karlidag H, Turan M, Dursun A, Goktepe F (2011) Growth, nutrient uptake, and yield promotion of broccoli by plant growth promoting rhizobacteria with manure. Hortscience 46:932–936

Azospirillum and Wheat Production 17

Mohammad Javad Zarea

Abstract

This review aims to elucidate the actual effect of *Azospirillum* spp. on wheat production under field condition and represent methods by which it can enhance the beneficial effect of *Azospirillum* on wheat. The bacterial genus *Azospirillum* is well known as a plant growth-promoting bacteria (PGPR). Wheat is the topmost important staple crop cultivated worldwide under contrast environment from arid land to wet area. Researchers evaluating the effect of *Azospirillum* inoculation on gramineous plants were back from the 1970s. Since then numerous researches have done, and investigating the performances of *Azospirillum* on various crops has been focused. Evidence indicates the beneficial impact of *Azospirillum* application on improved plant growth and economical yield. However, there are some studies showing no beneficial effects. For better wheat-*Azospirillum* management, reconsideration of the published paper dealing with *Azospirillum*-wheat association is necessary. In this review field research result from the 1980s until the present time is reviewed, and agronomical and technical management that affect *Azospirillum* performances from these published papers are presented. This review aimed to not repeat the traditional beneficial effect of *Azospirillum* on host plant since many reviews on this filed have been previously represented by considerable authors.

M.J. Zarea
Faculty of Agriculture, University of Ilam, Ilam, Iran

Faculty of Agriculture, University of Kurdistan, Sanandaj, Iran
e-mail: mj.zarea@ilam.ac.ir; mjzarea@ymail.com

V. Kumar et al. (eds.), *Probiotics in Agroecosystem*,
DOI 10.1007/978-981-10-4059-7_17

17.1 Introduction

The bacterial genus *Azospirillum* has been mentioned to be the most-investigated plant-associated bacteria (Bashan and de-Bashan 2010). *Azospirillum* is a gram-negative free-living nitrogen-fixing bacterium of the rhizosphere, blog to the group of bacteria known as associated bacteria with ability of plant growth-promoting (PGPR) properties. *Azospirillum* has been reported to be isolated from the rhizosphere of a wide range of many plants including grasses and cereals. Different species of the genus *Azospirillum* have been detected and isolated from tropical area and temperate climates. After elucidation of association relationship between plant and *Azospirillum*, interest in the application of *Azospirillum* species in monocotyledonous plants, especially grasses and grain crops, has been arisen (Sumner 1990).

Intensive researches under field condition and controlled condition have proven the potential benefits of *Azospirillum* on plant growth and crop yields of various crop including both leguminous and germanous plants. At present, sixteen species of *Azospirillum* have been described (Table 17.1). Among them *Azospirillum brasilense* gain the highest research attention in wheat production. Biological nitrogen fixation, ability of phytohormone production as plant growth *inducer* like auxin (Umali-Garcia et al. 1980) and help plant with nutrient abortion (Lin et al. 1983) are but no limited of main mechanisms by which beneficial effect of *Azospirillum* on plants excreted. Early researches (Jain and Patriquin 1984; Barbieri et al. 1986; Levanony and Bashan 1989) suggested that *Azospirillum* through induced modifications of host plant root system, enhancement of root development, improves mineral and water uptake by the inoculated plant compared to not inoculated plant (Sarig et al. 1988; Levanony and Bashan 1989; Pacovsky 1990). Cytokinins, gibberellins, and auxins are substances that *act* like phytohormone which has been claimed to be excreted by *Azospirillum*. These substances released by *Azospirillum* involve in the incitement of plant root system development (Barbieri and Galli 1993). Based on the worldwide results obtained from the field inoculation of *Azospirillum* sp. in different soils and climatic regions, it has been concluded that these bacteria are able to promote yield of staple important crops (Okon and Labandera-Gonzalez 1994). Research on the colonization of plant tissues by diazotrophic bacteria was initiated from 1985 to 1990, and consequently some aspects of the plant-bacteria interaction began to be elucidated (Baldani and Baldani 2005). However, positive effects of *Azospirillum* like hormonal effects, nitrogen assimilation, and biological nitrogen fixation have been confirmed, but negative effects also have been frequently detected (Boddey and Döbereiner 1982). Significant yield increase and better growth in wheat, a nontropical cereal, following inoculation with *Azospirillum* have been reposted in many related studies (Boddey et al. 1986; Mertens and Hess 1984; Millet et al. 1985; Rai and Gaur 1982; Reynders and Vlassak 1982a; Baldani et al. 1983, 1986; Díaz-Zorita and Fernández-Canigia 2009), carried out under nonlimiting condition and adverse condition such as salinity (Zarea et al. 2012) and drought stress (Creus et al. 1998).

Table 17.1 List of *Azospirillum* species and original of the isolation

Azospirillum	Original isolated	Country	References
A. lipoferum	Associated with wheat and maize	Brazil	Tarrand et al. (1978)
A. amazonense	Roots and rhizosphere soil of Gramineae	Amazon region, Brazil	Tarrand et al. (1978)
A. brasilense	Associated with wheat roots	Brazil	Baldani et al. (1986)
A. alopraeferens	Associated with Kallar grass	Saline soils of Pakistan	Reinhold et al. (1985)
A. irakense	Field-grown root rice plant	Grown in Iraq	Malik et al. (1994)
A. largimobile	Detected in water sample of a lake	Australia	Dekhil et al. (1997)
A. doebereinerae	grown *Pennisetum*	Germany	Eckert et al. (2001)
A. oryzae	Roots of the rice plant	Japan	Xie and Yokota (2005)
A. melinis	Molasses grass (*Melinis minutiflora* Beauv.) pasture and fodder grass	China	Peng et al. (2006)
A. canadense	Corn rhizosphere	Canadian soil	Mehnaz et al. (2007a)
A. zeae	Rhizosphere of corn	Canadian soil	Mehnaz et al. (2007b)
A. picis	From road tar abandoned by the marginal of a road in Taichung city area of Taiwan	Taiwan	Lin et al. (2009)
A. thiophilum	Sulfide spring shameless baths in Stavropol Krai, North Caucasus	Russia	Lavrinenko et al. (2010)
A. humicireducens	The anode biofilm microbial fuel cell	China	Zhou et al. (2012)
A. fermentarium	From a fermentative tank	Taiwan	Lin et al. (2013)
A. himalayense	Himalayan valley soil	India	Tyagi and Singh (2014)

17.2 Wheat

Wheat is the topmost important crop worldwide, providing the world's population with 20% of the calories. 2.5 billion people in less-developed countries receive 20% of daily protein from wheat (Braun et al. 2010). The modern wheat grower receives more return for money spent on nitrogen (N) than any other nutrient, and nitrogen application provides the largest net return of any nutrient used in wheat production. N has a vital role in the structures of protein. Proteins are the most important building substances that form protoplasm of every cell. Among all elements needed by plant, N is the first rank and needed at higher rate rather than other elements. Farmers worldwide annually apply considerable N to ensure a higher crop yield. N plays an important role in chlorophyll formation and structure. Transformation of energy from sunlight waves and converted into molecular enrich in energy (ATP and ADPH) is enabled by chlorophyll (photosynthetic activity). Therefore, N application ultimately affects leaf area and photosynthetic activity of plant as well as cell structure and size.

Wheat has two main growth stages, vegetative and reproductive growth stage. During vegetative growth stage, tiller number and spikelet number are formed, and any limiting growth factors like water and nutrient deficiency would negatively affect the yield potential of wheat. During reproductive growth stage, fertile grain and filling grain number and grain weight are determined. These traits are mostly affected by the limiting factors such as water availability, nutrient access, and other constraint environmental factors. The fertility of spike of tiller is most affected by two important nutrients N and P. Any limiting in these nutrients resulted in reduction of yield potential. In wheat tiller 1 and 2 are responsible for the 50% of the total grain yield and of the remaining main stem involved. Therefore, it would be important to match the initiation of *Azospirillum* benefits with demand of wheat according to the plant growth stage. Late beneficial excreted by *Azospirillum* would have a negligible effect on total yield since any sink that determines total grain yield previously is formed. However, it is seems that *Azospirillum* benefits would not be the same in spring wheat compared with winter wheat since growth stage in winter wheat occurs in prolonger time than that of spring wheat. Another opinion is about the environmental constraints that would affect the rate of benefits from *Azospirillum* inoculation excreted. Under environmental limiting *Azospirillum* performance may be aggressive. During recent years effect of *Azospirillum* under adverse condition has been a topic paid attention by some researchers.

17.3 Historical *Azospirillum* Background

Research on biological nitrogen fixation (BNF) associated with grasses was first begun by Johanna Döbereiner. Johanna Döbereiner joined the research team at the National Center of Education and Agricultural Research of the Ministry of Agriculture, located at Km 47 in the 1950s in Brazil (Baldani and Baldani 2005).

In the 1970s introduction of the acetylene reduction as a significant advance method helps researchers who focus in the study of BNF in grasses (Baldani and Baldani 2005). This advance was accompanied with the introduction of the semisolid nitrogen-free medium (NFb) that mimics the oxygen level which is found in soil niches. By introduction of NFb medium isolation of microaerophilic bacteria that fix nitrogen associated with plant roots was initiated. *Azospirillum lipoferum* and *Azospirillum brasilense* are the first bacterium species of *Azospirillum* genus isolated by the NFb medium. Another introduced semisolid medium named LGI medium allowed the isolation of upcoming new species, *A. amazonense* (Magalhães et al. 1983). By alteration of pH and carbon source of NFb medium, LGI medium was introduced (Baldani et al. 1984). Comparison of DNA-DNA homology of *Spirillum lipoferum* and *Spirillum brasilense* species resulted in the generation of a new genus with two species which are called *Azospirillum lipoferum* and *Azospirillum brasilense* (Tarrand et al. 1978). Following *Azospirillum lipoferum* and *Azospirillum brasilense description*, *Azospirillum* amazonense was later characterized (Magalhães et al. 1983). Species of *Azospirillum amazonense* have been isolated first from forage grass growing in the Amazon and in the state of Rio de Janeiro and later from different plants like rice, maize, and sorghum plants (Baldani 1984).

17.4 Adsorption to Wheat

Establishment of *Azospirillum* in the roots is a critical step toward an effective plant growth promotion (Bashan et al. 2004; Okon and Kapulnik 1986). Although *Azospirillum* are commonly noticed as rhizosphere bacteria and they mostly colonize the surface of the root, a few strains have been reported that have the ability to infect plant (Patriquin et al. 1983; Döabereiner et al. 1995). For example, strains of *Azospirillum* spp. that colonize the interior of root wheat have been reported by Baldani et al. (1986). *Azospirillum* can colonize different zones of wheat. *Azospirillum brasilense* Sp245 have been reported to colonize the regions of the lateral root and root hair (Vande Broek et al. 1993). Some *Azospirillum* species colonize further interior of the root. For example, in Brazilian wheat cultivar, *A. brasilense* Sp245 was observed in the root xylem, whereas *A. brasilense* Sp7 only occupied the root surface of wheat (Schloter et al. 1994). This new evident is in contrast to the report of Horemans et al. (1988). A field research work under field conditions in Belgium done by Horemans et al. (1988) showed no colonization of interior roots of barley, wheat, com, or grasses by *Azospirillum.* In this research *A. lipoferum* was detected only on the exterior root of corn and grass, whereas *A. brasilense* was detected on the root of all crop studied. Attachment of *Azospirillum* to the root system is not in a homogeneous pattern. *A. brasilense* Sp7 has been shown to attach preferentially to the root elongation rejoin of wheat seedlings (Kapulnik et al. 1985b), whereas in another study, this strain densely occupied the root hairs of pearl millet (Umali-Garcia et al. 1980). Pattern attachment of a specific *Azospirillum* strain is also varied among varieties of a certain crop (Murty and Ladha 1987). Environmental constraints can affect colonization of wheat roots by the *Azospirillum.* For example, movement of *Azospirillum* toward seedlings of wheat has been confirmed and was shown to be restricted by moisture content of soil (Bashan 1986). Saubidet et al. (2002) investigated the effect of inoculating a highly efficient *A. brasilense* strain on wheat plant grown under pot culture with three N regimes (0, 3 or 16 mMNO_3^-, 50 ml/pot once or twice a week) and in disinfected or non-disinfected soil. They found that at the booting stage, the inoculated roots in both soils showed a similar colonization by *Azospirillum* sp. that was not affected by N addition.

17.5 *Azospirillum* Field Research

Researchers evaluating the impact of *Azospirillum* inoculation on gramineous plants were from the 1970s. The results of these studies although were inconsistent demonstrated that biological nitrogen fixation by *Azospirillum* provides around 40% of the nitrogen requirement (Boddey and Döbereiner 1982). Field wheat inoculation experiments with *Azospirillum* strains isolated from sterilized roots of wheat (Sp 245, Sp 107 st) showed a consistent increase in total plant N; however, this was not the case for the heterologous strain Sp 7 (Baldani et al. 1983). Freitas et al. (1982) compared beneficial effect of homologous and heterologous strains, isolated from other plants, of *Azospirillum* with maize and confirmed superior advantageous of homologous strains of *Azospirillum* compared to heterologous strains (Freitas et al. 1982). During 1982 and 1983, Mertens and Hess (1984) conducted a field experiment to investigate the spring wheat response to *Azospirillum lipoferum* sp 108 with different soils, sands, and peat-clay mixtures containing 0.28% nitrogen. In this experiment sandy soil received phosphate, sodium, and in one assay supplied with N. However, in this conducted study, the most increase in grain yield (70%) obtained from sand received P and K only, but an increase in grain yield up to 32% was also observed in peat-clay soil containing 0.28% of total N. Further research showed that homologous *Azospirillum* strains of Sp245 and Sp107 preferentially colonize the interior roots of wheat and sorghum, whereas heterologous strains, Sp 7 and Cd, tend to establish on the outer surface of the root (Baldani et al. 1986). Boddey et al. (1986) claimed that positive influence of *Azospirillum* on wheat under field condition was due to the nitrate reductase activity of the bacteria within the root rather than biological nitrogen fixation of the bacteria. These results were later confirmed by Ferreira et al. (1987) who found that nitrate reductase activity of the bacteria was responsible for the positive influence of *Azospirillum* on wheat. To confirm this idea, Ferreira et al. (1987) compared the nitrate reductase activity of the negative mutants with wild species of *Azospirillum* under gnotobiotic conditions. In another field experiment conducted by Bhattarai and Hess (1993), feedback of different wheat cultivars (*Triticum aestivum* L.) yield to inoculations with a Nepal hilly region poor soil isolated *Azospirillum* was positive. Two hundred and ninety-seven experimental locations carried out under dry land cropping conditions of Argentina across the region of Pampas during 2002–2006 growing seasons were employed to determine the yield response of bread wheat to inoculation (seed inoculation) with a liquid formulation of *Azospirillum brasilense* (INTA Az-39 strain) (Díaz-Zorita and Fernández-Canigia 2009). Grain yield and harvested grain yield increased by about 6.1% and 8.0% (260 kg per ha), respectively, due to seed inoculation. Seventy percent of the experimental sites had positive response to the inoculation, regardless of the fertilization and other factors including management practices of soil and crop. A field trial conducted in Brazil by Hungria et al. (2010) was designed to test the response of wheat to elite strains of *A. brasilense.* Result indicated an increase of 18% in the grain yields of wheat due to *Azospirillum* inoculation. Millet and Feldman (1984) investigate the effect of *Azospirillum brasilense* on a common spring bread wheat yield response under four doses of chemical nitrogen fertilizer.

Table 17.2 Mechanisms through which *Azospirillum* positively affects wheat growth and yield

Mechanisms	References
Increase in the nitrate reductase (NR) activity of the bacteria in the roots	Boddey et al. (1986)
Stimulation of *b*-glucuronidase	Kapulnik et al. (1985a, b, 1987), Vande Broek et al. (1998)
Changes in root morphology	Kapulnik et al. (1985a, b)
Increase the nitrogen assimilation rate from the germination stage until the emergence of spikes	Ferreira et al. (1987), Rodrigues et al. (2000)
IAA (indole-3-acetic acid) production	Barbieri et al. (1986), Malik et al. (1994), Jain and Patriquin (1985)
Nitrogen fixation	Barbieri et al. (1986), Malik et al. (1994)
Greater nitrate reductase enzyme activity	Ferreira et al. (1987)
Production of auxins, gibberellins, and cytokinins	Hartmann and Zimmer (1994)
Nitrogenase activity	Elanchezhian and Panwar (1997)
Elastic adjustment under water stress	Creus et al. (2004)
Stimulating nitrogen uptake by the roots	Saubidet et al. (2002)

Results of this experiment showed that positive response of grain yield to inoculation was due to increase in grain spike^{-1}. At the higher doses of nitrogen fertilizer, these positive increases in yield in inculcated wheat were due to the higher number of spikes plant^{-1}. Irrespective of N doses used, a number of fertile spikelet main spike^{-1} increased by 0.5–1.4% due to *Azospirillum* inoculation. Further results of this study indicated no effect of inoculation on increased grain protein percentage. Mertens and Hess (1984) reported on the enhancement of spring wheat grain yield under field condition following *Azospirillum* inoculation. This increase was due to enhancement of thousand grain weight and total nitrogen. Based on the recent research, there is a positive correlation between wheat increased grain yield and the number of fertile tillers produced by the pant (Salantur et al. 2006). Saubidet et al. (2002) investigated the effect of *A. brasilense* on grain yield of wheat under sterilized soil. Positive influence of inculcation on total grain yield was due to the effect of bacteria which enhanced the tiller number in wheat. Therefore *Azospirillum* inoculation through enhancing the number of tillers in wheat plants (Saubidet et al. 2002) may result in increase in the total grain yield of wheat. Table 17.2 shows the mechanisms through which *Azospirillum* induces wheat growth and yield.

17.6 Factors Affecting Plant Promoting Effect of *Azospirillum* on Wheat

Response of crop to *Azospirillum* inoculation has not always been consistent, and the factors impacting the crop reply are not completely known (Okon and Labandera-Gonzalez 1994). Knowledge of *Azospirillum* extensive inoculation performances in large on-farm exploratory is still quite restricted (Díaz-Zorita and Fernández-Canigia 2009). Limited on-farm research is often justified by the reportedly low consistency of field results conducted in a context of more realistic crop management production conditions (Dobbelaere et al. 2001). The inconsistency in the outcomes could be related to the interaction of the inoculation methods with environmental conditions. Type of soil, water status, and crop management methods such as fertilization, chemical disease management, genotypes, etc. are mentioned factors affecting variability in the results of *Azospirillum* performances (Fages 1994). Complex interactions between the crop, the bacteria, and the environment have been assumed to affect wheat's grain yield response to inoculation with *Azospirillum* sp. and other diazotrophic bacteria varied among Brazilian locations (Sala et al. 2007).

17.7 Interactions Between Wheat Genotypes and *Azospirillum* Spp. Strains

Early research demonstrated the successful inoculation in wheat with homologous (strains isolated of the same crop) than nonhomologous strains (Sumner 1990). It seems that applying isolates of those strains that naturally colonize wheat plant has more advantages than strains isolated from a different crop. By using certain strains of *Azospirillum* that were isolated from wheat plant, it has been shown that those strains isolated from the outer disinfected part of wheat roots (Sp 245, Sp 107) were constantly superior in colonizing the internal root of wheat than those isolated from the rhizosphere soil of wheat (Baldani et al. 1983). From these results it seems that the beneficial effect of *Azospirillum* is more with those strains that penetrate the root. Research done by Ferreira et al. (1987) clearly showed that inoculation of wheat with homologous strains of *Azospirillum* (strain Sp 245) had more advantages than those heterologous strains due to the ability of homologous strains in colonizing the interior of wheat roots than those heterologous strains (Baldani et al. 1986, 1987). One of the main factors that may result in positive response of plant to inoculation is that the used strains can well penetrate the interior of the root and consistently colonize within the roots. In those experiments that inoculation resulted in positive yield responses, it is always shown that used strains have well settled in the interior of the root. The establishment and multiplication of *Azospirillum* especially under field conditions would be an important factor affecting plant growth promotion properties of the bacterium. In extensive field experimentation carried out by Baldani et al. (1986) in Brazil, *A. brasilense* strains Sp 107 and Sp 245 were shown to be established in all studied wheat assays and were the predominant strains in surface disinfected roots; however, strains Sp 7 and Cd showed poor establishment. Harris et al. (1989) reported on the weak establishment of *Azospirillum* in the rhizosphere of wheat following inoculation under temperate conditions. Reynders and Vlassak (1982a, b) reported on the superior of *A. brasilense* strain Sp BR14 than strain S-631 in enhanced grain yield of wheat. *A. brasilense* strain Sp BR14 and strain S-631 were isolated from wheat and maize, respectively. This result confirmed this idea that successful inoculation in wheat would be achieved with homologous strains isolated of the same crop than nonhomologous strains. Jagnow (1990) showed that there are not only interactions exists among cultivars and *Azospirillum* strains in rate of nitrogenase activity followed inoculation with *Azospirillum lipoferum*, but also that within a individual cultivar differences happen in response to various plants. A pervious study indicated that different strains of *A. brasilense* colonize variously the same cultivar of wheat (Saubidet and Barneix 1998). It has been shown that different *A. brasilense* strains colonize differently a single cultivar of wheat and that the individual strains colonize in various numbers when inoculated to various cultivars (Saubidet and Barneix 1998). Crop genotypes were also mentioned with the N fixation potential of *Azospirillum* strains (García de Salomone and Döbereiner 1996). Puente et al. (2005) tested wheat response to several *Azospirillum brasilense* strains and observed higher numbers of tiller, greater root dry matter, and increased total number of spikelets plant^{-1} obtained when only wheat crops were inoculated with the strains of INTA Az-8 or INTA Az-39. Strain of INTA Az-39, which has been reported to isolate from washed root wheat, came from Marcos Juárez, Córdoba Province, Argentina. Rodríguez Cáceres et al. (1996a, b) in a study conducted in a semiarid region of Argentina observed that wheat inoculation with this strain resulted in a significant higher grain yield. From these studies it seems that response of a crop to *Azospirillum* species is crop specific.

17.8 2,4-D and H_2O_2 Pretreatment

With consideration throughout the article published dealing with *Azospirillum* and wheat response, various agronomical management can be outlined that can enhance promoting effect of *Azospirillum* on wheat production. *Azospirillum* is a nitrogen-fixing microaerophilic bacterium with a noticeable aerotaxis and shows a considerable desire for very low oxygen tension (Patriquin et al. 1983). Nitrogenase activity of *Azospirillum* becomes very little under levels above 1 kPa (Patriquin et al. 1983). *Azospirillum* normally lacks the capability of nodules forming on host plants. 2,4-dichlorophenoxyacetic acid (2,4-D) has the ability to induce nodule-like structures or para-nodulates on plant roots (Tchan and Kennedy 1989). These nodule-like structures could be colonized by microorganisms including diazotrophs (rhizohia, *Azotobacter and Azospirillum).* 2,4-dichiorophenoxyacetic acid (2,4-D) has been shown to induce para-nodulates in root of wheat, and these para-nodulates are suitable colonization sites for *Azospirillum.* Zeman et al. (1992) observed higher rate of nitrogenase activity in para-nodulated wheat seedlings treated by 2,4-D than not treated with 2,4-D. An in vitro assay carried out by Elanchezhian and Panwar (1997) confirmed the positive effect of 2,4-dichiorophenoxyacetic acid on *Azospirillum brasilense* Sp7 performances and claimed that 2,4-D through induced para-nodulated structure or provide a suitable niche for *Azospirillum* in which higher nitrogenase activity occurred. Inside these nodule-like structures, *Azospirillum* protect from oxygen and higher nitrogenase activity in plants treated with 2,4-D confirmed this opinion. However, this technique after describing was not further considered in the forthcoming research. Recently a field experiment showed that pretreatment wheat with hydrogen peroxide can positively affect *Azospirillum* performances. Field study by and Jafariyan and Zarea (2016) was designed to elucidate if pretreatment wheat seed with H_2O_2 affects beneficial effect of *Azospirillum* on plant. In this study three concentrations of H_2O_2 (25%, 50%, and 80%) were used, and results elicit positive effect of H_2O_2 on *Azospirillum* performances toward wheat. Results from this study indicated that soaking wheat seed in 80% H_2O_2 and inoculated with *Azospirillum* improved total grain yield. This increased grain yield is accompanied with anatomic changes in leaf.

17.9 Nitrogen Application

The element nitrogen (N) has been assumed to be one of the topmost important macronutrients taken up in considerable amount than of any nutrient by the wheat. Based on the study's results, it seems that *Azospirillum* is not able to provide all N demented by the wheat and cannot completely replace N fertilizers for wheat. *Azospirillum* inoculation may result in improved grain yield or sustain a certain yield standard but with a reduction in the amounts of N fertilizers applied (Rothballer et al. 2003; Hungria et al. 2010; Venieraki et al. 2011; Piccinin et al. 2013). Rodríguez Cáceres et al. (1996b) claimed that the difference in wheat grain yield response to *Azospirillum brasilense* inoculation carried out in Entic Haplustolls from the Pampas region of Argentina was mainly related to variation in fertility of soil and availability of water. Minor contribution of N by *Azospirillum* to wheat plants under absence of exogenous carbohydrate condition in the plant rhizosphere was first reported by Lethbridge and Davidson (1983). Kapulnik et al. (1987) reported on negligible contribution of *Azospirillum* on the N_2 fixation in wheat plants. The early study by Millet and Feldman (1984) on response of the yield of a common spring wheat cultivar (*Triticum aestivum*) to inoculation with *Azospirillum brasilense* under four levels of application of N fertilization elucidated that increase in plant yield followed inoculation with *Azospirillum brasilense* obtained only from application of the medium and high levels of N fertilization and highest yield increase of approximately 8.0% achieved at the highest level N fertilization (1.0 g of pure N per plant). This study was a pot experiment conducted outdoors, and pots with volume of five liters were filled with a mixture ratio of 50:33:17 percent (v/v) of volcanic gravel, peat, and vermiculite and adequately irrigated. Mertens and Hess (1984), investigating the effect of *A. lipoferum* Sp 108 st, isolated from roots of maize in Brazil, on wheat found that yield response of wheat to this strain was positive irrespective of soil being fertilized with chemical nitrogen or not. Kapulnik et al. (1987) conducted field trails from 1980 to 1983 to investigate the response of eight commercial spring wheat cultivars (six *Triticum aestivum* and two *T. turgidum*) to *Azospirillum brasilense* inoculation with 40 and 120 kg N ha^{-1}. Increased plant dry matter under low level of N fertilization did not reflect into total grain yield except for cultivar "Miriam" which represented a positive and significant increase of grain yield. Further results of this study showed no significant impact of *Azospirillum* inoculation when high rate of N was applied. To confirm this obtained result, four cultivar responses of bread wheat (*T. aestivum*) to *Azospirillum* inoculation were investigated individually under four various locations with different N levels. Results of this experiment proved only positive and consistent response of grain yield of cultivar Miriam to inoculation with *Azospirillum*. Enhanced root development and branching was found to be the most striking impact of *Azospirillum* inoculation on cultivar Miriam. Spring wheat cultivar "Miriam" produced higher fertility number of tiller per unit area following *Azospirillum brasilense* inoculation, but this trait was significantly influenced by the dose of N applied (Kapulnik et al. 1987). Kapulnik et al. (1983) and Millet et al. (1985) also obtained the same result and found higher numbers of fertile tillers per area unit due to *Azospirillum* inoculation. With ^{15}N dilution technique, Kucey (1988) found that the contribution of *Azospirillum* to wheat through biological N_2 was 5–10%. In other experiments conducted by Kapulnik et al. (1985a) and Lethbridge and Davidson (1983), contribution of *Azospirillum* to wheat through fixed nitrogen was found to be negligible; however, with adding carbohydrate to the rooting medium, this contribution was increased. With using technique of ^{15}N dilution, Boddey et al. (1986) observed that increased grain yield and N content of wheat was not due to contribution of *Azospirillum* through N_2-fixation. Okon et al. (1988), through extrapolating of data from various $^{15}N_2$-incorporation and ^{15}N-dilution assessments, reported on an only fixation of 1 kg of N ha^{-1} in one growing season by *Azospirillum*. Galal et al. (2001), by evaluating the effects of bacterial inoculation (*Azospirillum brasilense* Sp 245) and N fertilizer doses on growth of wheat under field condition carried out in Egypt, showed that inoculation resulted in higher accumulation of dry matter in shoot and increased grain yield and N uptake by them with increasing N fertilizer doses up to 120 kg N ha^{-1}. Wheat grain yield is most affected by the N application, and among the nutrients N has the greatest impact on wheat yield. *Azospirillum* as a nitrogen-fixing bacteria cannot provide all N demand of the host plants. For example, Piccinin et al. (2013) investigate the yield response of wheat to two different forms of inoculants of *Azospirillum* (liquid inoculant and peat inoculant) under application of three levels of zero, 50, and 100 kg N ha^{-1}. Results indicated that inoculation of wheat seeds with *Azospirillum* spp. can substitute the half dose of N fertilizer applied. Results of this experiment also showed that nitrogen fertilization along with different forms of *Azospirillum* inoculation increased grain yield of wheat. These authors concluded that application of the half level of chemical N fertilizer associated with various forms of *Azospirillum* inoculants improved agronomic traits as well as grain yield. Therefore, it may be recommended using N fertilizer and *Azospirillum* inoculation together for the insurance of the most effective performances of *Azospirillum* on enhanced grain yield in wheat. However, the mineralization process of the associative N_2-fixing bacteria may share with input of supplemental N to the plants (Hungria 2011). Other studies have shown that *A. brasilense* inoculation resulted in increased N accumulated in the spike of wheat especially under application of N fertilizer (Santa et al. 2004). Zagonel et al. (2002) and Heinemann et al. (2006) also found considerable positive impact of levels of applied N on wheat grain yield. Dommelen et al. (2009), by studying the inoculation of *A. brasilense* in wheat, discovered that the major form of N assimilation occurs through activity of glutamine synthetase enzyme, which commonly releases ammonia into the medium.

The advantages induced by inoculation with *Azospirillum* associated to the use of half level of nitrogen contributed to the agronomic achievement of wheat. In contrast, results of some research indicated that application of higher levels of N fertilizer may negatively impact *Azospirillum* spp. performances because of rapid decreasing in the nitrogenase activity enzyme (Hartmann and Zimmer 1994). These negative effects were also observed in a field study conducted by Fukami et al. (2016). Numerous researches have shown that the inoculation of wheat with bacteria which belongs to the genus *Azospirillum* increased the root system of plant (Bashan and Levanony 1990; Didonet 1993; Didonet and Magalhães 1993). By improving both volume and number of root and subsequently increase in mineral and water uptake (Lin et al. 1983), grain yield would increase (Okon 1985; Okon and Labandera-Gonzalez 1994). However, the domination of *Azospirillum* within the roots is reported to occur when plant is at anthesis growth stage (Magalhães et al. 1979). At this wheat growth stage, noticeable portions of grain (sinks) have already been formed. In wheat grain yield is more restricted by the formed sinks (Fischer 1985; MacManey et al. 1986). Therefore, efficiency of bacteria on grain yield becomes negligible. However, bacterial activity results in increase content of N (Baldani et al. 1987) but that increased is out of sync with wheat demand, absorbs after anthesis (Rodrigues et al. 2000), and therefore has negligible impact on grain yield. For example, Rodrigues et al. (2000) conducted a field study to elucidate the role of inoculation with two strains of *Azospirillum brasilense*, strains 245 and JA 04, in assimilates and productivity of nitrogen and assimilates of bread wheat cultivar BR-23 under field condition. Results of this experiment showed no detectable variation in seed yield and in the assimilation translocation at anthesis and maturity resulting from a peat inoculation of strains. But differences occurred when N was added. Although inoculation treatments in the absence of N application significantly resulted in increase in content of nitrogen, but that increased had no effect in grain yield. Piccinin et al. (2013) also claimed that under lower levels of applying N, there was no stimulation effect of *Azospirillum* on wheat. However, this research has been done under not limited environmental recourses such as water deficiency or salinity stress, while environmental constraints may impact this effect of *Azospirillum* on the wheat.

17.10 Strains Used

The strains used as *Azospirillum* inoculants for wheat would be important since the root exudate of wheat attracts specific strains of *Azospirillum*. Earliest studies showed chemotactic response of *Azospirillum* to a range of amino acids, mono- and disaccharides, and organic acids which have been previously shown. Chemotactic response of *Azospirillum* is irrespective of its ability to metabolize the attractant (Barak et al. 1983; Heinrich and Hess 1985; Reinhold et al. 1985). Chemotaxis has been reported to be strain specific. For example, strains of *Azospirillum*, those that were isolated from plants with exhibiting C_4 pathway like Kallar grass and maize, had substantial appeal to malate but not to oxalate. Malate has been reported to be the most dominating organic acid exuded by Kallar grass and presumably also by maize, while oxalate is hardly detectable in these crops (Michiels et al. 1989). Reinhold et al. (1985) demonstrated that *Azospirillum* which were isolated from wheat were more attractive to oxalate compared with malate. Oxalate is not the major organic acid in the root exudates of C_4 plants (like maize or Kallar grass). In C_3 plant such as wheat, oxalate is assumed to be one of the dominating organic acid released from root. Early research showed that *A. lipoferum* colonize roots of C_4 plants (like sorghum and maize), whereas *A. brasilense* preferentially colonize the root of C_3 plant such as wheat (Döbereiner and De-Polli 1980; Rocha et al. 1981; Mandimba et al. 1986). It is assumed that there is *Azospirillum* strain-specific chemotaxis. Michiels et al. (1989) suggested that chemotactic may play a major role in determining host-specificity in colonization and attraction of certain species of *Azospirillum* toward root of certain plant.

17.10.1 Inoculation Method and Dose

Inoculation methods may effect on the wheat performances to *Azospirillum* inoculation. Early work done by Barbieri et al. (1988) reported on the changing influence of *Azospirillum brasilense* on root system by inoculums dose used. These authors claimed that inoculum concentration of 10^6 colony form unit ml^{-1} resulted in a greatest enhancement of the both number and length of lateral roots, whereas 10^9 colony form unit ml^{-1} inhibited development of root system in *Triticum durum* var. *Appula*. Creus et al. (1996) investigated two inoculation methods in bread wheat (*Triticum aestivum*). In this regard pre-germinated wheat seeds were inoculated with *Azospirillum* Sp 245 during imbibitions of seed. In this experiment seeds of wheat were immersed for 3 h in water and in media culture of *Azospirillum* cells mL^{-1} for 3 h. Results of this experiment showed that inoculation of seeds through immersing and then dried to lower level of water content (14%) caused the seeds to maintain a viable number of 10×3.7^6 cells•g^{-1} dry weight up to 27 days. Thirty days of stored seeds were able to germinate and retain up to 10^6 cells g^{-1} fresh weight within roots after 7 days of growth. Fukami et al. (2016) designed an experiment to test substitute practices for seed inoculation of two crops including maize and wheat in order to deter the direct connection of bacteria with pesticides. In this experiment *Azospirillum brasilense* was applied in a furrow, and its effectiveness were tested in comparison with seed inoculation. *Azospirillum brasilense* were sprayed at seeding time and after emergence of seedlings through the leaf spraying. Experiments are carried out under controlled environmental and field condition at contrast locations in Brazil. Results of the study showed that colonization rate of leaves was higher in plants that were foliar sprayed, whereas in soil inoculation method, the rate of colonization was higher in the root and rhizosphere. In this study *A. brasilense* resulted in a 25% reduction of need for N fertilizer. An inoculant dose is another important factor that must be considered. The higher doses of *Azospirillum* population within the rhizosphere of root may increase the release of plant hormones. Higher plant hormone secretion inhibits the growth of root. Negative impacts of high doses of *Azospirillum* have been previously observed in wheat (Bashan 1986). Hungria et al. (2010) have also observed benefits for soybean and common bean only at minor concentration of inoculation and growth restriction with higher concentration of *Azospirillum*. Fukami et al. (2016) also observed inhibitory effect of *Azospirillum* inoculation on growth of wheat when higher doses of *Azospirillum* were used.

17.11 Environmental Factors Affecting *Azospirillum* N_2 Fixation

However, N_2 fixation by *Azospirillum* is not very considerable, and other aspects aside from N_2 fixation have been proposed to define the increased yield of crops due to inoculation. N_2 fixation of *Azospirillum* in inoculated grasses has been shown to be negligible and can provide a minimum portion of the N requirement for the plant (Boddey and Dobereiner 1988) but always would be important in order to reduce chemical N fertilizer application under wheat production. It would be important to note that N_2 fixation occurs only under lower levels of O_2 and with low or moderate doses of N fertilizer. Several environmental factors and agronomical management influence N2 fixation in *Azospirillum*. It has been shown that destitution of O_2 can enhance denitrification in *Azospirillum*-wheat associations, resulting in the loss of N loss in the soil (Neuer et al. 1985). Under a severe O_2 deprivation, many strains of A. *lipoferum* and *A. brasilense* have been shown to dissimilate nitrate and nitrite (Eskew et al. 1977; Magalhaes et al. 1983). *A. amazonense* has not been shown to dissimilate nitrate and nitrite (Eskew et al. 1977; Magalhaes et al. 1983). In a process called denitrification, some of the strains are responsible for the more reduction of nitrite to nitrate or nitrogen (Krieg and Döbereiner 1984). Early studies demonstrated that the denitrification process carried out by A. *brasilense* in wheat takes place synchronously with nitrogen fixation (Neuer et al. 1985). High range of soil pH, high temperature, and O_2 limitation are some of the unfavorable conditions that enhance the process of denitrification. Neuer et al. (1985) proposed that the net result of both processes, N_2 fixation and denitrification process, may be a loss of N.

Conclusion

Knowledge about the performances of extensive inoculation with *Azospirillum* under large on-farm experiments are still actually restriction (Díaz-Zorita and Fernández-Canigia 2009). Relationships of crop, *Azospirillum*, and environment are complex, and the environment is a major factor that strongly impacts the grain yield response of wheat to inoculation with *Azospirillum* species (Sala et al. 2007). With consideration throughout the article published dealing with *Azospirillum* and wheat response, various management methods can be outlined that can enhance promoting effect of *Azospirillum* on wheat production:

1. Any physical or chemical treatments that the increased *Azospirillum* numbers in soil would be important factors.
2. Strain selection is important. The strains used as *Azospirillum* inoculants for wheat would be important since the root exudate of wheat attracts specific strains

of *Azospirillum*. The same strains of *Azospirillum* that are isolated from the disinfected roots of the same plant have been mentioned to be much effective compared to nonhomologous strains. Strains that establish within the root of crop may be successfully resulted in better yield positive responses of crop.
3. Inoculation method may effect on the wheat performances to *Azospirillum* inoculation.
4. *Azospirillum* cannot be a full replacement agent for chemical N fertilizers for wheat cropping. *Azospirillum* inoculation may result in increase yield and maintain actual yield standards when accompanied with N fertilizers applied.
5. Chemical agent like 2,4-dichlorophenoxyacetic acid (2,4-D) has been reported to enhance *Azospirillum* beneficial performances toward wheat plant through nodule-like structures inducing.
6. Wheat seed priming or exogenous application of wheat seed with hydrogen peroxide as a reactive oxygen species have been recently shown to enhance promoting effect of *Azospirillum* on increased wheat production under dry land farming.

References

Baldani JI (1984) Ocorrência e caracterização de *Azospirillum amazonense* em comparação com outras species deste gênero em raízes de milho, sorgo e arroz, M.Sc. Thesis, Universidade Federal Rural do Rio de Janeiro, RJ, Brasil

Baldani JI, Baldani VLD (2005) History on the biological nitrogen fixation research in graminaceous plants: special emphasis on the Brazilian experience. An Acad Bras Cienc 77:549–579

Baldani VLD, Baldani JI, Döbereiner J (1983) Effects of *Azospirillum* inoculation on the root infection and nitrogen incorporation in wheat. Can J Microbiol 29:433–439

Baldani JI, Baldani VLD, Sampaio MJAM, Dobereiner J (1984) A fourth Azospirillum species from cereal roots. An Acad Bras Cien 56:365

Baldani VLD, Alvarez MA, Baldani JA, Döbereiner J (1986) Establishment of inoculated *Azospirillum* spp. in the rhizosphere and in roots of field grown wheat and sorghum. Plant Soil 90:35–46

Baldani VLD, Baldani JL, Döbereiner J (1987) Inoculation of field grown wheat *(Triticum aestivum)* with *Azospirillum* spp. in Brazil. Biol Fertil Soils 4:37–40

Barak R, Nur I, Okon Y (1983) Detection of chemotaxis in *Azospirillum brasilense*. J Appl Bacteriol 54:399–403

Barbieri P, Galli E (1993) Effect on wheat root development of inoculation with an *Azospirillum brasilense* mutant with altered indole-3-acetic acid production. Res Microbiol 144:69–75

Barbieri P, Zanelli T, Gaili E, Zanetti G (1986) Wheat inoculation with *Azospirillum brasilense* Sp6 and some mutants altered in nitrogen fixation and indole-3-acetic acid production. FEMS Microbiol Lett 36:87–90

Barbieri P, Bernardi A, Galli E, Zanetti G (1988) Effects of inoculation with different strains of azospirillum brasilense on wheat roots development. In: Klinqmüller W (ed) *Azospirillum* IV: genetics, physiology, ecology. Springer, Berlin/Heidelberg, pp 181–188

Bashan Y (1986) Migration of the rhizosphere bacteria Azospirillum brasilense and Pseudomonas £uorescens towards wheat roots in the soil. J Gen Microbiol 132:3407–3414

Bashan Y, de-Bashan LE (2010) How the plant growth-promoting bacterium *Azospirillum* promotes plant growth – a critical assessment. Adv Agron 108:77–136

Bashan Y, Levanony H (1990) Current status of *Azospirillum* inoculation technology: *Azospirillum* as a challenge for agriculture. Can J Microbiol 36:591–608

Bashan Y, Holguin G, de-Bashan LE (2004) *Azospirillum*-plant relationships: physiological, molecular, agricultural, and environmental advances (1997–2003). Can J Microbiol 50(8):521–577

Bhattarai T, Hess D (1993) Yield responses of Nepalese spring wheat (*Triticum aestivum* L.) cultivars to the inoculation with *Azospirillum* spp of Nepalese origin. Plant Soil 151:67–76

Boddey RM, Döbereiner J (1982) Association of *Azospirillum* and other diazotrophs with tropical gramineae. In: International congress of soil science, 12th edn. New Delhi, Non-symbiotic nitrogen fixation and organic matter in the tropics, pp 28–47

Boddey RM, Döbereiner J (1988) Nitrogen fixation associated with grasses and cereals: recent results and perspectives for future research. Plant Soil 108:53–65

Boddey RM, Baldani VLD, Baldani JI, Döbereiner J (1986) Effect of inoculation of *Azospirillum* spp. on nitrogen accumulation by field-grown wheat. Plant Soil 95:109–121

Braun HJ, Atlin G, Payne T (2010) Multi-location testing as a tool to identify plant response to global climate change. In: Reynolds MP (ed) Climate change and crop production. CABI Climate Change Series, Surrey, pp 115–138

Creus CM, Sueldo RJ, Barassi CA (1996) A*zospirillum* inoculation in pregerminating wheat seeds. Can J Microbiol 42:83–86

Creus CM, Sueldo RG, Barassi CA (1998) Water relations in *Azospirillum*-inoculated wheat seedlings under osmotic stress. Can J Bot 76:238–244

Creus C, Sueldo RJ, Barassi CA (2004) Water relations and yield in *Azospirillum*-inoculated wheat exposed to drought in the field. Can J Bot 82:273–281

Dekhil S, Cahill M, Stackebrandt E et al (1997) Transfer of *Conglomeromonas largomobilis* subsp. *largomobilis* to the genus *Azospirillum* as *Azospirillum largomobile* comb. nov., and elevation of *Conglomeromonas largomobilis* subsp. *parooensis* to the new type species of *Conglomeromonas*, *Conglomeromonas parooensis* sp. nov. system. Appl Microbiol 20:72–77

Díaz-Zorita M, Fernández-Canigia MV (2009) Field performance of a liquid formulation of Azospirillum brasilense on dryland wheat productivity. Eur J Soil Biol 45:3–11

Didonet AD (1993) Aspectos do mecanismo de ação fisiológica associados à promoção do crescimento radicular de trigo (*Triticum aestivum* L.) por bactérias do gênero *Azospirillum*. Campinas: Unicamp, 68p. Tese de Doutorado

Didonet AD, Magalhães AC (1993) The role of auxin-like compounds in plant growth promoting rhyzobacteria: the wheat-*Azospirillum* association. Revista Brasileira de Fisiologia Vegetal, Londrina 5:179–183

Döabereiner J, Baldani VLD, Reis VM (1995) Endophytic occurrence of diazotrophic bacteria in non-leguminous crops. In: Fendrik I, Del Gallo M, Vanderleyden J, de Zamaroczy M (eds) *Azospirillum* VI and related microorganisms. Springer, Berlin, pp 3–14

Dobbelaere S, Croonenborghs AT, Thys A, Ptacek D, Vanderleyden J, Dutto P, Labandera-González C, Caballero MJ, Aguirre JF, Kapulnik Y, Brener S, Burdman S, Kadouri D, Sarig S, Okon Y (2001) Responses of agronomically important crops to inoculation with Azospirillum. Aust J Plant Physiol 28:871–879

Döbereiner J, De-Polli H (1980) Diazotrophic rhizocoenoses. In: Stewart WDP (ed) Nitrogen fixation. Academic Press, London, pp 301–333

Dommelen AV, Croonenborghs A, Spaepen S, Vanderleyden J (2009) Wheat growth promotion through inoculation with an ammonium-excreting mutant of Azospirillum brasilense. Biol Fertil Soils 45:549–553

Eckert B, Weber OB, Kirchhof G, Halbritter A, Stoffels M, Hartmann A (2001) Azospirillum doebereinerae sp. nov., a nitrogen fixing bacterium associated with the C4-grass Miscanthus. Int J Syst Evol Microbiol 51:17–26

Elanchezhian R, Panwar JDS (1997) Effects of 2,4-D and *Azospirillum brasilense* on nitrogen fixation, photosynthesis and grain yield in wheat. Agron crop sci 178:129–133

Eskew DL, Focht DD, Ting IP (1977) Nitrogen fixation, denitrification and pleomorphic growth in a highly pigmented *Spirillum lipoferum*. Appl Environ Microbiol 34:582–585

Fages J (1994) *Azospirillum* inoculants and field experiments. In: Okon Y (ed) *Azospirillum*/plant associations. CRC Press Inc., Boca Raton, pp 87–109

Ferreira MCB, Fernandes MS, Döbereiner J (1987) Role of Azospirillum brasilense nitrate reductase in nitrate assimilation by wheat plants. Biol Fertil Soils 4:47–53

Fischer RA (1985) Number of kernels in wheat crops and influence of solar radiation and temperature. J Agr Sci, Cambridge, Great-Britain 105:447–461

Freitas JLM, Rocha REM, Pereira PAA, Döbereiner J (1982) Matéria orgânica e inoculação de *Azospirillum* na incorporação de N pelo milho. Pesq Agropec Bras 17:1423–1432

Fukami J, Nogueira MA, Araujo RS, Hungria M (2016) Accessing inoculation methods of maize and wheat with *Azospirillum brasilense*. AMB Expr 6:12–13

Galal YGM, El-Ghandour IA, El-Akel EA (2001) Stimulation of wheat growth and N fixation through *Azospirillum* and *Rhizobium* inoculation: a field trial with techniques Food security and sustainability of agro-ecosystems. In: Horst WJ (ed). Plant nutrition-food security and sustainability of agroecosystems. Kluwer Academic, Netherlands, pp 666–667

García de Salomone I, Döbereiner J (1996) Maize genotype effects on the response to Azospirillum inoculation. Biol Fertil Soils 21:193–196

Harris JM, Lucas JA, Davey MR, Lethbridge G, Powell KA (1989) Establishment of *Azospirillum* inoculant in the rhizosphere of winter wheat. Soil Biol Biochem 21:59–64

Hartmann A, Zimmer W (1994) Physiology of *Azospirillum*. In: Okon Y (ed) *Azospirillum*/plant associations. CRC Press, Boca Raton, pp 15–39

Heinemann AB, Stone LF, Didonet AD, Trindade MG, Soares BB, Moreira JAA, Cánovas AD (2006) Eficiência de uso da radiac¸ ão solar na produtividade do trigo decorrente da adubac¸ ão nitrogenada. Rev Bras Eng Agríc Ambiental 10:352–356

Heinrich D, Hess D (1985) Chemotactic attraction of *Azospirillum lipoferum* by wheat roots and characterization of some attractants. Can J Microbiol 31:26–31

Horemans S, De Coninck K, Vlassak K (1988) Aspects of the ecology of Azospirillum sp. in Belgian soils. In: Klingmuller W (ed) *Azospirillum* N: genetics, physiology, ecology. Springer, Berlin

Hungria M (2011) Inoculac¸ ão com Azospirillum brasilense: inovac¸ ão em rndimento a baixo custo. Embrapa Soja, Londrina, 36 p

Hungria M, Campo RJ, Souza EM, Pedrosa FO (2010) Inoculation with selected strains of *Azospirillum brasilense* and *A. lipoferum* improves yields of maize and wheat in Brazil. Plant Soil 331:413–425

Jafariyan T, Zarea MJ (2016) Hydrogen peroxide affects plant growth promoting effects of *Azospirillum*. Crop Sci Biotech 19:167–175

Jagnow G (1990) Differences between cereal crop cultivars in root associated nitrogen fixation, possible causes of variable yield response to seed inoculation. Plant Soil 123:255–259

Jain DK, Patriquin DG (1984) Root hair deformation, bacterial attachment, and plant growth in wheat-*Azo,pirillum* associations. Appl Environ Microbiol 48:1208–1213

Jain DK, Patriquin DG (1985) Characterization of a substance produced by *Azospirillum* which causes branching of wheat root hairs. Can J Microbiol 31:206–210

Kapulnik Y, Sarig S, Nur I, Okon Y (1983) Effect of *Azospirillum* inoculation on yield of field grown wheat. Can J Microbiol 29:895–899

Kapulnik Y, Feldman M, Okon Y, Henis Y (1985a) Contribution of nitrogen fixed by *Azospirillum* to the N nutrition of spring wheat in Israel. Soil Biol Biochem 17:509–515

Kapulnik Y, Okon Y, Henis Y (1985b) Changes in root morphology caused by *Azospirillum* inoculation. Can J Microbiol 31:881–887

Kapulnik Y, Okon Y, Henis Y (1987) Yield response of spring wheat cultivars (*Triticum aestivum* and *T. turgidum)* to inoculation with *Azospirillum brasilense* under field conditions. Biol Fertil Soils 4:27–35

Krieg NR, Döbereiner J (1984) Genus *Azospirillum*. In: Holt JG, Krieg NR (eds) Bergey's manual of systematic bacteriology, 9th edn. Williams & Wilkins, Baltimore, pp 94–104

Kucey RMN (1988) Alteration of size of wheat root systems and nitrogen fixation by associative nitrogen-fixing bacteria measured under field conditions. Can J Microbiol 34:735–739

Lavrinenko K, Chernousova E, Gridneva E et al (2010) *Azospirillum thiophilum* sp. nov., a novel diazotrophic bacterium isolated from a sulfide spring. Int J Syst Evol Microbiol 60:2832–2837

Lethbridge G, Davidson MS (1983) Root-associated nitrogen-fixing bacteria and their role in the nitrogen nutrition of wheat estimated by 15N isotope dilution. Soil Biol Biochem 15:365–374

Levanony H, Bashan Y (1989) Enhancement of cell division in wheat root tips and growth of root elongation zone induced by *Azospirillum brasilense* Cd. Can J Bot 67:2213–2216

Lin W, Okon Y, Hardy RWF (1983) Enhanced mineral uptake by *Zea mays* and *Sorghum bicolor* roots inoculated with *Azospirillum brasilense*. Appl Environ Microbiol 45:1775–1779

Lin SY, Young CC, Hupfer H, Siering C, Arun AB, Chen WM, Lai WA, Shen FT, Rekha PD, Yassin AF (2009) *Azospirillum picis* sp. nov., isolated from discarded tar. Int J Syst Evol Microbiol 59:761–765

Lin S-H, Liu Y-C, Hameed A et al (2013) *Azospirillum fermentarium* sp. nov., a novel nitrogenfi xing species isolated from a fermenter in Taiwan. Int J Syst Evol Microbiol 63:3762–3768

Macmaney M, Diaz M, Simon C, Gioia A, Slafer GA, Andrade FH (1986) Respuesta a la reducción de la capacidad fotosintética durante el llenado de granos en trigo. In: Congreso Nacional Del Trigo, Pergamino. AINBA, Pergamino, pp 178–190

Magalhães FMM, Patriquin D, Döbereiner J (1979) Infection of field grown maize with *Azospirillum* spp. Revista Brasileira de Biologia, Rio de Janeiro 39:587–596

Magalhaes FMM, Baldani JI, Souto SM, Kuykendall JR, DObereiner J (1983) A new acid-tolerant *Azospirillum* species. Ann Acad Bras Cian 55:417–430

Malik KA, Rasul G, Hassan U, Mehnaz S, Ashraf M (1994) Role of N_2-fixing and growth hormones producing bacteria in improving growth of wheat and rice. In: Hegazi NA, Fayez M, Monib M (eds) Nitrogen fixation with non-legumes. American University in Cairo Press, Cairo, pp 409–422

Mandimba G, Heulin T, Bally R, Guckert A, Balandreau J (1986) Chemotaxis of free-living nitrogen-fixing bacteria towards maize mucilage. Plant Soil 90:129–139

Mehnaz S, Weselowski B, Lazarovits G (2007a) *Azospirillum canadense* sp. nov., a nitrogen-fixing bacterium isolated from corn rhizosphere. Int J Syst Evol Microbiol 57:620–624

Mehnaz S, Weselowski B, Lazarovits G (2007b) *Azospirillum zeae* sp. nov., a diazotrophic bacterium isolated from rhizosphere soil of *Zea mays*. Int J Syst Evol Microbiol 57:2805–2809

Mertens T, Hess D (1984) Yield increases in spring wheat (*Triticum aestivum* L.) inoculated with *Azospirillum lipoferum* under greenhouse and field conditions of a temperature region. Plant Soil 82:87–99

Michiels K, Vanderleyden J, Van Gool A (1989) *Azospirillum* – plant root associations: a review. Biol Fertil Soils 8:356–368

Millet E, Feldman M (1984) Yield response of a common spring wheat cultivar to inoculation with *Azospirillum brasilense* at various levels of nitrogen fertilization. Plant Soil 80:255–259

Millet E, Avivi Y, Feldman M (1985) Effects of rhizospheric bacteria on wheat yield under field conditions. Plant Soil 86:347–355

Murty MG, Ladha JK (1987) Differential colonization *of Azospirillum lipoferum* on roots of two varieties of rice *(Oryza sativa* L). Biol Fertil Soils 4:3–7

Neuer G, Kronenberg A, Bothe H (1985) Denitrification and nitrogen fixation by *Azospirillum*. Arch Microbiol 141:364–370

Okon Y (1985) *Azospirillum* as a potential inoculant for agriculture. Trends Biotechnol 3:223–228

Okon Y, Kapulnik Y (1986) Development and function of Azospirillum-inoculated roots. Plant Soil 90:3–16

Okon Y, Labandera-gonzalez CA (1994) Agronomic applications of *Azospirillum*: an evaluation of 20 years of worldwide field inoculation. Soil Biol Biochem 26:1591–1601

Okon Y, Fallik E, Sarig S, Yahalom E, Tal S (1988) Plant growth promoting effects of *Azospirillum*. In: Bothe H, Bruijn Newton WE (eds) Nitrogen fixation: hundred years after. Proceeding of the 7th international congress on nitrogen fixation, Colgne, FRG. Gustav Fischer, Stuttgart, pp 741–746

Pacovsky RS (1990) Development and growth effects in the *Sorghurn-Azospirilium* association. J Appl Bacteriol 68:555–563

Patriquin DG, Dobereiner J, Jain DK (1983) Sites and processes of association between diazotrophs and grasses. Can J Microbiol 29:900–915

Peng G, Wang H, Zhang G et al (2006) *Azospirillum melinis* sp. nov., a group of diazotrophs isolated from tropical molasses grass. Int J Syst Evol Microbiol 56:1263–1271

Piccinin GG, Braccini AL, Dan LGM, Scapim CA, Ricci TT, Bazo GL (2013) Efficiency of seed inoculation with Azospirillum brasilense on agronomic characteristics and yield of wheat. Ind Crop Prod 43:393–397

Puente M, Montecchia MS, Perticari A (2005) Evaluation of Azospirillum inoculant strains in wheat. In: SAGPyA-INTA (ed) 7th International Wheat Congress, Mar del Plata, SAGPyA-INTA, p. en CD

Rai SN, Gaur C (1982) Nitrogen fixation by *Azospirillum* spp and effect of *Azospirillum lipoferum* on the yield and N-uptake of wheat crop. Plant Soil 69:233–238

Reinhold B, Hurek T, Fendrik I (1985) Strain-specific chemotaxis of *Azospirillum* spp. J Bacteriol 162:190–195

Reynders L, Vlassak K (1982a) Use of *Azospirillum brasilense* as biofertilizer in intensive wheat cropping. Plant Soil 66:217–273

Reynders L, Vlassak K (1982b) Physio-ecological aspects and agricultural importance of *Azospirillum-plant* root associations. In: Klingmuller W (ed) *Azospirillum:*genetics, physiology, ecology. Birkhauser, Basel

Rocha REM, Baldani JI, DObereiner J (1981) Specificity of infection by *Azospirillum* spp. in plants with C 4 photosynthetic pathway. In: Vose PB, Ruschel AP (eds) Associative N 2 fixation, vol II. CRC Press, Boca Raton, pp 67–69

Rodrigues S, Didonet AD, Gouveia JA, De Cássia SR (2000) Nitrogen translocation in wheat inoculated with *azospirillum* and fertilized with nitrogen. Pesq Agrop Brasileira 35:1473–1481

Rodríguez Cáceres EA, Di Ciocco C, Pacheco Basurco JC, de la inoculaci6n con I (1996a) Azospirillum brasilense entrigo cultivado en suelos de la provincia de La Pampa, Argentina. Ciencia del Suelo 14:110–112

Rodríguez Cáceres EA, Gonzålez Anta G, L6pez JR, Di Ciocco CA, Pacheco Basurco JC, Parada JL (1996b) Response of field grown wheat to inoculation with *Azospirillum brasilense* and Bacillus polymyxa in the semiarid region of Argentina. Arid Soil Res Rehabil 10:13–20

Rothballer M, Schmid M, Hartmann A (2003) *In situ* localization and PGPB effect of *Azospirillum brasilense* strains colonizing roots of different wheat varieties. Symbiosis 34:261–279

Sala VMR, Cardoso EJBN, De Freitas JG, Da Silveira APD (2007) Wheat genotypes response to inoculation of diazotrophic bacteria in field conditions. Pesqui Agropecu Bras 42:833–842

Salantur A, Ozturk R, Akten S (2006) Growth and yield response of spring wheat (*Triticum aestivum* L.) to inoculation with rhizobacteria. Plant Soil Environ 52:111–118

Santa ORD, Hernadez RF, Alvarez GLM, Ronzelli P, Soccol CR (2004) Azospirillum sp. inoculation in wheat, barley and oats seeds greenhouse experiments. Braz Arch Biol Technol 47:843–850

Sarig S, Btum A, Okon Y (1988) Improvement of water status and yield of field-grown grain sorghum *(Sorghum bicotor) by* inoculation with *Azospirillum brasilense.* J Agric Sci 110:271–277

Saubidet MI, Barneix AJ (1998) Growth stimulation and nitrogen supply to wheat plants inoculated with *Azospirillum brasilense.* J Plant Nutr 21:2565–2577

Saubidet MI, Fatta N, Barneix AJ (2002) The effect of inoculation with *Azospirillum brasilense* on growth and nitrogen utilization by wheat plants. Plant Soil 245:215–222

Schloter M, Kirchhof G, Heinzmann J, Döabereiner J, Hartmann A (1994) Immunological studies of the wheat-root-colonization by the Azospirillum brasilense strains Sp7 and Sp245 using strain specific monoclonal antibodies. In: Hegazi NA, Fayez M, Monib M (eds) Nitrogen fixation with non-legumes. American University in Cairo Press, Cairo, pp 291–297

Sumner ME (1990) Crop responses to *Azospirillum* inoculation. In: Stewart A (ed) Advances in soil science. Springer, New York

Tarrand JJ, Krieg NR, Döbereiner J (1978) A taxonomic study of the *Spirillum lipoferum* group, with the descriptions of a new genus, *Azospirillum* gen. nov. and two species *Azospirillum lipoferum* (Beijerinck) comb. nov. and *Azospirillum brasilense* sp. nov. Can J Microbiol 24:967–980

Tchan YT, Kennedy IR (1989) Possible Nz-fixing root nodules induced in non-legumes. Agric Sci 2:57–59
Tyagi S, Kumar Singh D (2014) *Azospirillum himalayense* sp. nov., a *nifH* bacterium isolated from Himalayan valley soil, India. Ann Microbiol 64:259–266
Umali-Garcia M, Hubbell DH, Gaskins MH, Dazzo FB (1980) Association of *Azospirillum* with grass roots. Appl Environ Microbiol 39:219–226
Vande Broek A, Michiels J, Van Gool A, Vanderleyden J (1993) Spatial-temporal colonization patterns of Azospirillum brasilense on the wheat root surface and expression of the bacterial nifH gene during association. Mol Plant-Microbe Interact 6:592–600
Vande Broek A, Lambrecht M, Vanderleyden J (1998) Bacterial chemotactic motility is important for the initiation of wheat root colonization by Azospirillum brasilense. Microbiol 144:2599–2606
Venieraki A, Dimou M, Pergalis P, Kefalogianni I, Chatzipavlidis I, Katinakis P (2011) The genetic diversity of culturable nitrogen-fixing bacteria in the rhizosphere of wheat. Microb Ecol 61:277–285
Xie CH, Yokota A (2005) *Azospirillum oryzae* sp. nov., a nitrogen-fixing bacterium isolated from the roots of the rice plant *Oryza sativa*. Int J Syst Evol Microbiol 55:1435–1438
Zagonel J, Venâncio WS, Kunz RP, Tanamati H (2002) Doses de nitrogênio e densidades de plantas com e sem regulador de crescimento afetando o trigo, cultivar OR-1. Ciênc Rural 32:25–29
Zarea MJ, Hajinia S, Karimi N, Mohammadi Goltapeh E, Rejali F, Varma A (2012) Effect of Piriformospora indica and Azospirillum strains from saline or non-saline soil on mitigation of the effects of NaCl. Soil Biol Biochem 45:139–146
Zeman AMM, Tchan YT, Elmerich C, Kennedy IR (1992) Nitrogenase activity in wheat seedlings bearing *para-nodules* induced by 2,4-dichlorophenoxyacetic acid (2,4-D) and inoculated with *Azospirillum*. Res Microbiol 143:847–855
Zhou S, Han L, Wang Y et al (2012) *Azospirillum humicireducens* sp. nov., a nitrogen-fixing bacterium isolated from a microbial fuel cell. Int J Syst Evol Microbiol 63(pt 7):2618–2624

Current Scenario of Root Exudate–Mediated Plant-Microbe Interaction and Promotion of Plant Growth

18

Kanchan Vishwakarma, Shivesh Sharma, Vivek Kumar, Neha Upadhyay, Nitin Kumar, Rohit Mishra, Gaurav Yadav, Rishi Kumar Verma, and Durgesh Kumar Tripathi

Abstract

Over the last few years, a boom has been witnessed in the area of soil ecology which has produced numerous data on interactions between plant and rhizospheric microbes. The plant-microbe interactions in the rhizospheric niche have proved to be crucial for the advancement of sustainable farming practices which decrease the usage of chemical fertilizers and pesticides. Root exudates are substances released by plant roots that show a significant role in mediating the plant-microbe interactions in soil. These root exudates send chemical signals to microbes which in response are attracted towards the roots and influence growth of plants, soil properties, and microbial community. This chapter is focussed on

K. Vishwakarma • N. Upadhyay • N. Kumar • R.K. Verma
Department of Biotechnology, Motilal Nehru National Institute of Technology Allahabad, Allahabad, Uttar Pradesh, India

S. Sharma (✉)
Department of Biotechnology, Motilal Nehru National Institute of Technology, Allahabad, Uttar Pradesh, India
e-mail: ssnvsharma@gmail.com

V. Kumar
Himalayan School of Biosciences, Swami Rama Himalayan University, Jolly Grant, Dehradun, Uttarakhand, India
e-mail: vivekbps@gmail.com

R. Mishra • D.K Tripathi
Centre for Medical Diagnostic and Research, MNNIT Allahabad, Allahabad, Uttar Pradesh, India

G. Yadav
Department of Biotechnology, Motilal Nehru National Institute of Technology Allahabad, Allahabad, Uttar Pradesh, India

Centre for Medical Diagnostic and Research, MNNIT Allahabad, Allahabad, Uttar Pradesh, India

V. Kumar et al. (eds.), *Probiotics in Agroecosystem*,
DOI 10.1007/978-981-10-4059-7_18

recent advancements in the utilization of root exudates in plant-microbe interactions to enhance plant growth promotion. The plant-microbe interactions are categorized as beneficial or detrimental depending upon the characteristics of root exudates. This chapter also covers different types of root exudates and their function in modifying the exchanges between rhizospheric microbes and plants for the betterment of soil health and sustainable ecosystems.

18.1 Introduction

Soil, also known as a "black box," inhabits diverse macro- and micro-community structures, and the rhizosphere is where contact between soil and roots is established. Biological, physical, and chemical activities are affected by the compounds that the root exudes and microbes feed on (Kamilova et al. 2006; Kumar et al. 2007).

There is a group of chemical substances and signaling compounds produced by plants that help the plant to impart defensive mechanisms against pathogens and attract beneficial microbes (Haichar et al. 2008). The rhizosphere is representative of an extremely lively base of communication between roots, soil microorganisms, invertebrates, and neighboring root systems of competitor plants (Hirsch et al. 2003). There are usually two sub-divisions to the rhizosphere: First is the endo-rhizosphere which contains the root cortex, epidermis, and root hairs. The second is the ecto-rhizosphere with root-connected soil compartments up to 5 mm in length (Fig. 18.1). Hence, this area has made an exciting zone for exploring associations between microorganisms and plants.

The prime role of the root is to provide anchorage and backing for the plant and to facilitate absorption and conduction of water and nutrients (Abbott and Murphy 2003). There is production of root hairs over epidermal cells in tap root and lateral root systems. They are specialized for absorbing water and nutrients from soil. Even though such root functions have been known for a long time, the diversity of root exudates present in the rhizosphere and their exact role in influencing microbial behavior is still unclear (Narula et al. 2009). Plant root exudates are essential parameters for structuring the bacterial community in the rhizosphere (Walker et al. 2003a), performing key roles such as defending against pathogens (Abbott and Murphy 2003), and forming a basis for chemotaxis to mediate attraction and repulsion among particular microbial species and communities (Kumar et al. 2007). Root exudates also maintain moisture and wetting of the soil, mobilize minerals and nutrients, change the chemical characteristics of the soil, provide stability to soil-aggregates surrounding the roots, and inhibit the development of neighboring competitor plants (Narula et al. 2009).

Plant roots continuously release a massive array of significantly beneficial low and high–molecular weight components in the rhizosphere like ions, oxygen, enzymes, water, mucilage, and a number of carbon-containing compounds as well as secondary metabolites (Nardi et al. 2000; Vishwakarma et al. 2016) which helps in complex biological and physico-chemical interactions occurring amongst plant roots and the near-by soil environment. The associations that involve rhizospheric

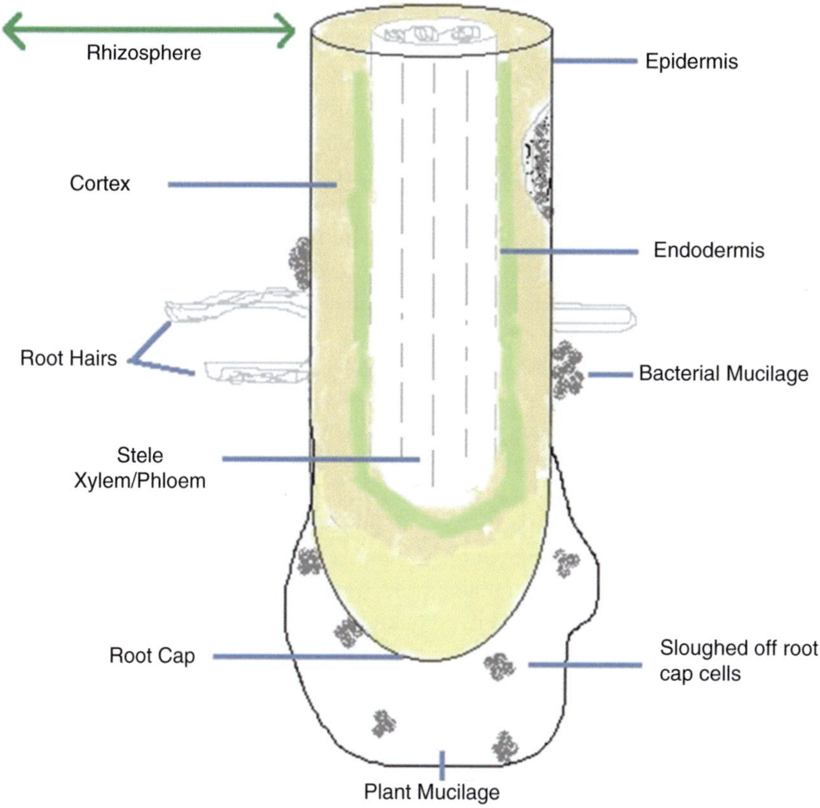

Fig. 18.1 Sub-division of the rhizosphere showing different sections of roots

roots consist of root-microbe, root-invertebrate, and root-root interactions (Bais et al. 2001; Gleba et al. 1999).

Root exudates are divided into two separate categories of compounds, namely, Low and High molecular weight compounds.

1. Low–molecular weight exudates consists sugars, organic acids, secondary metabolites, phenols, and the amino acids. This category comprises most of the diversity of root exudates (Rougier 1981).
2. High–molecular weight compounds include proteins and polysaccharides (mucilage) and form less molecular diversity than the low–molecular weight category but they make up a huge percentage of total exudates (Abbott and Murphy 2003; Walker et al. 2003b).

Numerous phytotoxic components exuded from roots have been characterised like 7,8-benzoflavone from *Acroptilon repens* (Russian knapweed) (Stermitz et al. 2003), (±)-catechin from *Centaurea maculosa* (spotted knapweed) (Bais et al.

2002a), DIMBOA and DIBOA from *Triticum aestivum* (wheat) (Wu et al. 2000), juglone from *Juglans nigra* (black walnut) (Jose and Gillespie 1998), 8- hydroxy-quinoline from *Centaurea diffusa* (diffuse knapweed) (Vivanco et al. 2004), sorgoleone from *Sorghum* spp. (Nimbal et al. 1996), and 5,7,4′-trihydroxy-3′,5′-dimethoxyflavone from *Oryza sativa* (rice) (Kong et al. 2004). A study by Bertin et al. (2003) have shown that differential expression of proteins that are exuded from the roots of the cowpea is in response to environmental conditions with their composition varying with respect to pH and ionic level within the root environment. In this research, western blotting and specific enzyme tests have been used to investigate the presence of defense proteins chitinase, LTPs, and β-1,3-glucanase in imbibed root exudates.

The root exudation process was earlier considered a passive process, but evidence is now available for ATP-binding cassette (ABC) transporters forming the basis for phytochemical translocation in the rhizosphere. The evidence supports the fact that the plant is dynamically releasing metabolites in the surrounding soil (Loyola-Vargas et al. 2007; Badri et al. 2008). There are a number of factors on which constituents of root exudates depend; these include plant species and cultivar, phase of development, substrate to grow plant, and stress parameters (Uren 2000). For instance, root exudates of tomato, sweet pepper, and cucumber were shown to contain more organic acids (mainly succinic, citric, and malic acid) than sugars (fructose and glucose) when grown under gnotobiotic conditions (Kamilova et al. 2006). There is accumulation of ubiquitous phenylpropanoids together with phylogenetically restricted glucosinolates in *A. thaliana* roots (Bednarek et al. 2005). There is also root-derived secretion and accumulation of many flavonoids, triterpene saponins, and isoflavonoids in the cell cultures of *M. trunculata* (Farag et al. 2007; Huhman et al. 2005).

18.2 Importance and Function of Root Exudates

The major function that roots perform includes providing support and anchorage to the plant and facilitating conduction and absorption of water and nutrients (Abbott and Murphy 2003). There is production of root hairs over epidermal cells besides tap root and lateral root systems. They are specialized for absorbing water and nutrients from soil.

The rhizosphere is considered to be supportive for a wide range of bacteria capable of accelerating the plant growth. These plant growth-promoting rhizobacteria (PGPR) work through a number of mechanisms that have N_2-fixation or phytohormone production (Barea et al. 2005).

18.2.1 Against Pathogenic Microorganisms

Plant-microbe communication is one of the most important interactions describing the below-ground zone. Root microbe interactions are influenced by some of the chemicals recognized in the root exudates; these include the flavonoid signals observed in the exudates of legumes, which trigger the genes of *R. melitoti* responsible for nodulation. Such compounds can also be accountable for foundation of vesicular arbuscular mycorrhizal (VAM) (Becard et al. 1995). For instance, extraction of the *Phytophthora cinnamoni* cell wall triggers the release of a multifunctional caffeic acid ester i.e. rosmarinic acid (RA), identified in the root exudates of sweet basil cultures (Bais et al. 2006). Similarly, in a report by Brigham et al. (1999), *Lithospermum erythrorhizon* was shown to produce cell-specific pigmented naphthoquinones.

The physiological state of plants is reflected by the constituents of their root exudates, which are further influenced by both abiotic and biotic parameters (Vishwakarma et al., 2017). For instance, energy is required for inducing expression of the plant defense. This demand for energy is well established for Arabidopsis, i.e., it requires substantial fitness costs to activate the defence responses (Van Hulten et al. 2006). As predicted by local reductions in photosynthetic activity, actuated plant defence need energy to form defence-related compounds (Berger et al. 2007; Bolton 2009). Salicylic acid treatment has been given to the roots of Arabidopsis, and a number of secondary metabolites were secreted, such as butanoic, ferulic and 3-indolepropanoic acid. All of these show antibacterial activity *in vitro* at the levels identified in exudates against pathogens like *Erwinia* spp., *Xanthomonas campestris*, and *P. syringae* (Walker et al. 2003a, b). On the other hand, *P. fluorescens*, which is non-pathogenic, was found to be less sensitive to those exudates.

A report by Mavrodi et al. (2012) shows the precise choice of plant growth protecting rhizobacteria under pathogen attack. It is supported by the results that DAPG-forming *Pseudomonas* were recruited by the wheat rhizosphere during irrigation, while dry conditions supported recruitment of phenazine producing *Pseudomonas*. *G. graminis* var. *tritici* is the chief soil-borne pathogen during irrigation conditions of wheat, while *Rhizoctonia solani* pose more threat during dry situations. Of note, *G. graminis* var. *tritici* shows more sensitivity to DAPG and *R. solani* has more sensitivity towards phenazines. Hence, in circumstances favoring particular pathogens, plants select those antagonists that show more efficiency against those pathogens (Walker et al. 2003b).

18.2.2 Remediation of Heavy Metals

A significant proportion of soil contamination is represented by heavy metals. The main cause of metal pollution in soil is due to anthropogenic activities such as the use of fertilizers and pesticides containing metal and accumulation of industrial waste (Vishwakarma et al. 2017). The microorganisms residing in rhizospheric niche possess high microbial activity and therefore their utilization of the transformation of organic pollutants or removal of contaminants from soil need to be considered (Kothe et al. 2005). The remediation process of pollutants in soil and water takes place exterior to the plant roots. Many organic pollutants are converted into nontoxic compounds by enzymatic actions of microorganisms, whereas several synthetic compounds known as being "recalcitrant" are resistant to any kind of biological degradation.

Bioaccumulation of several metals have been studied. Reports are available on the accumulation of metals such as cadmium, copper, and nickel on Streptomycetes, which denotes a group of Gram-negative bacteria found predominantly in poor and contaminated soils (Albarracín et al. 2008; Schmidt et al. 2005; Sineriz et al. 2009).

It was previously reported that the plant-microbe interaction in soils is greatly influenced by the presence of secondary metabolites existing in the root exudates. These secondary metabolites categorize the association between the individual microbe and plant as mutualistic, associative, or pathogenic. For example, *Rhizobium* spp. form a symbiotic association with legumes and are responsible for symbiotic nitrogen fixation. The interaction between the two organisms is facilitated in part through root-secreted flavones (Redmond et al. 1986).

Plant root exudates have a variety of functions that influence plant growth and also enhance the degradation process of soil contaminants. Some of the important functions of root exudates in the soil include influence on the nutrient cycling processes, enhancement of the degradation organic matter, inhibition of the soil-nitrification process, and interference with the bacterial quorum-sensing response.

18.2.3 Availability of Soil Resources

Availability of phytosiderophores and plant nutrients: Some compounds that are found in the rhizosphere as a part of root exudates function as metal chelators and enhance the availability of metals such as iron, copper, zinc, and manganese for plant uptake (Lambers et al. 2009). It has also been shown that plants use metal chelators in root exudates to enhance nutrient availability for plant growth, for example, the function of graminoid phytosiderophores to transform Fe (III) to form Fe (III)-phytosiderophores, which is taken up by grasses more efficiently than other chelated forms of iron (Doornbos et al. 2012).

Organic acids and phosphorus availability: Organic acids for example citric, malic, and oxalic acid are an important part of root exudates, and also function as metal-chelators in the rhizosphere. More specifically, they are responsible for the solubilization of insoluble forms of phosphate rather than enhancing micronutrient availability (Bais et al. 2006). They form complexes with aluminum or iron in aluminum or ferric phosphates and release phosphates in the form that is taken up by plants (Dakora and Phillips 2002). The mechanism of phosphate solubilization is also reported where plants increase the secretion of carboxylate in the limiting condition of P to solubilize adsorbed phosphate (Vance et al. 2003). Root exudates also play an important role in enhancing the activity of phosphate-solubilizing bacteria to increase the supply of P to the plant (Richardson et al. 2009, 2011).

C- and N- Bio availability: Microbial population in the rhizosphere is highly influenced by those plants that release different types of nutrients from their roots, which are then utilized by the microbes for their growth (Haichar et al. 2012; Baetz and Martinoia 2014). A compound named aminocyclopropane-1-carboxylic acid (ACC) that is secreted by plants is a precursor of ethylene formation and consumed as a C and N source by rhizospheric bacteria (Haichar et al. 2014).

During nitrogen fixation, a less mobile NH_4^+ is transformed to the highly mobile NO_3^- during the process of nitrification. It ultimately influences plant nitrogen uptake because the NO_3^- is vulnerable to loss by the denitrification process from the root surface (Subbarao et al. 2007). To improve nitrogen recovery in soils, it is important to regulate the nitrification step. An example of a nitrification inhibitor was discovered in the root exudates of forage grass (Subbarao et al. 2009).

18.2.4 Regulation of Chemotaxis

Chemotaxis, a phenomenon possessed by most of the motile bacteria, is defined as the process where a microorganism moves in response to chemical gradients. Chemotaxis is a well-defined process in the application of plant-microbe interactions, where soil microbes get attracted towards plant roots (Kumar et al. 2007). The chemotactic response of microorganisms to root exudates performs a significant ecological function in plant-linked bacteria and constitutes the initiation of the interaction between roots and microorganisms. A study of *Pseudomonas putida* showing positive chemotaxis to maize-derived aromatic metabolites was also reported (Neal et al. 2012).

18.3 Plant Root-Microbe Interaction

Plant-microbe interactions show significant communications that characterize the different levels of soil. Root exudates represent an essential constituent mainly helps in maintaining communication between root-associated bacteria and plants. A broad range of chemical compounds and signaling molecules are produced by plants. Approximately 100,000 types of different substrates were produced by plants, which serve as chemotactic agents for microorganisms in plant systems (Bais et al. 2004b).

The unique bacterial communities in the rhizospheric region varies among plant species and through time (Baudoin et al. 2002). Separate root zones in a single plant can assist different bacterial populations, reflecting quantitative and qualitative variations in root exudation (Yang and Crowley 2000). Plants modify the rhizobacterial population by secreting different compounds, which range from single carbohydrate molecules to complex aromatic compounds (Kamilova et al. 2006; Cheng et al. 2014). The reaction of fungal communities to plants is not well documented, although many reports are available targeting mycorrhizal fungi. Similar to bacteria, root exudates are also utilized in maintaining fungal diversity and community structure (Innes et al. 2004). Therefore, it is concluded that the soil ecosystem helps in determining and modifying the rhizospheric microbiome (Garbeva et al. 2008; Lundberg et al. 2012).

A study presented by Broeckling et al. (2008) showed that root exudates from distinctive plant species such as arbuscular mycorrhizal fungi (AMF), nodulating legumes, and a nonmycorrhizal Brassicaceae may affect the diversity of fungal communities of various types within the intact root biomass. The results obtained from the experiment carried out by Broeckling et al. (2008) signify that the regulation of fungal communities by plant root exudates are completed through two different mechanisms that target specific fungal phylotypes. The first mechanism is involved in reducing the relative fungal quantity via an antifungal property of the exudate or a chemical signaling mechanism that restricts their growth. The second mechanism positively regulates abundance either by enhancing chemical signals or by supplying appropriate nutrients for their growth.

18.3.1 Positive Impact

Plant-microbe soil interaction processes are considered to have a significant effect on plant growth by enhancing the supply of nutrients, fixing atmospheric nitrogen, increasing tolerance towards stress conditions, and enhancing resistance against plant pathogens by different classes of endophytic microbes and PGPR (Gray and Smith 2005). Some bacteria can also produce antibiotics and form biofilms that protect them from potential phytopathogens, or degrade toxic compounds produced by either plants or microbes (Bais et al. 2004a).

Plant root exudates perform several functions and play a significant role in root-microbe interactions. Activation of *Rhizobium meliloti* genes initiates nodulation process with the help of compounds like, flavonoids present in exudates of leguminous plants. The colonization of VAM is also influenced by the composition of root exudates (Becard et al. 1995). The root exudates contain several phytoalexins and many defense proteins, which always protect the survival and growth of the delicate and unprotected root cells from pathogenic microorganisms (Flores et al. 1999).

18.3.1.1 Nitrogen Fixation

Gram-negative nitrogen-fixing bacteria have important plant-microbe interactions with leguminous plants (Morgan et al. 2005). Symbiotic relationships between rod shaped proteobacteria like *Rhizobium, Bradyrhizobium, Sinorhizobium, Mesorhizobium, Azorhizobium,* and *Photorhizobium* and infected legume plants are prominent in nature (Bloemberg and Lugtenberg 2001; Madigan and Martinko 2006). Root nodules are formed when a suitable species of one of these rhizobia infects roots (Heidstra and Bisseling 1996), which is why leguminous crops around the world are being treated by these biofertilizers (Deaker et al. 2004). Nod factor, which is a bacterial substance, induces the penetration of rhizobia into root hairs (Perret et al. 2000). These bacteria follow root hair curlings to enter the plant root hairs and then trigger formation by the plant of a cellulosic tube known as an infection thread. These infection threads are then used as a way to spread all over the root hairs and infect adjacent root cells, which helps in triggering plant cell division. This continuous plant cell division later forms root nodules.

This rhizobia-legume communication is very specific, allowing only particular rhizobia strains to nodulate with particular host legumes. By using the lac*z* reporter gene, Zhu et al. (1997) have shown in their study that *Sinorhizobium meliloti* effectively nodulated species of the *Trigonella*, *Medicago* and *Melilotus genera*, while *Rhizobium leguminosarum bv viciae* produces nodulation in the *Lathyrus*, *Lens*, *Vicia* and *Pisum genera* via border cells.

In a study on non-legume *Parasponia andersonii*, a elm family member was nodulated by *Rhizobium* strain NGR234, which also nodulates 232 species of legumes (Pueppke and Broughton 1999). However, not every member of the legume family forms root nodules. For example, *Caesalpinoideae* are one of three non-nodulating legume sub-families: *Caesalpinoideae*, *Mimosoideae*, and *Papilionoideae*. Nod genes are induced in *Mesorhizobium loti* by aldonic, erythronic, and tetronic acid, which are all exuded by *Lotus corniculatus* (Morgan et al. 2005).

Studies have shown that bacteria have the potential to recognize plant-derived compounds (flavonoids) that help in rhizobia-legume interactions through many molecular signaling pathways (Nagahashi and Douds 2003). Flavonoids can be both antagonistic and agonistic for rhizobia species (Cooper 2007). For example, *Bradyrhizobium japonicum* nod genes are induced by daidzein and genistein, isoflavonoids produced by soybean (*Glycine max*), but inhibit *S. meliloti* nod gene expression, whereas luteolin is an inducing agent in the case of *S. meliloti* nod genes (Peters et al. 1986).

These specificity powers allow rhizobia to differentiate other legumes from their hosts. Rhizobial chemotaxis as well as *nod* gene expression are induced by a specific flavonoid (Peck et al. 2006; Wang et al. 2012). RNA interference was used by Wasson et al. (2006) to silence chalcone synthase (CHS), which then inhibited nodule formation. In *M. truncatula,* this enzyme catalyzes the initial dedicated step of the flavonoid pathway, which activates nod genes in *S. meliloti*. To achieve flavonoid accumulation and nodule development in plants, a supplementation of the flavonoid precursors liquiritigenin and naringenin could be used as an inducer.

Another nitrogen-fixing association is found with tree roots and the Gram-positive, filamentous actinobacterial genus *Frankia,* which forms an intracellular nitrogen-fixing symbiosis with over 200 angiosperm species belonging to eight families. The best-known association of *Frankia* is with *Alnus* (Daniel et al. 2007).

This interaction between plants that belong to eight dicotyledonous families and the actinobacterium *Frankia* is known as "actinorhizal symbiosis," collectively called "actinorhizal" (Wall 2000). *Frankia* was inoculated in *A. glutinosa* to check the expression of genes coding for chalcone synthase (chs) and phenyl-ammonia lyase (pal), which are involved in the biosynthesis of flavonoids (Hammad et al. 2003; Kim et al. 2003). A study analyzing the expressed sequence tag (EST) database of a nodule and a *Casuarina glauca* root led to the detection of eight genes responsible for enzymes involved in the biosynthesis of the flavonoid pathway (Auguy et al. 2011). Further study is needed to enhance the understanding of actinorhizal symbiosis and its relationship with *Frankia* cultivability. Such a study may provide important data on developmental biology and plant-microbe interactions.

18.3.1.2 Mycorrhizal Interactions

The symbiotic relationship between plant roots, fungi and soil is known as "mycorrhizae." Greater than 80% of terrestrial plants form an association with AMF. AMF are obligate symbionts, which means they are incapable of finishing their life cycle with a lack of host roots. Highly crossed web-like structures are formed by fungi after penetrating plant roots. These branching structures are known as "arbuscules," and they are thought to be the primary sites of nutrient interchange between roots and fungi (Akiyama and Hayashi 2006).

This branch formation by fungus is due to an inducing factor, which is a plant-signaling molecule that activates hyphal morphogenesis leading to successful root colonization (Buee et al. 2000; Giovannetti et al. 1996). Studies have shown that all the mycotrophic plants have this branch-inducing factor present in their root exudates, but it was absent in non-host plants. A study on root exudate sesquiterpene identified from *Lotus japonicas,* a mycotrophic plant, has confirmed its role in activating hyphal branching in dormant mycorrhizal fungi (Akiyama et al. 2005). Ecto and endomycorrhiza are differentiated by their expansion around or inside root cells. Endomycorrhizal AMF are constantly found in relationship with roots and are therefore considered as obligate parasites. Unlike AMF, ectomycorrhiza belonging mainly to the basidiomycetes and not often to asco- or zygomycetes, are not obligate biotrophs but can live saprophytically in soil.

VAM forms a tight and communally useful association between mycorrhizal fungus and plant roots. Metabolic changes are undergone by both host and AMF to meet each other's needs. As a result, root exudation induced by VAM will enhance microbial community growth in the rhizosphere and have growth-promoting effects on plants. Mycorrhizal fungi get carbon from the host root while inorganic nutrients are taken up from the soil around the plant root, creating one of the best examples of a plant-microbe symbiotic relationship where both benefit. The fungus benefits through the continuous supply of organic nutrients from the plant while the plant functions physiologically well and competes effectively with other plant communities (Bago et al. 2003).

This fungus-plant association has some other advantages such as a phosphate solubilization property that makes phosphorous available to plants. This is why VAM is widely used as a biofertilizer in the field (Behl et al. 2007). AMF may distinguish the occurence of a well-matched host through their root exudates, similar to the method of recognition used by rhizobia (Nagahashi and Douds 2003). Also, AMF can provide phytohormones, which enhance plant growth. The main step in AMF maturation is the development of extraradical hyphae induced by signal molecules secreted by plants. These lead to the beginning of AMF-induced symbiosis (De Carvalho-Niebel et al. 2002). These signal molecules are known as "strigolactones" (SLs), which acts as plant hormones (Koltai 2013). Currently, it is known that carotenoids are used to derive terpenid lactones like SLs (Matusova et al. 2005). Studies have shown that SLs are found in a broad different type of plant species, including primitive plants, dicots and monocots (Liu et al. 2009; Xie et al. 2010; Proust et al. 2011).

18.3.1.3 Plant Growth Promotion

Number of studies are focussed on a new cluster of microbes due to their involvement in increasing the productivity and health of crops. These microbes are called the PGPRs, they and affect crop yield and growth by releasing necessary compounds that stimulate growth (Bloemberg and Lugtenberg 2001). The establishment of bacteria in the roots of plants takes place due to signals given by root exudates. These exudates, for example- sugars, amino acids, etc., further rouse PGPR chemotaxis over the surface of roots and impact the motility of flagella in some bacteria (Somers et al. 2004).

Phytostimulators are secreted by some rhizobacteria that directly add to the growth of the plant. *Azospirillum* sp., despite its nitrogen-fixing properties, also releases phyto-hormones like auxins, gibberellins, and cytokinins (Steenhoudt and Vanderleyden 2000). Root exudates help PGPRs by supplying the precursors for biotransformation. For example, the root exudate tryptophan is a precursor for an important auxin, i.e., IAA (indole acetic acid), which is exploited by rhizobacteria (Cooke et al. 2002). A study also reported the presence of sugars and amino acids in *Avena barbata* root exudates.

There also exists an indirect mechanism for plant growth through suppression of the capacity of phyto-pathogens. This mechanism involves the capability of bacteria to produce siderophores (Parez-Miranda et al. 2007). Siderophores chelate iron hence making it unavailable to pathogens. Other mechanisms include production of anti-fungal metabolites, enzymes to degrade the cell wall, and hydrogen cyanide (HCN) to retard the growth of fungal pathogens.

Apart from indirect mechanisms, there are also certain direct ones. These include atmospheric nitrogen fixing to make it available to plants, formation of siderophores, phosphate solubilization and mineralization, and production of phytohormones (Page 1987; Guan and Kamino 2001).

18.3.2 Negative Impact

A number of secondary metabolites have been identified, and their properties with respect to the rhizosphere have been elucidated.

18.3.2.1 Inhibition of Pathogenicity by Secreting Antimicrobials

Soil microbial communities including pathogens are thought to be attracted by plants when compounds are exuded from plant roots. There is a wide variety of chemo-diversity in root exudates and a continuous search is on for the identification of suitable antimicrobials.

For instance, an extract of the cell wall from *Phytophthora cinnamoni* leads to the precipitation of release of rosmarinic acid (RA), a multifunctional caffeic acid ester, identified in the root exudates of sweet basil cultures (Bais et al. 2006). Root cultures of basil were also shown to exude RA when challenged *in situ* with *Pythium ultimum*, which further demonstrates antimicrobial activity against a number of soil-borne pathogenic microbes like *Pseudomonas aeruginosa* (Bais et al. 2002b). A similar study carried out by Brigham et al. (1999) showed that upon elicitation, pigmented naphthoquinones were released by hairy root cultures of *Lithospermum erythrorhizon* followed by many other biological activities against pathogens. Knowing the demonstrated antimicrobial activities of RA and naphthoquinones, root exudates can signify their importance in protecting the rhizosphere from pathogenic microbes.

18.3.2.2 Antimicrobials

There are certain compounds to which the bacterial pathogenic microbes causing disease and infection in roots were known to be resistant. Such compounds can find their role in providing defense against non-host pathogens (Haichar et al. 2012). For instance, phenyl propanoid is one such compound that has been shown to be considerably higher in roots challenged by non-host-pathogenic bacteria, i.e., non-host *Pseudomonas syringae* strains, in comparison to host-pathogenic bacteria, i.e., *P. syringae* pv. *tomato*DC3000. It is released in reaction to attack by pathogens.

Lanoue et al. (2010) observed that the root system of Barley (*Hordeum vulgare*) released some phenol compounds that have antimicrobial activity when subjected to infection with *Fusarium graminearum*. Earlier, in a report by Vaughan et al. (2013), it was observed that uninfected *A. thaliana* roots were constitutively producing and releasing a diterpene called "rhizathalene A." It was also shown that plants that are not producing this compound are found to be more vulnerable to be attacked by herbivorous insects.

The synthetic analogue of strigolactone, i.e. GR24, has been found to inhibit the growth of a wide varieties of phytopathogenic fungi in the growth medium (Dor et al. 2011). This indicates that released strigolactones can have either direct or indirect effects on their natural enemies by altering hormonal defense pathways along with contributing to the below-ground biotic stress response to plants (Dor et al. 2011; Torres-Vera et al. 2013; Baetz and Martinoia 2014).

Another set of compounds that can show antimicrobial activity towards a number of organisms are biosurfactants. They show antimicrobial activity against pathogenic oomycetes *Pythium* and *Phytophthora*, the fungus *Rhizoctonia*, in addition to various Gram-positive and Gram-negative bacteria that pose pathogenicity to humans like *Staphylococcus aureus* and *Proteus vulgaris* (Raaijmakers et al. 2006; Das et al. 2008). There are a number of secondary metabolites secreted by the roots of Arabidopsis when given treatment with salicylic acid (Walker et al. 2003a, b). Such compounds involve butanoic acid, ferulic acid, and 3-indolepropanoic acid. All of them were shown to display *in vitro* antibacterial activity against the pathogens *Erwinia*, *Xanthomonas campestris*, and *P. syringae* in the amounts found in root exudates.

18.3.2.3 Quorum Sensing

Quorum sensing (QS) is the capability of bacteria to develop communication and coordination of behavior through signaling molecules. It is a controlling process through which bacteria examine their growth. During growth, signal molecules are subsequently released by bacteria (Xuesong et al. 2003). Quorum-sensing systems are possessed by both Gram-negative and Gram-positive bacteria. They comprise significant plant pathogenic bacteria, namely *Erwinia* spp., *Pseudomonas* spp., and *Agrobacterium* spp., and control the expression of the number of genes that are necessary for creating pathogenesis (Fray 2002). QS is the phenomena by which production and release of the virulence parameters are regulated in several bacterial pathogens.

There is a difference in the detection of chemical signals in Gram-positive and Gram-negative bacteria, i.e. α-homoserine lactones (AHLs) are for Gram–negative bacteria whereas peptide auto-inducers are for Gram–positive bacteria. The mechanism of QS was first defined in *Vibrio fischeri* (an aquatic bacteria) through the signal-regulated induction of *lux* genes responsible for bioluminescence. This process is dependent on density. Normally there is constitutive synthesis of a basic level of AHLs until the levels reach a threshold value. This threshold is a point at which such molecules start acting as ligands for the global transcription regulator LuxR/LuxR-like proteins. These proteins are thought to activate many genes controlled by QS involving virulence factors. The rhizosphere has high levels of AHL-forming bacteria in comparison to bulk soil, showing their important role in colonization (Elasri et al. 2001).

The very first defined examples of QS mimicking plant-secreted compounds were halogenated furanones produced by *Delisea pulchra* (marine red algae) (Givskov et al. 1996). Such compounds are shown to be structurally similar to α-homoserine lactones. Certain bioactive components were found to be present in the root exudates of *Pisum sativum* (pea) that mimic AHL signaling in fully characterized reporter strains of bacteria. Through the regulation of AHLs these components can help with stimulation of behaviors in some strains and cause the inhibition of behaviors in others (Teplitski et al. 2000).

In a study carried out by Fray (2002), it was demonstrated that pathogenicity was re-established in AHL-producing transgenic tobacco plants to an avirulent AHL-deficient *Erwinia carotovora* mutant. In addition to the release of AHL-mimicking compounds, Rasmussen et al. (2005) revealed certain quorum sensing inhibitors (QSIs) in garlic extracts that are specific to the QS-controlled virulent genes in *P. aeruginosa* analyzed by gene chip-based transcriptomics.

Hence, it is likely that roots have the capability to develop defense strategies with the help of several secreted molecules in the rhizosphere. These secreted molecules obstruct the QS responses of bacteria like mimicking, blocking the signals, and secreti ng enzymes to degrade signals and thus induce chemical-attenuation of pathogens (Rasmussen and Givskov 2006; Defoirdt et al. 2010).

18.4 Root–Invertebrate (Nematode) Interaction

Root exudates are considered to be a well-known source of carbon for microorganisms residing in soil allowing profound populations to exist in the rhizosphere. The value and amount of carbon and other nutrients released in the rhizosphere, the microbial community structure surrounding the roots, and the effects on microorganism-nematode association, all are significantly influenced by the species of plant and the environmental conditions. The interaction between plants and invertebrates as facilitated by chemical signals has mainly been studied in leaves and stems, whereas the interaction between roots and invertebrates has just begun to be explored.

By utilizing a ^{14}C pulse-labelling practice, Yeates (1999) observed that there was a substantial upsurge in fixing of label C via photosynthesis in the soil microbial biomass after infecting the roots of white clover (*Trifolium repens*) with *Heterodera trifolii* and numerous other nematodes (Nobili et al. 2001). This outcome suggests that infection by parasitic nematodes in white clover plant liberates additional organic compounds in the rhizosphere in general. Similarly, *M. incognita* infection in the roots of tomatoes led to an increased amount of water-soluble ^{14}C and metal ions in its exudates as compared to healthy plants.

The majority of the information about communication between microorganisms and nematodes was the result of research done on rhizobia, mycorrhizal fungi, and plant pathogens (Baetz and Martinoia 2014). Such studies evidently demonstrate the complex tri-trophic webs wherein competition, addition, and synergistic interactions take place between nematodes and microbes in order to affect the plant host.

18.5 Root- Root Interaction

Even though there have been significant improvements in the understanding of root functions in the past decade, the complex associations arising at the interface of root and soil involving root exudation are still beginning to be investigated.

There are three modes of interference,namely resource competition, chemical interference, and/or parasitism, and the root exudates have the ability to affect all three of them. Root exudates have been shown to exhibit the properties of phytotoxins in order to mediate chemical interference for several species of plant. Natural compounds derived from plants to facilitate plant defense are termed allelochemicals; allelopathy is described as the phenomena in which bioactive secondary compounds are produced and released by the plants in order to affect the development of neighboring plant species (Weston et al. 2012). Allelochemicals that are secreted as root exudates are shown to penetrate the rhizosphere just after their secretion (Inderjit 2001). These chemicals are thought to be liberated in bulk but are subjected to sorption (physical), metal oxidation (chemical), and microbial degradation (biological) within the rhizosphere (Narula et al. 2009).

The allelochemicals released by roots are shown to reduce the growth of neighboring plants as well as suppress pathogenic microbes, insects, and herbivores. Nowadays, it is possible to characterize very small quantities of secondary bioactive compounds in the rhizosphere as well as study their metabolism and secretion in the soil (Mohney et al. 2009).

In one study, *Sorghum* spp., including johnson grass (*Sorghum halpense* L. Pers.) and sorghum sudan grass hybrid (*Sorghum bicolor* × *Sorghum sudanese*) produced ample amounts of potential allelochemicals in their exudates. Sorghum exudates when chemically characterized reveal many related long-chain hydroquinones that include sorgoleone with its resorcinol-analogue. These compounds are found to deter the growth of neighboring plants by inhibiting processes like photosynthesis and respiration (Czarnota et al. 2003; Dayan et al. 2009). Similarly, limited growth of weeds was observed in agricultural systems with *Triticum aestivum* (Wu et al. 2000) and *Oryza sativa* (Kong et al. 2004) in the presence of DIBOA and 5,7,4′-trihydroxy-3′,5′-dimethoxyflavone, respectively. In the maize rhizosphere, there was enhancement in the population of *P. putida* with valuable and positive characteristics gained as a result of exudation of benzoxazinone DIMBOA (2,4-dihydroxy-7-methoxy-1,4- benzoxazin-3-one) (Neal et al. 2012).

A secondary metabolite released from the roots of *Centaurea maculosa* (knapweed) sets up a standard example of an exudate compound displaying negative root-root interaction in the rhizosphere. The intrusive behavior of knapweed into the rhizosphere was due to the phytotoxin released by roots, i.e. (±)-catechin. Interestingly, (−)-catechin was found to show allelochemical activity, while (+)-catechin was shown to inhibit soil-borne bacteria (Perry et al. 2005). This study clearly shows that one root exudate can exhibit different properties such as autotoxicity and allelopathy in a plant species.

There are a few plants that prevent the inhibition of phytotoxin by changing their structure chemically. For instance, *N*-glucosylation is the pathway on which *Zea mays* (corn) depends in order to nullify the influence of DIMBOA, DIBOA, and BOA, phytotoxins that are released in the rhizosphere by *Triticum aestivum* (wheat) and many other grasses.

Root exudates are crucial in advancing the interaction between a parasitic plant and its host, where the interaction is considered negative for host and positive for parasite (Weston et al. 2012). The secondary metabolites released by the roots act as chemical messengers and are often utilized by plants for initiating the growth of invasive organs like haustoria needed for heterotrophic progression (Walker et al. 2003a).

Well-established physical associations between host and parasite are known for many obligate parasites such as *Striga* spp., witchweed, *Orobanche* spp., and broomrape (Palmer et al. 2004). A number of major food crops are parasitized by plants belonging to *Scrophulariaceae,* for example maize (*Zea mays*), sorghum (*Sorghum bicolor*), millet (*Panicum milaceum*), rice (*Oryza sativa*), and legumes. This family is known for invading the roots of surrounding plants in order to parasitize them for water, minerals, and other important nutrients (Yoder 2001).

Many exudates show positive responses in defense of neighboring plants in order to diminish the population of herbivores by drawing them indirectly towards aberrant plants. For example, infection of *V. faba* plants led to secretion of root exudates that are shown to regulate the green-leafy volatile formation in uninfected *V. faba* plants, thereby attracting the aphid parasitoids to the already infested *V. faba* (Du et al. 1998).

18.6 Conclusion and Future Prospects

Current studies are still at the learning phase about the multifaceted communication between root of the plants and their extremely varied and active micro-flora. It's capability that enables plants to respond differently to pathogenic and beneficial microbes is of utmost importance for their survival. Sorting out the molecular chemistry of plant and microbial relations will not only give the opportunity to make changes in plant defense for human advantage, but also to promote establishment of helpful rhizospheric microbes. Therefore, modelling novel techniques and procedures in order to investigate rhizospheric ecological parameters under inherent conditions is urgently needed. Finally, capturing knowledge about plant growth-promoting rhizobacteria and root exudation, from genetic to the ecosystem level, will actually help in improvement of plants in terms of absorption of nutrients, detoxification of soils, and protection against invasive weeds and microbial pathogens.

Acknowledgement The authors are thankful to Director MNNIT Allahabad for providing necessary facilities and "Design and Innovation Centre" a project sponsored by Ministry of Human Resource Development, Government of India, for supporting the execution of the study.

References

Abbott L, Murphy D (2003) Soil biology fertility: a key to sustainable land use in agriculture. Kluwer Academic Publishers, Dordrecht, pp 187–203

Akiyama K, Matsuzaki K, Hayashi H (2005) Plant sesquiterpenes induce hyphal branching in arbuscular mycorrhizal fungi. Nature 435:824–827

Akiyama K, Hayashi H (2006) Strigolactones: chemical signals for fungal symbionts and parasitic weeds in plant roots. Ann Bot Lond 97:925–931

Albarracín VH, Winik B, Kothe E, Amoroso MJ, Abate CM (2008) Copper bioaccumulation by the actinobacterium *Amycolatopsis* sp. AB0. J Basic Microbiol 48:323–330

Auguy F, Abdel-Lateif K, Doumas P, Badin P, Guerin V, Bogusz D, Hocher V (2011) Activation of the isoflavonoid pathway in actinorhizal symbioses. Funct Plant Biol 38:690–696

Badri DV, Loyola-Vargas VM, Broeckling CD, De-la-Pena C, Jasinski M, Santelia D, Martinoia E, Sumner LW, Banta LM, Stermitz F, Vivanco JM (2008) Altered profile of secondary metabolites in the root exudates of Arabidopsis ATP-binding cassette transporter mutants. Plant Physiol 146:762–771

Baetz U, Martinoia E (2014) Root exudates: the hidden part of plant defense. Trends Plant Sci 19:90–98

Bago NB, Pfeffer PE, Abubakar J, Jun J, Allen JW (2003) Carbon export from arbuscular mycorrizal roots involves the translocation of carbohydrate as well as lipid. Plant Physiol 13:1496–1507

Bais HP, Fall R, Vivanco JM (2004a) Biocontrol of *Bacillus subtilis* against infection of *Arabidopsis* roots by *Pseudomonas syringae* is facilitated by biofilm formation and surfactin production. Plant Physiol 134:307–319

Bais HP, Loyola-Vargas VM, Flores HE, Vivanco JM (2001) Root-specific metabolism: the biology and biochemistry of underground organs. In Vitro Cell Dev Biol Plant 37:730–741

Bais HP, Park SW, Weir TL, Callaway RM, Vivanco JM (2004b) How plants communicate using the underground information superhighway. Trends Plant Sci 9:26–32

Bais HP, Walker TS, Schweizer HP, Vivanco JM (2002a) Root specific elicitation and antimicrobial activity of rosmarinic acid in hairy root cultures of sweet basil (*Ocimum basilicum* L.) Plant Physiol Biochem 40:983–995

Bais HP, Walker TS, Stermitz FR, Hufbauer RA, Vivanco JM (2002b) Enantiomeric dependent phytotoxic and antimicrobial activity of (±)-catechin. A rhizosecreted racemic mixture from spotted knapweed. Plant Physiol 128:1173–1179

Bais HP, Tiffony L, Weir LT, Perry LG, Gilroy S, Vivanco JM (2006) The role of root exudates in rhizosphere interactions with plant and other organisms. Annu Rev Plant Biol 57:233–266

Barea JM, Pozo MJ, Azcon R, Azcon-Aguilar C (2005) Microbial cooperation in the rhizosphere. J Exp Bot 56:1761–1778

Baudoin E, Benizri E, Guckert AV (2002) Impact of growth stage on the bacterial community structure along maize roots, as determined by metabolic and genetic fingerprinting. Appl Soil Ecol 19:135–145

Becard G, Taylor LP, Douds DD, Pfeffer PE, Doner LW (1995) Flavonoids are not necessary plant signal compounds in arbuscular mycorrhizal symbiosis. Mol Plant-Microbe Interact 8:252–258

Bednarek P, Schneider B, Svatos A, Oldham NJ, Hahlbrock K (2005) Structural complexity, differential response to infection, and tissue specificity of indolic and phenylpropanoid secondary metabolism in *Arabidopsis* roots. Plant Physiol 138:1058–1070

Behl RK, Ruppel S, Kothe E, Narula N (2007) Wheat × Azotobacter × VA mycorrhiza interactions towards plant nutrition and growth – a review. J Appl Bot Food Qual 81:95–109

Berger S, Sinha AK, Roitsch T (2007) Plant physiology meets phytopathology: plant primary metabolism and plant-pathogen interactions. J Exp Bot 58:4019–4026

Bertin C, Yang X, Weston LA (2003) The role of root exudates and allelochemicals in the rhizosphere. Plant Soil 256:67–83

Bloemberg GV, Lugtenberg BJ (2001) Molecular basis of plant growth promotion and biocontrol by rhizobacteria. Curr Opin Plant Biol 4:343–350

Bolton MD (2009) Primary metabolism and plant defense—fuel for the fire. Mol Plant-Microbe Interact 22:487–497

Brigham LA, Michaels PJ, Flores HE (1999) Cell-specific production and antimicrobial activity of naphthoquinones in roots of *Lithospermum erythrorhizon*. Plant Physiol 119:417–428

Broeckling CD, Broz AK, Bergelson J, Manter DK, Vivanco JM (2008) Root exudates regulate soil fungal community composition and diversity. Appl Environ Microbiol 74:738–744

Buee M, Rossignol M, Jauneau A, Ranjeva R, Becard G (2000) The pre-symbiotic growth of arbuscular mycorrhizal fungi is induced by a branching factor partially purified from plant root exudates. Mol Plant-Microbe Interact 13:693–698

Cheng W, Parton WJ, Gonzalez-Meler MA, Phillips R, Asao S, McNickle GG, Brzostek E, Jastrow JD (2014) Synthesis and modelling perspectives of rhizosphere priming. New Phytol 201:31–44

Cooke TJ, Poli D, Sztein AE, Cohen JD (2002) Evolutionary patterns in auxin action. Plant Mol Biol 49:319–338

Cooper JE (2007) Early interactions between legumes and rhizobia: disclosing complexity in a molecular dialogue. J Appl Microbiol 103:1355–1365

Czarnota MA, Paul RN, Weston LA, Duke SO (2003) Anatomy of sorgoleone-secreting root hairs of *Sorghum* species. Int J Plant Sci 164:861–866

Dakora FD, Phillips DA (2002) Root exudates as mediators of mineral acquisition in low-nutrient environments. Plant Soil 245:35–47

Daniel G, Jaffre T, Prin Y (2007) Abundance of *Frankia* from *Gymnostoma* spp. in the rhizosphere of *Alphitonia neocaledonica*, a nonnodulated *Rhamnaceae endemic*to New Caledonia. Eur J Soil Biol 36:169–175

Das P, Mukherjee S, Sen R (2008) Antimicrobial potential of a lipopeptide biosurfactant derived from a marine Bacillus circulans. J Appl Microbiol 104:1675–1684

Dayan FE, Howell JE, Weidenhamer JD (2009) Dynamic root exudation of sorgoleone and its in planta mechanism of action. J Exp Bot 60:2107–2117

De Carvalho-Niebel F, Timmers AC, Chabaud M, Defaux P, Abarkar DG (2002) The nod factor-elicited annexin MtAnn1 is preferentially localized at the nuclear periphery in symbiotically activated root tissues of *Medicago truncatula*. Plant J 32:343–352

Deaker R, Roughley RJ, Kennedy IR (2004) Legume seed inoculation technology – a review. Soil Biol Biochem 36:1275–1288

Defoirdt T, Boon N, Bossier P (2010) Can bacteria evolve resistance to quorum sensing disruption? PLoS Pathog 6:e1000989

Doornbos RF, van Loon LC, Bakker PAHM (2012) Impact of root exudates and plant defense signalling on bacterial communities in the rhizosphere. A review. Agron Sustain Dev 32:227–243

Dor E, Joel DM, Kapulnik Y, Koltai H, Hershenhorn J (2011) The synthetic strigolactone GR24 influences the growth pattern of phytopathogenic fungi. Planta 234:419–427
Du YJ, Poppy GM, Powell W, Pickett JA, Wadhams LJ, Woodcock CM (1998) Identification of semiochemicals released during aphid feeding that attract parasitoid *Aphidius ervi*. J Chem Ecol 24:1355–1368
Elasri M, Delorme S, Lemanceau P, Stewart G, Laue B et al (2001) Acyl-homoserine lactone production is more common among plant-associated *Pseudomonas* spp. than among soil borne *Pseudomonas* spp. Appl Environ Microbiol 67:1198–1209
Farag MA, Huhman DV, Lei ZT, Sumner LW (2007) Metabolic profiling and systematic identification of flavonoids and isoflavonoids inroots and cell suspension cultures of *Medicago truncatula* using HPLC-UVESI- MS and GC-MS. Phytochemistry 68:342–354
Flores HE, Vivanco JM, Loyola-Vargas VM (1999) "Radicle" biochemistry: the biology of root-specific metabolism. Trends Plant Sci 4:220–226
Fray RG (2002) Altering plant microbe interaction through artificially manipulating bacterial quorum sensing. Ann Bot 89:245–253
Garbeva PV, Van Elsas JD, Van Veen JA (2008) Rhizosphere microbial community and its response to plant species and soil history. Plant Soil 302:19–32
Giovannetti M, Sbrana C, Silvia A, Avio L (1996) Analysis of factors involved in fungal recognition response to host-derived signals by arbuscular mycorrhizal fungi. New Phytol 133:65–71
Givskov M, Nys RD, Manefield M, Gram L, Maximilien R, Eberl L et al (1996) Eukaryotic interference with homoserine lactone-mediated prokaryotic signaling. J Bacteriol 178:6618–6622
Gleba D, Borisjuk NV, Borisjuk LG, Kneer R, Poulev A, Skarzhinskaya M et al (1999) Use of plant roots for phytoremediation and molecular farming. Proc Natl Acad Sci U S A 25:5973–5977
Gray EJ, Smith DL (2005) Intracellular and extracellular PGPR: commonalities and distinctions in the plant-bacterium signaling processes. Soil Biol Biochem 37:395–410
Guan LL, Kamino K (2001) Bacterial response to siderophore and quorum sensing chemical signals in the seawater microbial community. BMC Microbiol 1:27
Haichar FZ, Santaella C, Heulin T, Achouak W (2014) Root exudates mediate interactions belowground. Soil Biol Biochem 77:69–80
Haichar FZ, Marol C, Berge O, Rangel-Castro J, Prosser JI, Balesdent J, Heulin T, Achouak W (2008) Plant host habitat and root exudates shape soil bacterial community structure. ISME J 2:1221–1230
Haichar FZ, Roncato MA, Achouak W (2012) Stable isotope probing of bacterial community structure and gene expression in the rhizosphere of *Arabidopsis thaliana*. FEMS Microbiol Ecol 81:291–302
Hammad Y, Nalin R, Marechal K, Fiasson K, Pepin R, Berry AM, Normand P, Domenach AM (2003) A possible role for phenylacetic acid (PAA) in *Alnus glutinosa* nodulation by *Frankia*. Plant Soil 254:193–205
Heidstra R, Bisseling T (1996) Nod factor induced host responses and mechanisms of Nod factor perception. New Phytol 133:25–43
Hirsch AM, Bauer WD, Bird DM, Cullimore J, Tyler B, Yoder JI (2003) Molecular signals and receptors: controlling rhizosphere interactions between plants and other organisms. Ecology 84:858–868
Huhman DV, Berhow MA, Sumner LW (2005) Quantification of saponins in aerial and subterranean tissues of *Medicago truncatula*. J Agric Food Chem 53:1914–1920
Inderjit (2001) Soil: environmental effect on allelochemical activity. Agron J 93:79–84
Innes L, Hobbs PJ, Bardgett RD (2004) The impacts of individual plant species on rhizosphere microbial communities in soils of different fertility. Biol Fertil Soils 40:7–13
Jose S, Gillespie AR (1998) Allelopathy in black walnut (*Juglans nigra* L.) alley cropping. I. Spatio-temporal variation in soil juglone in a black walnut-corn (*Zea mays* L.) alley cropping system in the midwestern USA. Plant Soil 203:191–197
Kamilova F, Kravchenko LV, Shaposhnikov AI, Azarova T, Makarova N, Lugtenberg B (2006) Organic acids, sugars, and Ltryptophane in exudates of vegetables growing on stonewool and their effects on activities of rhizosphere bacteria. Mol Plant-Microbe Interact 19:250–256

Kanchan Vishwakarma, Neha Upadhyay, Nitin Kumar, Gaurav Yadav, Jaspreet Singh, Rohit K. Mishra, Vivek Kumar, Rishi Verma, R. G. Upadhyay, Mayank Pandey, Shivesh Sharma, (2017) Abscisic Acid Signaling and Abiotic Stress Tolerance in Plants: A Review on Current Knowledge and Future Prospects. Frontiers in Plant Science 08

Kim HB, Oh CJ, Lee H, Sun An C (2003) A type-i chalcone lsomerase mRNA is highly expressed in the root nodules of *Elaeagnus umbella*. J Plant Biol 46:263–270

Koltai H (2013) Strigolactones' ability to regulate root development may be executed by induction of the ethylene pathway. Plant Signal Behav 6:1004–1005

Kong CH, Liang WJ, Xu XH, Hu F, Wang P, Jiang Y (2004) Release and activity of allelochemicals from allelopathic rice seedlings. J Agric Food Chem 52:2861–2865

Kothe E, Bergmann H, Büchel G (2005) Molecular mechanisms in bio-geo-interactions. Chem Erde 65(S1):7–27

Kumar R, Bhatia R, Kukreja K, Behl RK, Dudeja SS, Narula N (2007) Establishment of *Azotobacter* on plant roots: chemotactic response, development and analysis of root exudates of cotton (*G. hirusitum* L.) and wheat (*T. aestivum* L.) J Basic Microbiol 47:436–439

Lambers H, Mougel C, Jaillard B, Hinsinger P (2009) Plant-microbe-soil interactions in the rhizosphere: an evolutionary perspective. Plant Soil 321:83–115

Lanoue A, Burlat V, Henkes GJ, Koch I, Schurr U, Rose USR (2010) De novo biosynthesis of defense root exudates in response to *Fusarium* attack in barley. New Phytol 185:577–558

Liu J, Magalhaes JV, Shaff J, Kochian LV (2009) Aluminum-activated citrate and malate transporters from the MATE and ALMT families function independently to confer Arabidopsis aluminum tolerance. Plant J 57:389–399

Loyola-Vargas VM, Broeckling CD, Badri D, Vivanco JM (2007) Effect of transporters on the secretion of phytochemicals by the roots of *Arabidopsis thaliana*. Planta 225:301–310

Lundberg DS, Lebeis SL, Herrera Paredes S, Yourstone S, Gehring J, Malfatti S et al (2012) Defining the core *Arabidopsis thaliana* root microbiome. Nature 488:86–90

Madigan MT, Martinko JM (2006) Brock: biology of microorganisms, 655–667. Pearson Prentice Hall, New Jersey

Matusova R, Rani K, Verstappen FWA, Franssen MCR, Beale MH, Bouwmeester HJ (2005) The strigolactone germination stimulants of the plant parasitic *Striga* and *Orobanche* spp. are derived from the carotenoid pathway. Plant Physiol 139:920–934

Mavrodi OV, Mavrodi DV, Parejko JA, Thomashow LS, Weller DM (2012) Irrigation differentially impacts populations of indigenous antibiotic-producing *Pseudomonas* spp. in the rhizosphere of wheat. Appl Environ Microbiol 78:3214–3220

Mohney BK, Matz T, LaMoreaux J, Wilcox DS, Gimsing AL, Mayer P, Weidenhamer JD (2009) In situ silicone tube microextraction: a new method for undisturbed sampling of root exuded thiophenes from marigold (*Tagetes erecta* L.) in soil. J Chem Ecol 35:1279–1287

Morgan JA, Bending W, White PJ (2005) Biological costs and benefits to plant microbe interactions in the rhizosphere. J Exp Bot 56:1729–1739

Nagahashi G, Douds DD Jr (2003) Action spectrum for the induction of hyphal branches of an arbuscular mycorrhizal fungus: exposure sites versus branching sites. Mycol Res 107:1075–1082

Nardi S, Concheri G, Pizzeghello D, Sturaro A, Rella R, Parvoli G (2000) Soil organic matter mobilization by root exudates. Chemosphere 5:653–658

Neal AL, Ahmad S, Gordon-Weeks R, Ton J (2012) Benzoxazinoids in root exudates of maize attract *Pseudomonas putida* to the rhizosphere. PLoS One 7:e35498

Narula N, Kothe E, Behl RK (2009) Role of root exudates in plant-microbe interactions. J Appl Bot Food Qual 82:122–130

Nimbal CI, Pedersen JF, Yerkes CN, Weston LA, Weller SC (1996) Phytotoxicity and distribution of sorgoleone in grain sorghum germplasm. J Agric Food Chem 44:1343–1347

Nobili M, Contin M, Mondini C, Brookes PC (2001) Soil microbial biomass is triggered into activity by trace amounts of substrate. Soil Biol Biochem 33:1163–1170

Page WJ (1987) Iron dependent production of hydroxamate by sodium dependent *Azotobacter chroococcum*. Appl Environ Microbiol 53:1418–1424

Palmer AG, Gao R, Maresh J, Erbil WK, Lynn DG (2004) Chemical biology of multihost/ pathogen interactions: chemical perception and metabolic complementation. Annu Rev Phytopathol 42:439–464

Parez-Miranda S, Cabirol N, George-Tellez R, Zamudio-Rivera LS, Fernandez FJ (2007) O-CAS, a fast and universal method for siderophore detection. J Microbiol Meth 70:127–131

Peck MC, Fisher RF, Long SR (2006) Diverse flavonoids stimulate NodD1 binding to nod gene promoters in *Sinorhizobium meliloti*. J Bacteriol 188:5417–5427

Perret X, Staehelin C, Broughton WJ (2000) Molecular basis of symbiotic promiscuity. Microbiol Mol Biol Rev 64:180–201

Perry LG, Thelen GC, Ridenour WM, Weir TL, Callaway RM et al (2005) Dual role for an allelochemical: (±)-catechin from *Centaurea maculosa* root exudates regulates conspecific seedling establishment. J Ecol 93:1125–1136

Peters NK, Frost JW, Long SR (1986) A plant flavone, luteolin, induces expression of *Rhizobium meliloti* nodulation genes. Science 233:977–980

Proust H, Hoffmann B, Xie X, Yoneyama K, Schaefer DG, Yoneyama K, Nogue F, Rameau C (2011) Strigolactones regulate protonema branching and act as a quorum sensing-like signal in the moss *Physcomitrella patens*. Development 138:1531–1539

Pueppke SG, Broughton WJ (1999) *Rhizobium* sp. strain NGR234 and R. fredii USDA257 share exceptionally broad, nested host ranges. Mol Plant-Microbe Interact 12:293–318

Raaijmakers JM, De Bruijn I, De Kock MJD (2006) Cyclic lipopeptide production by plant-associated pseudomonas spp.: diversity, activity, biosynthesis, and regulation. Mol Plant-Microbe Interact 19:699–710

Rasmussen TB, Givskov M (2006) Quorum-sensing inhibitors as anti-pathogenic drugs. Int J Med Microbiol 296:149–161

Rasmussen TB, Skindersoe ME, Bjarnsholt T, Phipps RK, Christensen BK, Jensen PO (2005) Identity and effects of quorum-sensing inhibitors produced by *Penicillium* species. Microbiology 151:1325–1340

Redmond JW, Batley M, Djordjevic MA, Innes RW, Kuempel PL, Rolfe BG (1986) Flavones induce expression of nodulation genes in Rhizobium. Nature 323:632–635

Richardson A, Barea JM, McNeill A, Prigent-Combaret C (2009) Acquisition of phosphorus and nitrogen in the rhizosphere and plant growth promotion by microorganisms. Plant Soil 321:305–339

Richardson AE, Lynch JP, Ryan PR, Delhaize E, Smith FA, Smith SE, Harvey PR, Ryan MH, Veneklaas JE, Lambers H, Oberson A, Culvenor RA, Simpson RJ (2011) Plant and microbial strategies to improve the phosphorus efficiency of agriculture. Plant Soil 4:59–61

Rougier M (1981) Secretory activity at the root cap. In: Tanner W, Loews FA (eds) Encyclopidia of plant physiology, B plant carbohydrates II, vol 13. Springer, Berlin, pp 542–574

Schmidt A, Haferburg G, Sineriz M, Merten D, Büchel G, Kothe E (2005) Heavy metal resistance mechanisms in actinobacteria for survival in AMD contaminated soils. Chem Erde 65:131–144

Sineriz ML, Kothe E, Abate CM (2009) Cadmium biosorption by *Streptomyces* sp. F4 isolated from former uranium mine. J Basic Microbiol 1:55–62

Somers E, Vanderleyden J, Srinivasan M (2004) Rhizosphere bacterial signalling: a love parade beneath our feet. Crit Rev Microbiol 30:205–235

Steenhoudt O, Vanderleyden J (2000) *Azospirillum*, a free-living nitrogen-fixing bacterium closely associated with grasses: genetic, biochemical and ecological aspects. FEMS Microbiol Rev 24:487–506

Stermitz FR, Bais HP, Foderaro TA, Vivanco JM (2003) 7, 8-benzoflavone: a phytotoxin from root exudates of invasive Russian knapweed. Phytochemistry 64:493–497

Subbarao GV, Nakahara K, Hurtado MP, Ono H, Moreta DE, Salcedo AF, Yoshihashi AT, Ishikawa T, Ishitani M, Ohnishi-Kameyama M, Yoshida M, Rondon M, Rao IM, Lascano CE, Berry WL, Ito O (2009) Evidence for biological nitrification inhibition in Brachiaria pastures. Proc Natl Acad Sci U S A 106:17302–17307

Subbarao GV, Rondon M, Ito O, Ishikawa T, Rao IM, Nakahara K, Lascano C, Berry WL (2007) Biological nitrification inhibition (BNI) – is it a widespread phenomenon? Plant Soil 294:5–18

Teplitski M, Robinson JB, Bauer WD (2000) Plants secrete substances that mimic bacterial N-acyl homoserine lactone signal activities and affect population density-dependent behaviors in associated bacteria. Mol Plant-Microbe Interact 13:637–648

Torres-Vera R, Garcia JM, Pozo MJ, Lopez-Raez JA (2013) Do strigolactones contribute to plant defence? Mol Plant Pathol 15:211–216

Uren NC (2000) Types, amounts and possible functions of compounds released into the rhizosphere by soil grown plants. In: Pinton R, Varanini Z, Nannipieri P (eds) The rhizosphere: biochemistry and organic substances at the soil interface. Marcel Dekker, New York, pp 19–40

Vance CP, Uhde-Stone C, Allen DL (2003) Phosphorus acquisition and use: critical adaptations by plants for securing a non-renewable source. New Phytol 157:423–447

Van Hulten M, Pelser M, Van Loon LC, Pieterse CMJ, Ton J (2006) Costs and benefits of priming for defense in Arabidopsis. Proc Natl Acad Sci U S A 103:5602–5607

Vaughan MM, Wang Q, Webster FX, Kiemle D, Hong YJ, Tantillo DJ, Coates RM, Wray AT, Askew W, O'Donnell C, Tokuhisa JG, Tholl D (2013) Formation of the unusual semivolatile diterpene rhizathalene by the Arabidopsis class I Terpene synthase TPS08 in the root stele is involved in defense against belowground herbivory. Plant Cell 25:1108–1125

Vishwakarma K, Sharma S, Kumar N, Upadhyay N, Devi S, Tiwari A (2016) Contribution of Microbial Inoculants to Soil Carbon Sequestration and Sustainable Agriculture. In: Microbial Inoculants in Sustainable Agricultural Productivity. Springer India, pp. 101–112.

Vivanco JM, Bais HP, Stermitz FR, Thelen GC, Callaway RM (2004) Biogeographical variation in community response to root allelochemistry: novel weapons and exotic invasion. Ecol Lett 7:285–292

Walker TS, Bais HP, Halligan KM, Stermitz FR, Vivanco JM (2003a) Metabolic profiling of root exudates of *Arabidopsis thaliana*. J Agric Food Chem 51:2548–2554

Walker TS, Bais HP, Grotewold E, Vivanco JM (2003b) Root exudation and rhizosphere biology. Plant Physiol 132:44–51

Wall LG (2000) The actinorhizal symbiosis. J Plant Growth Regul 19:167–182

Wang E, Schornack S, Marsh JF, Gobbato E, Schwessinger B, Eastmond P, Schultze M, Kamoun S, Oldroyd GE (2012) A common signaling proces that promotes mycorrhizal and oomycete colonization of plants. Curr Biol 22:2242–2246

Wasson AP, Pellerone FI, Mathesius U (2006) Silencing the flavonoid pathway in *Medicago truncatula* inhibits root nodule formation and prevents auxin transport regulation by rhizobia. Plant Cell 18:1617–1629

Weston LA, Ryan PR, Watt M (2012) Mechanism for cellular transport and release of allelochemicals from plant roots into rhizosphere. J Exp Bot 63:1–10

Wu HW, Haig T, Pratley J, Lemerle D, An M (2000) Allelochemicals in wheat (*Triticum aestivum L.*): variation of phenolic acids in root tissues. J Agric Food Chem 48:5321–5325

Xie X, Yoneyama K, Yoneyama K (2010) The strigolactone story. Annu Rev Phytopathol 48:93–117

Xuesong H, Williams C, Deanne LP, Laura OS, Wagner J, Clay F (2003) Quorum sensing in *Rhizobium* sp. strain NGR234 regulates conjugal transfer (*tra*) gene expression and influences growth rate. J Bacteriol 185:809–822

Yang CH, Crowley DE (2000) Rhizosphere microbial community structure in relation to root location and plant iron nutritional status. Appl Environ Microbiol 66:345–351

Yeates GW (1999) Effects of plants on nematode community structure. Annu Rev Phytopathol 37:127–149

Yoder JI (2001) Host-plant recognition by parasitic *Scrophulariaceae*. Curr Opin Plant Biol 4:359–365

Zhu Y, Pierson LS, Hawes MC (1997) Induction of microbial genes for pathogenesis and symbiosis by chemicals from root border cells. Plant Physiol 115:1691–1698

Mycorrhiza: An Alliance for the Nutrient Management in Plants

19

Aisha Sumbul, Irshad Mahmood, Rose Rizvi,
Rizwan Ali Ansari, and Safiuddin

Abstract

Mycorrhizal fungi are a wide ranging group of heterogeneous fungal taxa found to be allied with the roots of over 90% of all plant species. Among several types of mycorrhizal associations, two types are of high ecological and economic importance, i.e. arbuscular and ectotrophic mycorrhizal interactions. We have given a brief account on habitat, host specificity, and structural components of these mycorrhizal groups. An elaborated discussion on mineral absorption, different absorption pathways and the mechanisms involved has been presented in this chapter. Besides improving plant uptake of mineral nutrients already present in soil, many mycorrhizal fungi play a significant role in mobilizing nutrients either from organic substrate, mineral particles or rock surface. Mycorrhizal fungi take on several mechanisms to accomplish the function successfully, such as enhanced absorbing area of plant roots, release of biochemicals and consortium with other microorganisms. In addition to mobilizing nutrients, mycorrhizal fungi also serves as an important C sink in the soil, thus having an important influence on the cycling of these elements. The contributions of each partner in a mycorrhizal association are starting to be revealed by the use of molecular and genetic tools, coupled to high-throughput sequencing and advanced microscopy. Signalling pathways between plants and fungi have now been marked out, and the recognition of various novel nutrient transporters has unveiled some of the cellular processes that are fundamental to the mycorrhizal symbiosis. Different transporters, especially proton-coupled phosphate transporters, have been recognized on both the plant and fungal membranes and contribute to delivering phosphate from fungi to plants. Although much work has been previously done on several aspects of such symbioses, the extent to which they are functionally important in agriculture remains unclear. We are in urgent need to focus on the questions, the answers of which will give the new perspectives on mycorrhizal function.

A. Sumbul (✉) • I. Mahmood • R. Rizvi • R.A. Ansari • Safiuddin
Section of Plant Pathology and Nematology, Department of Botany, Aligarh Muslim University, Aligarh, UP, India
e-mail: aishasumbul92amu@gmail.com

V. Kumar et al. (eds.), *Probiotics in Agroecosystem*,
DOI 10.1007/978-981-10-4059-7_19

19.1 Introduction

Little things run the world especially when the case is of getting plants established. Under natural environment, plants remain closely associated with soil organisms called mycorrhizal fungi. These fungi colonize plant roots and stretch the root system into the surrounding soil. Surprising amounts of mycorrhizal filaments are found to be present in healthy soil. A very small portion of soil associated with vigorously growing plants may contain several miles of fungal filaments. The relationship is advantageous because the plant enjoys improved nutrient and water uptake, disease resistance and superior survival and growth (Fig. 19.1).

The term 'mycorrhiza' was proposed by a German scientist, A. B. Frank, more than 100 years ago. Literally the word 'mycorrhiza' means fungus root, and it describes the mutualistic association existing between a group of soil fungi and higher plants (Habte 2000). The association is based on symbiotic interaction taking into account a bidirectional trade of resources across the mycorrhizal interface. The mycorrhizal fungus supplies the host plant with nutrients, such as phosphate and nitrogen, and increases the abiotic (drought, salinity, heavy metals) and biotic (root pathogens) stress resistance of the host, and in turn, the host plant transfers between 4% and 20% of its photosynthetically fixed carbon to the mycorrhizal fungus (Wright et al. 1998). Fossil records point out that mycorrhizal interactions developed about 400–450 million years ago (Smith and Read 2008) and that they played a critical role in the colonization of land by plants. Although mycorrhizal associations came to light over 100 years ago, their importance in plant productivity did not receive due recognition until the past 50 years, molecular biology got advanced and gave insight into the mechanism of action of mycorrhizal fungi. At present, thousands of scientists all over the world are involved in the study of mycorrhizal associations, and any discussion of plant productivity that does not include mycorrhizal associations can barely be regarded as complete (Habte 2000). Approximately 90% of all known land plant species form mycorrhizal association with ubiquitous soil fungi (Bonfante and Genre 2010). In contrast to mutually beneficial mycorrhizal association, some mycoheterotrophic plants (approximately 400 plant species from

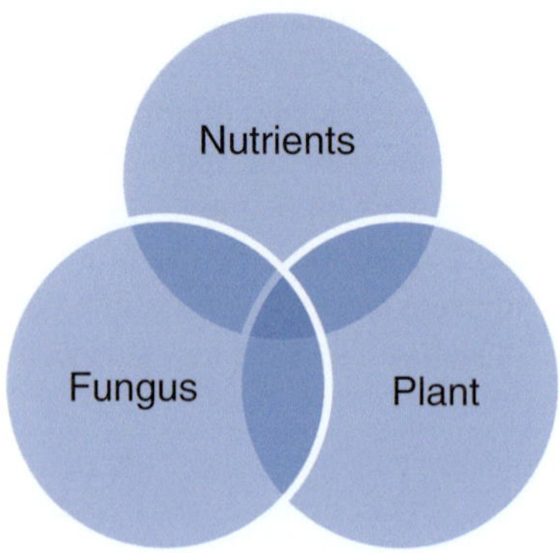

Fig. 19.1 Diagrammatic representation of association between plant and fungus and their combined sharing of nutrients

different plant families, such as bryophytes, pteridophytes, and angiosperms) depend on mycorrhizal fungi for their carbon supply. These plants have lost their photosynthetic potential and parasitize mycorrhizal fungi associated with nearby autotrophic plants (Bücking et al. 2012).

In this chapter, the main emphasis is being given on mutually beneficial arbuscular and ectotrophic mycorrhizal interactions, as they are of high economic and ecological significance (Marschner and Dell 1994). Arbuscular mycorrhizal fungi can be considered as 'biofertilizers and bioprotectors' in environmentally sustainable agriculture due to their ability to colonize and benefit a wide variety of food and cash crops. Ectomycorrhizal fungi, on the other hand, colonize a fewer number of plant species, but act as symbiotic partners of tree and shrub species and play a leading role in forest ecosystems (Finlay 2008) and could be a pivotal component in phytoremediation and/or revegetation applications (Bücking 2011; Giri et al. 2005).

19.2 Occurrence and Host Specificity of Mycorrhizal Fungi

AM fungi belong to six genera within the azygosporous zygomycetes. On the other hand, most ectomycorrhizal fungi belong to several genera within the class Basidiomycetes, while some belong to the zygosporic zygomycetes and ascomycetes. AM fungi are very effective in taking up inorganic phosphorus (P) and thus dominant in warm, dry climates where P is often a limiting factor. AM associations exist in a wide range of tropical and temperate tree species. They are found to have little specificity of association with host species (Bücking et al. 2002). These associations are known not to exist only in a few plants, namely, members of the families Amaranthaceae, Pinaceae, Betulaceae, Brassicaceae, Chenopodiaceae, Cyperaceae, Juncaceae, Proteaceae and Polygonaceae. EM fungi are more efficient in taking up N than AM, and they are the most common in the boreal zone and the humid parts of the temperate zone, where the low temperatures and high humidity promote the accumulation of organic matter, decreasing pH and low N availability (Kilpeläinena et al. 2016). Ectomycorrhizal fungi associates with a comparatively lower portion of all plant species, perhaps only about 3%, but this 3% represents the majority of tree of the temperate and boreal forests (particularly the plant families Fagaceae and Pinaceae), so, in terms of land area, the majority of the earth's forests are reliant on ectomycorrhizal fungi (Habte 2000; Smith and Read 2008; Bonfante and Genre 2010).

19.3 Structural Variability Among Mycorrhizal Fungi

Of the many types of mycorrhizal association, two are of major economic and ecological importance: endomycorrhizal association of the arbuscular (AM) type and ectomycorrhizal associations. Arbuscular mycorrhizal (AM) and ectomycorrhizal (ECM) interactions vary in their structural attributes and also in the plant and fungal species they include.

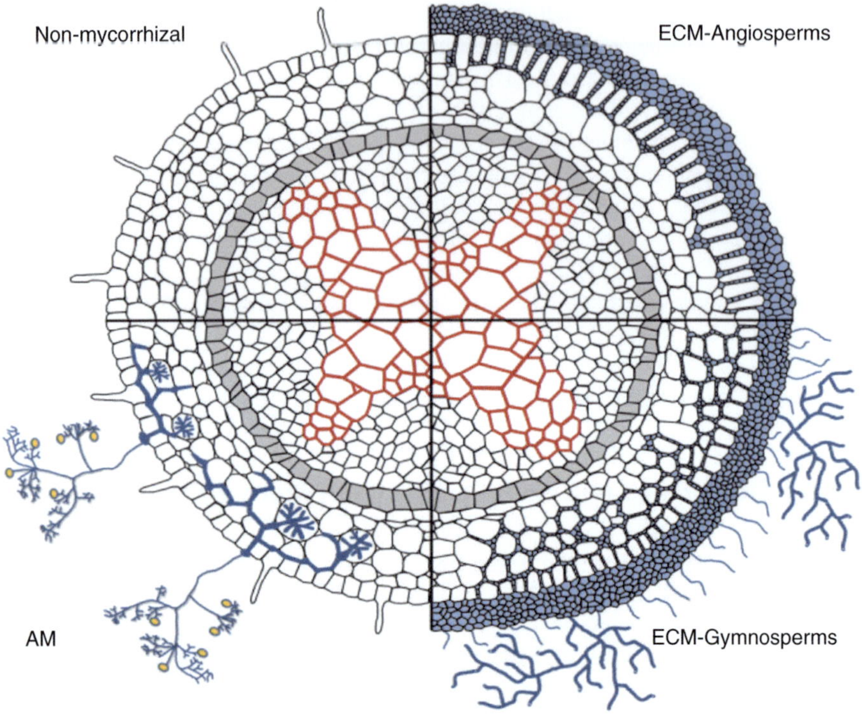

Fig. 19.2 Main structural differences between AM and ECM associations of angiosperms or gymnosperms (Adapted from: © Bücking et al. 2012. Published in [short citation] under CC BY 3.0 license. Available from: http://dx.doi.org/10.5772/52570)

In ECM, the fungi intrude the cortical portion of the host root without penetrating cortical cells. They are also known as 'sheathing mycorrhiza' due to the formation within the root of a hyphal network known as the 'Hartig net' around cortical cells and a thick layer of hyphal mat on the root surface known as sheath or mantle, which covers feeder roots. Infection of host plants by ectomycorrhizal fungi usually leads to alterations in feeder roots that are visible to the naked eye (Genre 2010). Feeder roots colonized by the fungi are thicker and more branched than uncolonized roots and also are differently coloured. Generally, ectomycorrhizas are formed between fine roots and dikaryotic mycelia originating from the fusion of two different monokaryotic hyphae germinated from spores. The characteristic fungal sheath (the mantle) adheres to the root surface and consists of aggregated hyphae. This mycelium is linked to extramatrical hyphae that explore the substrate and are responsible for the mineral nutrition and water uptake of the symbiotic tissues. From the inner zone of the mantle, some hyphae penetrate between the root cells to form an interface called the Hartig net, where metabolites are exchanged. The hyphae always remain apoplastic and can colonize the epidermal (angiosperms) and the cortical cell (gymnosperms) layers. Root cells surrounded by hyphae are still alive, as in arbuscules (Barker et al. 1998).

In AM fungi, the fungi penetrate the cortical cells and form clusters of finely divided hyphae known as arbuscules in the cortex. In some cases (called as VAM), they also form vesicles, which are membrane-bound organelles of diverse shapes, inside or outside the cortical cells. Arbuscules are believed to be the sites where materials are exchanged between the host plant and the fungi. Vesicles have dual function, they generally serve as storage structures, and when they are old, they can act as reproductive structures. Vesicles and arbuscules, together with large spores, comprise the characteristic features of the VA mycorrhizas. Vesicles are not always visible in such types of mycorrhizal associations; some scientists propose the designation arbuscular mycorrhiza (AM), more preferable over the term vesicular-arbuscular (VA) mycorrhiza. Both AM fungi and ectomycorrhizal fungi extend hyphae from the root into the soil, and these external (or extraradical) hyphae are responsible for translocating nutrients from the soil to the root (Fig. 19.2).

19.4 Nutrient Uptake Pathways in Mycorrhizal Roots

There can be two pathways through which plants absorb nutrients from the soil (Smith et al. 2011): 'plant pathway' that includes the unmediated uptake of nutrients from the soil by the root epidermis and its root hairs and the 'mycorrhizal pathway' that includes the uptake of nutrients through extraradical mycelium of the fungus and the transfers to the Hartig net in ECM association or to the intra-radical mycelium in AM association and the uptake by the plant from the interfacial apoplast (Harrison et al. 2002). The uptake of nutrients from the soil via the plant pathway, however, is often restricted by the low mobility of nutrients in the soil (Bücking and Kafle 2015). AM and ECM roots vary in their structural attributes, and this difference has connection with slightly different mode of nutrient uptake in AM or ECM plants.

AM roots do not form a fungal sheath and can presumably utilize both pathways for nutrient uptake (Bücking et al. 2012). It has previously been suggested that in the AM symbiosis, both uptake pathways act combined (Bücking and Kafle 2015). This led to the assumption that the uptake via the mycorrhizal pathway can be avoided when nutrient availability in the soil is high and mycorrhizal plants not always show a positive growth response. This view, however, has now become controversial (Smith and Read 1997; Smith et al. 2009, 2011), and it has been asserted that the mycorrhizal pathway can dominate the total P uptake and that the true contribution of the mycorrhizal pathway to total P uptake can be 'hidden' (Smith et al. 2003; Nagy et al. 2009). Plant P transporters that are involved in the uptake via the plant pathway are down-regulated in response to the AM symbiosis (Harley and Smith 1983; Chiou et al. 2001; Grunwald et al. 2009), while mycorrhiza-specific transporters that are involved in the P uptake from the mycorrhizal interface are induced (Xu et al. 2007; Paszkowski et al. 2002).The benefaction of both the pathways to total P uptake also relies on the plant and fungal species. Zhang et al. (2015) demonstrated that *Rhizophagus irregularis* was more efficient in P absorption than *Acaulospora longula* and *Gigaspora margarita* in *Lotus japonicas*. Grunwald et al. (2009) has demonstrated that *Glomus intraradices* has the highest ability to suppress the expression of plant P transporters of the plant pathway, while *G. mosseae* had the least effect. Such facts advocate that the contribution of the mycorrhizal pathway to nutrient acquisition also relies on the efficacy with which both partners interact and exchange nutrients across the mycorrhizal interface (Bücking et al. 2012).The suppression of the plant pathway by AM fungi can result in growth reductions in mycorrhizal plants when the mycorrhizal pathway does not repay for the depressed uptake by the plant pathway (Smith and Smith 2011).It has been hypothesized that the AM fungus could use the downregulation of the plant pathway to enhance its C availability. A greater dependency on the mycorrhizal pathway for nutrient acquisition has been shown to motivate the C distribution to the root system (Nielsen et al. 1998; Postma and Lynch 2011) (Fig. 19.3).

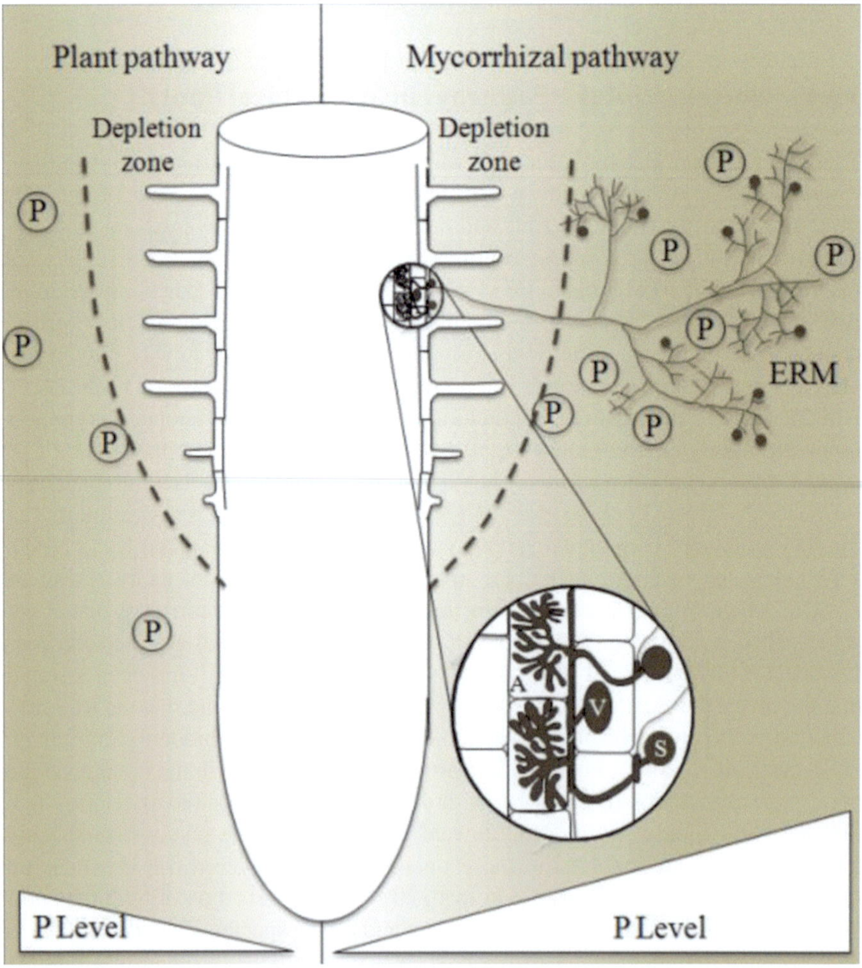

Fig. 19.3 Nutrient uptake pathways in mycorrhizal roots (Adapted from: © Bücking et al. 2012. Published in [short citation] under CC BY 3.0 license. Available from: http://dx.doi.org/10.5772/52570)

In ectomycorrhizal tree species, most of the root surface is composed of the areas that are not useful in nutrient acquisition, and the zones that contribute in nutrient acquisition such as the non-mycorrhizal white or ECM roots represent only 2% and 16% of the total root length, respectively (Taylor and Peterson 2002). In such cases the impact of fungal mantle that surrounds root tips is most critical (Taylor and Peterson 2005). If the fungal mantle does not allow the nutrient ions to permeate through it, the underlying root tissue would be detached from the soil solution, and these roots would be solely dependent upon the mycorrhizal pathway for nutrient acquisition. ECM fungal species and the structure and properties of the mantle decide if the fungal sheath represents an apoplastic barrier. Taylor and Peterson (2005) conducted an experiment related to the assessment of the permeability of *Pinus banksiana*/*Hebeloma cylindrosporum* fungal mantle to both berberine and radioactive sulphate ions. They found that the fungal mantle was completely impermeable to tracer dye. The fungal mantle also proved to be impermeable to sulphate over a 24-h exposure period. These results suggested that the plant may be highly dependent on the fungus to supply mineral nutrients as there is little plant tissue capable of nutrient absorption outside the fungal mantle. Some other fungi have been shown to release hydrophobins during ECM development (Coelho et al. 2010). Hydrophobins, the small hydrophobic proteins that are involved in the fastening of hyphae to surfaces, can also enhance the water impermeability of the fungal sheath (Unestam 1991; Unestam and Sun 1995). Due to the fact that only 2% of the root surface of pines is non-mycorrhizal and that the ERM of an ECM fungus can represent up to 99% of the nutrient-absorbing surface length of pine roots (Rousseau et al. 1992), ECM tree species such as pines are considered to be highly dependent on their fungal associates (Ouahmane et al. 2009; Brundrett 2002), and it can be considered that the mycorrhizal pathway plays an even more important role for nutrient uptake in ECM root systems than in AM root systems (Bücking et al. 2012).

19.5 Possible Mechanisms of Nutrient Acquisition by Mycorrhizal Fungi

Mycorrhizal fungi are able to absorb and transport all of the 15 major macro and micronutrients essential for plant growth. Mycorrhizal fungi produce powerful chemicals into the soil that dissolve hard to arrest nutrients such as phosphorous, iron and other 'tightly bound' soil nutrients. This whole process is immensely important in plant nutrition and gives the idea why non-mycorrhizal plants require high levels of fertility to maintain their health. Mycorrhizal fungi form an intricate web that captures and assimilates nutrients conserving the nutrient capital in soils. In non-mycorrhizal conditions, much of this fertility is wasted or lost from the system. Mycorrhizal associations may affect the mineral nutrition of the host plant directly by improving plant growth through nutrient acquisition by the fungus, or indirectly by modifying transpiration rates and composition of rhizosphere microflora (Marschner and Dell 1994), by nutrient

Table 19.1 AM and ECM responsible for transport of specific class of nutrients

Nutrient transported	VAM	ECM
P	+	+
NH_4^+	+	+
NO_3	–	+
K	+	+
Ca	+	–
SO_4^+	+	–
Cu	+	–
Zn	+	–
Fe	–	+

Adapted and modified from Marschner and Dell (1994)

mobilization from organic substrates (Finlay 2008), by enhancing fertilizer use efficiency (Jeff and Taylor 2005) or by beneficial association with other micro-organisms (Finlay 2008) (Table 19.1).

Two major steps are involved in nutrient acquisition and delivery of nutrients via mycorrhizal association:

1. Mobilization and absorption by fungal mycelia
2. Transfer of mobilized nutrients across fungus–root interface

19.6 Mobilization and Absorption of Nutrients

Besides the hyphae in direct contact with the root surface, all mycorrhizal fungi develop mycelium (extramatrical mycelium) which extends from the infected root surface into adjoining soil. Both arbuscular mycorrhizal and ectomycorrhizal fungi produce enormous amount of extramatrical mycelium, with arbuscular mycorrhizal mycelia extending several centimetres from the infected root surface and ectomycorrhizal mycelium potentially spreading for up to several meters (Goltapeh et al. 2008). In either case, the mycelium spread well beyond the nutrient depletion zone for immobile nutrients around individual roots and exhibit a complex framework that renders it an effective nutrient collecting network (Schachtman et al. 1998; Bücking and Heyser 2001; Goltapeh et al. 2008). Extramatrical mycelium is an element of mycorrhiza which efficiently excavate bulk soil for scanty nutrients and translocates acquired nutrients to the fungus root interface where transfer to the host plant is affected (Bücking and Kafle 2015). Extramatrical mycelium of many ecto-mycorrhizal fungi spread as a diffuse mat of individual hyphae forming a complex linear multi-hyphal structures known as rhizomorphs. Hyphae up to 35 μm in diameter at the centre of rhizomorphs lack cell walls and perform an important function in transport of inorganic nutrients or photoassimilates. On the other hand, diffuse hyphae (diameter 1–5 μm) at the growing front of arbuscular mycorrhizas provide an extensive surface area for nutrient absorption, while larger diameter hyphae (up to 10 μm) accounts for an excellent translocatory infrastructure for effectively transferring solutes from bulk soil through the rhizosphere to root surfaces (Ravnskov and Jakobsen 1995). In addition to improving plant uptake of mineral nutrient already present in soil, many mycorrhizal fungi may play a significant role in mobilizing nutrients either from organic substrate (Hodge and Fitter 2010), mineral particles or rock surface (Finlay and Rosling 2006).

Many mycorrhizal fungi may play a valuable role in the mobilization of nutrients such as N and P from structural and other polymers which are otherwise unavailable to plant roots. Extraction of N and P by mycorrhizal fungi from a range of organic substrates such as pollen (Perez-Moreno and Read 2001a; Finlay 2008), dead nematodes (Perez-Moreno and Read 2001b), Collembola (Klironomos and Hart 2001) and saprotrophic mycelia (Lindahl et al. 1999) has been evidenced by several scientists. Involvement of mycorrhizal fungi in microbial mobilization–immobilization cycles leads to mobilization of N and P from microbial, microfaunal, meso-faunal and plant litter, allowing the distinctive plant communities to bloom along the altitudinal or latitudinal gradients (Smith et al. 2003, 2009).

Ectomycorrhizal fungi colonizing boreal forest ecosystems are the suitable examples of such incidents. In these ecosystems, N and P are available in organic forms that are not readily accessible to autotrophs. Here the dominant plant species are highly dependent on mycorrhizal symbionts to satisfy their nutrient needs. Ectomycorrhizal fungi have the ability to directly attack the structural polymers which may render nutrients unavailable and in mobilization of N and P from the organic polymers (Read and Perez-Moreno 2003). The observations made by Lindahl et al. (2007) suggest that saprotrophs with a full complement of litter-degrading enzymes are required during the initial stages of decomposition and that N mobilized by these fungi is retained within their mycelia. As the C:N ratio of the litter declines, the saprotrophs are considered to become less competitive in relation to mycorrhizal species which are directly supplied with host assimilates (Hodge et al. 2000). The prevalence of ectomycorrhizal fungi in the lower, well-degraded litter and humus gives indication that mycorrhizal hyphae play a considerable role in mobilizing N from well-decomposed organic matter in boreal forest soils and that unstable C entering the soil through roots and associated mycorrhizal fungi may play a pivotal role in carrying out mobilization of this N. Ectomycorrhizal fungi produce extracellular proteinases and peptidases that efficiently hydrolyse organic nitrogen sources to release amino acids which can be sucked up by the fungi. Ectomycorrhizal fungi also secrete extracellular phosphomonoesterases and phosphodiesterases. The phosphodiesterases are able to mediate the mobilization of phosphorus sequestered within nucleic acids. Some ectomycorrhizal fungi produce hydrolytic enzymes within the cellulase, hemicellulase and lignase families that may encourage hyphal entry to dead and decaying plant material in soil and reach to mineral nutrients sequestered therein. By such ways ectomycorrhizal fungi reduce the conventional nutrient cycles, liberating nutrients from soil organic matter, free from the involvement of saprotrophic organisms. Ectomycorrhizal fungi have also been reported to be able to produce siderophores capable of complexing iron and oxalate to improve potassium uptake. Reducing agents produced by ectomycorrhizal fungi increase ion acquisition from stable oxides (e.g. MnO_2), thus helping in improving plant nutrition (Lindahl et al. 2001, 2007).

The obligate biotrophic nature of AM fungi has suggested that these fungi are not able to use organic N sources (Bücking and Kafle 2015); however, several studies demonstrate that hyphae of AM fungi grow on organic substrates and transfer N to their host plants (Leigh et al. 2009; Hodge and Fitter 2010) that leads to higher plant nitrogen content in mycorrhizal plants (Thirkell et al. 2015). Reynolds et al. (2005) found no evidence that AM fungi promote plant N acquisition and growth of old field perennials under conditions of low N supply, but AM fungi may be associated with decaying organic matter in some ecosystems. Hodge et al. (2001) demonstrated enhanced decomposition and N capture from decaying grass leaves in the presence of AM fungi. Leigh et al. (2009) confirmed that AM fungi do not have saprophytic ability and the fungus absorbs N from the organic substrates most probably as decomposition product. However, AM fungus speeds up the N mobilization from organic matter (Atul-Nayyar et al. 2009) and influence the C flow through soil microbial communities during decomposition (Herman et al. 2012). However, further research is still needed to distinguish between the direct capacity of AM fungi to mobilize organic substrates and their possible, indirect effects on decomposition and plant nutrient uptake, caused by stimulation of decomposers and subsequent uptake of their decomposition products by mycorrhizal hyphae (Li et al. 2006; Finlay 2008).

Other than organic matter, mycorrhizal fungi (either by themselves or in association with bacteria or other fungi) has been reported to actively mobilize nutrients from mineral particles and rock surfaces through weathering (Landeweert et al. 2001; Finlay and Rosling 2006; Wallander 2006; Finlay 2008). The role of arbuscular mycorrhizal (AM) fungi in mineral weathering is conflicting, and there are only a few evidences suggesting enhanced utilization of relatively insoluble forms of inorganic P such as rock phosphate by AM fungi. These effects could depend upon synergistic interactions of AM fungi with other P-solubilizing microorganisms. Wallander (2006) reported a significant mycorrhizal contribution to mineral weathering in forest soils. Ectomycorrhizal fungi are reported to produce low-molecular-weight (LMW) organic acids that are being implicated in weathering of minerals (Ahonen-Jonnarth et al. 2000). Breemen et al. (2000) observed that numerous open, tubular pores, 3–10 gum in width, were present in weatherable minerals in all podzol surface soils and shallow granitic rocks under European coniferous forests, and they hypothesized these pores were formed by complex forming, low-molecular-weight organic acids released by or formed in association with mycorrhizal fungi. The mycelium of ectomycorrhizal fungi is able to penetrate and most probably create microsites which are beyond the reach of plant roots and isolated from bulk soil solution phenomena. Dissolved products could be transferred to the host plant roots, avoiding the soil solution with often toxic concentration of $A1^{3+}$ from acid rain (Clark 1997) and also bypassing competition for nutrient uptake by other organisms.

19.7 Movement of Carbon and Nutrients Across the Fungus–Root Interface

Whatever the type of mycorrhizal fungi or the way they adopt to mobilize, the nutrients reach at the fungus root interface within the symplasm of the fungus. Transfer of the nutrients to the host plant involves efflux across the fungal plasma membrane followed by absorption from the apoplasm of the interface across the plasma membrane of the host root cells (Cairney and Burke 1996). Runaway of the substrates from the interface is reduced by complicated fungal structures. Impermeable extracellular materials get accumulated between hyphae of the mantle in some ectomycorrhizas and at the points of hyphal entry into cells in arbuscular mycorrhizas. Ectomycorrhizas form a specific apoplasmic chamber. This prevents runoff of nutrients from the interface apoplasm. This suggests that local chemical and physical conditions can be managed by the activities of both the associates in the symbiosis. Mycorrhizal fungi obtain carbon for growth and metabolism from host roots, mainly as photoassimilate (Smith and Read 2008; Bonfante and Genre 2010). In contrast to phytopathogenic fungi or ericoid mycorrhizal fungi, AM and ECM fungi are not able to use sucrose as a carbon source, and they take up simpler sugars, such as glucose or fructose from the mycorrhizal interface. The presence of invertase genes in fungal genomes is correlated with the nutritional mode and in contrast to other plant associated fungi, such as pathogens or endophytes, there are no indications that AM or ECM fungi possess invertase genes (Parrent et al. 2009; Bonfante and Genre 2010; Wahl et al. 2010) or have invertase activity (Salzer and Hager 1996). Consequently, mycorrhizal fungi rely on the invertase activity of the host in the interfacial apoplast for sucrose hydrolysis. Sucrose hydrolysis makes the hexoses glucose and fructose available for the fungus, and it has been suggested that glucose is mainly taken up by hyphae of the Hartig net and fructose mainly by hyphae of inner mantle layers (Nehls et al. 2001). Several transporters have been identified on both the plant and fungal membranes and contribute to delivering nutrients from fungi to plants. In context of ectomycorrhizal association, the high affinity NH4+ importer AmAMT2 of *Amanita muscaria* is upregulated in the extraradical mycelium, but downregulated in Hartig net and the fungal sheath (Willmann et al. 2007; Martin and Nehls 2009). The high expression of this transporter in the ERM suggests a high capability of the ERM for NH4+ uptake. The low expression level in the Hartig net on the other hand indicates that NH4+ can serve as a potential nitrogen source that is delivered by the mycorrhizal fungus to the host. A low expression level of this NH4+ importer in the Hartig net would reduce the reabsorption of NH4+ by the fungus from the interfacial apoplast and increase the net transport of NH4+ to the host. The potential transport of NH4+ across the ECM interface is also supported by the presence and upregulation of plant high affinity NH4+ importers in ECM roots (Selle et al. 2005; Couturier et al. 2007). Wang and Qiu (2006) observed in their study on the rice and *Medicago truncatula* that plasma membrane H^+-ATPases that are specifically induced in arbuscule containing cells are required for enhanced proton pumping activity in membrane vesicles from AM colonized roots (Harrison et al. 2002). Mutation of the H^+-ATPase decreased arbuscule size

Table 19.2 Comparison between AM and ECM associations

Features	AM association	ECM association
Nutrients transported to plant	Specifically important for P transport, also contribute in N transport	Specifically important for N transport but also have significant contribution in P transport
Occurrence	Mainly in warm and dry climates where P availability is low	Climates with low temperature and high humidity, where N availability is low
Host range	Associates with a very wide range of hosts	Associates with comparatively lower portion (3%) of all plant species
Mode of fungal nutrition	Obligate biotroph	Facultative saprotroph
Structural elements	Arbuscules, ERM and vesicles in Some types	Mantle, Hartig net and ERM
Mode of penetration	Both inter as well as extracellular	Only intercellular
Nutrient uptake pathway	Both plant and mycorrhizal pathway	Mainly mycorrhizal pathway

Modified after Bücking et al. (2012)

and hindered nutrient uptake by the host plant through the mycorrhizal symbiosis. Overexpression of the H^+-ATPase Os-HA1 enhanced both phosphate uptake and the plasma membrane potential, demonstrating that this H^+-ATPase plays a lead role in energizing the peri-arbuscular membrane, thereby facilitating nutrient exchange in arbusculated plant cells. Another transporter, Pt4, a high affinity phosphate (P) transporter, is exclusively manifested in mycorrhizal roots and is implicated in the acquisition of P delivered by the fungus (Xu et al. 2007). A high affinity ammonium transporter (AMT2;2) is also found to be situated in peri-arbuscular membrane (Guether et al. 2009), and the presence of mycorrhiza inducible sulphate transporters in AM roots suggests that sulphate is also transferred from the AM fungus to the host across the mycorrhizal interface (Casieri et al. 2012; Allen and Shachar-Hill 2009; Helber et al. 2011) (Table 19.2).

Conclusions

Mycorrhizal fungi are found to be associated with majority of higher plants. These symbiotic associations vary widely in their structure and function. Out of several types of mycorrhizal fungi, AM and ECM fungi play significant role in nature. Both types of mycorrhizal fungi not only help in plant uptake of major nutrients like P and N but also assist in taking up micronutrients such as Zn, Cu, Fe, etc. Mycorrhizal fungi adopt several mechanisms to accomplish the task successfully, including enhanced absorbing area of plants, release of biochemical and association with other microorganisms in surroundings. In addition to mobilizing nutrients, mycorrhizal fungi also serve as an important C sink in the soil, thus having an important impact on the cycling of these elements. Therefore, mycorrhiza has proved to be an important alliance for the nutrient management in ecosystem.

References

Ahonen-Jonnarth U, Van Hees PAW, Lundström US, Finlay RD (2000) Production of organic acids by mycorrhizal and non-mycorrhizal *Pinus sylvestris* L. seedlings exposed to elevated concentrations of aluminium and heavy metals. New Phytol 146:557–567

Allen JW, Shachar-Hill Y (2009) Sulfur transfer through an Arbuscular mycorrhiza. Plant Physiol 149(1):549–560

Atul-Nayyar A, Hamel C, Hanson K, Germida J (2009) The arbuscular mycorrhizal symbiosis links N mineralization to plant demand. Mycorrhiza 19:239–246

Barker SJ, Stummer B, Gao L, Dispain I, O'Connor PJ, Smith SE (1998) A mutant in Lycopersicon esculentum mill, with highly reduced VA mycorrhizal colonization: isolation and preliminary characterization. Plant J 15:791–797

Bonfante P, Genre A (2010) Mechanisms underlying beneficial plant–fungus interactions in mycorrhizal symbiosis. Nat Commun 1:48. doi:10.1038/ncomms1046

Breemen NV, Finlay R, Lundstrom U, Jongmans AG, Giesler R, Olsson M (2000) Mycorrhizal weathering: a true case of mineral plant nutrition? Biogeochemistry 49:53–67

Brundrett MC (2002) Tansley review no. 134: Coevolution of roots and mycorrhizas of land plants. New Phytol 154:275–304

Bücking H (2011) Ectomycoremediation: an eco-friendly technique for the remediation of polluted sites. In: Rai M, Varma A (eds) Diversity and biotechnology of ectomycorrhizae. Soil biology. Springer, Heidelberg, pp 209–229

Bücking H, Heyser W (2001) Microautoradiographic localization of phosphate and carbohydrates in mycorrhizal roots of *Populus tremula* x *Populus alba* and the implications for transfer processes in ectomycorrhizal associations. Tree Physiol 21(2):101–107

Bücking H, Kafle A (2015) Role of Arbuscular Mycorrhizal fungi in the nitrogen uptake of plants: current knowledge and research gaps. Agronomy 5:587–612

Bücking H, Kuhn AJ, Schröder WH, Heyser W (2002) The fungal sheath of ectomycorrhizal pine roots: an apoplastic barrier for the entry of calcium, magnesium, and potassium into the root cortex? J Exp Bot 53:1659–1669

Bücking H, Liepold E, Ambilwade P (2012) The role of the mycorrhizal symbiosis in nutrient uptake of plants and the regulatory mechanisms underlying these transport processes In: Dhal NK, Sahu SC (eds) Plant science, ISBN 978-953-51-0905-1

Cairney JWG, Burke RM (1996) Physiological heterogeneity within fungal mycelia: an important concept for a functional understanding of the ectomycorrhizal symbiosis. New Phytol 134:685–695

Casieri L, Gallardo K, Wipf D (2012) Transcriptional response of *Medicago truncatula* sulphate transporters to arbuscular mycorrhizal symbiosis with and without Sulphur stress. Planta. doi:10.1007/s00425-012-1645-7

Chiou TJ, Liu H, Harrison MJ (2001) The spatial expression patterns of a phosphate transporter (MtPT1) from *Medicago truncatula* indicate a role in phosphate transport at the root/soil interface. Plant J 25(3):281–293

Clark RB (1997) Arbuscular mycorrhizal adaptation, spore germination, root colonization, and host plant growth and mineral acquisition at low pH. Plant Soil 192:15–22

Coelho ID, de Queiroz MV, Costa MD, Kasuya MCM, de Araujo EF (2010) Identification of differentially expressed genes of the fungus *Hydnangium*sp during the pre-symbiotic phase of the ectomycorrhizal association with *Eucalyptus grandis*. Mycorrhiza 20(8):531–540

Couturier J, Montanini B, Martin F, Brun A, Blaudez D, Chalot M (2007) The expanded family of ammonium transporters in the perennial poplar plant. New Phytol 174:137–150

Finlay RD, Rosling A (2006) Integrated nutrient cycles in forest ecosystems, the role of ectomycorrhizal fungi. In: Gadd GM (ed) Fungi in biogeochemical cycles. Cambridge University Press, Cambridge, pp 28–50

Finlay RD (2008) Ecological aspects of mycorrhizal symbiosis: with special emphasis on the functional diversity of interactions involving the extraradial mycelium. J Exp Bot 59(5):1115–1126

Genre A (2010) Mechanisms underlying beneficial plant-fungus interactions in mycorrhizal symbiosis. Nat Commun 1:48–52

Giri B, Giang PH, Kumari R, Prasad R, Sachdev M, Garg AP, Oelmüller R, Varma A (2005) Mycorrhizosphere: strategies and functions. Soil Biol 3:213–252

Goltapeh EM, Danesh YR, Prasad R, Varma A (2008) Mycorrhizal fungi: What we know and what should we know? In Mycorrhiza Springer, Berlin, Heidelberg, pp. 3–27

Grunwald U, Guo W, Fischer K, Isayenkov S, Ludwig-Müller J, Hause B, Guo W, Yan X, Frankene P (2009) Overlapping expression patterns and differential transcript levels of phosphate transporter genes in arbuscular mycorrhizal, P_i-fertilised and phytohormone-treated *Medicago truncatula* roots. Planta 229:1023–1034

Guether M, Neuhauser B, Balestrini R, Dynowski M, Ludewig U, Bonfante P (2009) A mycorrhizal-specific ammonium transporter from *Lotus japonicus* acquires nitrogen released by arbuscular mycorrhizal fungi. Plant Physiol 150(1):73–83

Habte M (2000) Mycorrhizal fungi and plant nutrition In: Silva JA and Uchida R (eds) Plant nutrient management in Hawaii's soils, approaches for tropical and subtropical agriculture, College of Tropical Agriculture and Human Resources, University of Hawaii at Manoa, US.

Harley JL, Smith SE (1983) Mycorrhizal symbiosis. Academic Press, Toronto, pp 112–115

Harrison MJ, Dewbre GR, Liu J (2002) A phosphate transporter from *Medicago truncatula* involved in the acquisition of phosphate released by arbuscular mycorrhizal fungi. Plant Cell 14:2413–2429

Helber N, Wippel K, Sauer N, Schaarschmidt S, Hause B, Requena N (2011) A versatile monosaccharide transporter that operates in the arbuscular mycorrhizal fungus *Glomus* sp. Is crucial for the symbiotic relationship with plants. Plant Cell 23:3812–3823

Herman DJ, Firestone MK, Nuccio E, Hodge A (2012) Interactions between an arbuscular mycorrhizal fungus and a soil microbial community mediating litter decomposition. FEMS Microbiolol Ecol 80:236–247

Hodge A, Campbell CD, Fitter AH (2001) An arbuscular mycorrhizal fungus accelerates decomposition and acquires nitrogen directly from organic material. Nature 413:297–299. doi:10.1038/35095041

Hodge A, Fitter AH (2010) Substantial nitrogen acquisition by arbuscular mycorrhizal fungi from organic material has implications for N cycling. Proc Nat Acad Sci USA 107:13754–13759

Hodge A, Robinson D, Fitter AH (2000) An arbuscular mycorrhizal inoculum enhances root proliferation in, but not nitrogen capture from, nutrient-rich patches in soil. New Phytol 145:575–584

Jeff H, Taylor PCA (2005) Ectomycorrhizal impacts on nutrient uptake pathways in woody roots. New For 30:203–214

Kilpeläinena J, Vestbergb M, Repoc T, Lehtoa T (2016) Arbuscular and ectomycorrhizal root colonisation and plant nutrition in soils exposed to freezing temperatures. Soil Biol Biochem 99:85–93

Klironomos JN, Hart MM (2001) Animal nitrogen swap for plant carbon. Nature 41:651–652

Landeweert R, Hoffland E, Finlay RD, Kuyper TW, van Breemen N (2001) Linking plants to rocks: ectomycorrhizal fungi mobilize nutrients from minerals. Trends Ecol Evol 16: 248–254

Leigh J, Hodge A, Fitter AH (2009) Arbuscular mycorrhizal fungi can transfer substantial amounts of nitrogen to their host plant from organic material. New Phytol 181:199–207

Li HY, Smith SE, Holloway RE, Zhu YG, Smith FA (2006) Arbuscular mycorrhizal fungi contribute to phosphorus uptake by wheat grown in a phosphorus-fixing soil even in the absence of positive growth responses. New Phytol 172:536–5343

Lindahl BD, Ihrmark K, Boberg J, Trumbore SE, Hogberg P, Stenlid J, Finlay RD (2007) Spatial separation of litter decomposition and mycorrhizal nitrogen uptake in a boreal forest. New Phytol 173:611–620

Lindahl B, Olsson S, Stenlid J, Finlay RD (2001) Effects of resource availability on mycelial interactions and 32P–transfer between a saprotrophic and an ectomycorrhizal fungus in soil microcosms. FEMS Microbiol Ecol 38:43–52

Lindahl B, Stenlid J, Olsson S, Finlay RD (1999) Translocation of 32P between interacting mycelia of a wood decomposing fungus and ectomycorrhizal fungi in microcosm systems. New Phytol 44:183–193
Marschner H, Dell B (1994) Nutrient uptake in mycorrhizal symbiosis. Plant Soil 159:89–102
Martin F, Nehls U (2009) Harnessing ectomycorrhizal genomics for ecological insights. Curr Opi Biotechnol 12:509–515
Nagy R, Drissner D, Amrhein N, Jakobsen I, Bucher M (2009) Mycorrhizal phosphate uptake pathway in tomato is phosphorus-repressible and transcriptionally regulated. New Phytol 181:950–959
Nehls U, Mikolajewski S, Magel E, Hampp R (2001) The role of carbohydrates in ectomycorrhizal functioning: gene expression and metabolic control. New Phytol 150:533–541
Nielsen KL, Bouma TJ, Lynch JP, Eissenstat DM (1998) Effects of phosphorus availability and vesicular–arbuscular mycorrhizas on the carbon budget of common bean (*Phaseolus vulgaris*). New Phytol 139:647–656
Ouahmane L, Revel JC, Hafidi M, Thioulouse J, Prin Y, Galiana A, Dreyfus B, Duponnois R (2009) Responses of *Pinus halepensis* growth, soil microbial catabolic functions and phosphate-solubilizing bacteria after rock phosphate amendment and ectomycorrhizal inoculation. Plant Soil 320(1–2):169–179
Parrent JL, James TY, Vasaitis R, Taylor AFS (2009) Friend or foe? Evolutionary history of glycoside hydrolase family 32 genes encoding for sucrolytic activity in fungi and its implications for plant-fungal symbioses. BMC Evol Biol 9:148–154
Paszkowski U, Kroken U, Roux C, Briggs SP (2002) Rice phosphate transporters include an evolutionarily divergent gene specifically activated in arbuscular mycorrhizal symbiosis. Proc Nat Acad Sci USA 99(20):13324–13329
Perez-Moreno J, Read DJ (2001a) Exploitation of pollen by mycorrhizal mycelial systems with special reference to nutrient cycling in boreal forests. Proc Royal Soc B 268:1329–1335
Perez-Moreno J, Read DJ (2001b) Nutrient transfer from soil nematodes to plants: a direct pathway provided by the mycorrhizal mycelial network. Plant Cell Environ 24:1219–1226
Postma JA, Lynch JP (2011) Root cortical aerenchyma enhances the growth of maize on soils with suboptimal availability of nitrogen, phosphorus, and potassium. Plant Physiol 156:1190–1201
Ravnskov S, Jakobsen I (1995) Functional compatibility in arbuscular mycorrhizas measured as hyphal p transport to the plant. New Phytol 129:611–618
Read DJ, Perez-Moreno J (2003) Mycorrhizas and nutrient cycling in ecosystems: a journey towards relevance? New Phytol 157:475–492
Reynolds HL, Hartley AE, Vogelsang KM, Bever JD, Schultz PA (2005) Arbuscular mycorrhizal fungi do not enhance nitrogen acquisition and growth of old-field perennials under low nitrogen supply in glasshouse culture. New Phytol 167:869–880
Rousseau JVD, Reid CPP, English RJ (1992) Relationship between biomass of the mycorrhizal fungus *Pisolithus tinctorius* and phosphorus uptake in loblolly pine seedlings. Soil Biol Biochem 24(2):183–184
Salzer P, Hager A (1996) Sucrose utilization of the ectomycorrhizal fungi *Amanita muscaria* and *Hebeloma crus-tuliniforme* depends on the cell wall-bound invertase activity of their host *Picea abies*. Botanica Acta 104:439–445
Schachtman DP, Reid RJ, Ayling SM (1998) Phosphorus uptake by plants: from soil to cell. Plant Physiol 116(2):44–53
Selle A, Willmann M, Grunze N, Gessler A, Weiss M, Nehls U (2005) The high-affinity poplar ammonium importer PttAMT1.2 and its role in ectomycorrhizal symbiosis. New Phytol 168(3):697–706
Smith FA, Grace EJ, Smith SE (2009) More than a carbon economy: nutrient trade and ecological sustainability in facultative arbuscular mycorrhizal symbioses. New Phytol 182:347–358
Smith SE, Smith FA, Jakobsen I (2003) Mycorrhizal fungi can dominate phosphate supply to plants irrespective of growth responses. Plant Physiol 133:16–20
Smith SE, Jakobsen I, Grønlund M, Smith FA (2011) Roles of arbuscular mycorrhizas in plant phosphorus nutrition: interactions between pathways of phosphorus uptake in arbuscular

mycorrhizal roots have important implications for understanding and manipulating plant phosphorus acquisition. Plant Physiol 156:1050–1057
Smith SE, Read DJ (1997) Mycorrhizal symbiosis. Academic Press, London
Smith SE, Read DJ (2008) Mycorrhizal Symbiosis, 3rd edn. Academic Press, London, p 787
Smith SE, Smith FA (2011) Roles of arbuscular mycorrhizas in plant nutrition and growth: new paradigms from cellular to ecosystem scales. Ann Rev Plant Biol 62:227–250
Taylor JH, Peterson CA (2002) Morphometric analysis of *Pinus banksiana* lamb. Root anatomy during a 3-month field study. Trees 14:239–247
Taylor JH, Peterson CA (2005) Ectomycorrhizal impacts on nutrient uptake pathways in woody roots. New For 30(2–3):203–214
Thirkell JD, Cameron DD, Hodge A (2015) Resolving the "nitrogen paradox" of arbuscular mycorrhizas: fertilization with organic matter brings considerable benefits for plant nutrition and growth. Plant Cell Environ. doi:10.1111/pce.12667
Unestam T (1991) Water repellency, mat formation, and leaf-stimulated growth of some ectomycorrhizal fungi. Mycorrhiza 1:13–20
Unestam T, Sun YP (1995) Extramatrical structures of hydrophobic and hydrophilic ectomycorrhizal fungi. Mycorrhiza 5:301–311
Wahl R, Wippel K, Goos S, Kämper J, Sauer N (2010) A novel high-affinity sucrose transporter is required for virulence of the plant pathogen *Ustilago maydis*. PLoS Biol 8:1000303. doi:10.1371/journal.pbio.1000303
Wallander H, Bonfante P, Wickman T, Jacks G (1997) Apatite as a source of mycorrhizal and non-mycorrhizal *Pinus sylvestris*. Plant Soil 196:123–131
Wallander H (2006) Uptake of P from apatite by *Pinus sylvestris* seedlings colonized by different ectomycorrhizal fungi. Plant Soil 218:249–256
Wang B, Qiu YL (2006) Phylogenetic distribution and evolution of mycorrhizae in land plants. Mycorrhiza 16:299–363
Willmann A, Weiss M, Nehls U (2007) Ectomycorrhiza-mediated repression of the high affinity ammonium importer Gene AmAMT2 in *Amanita muscaria*. Curr Genet 51(2):71–78
Wright DP, Read DJ, Scholes JD (1998) Mycorrhizal sink strength influences whole plant carbon balance of *Trifolium repens* L. Plant Cell Environ 21:881–891
Xu GH, Chague V, Melamed-Bessudo C, Kapulnik Y, Jain A, Raghothama KG, Levy AA, Silber A (2007) Functional characterization of LePT4: a phosphate transporter in tomato with mycorrhiza-enhanced expression. J Expt Bot 258(10):2491–2501
Zhang X, Chen B, Ohtomo R (2015) Mycorrhizal effects on growth, P uptake and Cd tolerance of the host plant vary among different AM fungal species. Soil Sci Plant Nutri 61(2):359–368

Sustainable Management of Waterlogged Areas Through a Biodrainage and Microbial Agro-ecosystem

20

Kumud Dubey, Alok Pandey, Praveen Tripathi, and K.P. Dubey

Abstract

Irrigation potential has been increased in recent years to achieve greater and sustained yield of agricultural products. The introduction of canal irrigation has caused a rise in the ground water table leading to waterlogging and secondary salinization. Management of this high water table is a major challenge globally as well as in India. Globally, about one-third of irrigated land is presently facing the threat of waterlogging and associated soil salinization. Approximately 4,981.43 square kilometers in the state of Uttar Pradesh, India are suffering from waterlogging and soil salinization, resulting in reduced agricultural productivity. The ability of vegetation with prolific transpiration characteristics may be utilized to reduce the water table markedly, combating this problem in a less expensive and more environmentally friendly manner. This drainage of excess of ground water through vegetation is termed biodrainage and appears promising for the management of a high water table and resultant soil salinity problems. This biodrainage technology may be more efficient in combination with specific beneficial microbes with pH-reducing properties allowing for the amelioration of soil characteristics. The present review discusses the application of biodrainage vegetation in combination with beneficial microbes for the sustainable management of waterlogged/high ground water table area.

K. Dubey (✉) • A. Pandey • P. Tripathi
Center for Social Forestry and Eco-Rehabilitation, Allahabad, Uttar Pradesh, India
e-mail: dkumud@yahoo.com

K.P. Dubey
Uttar Pradesh Forest Corporation, Allahabad, Uttar Pradesh, India

V. Kumar et al. (eds.), *Probiotics in Agroecosystem*,
DOI 10.1007/978-981-10-4059-7_20

20.1 Introduction

The total irrigated area in the world is 255 million hectares (Mha), of which more than two-thirds lies in Asia. About 20% of the irrigated land has been rendered saline due to waterlogging. Each year, an additional area of approximately 1.5 Mha of irrigated land is affected by secondary salinization due to waterlogging thereby losing its productivity. In Uttar Pradesh, India, waterlogging has emerged as a serious problem, especially in the Gangetic plains, where it is adversely affecting agriculture productivity. An area of approximately 4981.43 km^2 area in Uttar Pradesh is suffering from waterlogging resulting in reduced productivity.

Attempts are being made by various departments, organizations, and agencies for the prevention of waterlogging and soil salinization and the reclamation of such degraded lands, without much success. The problems of waterlogging and salinization can be effectively tackled by conventional engineering-based subsurface drainage systems, provided these are properly designed, installed, maintained, and operated. Conventional subsurface drainage systems consist of two types, vertical (tube wells) and horizontal (drain pipe). When properly designed, installed, and maintained, these systems are efficient in lowering the water table and preventing waterlogging and salinization (Heuperman and Kapoor 2002), but are more expensive and cause environmental problems. Moreover, they require periodic maintenance and are expensive. Under such a situation biodrainage is envisioned as a benign and cost-effective technology for lowering the rising water table so as to take it well below (>1.5 m) the root zone of crop plants (Angrish et al. 2006). Biodrainage is the vertical drainage of soil water through evapotranspiration by vegetation. The term *biodrainage* is relatively new, although the use of vegetation to dry out soil profiles has been known for a long time. Biodrainage is economically attractive because it requires only an initial investment for planting the vegetation, and when established, the system can produce economic returns by means of its harvested byproduct like fodder, timber, fuel wood, or fiber (Heuperman et al. 2002). There is consensus that biodrainage, when properly implemented, can lower the water table. It could solve problems associated with waterlogged areas and canal seepage. It has been demonstrated that under ideal conditions, a tree canopy may lower the water table by 1–2 m over a time period of 3–5 years (Kapoor 2002). Fast growing tree species may perform as effective biodrainage systems. The deep rooting characteristics of these trees make them extremely efficient users of water compared to the crop plants. Fast growing species like *Eucalyptus*, known for extensive water consumption under excess soil moisture conditions, are suitable for biodrainage. Other suitable species for biodrainage may be *Acacia nilotica*, *Casuarina glauca*, *Terminalia arjuna*, *Pongamia pinnata*, and *Syzygium cuminii*, etc. These species can be planted in blocks in the form of farm forestry or along the field boundary in the form of agroforestry (Dubey 2012, 2016). Under an Indian Council of Forestry Research and Education, Dehradun (ICFRE, Dehradun)-sponsored project a study was conducted to phytoremediate the waterlogged area through planting of a *Eucalyptus* hybrid, *Terminalia arjunaa*, *Trewia nudifora*, *Acacia nilotica*, and *Syzygium cumini* as biodrainage species with microbial amendments to ameliorate soil conditions. These selected tree species are common and prevalent species of the area and may be sustained in waterlogged conditions. Biodrainage is an emerging concept in India for land reclamation and the biodrainage potential of Eucalyptus is well established. The suitability of other tree species for biodrainage vegetation was compared with *Eucalyptus* sp.

20.2 Waterlogging and Soil Salinization

Waterlogging is the condition of the soil in which excess water limits gaseous transmission (Setter et al. 2002). The causes of waterlogging can be natural and anthropogenic. Main causes are heavy rainfall, poor water management systems, high water table, floods, over-irrigation, and seepage from canals and dams (Bilal et al. 2014). In water-logged situations, the ground water table gets allied to the crop root zone for most of the year (Michael and Ojha 2006). In such conditions, the soil becomes saturated with water and since the space between the soil particles is occupied by water instead of air, it suffocates the plant and is not suitable for the proper growth of plants. A rise in the groundwater level followed by waterlogging and secondary salinization has become a serious problem in canal-irrigated areas, located in arid and semi-arid regions of the world (Singh 2013). Commonly cultivated agricultural crop plants such as cereals, pulses, oilseeds, vegetables, and cash crops are susceptible to excess salts in soil or in irrigation water (Blumwald et al. 1983). High salinity in soil or irrigation water reduces the plant's capacity to extract water and nutrients, which affects the agricultural crop production (Dash et al. 2005). Plants resistant to waterlogging may thrive only in such soils. A rise in groundwater level followed by waterlogging and secondary salinization has become a serious problem in canal-irrigated areas located in arid and semi-arid regions of the world and has an adverse impact on crop productivity (Bilal et al. 2014; Mohamedin et al. 2010; Zhen et al. 2008, 2009).

These problems have been caused mainly by people in irrigation areas. Increasing pressure on land resources caused by rising populations has required an emergent need to produce more food, fuel, fodder, and fiber. This will necessitate effective utilization of degraded lands under crop cultivation and under canal irrigation. Irrigated agriculture, covering about 17% of the total cropped area of the world, contributes 40% of global food production (INCID 2003). Also in India, only one-third of the area under irrigation produces two-thirds of the food grains. However, the introduction of canal irrigation has caused a rise in the ground water table leading to waterlogging and secondary salinization. Presently, about one-third of the world's irrigated area faces the threat of waterlogging, about 60 Mha has already become waterlogged and 20 Mha salt affected (Heuperman et al. 2002). As per the estimate of Ministry of Water Resources, Government of India, in canal command areas of the country 2.46 Mha is waterlogged and 3.30 Mha salt affected (MOWR 1991; Ram et al. 2008). India has the largest irrigated area in the world with an ultimate irrigation potential of 139.91, of which 98.84 Mha has been utilized by the end of the tenth plan (India, 2007).

However, unscientific use of water along with other natural and man-made causes leads to waterlogging, soil salinity, and, consequently, suboptimal agricultural production (Sarangi and Bundela 2011). Currently at least 20% of the world's irrigated land is salt affected and/or irrigated with waters containing elevated levels of salts. Several major irrigation schemes have suffered from the problems of salinity and sodicity, reducing their agricultural productivity and sustainability (Ghassemi et al. 1995; Qadir et al. 2008). Several major irrigation schemes throughout the world have suffered from salinity problems (Gupta and Abrol 2000; Herczeg et al. 2001; Cai et al. 2003; Sarraf 2004). Introduction of canal irrigation has led to a rising water table and consequent waterlogging and salinity problems (Kumar 2004). The Ministry of Agriculture estimated in 1984–1985 that an area of 8.53 Mha was suffering from waterlogging including both irrigated and non-irrigated areas. Waterlogging is a widespread problem in non-irrigated areas, where low-lying depressions serve as discharge areas, and on irrigated lands. In arid and semi-arid climates, in addition to waterlogging the major problem associated with irrigation, in the absence of drainage, is salinization. In India too, salinization/alkalization and waterlogging have rendered a sizeable area of arable lands unproductive (Dwivedi 2006). A total of 22.69 lakh ha or 9.40% of geographical area of Uttar Pradesh is covered by wastelands. About 4.913 lakh ha in Uttar Pradesh is suffering from waterlogging, resulting in reduced productivity. Waterlogging is mainly due to seepage from canal irrigation, water stagnation on the surface, and/or shallow groundwater levels. High rainfall in low land gradient areas also contributes to waterlogging. In Uttar Pradesh waterlogging is associated with alkalinity. The problem is acute in the districts of the north eastern plains, eastern plains, Gangetic plains, eastern tarai, and their periphery (Chaudhary et al. 2005).

20.3 Bio-drainage

Although the term biodrainage is relatively recent, the concept is not new. Biodrainage methods consist of the strategic planting of trees with high transpirative capacity (Heuperman et al. 2002). Biodrainage relies on vegetation, rather than mechanical means, to remove excess water. The driving force behind the biodrainage concept is the consumptive water use of plants. Biodrainage may be defined as "*Draining out of excess soil water in atmosphere through deep-rooted plants using their bio-energy*" (Chauhan et al. 2012; Ram et al. 2008; Dubey 2012). It consists of the planned planting of trees with a high transpiration rate (Khamzina et al. 2005; Akram et al. 2010; Dubey 2012). The biodrainage system consists of fast growing tree species, which absorb water from the capillary fringe located above the ground water table. The absorbed water is translocated to different parts of plants and finally more than 98% of the absorbed water is transpired into the atmosphere mainly through the stomata. This combined process of absorption, translocation, and transpiration of excess ground water into the atmosphere by deep-rooted vegetation conceptualizes biodrainage technology (Ram et al. 2008). Biodrainage utilizes transpiration from trees to achieve water balance in groundwater, and to check the rise of the water table. This enables control of waterlogging and salinization of soils (Jain 2006; Kapoor 2000). In developing countries like India, farmers have small holdings and cannot afford to put their entire piece of land under tree plantations, thus Agroforestry can be a viable and remunerative option that provides additional income as tree products (timber, fuel wood, etc.), in addition to regular income from agricultural crop produce. Many workers have recommended rehabilitation of such salt-affected, waterlogged lands through tree plantations with biodrainage qualities (Dash et al. 2005, 2008; Dhyani et al. 2007; Ram et al. 2008, 2011; Angrish et al. 2009; Bala et al. 2009, 2014; Roy Chowdhury et al. 2011; Fanish and Priya 2013; Bilal et al. 2014; Singh et al. 2014; Dubey 2016).

In addition to the lowering of the groundwater table, biodrainage plantations may offer other advantages such as combating wind erosion (Thorburn and George 1999) and provision of construction material, fodder, and fuelwood (Heuperman et al. 2002). Thornburn and George (1999) reported that evaporation from the soil takes place up to a depth of 4 m. Therefore, soil management should be planned to keep this 4 m soil depth free from waterlogging to minimize the process of secondary salinization of soils and to sustain crop productivity. The biodrainage technique can be applied in two contexts, namely curative (for waterlogged areas) and preventive (for potentially waterlogged areas and shallow a water table).

20.4 Application of *Eucalyptus* Species as Biodrainage Vegetation

Eucalyptus has over 700 species distributed throughout the world. It is an important species for Agroforestry. The growth of Eucalyptus species is fast and produces large quantities of biomass. Mostly Eucalyptus species are introduced to provide various products such as fuel wood, pulp, and paper, sawn timber, essential oils, e.g. for medicine and perfumes, and services including reclamation of degraded lands, saline areas, and drainage of waterlogged areas (Munishi 2006; Bilal et al. 2014). They have a special rooting system consisting of a shallow rooting system just beneath the soil surface, and taproots that penetrate deep into the soil reaching the water table. The shallow roots extend horizontally to more than 3–5 m, these roots are used to absorb surface soil moisture but they are not very dense. The tap roots can grow up to 9 m into deeper soil layers. They are used to take up groundwater from aquifers that are more permanently available than surface soil moisture (Fritzsche et al. 2006). Due to such features, they rapidly draw down the water table without affecting the water availability in the agricultural crop root zone. Due to fast growth rates, they utilize more water, which makes them a water pumper in waterlogged areas. They have an inbuilt mechanism to utilize water in great amounts (Tushar 2002; Varghese et al. 2002). Depending upon the genetic makeup, they are tolerant to salinity, waterlogging, etc. (Bilal et al. 2014). This property makes Eucalyptus a suitable biodrainage species for sustainable management of waterlogged agro-ecosystems.

The biodrainage potential of *Eucalyptus tereticornis* for reclamation of shallow water table areas in north-west India was studied by Ram et al. (2007). They studied the groundwater table levels under 18-year-old plantations of *Eucalyptus tereticornis* (Mysure gum) and 350 m apart from the plantations at Dhob-Bhali research plot located in Rohtak district of Haryana state (north-west India). Throughout the study, the ground water table underneath the plantations remained lower than the ground water table in the adjacent fields. According to them, in shallow ground water table areas of semi-arid regions with alluvial sandy loam soils, the plantations of *E. tereticornis* were acting as bio-pumps and, therefore, they recommended closely spaced parallel strip plantations of this species for the reclamation of waterlogged areas.

According to Singh et al. (2014), biodrainage technology was applied for controlling seepage, waterlogging, and salt accumulation in the root zone due to its excessive evapo-transpirative (ET) demand. Eucalyptus is most suitable for establishing a biodrainage belt. Farmers are also growing Eucalyptus as a sole plantation crop for meeting timber demands. Eucalyptus wood has different uses in different sectors. Eucalyptus logs are most commonly used for preparing shuttering of building construction. Its plies in boxes are used for packing fruits. It is also used in the paper and pulp industry. Its chips are used in the making of particle board. It can be successfully grown on marginal, saline, sodic, and waterlogged land. They are fast growing trees and their economic return depends on height and girth. Use of Eucalyptus plants for biodrainage of waterlogged soil is based on the associated high evapo-transpiration rate.

Chhabra and Thakur (2006) conducted water balance studies in Karnal, Haryana, India for 5 years in big lysimeters showing that *Eucalyptus tereticornis* plants can biodrain 5.03, 5.14, 6.96, and 8.01 times the potential evaporation in the second, third, fourth, and fifth years respectively. They stated Eucalyptus as an excellent species for removing excess water and controlling water stagnation in land locked low-lying areas and for disposal of waste waters through land application.

A study was carried out on a 4-ha area of 6-year-old Eucalyptus plants near Bahawalnagar, Pakistan by Chaudhry et al. (2000). The role of a Eucalyptus plantation in the biological control of waterlogging and its impact on soil salinity was studied. It was reported that under the Eucalyptus, the water-table rose away from the plantation, with a maximum rise of over 30%. Salinity was maintained in the area under plantation.

20.5 Other Multipurpose Species as Biodrainage Vegetation

Khamzina et al. (2006) evaluated the potential of nine multipurpose tree species like *Prunus armeniaca*, *Populus nigra var. pyramidalis*, *Salix nigra*, *Catalpa bignonioides*, *Elaeagnus angustifolia*, *Fraxinus pennsylvanica*, *Morus alba*, *Populus euphratica*, and *Ulmus pumila* for afforestation of degraded land in the Khorezm region, Central Asia (Uzbekistan), particularly their suitability for biodrainage, i.e., lowering the elevated groundwater table through the transpirative capacity of plantations. For this purpose, water use, water use efficiency, and tree physiological factors influencing transpiration were assessed during two consecutive years. *Elaeagnus angustifolia*, *Populus* sp., and *Ulmus pumila* were found suitable for biodrainage.

Khamzina et al. (2005) evaluated young and adult tree plantations for biodrainage management in the Lower Amudarya River Region, Uzbekistan. They compared leaf transpiration rates of nine tree species, viz. the apricot tree *(Prunus armeniaca* L.*)*, black poplar *(Populus nigra var. pyramidalis* (Rozan) Spach*)*, black willow *(Salix nigra* Marshall*)*, Eastern catalpa *(Catalpa bignonioides* Walter*)*, Euphrates poplar *(Populus euphratica* Olivier*)*, Russian olive *(Elaeagnus angustifolia*L.*)*, Siberian elm *(Ulmus pumila* L.*)*, swamp ash *(Fraxinus pennsylvanica* Marshall), and white mulberry *(Morus alba)*, to identify the most water-consuming species for use in biodrainage plantations. The assortment highlighted the potential of *E. angustifolia*, which combined high transpiration, salinity tolerance, fast growth, and production of nutritious feed. Performance of *Populus* sp. and *Ulmus pumila* was less consistent, but promising enough to make them potentially suitable candidates.

Toky et al. (2011) studied the role of tree plantation of *Callistemon lanceolatus*, *Eucalyptus* hybrid, *Melia azedarach*, *Pongamia pinnata*, *Prosopis juliflora*, *Tamarix aphylla*, and *Terminalia arjuna* to bioremediate the water table through efficient biodrainage (evapo-transpiration) and for the development of a farmer's agroforestry model on an abandoned waterlogged area. A decline in the water table was observed on the entire site over this period, making the agricultural land arable.

20.6 Microbial Application for Management of Waterlogged Area

Associated soil salinity is a pervasive problem of waterlogged areas. Microbes are involved in primary production, decomposition, nutrient recycling, and other associated processes in agro-ecosystems. Therefore, microbe assisted management holds promise for *in situ* treatment of such problematic soils. These beneficial microbes, also termed biofertilizers, are involved in nitrogen (N) fixation, carbon fixation, and in improving the nutrient availability to the crop in the soil. Biofertilizers help in improving soil fertility and enhance nutrient uptake by plants in deficient soils, thereby aiding in better establishment and growth of crop plants. They also secrete growth substances and antifungal chemicals, as well as improve seed germination and root growth. In the present study, cyanobacteria and phosphate solubilizing bacteria have been used to phytoremediate the waterlogged area. Anand et al. (2015) applied cyanobacteria to phytoremediate alkaline soil. These soils have very low total nitrogen, sulphate, and phosphate content. Cyanobacteria, also commonly known as blue green algae, act as good fertilizer and could be used to remediate such soil. Cyanobacteria are considered as an important group of micro-organisms having the ability to carry out both photosynthesis as well as nitrogen fixation non-symbiotically. Fixation of nitrogen in blue green algae takes place in specialized cells called "Heterocysts." Heterocystous filamentous forms increase nitrogen content of soil and are capable of solubilizing microbial nutrients. They also ameliorate the soil by the addition of organic carbon and produce growth-promoting substances like Vitamin B-12, auxin, and ascorbic acid etc., which stimulate the growth of crop plants (Subbarao 1997). In waterlogged conditions, cyanobacteria can easily grow. Cyanobacteria can be used by farmers to make their agriculture land nitrogen-rich and fertile economically and naturally, in a sustainable manner.

Phosphorous is an important plant nutrient, which is referred to as the master key element in crop production. Phosphorous is found in soil in various organic and inorganic combinations, most of which is unavailable to plants. Plants take phosphorous in the form of soluble orthophosphate ions. The most important aspect of the phosphorous cycle is microbial mineralization and solubilization, otherwise the extraction of phosphorous is not handy to plant roots and mobilization. Microbial solubilization of inorganic phosphate compounds is of great economic importance in plant nutrition. Phosphate-solubilizing microorganisms (PSMs) offer a biological rescue system capable of solubilizing the insoluble inorganic P of soil and make it available to the plants (Khan et al. 2006). PSMs consist largely of bacteria and fungi. Such microbes not only assimilate P but a large portion of the soluble phosphate is released in quantities in excess of their own requirement (Gaur 1990). They reduce the pH of the soil by producing various types of organic acids. In such conditions, insoluble fixed phosphate changes into soluble free phosphate, which is readily available for plants.

The uses of these microbial biofertilizers also promote the plant growth. The mechanisms of plant growth promotion by microbial biofertilizers have not been completely elucidated but the important mechanisms may include phytohormone production, plant disease suppression, enhancement of plant nutrient availability and their effective absorption and enhancement of other beneficial microorganisms (Gerhardson and Wright 2002; Jeon et al. 2003). Production of plant growth stimulating compounds like vitamins, gibberellins, auxins, Vitamin B_{12}, GA_3, and IAA by PSMs has been reported by several workers (Subbarao 1997). These growth-promoting substances stimulate plant growth and thus produce greater plant biomass.

They also support the growth of other beneficial microorganisms and have synergistic effects. The positive effects of microbial biofertilizer on plant growth are generally associated with remarkable changes in root morphology, namely increased lateral root length and root hair numbers and length. It is generally assumed that these developmental responses are triggered by phytohormones produced by the bacteria. Among the plant growth regulators, auxin may play a major role. Consistent with the hypothesis of an auxin-mediated effect of these bacterial inoculations, it had a positive impact on rooting. Other phytohormones including cytokinins and gibberellins may be involved in the effect of microbial biofertilizer on root morphogenesis. Furthermore, some microbial biofertilizers have been shown to have an aminocyclopropane carboxylatedeaminase, an enzyme that hydrolyses amino cyclopropane carboxylate (ACC). Such bacteria are likely to divert ACC, the precursor of ethylene, from the plant root, which has the effect of reducing the inhibition of root growth by ethylene. This positive effect of microbial biofertilizer on root systems may also improve the biodrainage capacity of trees (Subbarao 1997).

Cyanobacteria also affects the protein pattern and metabolic activities of plants and hence plant growth (Haroun and Hussein 2003). Adam (1999) reported that nitrogen fixer Cyanobacterium inoculation led to a significant increase in growth parameters as well as nitrogenous compounds in crop plants wheat, maize, sorghum, and lentils. This promotion could be attributed to the nitrogenase as well as nitrate reductase activities of the alga associated with the surface of plants or the amino acids and peptides produced in the algal filtrate and/or other compounds that stimulate growth of crop plants (Adam 1999). The plant growth-promoting effect of PSMs may be caused by phytohormone production and their capability to solubilize insoluble phosphates (Jeon et al. 2003; Krishnan et al. 2004). The increase in growth after inoculation with PSMs as compared to control might be attributed to the protocooperative and synergistic effect of phosphate solubilization and phytohormone production (Dutta et al. 2002).

In the present study, *Oscillatoria sp.* and *Aulosira sp.* prevalent native species of Cyanobacteria in the area; and *Bacillus subtilis* as phosphate solubilizing bacteria (PSB) have been used.

20.7 Methodology

20.7.1 Mass Propagation of Cyanobacteria

Oscillatoria and *Aulosira* sp. were cultured and propagated on BG11 medium. Mass production of cyanobacteria was done in tanks in outdoor conditions. Mixtures of both cyanobacteria species were used as starter inoculums for bulk production. About 10 kg of farm soil was taken and spread in a tank of about 1 m × 1 m × 0.5 m in dimension. 100 g of superphosphate was added and watered the tank to about 10 cm height. The pH was adjusted to 7 by mixing lime. Two millitres of insecticide, e.g. malathion, was added to protect the culture from mosquitoes and insects. The mixture was mixed well and the soil particles were allowed to settle down. When the water became clear, 100 g of starter inoculums were sprinkled on the water surface. When the temperature remained between 35° and 40° during summer, optimum growth of cyanobacteria was achieved. The water level was maintained to about 10 cm during this period. After drying, the algal mat got separated from the soil and formed flakes.

20.7.2 Site Preparation for the Study

The site selected for the study was located in Badshahpur, Jaunpur District in a canal-irrigated area. The site had a high ground water table that varied from 0 to 2 m throughout the year. This adversely affects crops by virtue of saturating the root zone due to capillary rise. Such areas are potentially threatened by surface waterlogging in the due course of time if water accumulation continues (Annon 2014). The site remained waterlogged from July to November and during canal-running conditions suffered from seepage problems. Phosphate solubilizing bacteria, *Bacillus subtilis* (CFU 1 × 10^8) was procured from IFFCO, Fulpur in the form of liquid culture. Both microbes are amended to the site (cyanobacteria at 5 kg/ha and PSB at 1 L/ha). The parameter, pH, and organic matter (%) were monitored regularly.

20.7.3 Establishment of a Plantation Trial of Biodrainage Species

Quality seeds of selected species viz. *Eucalyptus* hybrid, *Terminalia arjuna*, *Trewia nudifora*, *Acacia nilotica*, and *Syzygium cumini* were procured. Nursery plants were raised. For nursery raising seeds were sown in polybags filled with rooting media consisting of FYM:Sand:Soil (1:1:1). Regular maintenance and management of the nursery was carried out. The experiment was statistically designed as per the land availability at farmer's field. The plantation trial was established on raised bunds at a selected site in the month of July, after the site treatment. These raised bunds will provide a comfortable root zone for young planted seedlings of selected species by facilitating soil aeration. Maintenance and management of the trial was done as per requirement. Growth data from the experimental trial was recorded regularly. Soil moisture content (%) at 1 ft depth was also monitored to study the biodrainage potential of the planted species in field conditions in April.

20.8 Results and Discussion

Soil pH and Organic Matter (OM) % was monitored. The observations are depicted in Fig. 20.1.

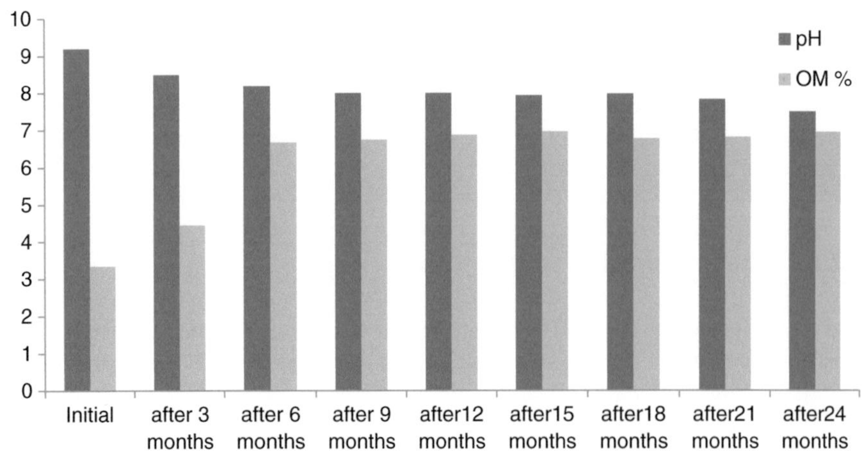

Fig. 20.1 Soil pH and organic matter of soil from planted waterlogged site

A rapid decrease in pH was initially observed after just 3 months of treatment of the site with microbes after that the decline was not significant and maintained at same level. As far as organic matter was concerned, this increased after treatment but after 6 months the enhancement was not significant. It was observed that treatment ameliorated the soil characteristics by lowering the pH and increasing the OM of the soil after the treatment (Fig. 20.1). The microbial community influences soil fertility through soil processes viz. decomposition, mineralization, and storage/release of nutrients (Alexander and Asea 1997). Treatment with *Bacillus subtilis*, a PSB, enriches the site conditions and reduces pH through excreting organic acids that dissolve phosphatic minerals and release phosphorous into solution (Khan et al. 2009). *Bacillus subtilis* can grow on saline soil and was reported to tolerate up to 2.2% NaCl concentration. The salt tolerance ability of *B. subtilis* can help to serve a suitable bio-fertilizer for saline-alkali soil- based agriculture (Patil 2014). Growth of P-solubilizing microorganisms is generally accompanied by a decrease in pH of the medium. Bacteria play a role in phosphorus nutrition by enhancing its availability to plants through release from inorganic and organic soil P pools by solubilization and mineralization. The principal mechanism in soil for mineral phosphate solubilization is lowering of soil pH by microbial production of organic acids, which include citric, gluconic, fumaric, malic, oxalic, lactic, 2- ketogluconic, malonic acids, etc. and mineralization of organic P by acid phosphatase (Mohammadi 2012). A positive relationship between PSB and plants is synergistic in nature as bacteria provide soluble phosphate and plants supply root-borne carbon compounds (mainly sugars), that can be metabolized for bacterial growth. PSB plays a vital role in P solubilization by producing organic acids. Organic acids perform many functions in the soil, such as root nutrient acquisition, mineral weathering, microbial chemotaxis, and metal detoxification. They play an important role in the mobilization of soil P and enhance P bioavailability with decreasing P adsorption and dissolution of insoluble P compounds such as Ca, Fe, and Al phosphates and decrease the pH in basic soils (Panhwar et al. 2013). Lowering the pH of soil is mainly through organic acid production, acid phosphatase secretion, and production of low molecular weight organic acids, mainly gluconic and keto gluconic acids; in addition the pH of rhizosphere is lowered through biotical production of proton/bicarbonate release (anion/cation balance) and gaseous (O_2/CO_2) exchanges. Although a high buffering capacity of soil reduces the effectiveness of PSB in lowering the pH and releasing P from bound phosphates, enhancing microbial activity through PSB inoculants may contribute considerably to plant P uptake and to reducing the rhizospheric soil pH. Phosphorus-solubilizing bacteria mainly *Bacillus*, *Pseudomonas*, and *Enterobacter* are very effective for increasing the plant-available P in soil as well as the growth and yield of crops (Mohammadi 2012). PSB application had a positive impact on root growth. The root development and plant biomass were correlated with higher availability of P; moreover, PSB application may also have some other beneficial effects like phytohormones production (Panhwar et al. 2013). This positive effect of PSB application on root growth may enhance biodrainage potential.

Cyanobacteria may potentially be used for remediating such soil suffering from waterlogging and secondary soil salinity problems. They tolerate salinity and extensively grow on the soil surface. Their tolerance to salinity is due to accumulation of inorganic ions, organic compounds (sugar, polyols, quaternary amines), and osmoregulators (Rao and Burns 1991; Pade and Hagemann 2015). The inoculation of cyanobacteria to such soils supplements the soil nutrients and improves the soil quality by making it arable through bringing about a decrease in pH, exchangeable sodium, Na/Ca, and an overall increase in N, P, organic matter, and the water-holding capacity of soil. Excreted extracellular polysaccharide by cyanobacteria can improve soil structure by increasing soil binding property (Rogers and Burns 1994). The nutrient content of saline soil was enhanced by the application of cyanobacteria in the form of organic matter (Apte and Thomas 1997). Available phosphorous and sulphur increased in soil in response to cyanobacterial application (Hashem 2001; Aziz and Hashem 2003). Cyanobacteria application to saline soil reduces electrical conductivity (Elayarajan 2002; Prabu and Udayasoorian 2007; Rai 2015). Cyanobacteria not only grow in saline ecosystems but also improve the physicochemical properties of soil by enriching them with carbon, nitrogen, and available phosphorous (Antarikanonda and Amarit 1991). The mucilaginous sheath of the cyanobacteria utilizes excess water for its multiplication and forms a thick mat, which adds organic matter, organic N, and organic P to the soil and is responsible for binding soil particles, thus improving soil permeability, texture, and aeration, which is a major problem of such sites (Singh 1961; Pandey et al. 2005; Ibraheem 2007; Anand et al. 2015; Rai 2015; Singh and Singh 2015). They stimulate plant growth including biofertilization (increasing the supply of mineral nutrients to the plant), biological control (elimination of the plant enemies including microbial pathogens, insects, and weeds) and direct plant growth production by delivering plant growth hormones (Lugtenberg et al. 1991; Abdel-Raouf et al. 2012). This stimulation in overall plant growth may lead to an enhanced biodrainage potential of tree vegetation.

The co-inoculated cyanobacteria fix carbon and act as a carbon source for PSB and synergistically for PSB. Cyanobacteria also fix nitrogen and it has been reported that co-inoculation of PSB with N_2 fixers has a positive effect on plant growth, thereby enhancing productivity and plant biomass (Krishnaveni 2010; Mohammadi 2011; Rathi and Gaur 2016). This excreted extracellular polysaccharide by cyanobacteria may also act as carbon source for co-inoculated PSB and *Bacillus subtilis* and have synergistic effects.

The water table was also observed in an observation well located in an area nearby to the plantation trial. A slight decline in the water table pattern was observed in the observation well 2 years after plantation (Fig. 20.2).

It was reported that trees could be used to manage the rising water table and salinity problems. The clear impact of the lowering water table was reported after the planting of biodrainage tree species. The approach was found to be relatively cheap, sustainable, and ecologically compatible, relying on the natural capability of high transpiration potential of tree vegetation (Kapoor 2002). In an observation, it was reported that the tree growth caused an increase in the depth of the water table.

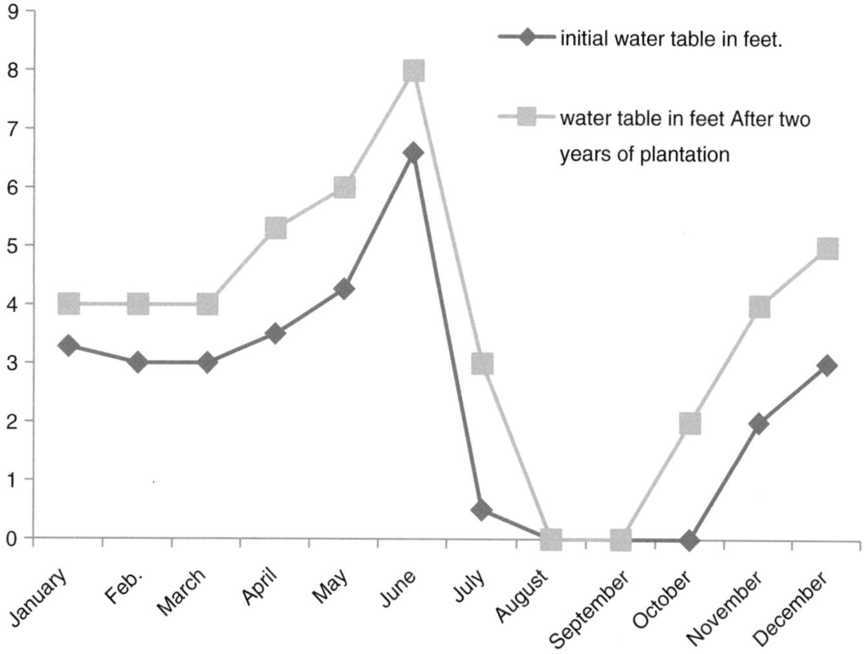

Fig. 20.2 Water table in observation well located near plantation trial

The rate of fall of the water table doubled with the development of the trees (Rodríguez-Suárez et al. 2011). Ahmad et al. (2007) studied a Eucalyptus plantation for intercepting canal seepage and controlling the water table. A "cone of depression" in the water table immediately beneath the strip of tree plantations of *Callistemon lanceolatus*, *Eucalyptus hybrid*, *Melia azedarach*, *Pongamia pinnata*, *Prosopis juliflora*, *Tamarix aphylla*, and *Terminalia arjuna* was also observed by Toky et al. (2011).

Growth data of the biodrainage plantation is depicted in Figs. 20.3 and 20.4 for height and girth, respectively.

From the observations, it was concluded that as far as growth was concerned, the *Eucalyptus hybrid* performed best followed by *Terminalia arjunaa* and *Acacia nilotica. Syzygium cumini* and *Trewia nudifora* performed almost at par. An almost six times increment in height and eight times increment in girth was observed in the *Eucalyptus hybrid.* The growth behavior, biomass accumulation by the plants, and physiological parameters suggested that Eucalyptus has high potential to be used as an efficient biodrainage species, similarly to the report by Bala et al. (2009). *Eucalyptus tereticornis* and *Eucalyptus hybrid* were reported as fast bio-drainers primarily due to their ability to display large leaf area (Angrish et al. 2009). Due to its high evapo-transpiration rate, its suitability to all soil types, adoptability to varying climatic conditions and tolerance to waterlogging, salinity, and sodicity, its timber value, fast growth, and other multipurpose values such as shuttering in building

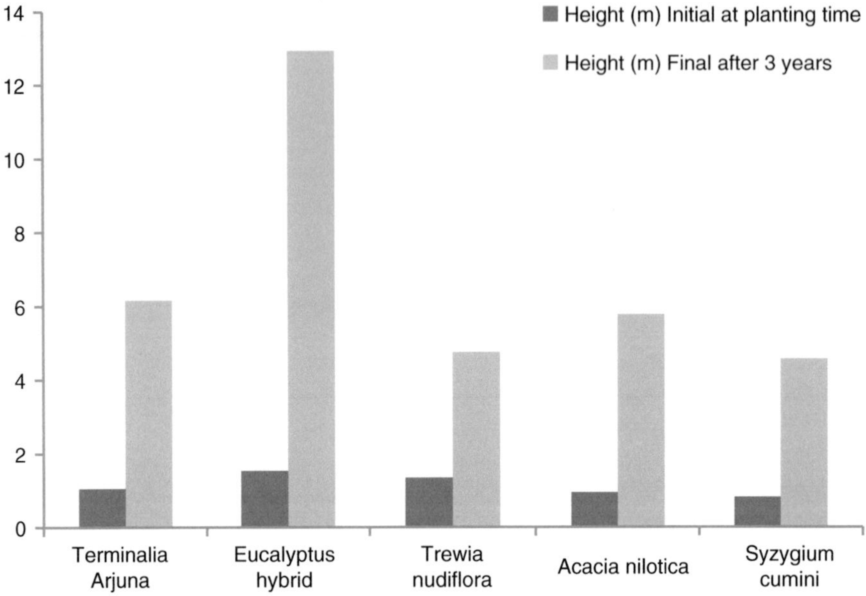

Fig. 20.3 Height of planted biodrainage species at waterlogged site

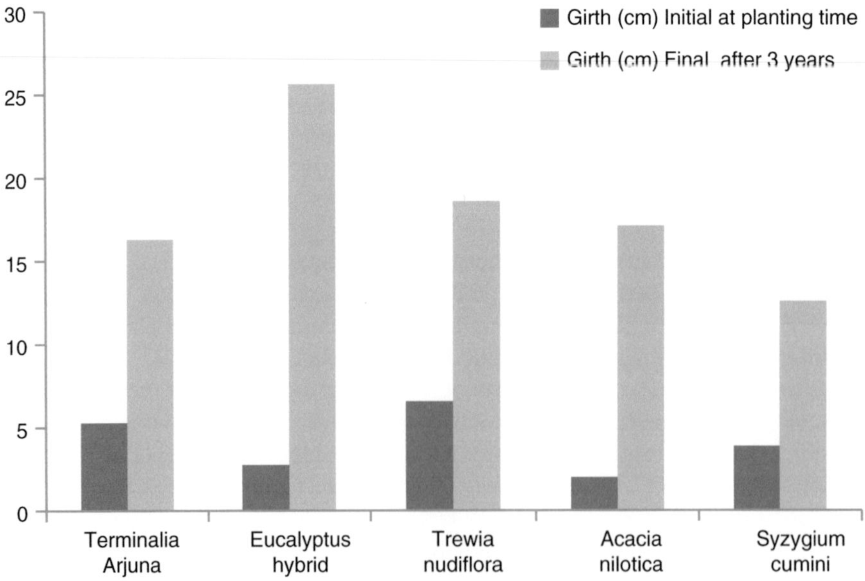

Fig. 20.4 Girth of planted biodrainage species at waterlogged site

construction, in paper, pulp, and particles board industry etc., Eucalyptus is the most preferred biodrainage species by farmers. Most of the farmers of the eastern UP area are marginal farmers and they plant Eucalyptus as a sole crop for establishing a biodrainage belt in canal-irrigated command areas for combating the high water table-associated problems (Singh et al. 2014). In waterlogged areas, Eucalyptus can be successfully grown by ridge planting. The world's Eucalyptus plantation area has increased to 19 Mha because of its fast growth rate, good wood properties, and carbon sequestration, and thus seems to be a good option for biodrainage (Iglesias and Wilstermann 2009). Ram et al. (2011) reported that Eucalyptus plantations generated 46.6 tons per ha fresh biomass with a benefit cost ratio of 3:5 and also sequestered 15.5 tons carbon per ha. Lowering of the water table and associated soil improvement by Eucalyptus plantations increased the wheat grain yield by 3.4 times and resulted in remediation of waterlogged areas.

Soil Moisture content (%) at a depth of 1 ft of planted tree species is described in Fig. 20.5.

It has been shown from observations that soil moisture content was reduced the most in the cases of *Eucalyptus* hybrids in comparison to other planted species, which illustrates its high water consumption rate in comparison to other planted species. This made it a potential biodrainage species. Similar observations were also reported by Zahid et al. (2010), who found Eucalyptus species used more water than other tree species. Soil moisture depletion rates were higher under Eucalyptus trees in dry seasons and were lower under the teak and jackfruit. The roots of the

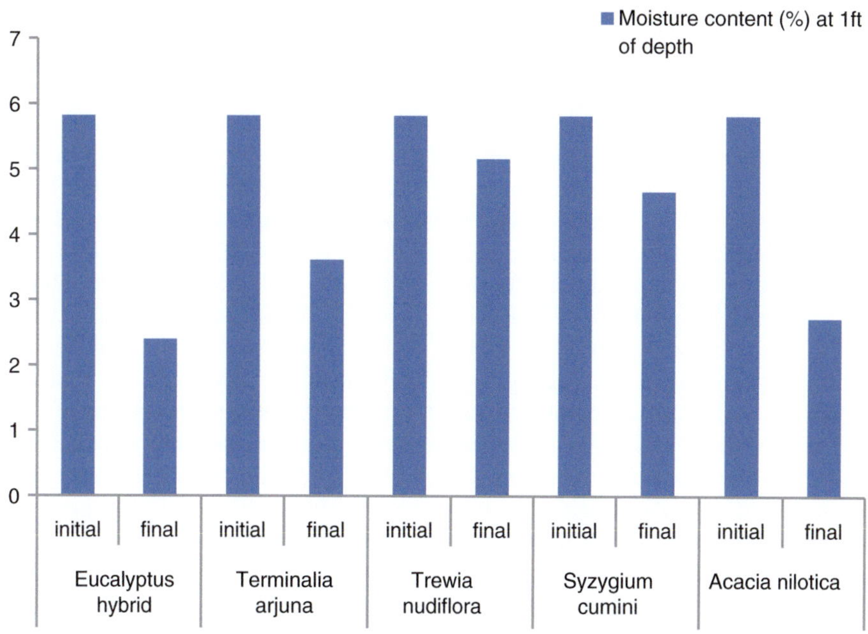

Fig. 20.5 Moisture content in 1 ft of planted biodrainage species at waterlogged site in April

Eucalyptus penetrate into the soil at 2.5 m per year and absorb water from the ground reservoir. Eucalyptus has a tendency to utilize ground water as well as water from the upper zone (Calder et al. 1997).

Conclusion

Agriculture around the globe is confronted with the serious problem of waterlogging and salinization. Management of this excess water in agricultural fields is a major challenge globally, including in India, especially in canal-irrigated command areas where seepage-associated high ground water tables and soil salinity have already thrown a serious challenge for the sustainability of irrigated agriculture. In combating a problem of such an extent, the application of biodrainage technology has proved less expensive and more environmentally friendly for its sustenance. These species can be planted in blocks in the form of farm forestry or along the field boundary in the form of agroforestry (Dagar et al. 2009). Biodrainage can be a feasible option for controlling waterlogging and salinity in irrigated lands (Kapoor 2002). The most common tree species recommended for biodrainage in the Indian subcontinent and other parts of the world is *Eucalyptus* due to its high transpiration rate and adaptability to the varying soil conditions, e.g. wetness and salinity (Singh et al. 2014). In the current study *Eucalyptus* species also performed best as far as growth and water consumption were concerned in waterlogged conditions and were found to be more effective in combating waterlogging. Eucalyptus can provide substantial yields of biomass and can reduce greenhouse gas emissions from fossil fuel consumption. By utilizing high water-uptake trees like the *Eucalyptus* species, biodrainage may be a viable alternative to conventional engineering-based techniques. Biodrainage is economical because it requires only initial investment for planting the vegetation, and when established, the system provides economic returns by means of fodder, wood, or fiber harvested and additionally sequesters carbon in the timber. For effective application of this technology, suitability of multipurpose tree species for different agro-climatic zones has to be studied. Application of cyanobacteria and PSB to these plantations ameliorated the soil by adding nutrients and reducing pH and promoted the plant growth by creating favorable soil conditions. Cyanobacteria, because of their dual capacity for photosynthesis and N_2 fixation, are capable of contributing to productivity in different situations. The invasion of cyanobacteria in the plantation promotes soil genesis, adds humus, dissolves certain minerals, absorbs moisture in its mucilaginous sheath, increases polysaccharide content, reduces soil loss, and improves texture. Cyanobacteria is an important part of wet agro-ecosystems, is easily available, and serves as the cheapest source of natural biofertilizers for such waterlogged sites. PSB application also has synergistic beneficial effects in remediation of such soils. There are several plant growth-promoting bacteria (PGPB) present in the soil rhizosphere. These PGPBs also have positive effects on plant growth and soil fertility. The effect of these PGPBs also has to be investigated in combination with above microbial biofertilizers in the plantation of biodrainage tree species.

It may be concluded from the above study that an integrated approach of microbial application with biodrainage technology may phytoremediate agricultural land suffering from waterlogging associated with a high water table and soil salinization, more promptly and at proven lower cost, higher return, in a socially acceptable and environmentally friendly method.

References

Adam MS (1999) The Promotive effect of the Cyanobacterium *Nostoc muscorum* on the growth of some crop plants. Acta Mirobiol 48(2):163–171

Abdel-Raouf N, Al-Homaidan AA, Ibraheem IBM (2012) Agricultural importance of algae. Afr J Biotechnol 11(54):11648–11658

Ahmad S, Mohyuddin J, Siddiqui SM et al (2007) Tree plantation for intercepting canal seepage and controlling Watertable. Pakistan J Water Res 11(2):35

Akram S, Abkavosh S, Liaghat H et al (2010) World Congress of the International Commission of Agricultural and Biosystems Engineering (CIGR), p 1–9

Alexander M, Asea PEA (1997) Introduction to soil microbiology. Wiley, New York, pp 33–399

Annon (2014) Salt-affected and waterlogged areas of India. Technical Report by National Remote Sensing Centre, Balanagar, Hyderabad, Indian Space Research Organization, p 10

Anand AK, Prasad V, Alam M (2015) Physico-chemical characterization of Usar soil and its natural reclamation by cyanobacteria. J Chem Cheml Sci 5(4):145–152

Antarikanonda P, Amarit P (1991) Influence of blue-green algae and nitrogen fertilizer on rice yield in saline soils Kasctsart. J Nat Sci 25:18–25

Angrish R, Toky OP, Datta KS (2006) Biological water management: biodrainage. Curr Sci 90(7):897

Angrish R, Datta C, Rani VS (2009) Comparative bio-drainage potential of some tree species. Asian Regional Conference, New Delhi, India, 6–11 Dec 2009

Apte SK, Thomas J (1997) Possible amelioration of coastal soil salinity using halo tolerant nitrogen fixing cyanobacteria. Plant Soil 189:205–211

Aziz MA, Hashem MA (2003) Role of cyanobacteria in improving fertility of saline soil. Pakistan J Biol Sci 6(20):1751–1752

Bala N, Singh G, Bohra NK et al (2009) Increasing productivity of waterlogged zone of canal command area in Indian Desert. 5th International Executive Council Meeting & Asian Regional Conference, New Delhi, India, 6–11 Dec 2009

Bala N, Singh G, Bohra NK (2014) Biodrainage for restoration of canal command waterlogged area in Indian desert. Indian Forester 140(5):462–467

Bilal H, Ali SS, Kim KM (2014) Potential of eucalyptus in the remediation of environmental problems: a review. Int J Innov Sci Res 4(2):136–144

Blumwald E, Mehlhorn RJ, Packer L (1983) Studies of osmoregulation in salt adaptation of cyanobacteria with ESR spin-probe techniques. Proc Natl Acad Sci U S A 80(9):2599–2602

Calder I, Paul R, Roiser TW (1997) Eucalyptus water use greater than rainfall input. Hydrol Earth Syst Sci 1(2):249–256

Cai X, McKinney DC, Rosegrant MW (2003) Sustainability analysis for irrigation water management in the Aral Sea region. Agric Syst 76:1043–1066

Chhabra R, Thakur NP (2006) Evaluation of transpiration capacity of eucalyptus (*Eucalyptus tereticornis*) and bamboo (*Bambusa arundinacea*) for biodrainage of surface waters. Indian J For 29(1):1–8

Chaudhry MR, Chaudhry MA, Subhani AKM (2000) Biological control of waterlogging and impact on soil and environment. International Waterlogging and Salinity Res. Inst. 13 West Wood Colony, Thoka Niaz Baig, Lahore-53700, Pakistan

Chaudhary HP, Singh PK, Upadhyay KD (2005) Possibilities of enhancing green cover in the areas other than reserve forest & need strong joint attention of government, pias& people. In Proceeding of Workshop on National Afforestation plan and Forest development agencies, Lucknow, India, Sept 6–7, 2005

Chauhan MK, Ram J, Dagar JC (2012) Biodrainage and carbon sequestration. Lambert Academic Publishing, Germany, (ISBN-10: 3659147095, ISBN-13: 978-3659147098) pp 304

Dagar JC, Singh G, Ram J (2009) Bio-drainage: an eco-friendly technique for combating water logging and salinity. 5th International Executive Council Meeting & Asian Regional Conference, New Delhi, India, 6–11 Dec 2009

Dash CJ, Sarangi A, Singh AK (2005) Bio-drainage: an alternate drainage technique to control waterlogging and salinity. J Soil Water Conserv India 4(3&4):149–155

Dash CJ, Sarangi A, Singh AK (2008) A decision support system on biodrainage for land reclamation. J Soil Water Conserv 7(1):33–37

Dhyani SK, Samra JS, Ajit AK (2007) Forestry to support increased agricultural production: focus on employment generation and rural development. Agric Econ Res Rev 20:179–202

Dubey K (2012) SWOT analysis for the application of biodrainage technology to phytoremediate water logged sites. Int J Soc Forest 5(2):47–59

Dubey K (2016) Phyto-remediation of waterlogged waste land through biodrainage and soil amendments, Project Report-2016, Indian Council Forestry Res Education, Dehradun, India

Dutta M, Banik S, Dhimak KR (2002) Efficacy of Phospobacterium (*Basillus firmus*) in combination with phosphates and organics on rice productivity in acid soils. 17th World Congress of Soil Science, Thailand, 14–21 Aug 2002

Dwivedi RS (2006) Study of salinity and waterlogging in Uttar Pradesh (India) using remote sensing data. Land Degrad Dev 5(3):191–199

Elayarajan M (2002) Land application of treated paper board mill effluent on soil- water-plant ecosystem (Soil Science). Coimbatore, TNAU, PhD Thesis

Fanish SA, Priya RS (2013) Review on benefits of agro forestry system. Int J Educ Res 1(1):1–12

Fritzsche F, Abate A, Fetene M (2006) Soil-plant hydrology of indigenous and exotic trees in an Ethiopian montane forest. Tree Physiol 26(8):1043–1054

Gaur AC (1990) Phosphate solubilizing microorganisms as biofertilizers. Omega Scientific Publishers, New Delhi, p 176

Gerhardson B, Wright S (2002) Bacterial associations with plants: beneficial, non N-fixing interactions. In: Sivasithamparam K, Dixon KW, Narrett RL (eds) Microorganism in plant conservation and biodiversity. Kluwer Academic Press, London, pp 79–103

Ghassemi F, Jakeman AJ, Nix HA (1995) Salinisation of land and water resources: human causes, extent, management and case studies. CABI Publishing, Wallingford

Gupta RK, Abrol IP (2000) Salinity build-up and changes in the rice-wheat system of the Indo-Gangetic Plains. Exp Agric 36:273–284

Haroun SA, Hussein MH (2003) The Promotive effect of algal biofertilizers on growth, protein pattern and some metabolic activities of *Lupinus termis* plants grown in siliceous soil. Asian J Plant Sci 2(13):944–951

Hashem MA (2001) Role of blue-green algal inoculum for improving soil fertility and reclaiming salinity of soil. Research Report. BARC. Dhaka, Bangladesh, pp 2

Herczeg AL, Dogramaci SS, Leany FWJ (2001) Origin of dissolved salts in a large, semi-arid groundwater system: Murray Basin, Australia. Mar Freshw Res 52:41–52

Heuperman AF, Kapoor AS (2002) Bio-drainage: principal experiences and applications. IPTRID, FAO, Rome, pp 1–79

Heuperman AF, Kapoor AS, Denecke HW (2002) Biodrainage – principles, experiences and applications. International Programme for Technology and Research in Irrigation and Drainage (IPTRID).Food and Agriculture Organization of the United Nations (FAO) publication, Italy, pp 79

Ibraheem IBM (2007) Cyanobacteria as alternative biological conditioners for bioremediation of barren soil. Egyptian J Phycol 8:99–116

Iglesias TG, Wilstermann D (2009) *Eucalyptus universalis* Global cultivated eucalypt forests map, In GIT forestry consulting's eucalyptologics: information resources on Eucalyptus cultivation worldwide. Retrieved from http://www.git-forestry.com/; 19 Jan 2009

INCID (2003) Bio-drainage – status in India and other countries. Publication of the Indian National Committee on Irrigation and Drainage, New Delhi

Jain AK (2006) The concept of bio-drainage in flood prevention and the avoidance of water scarcity. Int J Environ Stud 63(1):39–48

Jeon JS, Lee SS, Kim HY (2003) Plant growth promotion in soil by some inoculated microorganisms. J Microbiol 41(4):271–276
Kapoor AS (2000) Bio-drainage: to control waterlogging and salinity in irrigated lands. Challenges facing irrigation and drainage in the new millennium. Proceedings US Committee on Irrigation and Drainage, Fort Collins, Colorado, USA, p 217–235, Jun 2000
Kapoor AS (2002) Bio-drainage: a biological option for controlling water logging and salinity. Tata McGraw Hill, New Delhi, pp 1–332
Khamzina A, Lamers JPA, Wickel B (2005) Evaluation of young and adult tree plantations for biodrainage management in the lower Amudarya River Region, Uzbekistan. ICID 21st European Regional Conference, 15–19 May 2005 – Frankfurt (Oder) and Slubice – Germany and Poland, pp 1–11
Khamzina A, Lamers JPA, Martius C (2006) Potential of nine multipurpose tree species to reduce saline groundwater tables in the lower Amu Darya River region of Uzbekistan. Agrofor Syst 68(2):151–165
Khan MS, Zaidi A, Wani PA (2006) Role of phosphate-solubilizing microorganisms in sustainable agriculture – a review. Agron Sustain Dev 27:29–43. doi:10.1051/agro:2006011
Khan AA, Jilani G, Akhtar MS et al (2009) Phosphorus solubilizing bacteria: occurrence, mechanisms and their role in crop production. J Agric Biol Sci 1(1):48–58
Krishnaveni MS (2010) Studies on phosphate solubilizing bacteria (PSB) in rhizosphere and nonrhizosphere soils in different varieties of foxtail millet (*Setariaitalica*). Int J Agric Food Sci Technol 1(1):23–39
Krishnan PR, Rajapandian SJ, Selvi TK (2004) Influence of inoculation of biofertilizers on growth and biomass productivity of Simarouba Glauca seedlings. My Forest 40(2):197–202
Kumar R (2004) Groundwater use in north- West India (eds Abrol I P et al). Centre for Advancement of Sustainable Agriculture, New Delhi, pp 1–26
Lugtenberg BJJ, Weger LA, de Bennett JW (1991) Microbial stimulation of plant growth and protection from disease. Curr Opin Biotechnol 2(3):457–464
Michael AM, Ojha TP (2006) Principles of agricultural engineering, vol vol. II, 5th edn. Jain Brothers, New Delhi, pp 391–457
Mohamedin AAM, Awaad MS, Ahmed AR (2010) The negative role of soil salinity and waterlogging on crop productivity in the northeastern region of the Nile Delta, Egypt. Res J Agric Biol Sci 6(4):378–385
Mohammadi K (2011) Soil, plant and microbe interaction. Lambert Academic Publication. pp 120
Mohammadi K (2012) Phosphorus solubilizing bacteria: occurrence, mechanisms and their role in crop production. Resour Environ 2(1):80–85
MOWR (1991). Ministry of Water Resources, Govt. of India. Report of the working group on waterlogging, soil salinity and alkalinity (mimeograph).
Munishi PKT (2006) The eucalyptus controversy in Tanzania. Department of Forest Biology, Sokoine University of Agriculture, Morogoro
Pade N, Hagemann M (2015) Review salt acclimation of cyanobacteria and their application in biotechnology. Life 5:25–49
Pandey KD, Shukla PN, Giri DD (2005) Cyanobacteria in alkaline soil and the effect of cyanobacteria inoculation with pyrite amendments on their reclamation. Biol Fertil Soils 41(6):451–457
Panhwar QA, Jusop S, Naher UA et al (2013) Application of potential phosphate-solubilizing bacteria and organic acids on phosphate solubilization from phosphate rock in aerobic rice. Sci World J 2013(2013):272409, 10 pages. http://dx.doi.org/10.1155/2013/272409
Patil VS (2014) *Bacillus subtilis*: a potential salt tolerant phosphate solubilizing bacterial agent. Int J Life Sci Bt PharmRes 3(2):141–145
Prabu PC, Udayasoorian C (2007) Native cyanobacteria *Westiellopsis* (TL-2) sp for reclaiming paper mill effluent polluted saline sodic soil habitat of India. EJEAFChe 6(2):1775–1786
Qadir M, Tubeileh A, Akhtar J (2008) Productivity enhancement of salt-affected environments through crop diversification. Land Degrad Develop 19:429–453
Rai A (2015) Salt tolerance by cyanobacteria and reclamation of usar soil. Indian J Plant Sci 4(2):59–62

Ram J, Garg VK, Toky OP (2007) Biodrainage potential of Eucalyptus tereticornis for reclamation of shallow water table areas in north-west India. Agrofor Syst 69(2):147–165

Ram J, Dagar JC, Singh Gurbachan, Lal Khajanchi, Tanwar VS, Shoeran SS, Kaledhonkar MJ, Dar SR and Kumar Mukesh (2008). Biodrainage: EcoFriendly Technique for Combating Waterlogging & Salinity. Technical Bulletin: CSSRI / Karnal / 9 / 2008, Central Soil Salinity Research Institute, Karnal, India, pp 24

Ram J, Dagar JC, Lal K (2011) Bio-drainage to combat waterlogging , increase farm productivity and sequester carbon in canal command areas of northwest India. Curr Sci 100(11):1673–1680

Rao DLN, Burns RG (1991) The influence of blue green algae on the biological amelioration of alkali soils. Biol Fertil Soil 11:306–312

Rathi M, Gaur N (2016) Phosphate solubilizing bacteria as biofertilizer and its applications. J Pharma Res 10(3):146–148

Rodríguez-Suárez JA, Soto B, Perez R et al (2011) Influence of *Eucalyptus globulus* plantation growth on water table levels and low flows in a small catchment. J Hydrol 396(3):321–326

Rogers SL, Burns RG (1994) Changes in aggregate stability nutrient status, indigenous microbial populations, and seedling emergence following inoculation of soil with Nostocmuscorum. Biol Fertil Soil 18:209–215

Roy Chowdhury S, Kumar A, Brahmanand PS (2011) Application of biodrainage for reclamation of waterlogged situations in deltaic Orissa. Research bulletin no. 53. Directorate of Water Management (Indian Council of Agricultural Research). Chandrasekharpur, Bhubaneswar. pp 32

Sarraf M (2004) Assessing the costs of environmental degradation in the Middle East and North Africa countries. Environmental strategy notes (no. 9). Environment Department, World Bank, Washington, DC

Sarangi A, Bundela DS (2011) Decision support Systems in Water Resources Management – a REVIEW. Technical bulletin-2011. Water Technology Centre. Indian Agricultural Research Institute, New Delhi, p 50

Setter TL, Waters I, Sharma SK (2002) Water use by contour-planted belts of trees comprised of four eucalyptus species. Agric Water Manag 53:133–152

Singh RN (1961) Role of blue-green algae in nitrogen economy of Indian agriculture. Indian Council of Agricultural Research, New Delhi

Singh A (2013) Groundwater modelling for the assessment of water management alternatives. J Hydrol 481:220–229

Singh V, Singh DV (2015) Cyanobacteria modulated changes and its impact on bioremediation of saline-alkaline soils. Bangladesh J Bot 44(4):653–658

Singh SK, Verma CL, Sharma DK (2014) Plant height model for eucalyptus plantations for biodrainage use. Int J Res Eng Technol 3(6):250–259

Subbarao NS (1997) Biofertilizers in agriculture and forestry. Oxford & IBH publishing Co. Pvt. Ltd, New Delhi

Thorburn PJ and George RJ (1999).Interim guidelines for re-vegetating areas with shallow saline water tables.Agroforestry over shallow water tables.Water and Salinity Issues in Agroforestry No.4.RIRDC Publication No. 99/36.RIRDC Project No.WS 967–7. Pp 13

Toky OP, Angrish R, Datta KS et al (2011) Biodrainage for preventing waterlogging and concomitant wood yields in arid agro-ecosystems in north-western India. J Sci Ind Res 70(08):639–644

Tushar (2002) Eucalyptus – falsely cursed. Farmers Eorums 2(3):17

Varghese MA, Nicodemus B, Nagarajan N et al (2002) A breeding programmed for improving productivity of Eucalyptus camaldulensis and Eucalyptus tereticornis in India. In: Bagshi SK, Varghese M, Siddappa (eds) Recent eucalyptus research in India. Indian institute of forest genetics and tree breeding, Coimbatore, pp 19–28

Zahid DM, Shah FR, Majeed A (2010) Planting *Eucalyptus camaldulensis* in arid environment is it useful species under water deficit system. J Bot 42(3):1733–1744

Zhen C, Jiang D, Dai T et al (2009) Effects of salt and waterlogging stresses and their combination on leaf photosynthesis, chloroplast ATP synthesis, and antioxidant capacity in wheat. Plant Sci 176:575–582

Zhen L, Hezhong D, Weijiang L et al (2008) Individual and combined effects of salinity and waterlogging on Cry1Ac expression and insecticidal efficacy of Bt cotton. Crop Prot 27:1485–1490

Traditional Ecological Knowledge-Based Practices and Bio-formulations: Key to Agricultural Sustainability

21

Seema B. Sharma

Abstract

The ever-increasing population and the pressure it exerts on the food systems have raised the demands on agriculture in the form of higher yields in shorter time. The high rate of production with limited arable land has caused devastating effects on the entire food chain. The soil is degraded and water is polluted. Chemical intensive systems with indiscriminate use of fertilizers and pesticides coupled with the global climate change have geared the agricultural scenario into an era of irreparable damage. We need to realize that such growth comes with a big price tag not only for humanity but also for the natural environment that sustains life and its various forms. The use of chemical-based formulations in agriculture has already raised questions on their sustainability in the long term. It has crippled not only the economic conditions of the farmers in developing countries and regions with harsh climatic conditions, but also their esthetic values are now being questioned. Our age-old systems, which withstood the test of time, always seemed to have an answer to such unforeseen situations, incurred due to governmental policies that were not futuristic enough to see the damage that might occur in the long run. Subsidized prices of chemical fertilizers, hybrid seeds, and farming policies framed to propagate them are devoid of the issues at regional and local demands. Traditional Ecological Knowledge (TEK) is the indigenous knowledge of local people in any area of the world. It is framed and processed by inhabitants and carefully passed on to future generations. This knowledge is generated keeping in mind needs as well as esthetic values. Each string is woven delicately into the life-sustaining system so intricately that it becomes an inseparable part of the community in particular and the ecosystem in general. This TEK has its roots in various life-sustaining activities such as customs, rituals, health,

S.B. Sharma
Department of Earth and Environmental Science, KSKV Kachchh University,
Mundra Road, Bhuj, Gujarat, India
e-mail: seemabhargavsharma@gmail.com

V. Kumar et al. (eds.), *Probiotics in Agroecosystem*,
DOI 10.1007/978-981-10-4059-7_21

dairy, and agro-ecosystems. In agricultural systems the TEK-based adaptations are gaining relevance due to their long-term sustainable solutions. Bioformulations and practices from these TEK systems are being revised and adapted in the present scenario to revive agri-management systems in a sustainable manner.

21.1 Introduction

Ensuring sustainability across the entire agri-food chain is a major focus not just for farmers, researchers, and policy makers, but also for processors and retailers in the food and beverage industry. Our ancient methods in agri-system management are future proof. The evolution of mankind has been a witness to several changes, finally evolving into present-day civilized and settled man. From early nomadic wanderers who were hunters and food gatherers to present-day mechanized and high input–based agriculture systems, mankind has been a witness to drastic and rapid changes in food consumption as well as in production techniques. With the dawn of early agriculture systems and domestication of animals for sustenance, humans realized the long-term benefits of agriculture. Innovative and adaptive practices in agriculture led to the present day forms. One can very correctly say that agriculture provided the stepping stones towards a more organized society and hence is the very basis of human existence.

An ever-expanding world population is putting tremendous pressure on cultivable lands around the world. To meet such challenges, a continuous expansion of food-producing ecosystems is required. In every agri-production system, crops must be provided with major nutrients for better growth and substantial yields (Rengel 2008). There has been an indiscriminate use of chemical fertilizers in agriculture worldwide to provide nutrients to support plant growth and consequently to boost crop productivity. Undoubtedly, chemical fertilizers have offered benefits to modern cropping systems, but their overuse has resulted in the deteriorating health of agricultural soils leading to both a decrease in production and ecosystem degradation. Scientists are, therefore, currently interested in developing alternative technologies to minimize the dependence on chemical fertilizers. Various environmental implications and health hazards coupled with socio-economic problems stem from the indiscriminate use of chemical fertilizers, pesticides, and hybrid seeds. Though agricultural production overall continues to increase, the rate of yield per hectare is declining.

Indigenous knowledge is a synchronized body of knowledge acquired by local people through informal experiments, accumulation of experiences, and an indepth understanding of the environment. The traditional methods in agriculture have helped sustain the environment over the centuries. However, increased modernization and high input-based agri-management systems have diminished these age-old methodologies and this knowledge is now surviving only in bits and pieces. Traditional Ecological Knowledge or TEK is an accumulating body of knowledge, beliefs, and practice that evolves through adaptive processes and is handed down through generations by cultural transmission, and deals with the relationship of living beings (human and non-human) with one another and with the environment. TEK may be defined as "a cumulative body of knowledge, practice and belief evolving by adaptive processes and handed down through generations by cultural transmission, about the relationship of living beings (including humans) with one another and with their environment" (Berkes 1999). TEK is known by different names in different parts of the world, e.g. indigenous knowledge, local knowledge, etc., but in essence they all have similar meanings.

21.2 TEK in Agri-Management Systems: World Scenario and the Indian Context

All over the world different groups of people may perceive and interact with nature in a different manner but history has withstood the testimony that the only sustainable systems are those able to maintain the natural flow of materials to and from the system. In the quest for an ecologically sustainable society, indigenous people and traditional ecological knowledge have been shaped so that the systems can sustain themselves in the long run.

The rich cultural heritage of the indigenous knowledge in the agriculture sector of Asia, Africa, America, Australia, and many other parts of the world is an immense source of systems and adaptive methodologies that are suited to the microenvironment, but have strong chances of replicability in similar climatic zones.

The Globally Important Agricultural Heritage Systems (GIAHS) are good examples of evolutionary adapted socio-ecosystems in human history (CelaCruz and Koohafkan 2009) and TEK is an important component of this. These GIAH systems, due to excellent local traditional knowledge and practices, have been passed down for generations. The Hani Rice Terraces System, located in China's south-western Yunnan Province, represents a living example of GIAHS (Yuan et al. 2014). It has existed for more than 1,000 years, following indigenous knowledge related to cultivation and natural resources management (NRM), which was collected and practiced continually. Over this long time period, TEK has enabled the Hani people to manage their terraces and related natural resources in a sustainable way (Yuan et al. 2014).

The indigenous "Maya" people of south-eastern Mexico and central America have adapted an agriculture system known as "Milpa" (to the field) agricultural systems for almost three millennia (Flores-Delgadillo et al. 2011). They used slash-and-burn methods to manage agriculture of different varieties like beans, maize, and squash, among other plants, for food and medicinal purposes; they also used the method of terracing in their agricultural fields and manipulated wetland systems for agricultural production (Flores-Delgadillo et al. 2011). To make their agricultural system effective the Maya used a different cultivation method, instead of traditional tillage, which included site-specific crop management by planting crops along with perennial plants in soil-filled cavities of limestone bedrock (Flores-Delgadillo et al. 2011).

Sacred forests, or "kayas" as they are popularly known in coastal Kenya, conserve faunal and floral biodiversity and are a valuable source of germplasm for species that are tolerant to extreme weather and soil conditions. Unfortunately, the so-called "Green Revolution" and the urge to use modern agriculture to improve food production and security, has meant a huge proportion of farmers were lured into growing modern monoculture crop varieties. Studies from regions like China, Bolivia, and Kenya all recognize the need to support local initiatives such as local seed production, community-based landrace conservation, and local seed banks to increase self-reliance in the farming community.

The above-mentioned strategic methodologies suited to different indigenous groups in different parts of the world have a very strong foundation of living in harmony with nature. These systems were not driven by greed but the underlying principle of respect and love for the whole ecosystem that helped man and nature to flourish intricately. However, the era of industrial revolution, an unprecedented boom in population, depleting resources, and above all the driving greed of man ushered the whole agri-management system into the present state of failure and distress.

In India, after the launch of the "Green Revolution" in the mid-1960s, there was a substantial increase in the production of food grains which was achieved through the use of chemical fertilizers and high-yielding hybrid varieties of crops. It has now been realized that this increase in production was achieved at a heavy cost to the environment, in terms of the depletion of soil fertility, salinization, irreparable damage to soil structure, and its water-holding capacity. There was an indiscriminate killing of useful insects, microorganisms, and predators which has removed the privilege of keeping a natural biological check on the insect pests. Chemical pollution in the form of fertilizers and pesticides has not only endangered the health of the farming communities who produce them but has also poisoned the produce with highly toxic residues, crippling the overall health of the present as well as future generations. It has become imperative to adopt a strategy wherein high yield can be achieved to meet the demands of an increasing population along with the maintenance of overall soil health for future generations. Sustainability of any agricultural system depends upon the consistent fertility of the soil as a factor of time. The cases of farmer suicide in India substantiate the fact that what the policy makers and researchers had planned contained a big loophole and we need to relook at our agriculture policies, whcih are system oriented and not yield centric.

21.3 The "Vedic Krishi" System of India: The World Heritage

India is a land of rich cultural heritage. In today's scenario when we talk of the whole world as a "Global Village," we as global citizens cannot segregate one knowledge system and its useful implications as an intellectual property of a single nation or a cult. It is a knowledge system of humanity, for humanity, and by humanity. The ancient Indian scriptures The Vedas and Upanishads are the collection of knowledge and experiences of the intellectual ancient Indian civilizations. Each hymn in these scriptures has described various diversified aspects of mankind from astrology to spirituality and from medicine to agriculture; all aspects have been covered in a very subtle way and are very much relevant today. "Rig Veda and Krushi Parashar" finely describe crops and their cultivation techniques, rain forecasting, and soil adaptations. All the diversified aspects related to agro-management techniques have been dealt with in detail in these valuable treatises.

21.3.1 Broad Principles

For centuries the tradition of holistic agriculture has been a way of life in India that has shaped the outlook, the thought process, the culture, and the economic life of its people.

Below are the broad principles that have been part and parcel of this Vedic Krishi or ancient agriculture system:

1. Restoration of soil health by incorporation of the organic matter that holds its fertility.
2. Establishing and maintaining the soil as a living system.
3. Skilful applications of factors that maintain soil health.
4. In crop growth all the five elements have to be balanced, they are: earth, wind, rain, fire, and space.

21.3.2 Important Aspects of Vedic Krishi

Based on the broad principles following practices which frame this agro-management system:

1. Harnessing energies from the cosmos: A biodynamic calendar and its use is very much prevalent in India as well as in other parts of the world. During the particular time and phase of month and year, the cosmic influences are most supportive for the growth and overall development of plants. The ancient sages had an in-depth knowledge of this aspect and they very prominently used it in different phases of crop growth, right from sowing up until the harvesting phase.
2. Harnessing energies from the cow: The cow has occupied a pivotal place in social, economic, and spiritual spheres of the Indian civilization. The Vedic literature rightly describes in detail its importance. Different products from the cow like cow dung and urine are an important constituent of traditional bio-formulations.
3. Agnihotra therapy: Chanting of particular "Mantras" at a particular time of the day with simple offerings in fire help to fine-tune the cosmic energies and utilize them for beneficial aspects and purifying the environment.

4. Bio-formulations from locally available materials: Compost and farm yard manure along with locally and readily available materials are the basic constituents of traditional bio-formulations. Some of them are as follows:
   - Panchgavya: This is a special preparation made from five by-products of the cow along with other ingredients (coconut water, sugarcane juice) that are incubated in earthen pots and are used as liquid manure.
   - Jivamrit: This is a beautiful fermented concoction of cow dung, cow urine, jiggery, and gram flour along with some soil. The microbial population grows rapidly with the availability of carbohydrate (jaggery) and protein (gram flour), and after 5 days this bio-formulation exerts wonderful effects on the soil and crops.
   - Vermicomposting: The farm yard manure or any organic waste is composted with the help of earthworms.

     The ultimate aim of these biofertilizers is to maintain and propagate the soil microcosm wherein different beneficial microorganisms help in different processes, like the phosphate-solubilizing microbes (PSMs) that help to solubilize and mineralize locked forms of phosphorus into bioavailable forms (Sharma et al. 2013, 2014b).
5. Beej Sanskar: Seed treatment. Prior to sowing, seeds are treated with organic materials like cow urine and then sown.
6. Bhramastra: Natural pesticide. Prepared from leaves of local tree species like guava, papaya, pomegranate, custard apple, etc. with cow urine as the basic ingredient.
7. Acchadan: Mulching technique. The soil is covered with dried vegetation like sugarcane waste and this prevents evaporation of moisture from the soil.

TEK-based agro-systems serve as natural "Gene Banks" because farmers through their traditional methods of sowing the native seeds and using local species are able to maintain a genetic pool that is indigenous as well as suited to the local environment. These local varieties are more sturdy and resilient and also are able to withstand environmental stresses. Modern or hybrid varieties on the contrary lack these qualities. Moreover, these modern varieties cripple the farming community in more ways than one as they have to be procured from the sellers in each season, are subject to market availability and quality, and are often intellectual property right (IPR) protected, which can restrict their use. They also require costly inputs of fertilizers and pesticides. The overall goal of this knowledge in agriculture is to assimilate various components in the holistic fabric of ecosystem integrity so that overall biodiversity is preserved and biological efficiency is improved. This ultimately leads to agro-ecosystem productivity and maintenance of self-regulating capacity. The central idea is to maintain an agro-ecosystem that closely mimics the function and structure of local natural ecosystems.

21.4 TEK-Based Agricultural Practices: The Indian Scenario

Traditional methods of cultivation have revealed several interesting facts about the assets of traditional agriculture practices in India. The indigenous skills of the farmers of remote tribal villages are still untouched by the official extension and development programs. These existing indigenous practices of cultivation have emerged after centuries of experience, based on the policy of trial and error, and as such they have a sound base for their wider acceptance. Traditionally, man, animals, forests, and agricultural fields were inseparable and harmonious entities of a single system.

As an example, there are biodiversity hot spots, with forests (Rai 2007) and even agricultural systems (Barooah and Pathak 2009) in major areas of north-eastern India that are home to large pools of genetic biodiversity. Bari is a common farming system prevalent in north-east India; it is an area on a homestead where different crops are grown with poultry, livestock, and fish next to the main household. This type of system, in conjunction with the taboos related to cultural/religious beliefs, have contributed to management and conservation of biodiversity and allowed the ethnic group Thengal-Kacharis to survive sustainably (Barooah and Pathak 2009).

Bamboo was found almost everywhere and in every Bari system (Barooah and Pathak 2009). Sustainable harvesting of bamboo is carried out so that the people can continue to benefit from the resources it provides. Religious beliefs permit harvesting only on certain days and on every new moon. Efforts for conservation and preservation of certain species stem from the cultural and religious beliefs of "Thengal-Kacharis" (Barooah and Pathak 2009). Dhekia (*Diplazium esculentum*), a leafy vegetable, is a taboo during the fall season because that is when the plant sporulates and propagates. Many plants are also considered taboo and preserved for religious reasons like the holy tree, Sijou Goch (*Euphorbia nerifolia*) (Barooah and Pathak 2009). The women of this community have an extensive traditional knowledge on how to store food and seeds. Seeds from brinjal, ladyfinger, and chillies are stored and dried in the fireplace (dhua chang) to prevent attack by pathogens (Barooah and Pathak 2009). Pumpkin, sesame, ash gourd, and cucumber seeds are removed, sun dried, and stored in bamboo for the next growing season. Medicinal plants like amla (*Emblica officinalis*) and myrobalan (*Terminalia chebula*) are also dried and stored (Barooah and Pathak 2009). *Areca catechu* nuts are stored in pits lined with palm or banana leaves because they help preserve the nuts for consumption until the next growing season.

A comparative microbial diversity and nutrient availability analysis of TEK-based systems versus chemical integrated agri-amendment systems was carried out in the earmarked fields of Kachchh, western India by Sharma et al. (2014a). The organic farms were subjected to various practices, viz.: Acchadan, seed treatment, Jivamrit application and use of natural pesticides as "Brahmastra" that are a traditional wealth of the age old "Vedic Krishi" system from ancient India. On the other hand, the synthetic input-based fields were subjected to application of chemical fertilizers and pesticides. Samples were collected from the crop rhizosphere up to a depth of 12 cm using standard soil sampling procedure. Samples were analyzed in triplicate for physical-chemical characteristics and microbiological analysis. It was observed that organic-based farming systems provide the major plant nutrients to the crops at the required timings. Organic amendments not only increase the organic matter content of the soil but also provide nutrients that are seldom applied by farmers (e.g., manganese, zinc, and sulphur) as insurance against potential yield limitations. Furthermore, nutrients that are normally applied in ommercial fertilizers (e.g.

potassium) and liming sources (i.e. magnesium and calcium) are supplemented in organic amendments and permitted to accrue in the soil. Hence these TEK-based systems help to build drought-resilient soils (Sharma and Thivakaran 2016).

21.5 Conclusions and Future Prospects

Before the advent of modern agricultural technologies, traditional agricultural systems have sustained populations of various indigenous people in various parts of the world. Agriculture and the associated traditional knowledge involve two major aspects: the characteristics of crops and the management of those crops (Wilken 1987). Management of biodiversity and mixed inter-cropping has helped sustain agriculture practices of indigenous and local communities. The agro-ecological evidence suggests that the cultural adaptations developed by the farming community in India like the mix-cropping system helps maintain a wide genetic resource base. Similarly domestication of cattle especially the local breeds of cow provide different manures. This has helped to sustain farming practices in harsh climatic conditions and terrains. TEK systems are major contributors in sustaining regional as well as local biodiversity and ecosystem services and are important in building systems that are resilient in the face of global environmental change. This knowledge in the agriculture sector must be a part of policies adapted by governments, institutions, and Non-Governmental Organizations (NGOs) when designing and implementing climate change and agriculture-related policies. TEK has the potential to provide valuable information and useful models that can be adapted for resource management in today's scenario where biodiversity is now becoming synonymous with sustainable development and human survival. Agricultural techniques and products based on indigenous knowledge are now being widely adapted.

Many developing countries including India have a rich cultural tradition of their own. The so-called "Modern Progressive Agriculture" that relies on high-yielding hybrid crop varieties and ever-increasing and indiscriminate doses of chemical fertilizers and pesticides have crippled the whole food system and indiscriminately uprooted the time-honored agro-management practices. Before this enormous wealth of TEK in the agri-sector is lost forever, scientists and researchers should systematize and try to give it a scientific standing, given that these practices are rapidly diminishing due to socio-economic and political pressures. The decline of the TEK-based agriculture systems can be summarily attributed to the erosion of cultural values and weakening of traditional authorities and governmental policies promoting modern techniques at the expense of indigenous knowledge based practices.

Integration of scientific and indigenous knowledge systems is vital for agricultural sustainability. This integration of knowledge systems is important because the indigenous knowledge system is far more accurate with regard to the nomenclature for identifying soils and their suitability for varied uses. However, in order to render it suitable for varied environments and regions it needs to be provided with a scientific standing, as mentioned above. TEK is an important component of local adaptive capacity. There is an immediate need to tackle the research/extension systems, agricultural policies, and IPR regimes that drive the loss of local crop varieties. It is high time that we realized the huge importance of this traditional knowledge system and its invaluable implications in the agri-sector and sustainable ecosystems.

References

Berkes F (1999) Sacred ecology. In: Traditional ecological knowledge and resource management. Taylor & Francis, Philadelphia

Barooah M, Pathak A (2009) Indigenous knowledge and practices of *Thengal Kachari* women in sustainable management of *bari* system of farming. Indian J Tradit Knowl 8(1):35–40

CelaCruz M, Koohafkan P (2009) Globally important agricultural heritage systems: a shared vision of agricultural, ecological and traditional societal sustainability. Resour Sci 31:905–913

Flores-Delgadillo L, Fedick SL, Solleiro-Rebolledo E, Palacios-Mayorga S, Ortega-Larrocea P, Sedov S, Osuna-Ceja E (2011) A sustainable system of a traditional precision agriculture in a Maya homegarden: soil quality aspects. Soil Tillage Res 113:112–120

Rai SC (2007) Traditional ecological knowledge and community-based natural resource management in Northeast India. J Mt Sci 4:248–258

Rengel Z (2008) Bioavailability of phosphorus and micronutrients in the soil-plant-microbe continuum. K-8, Fifth International Symposium ISMOM 2008, Pucon, Chile, 24–28

Sharma SB, Trivedi MH, Sayeed RZ, Thivakaran GA (2014a) Study on the status of soil phosphorus in context with phosphate solubilizing microorganisms in different agricultural amendments in Kachchh, Gujarat, Western India. Annu Res Rev Biol 18:2901–2909

Sharma SB, Trivedi MH, Thivakaran GA (2014b) Phosphate solubilization by bacteria isolated from agric-soils of Kachchh, Western India. Int J Biotechnol Biosci 4(2):149–151

Sharma SB, Thivakaran GA (2016) Impact of drought on soil and microbial diversity in different agro ecosystems of the semi-arid zones. In: Hakeem KR, Akhtar MS, Abdullah SNA (eds) Plant, soil and microbes - interactions and implications in crop science. Springer International, Cham

Sharma SB, Sayeed RZ, Trivedi MH, Thivakaran GA (2013) Phosphate solubilising microbes: sustainable approach for managing phosphorus deficiency in agricultural soils. Springer Plus 2:587–591

Wilken GC (1987) Good farmers: traditional agricultural resource management in Mexico and Central America. University of California Press, Berkeley/Los Angeles

Yuan Z, Lun F, He L, Cao Z, Min Q, Bai Y, Liu M, Cheng S, Li W, Fuller AM (2014) Exploring the state of retention of traditional ecological knowledge (TEK) in a Hani rice terrace village, Southwest China. Sustainability 6:4497–4513

Influence of Arbuscular Mycorrhizal Fungal Effect and Salinity on *Curcuma longa*

22

B. Sadhana and S. Muthulakshmi

Abstract

The endophytic Arbuscular Mycorrhizal (AM) fungi is mutually associated with root systems of higher plants. This fungus colonizes both inter- and intracellularly in the roots of plants. *Curcuma longa* is a herbaceous perennial plant commonly called turmeric, belonging to the Zingiberaceae family and native to southern Asia, particularly India. Our country is a leading producer and exporter of turmeric in the world. It is used as a condiment, dye, cosmetic, and medicine, and is also used in religious ceremonies. The present study focuses on the influence of AM fungal effect and salinity on *Curcuma longa* plants grown under greenhouse conditions. This investigation reported that lower concentrations of sodium chloride do not show drastic effects on plant growth when they are treated with AM fungi compared to non-AM fungi-inoculated control plants. Thus AM fungi improved the salt tolerance in *Curcuma longa* plants at lower concentrations of sodium chloride.

22.1 Introduction

Arbuscular Mycorrhiza (AM) fungi are beneficially associated with root systems of most higher plants. Turmeric (*Curcuma longa* L) is an ancient and sacred spice of India, which is commonly known as "Indian saffron." It is one of the most important commercial spice crops grown in India. Andhra Pradesh, Tamil Nadu, Orissa, Karnataka, West Bengal, Gujarat, Meghalaya, Maharashtra, and Assam are

B. Sadhana (✉) • S. Muthulakshmi
PG and Research Centre, Department of Botany, Thiagarajar College, Madurai, Tamil Nadu, India
e-mail: sadhanakarthik2004@yahoo.co.in

V. Kumar et al. (eds.), *Probiotics in Agroecosystem*,
DOI 10.1007/978-981-10-4059-7_22

important states which cultivate turmeric. Among these Andhra Pradesh alone occupies 35.0% of the area and represents 47.0% of production.

Turmeric is used in various forms as a condiment, flavoring, and coloring agent and as a principal ingredient in Indian culinary curry powder. It has anti-cancer and anti-viral properties and hence has use in the drug and cosmetic industry. "Kum-kum," popular with Hindu temple pooja on religious and ceremonial occasions, is also a by-product of turmeric. There is increasing demand for natural products as food additives like turmeric, which acts as a natural food colorant. This study deals with the analysis of different concentrations of sodium chloride on the influence of vegetative growth, physiological tolerance, and uptake of nutrients in *Curcuma longa* under greenhouse condition.

22.2 Methods

22.2.1 Sample Collection

The AM fungal inocula were collected from the Tamil Nadu Agricultural University, Madurai, Tamil Nadu, India for the present study. The rhizome of *Curcuma longa* was collected from cultivation area of Madurai.

22.2.2 Potting and Seeding

One hundred grams of collected soil samples (inoculum) were layered over pot mixture (1:3 of sterilized soil and sand; about 3 kg) in earthen pots (height: 20 cm, diameter: 25 cm). These pots were used in the study. The control pot was not inoculated with AM fungi. The *Curcuma longa* rhizome 2 cm in size was propagated at a depth of 5 cm in pot soil. The pot was assigned for the following treatments: T1 – 0.1, T2 – 0.2, T3 – 0.3, T4 – 0.4, and T5 – 0.5% sodium chloride concentration for a 7-day period after the 90th day of vegetative growth.

22.2.3 Growth Determination

The plant growth was measured at regular intervals of 30 days for both AM fungi-inoculated and control plants.

22.2.4 Estimation of Chlorophyll

The chlorophyll content of leaf tissue was estimated by the method of Arnon (1949). Fifty milligrams of leaf tissue were ground in 80% acetone using a mortar and pestle and centrifuged. The pellet was extracted again with acetone and centrifuged. This was repeated until the pellet turned non-green. The supernatants were collected and the optical density of the extract was read at 645 nm and 663 nm.

The chlorophyll content was calculated on a fresh weight basis using the following formula:

$$\text{Total chlorophyll}\left(\mu g/\text{g } fr.wt\right) = \frac{22.4\times \text{A}645 + 8.02\times \text{A}663}{1\times 1000\times W}\times V$$

$$\text{Chlorophyll a}\left(\mu g/\text{g } fr.wt\right) = \frac{22.9\times \text{A}663 - 2.69\times \text{A}645}{1\times 1000\times W}\times V$$

$$\text{Chlorophyll b}\left(\mu g/\text{g } fr.wt\right) = \frac{22.9\times \text{A}645 - 4.68\times \text{A}663}{1\times 1000\times W}\times V$$

Where

l = path of light length in cm (1 cm)
V = Volume of the extract in ml and
W = Fresh weight of the sample in g.

22.2.5 Estimation of Carotenoid

The optical density of the acetone extract of leaves was read at 480, 645, and 663 nm and the amount of carotenoids were estimated (Ridley 1977) using the formula:

A480 + (0.114 × A663) – (0.638 × A645) and the extinction coefficient of 100 mM^{-1} cm^{-1}.

22.2.6 Estimation of Proline

The proline content of the leaf is determined by the method of Bates et al. (1973). About 500 mg of leaf sample was ground in a mortar with 10 ml of 3% aqueous sulfosalicylic acid. The homogenate was filtered through Whatman No: 2 filter paper. The extraction was repeated and the filtrates were combined. Two milliliters of the filtrate were taken in a test tube and 2 ml of acid ninhydrin and 2 ml of glacial acetic acid were added. The mixtures were incubated at 100 °C in a hot water-bath for 1 h. Then the tubes were transferred to an ice-bath to terminate the reaction. Four milliliters of toluene was added and the contents mixed vigorously for 15–20 s. The optical density of the upper toluene layer containing proline was measured at 520 nm against a reagent blank by using UV-VIS Spectrophotometer-Systronics-119. Authentic proline was used as a standard for calculating the leaf proline content.

22.2.7 Total Nitrogen Estimation

The dried biomass of each sample was ground in a porcelain mortar and pestle and the total nitrogen content was estimated by modified micro kjeldahl method (Umbreit et al. 1972). Ten milligrams of powdered biomass were taken in a micro kjeldahl flask. A pinch (about 50 mg) of catalyst (Humphries 1956) and 0.5 ml of concentrated sulphuric acid were added into the kjeldahl flask. The flask was gently heated in a digestion rack until the fumes of sulfuric acid evolved. Then it was heated strongly until the digest turned to an apple green color. After cooling, the digest was made up to 20 ml with double distilled water. To 1 ml of the above diluted digest, 2 ml of water, 2 ml of color reagent, and 3 ml of 2 N sodium hydroxide were added in series. After 15 min, the optical density of the solution was read at 490 nm against a reagent blank by using UV-VIS Spectrophotometer Systronics-119. The nitrogen content of the sample was determined with reference to a standard graph prepared using authentic ammonium chloride.

22.2.8 Total Phosphorus Estimation

The dried biomass of each sample was ground in a porcelain mortar and pestle and the total phosphorus content was estimated by the micro kjeldahl method (Bartlett 1959). 100 mg of powdered biomass were ground in ice-cold 0.2 N perchloric acid, and the extract was held in ice for 15 min prior to clearing by centrifugation. The extraction by perchloric acid was repeated at least three times and the extracts were pooled and analyzed for total phosphorus. To 1 ml of the extract in a tube, 1 ml of 60% TCA was added and the contents were digested at 160–180 °C. After digestion, each sample received 4.5 ml of 2.5% ammonium molybdate and 0.2 ml of 0.25% 1-amino 2- naphthol 4- sulphuric acid (ANSA). The contents were mixed well and heated in a boiling water bath for 10 min. After cooling, the volume was diluted to 10 ml with distilled water. The blue color developed was measured at 650 nm against blank by using UV-VIS Spectrophotometer-Systronics-119. Potassium dihydrogen phosphate was used as a standard for calculating total phosphorus of the sample.

22.2.9 Estimation of Proteins

The total protein content of fresh leaf tissue was estimated by Lowrey's method (Lowrey et al. 1951). Fifty milligrams of leaf tissue was extracted in hot 80% ethanol and macerated in a mortar with pestle. The homogenate was collected and the supernatant discarded. The pellet was suspended in a suitable volume of 5% TCA in an ice-bath for 15 min. The pellet was re-extracted once in hot absolute ethanol and twice with an ethanol–ether mixture, each time discarding the supernatants after centrifugation. This pellet contained proteins and nucleic acids. The protein sample was placed in 1 ml of sodium hydroxide at 100 °C for 4–5 min. Thereafter, 5 ml of alkaline copper reagent were added and allowed to incubate at room temperature for 10 min. Folin phenol reagant 0.5 ml was added rapidly and mixed immediately. After 30 min, the optical density was measured at 750 nm by using UV-VIS Spectrophotometer-Systronics-119. The bovine serum albumin was used as standard for calculating the protein content of the sample.

22.2.10 Statistical Analysis

The measured and collected data of this study were analyzed by analysis of variance (ANOVA) and means comparison was carried out using Duncan's multiple range test (DMRT) (Little and Hills 1978).

22.3 Results and Discussion

One of the chief abiotic stresses is the soil salinity, which negatively influences plant growth and crop productivity all over the world. Valuable and advantageous microorganisms such as fungi and bacteria can enhance plant growth under several stress conditions and also increase crop yield (Evelin et al. 2009). The salt stress or salinity not only negatively influences the host plant but also affects the inoculated beneficial microbiomes in soil. Inoculation of AM fungi improves the root colonization capacity, spore germination, and growth of fungal hyphae (Hirrel 1981; Estaun 1989; McMillen et al. 1998; Jahromi et al. 2008). There are several scientific reports that exhibited the root colonization by AM fungi in host plants decreasing under saline conditions or in a saline environment (Hirrel and Gerdemann 1980; Ojala et al. 1983; Duke et al. 1986; Poss et al. 1985; Rozema et al. 1986; Menconi et al. 1995; Juniper and Abbott 2006; Giri et al. 2007; Sheng et al. 2008). Juniper and Abbott (2006)) demonstrated that the saline environment repressed the growth AM fungi and its spore formation, which leads to poor plant growth (Tian et al. 2004; Sheng et al. 2008).

In the present study the AM fungi-inoculated *Curcuma longa* (turmeric) plant exhibited a significant ($P \leq 0.05$) difference in chlorophyll *a* and *b* pigments ($0.7532^{b} \pm 0.012$ μg; $0.2394^{b} \pm 0.026$ μg), which was noticed at the 90th day of growth period compared to the control plant ($0.3212^{c} \pm 0.012$ μg and $0.0095^{d} \pm 0.023$ μg). A gradual increase in chlorophyll pigment contents was observed in AM fungi-inoculated plants compared to the control plants. Pigment contents were decreased under salinity stress conditions (0.1–0.5% of NaCl) in all plants. Therefore, during regular watering of the plants, AM fungi-inoculated *Curcuma longa* plants under saline stress of 0.1% and 0.2% NaCl recovered quickly in comparison to the uninoculated or control plants. The maximum chlorophyll *a* and *b* was observed ($8.1253^{f} \pm 0.140$ μg and $0.6853^{e} \pm 0.023$ μg) in 0.1% NaCl-stressed + AM fungi-treated plants compared to control and other higher sodium chloride-stressed AM fungi-inoculated plants. Chlorophyll *a* and *b* were decreased gradually during harvesting period in all treatments (Tables 22.1 and 22.2).

Hayat et al. (2010) reported that the photosynthetic activities of plants were affected by salinity stress. This salinity stress has numerous adverse effects on plant physiological activities such as ion toxicity changes in growth, increased respiration rate, unequal mineral distribution in plant, membrane instability resulting from calcium and potassium displacement by sodium (Grattan and Grieve 1992), membrane permeability (Gupta et al. 2002), and lowering of photosynthetic efficacy (Ashraf and Shahbaz 2003; Kao et al. 2003; Sayeed 2003).

Table 22.1 Influence of Arbuscular Mycorrhizal fungal effect and salinity on the chlorophyll *a* (μg/g fresh leaf) content of *Curcuma longa*

Sl No	Treatments	60th Day	90th day	97th day	120th day	150th day	180th day	210th day	240th day
1	Control	$0.1132^{a} \pm 0.020$	$0.3212^{c} \pm 0.012$	$0.1054^{a} \pm 0.010$	$0.2104^{b} \pm 0.102$	$0.5261^{e} \pm 0.027$	$0.4203^{d} \pm 0.020$	$0.4015^{d} \pm 0.020$	$0.3209^{c} \pm 0.010$
2	T1	$0.3845^{a} \pm 0.040$	$0.7532^{b} \pm 0.012$	$0.7911^{b} \pm 0.020$	$1.1043^{c} \pm 0.116$	$3.0241^{d} \pm 0.020$	$5.0241^{e} \pm 0.200$	$8.1253^{f} \pm 0.140$	$5.0126^{e} \pm 0.122$
3	T2	$0.3845^{a} \pm 0.040$	$0.7532^{b} \pm 0.032$	$0.7221^{b} \pm 0.011$	$0.8242^{c} \pm 0.113$	$1.1023^{d} \pm 0.103$	$3.0221^{f} \pm 0.110$	$5.0012^{g} \pm 0.520$	$2.0134^{e} \pm 0.310$
4	T3	$0.3845^{e} \pm 0.040$	$0.7532^{b} \pm 0.021$	$0.6713^{b} \pm 0.023$	$0.8013^{c} \pm 0.025$	$1.0242^{d} \pm 0.102$	$3.1325^{f} \pm 0.053$	$4.1370^{g} \pm 0.240$	$2.0143^{e} \pm 0.126$
5	T4	$0.3845^{a} \pm 0.040$	$0.7532^{c} \pm 0.010$	$0.6201^{b} \pm 0.031$	$0.7804^{c} \pm 0.031$	$1.1432^{d} \pm 0.028$	$3.4572^{f} \pm 0.125$	$4.0321^{g} \pm 0.210$	$2.1187^{e} \pm 0.031$
6	T5	$0.3845^{a} \pm 0.040$	$0.7532^{c} \pm 0.040$	$0.5023^{b} \pm 0.025$	$0.7724^{c} \pm 0.009$	$1.2314^{d} \pm 0.027$	$3.2144^{f} \pm 0.052$	$4.1452^{g} \pm 0.220$	$2.1658^{e} \pm 0.113$

Values are mean of five replicates ± SD The mean difference is significant at a *P* value of ≤ 0.05

Table 22.2 Influence of Arbuscular Mycorrhizal fungal effect and salinity on the chlorophyll *b* (μg/g fresh leaf) content of *Curcuma longa*

Sl No	Treatments	60th day	90th day	97th day	120th day	150th day	180th day	210th day	240th day
1	Control	$0.0072^{c} \pm 0.010$	$0.0095^{d} \pm 0.023$	$0.0064^{b} \pm 0.002$	$0.0070^{c} \pm 0.002$	$0.0061^{b} \pm 0.017$	$0.0103^{e} \pm 0.030$	$0.0100^{e} \pm 0.021$	$0.0009^{a} \pm 0.003$
2	T1	$0.0934^{a} \pm 0.022$	$0.2394^{b} \pm 0.026$	$0.2441^{b} \pm 0.007$	$0.3233^{c} \pm 0.106$	$0.4241^{d} \pm 0.030$	$0.6231^{e} \pm 0.030$	$0.6853^{e} \pm 0.023$	$0.4023^{d} \pm 0.008$
3	T2	$0.0934^{a} \pm 0.022$	$0.2394^{b} \pm 0.026$	$0.2100^{b} \pm 0.011$	$0.2205^{b} \pm 0.013$	$0.3123^{c} \pm 0.011$	$0.4125^{d} \pm 0.033$	$0.5002^{e} \pm 0.010$	$0.3032^{c} \pm 0.010$
4	T3	$0.0934^{a} \pm 0.022$	$0.2394^{c} \pm 0.026$	$0.2101^{b} \pm 0.063$	$0.2113^{b} \pm 0.125$	$0.2245^{b} \pm 0.020$	$0.4226^{e} \pm 0.062$	$0.5200^{f} \pm 0.020$	$0.3140^{d} \pm 0.024$
5	T4	$0.0934^{a} \pm 0.022$	$0.2394^{d} \pm 0.026$	$0.2013^{c} \pm 0.004$	$0.1808^{b} \pm 0.021$	$0.2232^{c} \pm 0.018$	$0.3802^{e} \pm 0.015$	$0.4821^{f} \pm 0.040$	$0.3017^{d} \pm 0.024$
6	T5	$0.0934^{b} \pm 0.022$	$0.2394^{e} \pm 0.026$	$0.0900^{a} \pm 0.025$	$0.1621^{c} \pm 0.012$	$0.2114^{d} \pm 0.017$	$0.3704^{f} \pm 0.022$	$0.3955^{g} \pm 0.032$	$0.2057^{d} \pm 0.015$

Values are mean of five replicates ± SD The mean difference is significant at a *P* value of ≤ 0.05

Allakhverdiev et al. (2000) found that major abiotic stresses such as cold, heat, water stress, and salinity adversely influence plant growth and productivity, but compared to other stresses salt stress exerts more severe and extreme effects, which lead to low productivity (Munns 2002). According to Munns (1993), a high salt content decreased plant growth and crop production by influencing the significant physiological courses such as ion balance modification, water status, mineral nutrition and distribution, stomatal behavior, and photosynthetic efficiency.

In another study, AM fungi-inoculated plants showed augmented vegetative growth, total chlorophyll content, and uptake of nutrients like nitrogen, phosphorus, potassium, calcium, and magnesium in maize plants (Sitaramaiah et al. 1998). Goicoechea et al. (Goicoechea et al. 1997) reported that VAM-inoculated alfalfa plants were better adapted to water-deficit conditions and increased concentration of proline compared to the non-mycorrhizal plants.

In another study, chlorophyll content was reduced by increasing salinity (Sheng et al. 2008), owing to the suppression of specific enzymes that are responsible for the synthesis of photosynthetic pigments (Murkute et al. 2006). The uptake of important minerals needed for chlorophyll synthesis and concentrations in leaves was also reduced (El-Desouky and Atawia 1998).

The carotenoid contents in all salt-treated concentrations of AM fungi-inoculated *Curcuma longa* exhibited higher increases compared to the control, and were observed to be highest during the harvesting period (Fig. 22.1). Yancey et al. (Yancy

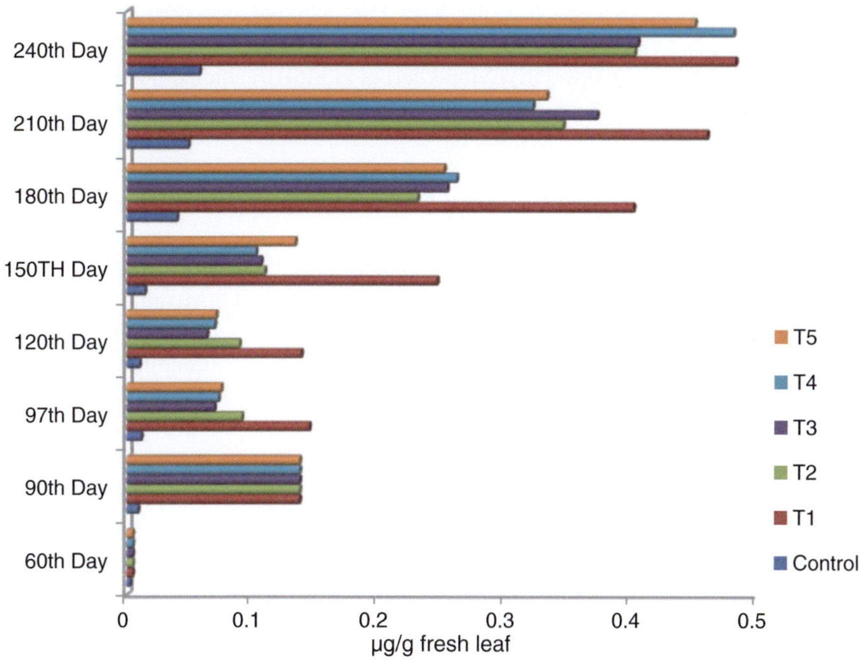

Fig. 22.1 Influence of Arbuscular Mycorrhizal fungal effect and salinity on the carotenoid content (μg/g fresh leaf.) of *Curcuma longa*

et al. 1982) suggested that plants resort to many adaptive techniques in response to various abiotic and environmental stresses such as dehydration and unnecessary osmotic pressure. Such adaptive techniques in plants include changes in physiological and biochemical processes, which are helped by beneficial microbes. Accumulation of proline and carotenoids during stress conditions with the help of beneficial microbes could be due to the direct impact of a reduced photosynthetic rate since a positive correlation between photosynthetic rate and chlorophyll content was found in many instances (Tester et al. 1986; Sivaprasad and Rai 1987). The reduction in photosynthetic activity might also influence the nitrogen fixation by legumes under salt stress (Georgiev and Atkias 1993). The dehydration and turgor pressure loss occur as a result of the higher ratios of toxic salts in leaf apoplasm, which leads to death of leaf cells and tissues (Marschner 1995).

Moisender et al. (2002) and Sheekh-El and Omar (2002) reported that the salt-sensitive plants had either a relatively low salt tolerance or strongly repressed growth at low salt concentration or may exhibit severely inhibited growth at low salinity levels, so differ in the growth response to various levels of salinity. Murphy and Durako (2003) suggested that this salt stress influenced the physiological activities at the whole plant level and also at cellular levels through osmotic and ionic stress. The proline buildup in plants is also a part of the stress signal affecting the adaptive mechanism (Maggio et al. 2002). This metabolic activity has been studied in plants in response to various osmotic stresses (Verbruggen and Heramns 2008). Mohammadkhani and Heidari (2008) reported induced accumulation of soluble sugars and proline in two varieties of maize. This accumulation of sugars and proline in plants has been a parameter for selection for stress tolerance (Yancy et al. 1982; Jaleel et al. 2007). Similarly, Mafakheri et al. (2010) studied the influence of drought stress on crop yield, proline accumulation, and chlorophyll contents in three chickpea varieties.

The present study showed the proline content was maximum in AM fungi-inoculated *Curcuma longa* plants stressed with 0.1–0.5% NaCl and was minimum in control plants at all stages of the plant. It was three- to fourfold higher during salt stress in AM fungi-inoculated plants when compared to control. After the salinity stress period, its level declined sharply in both control and AM fungi-inoculated plants (Fig. 22.2). This proline accumulation was higher in the *Curcuma longa* plant and was stimulated by the AM fungi under mild salt-stressed conditions. Further, the diffusion of proline after regular hydration of plants might be taken to indicate that proline served as a storage compound during the stress period.

The accumulation of amino acid proline has been one of the most frequently induced components by drought and salinity stress in plants. There are many plants that accumulate proline as a nontoxic and protective osmolyte to maintain osmotic balance, which implies low water potential (Jain et al. 2001; Parida et al. 2002; Ashraf and Foolad 2007; Sannazzaro et al. 2007) under various salinity conditions. Goas et al. (1982) suggested that microbial infibulation helps in acting as a reservoir of energy and nitrogen for utilization during salt stress.

Jindal et al. (1993) reported that mycorrhizal-inoculated mung bean (*Vigna radiata*) plants had a greater proline content compared to the control plants at 12.5 and

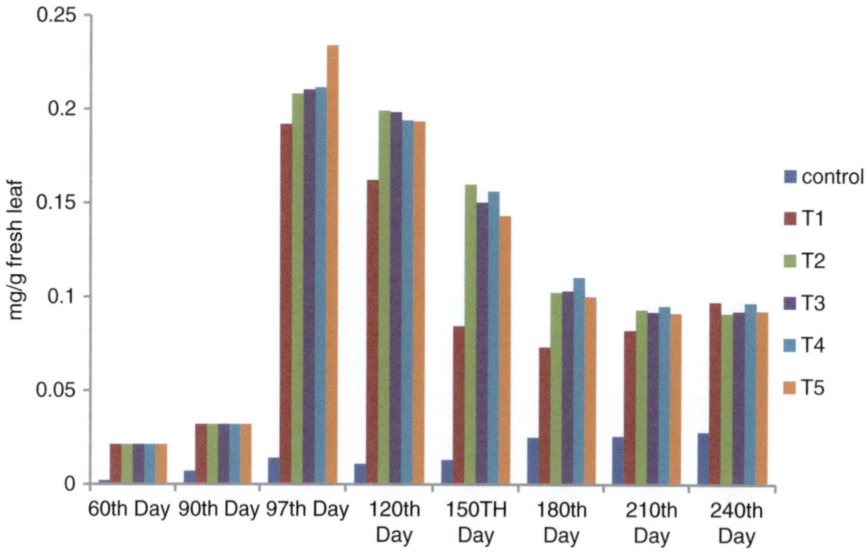

Fig. 22.2 Influence of Arbuscular Mycorrhizal fungal effect and salinity on the proline content (mg/g fresh leaf.) of *Curcuma longa*

25 mM sodium chloride at the 40th and 60th days after sowing. There was a higher proline concentration in AM soybean than control plants at different salinity levels like 0, 50, 100, 150, and 200 mM sodium chloride (Sharifi et al. 2007). The accumulation of proline in plants could be a symptom of stress tolerance in less salt-tolerance species and its contribution to osmotic modification was seemingly insignificant compared to the potassium ions (Wang et al. 2004.

The protein content of leaf was significantly ($P \leq 0.05$) increased gradually in AM fungi-treated *Curcuma longa* plants. This content declined during the salt-stressed period. After salt stress recovery, the maximum was observed as $1.825^{f} \pm 0.002$ mg and $1.420^{f} \pm 0.000$ mg on the 210th day of growth period of AM fungi-treated *Curcuma longa* under stressed conditions with 0.1% and 0.2% sodium chloride compared to control ($0.438^{e} \pm 0.002$ mg) and other salt-stressed plants (Table 22.3). Dumas et al. (1994) observed higher protein content in AM fungi-inoculated *Nicotianatabacum* and *Allium cepa* roots compared with control plants.

The enhancement of phosphate uptake and growth of leguminous plants by vesicular AM fungi (Ezawa et al. 2000; Arihara and Karasawa 2000; Meshram et al. 2000; Joner 2000; Mayoral et al. 2000; Guriqbal et al. 2001; Mamtha et al. 2002; Atimanav and Adholeya 2002) has been reported. Significant effects of mycorrhizae on nitrogen uptake percentage in grass species were studied. Alvey et al. (2001) showed strong evidence that cereal/legume rotations could enhance phosphorus nutrient uptake of cereal through arbuscular mycorrhizae. These results were supported by Johnny (1999), who reported that the legumes formed tripartite symbiosis with AM fungi and *Rhizobia,* which variably influenced plant productivity. Phosphorus nutrition is also an essential factor in the activation of nitrogen assimilation by legumes.

Table 22.3 Influence of Arbuscular Mycorrhizal fungal effect and salinity on the protein (mg/g fresh leaf) content of *Curcuma longa*

Sl No	Treatments	60th day	90th day	97th day	120th day	150th day	180th day	210th day	240th day
1	Control	$0.002^{a} \pm 0.001$	$0.025^{b} \pm 0.000$	$0.031^{b} \pm 0.001$	$0.099^{c} \pm 0.001$	$0.140^{c} \pm 0.002$	$0.219^{d} \pm 0.002$	$0.438^{e} \pm 0.002$	$0.177^{c} \pm 0.001$
2	T1	$0.002^{a} \pm 0.001$	$0.025^{a} \pm 0.000$	$0.405^{c} \pm 0.000$	$0.765^{c} \pm 0.010$	$0.933^{d} \pm 0.020$	$1.022^{e} \pm 0.000$	$1.825^{f} \pm 0.002$	$0.137^{b} \pm 0.001$
3	T2	$0.002^{a} \pm 0.001$	$0.025^{b} \pm 0.000$	$0.050^{b} \pm 0.000$	$0.075^{c} \pm 0.001$	$0.094^{c} \pm 0.020$	$0.907^{e} \pm 0.020$	$1.420^{f} \pm 0.000$	$0.102^{d} \pm 0.002$
4	T3	$0.002^{a} \pm 0.001$	$0.025^{b} \pm 0.000$	$0.025^{b} \pm 0.002$	$0.020^{b} \pm 0.010$	$0.062^{c} \pm 0.000$	$0.085^{d} \pm 0.001$	$0.099^{d} \pm 0.020$	$0.075^{c} \pm 0.003$
5	T4	$0.002^{a} \pm 0.001$	$0.025^{b} \pm 0.000$	$0.022^{c} \pm 0.003$	$0.028^{c} \pm 0.000$	$0.037^{d} \pm 0.001$	$0.195^{e} \pm 0.003$	$0.853^{f} \pm 0.000$	$0.800^{f} \pm 0.001$
6	T5	$0.002^{a} \pm 0.001$	$0.025^{b} \pm 0.000$	$0.029^{b} \pm 0.002$	$0.085^{d} \pm 0.003$	$0.058^{c} \pm 0.010$	$0.085^{d} \pm 0.002$	$0.196^{f} \pm 0.004$	$0.154^{e} \pm 0.002$

Values are mean of five replicates ± SD The mean difference is significant at P values of ≤ 0.05

Phosphate-solubilizing microorganisms (PSMs) also play a significant role in plant growth and metabolism and crop productivity. Raja et al. (2002) have reported enhanced availability and uptake of native soil phosphorus by converting the insoluble phosphorus to soluble forms by producing various organic acids. Root colonization of plants by AM fungi greatly augmented phosphorus and nitrogen nutrient uptake (Chen et al. 2005), similarly AM fungi also improved the uptake of nutrients by extra radical mycorrhizal hyphae (Ruiz-Lozano 2006).

In soil plant systems, the nutrient cycling is also influenced by AM fungal association and it has also enhanced the plant health through increased protection against many biotic and abiotic stresses, and also soil structure through aggregate formation (Bethelenfalvay and Linderman 1992; Gianinazzi et al. 2002; Turnau and Haselwandter 2002; Van der Heijden and Sanders 2002; Jeffries 1987; Barea et al. 2005; Turnau et al. 2005).

Generally, in legumes the AM fungi increased root nodulation and nitrogen fixation as a consequence of improved phosphorus nutrition (Athar 2005). The present investigation revealed that the *Curcuma longa* plant's total nitrogen and phosphorus contents were greater in AM fungi-inoculated plants than in the controls at all stages of growth. Salt stress did not affect the nitrogen and phosphorus uptake in AM fungi-inoculated plants. A significant ($P \leq 0.05$) increase in total nitrogen and phosphorus content ($33.55^{h} \pm 0.21$ mM; $0.350^{g} \pm 0.011$ mg and $33.05^{h} \pm 0.01$ mM; $0.220^{g} \pm 0.003$ mg) was found on the 210th day in the AM fungi-inoculated *Curcuma longa* plants stressed with 0.1% and 0.2% sodium chloride followed by the rest of plants, and was minimum in control plants ($16.45^{h} \pm 0.20$ mM and $0.155^{f} \pm 0.001$ mg). Those plants were tolerant to mild salt stress due to the induction of continuous nutrient mobilization by AM fungi. These contents were decreased during harvesting period (Tables 22.4 and 22.5).

The influence of VAM fungi on growth and mineral accumulation in *Bromus inermis* at 0, 1, 2, and 3 g sodium chloride/kg soil in a pot experiment was studied. The report concluded that the mechanism of VAM fungi in improving the salt tolerance of *Bromus inermis* is to increase the plant's nutritional condition by increasing the uptake of nitrogen and phosphorus while decreasing the relative concentration of sodium and chlorine in the plant (Feng et al. 1998.

Lioyd et al. (1989) reported that the greater salt concentrations in the root zone lessen availability of water in soil and lower water potential. This deficiency leads to cellular level dehydration and eventually osmotic stress also occurs. Hasegawa et al. (2000) also reported that extreme amounts of toxic ions such as Na^{+} ion and Cl^{-} ion create an ionic imbalance by lessening the uptake of helpful ions such as K^{+}, Ca^{2+} and Mn^{2+} ions. Shahid et al. (2011) reported that the Na^{+} and Cl^{-} ion buildup was connected with a decrease in concentration of K^{+} in leaves and roots.

The present study reported that employment of AM fungi in the *Curcuma longa* (turmeric) plant (Plate 22.1) causes a significant increase in vegetative plant growth along with an upsurge in the chlorophyll and carotenoid pigments. There was 90–100% of root colonization by AM fungi (Plate 22.2) and greater synthesis of proline in mycorrhizal plants compared to the control or uninoculated plants. Application of lower concentrations (0.1% and 0.2%) of sodium chloride does not

Table 22.4 Influence of Arbuscular Mycorrhizal fungal effect and salinity on the nitrogen (mM/g biomass) content of *Curcuma longa*

Sl No	Treatments	60th day	90th day	97th day	120th day	150th day	180th day	210th day	240th day
1	Control	$05.20^{a} \pm 0.05$	$07.75^{b} \pm 0.03$	$07.95^{c} \pm 0.1$	$08.45^{d} \pm 0.14$	$11.05^{e} \pm 0.00$	$13.75^{f} \pm 0.21$	$16.45^{h} \pm 0.20$	$15.20^{g} \pm 0.03$
2	T1	$05.20^{a} \pm 0.05$	$07.75^{b} \pm 0.03$	$08.53^{c} \pm 0.05$	$09.85^{d} \pm 020$	$16.85^{e} \pm 0.00$	$26.05^{f} \pm 0.43$	$33.55^{h} \pm 0.21$	$26.80^{g} \pm 0.02$
3	T2	$05.20^{a} \pm 0.05$	$07.75^{b} \pm 0.03$	$09.95^{c} \pm 0.02$	$14.95^{d} \pm 0.00$	$26.92^{e} \pm 0.03$	$31.05^{f} \pm 0.05$	$33.05^{h} \pm 0.01$	$30.10^{g} \pm 0.12$
4	T3	$05.20^{a} \pm 0.05$	$07.75^{b} \pm 0.03$	$09.95^{c} \pm 0.03$	$09.75^{c} \pm 0.21$	$14.08^{d} \pm 0.10$	$21.00^{e} \pm 0.02$	$25.80^{g} \pm 0.03$	$23.20^{f} \pm 0.03$
5	T4	$05.20^{a} \pm 0.05$	$07.75^{b} \pm 0.03$	$8.05^{c} \pm 0.11$	$09.75^{d} \pm 0.00$	$13.65^{e} \pm 0.21$	$14.00^{e} \pm 0.05$	$16.87^{g} \pm 0.00$	$15.20^{f} \pm 0.05$
6	T5	$05.20^{a} \pm 0.05$	$07.75^{b} \pm 0.03$	$8.55^{c} \pm 0.01$	$08.75^{c} \pm 0.03$	$10.02^{d} \pm 0.05$	$13.34^{e} \pm 0.02$	$15.75^{f} \pm 0.12$	$13.10^{e} \pm 0.00$

Values are mean of five replicates ± SD The mean difference is significant at P values of ≤ 0.05

Table 22.5 Influence of Arbuscular Mycorrhizal fungal effect and salinity on the phosphorus (mg/g biomass) content of *Curcuma longa*

Sl No	Treatments	60th day	90th day	97th day	120th day	150th day	180th day	210th day	240th day
1	Control	$0.003^{a} \pm 0.003$	$0.005^{a} \pm 0.002$	$0.005^{a} \pm 0.000$	$0.006^{b} \pm 0.001$	$0.014^{c} \pm 0.002$	$0.065^{d} \pm 0.011$	$0.155^{f} \pm 0.001$	$0.092^{e} \pm 0.010$
2	T1	$0.003^{a} \pm 0.003$	$0.005^{a} \pm 0.002$	$0.085^{b} \pm 0.002$	$0.099^{c} \pm 0.001$	$0.107^{d} \pm 0.020$	$0.202^{e} \pm 0.031$	$0.350^{g} \pm 0.011$	$0.247^{f} \pm 0.012$
3	T2	$0.003^{a} \pm 0.003$	$0.005^{a} \pm 0.002$	$0.093^{b} \pm 0.000$	$0.105^{c} \pm 0.002$	$0.139^{d} \pm 0.012$	$0.187^{e} \pm 0.002$	$0.220^{g} \pm 0.003$	$0.125^{f} \pm 0.010$
4	T3	$0.003^{a} \pm 0.003$	$0.005^{a} \pm 0.002$	$0.099^{b} \pm 0.021$	$0.109^{c} \pm 0.003$	$0.172^{d} \pm 0.010$	$0.180^{d} \pm 0.002$	$0.215^{f} \pm 0.000$	$0.205^{e} \pm 0.051$
5	T4	$0.003^{a} \pm 0.003$	$0.005^{a} \pm 0.002$	$0.083^{b} \pm 0.000$	$0.101^{c} \pm 0.002$	$0.154^{e} \pm 0.010$	$0.162^{e} \pm 0.030$	$0.145^{d} \pm 0.021$	$0.138^{d} \pm 0.002$
6	T5	$0.003^{a} \pm 0.003$	$0.005^{a} \pm 0.002$	$0.078^{b} \pm 0.002$	$0.100^{c} \pm 0.000$	$0.125^{d} \pm 0.003$	$0.127^{d} \pm 0.030$	$0.139^{f} \pm 0.021$	$0.131^{e} \pm 0.003$

Values are mean of five replicates ± SD The mean difference is significant at P values of ≤ 0.05

Plate 22.1 Influence of Arbuscular Mycorrhizal fungal effect and salinity on *Curcuma longa* (**a–d**)

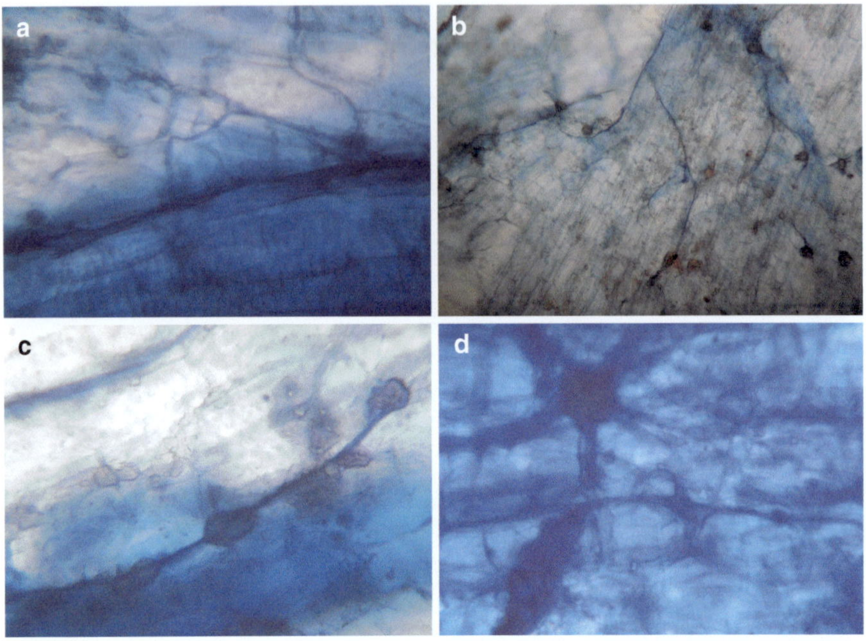

Plate 22.2 Arbuscular Mycorrhizal fungal infection in root tissues of *Curcuma longa* (**a**–**d**)

negatively affect the AM fungi- inoculated plant growth when grown in greenhouse conditions. Moreover, these salinity-stressed plants recovered faster compared to the other salt-induced plants and control plants. It can be concluded that the AM fungi-inoculated turmeric plant exhibited noteworthy vegetative growth and enhanced mild salt tolerance compared to control plants.

Smith and Read (2008) also reported that AM fungi are associated with the roots of over 80% terrestrial plant species and they have also been shown to encourage plant growth yield and salinity tolerance by many scientists. Application of AM fungi promotes salinity tolerance in plants by applying numerous mechanisms, such as enhancing nutrient uptake (Evelin et al. 2012), increasing plant growth–hormone production, and improving rhizospheric and soil conditions (Asghari et al. 2005), enhancement in photosynthetic activity and water use efficiency (Hajiboland et al. 2010), buildup of compatible solutes (Evelin et al. 2013), and manufacture of greater antioxidant enzymes, as reported by Manchanda and Garg (2011)).

The positive aspects of AM fungi inoculation with crop plants results in better survival of root-colonized plants in terms of better growth, the maintenance of plant biodiversity, the soil microflora improvement (Boer et al. 2005), resistance to various biotic (BØdker et al. 2002; Dalpe 2005) as well as abiotic pressures and stresses (Evelin et al. 2009; Neumann and George 2009), enhancement of the soil structure and lessening of chemical fertilizer and pesticide application (Strack et al. 2003).

Conclusion

The current study illustrates the effects of AM fungus and the low concentrations of NaCl (0.1 and 0.2%) on the *Curcuma longa* (turmeric) plant. Inoculation of the beneficial fungus boosted the vegetative growth by increasing the chlorophyll and carotenoid pigment contents along with tolerance towards applied salt stress for a few days. Application of higher concentrations of sodium chloride (above 0.2%) negatively influenced the growth pattern of *Curcuma longa* (turmeric) plants under greenhouse conditions. Consequently, application of AM fungus encouraged nutrient uptake, particularly N and P, and also provided minor salt tolerance in turmeric plants for the applied concentrations of NaCl. The results of this study suggest that using AM fungi soil application not only enhances the vegetative growth of the plant but also augments soil fertility and lessens the application of chemical fertilizers in the crop field, which becomes cost effective. AM fungus is generally known as an ecofriendly biofertilizer or bioinoculant that does not cause any pollution on or adverse effects to our ecosystem.

Acknowledgments The author would like to express her immense thanks to the President, Vice-President, Secretary, Principal, HOD of the Botany Department, Thiagarajar College, Madurai, Tamil Nadu, India for their encouragement and for providing the facilities for the successful completion of this research work. The author would also like to express her heartfelt thanks to her family and friends for support in the successful completion of this work.

References

Allakhverdiev SI, Sankamoto A, Nishijama Y, Inaba M, Murata N (2000) Ionic and osmotic effects of sodium chloride – induced inactivation of photosystems I and II in *Synechococcus sp*. Plant Physiol 123:1047–1056

Alvey S, Bagyaoko M, Neumann G, Buerkert A (2001) Cereal/legume rotations affect chemical properties and biological activities in two West African soils. Plant Soil 231:45–54

Arihara J, Karasawa T (2000) Effect of previous crops on arbuscular mycorrhizal formation and growth of succeeding maize. Soil Sci Plant Nutr 46(1):43–51

Arnon DI (1949) Copper enzymes in isolated chloroplasts. Polyphenol oxidase in *Beta vulgaris*. Plant Physiol 24:1–15

Asghari HR, Marschner P, Smith SE, Smith FA (2005) Growth response of *Atriplex nummularia* to inoculation with arbuscular mycorrhizal fungi at different salinity levels. Plant Soil 373:245–256

Ashraf M, Foolad MR (2007) Role of glycine betaine and proline in improving plant abiotic stress resistance. Environ Exp Bot 59:207–216

Ashraf M, Shahbaz M (2003) Assessment of genotypic variation in salt tolerance of early CIMMYT hexaploid wheat germplasm using photosynthetic capacity and water relation as selection criteria. Photosynthetica 41:273–280

Athar M (2005) Nodulation of native legumes in Pakistani range lands. Agric Conspec Sci 70:49–54

Atimanav G, Adholeya A (2002) AM inoculations of five tropical fodder crops and inoculum production in marginal soil amended with organic matter. Biol Fertil Soils 35:214–218

Barea JM, Azcón R, Azcón-Aguilar C (2005) Interactions between mycorrhizal fungi and bacteria to improve plant nutrient cycling and soil structure. In: Buscot F, Varma S (eds) Micro organisms in soils: roles in genesis and functions. Springer, Heidelberg, pp 195–212

Bartlett GR (1959) Phosphorus assay in column chromatography. J Biol Chem 234:466–468
Bates LS, Waldran RP, Teare ID (1973) Rapid determination of free proline for water stress studies. Plant Soil 39:205–208
Bethlenfalvay GJ, Linderman RG (1992) Mycorrhizae in sustainable agriculture. ASA Special Publication, Madison, pp 73–89
Boer W, Folman LB, Summerbell RC, Boddy L (2005) Living in a fungal world: Impact of fungi on soil bacteria niche development. FEMS Micro Biol Rev 29:795–811
BØdker L, KjØler R, Kristensen K, Rosendahl S (2002) Interactions between indigenous arbuscular mycorrhizal fungi and *Aphanomyceseuteiches* in field grown pea. Mycorrhiza 12:7–12
Chen X, Tang JJ, Zhi GY, Hu SJ (2005) Arbuscular mycorrhizal colonization and phosphorus acquisition of plants: effects of coexisting plant species. Appl Soil Ecol 28:259–269
Dalpe Y (2005) Les Mycorrhizes: un outil de protection desplantesmais non unepanace. Phytoprotection 86:53–59
Duke ER, Johnson CR, Koch KE (1986) Accumulation of phosphorus, dry matter and betaine during Sodium chloride stress of split-root citrus seedlings colonized with vesicular-arbuscular mycorrhizal fungi on zero, one or two halves. New Phytol 104:583–590
Dumas GE, Guillaume P, Tahiri AA, Gianinazzi-Pearson V, Gianinazzi S (1994) Changes in polypeptide patterns in tobacco roots by Glomus species. Mycorrhiza 4:215–221
El-Desouky SA, Atawia AAR (1998) Growth performance of citrus rootstocks under saline conditions. Alexandria J Agric Res 43:231–254
Estaun MV (1989) Effect of sodium chloride and mannitol on germination and hyphal growth of the vesicular-arbuscular mycorrhizal fungus *Glomus mosseae*. Agric Ecosyst Environ 29:123–129
Evelin H, Giri B, Kapoor R (2013) Ultrastructural evidence for AMF mediated salt stress mitigation in *Trigonella foenumgraecum*. Mycorrhiza 23:71–86
Evelin H, Giri B, Kapoor R (2012) Contribution of *Glomusintraradices* inoculation to nutrient acquisition and mitigation of ionic imbalance in Sodium chloride stressed *Trigonella foenu graecum*. Mycorrhiza 22:203–217
Evelin H, Kapoor R, Giri B (2009) Arbuscular mycorrhizal fungi in alleviation of salt stress: a review. Ann Bot 104:1263–1280
Ezawa T, Yamanoto K, Yoshida S (2000) Species composition and spore density of indigenous vesicular arbuscular mycorrhizal fungi under different condition of P- fertility as revealed by soybean trap culture. Soil Sci Plant Nutr 46(2):297
Feng GU, Yang MQ, Bai DS, Feng G, Yang MQ, Bai DS (1998) Influence of VAM Fungi on mineral elements concentration and composition in *Bromus inermis* under salinity stress. Acta-Prataculturae-Sinica 7:21–28
Georgiev GI, Atkias CA (1993) Effects of salinity on N_2 fixation, nitrogen metabolism and export and diffusive conductance of cowpea root nodules. Symbiosis 15:239–255
Gianinazzi S, Schüepp H, Barea JM, Haselwandter K (2002) Mycorrhizal technology in agriculture: from genes to bioproducts. BirkhäuserVerlag, Basel
Goas G, Goas M, Larher F (1982) Accumulation of free proline and glycine betaine in *Aster tripolium* subjected to a saline shock: a kinetic study related to light period. Physiol Plant 55:383–388
Goicoechea N, Antolín MC, Sánchez-Díaz M (1997) Influence of arbuscular mycorrhizaeand *Rhizobium* on nutrient content and water relations in drought-stressed alfalfa. Plant Soil 92:261–268
Giri B, Kapoor R, Mukerji KG (2007) Improved tolerance of *Acacia nilotica* to salt stress by arbuscular mycorrhiza, *Glomus fasciculatum*, may be partly related to elevated K^+/Na^+ ratios in root and shoot tissues. Microb Ecol 54:753–760
Grattan SR, Grieve CM (1992) Mineral element acquisition and growth response of plants grown in saline environment. Agric Ecosyst Environ 38:275–300
Gupta NK, Meena SK, Gupta S, Khandelwal SK (2002) Gas exchange, membrane permeability and ion uptake in two species of Indian Jujube differing in salt tolerance. Photosynthetica 40:535–539
Guriqbal S, Sekhon HS, Poonam S, Singh G, Sharma P (2001) Effect of *Rhizobium*, vesicular arbuscular mycorrhiza and phosphorus on the growth and yield of lentil (*Lensculinaris*) and field pea (*Pisumsativum*). Environ Ecol 19(1):40–42

Hajiboland R, Aliasgharzadeh N, Laiegh SF, Poschenreider C (2010) Colonization with arbuscular mycorrhizal fungi improves salinity tolerance of tomato (*Solanum lycopersicum* L.) plants. Plant Soil 331:313–327

Hasegawa PM, Bressnan RA, Zhu JK, Bohnert HT (2000) Plant cellular and molecular responses to high salinity. Annu Rev Plant Physiol Plant Mol Biol 51:463–499

Hayat S, Hasan SA, Yusuf M, Hayat Q, Ahmad A (2010) Effect of 28 homo brassinolide on photosynthesis, fluorescence and anti oxidant system in the presence or absence of salinity and temperature in *Vignaradiata*. Environ Exp Bot 69:105–112

Hirrel MC (1981) The effect of sodium and chloride salts on the germination of *Gigaspora margarita*. Mycologia 73:610–617

Hirrel MC, Gerdemann JW (1980) Improved growth of onion and bell pepper in saline soils by two vesicular-arbuscular mycorrhizal fungi. Soil Sci Soc Am J 44:654–665

Humphries EC (1956) Mineral components and ash analysis, determination of nitrogen. In: Peach K, Tracey MV (eds) Modern methods of plant analysis, vol 1. Springer, Berlin, pp 469–502

Jahromi F, Aroca R, Porcel R, Ruiz-Lozano JM (2008) Influence of salinity on the *invitro* development of *Glomus intraradices*and on the *in vivo* physiological and molecular responses of mycorrhizal lettuce plants. Microb Ecol 55(1):45–53

Jain M, Mathur G, Koul S, Sarin NB (2001) Ameliorative effects of proline on salt stress-induced lipid peroxidation in cell lines of groundnut (*Arachishypogaea* L.) Plant Cell Rep 20:463–468

Jaleel CA, Gopi R, Sankar B, Manivannan P, Kishore Kumar A, Sridharan R, Panneerselvam R (2007) Studies on germination, seedling vigour, lipid peroxidation and proline metabolism in *Cathranthusroseus* seedlings under salt stress. S Afr J Bot 73:190–195

Jeffries P (1987) Use of mycorrhiza in agriculture. Crit Rev Biotechnol 5:319–357

Jindal V, Atwal A, Sekhon BS, Singh R (1993) Effect of vesicular arbuscular mycorrhizae on metabolism of mung plants under sodium chloride salinity. Plant Physiol Biochem 3:475–481

Johnny LL (1999) Effects of interactions between arbuscular mycorrhizal fungi and *Rhizobium leguminosarum* on pea and Lentil. Ph.D.Thesis. The University of Saskatchwan, Canada-0780

Joner EJ (2000) The effect of long-term fertilization with organic and inorganic fertilizers on mycorrhiza mediated P uptake in subteraranean clover. Biol Fertil Soils 32:435–440

Juniper S, Abbott LK (2006) Soil salinity delays germination and limits growth of hyphae from propagules of arbuscular mycorrhizal fungi. Mycorrhiza 16:371–379

Kao WY, Tsai TT, Shih CN (2003) Photosynthetic gas exchange and chlorophyll *a* fluorescence of three wild soybean species in response to sodium chloride treatments. Photosynthetica 41:415–419

Lioyd J, Kriedemaan PE, Aspinall D (1989) Comparative sensitivity of Prior Lisbon lemon and Valencia orange trees to foliar sodium chloride concentrations. Plant Cell Environ 12:259–540

Little TM, Hills FC (1978) Agricultural experimentation. Wiley, New York

Lowrey OH, Rosenbrough NJ, Farr AL, Randall RJ (1951) Protein measurement with the folin phenol reagent. J Biol Chem 193:265–275

Mafakheri A, Siosemardeh A, Bahramneijad B, Struik PC, Sohrabi E (2010) Effect of drought stress on yield, proline and chlorophyll contents in three chickpea cultivars. Aust J Crop Sci 4(8):580–585

Maggio A, Miyazaki S, Veronese P, Fujita T, Ibeas JI, Damsz B, Narasimhan ML, Hasegawa PM, Joly RJ, Bressan RA (2002) Does proline accumulation play an active role in stress-induced growth reduction. Plant J 31:699–712

Mamtha G, Bagyarai DJ, Jaganath S (2002) Inoculation of field established mulberry and papaya with CAM fungi and mycorrhiza helper bacterium. Mycorrhiza 12:313–316

Manchanda G, Garg N (2011) Alleviation of salt-induced ionic, osmotic and oxidative stresses in *Cajanuscajan* nodules by AM inoculation. Plant Biosyst 145:88–97

Marschner H (1995) Mineral nutrition of higher plants, 2nd edn. Academic Press, San Diego

Mayoral MLI, Carballo O, Carreno L, de Mejia MG (2000) Effects of VA mycorrhizal inoculation on growth, yoeld, N and P nutrition of nodulating bean varieties in two soil substrates of contrasting fertility. J Plant Nutr 23(8):1117–1133

McMillen B, Juniper S, Abbott LK (1998) Inhibition of hypha growth of a vesicular-arbuscular mycorrhizal fungus in soil containing sodium chloride limits the spread of infection from spores. Soil Biol Biochem 30:1639–1646
Menconi M, Sgherri CLM, Pinzino C, Navari-izzo F (1995) Activated oxygen species production and detoxification in wheat plants subjected to a water deficit programme. J Exp Bot 46:1123–1130
Meshram AT, Jadhav AC, Konde BK, Wani PV (2000) Effect of VAM fungi and P-sources on nodulation, dry matter and yield of chickpea. J Maharashtra Agric Univ 25(1):99–101
Mohammadkhani N, Heidari R (2008) Drought-induced accumulation of soluble sugars and proline in two maize varieties. World Appl Sci J 3(3):448–453
Moisender PH, McClinton E, Paerl HW (2002) Salinity effects on growtphotosynthetic parameters and nitrogenase activity in estuarine planktonic Cyanobacteria. Microb Ecol 43:432–442
Munns R (2002) Comparative physiology of salt and water stress. Plant Cell Environ 25:239–250
Munns R (1993) Physiological processes limiting plant growth in saline soils: some dogmas and hypothesis. Plant Cell Environ 16:15–24
Murkute AA, Sharma S, Singh SK (2006) Studies on salt stress tolerance of citrus rootstock genotypes with arbuscular mycorrhizal fungi. Hortic Sci 33:70–76
Murphy KST, Durako MJ (2003) Physiological effects of short term salinity changes on Ruppiamaritima. Aquat Bot 75:293–309
Neumann E, George E (2009) The effect of arbuscular mycorrhizal root colonization on growth and nutrient uptake of two different cowpea (*Vignaunguiculata* L.) Walp. Genotypes exposed to drought stress. Emir J Food Agric 21:1–17
Ojala JC, Jarrell WM, Menge JA, Johnson ELV (1983) Influence of mycorrhizal fungi on the mineral nutrition and yield of onion in saline soil. Agron J 75:255–259
Parida A, Das AB, Das P (2002) Sodium chloride stress causes changes in photo synthetic pigments, proteins and other metabolic components in the leaves of a tree mangrove, *Bruguieraparviflora*, in hydroponic cultures. J Plant Biol 45:28–36
Poss JA, Pond EC, Menge JA, Jarrell WM (1985) Effect of salinity on mycorrhizal onion and tomato in soil with and without additional phosphate. Plant and Soil 88:307–319
Raja AR, Shah KH, Aslam M, Memon MY (2002) Response of phosphobacterial and mycorrhizal inoculation in wheat. Asian J Plant Sci 1(4):322–323
Ridley SM (1977) Interaction of chloroplast with inhibitors. Induction of chlorosis by diuron during prolonged illumination *in vitro*. Plant Physiol 59:724–732
Rozema J, Arp W, Van Diggelen J, Van Esbroek M, Broekman R, Punte H (1986) Occurrence and ecological significance of vesicular arbuscular mycorrhiza in the salt marsh environment. Acta Botanica Netherlandica 35:457–467
Ruiz-Lozano JM (2006) Physiological and molecular aspects of osmatic stressalleviation in arbuscularmycorrhizal plants. In: Rai M (ed) Hand book of microbial biofertilizers. Haworth Press, New York, pp 283–303
Sannazzaro AI, Echeverria M, Albertó EO, Ruiz OA, Menéndez AB (2007) Modulation of polyamine balance in *Lotus glaber* by salinity and arbuscular mycorrhiza. Plant Physiol Biochem 45:39–46
Sayeed OH (2003) Chlorophyll fluorescence as a tool in cereal crop research. Photosynthetica 41:321–330
Shahid MA, Pervez MA, Balal RM, Ahmad R, Ayyub CM, Abbas T, Akhtar N (2011) Salt stress effects on some morphological and physiological characteristics of okra (*Abelmoschusesculentus*, L.) Soil Environ 30(1):66–73
Sharifi M, Ghorbanli M, Ebrahimzadeh H (2007) Improved growth of salinity-stressed soybean after inoculation with pre-treated mycorrhizal fungi. J Plant Physiol 164:1144–1151
Sheekh-El MM, Omar HH (2002) Effect of high salt stress on growth and fatty acids content of unicellular green agae *Chlorella vulgaris*. Am J Microbiol 55:181–191
Sheng M, Tang M, Chan H, Yang B, Zhang F, Huang Y (2008) Influence of arbuscular mycorrhizae on photosynthesis and water status of maize plants under salt stress. Mycorrhiza 18:287–296

Sitaramaiah K, Rahulkhanna T, Trimuthulu N (1998) Effect of *Glomus fasciculatum* on growth and chemical composition of maize. Soil Microbe Plant Pathol 64:34–37
Sivaprasad P, Rai PV (1987) Mechanism of enhanced nodulation in vesicular – arbuscular mycorrhizal (VAM) pigeon-pea, *Cajanuscajan* (L) Millsp. Agric Res J Kerala 25:99–102
Smith SE, Read DJ (2008) Mycorrhizal symbiosis. Academic Press, San Diego
Strack D, Fester T, Hause B, Schliemann W, Walter MN (2003) Arbuscular mycorrhiza: biological, chemical and molecular aspects. J Chem Ecol 29:1955–1979
Tester M, Smith SE, Smith FA, Walker NA (1986) Effects of photon irradiation on the growth of shoots and roots, on the rate of initiation of mycorrhizal infection and on the growth of infection units in *Trifolium subterraneum*. New Phytol 103:375–390
Tian CY, Feng G, Li XL, Zhang FS (2004) Different effects of arbuscular mycorrhizal fungal isolates from saline or non-saline on salinity tolerance of plants. Appl Soil Ecol 26:143–148
Turnau K, Jurkiewicz A, Lingua G, Barea JM, Gianinazzi-Pearson V (2005) Role of arbuscular mycorrhiza and associated micro-organisms in phytoremediation of heavy metal polluted sites. In: Prasad MNV, Sajwan D, Ravi S (eds) Trace elements in the environment. Biogeochemistry biotechnology and bioremediation. New York, USA, CRC Press/Lewis Publishers. (in press)
Turnau K, Haselwandter K (2002) Arbuscular mycorrhizal fungi, an essential component of soil microflora in ecosystem restoration. In: Gianinazzi S, Schüepp H, Barea JM, Haselwandter K (eds) Mycorrhiza technology in agriculture: from genes to bioproducts. BirkhäuserVerlag, Basel, pp 137–149
Umbreit WW, Burris RH, Stauffer JF (1972) Method for nitrogen. In: Manometric and biochemical techniques, 5th edn. Burgess Publishing Company, Minnesota, pp 259–260
Van der Heijden MGA, Sanders IR (2002) Mycorrhizal ecology. Springer, Berlin/Heidelberg
Verbruggen N, Heramns C (2008) Proline accumulation in plants: a review. Amino Acids 35:753–759
Wang S, Wan C, Wang Y (2004) The characteristics of Na^+, K^+ and free proline distribution in several drought-resistance plants of the Alxa Desert, China. J Arid Environ 56:525–539
Yancy PH, Clark ME, Hand SC, Bowlus RD, Somero GN (1982) Living with water stress: evolution of osmolyte system. Science 217:1214–1122

Microbes and Crop Production

23

Priyanka Arora and Archana Tiwari

Abstract

Our environment includes a wide diversity of microbes that interact with plants in different ways. The range of interaction may be from two-partite symbiosis (nodule formation by legume-rhizobia interaction during N_2 -fixation) to multi-partite epiphytic as well as endophytic. Soil microbes find their major role in cycling of phosphorus, nitrogen, and other nutrients via production of certain exudates. Effective agronomic practices are associated with both beneficial and biocontrol mechanisms of microbes associated with symbiosis. To develop a microbe-based sustainable agriculture, fundamental knowledge of molecular biology, evolution, ecology, and genetics are needed. The employment of such relevant studies could lead to more prominent productivity as well as adaptive functions. The microbial-based crop production could lead to the replacement of hazardous chemicals with biofertilizers, thereby reducing the cost and enhancing the quality of agricultural products thus obtained.

23.1 Introduction

Plants are in continuous interaction with micro-organisms. Some interactions are beneficial while some are disadvantageous. The beneficiary role of micro-organisms still lacks exploration. This chapter deals with the exploration of various microbes with special reference to their role in enhancing crop production and hence agriculture, which may provide opportunities for future prospects.

Currently, there is a great need to provide proper nourishment to the continuously growing human population, and therefore an urgent need for techniques in agriculture that are eco-friendly as well as capable of enhancing agricultural products both quantitatively and qualitatively. Microorganisms can serve as an

P. Arora • A. Tiwari (✉)
School of Sciences, Noida International University, Greater Noida, India
e-mail: panarchana@gmail.com

V. Kumar et al. (eds.), *Probiotics in Agroecosystem*,
DOI 10.1007/978-981-10-4059-7_23

excellent alternative for traditional agricultural practices that are known to severely affect the agro-ecosystem balance, thereby damaging health severely. For instance, nitrogen fertilizers, which are used on a regular basis, lead to ground water contamination. Also, the chemicals used in crop protection persist in foods, which is a cause for great concern. The replacement of chemicals with micro-organisms has also been reported previously in agriculture (Dobbelaere et al. 2003; Burdman et al. 2000).

Some of the beneficial microorganisms include plant growth–promoting rhizobacteria (PGPR) and biological control agents (BCAs). These microbes play a key role in improving and enhancing agriculture (Whipps 1997; Raaijmakers et al. 2009; Hermosa et al. 2011). Microbes can be applied directly in the form of biofertilizers in order to protect and stimulate plant growth. Although it has been repeatedly stated that microbes such as fungi or bacteria play a major role in demolishing plant pathogens, thereby promoting plant growth, their extensive exploitation is still lacking in biotechnology as far as agriculture is concerned (Berg 2009).

The rhizosphere of plants serves as the major platform for soil-microbe interactions. The types of interaction that exist between the soil ecological environments may be symbiotic, parasitic, associative, or neutralistic. This depends upon the grade of plant nutrients available in soil, the defense mechanisms occurring in plants, and the kinds of microbes present in the soil near the rhizospheric-zone. As the microbes approach the epidermis the plant starts secreting some signal compounds as a means of protection against the rush of microbes that are continuously invading the root zone. This is the stage where microbes are differentiated into associative, pathogenic, or neutralistic forms in plants (Hayat et al. 2010).

Most commonly the signal molecules that are being produced as a response to microbial invasion involve flavones and flavonoids. These are secreted in the microbial rhizosphere. However, some of these compounds serve as antimicrobial agents by remaining attached to the cells of plants.

In the case of symbiosis occurring between the legume and *Rhizobium*, nitrogen-fixing nodules are induced in the roots of leguminous plants by *Rhizobium*, the rod-shaped soil bacterium. Various species of rhizobacteria, in particular *Pseudomonas* and *Bacillus*, have been found to play a major role in the colonization of plants and plant pathogen suppression (Table 23.1).

Dinitrogen is chemically inert and constitutes approximately 80% of the volume of the earth's atmosphere. Enzyme nitrogenase, which is secreted by bacteria, reduces dinitrogen to ammonia. To support the metabolism of proliferating microbial populations, plants provide a micro-aerobic environment for proper functioning of the nitrogenase enzyme, which is oxygen sensitive.

The bacteria in turn fix the atmospheric nitrogen and synthesize organic nitrogenous compounds as per the plant's biological needs. As this symbiotic association finds its importance in agriculture, continuous research has been aiming to increase the effectiveness of this symbiotic mechanism. Genetic manipulations are being continuously carried out so that this symbiosis can also be made applicable to other non-leguminous plants (Stacey et al. 1980; Fisher et al. 1985; Fisher and Long 1992).

Table 23.1 Important plant growth–promoting rhizobacteria (PGPR)

Plant growth promoting rhizobacteria (PGPR)	Agricultural crop	References
Comamonas acidovorans	Kiwi	Erturk et al. (2010)
Pseudomonas brassicacaerum, P. Marginali, P. oryzihabitans, P. putida, Alcaligenes xylosoxidans	Indian mustard and rape	Belimov et al. (2007)
Bacillus subtilis FB17	*Arabidopsis thaliana*	Rudrappa et al. (2008)
Bacillus cepacia strain OSU-7	Stored potatoes	Recep et al. (2009)
Bacillus subtilis, B. amyloliquefaciens IN 937, *Enterobacter cloaca*	*Arabidopsis* sp.	Ryu et al. (2003)
P. putida KD	Tomato and cucumber	Rezzonoco et al. (2005)
P. fluorescens CHA0	*Arabidopsis* sp.	Iavicoli et al. (2003)
P. fluorescens WCS 365	Tomato	Kamilova et al. (2006a)
Pseudomonas fluorescens PCL1606	Avocado	Cazorla et al. (2006)
Bradyrhizobium and PGPR	Mungbean	Shaharoona et al. (2006)
Bacillus cereus UW 85	Grain legumes	Vessey and Buss (2002)
Collimonas fungivorans	Tomato	Kamilova et al. (2008)
Agrobacterium amazonense	Rice	Rodrigues et al. (2008)
Pseudomonas BA-8, *Bacillus* OSU-142, *Bacillus M-3*	Strawberry	Pirlak and Kose (2009)

As per the research, several nodule-promoting rhizobacteria (NPR) and plant growth–promoting rhizobacteria (PGPRs) have been identified. These microbes are selected because of their capability to produce substances that promote plant growth. These microbes are found to produce siderophores which chelate cations that are insoluble, serving an associative function with plants. Also, these microbes induce the production of phytoalexins by the plants which create antibiosis for pathogenic forms prevailing in rhizosphere (Lifshitz et al. 1986; Halverson and Handelsman 1991).

The microbes involved in these types of interactions in particular include *Bacillus* and *Pseudomonas* species (Capper and Higgin 1993; Guaiquil and Luigi 1992; Parmar and Dadarwal 1997). This chapter focuses on beneficiary effects of bacteria prevailing in rhizospheric soil and also provides an insight into plant-microbial interactions. The chapter is aimed at covering different perspectives of soil beneficial microbes and different direct and indirect mechanisms involved in the promotion of plant growth. Further study on the mechanisms involved in such interactions will prove to be a key tool in enhancing sustainable agriculture in the near future.

23.2 Microbial Intervention

Microbial intervention refers to any action or process which leads to intervention of biological processes occurring in soil or in plants via microorganisms. Mostly this intervention is beneficial, leading to better availability of nutrients for plants, thereby enhancing plant growth and yield. Higher productivity could be attained

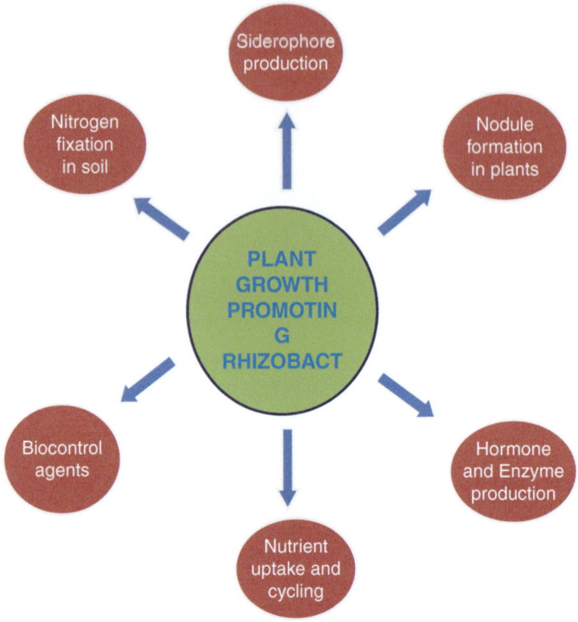

Fig. 23.1 Plant growth–promoting rhizobacteria (PGPRs) and their role in plants

through microbial intervention. As a result of this microbial intervention, various biological processes are enhanced, namely nutrient availability, atmospheric nitrogen availability, organic waste decomposition and recycling, and inorganic compounds leaching via microbes (Brierley 1985; Ehrlich 1990), repression of soil-borne pathogens, antibiotic production, heavy metal complexation to limit their uptake by plants, polysaccharide production to enhance the aggregation of soil, biodegradation of toxin-producing compounds like pesticides, simple organic compound production for uptake via plants, and many more (Fig. 23.1).

23.3 Plant Microbial Interaction

Microorganisms prevailing in soil continuously interact with roots of plants and constituents of the soil at the soil root junction. The heteromorphic biota present there obtain their nutrient requirements via root exudates and decaying plant biomass (Barea et al. 2005; Bisseling et al. 2009). Comparatively higher numbers of microbes are present in the rhizosphere (region in close vicinity of roots) and rhizoplane (root's external surface with soil particles and debris) than in soil devoid of plantation. The most probable reason for this may be lack of root exudates that serve as attractants for microbes. Germination of seeds leads to excretion of large amounts of nitrogen and carbon compounds, namely sugars, amino acids, vitamins, and organic acids, into the surrounding environment. A large number of microbial populations are attracted by these compounds, which may lead to vigorous competition between different species of microbes (Okon 1994). Also, the microbiomes prevailing in the rhizosphere differ between different species of plants (Bisseling et al. 2009).

Beneficial microbes are also known as biocontrol agents and growth promoters. These microbes adopt various modes of action for improving plant growth. These modes of action either directly or indirectly exert positive effects on plants.

The indirect positive effects of microbes on plants include the depression or alleviation of population dynamics, population density, and other metabolic activities of pathogens inhabiting the soil. This occurs mainly through competition, lysis, antibiosis, and hyperparasitism. At the root surface, competition takes place for nutrients and space. Some antagonistic microbes start producing lytic enzymes and a range of secondary metabolites, and serve as antimicrobial agents. In *Trichoderma*, hyperparasitism has been well cited involving the secretion of enzymes such as chitinase and cellulase, pathogen contact, hyphae coiling, digestion of the cell wall with the help of enzymes, and finally penetration.

However, rhizosphere microorganisms exert direct positive effects on plants through biofertilization of plants and phytostimulation. The major processes involved in this may be phytohormone production, increase in phosphate availability, and nitrogen fixation non-symbiotically (Burdman et al. 2000). PGPRs secrete several compounds that are toxic or harmful to pathogens like HCN, pyrrolnitrin, phenazines, and proluteosin. In addition to this, several antibiotics, enzymes, phytohormones, and metabolites that are toxic to pathogens are the means through which PGPRs act.

In a similar fashion, chemotaxis and quorum sensing are also essential for colonization of the rhizosphere (Castro-Sowinski et al. 2007; Ramette et al. 2011; Jousset et al. 2011). PGPRs produce siderophores, which are large molecular-weight compounds. Under conditions where iron availability is limited, these siderophores in a competitive way confiscate iron, thus making it unavailable for pathogenic fungi (Pedraza et al. 2007).

23.4 General Microbial Growth Promoters

Several microbes have the potential to promote plant growth. They help to enhance the vigor of plants as well as their establishment (Fig. 23.2, Table 23.2).

23.5 Biofertilizers for Specific Nutrients

Biofertilizers are microbes promoting growth of plants. These microbes have the capability to solubilize phosphate, fix N_2, and/or produce siderophores. These biofertilizers act via increasing the nutrient availability to plants (Fuentes-Ramirez and Caballero-Mellado 2006).

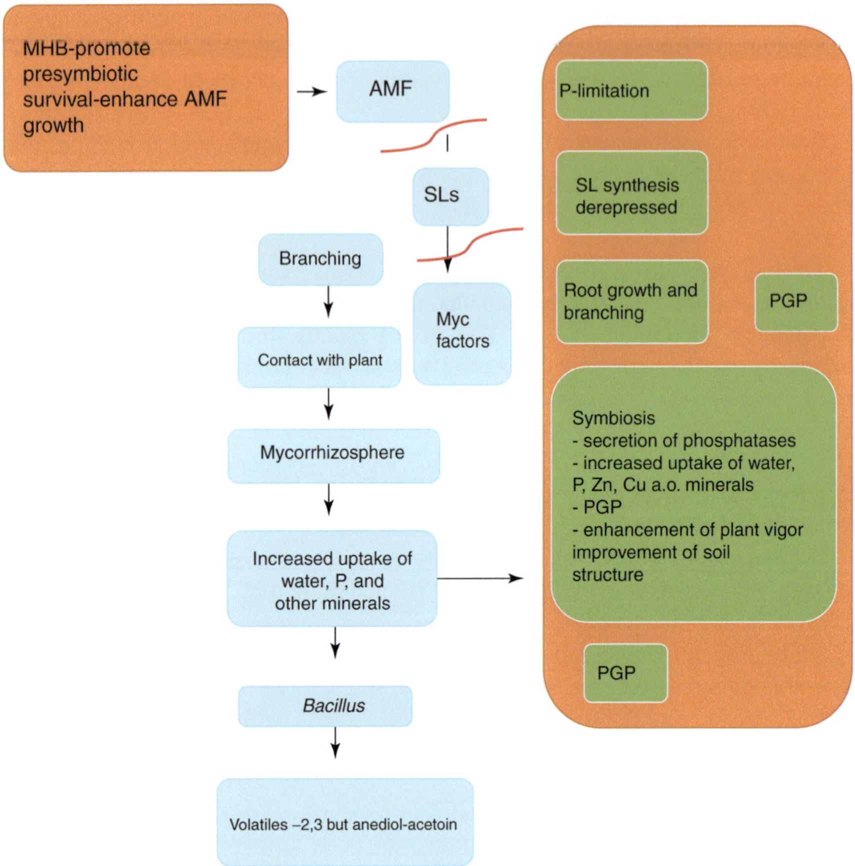

Fig. 23.2 Plant growth promotion and role of Arbuscular Mycorrhizal fungi

23.5.1 Nitrogen Fixation

Nitrogen is one of the most abundant elements present on earth but plants cannot take up nitrogen as such. It can only be utilized in the form of ammonium or nitrate ions.

Diazotrophs or biological nitrogen fixation refers to the conversion of atmospheric nitrogen to ammonium ions. Prokaryotes have the immense ability to fix N_2 (Dekas et al. 2009). These bacteria have the potential to establish symbiotic associations with plants. The symbiotic association of *Rhizobium* with leguminous plants that induces nodule formation is the best-studied example (Fig. 23.3). Genus *Parasponia* of the Rosales family is the only exception that is nodulated by rhizobia (Markmann and Parniske 2009). This symbiotic association is considered as the major source of nitrogen fixation in plants. It has been reported in the terrestrial ecosystem that per year more than 45 million metric tons of N are contributed by this symbiosis between *Rhizobia* and leguminous plants (Vance

Table 23.2 Plant growth–promoting (PGP) microbes and their significant PGP traits

Trait	Reference
(a) *Pseudomonas* spp.	
Cytokine producer	Garca' de Salome et al. (2001)
Auxin producer	Kamilova et al. (2006a)
ACC deaminase producer	Glick et al. (2007)
Phosphate solubilizer	Rodriguez et al. (2006)
Associative N-fixer	Dobbelaere et al. (2003)
Siderophore producer	Lemanceau et al. (2009)
(b) AMF (Arbuscular Mycorrhizal fungi)	
Promotion of growth of fungus and pre-symbiotic survival by Mycorrhiza helper bacteria (MHB)	Frey-Klett et al. (2007)
Stmulation of root growth by Mycorrhiza factors secreted by AMF	Maillet et al. (2011)
Protection against (a) biotic stresses and improvement of soil structure	Smith and Read (2008)
Uptake of P, Cu, water, Zn, and other nutrients	Clark and Zeto (2000)
(c) *Bacillus* spp.	
Release of Pi from phytate	Idriss et al. (2002)
N_2-fixers	Borriss (2011)
Potassium solubilizer	Wu et al. (2005)
Phosphate solubilizer	Borriss (2011), Rodriguez et al. (2006)
(d) *Trichoderma* spp.	
Auxin production	Contreras-Cornejo et al. (2009)
Degradation of phenolic compounds secreted by plants	Ruocco et al. (2009)
Can perform as endophyte; increases water and nutrient uptake; increases soil nutrient solubilization; nitrogen use efficiency, and plant vigor enhancement; development of above-ground plant parts and roots; root hair formation; causes deeper rooting; photosynthetic efficiency improvement; sucrose usage	Hermosa et al. (2012), Harman (2006), Shoresh et al. (2010), Lorito et al. (2010)
Seed germination acceleration	Mastouri et al. (2010)
The secondary metabolite harzianic acid promotes plant growth	Vinale et al. (2009)
Abiotic stress amelioration and alleviation of physiological stresses	Shoresh et al. (2010), Mastouri et al. (2010)
Increase of plant resistance, especially under suboptimal growth conditions	Lorito et al. (2010)

2001). Also, Hurek et al. (2002) reported that *Azoarcus sp.*, endophytic diazotrophic bacteria, have their beneficial effects directly associated with their ability to fix nitrogen. This is also case in sugarcane with *Acetobacter diazotrophicus* (Sevilla et al. 2001).

CROP	PASTURES and FODDER
Plant associated microbes	Plant associated microbes
• Legume- rhizobia (symbiotic)	• Cereal endophytic bacteria
• Cereal endophytic bacteria	• Cereal associative bacteria
• Cereal associative bacteria	• Legume- rhizobia (symbiotic)
• Azolla-cyanobacteria (symbiotic)	
Free living microbes	Free living microbes
• Heterotrophic bacteria	• Heterotrophic bacteria
• Cyanobacteria	• Cyanobacteria
• Autotrophic bacteria	• Autotrophic bacteria

Fig. 23.3 Biological nitrogen fixation and microbes

23.5.2 Phosphate Solubilization

Phosphorus is the third most important compound after water and nitrogen, and plays a key role in various metabolic activities of plants, including synthesis of nucleic acids, respiration, photosynthesis, generation of energy, and cellular signaling (Vance et al. 2003). Plants can only take up phosphorus in $H_2PO_4^-$ and HPO_4^{2-} ions. Although soil contains sufficient phosphorus needed for plant growth, the majority of forms available are not soluble and hence cannot be taken up by plants (Fig. 23.4). Also, the phosphorus supplied in the form of chemical fertilizers is rapidly converted to insoluble forms and hence is made unavailable to plants (Rodriguez and Frago 1999; Igual et al. 2001; Smyth 2011). P-limitation leads to rhizosphere acidification, which is the result of secretion of organic anion, particularly with citrate and oxalate, together with proton. The phosphorus is hence facilitated for further mobilization (Richardson et al. 2009). Some bacteria aid in phosphate solubilization and hence make it available to plants. These are known as phosphorus-solubilizing bacteria (Igual et al. 2001; Kim et al. 1998; Lipton et al. 1987).

Mineral phosphates serve as major factors for the release of phosphates via production of organic acids (Rodriguez et al. 2006). Vyas and Gulati (Vyas and Gulati 2009 reported that *Pseudomonas* spp., which play a major role in phosphate solubilization, have the potential to enhance growth parameters as well as phosphorus content in maize plants. Sundara et al. (2002) reported that the yield as

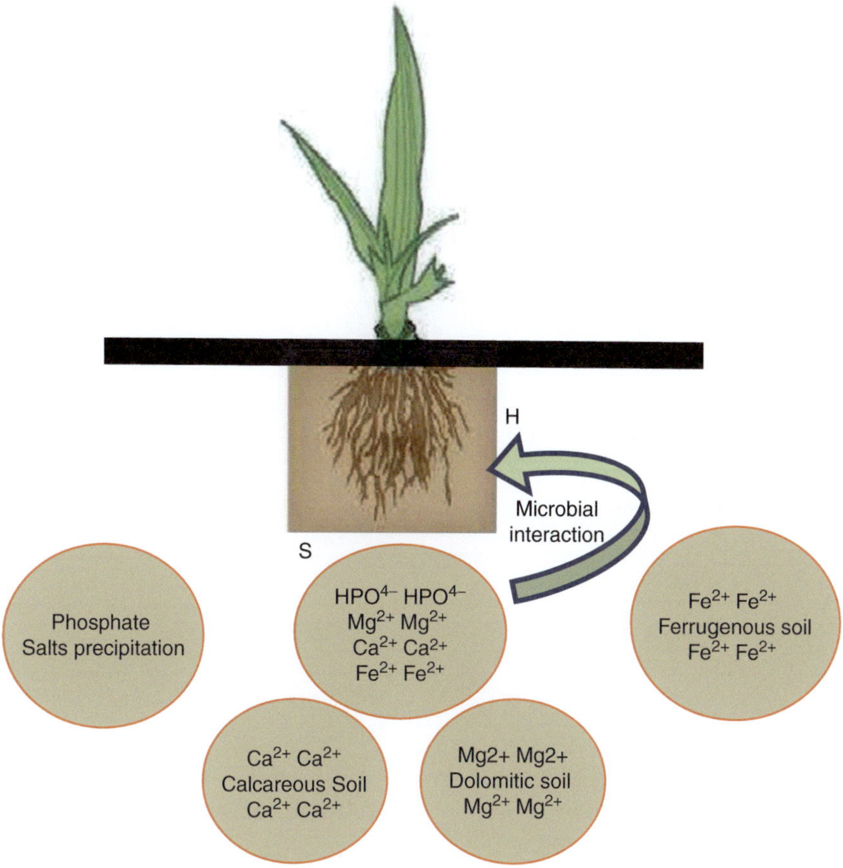

Fig. 23.4 Conventional chemistry of phosphate and microbes

well as available phosphorus in sugarcane was enhanced by *Bacillus megaterium*. An increase in yield was also observed in canola through the use of phosphorus-solubilizing *Bacillus* spp. (de Freitas et al. 1997).

23.5.3 Siderophores

Iron is a critical element in all living organisms. Although it is found in abundance in the earth's crust, it is mostly present in insoluble forms and therefore cannot be utilized by plants. In response to this, plants secrete siderophores, which are metal chelators. These siderophores bind to Fe^{3+} and are transported to the surface of roots where Fe^{3+} is reduced to Fe^{2+}, which can now be taken up by plants. By binding to siderophores, these Fe^{3+} ions can also be taken up by plants as Fe^{3+}-siderophore complex (Lemanceau et al. 2009; Fig. 23.5).

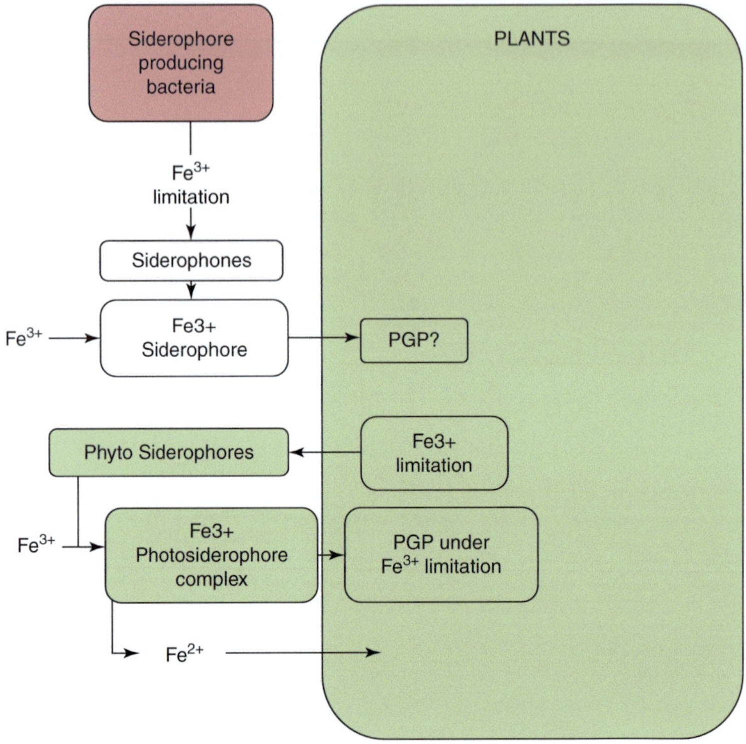

Fig. 23.5 Feasible contribution of microbes towards plant-Fe nutrition

Bacteria also have the potential to produce a variety of siderophores under low Fe^{3+} concentration. These siderophores bind to Fe^{3+} ions with high affinity. Plants can take up this bacterial Fe^{3+}- siderophore complex via absorption. However the significant role of this uptake is still unclear (Zhang et al. 2008).

23.5.4 Conclusions and Future Aspects

Interactions of rhizosphere microflora in ecosystems with plants vary or rely on establishment of close associations between the two associates. Research on a few of these associations, such as between the symbiotic rhizobia bacteria and leguminous plants, has established that this adjacent collaboration between bacteria and plant exhibit high levels of host specificity. Moreover, there is also a growing body of signals that suggest that many other relations between plants and microbiomes exhibit the same degree of specification, with dissimilar plant species, and even dissimilar and diverse cultivars of same plant species. This establishes a dissimilar and discrete microbial population in their rhizospheric zone when grown in the same soil. Formation of these societies depends on, at least in part, the activation of precise and explicit programs of gene expression or statement in the microbiome in response to the chemical signals excreted from plants. A relevant and appropriate example is the initiation of nodulation genes in receptive rhizobia, which are then triggered by manufacture and excretion of the specific flavonoids by plants in any ecosystem. In the case of the rhizobia legume plant communication, plants also respond to the bacteria-produced signals, and it is probable that this kind of chemical signal exchange is distinctive and characteristic of other plant-microbe communications.

We understand that the world's population is projected to double by end of 2033. The food demand in Asia is expected to surpass the capacity for production by the end of 2010. This poses a serious challenge to existing agricultural systems. Conventional farming tools and practices are becoming obsolete as they reach their limits of efficacy in intensifying agricultural productivity. As countries advance, people are demanding nutritionally rich, healthy, and fresh food. However, these burdens are burgeoned by decreasing farmland, growing labor costs, and scarcity of farm workforces. Microbial technology offers added tools and methods to augment the sustainability of the prevailing system to produce a higher, improved, and healthier quality of our agricultural produce. Potential reimbursements of microbial and plant technology are many and abundant, and also include providing endurance to crop pests and enhancing crop yield by minimizing chemical pesticide practice. Processing of foodstuff and food constituents using microbial technology provides an extensive and varied diversity of fermented food products and food constituents that are widely consumed. Microbial technologies employed in agriculture systems that ensured a "green revolution" in the middle of the 20th century led to the production of healthy food with low economic investment. Comprehensive and extensive application of potential microbes for crop production is a very important and significant step toward sustainable agriculture. Consequently, microbial-based technology and its applications in sustainable agriculture development and safe environmental health are achieving greater responsiveness. The purpose of this chapter was to additionally prioritize the significance and meaning in the scientific community among students and researchers about the importance of potential microbes in agriculture.

References

Barea JM, Pozo MJ, Azcón R, Azcón-Aguilar C (2005). Microbial cooperation in the rhizosphere. J Exp Bot 56:1761–1778

Belimov AA, Dodd IC, Safronova VI, Hontzeas N, Davies WJ (2007) *Pseudomonas brassicacearum* strain Am3 containing 1-aminocyclopropane-1-carboxylate deaminase can show both pathogenic and growth-promoting properties in its interaction with tomato. J Exp Bot 58:1485–1495

Berg G (2009) Plant–microbe interactions promoting plant growth and health: perspectives for controlled use of microorganisms in agriculture. Appl Microbiol Biotechnol 84:11–18

Bisseling T, Dangl JL, Schulze-Lefert P (2009) Next-Generation Communication. Science 324:691–691

Borriss R (2011) Use of plant-associated *Bacillus* strains as biofertilizers and biocontrol agents in agriculture. In: Maheshwari DK (ed) Bacteriain agrobiology: plant growth responses. Springer, Berlin/Heidelberg, pp 41–76

Brierley JA (1985) Use of microorganisms for mining metals. In: Halvorson HO, Pramer D, Rogul M (eds) Engineered organisms in the environment: scientific issues. ASM Press, Washington, DC, pp 141–146

Burdman S, Jurkevitch E, Okon Y (2000) Recent advance in the use of plant growth promoting rhizobacteria (PGPR) in agriculture. In: Subba Rao NS, Dommergues YR (eds) Microbial interaction in agriculture forestry, vol II. Science Publishers, Enfield, pp 229–250
Capper AL, Higgin KP (1993) Application of Pseudomonas fluorescens isolates to wheat as potential biological control agents against take-all. Plant Pathol 42:560–567
Castro-Sowinski S, Herschkovitz Y, Okon Y, Jurkevitch E (2007) Effects of inoculation with plant growth-promoting rhizobacteria on resident rhizosphere microorganisms. FEMS Microbiol Lett 276:1–11
Cazorla FM, Duckett SB, Bergstrom FT, Noreen S, Odik R et al (2006) Biocontrol of avocado Dematophora root rot by the antagonistic *Pseudomonas fluorescens* PCL 1606 correlates with the production 2-hexyl-5-propyl resorcinol. Mol Plant-Microbe Interact 19:418–428
Clark RB, Zeto SK (2000) Mineral acquisition by arbuscular mycorrhizal plants. J Plant Nutr 23:867–902
Contreras-Cornejo HA, Macias-Rodriguez L, Cortés-Penagos C, López-Bucio J (2009) *Trichoderma virens*, a plant beneficial fungus, enhances biomass production and promotes lateral root growth through an auxin-dependent mechanism in *Arabidopsis*. Plant Physiol 149:1579–1592
de Freitas JR, Banerjee MR, Germida JJ (1997) Phosphate-solubilising rhizobacteria enhance the growth and yield but not phosphorous uptake of canola (*Brassica rapus* L.) Biol Fertil Soils 24:358–364
Dekas AD, Poretsky RS, Orphan VJ (2009) Deep-sea archaea fix and share nitrogen in methane-consuming microbial consortium. Science 326:422–426
Dobbelaere S, Vanderleyden J, Okon Y (2003) Plant growth promoting effects of diazotrophs in the rhizosphere. Crit Rev Plant Sci 22:107–149
Ehrlich HL (1990) Geomicrobiology, 2nd edn. Dekker, New York, p 646
Erturk Y, Ercisli S, Haznedar A, Cakmakci R (2010) Effects of plant growth promoting rhizobacteria (PGPR) on rooting and root growth of kiwifruit (*Actinidia deliciosa*) stem cuttings. Biol Res 43:91–98
Fisher RF, Long SR (1992) Rhizobium-plant signal exchange. Nature 357:655–660
Fisher RF, Tu JK, Long SR (1985) Conserved nodulation genes in *Rhizobium meliloti* and *Rhizobium trifolii*. Appl Environ Microbiol 49:1432–1435
Frey-Klett P, Garbaye J, Tarkka M (2007) The mycorrhizae helper bacteria revisited. New Phytol 176:22–36
Fuentes-Ramirez L, Caballero-Mellado J (2006) Bacterial biofertilisers. In: Siddiqui ZA (ed) PGPR: biocontrol and biofertilization. Springer, Dordrecht, pp 143–172
Garcá de Salome IE, Hynes RK, Nelson LM (2001) Cytokinin production by plant growth promoting rhizobacteria and selected mutants. Can J Microbiol 47:404–411
Glick BR, Cheng Z, Czarny J, Duan J (2007) Promotion of plant growth by ACC deaminase-producing soil bacteria. Eur J Plant Pathol 119:329–339
Guaiquil VH, Luigi C (1992) Plant growth promoting rhizobacteria and their effect on rapeseed (*Brassica napus L.*) and potato seedlings. Microbiol Rev 23:264–273
Halverson LJ, Handelsman J (1991) Enhancement of soybean nodulation by *Bacillus cereus* UW85 in the field and in a growth chamber. Appl Environ Microbiol 57:2767–2770
Harman GE (2006) Overview of mechanisms and uses of *Trichoderma* spp. Phytopathology 96:190–194
Hayat R, Safdar Ali S, Amara U, Khalid R, Ahmed I (2010) Soil beneficial bacteria and their role in plant growth promotion: a review. Ann Microbiol 60:579–598
Hermosa R, Viterbo R, Chet I, Monte R (2012) Plant-beneficial effects of *Trichoderma* and of its genes. Microbiology 158:17–25
Hermosa R, Botella L, Alonso-Ramírez A, Arbona V, Gómez-Cadenas A, Monte E, Nicolás C (2011) Biotechnological applications of the gene transfer from the beneficial fungus *Trichoderma harzianum* spp. to plants. Plant Signal Behav 6(8):1235–1236
Hurek T, Handley LL, Reinhold-Hurek B, Piche Y (2002) Azoarcus grass endophytes contribute fixed nitrogen to the plant in an unculturable state. Mol Plant-Microbe Interact 15:233–242
Iavicoli A, Boutet E, Buchala A, Metraux JP (2003) Induced systemic resistance in Arabidopsis thaliana in response to root inoculation with *Pseudomonas fluorescens* CHA0. Mol Plant-Microbe Interact 16:851–858

Idriss EE, Makarewicz O, Farouk A, Rosner K, Greiner R, Bochow H, Richter T, Borriss R (2002) Extracellular phytase activity of *Bacillus amyloliquefaciens* FZB45 contributes to its plant-growth promoting effect. Microbiology 148:2097–2109

Igual JM, Valverde A, Cervantes E, Velaquez E (2001) Phosphate solubilizing bacteria as inoculants for agriculture: use of updated molecular techniques in their study. Agronomie 21:561–568

Jousset A, Rochat L, Lanoue A, Bonkowski M, Keel C, Scheu S (2011) Plants respond to pathogen infection by enhancing the antifungal gene expression of root-associated bacteria. Mol Plant-Microbe Interact 24:352–358

Kamilova F, Kravchenko LV, Shaposhnikov AI, Azarova T, Makarova N, Lugtenberg BJJ (2006a) Organic acids, sugars, and Ltryptophane in exudates of vegetables growing on stonewool and their effects on activities of rhizosphere bacteria. Mol Plant-Microbe Interact 19:250–256

Kamilova F, Kravchenko LV, Shaposhnikov AI, Makarova N, Lugtenberg B (2006b) Effects of the tomato pathogen *Fusarium oxysporum f.* sp. *radicis-lycopersici* and of the biocontrol bacterium *Pseudomonas fluorescens* WCS365 on the composition of organic acids and sugars in tomato root exudate. Mol Plant-Microbe Interact 19:1121–1126

Kamilova F, Lamers G, Lugtenberg B (2008) Biocontrol strain *Pseudomonas fluorescens* WCS365 inhibits germination of *Fusarium oxysporum* spores in tomato root exudate as well as subsequent formation of new spores. Environ Microbiol 10:2455–2461

Kim KY, Jordan D, McDonald GA (1998) *Enterobacter agglomerans*, phosphate solubilizing bacteria and microbial activity in soil: effect of carbon sources. Soil Biol Biochem 30:995–1003

Lemanceau P, Bauer P, Kraemer S, Briat JF (2009) Iron dynamics in the rhizosphere as a case study for analyzing interactions between soils, plants and microbes. Plant Soil 321:513–535

Lifshitz R, Kloepper JW, Scher FM, Tipping EM, Laliberte M (1986) Nitrogen-fixing pseudomonads isolated from roots of plants grown in the Canadian High Arctic. Appl Environ Microbiol 51:251–255

Lipton DS, Blanchar RW, Blevins DG (1987) Citrate, malate and succinate concentration in exudates from P-sufficient and P-stressed *Medicago sativa* L. seedlings. Plant Physiol 85:315–317

Lorito M, Woo SL, Harman GE, Monte E (2010) Translational research on *Trichoderma*: from 'omics' to the field. Annu Rev Phytopathol 48:395–417

Maillet F, Poinsot V, André O, Puech-Pages V, Haouy A, Guenier M, Cromer L et al (2011) Fungal lipochitooligosaccharide symbiotic signals in arbuscular mycorrhiza. Nature 469:58–63

Markmann K, Parniske M (2009) Evolution of root endosymbiosis with bacteria: how novel are nodules? Trends Plant Sci 14:77–86

Mastouri F, Björkman T, Harman GE (2010) Seed treatment with *Trichoderma harzianum* alleviates biotic, abiotic, and physiological stresses in germinating seeds and seedlings. Phytopathology 100:1213–1221

Okon Y, Labandera-Gonzales CA (1994) Agronomic application of Azospirillum: an evaluation of 20 years worldwide field inoculation. Soil Biol Biochem 26:1591–1601

Parmar N, Dadarwal KR (1997) Rhizobacteria from rhizosphere and rhizoplane of chick pea (*Cicer arietinum L.*) Indian J Microbiol 37:205–210

Pedraza R, Motok J, Tortora M, Salazar S, Díaz-Ricci J (2007) Natural occurrence of *Azospirillum brasilense* in strawberry plants. Plant Soil 295:169–178

Pirlak M, Kose M (2009) Effects of plant growth promoting rhizobacteria on yield and some fruit properties of strawberry. J Plant Nutr 32:1173–1184

Raaijmakers JM, Paulitz TC, Steinber C, Alabouvette C, Moënne-Loccoz Y (2009) The rhizosphere: a playground and battlefield for soilborne pathogens and beneficial microorganisms. Plant Soil 321:341–361

Ramette A, Frapolli M, Fischer-Le Saux M, Gruffaz C, Meyer JM, Défago G, Sutra L, Moënne-Loccoz Y (2011) *Pseudomonas protegens* sp. nov., widespread plantprotecting bacteria producing the biocontrol compounds 2, 4-diacetylphloroglucinol and pyoluteorin. Syst Appl Microbiol 34:80–88

Recep K, Fikrettin S, Erkol D, Cafer E (2009) Biological control of the potato dry rot caused by *Fusarium species* using PGPR strains. Biol Control 50:194–198

Rezzonoco F, Binder C, Defago G, Moenne-Loccoz Y (2005) The type III secretion system of biocontrol *Pseudomonas fluorescencs* KD targets the phytopathogenic chromista *Pythiu ultimum* and promotes cucmuber protection. Mol Plant-Microbe Interact 9:991–1001

Richardson AE, Barea J-M, McNeill AM, Prigent-Combaret C (2009) Acquisition of phosphorous and nitrogen in the rhizosphere and plant growth promotion by microorganisms. Plant Soil 321:305–339

Rodrigues EP, Rodrigues CS, de Oliveira ALM, Baldani VL, Teixeira da Silva JA (2008) *Azospirillum amazonense* inoculation: effects on growth, yield and N2 fixation of rice (*Oryza sativa L.*) Plant Soil 302:249–261

Rodriguez H, Fraga R, Gonzalez T, Bashan Y (2006) Genetics of phosphate solubilisation and its potential applications for improving plant growth-promoting bacteria. Plant Soil 287:15–21

Rodriguez H, Frago R (1999) Phosphate solubilising bacteria and their role in plant growth promotion. Biotechnol Adv 17:319–339

Rudrappa T, Czymmek KJ, Pare PW, Bais HP (2008) Root-secreted malic acid recruits beneficial soil bacteria. Plant Physiol 148:1547–1556

Ruocco M, Lanzuise S, Vinale F, Marra R, Turra D, Woo SL, Lorito M (2009) Identification of a new biocontrol gene in *Trichoderma viride*: the role of an ABC transporter membrane pump in the interaction with different plant–pathogenic fungi. Mol Plant-Microbe Interact 22:291–301

Ryu CM, Farag MA, Hu CH, Reddy MS, Wei HX, Pare PW, Kloepper JW (2003) Bacterial volatiles promote growth in Arabidopsis. Proc Natl Acad Sci U S A 100:4927–4932

Sevilla M, Burris RH, Gunapala N, Kennedy C (2001) Comparison of benefit to sugarcane plant growth and 15N2 incorporation following inoculation of sterile plants with *Acetobacter diazotrophicus* wild type and Nif-mutant strains. Mol Plant-Microbe Interact 14:358–366

Shaharoona B, Arshad M, Zahir ZA (2006) Effect of plant growth promoting rhizobacteria containing ACC-deaminase on maize (Zea *mays L.*) growth under axenic conditions and on nodulation in mung bean (*Vigna radiata L.*) Lett Appl Microbiol 42:155–159

Shoresh M, Harman GE, Mastouri F (2010) Induced systemic resistance and plant responses to fungal biocontrol agents. Annu Rev Phytopathol 48:1–23

Smith SE, Read DJ (2008) Mycorrhizal symbiosis, 3rd edn. Elsevier/Academic Press, Amsterdam

Smyth E (2011) Selection and analysis of bacteria on the basis of their ability to promote plant development and growth. PhD Thesis, University College Dublin, Dublin

Stacey G, Paau AS, Brill WJ (1980) Host recognition in the rhizobium-soybean symbiosis. Plant Physiol 66:609–614

Sundara B, Natarajan V, Hari K (2002) Influence of phosphorus solubilising bacteria on the changes in soil available phosphorus and sugarcane and sugar yields. Field Crop Res 77:43–49

Vance CP, Ehde-Stone C, Allan DL (2003) Phosphorous acquisition and use: critical adaptations by plants for screening a renewable resource. New Phytol 157:423–447

Vance CP (2001) Symbiotic nitrogen fixation and phosphorus acquisition: plant nutrition in a world of declining renewable resources. Plant Physiol 127:390–397

Vessey JK, Buss TJ (2002) Bacillus cereus UW85 inoculation effects on growth, nodulation, and N accumulation in grain legumes. Controlled environmental studies. Can J Plant Sci 82:282–290

Vinale F, Flematti G, Sivasithamparam K, Lorito M, Marra R, Skelton BW, Ghisalberti EL (2009) Harzianic acid, an antifungal and plant growth promoting metabolite from *Trichoderma harzianum*. J Nat Prod 72:2032–2035

Vyas P, Gulati A (2009) Organic acid production *in vitro* and plant growth promotion in maize under controlled environment by phosphate-solubilizing fluorescent *Pseudomonas*. BMC Microbiol 9:174–189

Whipps JM (1997) Ecological considerations involved commercial development of biological control agents for soil-borne disease. In: Van Elsas JD, Trevors JT, Wellington EMH (eds) Modern soil microbiology. Marcel Dekker, New York, pp 525–533

Wu SC, Cao ZH, Li ZG, Cheung KC, Wong MH (2005) Effect of biofertiliser containing N-fixer, P and K solubilizers and AM fungi on maize growth: a greenhouse trial. Geoderma 125:155–166

Yaacov Okon, Carlos A. Labandera-Gonzalez, (1994) Agronomic applications of azospirillum: An evaluation of 20 years worldwide field inoculation. Soil Biology and Biochemistry 26 (12):1591–1601

Zhang H, Sun Y, Xie X, Kim MS, Dowd SE, Pare PW (2008) A soil bacterium regulates plant acquisition of iron via deficiency-inducible mechanisms. Plant J 58:568–577

Probiotic Microbiome: Potassium Solubilization and Plant Productivity

24

Priyanku Teotia, Vivek Kumar, Manoj Kumar, Ram Prasad, and Shivesh Sharma

Abstract

The rhizosphere of plant roots supports a range of potassium-solubilizing microbes (KSMs). These KSMs solubilize the insoluble and unavailable potassium (K) to forms of K available for uptake and transport by the plant. Potassium is one of the unavoidable elements required for growth and yield. The specific rhizospheric microbes that perform the process of K solubilization include both bacteria and fungi, the foremost of which are: *Bacillus* sp. (*B. Mucilaginosus*, *B. megaterium*, *B. globisporus*, *B. edaphicus) Pseudomonas putida*, *Enterobacter hormaechei*, *Acidothiobacillus ferrooxidans*, *Paenibacillus* sp., and *Arthrobacter* sp.) *Aspergillus terresus*, *Fusarium oxysporum*, *Aspergillus fumigatus*, and *Aspergillus niger*. Agricultural soil particulates hold minerals such as illite, biotite, orthoclase, mica, and feldspar that contain potassium; however, this is not accessible to plants due to its immobilized form. In soil chemistry, after N and P, K is an important element; a major role is played by the rhizosphric microbes in mobilizing the inaccessible form of K to the roots of the plant. The rhizospheric K-solubilizing microbes such as *Bacillus*, *Pseudomonas*, and *Aspergillus* expel organic acids, which solubilize the insoluble K and make it available to plant roots. Most of the research work in this area has been conducted on nitrogen fixing and phosphate-solubilizing microbes. Solubilized K (quickly available) in

P. Teotia
Department of Botany, CCS University, Meerut, India

V. Kumar (✉)
Himalayan School of Biosciences, Swami Rama Himalayan University, Jolly Grant, Dehradun, Uttarakhand, India
e-mail: vivekbps@gmail.com

M. Kumar • R. Prasad
Amity Institute of Microbial Technology, Amity University, Noida, Uttar Pradesh, India
e-mail: manojjnu@gmail.com; rprasad@amity.edu

S. Sharma
Department of Biotechnology, Motilal Nehru National Institute of Technology, Allahabad, Uttar Pradesh, India
e-mail: ssnvsharma@gmail.com

V. Kumar et al. (eds.), *Probiotics in Agroecosystem*,
DOI 10.1007/978-981-10-4059-7_24

addition to the existing biofertilizers needs additional consideration at a profitable scale. The current chapter presents information to fill the knowledge gaps about K-solubilizing/mobilizing microorganisms in soil, and looks at the current and future facets of K-solubilizing microbes for enhanced crop production.

24.1 Introduction

The letter K is used to symbolize potassium; it is taken from the German word "Kalium." For a long time, locals used to prepare soap by burning organic matters and wood in vessels. The ashes left after burning wood and other materials were rinsed and the residue left behind after the evaporation of the rinse water consisted of potassium salts. The remainder was popularly acknowledged as potash or "pot ashes." The left-over salts were mixed with animal fat and boiled to manufacture soap. Samuel William Jackson, a botanist from Connecticut, examined the ash of burned organic matters and wood. It was observed that there was a great amount of potassium in various parts of plants, besides other minor and macro minerals.

Potassium (K) is an important and indispensable nutrient for the growth of plants. A vast amount of K is absorbed by the roots of the plant for growth and development and it is therefore classified as a macronutrient. K plays a key role in activation of enzymes, synthesis of protein, photosynthesis, and production. With the ever-increasing extent of rigorous and extensive agriculture, the K levels of soil have been depleted due to leaching, plant uptake, soil erosion, and water runoff (Li et al. 2006). Therefore, it is necessary to build up various alternative sustainable biological methods that can efficiently diminish the loss. K is a fundamental element that is allied with transportation of water, nutrients, and carbohydrates in plant tissues. If K is lacking or insufficient in the soil, the growth of plants is stunted and yield is reduced. Approximately 95% of K fertilizers are available in the form of muriate of potash, which is also known as potassium chloride. For crops that are unable to withstand it, chloride-free salts are used, such as potassium sulfate and potassium nitrate.

Diverse research shows that K encourages early growth, rise in protein production along with improved efficiency of water use, is essential for prolonged existence, affords winter hardiness, and increases resistance to diseases and insects. It is also essential for plant cells in high measure and possesses vital biochemical and physiological functions relating to cell osmotic regulation and activation of enzymes (Hasanuzzaman et al. 2013). The use of microbes as an option for biological processes to influence the release of K from rocks and minerals in the soil is an unconventional view (Rogers et al. 1998). There are a number of sets of microbes, for instance bacteria, fungi and yeast, which are capable in solubilizing unavailable K restricted in rock and soil minerals through mineralization (Sugumaran and Janarthanam 2007; Magri et al. 2012; Meena et al., 2014). The discharge of K from the soil and rock minerals is largely commenced by the release of organic acids which are produced by the microbes as they survive and proliferate in the rhizosphere. The organic acids produced by rhizospheric microbes include oxalic acid, malic acid, formic acid, and citric acid. The organic acids provide protons and form complexes with Ca^{2+} ions in soil, and thus enhance solubilization of the K ions in the soil system.

Sheng and He (2006) have revealed that organic compounds excreted by microbes, for example citrate, acetate, and oxalate, can enhance mineral solubilization in the soil. The complex formation between various metal ions like calcium, aluminum, and iron and organic acids also enhances K solubilization (Uroz et al. 2009).

Over the last few decades the knowledge of rhizosphere biology has increased greatly with the discovery of a significant and specific collection of microbes, acknowledged as plant growth–promoting microbes (PGPMs). The plant root system inhabits the PGPM, which implement valuable and affirmative effects on plant growth using various means (Ahemad and Kibret 2014). In addition, the use of K-mobilizing microbes (KMMs) as bioinoculants unaccompanied or accompanied by other microbes, has been shown to improve plant growth (Wu et al. 2005). In a phytotron growth chamber wheat and maize yields increased with the use of KMMs such as *Bacillus mucilaginosus*, *Azotobacter chroococcum*, and *Rhizobium* spp., as shown by Gupta et al. (2015). The outcome revealed that the assimilation of K was considerably enhanced by both maize and wheat by the application of KMMs, wherever waste mica was the only resource of K. Under abiotic or biotic stress and nutrient-imbalance conditions, KMMs have been found to be significant organisms for plant nourishment, root establishment, root escalation archetype, and competitiveness (Zhanga and Konga 2014). The use of KMMs in agriculture can significantly reduce the use of agrochemicals and maintain eco-friendly production of crops (Sheng et al. 2002; Pettigrew 2008). Diverse KMMs together with associative bacteria and fungi, for example *Paenibacillus*, *Azospirillum*, *Bacillus*, *Pseudomonas*, *Azotobacter*, *Enterobacter*, and *Aspergillus*, have been used for their favorable results on plant growth (Archana et al. 2013; Diep and Hieu 2013; Zhang et al. 2013). KMMs improve plant growth and development through a variety of means, but the exact and precise mechanisms involved are still not properly described (Shanware et al. 2014). The KMMs have been shown to candidly boost plant augmentation by diverse techniques like solubilization of minerals (Argelis et al. 1993; Valmorbida and Boaro 2007) and synthesis of phytohormones (Kumar and Narula 1999; Kumar et al. 2012). Direct enhancement of plant growth by solubilization and the mobilization of minerals due to amplification of the precise ion fluxes by KMMs present on the surface of roots has also been reported (Sheng et al. 2008; Meena et al. 2014). Microbes in soil and in plant rhizospheres play a key role in the natural K cycle and solubilizing of K (Zörb et al. 2014).

24.2 Soil and Potassium

Potassium is an essential, fundamental, and indispensable macronutrient found in soil. It has an important function in growth, metabolism, and the development of the plant. Plants with insufficient K will have poorly developed root systems, retarded growth, produce diminutive seeds, and have smaller yields. Although K comprises of about 2.52% of the top layer of the earth's crust, the tangible sum of this nutrient fluctuates from 0.04% to 3.0% in the soil (Blake et al. 1999; Lopo de SáI et al. 2014). The plants gain K from the soil of the rhizosphere and the accessibility of K depends upon the quantity present in the soil and its dynamics. There are usually three types of K in soil that are available to plants. Foremost is the readily unavailable form, minerals such as mica and feldspar contain most of the K, depending on soil type. These minerals are the basis of about 90–98% of K that exists in the soil (Memon et al. 1988; Zörb et al. 2014). The K is liberated at a slow rate to the more available forms as these break down. The second type is the slowly available K form, which make up 1–10% of K in soil and forms the colloidal portion of the interlayer of K in non-expanded clay minerals such as illite and lattice K. Potassium-feldspars in the soil contribute significantly to plant uptake (Sheng et al. 2008; Zörb et al. 2014). The slowly available K form is also recognized as "non-exchangeable" potassium; it cannot be reinstated by customary cation exchange processes. The third form is readily available K, which includes water-soluble K and transferable K in the soil. It is absorbed on the soil colloid surfaces and is freely available to plants (Maathuis and Sanders 1997; Blake et al. 1999). Despite this, higher plants obtain the majority of their K from the soil solution fraction. The liberation of unavailable K to the less available and easily available water-soluble form takes place when levels of exchangeable/available K or solution K are reduced via crop runoff, erosion, exclusion, or leaching, as shown in Fig. 24.1.

The quantity of accessible and inaccessible K differs from soil to soil and the active balance interaction between the different pools of K in soil. Thus, the fixation and discharge of K from mineral soils are influenced by numerous physical and chemical characteristics of the soil in addition to the plant interactions and soil microbial community. Potassium is taken up in greater measure by the plant than

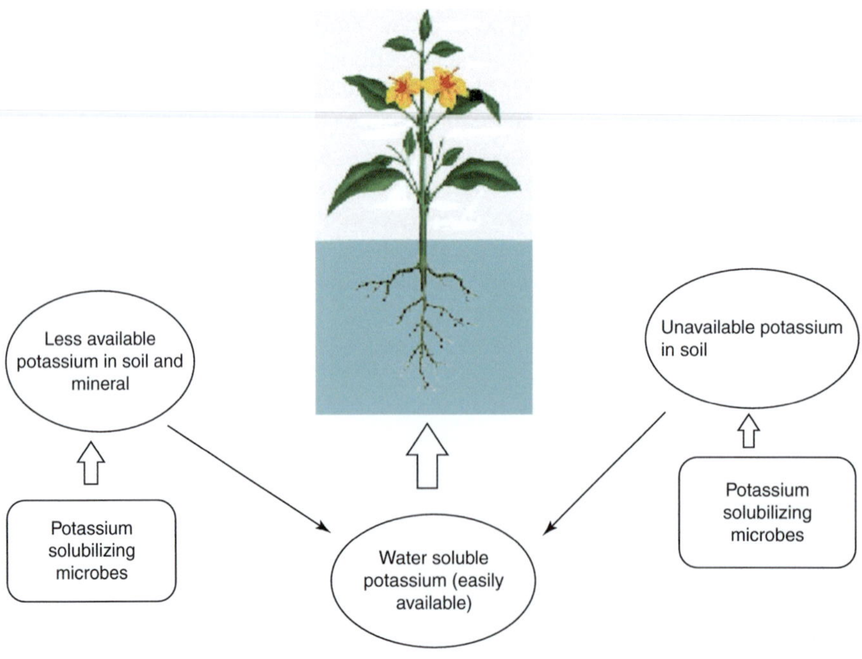

Fig. 24.1 The less available, water-soluble, and unavailable K in soil. Microbes play an important role in K mobilization

other vital elements except for N. The plant roots absorb the mineral nutrients that are dissolved in the water in soil. Nevertheless, the sum of nutrients in soil is always random and is not adequate for the growth of plants. Potassium comprises about 2.1% of the earth's crust and is therefore the seventh most plentiful element. Although the soil K treasury in the structural form is large in the soil (Sardans and Peñuelas 2015), vast areas of agricultural land in the world are still found to have poor availability of K, including 75% of the paddy soils in China and 66% of the wheat belt in southern Australia. This scarcity is due to the sluggish release compared to the requirement of K by the crop. Biofertilizer is a superior means of conveying this prime macronutrient to plants with the aid of K-solubilizing microorganisms (KSMs), which switches the inaccessible K to available K. With the employment of high-yielding varieties of crops, hybrids, and other progressive amplifications in agriculture, the K reserves of the soil are becoming exhausted at a quicker pace. Furthermore, K deficiency is also becoming one of the chief restraints in crop production due to excessive fertilizer application. This has placed emphasis on the search to unearth a remarkable indigenous supply of K for plant uptake and to maintain the K level in the soil for sustainable agriculture (Supanjani et al. 2006; Sardans and Peñuelas 2015). Soil microbes have been found to play a major role in natural K cycles and, thus, K-solubilizing microbes present in soil could provide a substitute system for formulating K available for plant uptake (Mikhailouskaya and Tcherhysh 2005). Therefore, detection of microorganisms that are able to solubilize K minerals quickly can safeguard the existing resources and shun environmental pollution perils caused by harmful application of chemical fertilizers.

Many soil bacteria such as *Acidothiobacillus ferrooxidans*, *Burkholderia*, *Pseudomonas*, *Bacillus mucilaginosus*, *Bacillus edaphicus*, *B. circulans*, and *Paenibacillus* sp. can liberate K from minerals and provide it to plants (Han et al. 2006; Andrist-Rangel et al. 2007). It is reported that the K-solubilizing bacteria leach out organic acids, siderophores, and hydrogen ions found in mobilizing K from minerals like illite, feldspar, and micas (Li 1994; Liu 2001; Lian et al. 2002; Liu et al. 2012). Some crops, for instance wheat (Mikhailouskaya and Tcherhysh 2005, Pettigrew 2008; Singh et al. 2010), sorghum (Basak and Biswas 2010; Gopalakrishnan et al. 2013) cucumber and pepper (Han et al. 2006), eggplant (Han and Lee 2005), soybean and cotton (Pettigrew 2008), rape and cotton (Xeng 2005), rice (Gopalakrishnan et al. 2013), and maize (Singh et al. 2010; Abou-el-Seoud and Abdel-Megeed 2012) have been boosted with K-solubilizing microbial isolates, and have generated motivating and convincing results. Likewise, a bacterium recognized as *Paenibacillus glucanolyticus* strain IISRBK2 that has huge potential to solubilize potash was isolated from the rhizosphere of the black pepper plant. This bacterial strain was also reviewed for growth and K uptake by black pepper in the soil. It was administered with 0.5, 1, and 1.5 g K kg^{-1} soil in pot, where the source of K was wood ash, which contained 53.1 g kg^{-1} K. In view of this, K-mobilizing microbes are being extensively engaged as bioinoculants in a number of countries where K is undersupplied or less accessible to plants in the agricultural soils (Gundala et al. 2013; Zarjani et al. 2013; Diep and Hieu 2013). Hence, application of K-mobilizing microbiomes as bioinoculants for superior crop production possibly will lessen the use of chemical fertilizers to uphold and prolong crop production (Sheng et al. 2002; Singh et al. 2010).

Presently, very little is known regarding KSMs and their effectiveness, mechanism of solubilizing K, making it accessible to plant roots, and lastly affecting plant growth structure in a range of agro-climatic conditions (Shanware et al. 2014). Sheng and Huang (2002) showed that pH, oxygen concentration, and the types of bacterial strains engaged influenced K expulsion from soil minerals. The efficacy of K solubilization by various microbes was observed to alter according to different environmental conditions and types of minerals. There is information regarding K solubilization by a *Bacillus* sp. in the liquid medium that showed that K mineral illite exhibited added growth as compared to feldspar (Sheng and He 2006). Therefore, there could be enormous potential for added and augmented crop production via application of K-bearing rock materials with K-solubilizing microbes as probiotic agents.

24.3 Potassium and Plant Productivity

Potassium is crucial for many plant processes. Its function encompasses the fundamental physiological and biochemical activities of plants. K is taken up by plants in larger amounts as compared to several other mineral elements apart from nitrogen and, in a number of cases, calcium (Bahadur et al. 2014). K is required in large quantities for a crop to attain its utmost yield. K assists in the building up of proteins and sugar, boosts photosynthesis, improves fruit quality, and reduces the incidence of diseases (Wang et al. 2013). It also encourages activation of the enzyme and nitrogen (N) utilization. K is supplied to plants with soil minerals, organic resources, and fertilizer. Plants are able to absorb K only through the soil or as water-soluble K. K deficiency in plants causes yellowing of leaf edges, giving the plant a burned facade. It could also be a reason for slow growth and for imperfect root growth of the plant. A plant growing in soil devoid of ample K produces small seeds and has smaller yields (Sparks and Huang 1987). Despite the fact that K is not a fundamental part of the chemical structure of plants, it plays many vital authoritarian roles in the natural growth and development of plants. To bring about every chemical reaction, enzymes act as the catalysts without being exploited and consumed in the reaction process. The element K "triggers" a minimum of 60 different enzymes that are involved in the overall plant growth, development, and yield. K modifies the physical characters of the enzyme molecule and exposes the chemically suitable efficient sites for the reaction. Diverse organic anions in addition to organic as well as inorganic compounds are too neutralized by K, inside the cells of a plant, resulting in stabilization of the pH of the plant cell between 7 and 8, which is most favorable for the majority of enzymatic reactions (Zörb et al. 2014). The amount of K present within the cell determines the number of enzymes that can be activated and the pace at which a chemical reaction can advance. Thus, the rate of a specific chemical reaction is administered by the swiftness at which the K ions enter or leave the cell cytoplasm. K plays a significant role opening and closing of the leaf stomata in plants (Armengaud et al. 2009). Appropriate functioning of stomatal opening and closing is obligatory for a lot of plant processes like photosynthesis, water transport, nutrient uptake, and also plant cooling. Upsurge of the K ions in the roots of a plant creates a gradient of the osmotic pressure so as to absorb water molecules into the roots. The insufficiency of K ions in plants leads to stress conditions and less water absorption (Sparks and Huang 1987; Wang et al. 2013).

During K deficiency in plants, the rate of generation of ATP molecules and photosynthesis is turned down, and the majority of the processes within the cell are ATP dependent. Consequently, the cellular activities are also slowed down. ATP is also required for the transportation of carbohydrates that are produced during the process of photosynthesis to different parts of the plant via phloem for usual growth, consumption, and storage (Bahadur et al. 2014). This transport system of plants utilizes energy in the shape of the ATP. If K is insufficient, not as much ATP is available, and the plant's transport system breaks down (Arquero et al. 2006). This leads the photosynthates to assemble in leaves, which reduces the pace of photosynthesis. Since K is a requisite for almost all key steps of protein synthesis like the "reading" of the genetic code in plant cells, which leads to the manufacturing of proteinaceous enzymes that control all growth and developmental processes, these processes would be unfeasible if the cells are deficient in K (Wang et al. 2013). Plants that are deficient in K are unable to synthesize proteins regardless of the presence of N in large amounts. Instead, amino acids, amides, and nitrates that are the "resources" or precursors of protein accumulate in the cells (Britzke et al. 2012).

In plants, K is a fairly mobile element and is transported from older to younger leaves. Consequently, K deficiency indications characteristically occur initially on the lower older leaves of the plant, and advance to the upper younger leaves, in accordance with the increasing severity of the K deficiency. The commonly prevalent and worldwide signs and indicators of K deficiency is yellow chlorosis or yellow scorching along the length of the leaf margin (Sparks and Huang 1987). In heightened and severe cases, the yellow and dried margins of the leaf may fold over. Conversely in crops with wide leaves such as cotton, soybeans, and banana, the entire leaf can be cast off, which results in untimely defoliation of the plant. Severe K deficiency in wheat and other cereal crops may cause a slowed growth rate, poorly developed roots, weak stalks, and undersized grain of poor quality. Death of frequent winter crops such as alfalfa and grasses may also occur in conditions of insufficient K (Dordas 2008) (Table 24.1).

K deficiency and K fertilizer deficiency in soil has become a vital limiting reason for the growth and sustainability of the agriculture system (Sheng et al. 2002). Escalating employment of chemical fertilizers in cultivation makes countries self reliant in food production but it depreciates the environment and ecosystem due to the harmful impacts on living organisms. Inadequate uptake of these chemical fertilizers by plants results in their leaching or discharge into water bodies through rain or irrigation water, which causes eutrophication in the water bodies and has a significant effect on living organisms together with plant growth–inhibiting microbes (Uroz et al. 2009). The excess use of chemical fertilizers in farming is expensive and also has a range of unfavorable effects on soils such as depleting water-holding capacity, poor soil fertility, and inconsistency in soil nutrients. For some time now, efforts have been made to build up various economical, effectual, and eco-friendly fertilizers which work without distressing effects on the ecosystem. Currently, various species of microbes are extensively employed that have exceptional assets in

Table 24.1 Percent nutrient content in potassium (K) fertilizers

Chemical formula	K_2O	Element	N	S
$K_2CO_3KHCO_3$	<68	Potassium carbonate		
$K_2SO_42MgSO_4$	22	Potassium magnesium sulfate		22
K_2SO_4	50–52	Potassium sulfate		18
$KNa(NO_3)_2$	14	Potassium sodium nitrate	15	
KPO_3	38	Potassium metaphosphate		
KCl	60–62	Potassium chloride		
KOH	83	Potassium hydroxide		
$K_4P_2O_7$	22–48	Potassium polyphosphate		
$KH_2PO_4K_2HPO_4$	30–50	Potassium orthophosphate		
KNO_3	44	Potassium nitrate	13	

terms of natural products, and provide the same results as a good chemical fertilizer. The surplus application of chemical fertilizer can increase expenses, reduce the effectiveness of K fertilizer, and eventually harm the environment (Zhang et al. 2013). A substitute for the synthetic K fertilizer is essential for the sustainable improvement of agriculture. It is anticipated that by the year 2020, to realize the target production of nearly 321 million tons of food grain, the nutrient requirement will be 28.8 million tons, whereas the accessibility will be just 21.6 million tons, i.e. a deficit of about 7.2 million tons (Swaminathan and Bhavani 2013). There is a growing concern about environmental hazards and increasing threat to sustainable agriculture because of diminishing soil fertility resulting in an amplifying gap between nutrient elimination and supplies. Above and beyond these issues, the extensive use of biofertilizers is not hazardous to the environment, is inexpensive and more efficient and productive, and is within reach for small-time farmers compared to chemical fertilizers (Subba Rao 2001).

24.4 Microbial Mechanism for Potassium Solubilization

The mechanism of K solubilization signifies the means through which the insoluble K and structural inaccessible forms of K complexes are mobilized and solubilized due to the excretion of a wide range of organic acids by microbes. These acids undergo a sequence of exchange reactions like acid lysis and complex lysis. In addition, these reactions are input processes which cause the alteration of insoluble forms of K into soluble forms. The KSMs produce organic acids, which enables them to dissolve K from insoluble minerals like micas, orthoclases, and illite (Shanware et al. 2014). These organic acids may either directly solubilize rock K or excrete the chelated silicon ions to turn K into a solution form that is available to plants. Microbial organic acids enhance the mobilization of K compounds by offering protons and also by chelating with Ca^{2+} ions present in the soil (Singh et al. 2010). Organic compounds produced by microorganisms such as oxalate, citrate, and acetate can improve mineral dissolution in soil (Sheng et al. 2003). In another study by Styriakova et al. (2003) it was reported that K solubilization ensues through the configuration of complexes among organic acids and metal ions such as iron, calcium, and aluminum. Microbial arbitrated organic chelates, ligands, and other metabolic derivatives like excreted enzymes and simple or complex molecules of organic acids enhance the solubilization of the aluminosilicate (usually quartz) minerals in *in vitro* and *in situ* environments (Zeng et al. 2012). The solubilization of K within feldspar and illite is enhanced by production of microbial organic acids like oxalic and tartaric acid (Sheng and He 2006). A study by Groudev (1987) revealed that solubilization of K by inorganic and organic acid production is also supported by the production of mucilaginous casing made up of exopolysaccharides formed by bacteria like *Bacillus* sp., *Clostridium* sp., and *Thiobacillus* sp. Sugumaran and Janarthanam (2007) have reported an analogous feasible method of K solubilization wherein *Bacillus mucilaginosus* was examined for K solubilization. During the period of bacterium inoculation, there was no decrease in pH of the medium, implying that *Bacillus* sp. did not excrete organic or inorganic acids and slime formation by bacterium could possibly be responsible for K solubilization. The soil microbiome involved in mineral weathering produces organic and inorganic acids, protons, chelates, siderophores, and ligands. Similar potential has been reported in fungal species like *Cladosporium*, *Aspergillus*, and *Pencillium*. These have been found to excrete enormous amounts of citric acid, gluconic acid, and oxalic acids in *in vitro* conditions, causing the mobilization of silicates, mica, and feldspar in soil (Lian et al. 2008). Similarly, Yang et al. (2014) reported the leaching of K from minerals containing K-rich shale was caused by the formation of biofilm by bacteria growing on it. These biofilms were made up of acids, protein, and polysaccharides produced by bacteria.

The ability of K-solubilizing microbes to solubilize insoluble K has been quantitatively examined by various researchers. The media generally used for quantitative assessment of solubilization of K by KSM is Aleksandrov medium. It contains 0.5% mica as a source of an insoluble form of K, besides 1% glucose, 0.05% $MgSO_4.7H_2O$, 0.0005% $FeCl_3$, 0.01% $CaCO_3$, 0.2% $CaPO_4$ and agar 3%, with pH adjusted to 6.5. The dishware are incubated at 28 ± 2 °C for 2–3 days and bacterial colonies displaying clear zones are selected and the diameter of the solubilization zone can then be measured (Kumar and Narula 1999; Prajapati et al. 2013).

$$\text{Ratio} = \text{Diameter of zone of clearance} \,/\, \text{Diameter of growth}$$

The quantitative evaluation of K solublization is done using flame photometry or atomic absorption spectrophotometry, in which the supernatant obtained after centrifugation of culture broth is used for precipitation of cobalt nitrite. Potassium chloride is employed as the standard for the quantification of K (Hu et al. 2006).

24.5 Contemporary State of K Solubilization and Crop Production

Potassium replenishment in soil depends heavily on the application of synthetic fertilizer; however, these products have a noteworthy negative effect on the environment. The use of K-solubilizing microbes (KSMs) as inoculants has potential, as these KSMs transform insoluble forms of K in the soil into a soluble form which is accessible to plants. This embodies a competent strategy for enhancement of K absorption by the plant, in addition to reducing the use of chemical fertilizer (Zhanga and Konga 2014). Many workers have reported the significant role played by potassic biofertilizers in agriculture, especially for enhancing soil fertility, yield-attributing characters, and thereby final yield (Basak and Biswas 2010; Archana et al. 2013; Mikhailouskaya and Tcherhysh 2005). Additionally, the application of potassic biofertilizers to soil improves the soil microbiota and reduces the soil compactness and application of chemical fertilizers. It is well recognized that, although the Indian soil is a rich source of K within secondary minerals, it is not easily accessible to the plant. This could be made possible by application of K-solubilizing microbes that could make K available to plants. Consequently, inoculatng soil with K-solubilizing bacteria and application of additional beneficial microbial inoculants become obligatory to reinstate and uphold soil fertility.

The use of an efficient microbiome for fixed K solubilization, mobilization, and accessibility of other micro- and macronutrients to obtain superior and sustainable yield. The growth-enhancing microbiome found in the rhizosphere of the plant implements beneficial effects on the uptake of nutrients and their mobilization together with various other means such as nitrogen fixation, siderophores, HCN and phytohormone production, and mobilization of micro- and macronutrients minerals like K, iron, phosphorous, copper, and zinc. The total K amount in Indian soil is by and large adequate to sustain the crop production and growth. The reachable amount and allocation pattern of this mineral element varies in different types of soil in various regions, which results in K unavailability for plant uptake (Zörb et al. 2014). The microbes replenish the root zone environment by liberating accessible and transferable forms of K, which accomplish the basic requirements of a plant. The restoration of mineral nutrients in the rhizospheric zone is vital and a necessity for higher crop yields, and the deliberate inoculation of microbes or those naturally present play a very important part (Sheng et al. 2002). Researchers have established that KSMs performed nicely with several crops under a diverse agro-climatic environment. Supanjani et al. (2006) illustrated that K-solublizing bacteria amplified photosynthesis by 16% in hot pepper *Capsicum annuum* L. along with increasing leaf area by 35%, in contrast to the control plants. Moreover, biomass and fruit production of the treated plants were also enhanced by 23% and 30%, respectively, compared to control plants. Researchers also established that there is a similar effect on plants when treated with either phosphorus or K rocks along with phosphorus/potassium-solubilizing bacterial strains or with a usual, soluble fertilizer. Sheng (2005) studied the effect of *Bacillus edaphicus* NBT (K-releasing bacterial strain) on cotton and rape for plant growth–promoting effects and nutrient uptake by plants in K-deficient pot soil. The experiment showed that inoculation with the bacterial strain *B. edaphicus* NBT increased the root and shoot growth of cotton and rape plants. Strain NBT was also capable of mobilizing K competently in both crop plants after the addition of illite to the soil. There was also an increase in the K content by 30% and 26% in cotton and rape, respectively, when grown in soil treated with insoluble K and inoculated with the bacterial strain NBT. The inoculation also resulted in elevated N and P contents of above-ground parts of the plant. Furthermore, the bacterial strain was also able to inhabit and proliferate in the rhizosphric soil of cotton and rape after root inoculation.

Basak and Biswas (Basak and Biswas 2008) investigated the effectiveness of K-solubilizing bacteria (*Bacillus mucilaginosus*) on Sudan grass (*Sorghum vulgare* Pers.) var. Sudanensis as a test crop grown in two Alfisols. Results demonstrated that the use of mica appreciably improved biomass yield, mineral uptake, and percent K recoveries by Sudan grass as compared to control plants (without KSM). Furthermore, when the mica was inoculated with the bacterial strain in both the soils, there was an additional boost in biomass yield, K uptake, and percent K recoveries in contrast to those soils without the application of KSM-inoculated mica. Another study conducted by Zhanga and Konga (Zhanga and Konga 2014) on 27 K-solubilizing strains revealed that among them, 17 strains were from *Klebsiella variicola*, two strains each from *Enterobacter cloacae* and *Enterobacter asburiae*, and the other six strains belonged to *Agrobacterium tumefaciens*, *Enterobacter aerogenes*, *Microbacterium foliorum*, *Pantoea agglomerans*, *Burkholderia cepacia*, and *Myroides odoratimimus*, respectively. *Klebsiella variicola* showed the highest frequency of occurrence with 17 strains. A greenhouse pot experiment was conducted using four K-solubilizing bacterial isolates, GL7, JM3, XF4, and XF11, for determination of the K-solubilizing capabilities. The tobacco seedlings were treated with the four KSM strains to observe the effectiveness of K-solubilizing isolates; it was found that the treatment significantly enhanced K and N uptake and plant dry weight. This increase was further elevated with the application of a combination of K-solubilizing bacterial inoculation along with feldspar powder. Isolate XF 11 exhibited the most prominent and advantageous effect on tobacco seedling plant growth and nutrient (K and N) uptake. Therefore, a potential substitute to commercial chemical K fertilizer that will possibly help to maintain the viability of soil nutrients could be a combination of KSM with the addition of K feldspar powder.

Recently, Prajapati et al. (2013) studied the effects of KSMs *Enterobacter hormaechei* and *Aspergillus terreus* (a fungal strain) on Okra (*Abelmoscus esculantus*) grown in pot soil deficient in K. Results showed that the *Enterobacter hormaechei* enhanced shoot and root growth of the plant. Furthermore, with the application of feldspar into the pot soil, both the microbes were able to mobilize K in the Okra plant. The K content was increased in Okra plants when the pot soils were modified with insoluble K and inoculated with *Enterobacter hormaechei* and *Aspergillus terreus*. Likewise, Han et al. (2006) considered the outcome of bacterial KSM *Bacillus mucilaginosus* on pepper and cucumber plants. The experiment confirmed that coinoculation of phosphate-solubilizing bacteria (PSB) and the KSM resulted in elevated P and K availability in contrast to the control which did not have bacterial inoculum and rock fertilizer. The inoculation of PSB with phosphorus incorporated-rock enhanced the accessibility of P and K in the soil, boosted N, P, and K uptake by shoots and root, and increased the biomass of pepper and cucumber plants. Comparable but less prominent results were attained when rock K and KSM were applied concurrently. Hassan et al. (2010) measured the efficacy of *Bacillus circulans*, a KSM, on *Ammi visnaga* (Khella) augmentation. The plant growth parameters were enhanced by the inoculation of KSMs in conjunction with feldspar. Sugumaran and Janarthanam (2007) measured the K solubilization efficiency of isolated K-solubilizing bacteria (*Bacillus mucilagenosus*). The maximum K solubilization was found to be 4.29 mg l^{-1}. Another striking research effort by Bagyalakshmi et al. (2012) exhibited an improvement in productivity and nutrient uptake in tea plants inoculated with K-mobilizing bacteria (*Pseudomonas putida*). Tea excellence factors like aflavin, arubigin, highly polymerized substances, sum liquor color, caffeine, vigor, color of leaf, and flavor indexes were enhanced to a great extent in plants treated with K-solubilizing bacteria.

24.6 Future Aspects of K Mobilizing Microbes

The prime nutrients for the development and growth of crop plants include minerals like N, P, and K. Haphazard employment of chemical fertilizers for the sustenance of crop plants is a major cause of contamination and infertile soil, in addition to causing pollution of water basins and destruction of microbes, which results in poor soil health. On the other hand, application of biofertilizers is an eco-friendly approach for the replenishment of nutrients to the soil for the sustainable growth of plants. Complex and elaborate transactions between the KSM, potential rhizospheric microorganisms, the roots of a plant, and the surrounding ecosystem are responsible for the mobilization of rock K and its inconsistency in uptake have a large effect on plant growth and development. Potential and probable approaches include cloning of genes liable for K solubilization in the genome of those microbes that have additional advantageous functions, for instance, exceptional proliferation and endurance in the rhizosphere, nitrogen-fixing ability, phosphate-solubilizing capacity, and production of biocontrol metabolites and phytohormones. Furthermore, the effectiveness of the KSMs can be amplified by development of superior culture techniques and deliverance protocols that sustain their existence in the rhizospheric zone. The amalgamation and utilization of plant growth endorsing microbes with varied beneficial functions embracing the ability for K and P solubilization is an ecologically favorable and enhanced approach as it may result in improved endurance, propagation, and superior adaptation to varied agroclimatic fluctuations that occur throughout the plantation period. An additional beneficial prospect could be the recent approaches incorporating application of molecular biology along with techniques for exploitation of useful microbial functions that would enhance K-solubilization capability. Moreover, the commercial utilization of these superior microbes as K bioinoculants will further increase crop productivity and growth for feasible, sustainable, and enduring agriculture. Further and extended studies that focus on such comparable issues related to other existing micro- and macronutrient elements in soil, particularly in rhizospheric soil, and an account of the microbial molecular means of plant nutrition uptake will definitely assist in increasing our knowledge about the development of improved microbial inoculants to augment the K requirement of the plant.

Conclusions

Evidently, the application of synthetic fertilizers and organic manures cannot be diminished radically or eradicated at this stage without a considerable decrease in food production. Concurrently, there are toxic after-effects on the environment from the use of chemical fertilizers like the increasing dead zones in marine ecosystems throughout the world. These cannot be ignored in the long term, as this will result in devastating effects on the ecological balance. Therefore, there is an urgent need for an integrated approach of nutrient management that would endeavor to reduce agricultural inputs along with decreasing the adverse and objectionable environmental side effects of synthetic or organic agricultural fertilizer use and production. It is very important to have an advanced understanding about the intricate relationships between microbes, fertilizers, and plants. There is a call for additional information, besides the techniques mentioned earlier, the application of which is also bi-pronged. First, there is a need for the introduction of added applied K nutrients into the plant through microbial inoculants, since not as much K nutrient is lost to the environment for the period during and after crop production. Secondly, loss of fertilizer could be curtailed by escalating the efficiency of the plant's nutrient uptake. This may possibly be accomplished by application of K solubilizers. In either case, there would be a huge drop in agricultural environmental pollution caused by the unsystematic use of synthetic fertilizers. The results show that inoculating the rhizosphere with PGPMs along with microbial strains of KSMs have greatly enhanced crop production. Consequently, the utilization of this arrangement will be a healthier approach, utilizing a mapping system that puts together the consortium of microbial strains. In the meantime, several related areas need to be better understood, such as where K solubilization under *in vitro* and in field conditions is required. However, there are no apparent information/data available about the amount of K solubilization and absorbtion by plants, either *in vitro* or under field conditions, besides the consumption of K by the microbes for their individual growth and metabolic activities. The present study along with other related information will undoubtedly assist in understanding as well as determining the status of insoluble K, and the use of bioinoculants may possibly be required for a realistic approach in an actual field situation. In the meantime, it is essential to measure the solubilized K as there are many apparent factors that may possibly influence K solubilization and uptake by the plant, among them predominantly the K needs of microbes, root exudation by each plant, and the soil environment, such as levels of pH, total dissolved solids, and total and available K.

These outcomes show that plant type influences the root colonization of inoculated strains. Research has illustrated that effectual plant-growth-endorsing bacterium-plant synchronization ought to be tested and recognized in controlled floral experimental designs with defined ecological site conditions and practical applications, such as the soil and plant type. Alternatively, besides the plant growth-enhancing capability of commercially used microbes, the amount of stimulus of crop plants in addition to their perseverance in the rhizosphere remains uncertain and indistinct under real field conditions. As a result, experiments pertaining to the stimulation of cotton and rape should be pursued by examination under authentic field conditions. Currently the application of K-solubilizing microbes in our agricultural system in soils that are K deficient where K is lacking or undersupplied will definitely help to resolve the K element quandary and advance research in this field. Aiming towards development of potential K solubilizers may perhaps help lead Indian agriculture to an unconventional means of K nutrition enrichment for use in our cropping system.

References

Abou-el-Seoud II, Abdel-Megeed A (2012) Impact of rock materials and biofertilizations on P and K availability for maize (*Zea Maize*) under calcareous soil conditions. Saudi J Biol Sci 19(1):55–63

Ahemad M, Kibret M (2014) Mechanisms and applications of plant growth promoting rhizobacteria: Current perspective. J King Saud Univ Sci 26(1):1–20

Andrist-Rangel Y, Edwards AC, Hillier S, Oborn I (2007) Long-term K dynamics in organic and conventional mixed cropping systems as related to management and soil properties. Agric Ecosyst Environ 122:413–426

Archana DS, Nandish MS, Savalagi VP, Alagawadi AR (2013) Characterization of potassium solubilizing bacteria (KSB) from rhizosphere soil. Bioinfolet 10:248–257

Argelis DT, Gonzala DA, Vizcaino C, Gartia MT (1993) Biochemical mechanism of stone alteration carried out by filamentous fungi living in monuments. Biogeo Chem 19:129–147

Armengaud P, Sulpice R, Miller AJ, Stitt M, Amtmann A, Gibon Y (2009) Multilevel analysis of primary metabolism provides new insights into the role of potassium nutrition for glycolysis and nitrogen assimilation in Arabidopsis roots. Plant Physiol 150:772–785

Arquero O, Barranco D, Benlloch M (2006) Potassium starvation increases stomatal conductance in olive trees. Hortscience 41:433–436

Bagyalakshmi B, Ponmurugan P, Marimuthu S (2012) Influence of potassium solubilizing bacteria on crop productivity and quality of tea (*Camellia sinensis*). Afr J Agric Res 7(30):4250–4259

Bahadur I, Meena VS, Kumar S (2014) Importance and application of potassic biofertilizer in Indian agriculture. Int Res J Biol Sci 3(12):80–85

Basak BB, Biswas DR (2008) Influence of potassium solubilizing microorganism (*Bacillus mucilaginosus*) and waste mica on potassium uptake dynamics by sudan grass (*Sorghum vulgare* Pers.) grown under two Alfisols. Plant Soil 317:235–255

Basak BB, Biswas DR (2010) Co-inoculation of potassium solubilizing and nitrogen fixing bacteria on solubilization of waste mica and their effect on growth promotion and nutrient acquisition by a forage crop. Biol Fertil Soils 46(6):641–648

Blake L, Mercik S, Koerschens M, Goulding KWT, Stempen S, Weigel A, Poulton PR, Powlson DS (1999) Potassium content in soil, uptake in plants and the potassium balance in three European long-term field experiments. Plant Soil 216(1):1–14

Britzke D, da Silva LS, Moterle DF, Rheinheimer D, Bortoluzzi EC (2012) A study of potassium dynamics and mineralogy in soils from subtropical Brazilian lowlands. J Soils Sediments 12:185–197

Diep CN, Hieu TN (2013) Phosphate and potassium solubilizing bacteria from weathered materials of denatured rock mountain, Ha Tien, Kiên Giang province, Vietnam. Am J Life Sci 1(3):88–92

Dordas C (2008) Role of nutrients in controlling plant diseases in sustainable agriculture. A review. Agron Sustain Dev 28:33–46

Gopalakrishnan S, Srinivas V, Vidya MS, Rathore A (2013) Plant growth-promoting activities of Streptomyces spp. in sorghum and rice. Springerplus 2:574

Groudev SN (1987) Use of heterotrophic micro-organisms in mineral biotechnology. Acta Biotechnol 7:299–306

Gundala PB, Chinthala P, Sreenivasulu B (2013) A new facultative alkaliphilic, potassium solubilizing, *Bacillus* Sp. SVUNM9 isolated from mica cores of Nellore District, Andhra Pradesh, India. Res Rev J Microbiol Biotechnol 2(1):1–7

Gupta G, Parihar SS, Ahirwar NK, Snehi SK, Singh V (2015) Plant growth promoting rhizobacteria (PGPR): Current and future prospects for development of sustainable agriculture. J Microb Biochem Technol 7:96–102

Han HS, Lee KD (2005) Phosphate and potassium solubilizing bacteria effect on mineral uptake, soil availability and growth of eggplant. Res J Agric Boil Sci vol 1(2):176–180

Han HS, Supanjani K, Lee KD (2006) Effect of co-inoculation with phosphate and potassium solubilizing bacteria on mineral uptake and growth of pepper and cucumber. Plant Soil Environ 52:130–136

Hasanuzzaman M, Nahar K, Alam MM, Roychowdhury R, Fujita M (2013) Physiological, biochemical, and molecular mechanisms of heat stress tolerance in plants. Int J Mol Sci 14(5):9643–9684

Hassan EA, Hassan EA, Hamad EH (2010) Microbial solubilization of phosphate – potassium rocks and their effect on khella (*Ammi visnaga*) growth. Ann Agric Sci (Cairo) 55:37–53

Hu X, Chen J, Guo J (2006) Two phosphate- and potassium-solubilizing bacteria isolated from Tianmu Mountain, Zhejiang, China. World J Microbiol Biotechnol 22(9):983–990

Kumar V, Narula N (1999) Solubilization of inorganic phosphates and growth emergence of wheat as affected by Azotobacter chroococcum mutants. Biol Fertil Soils 28(3):301–305

Kumar P, Dubey RC, Maheshwari DK (2012) *Bacillus* strains isolated from rhizosphere showed plant growth promoting and antagonistic activity against phytopathogens. Microbiol Res 167:493–499

Li YF (1994) The characteristics and function of silicate dissolving bacteria fertilizer. Soil Fertil 2:48–49

Li FC, Li S, Yang YZ, Cheng LJ (2006) Advances in the study of weathering products of primary silicate minerals, exemplified by mica and feldspar. Acta Petrol Mineral 25:440–448

Lian B, Fu PQ, Mo DM, Liu CQ (2002) A comprehensive review of the mechanism of potassium release by silicate bacteria. Acta Mineral Sinica 22:179–183

Lian B, Wang B, Pan M, Liu C, Teng HH (2008) Microbial release of potassium from K-bearing minerals by thermophilic fungus *Aspergillus fumigatus*. Geochim Cosmochim Acta 72:87–98

Liu GY (2001) Screening of silicate bacteria with potassium releasing and antagonistic activity. Chin J Appl Environ Biol 7:66–68

Liu D, Lian B, Dong H (2012) Isolation of *Paenibacillus* sp. and assessment of its potential for enhancing mineral weathering. Geomicrobiology J 29:413–421

Lopo de SáI AF, Valeri SV II, Pessoa da Cruz IIMC, Carlos Barbosa IIJ, Rezende GM II, Teixeira MP (2014) Effects of potassium application and soil moisture on the growth of Corymbia citriodora plants. Cerne 20(4):645–651

Maathuis FJM, Sanders D (1997) Regulation of K^+ absorption in plant root cells by external K^+: interplay of different plasma membrane K^+ transporters. J Exp Bot 48:451–458

Magri MMR, Avansini SH, Lopes-Assad ML, Tauk-Tornisielo SM, Ceccato-Antonini SR (2012) Release of potassium from rock powder by the yeast *Torulaspora globosa*. Braz Arch Biol Technol 55(4):577–582

Meena VS, Maurya BR, Verma JP (2014) Does a rhizospheric microorganism enhance K^+ availability in agricultural soils? Microbiol Res 169(5–6):337–347

Memon YM, Fergus IF, Hughes JD, Page DW (1988) Utilization of non-exchangable soil potassium in relation to soil types, plant species and stage of growth. Aust J Soil Res 26:489–496

Mikhailouskaya N, Tcherhysh A (2005) K-mobilizing bacteria and their effect on wheat yield. Latnian J Agron 8:154–157

Pettigrew WT (2008) Potassium influences on yield and quality production for maize, wheat, soybean and cotton. Physiol Plant 133:670–681

Prajapati K, Sharma MC, Modi HA (2013) Growth promoting effect of potassium solubilizing microorganisms on okra (*Abelmoscus Esculantus*). Int J Agric Sci 3:181–188

Rogers JR, Bennett PC, Choi WJ (1998) Feldspars as a source of nutrients for microorganisms. Am Mineral 83:1532–1540

Sardans J, Peñuelas J (2015) Potassium: a neglected nutrient in global change. Glob Ecol Biogeogr 24:261–275

Shanware AS, Kalkar SA, Trivedi MM (2014) Potassium solublisers: Occurrence, mechanism and their role as competent biofertilizers. Int J Curr Microbiol Appl Sci 3(9):622–629

Sheng XF (2005) Growth promotion and increased potassium up-take of cotton and rape by a potassium releasing strain of *Bacillus edaphicus*. Soil Biol Biochem 37:1918–1922

Sheng XF, He LY (2006) Solubilization of potassium bearing minerals by a wild type strain of *Bacillus edaphicus* and its mutants and increased potassium uptake by wheat. Can J Microbiol 52:66–72

Sheng XF, Huang WY (2002) Study on the conditions of potassium release by strain NBT of silicate bacteria. Sci Agric Sin 35:673–677

Sheng XF, He LY, Huang WY (2002) The conditions of releasing potassium by a silicate dissolving bacterial strain NBT. Agric Sci China 1:662–665

Sheng XF, Xia JJ, Chen J (2003) Mutagenesis of the *Bacillus edphicaus* strain NBT and its effect on growth of chili and cotton. Agric Sci China 2:40–41

Sheng XF, Zhao F, He LY, Qiu G, Chen L (2008) Isolation and characterization of silicate mineral solubilizing *Bacillus globisporus* Q12 from the surfaces of weathered feldspar. Can J Microbiol 54(5):1064–1068

Singh G, Biswas DR, Marwah TS (2010) Mobilization of potassium from waste mica by plant growth promoting rhizobacteria and its assimilation by maize (Zea mays) and wheat (*Triticum aestivum* L.) J Plant Nutr 33:1236–1251

Sparks DL, Huang PM (1987) Physical chemistry of soil potassium. In: Munson RD (ed) Potassium in agriculture. American Society of Agronomy, Madison, pp 201–276

Styriakova I, Styriak I, Galko I, Hradil D, Bezdicka P (2003) The release of iron-bearing minerals and dissolution of feldspar by heterotrophic bacteria of Bacillus species. Acta Pedol Sin 47(1):20–26

Subba Rao NS (2001) An appraisal of bio fertilizers in India. In: Kannian S (ed) Biotechnology of biofertilizers. Narosa Publication House, New Delhi

Sugumaran P, Janarthanam B (2007) Solubilization of potassium containing minerals by bacteria and their effect on plant growth. World J Agrl Sci 3:350–355

Supanjani HHS, Jung JS, Lee KD (2006) Rock phosphate-potassium and rock-solubilising bacteria as alternative, sustainable fertilizers. Agron Sustain Dev 26:233–240

Swaminathan MS, Bhavani RV (2013) Food production & availability – essential prerequisites for sustainable food security. Indian J Med Res 138(3):383–391

Uroz S, Calvaruso C, Turpault MP, Freyklett P (2009) Mineral weathering by bacteria: ecology, actors and mechanisms. Trends Microbiol 17:378–387

Valmorbida J, Boaro CSF (2007) Growth and development of *Mentha piperita* L. in nutrient solution as affected by rates of potassium. Braz Arch Biol Technol 50:379–384

Wang M, Zheng Q, Shen Q, Guo S (2013) The critical role of potassium in plant stress response. Int J Mol Sci 14(4):7370–7390

Wu SC, Cao ZH, Li ZG, Cheung KC, Wong MH (2005) Effects of biofertilizer containing N-fixer, P and K solubilizers and AM fungi on maize growth: A greenhouse trial. Geoderma 125:155–166

Xeng XF (2005) Growth promotion and increased potassium uptake of cotton and rape by a potassium releasing strain of *Bacillus edaphicus*. Soil Biol Biochem 37:1918–1922

Yang ML, Yan CX, Si SD (2014) Effect of potassium-solubilizing bacteria-mineral contact mode on decomposition behavior of potassium-rich shale. Chin J Nonferrous Metals 24:48–52

Zarjani JK, Aliasgharzad N, Oustan S, Emadi M, Ahmadi A (2013) Isolation and characterization of potassium solubilizing bacteria in some Iranian soils. Arch Agron Soil Sci 59(12):1713–1723

Zeng X, Liu X, Tang J, Hu S, Jiang P, Li W, Xu L (2012) Characterization and potassium solubilizing ability of *Bacillus circulans* Z1-3. Adv Sci Lett 10:173–176

Zhang A, Zhao G, Gao T, Wang W, Li J, Zhang S (2013) Solubilization of insoluble potassium and phosphate by Paenibacillus kribensis a soil microorganism with biological control potential. Afr J Microbiol Res 7(1):41–47

Zhanga C, Konga F (2014) Isolation and identification of potassium-solubilizing bacteria from tobacco rhizospheric soil and their effect on tobacco plants. Appl Soil Ecol 82:18–25

Zörb C, Senbayram M, Peiter E (2014) Potassium in agriculture – status and perspectives. J Plant Physiol 171(9):656–669

Earthworms and Associated Microbiome: Natural Boosters for Agro-Ecosystems

25

Khursheed Ahmad Wani, Mamta, Razia Shuab, and Rafiq A. Lone

Abstract

Nature has bestowed every living creature with unique qualities for maintaining an ecological balance. Earthworms are equipped with wonderful machinery, absolutely different from other organisms, which allow them to nurture the soil beautifully, having a direct impact on the production and quality of crops. Worms act as natural boosters when organic matter is converted into vermicompost and as soil conditioners bringing beneficial microbial activity to plants for growth and development. Microbial stimulation in the presence of earthworms may be due to the utilization of additional nutritive substances (secretion and excretion products) that they provide. Vermicomposting is highly nutritive and a growth promoter as compared to conventional compost. The process of vermicomposting has been well studied by earlier researchers, covering almost every aspect, but scant scientific literature is available on the relationship of earthworms with microbial diversity in different ecosystems. This chapter investigates how earthworms are natural boosters for agro-ecosystems and the role earthworms play in activating different microbes in agriculture fields.

K.A. Wani
Department of Environmental Science, ITM University, Gwalior, MP, India

Mamta
Department of Environmental Science, Jiwaji University, Gwalior, MP, India

R. Shuab
School of Studies in Botany, Jiwaji University, Gwalior, MP, India

R.A. Lone (✉)
School of Studies in Botany, Jiwaji University, Gwalior, MP, India

Department of Natural Sciences, SBBS University, Khiala (Padhiana), Jalandhar, Punjab, India
e-mail: rafiqlone@gmail.com

V. Kumar et al. (eds.), *Probiotics in Agroecosystem*,
DOI 10.1007/978-981-10-4059-7_25

25.1 Introduction

Microorganisms transform organic species (proteins, nucleic acids, fats, carbohydrates, etc.) in their digestive system into more stable products in the process of vermicomposting (Bianchina 2009). Earthworms are also able to clean up various pollutants from the soil as they accumulate in the worms' bodies. Hence, earthworms offer a solution to the tons of organic agro-waste that are being burned by farmers by recycling, by reusing this refuse to promote agriculture in a more efficient, economic, and environmentally friendly manner. The potential role of earthworms in organic solid waste management has been well established since the time of Darwin (1881), and has flourished to process waste to produce an efficient bioproduct, vermicompost (Kale and Bano 1986; Ismail 1995, Ismail 2005). Epigeic earthworms like *Perionyx excavatus*, *Eisenia fetida*, *Lumbricus rubellus*, and *Eudrilus eugeniae* are used for vermicomposting but local species like *Eisenia fetida* have been shown to be efficient for composting in tropical or subtropical conditions (Ismail et al. 1993). The method of vermicomposting involving a combination of local epigeic and anecic species of earthworms is called vermitech (Ismail et al. 1993; Ismail 2005). The nutrient content of vermicompost greatly depends on the input material. It usually contains higher levels of most of the mineral elements, which are in more available forms than the parent material (Edwards and Bohlen 1996).

Fungi are heterotrophic organisms that are totally different from bacteria. They have colonized a highly disturbed environment. The most important factor in the process of this colonization is the widespread mycelium composed of hyphae, which are often branched (Bardgett et al. 1993) and penetrate organic residues for the adsorption of nutrients. The production of different enzymes enables them to degrade the complex organic substances into simpler compounds. They can degrade complex organic substances like lignin and cellulose making them the best competitors for the decomposition of plant residues (Harley 1971). The process of mycorrhiza is highly beneficial in terms of soil fertility, and is possible due to fungal association, which may act as a parasitic or symbiotic process depending on the nutrient availability of the soil (Farrell et al. 2006). The fertility of soil, on the other hand, is increased by the application of different types of synthetic and organic fertilizers. Synthetic fertilizers increase the fertility of soil very rapidly but cause a negative impact on the ecosystem. Hence, organic fertilizers are being widely used and are highly recommended by experts. Of the organic fertilizers, vermicomposting is being applied almost everywhere and is considered a simple and viable fertilizer by the farming community.

Diverse microflora are present in different soil ecosystems as soil is regarded as the soul of infinite life. Earthworms are regarded as farmers' friends and influence the microbial community and the physical and chemical properties of soil (Pathma et al. 2011). Microbial species are activated as soon as organic waste is degraded into valuable vermicompost by earthworms. Of interest is the fact that earthworm activity increases beneficial microflora and suppresses harmful pathogenic microbes. Vermicomposting increases soil fertility, enhances plant growth, repels pests, and influences microbial activity in the soil. The abundance of nutrients in vermicompost and the presence of different microbial enzymes are basic elements that help in maintaining the fertility of soil (Maboeta and Van Rensburg 2003). The decomposition pathway from earthworm activity is probably due to the contrasting effects on bacterial and fungal populations as both the decomposers have different resource requirements. Fungal species, with their hyphal network, can immobilize great quantities of nutrients and the exploitative strategy of nutrient use of bacteria produces unstable substrates during vermicomposting (Bardgett 2005).

Soil microbes are very beneficial for soil health. Earthworms are the important visible species in the soil as their activity helps in soil nutrient cycling through rapid incorporation of dead organic matter into microbial soil. Mucous produced by the earthworm gut provokes the activity of other organisms. They increase the nitrogen, phosphorus, potassium, and calcium concentration in the soil and cause increased nitrogen mineralization through direct and indirect effects on the microbial community (Bhadauria and Saxena 2007).

25.2 The Choice of Food

Most of the organisms ingested by the earthworm during the process of feeding as microorganisms are considered an unavoidable constituent of their natural diet (Edwards and Bohlen 1996). Earthworms are selective feeders for certain fungal and bacterial species (Satchell 1967; Doube et al. 1997; Boag 2003). Although the nature of food preference by earthworms is still a mystery, there is an indication that the presence of different microorganisms brings changes that may be helpful in the growth of earthworms. The presence of fungal growth on food substances increases the availability of carbohydrate and nitrogen compounds for earthworms. Jayasinghe and Parkinson (2009) showed that earthworms prefer organic matter inoculated with different species of actinomycetes with evidence for chemoreception for the selection of food material. Chemoreception in earthworms is associated with sensory modulates based on the principle of neural organization that is utilized for the detection of food and reaction to different exogenous chemicals (Hildebrand 1995). Perhaps this application is used by agriculturists in order to use different types of pesticides for different types of pests.

The earthworm's gut is home to a large group of organisms, and their survival in the gut depends upon their capacity to resist both intra- and intercellular digestive enzymes, mucous of the intestine, calcium carbonate, and microbial substances (Brown 1995). Apart from grass fragments and other plant material, the digestive tract of earthworms contains a number of organisms and there is evidence for the possible existence of an ecological group of specific gut microbiota in earthworms (Lavelle and Spain 2001). This is further strengthened by the food quality and type of microorganisms. Automated image analysis showed that some microorganisms of the soil, namely *Pseudomonas* spp., increase in abundance through the gut track of *L. rubellus*. The earthworms' digestive tract is considered a suitable habitat for N_2O-producing bacteria as earthworms activate these microorganisms during gut passage (Horn et al. 2003).

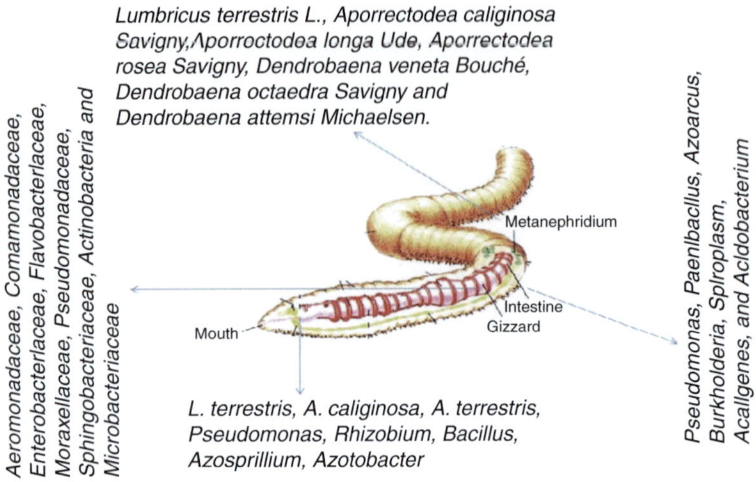

Fig. 25.1 Microbes in different parts of the earthworm

Among the microflora earthworms were seen to predate on fungal species, namely *Alternaria solani*, *Blastomyces* spp., *Botryotrichus* spp. *Chaetomium glabrum*, *Chaetomium* spp., *Cunninghamella echinulate*, *Curvularia* spp., *Fusarium oxysporum*, *Helminthosporium* spp., *Mucor hiemalis*, *Neocusmospora vasiinfecta*, *Nigrospora sphaerica*, *Rhizopus nigrican*, and *Trichoderma viridi* (Cooke and Luxton 1980), are also bacterial species such as *Bacillus cereus*, *mycoides*, *Serratia marcescens*, *E. coli*, and *Enterobacter cloacae* (Pedersen and Hendriksen 1993), yeast, *Candida famata* (Byzov et al. 2009) and *Lesquereusia spirali,s* which is a protozoa (Mukherjee and Julka 1984). The earthworm and its associated microbiome is shown in Fig. 25.1.

25.3 Soil Ecosystems, Earthworms, and Microbes

The soil ecosystem harbors different microbial communities that comprise bacteria, fungi, and protozoa. Actinomycetes, algae, and the microbial biomass mainly consist of bacteria and fungi. The compositions of microbial communities are dependent upon various key environmental factors and most of them either promote or inhibit their growth in agricultural ecosystems. However, every organism has a different role to play under various circumstances for the growth and development of the soil ecosystem. Most of the microorganisms like bacteria are ubiquitously distributed everywhere and have been reported to be found in hot springs, volcanoes, ice caps, etc. They have the ability to withstand erratic and extreme environmental conditions. However, the distribution of bacteria is dependent on energy resources. Bacterial species are autotrophic, photoautotrophic, heterotrophic, and chemoautotrophic. Furthermore, the abundance of bacteria in soil increases fertility as it is estimated that 1 g of soil contains above 10^{11} bacteria.

Earthworm species actively modified the structure of the microbial community in soil samples which were incubated with different plant materials (Clapperton et al. 2001). The activity of earthworms (*E. andrei*) had a great impact on the structure of the grape marc microbial community as revealed by phospho lipid fatty acid analysis. Soil-dwelling endogenic, anecic, and epigenic earthworms have been reported to modify the structure of the microbial community and interact very closely with different types of microorganisms (Lores et al. 2006), with epigenic worms having the lowest number of bacterial and fungal species. Only a few studies have shown very low (Enami et al. 2001) or scant changes (Marhan et al. 2007). Furthermore, the abundance of different microorganisms associated with earthworms are dependent on the kind of food source (Knapp et al. 2008), worm density, and substrate (Aira et al. 2008). A decrease in the activities of protease and cellular enzymes was also reported as a result of the presence of earthworms, with no such reduction in the control mesocosm. Correlations between microbial biomass and enzymes like protease and cellulose have been observed by Aira et al. (2007), indicatings that microorganisms play a vital role in shaping the patterns of enzymes present during the process of vermicomposting, which is attributed to a lower microbial biomass due to earthworm activity (Benitez et al. 2005). Earthworms modify the carbon and nitrogen pool during the process of vermicomposting, which in turn activates different microorganisms like nitrogen-fixing bacteria. The interaction between microbes and earthworms is essential for many soil processes in agro-ecosystems which are enhanced by the assimilation of labile carbon (Edwards and Bohlen 1996; Lee 1985). The influence of earthworms on soil microflora facilitates carbon and nitrogen transformations (Martin 1991; Scheu 1987).

The earthworm microbial interactions enhance soil carbon evolution, soil nutrient availability, and microbial availability, but can reduce microbial biomass (Edwards 1995). The earthworm *Lumbricus terrestris*, which occurs commonly in northern temperate agro-ecosystems, forms permanent or semi-permanent vertical burrows with small patches of plant litter and casts called middens gathered around the burrow entrance. These middens are dominant in agro-ecosystems and large numbers of *L. terrestris* affect the breakdown of crop residues and the spatial heterogenicity of residue microenvironments on the soil surface (Bohlen and Edwards 1995). The earthworm middens are characterized by higher microbial activity than the surrounding environment with greater diversity in forest ecosystems. Greater turnover of microbial populations comes with alteration of the quantity of plant litter in the midden environment. The soil is composed of organic and inorganic components, and promotes different types of microorganisms. Some bacteria like *Bacillus*, *Pseudomonas*, and *Streptomyces* induce secondary metabolites that act against secondary phytopathogenic fungi and human pathogenic bacteria (Pathma et al. 2011). Most edaphic organisms influence the properties of soil and activate the microbial community, thereby changing the soil's physical and chemical properties. Earthworms modify soil particles and decompose leaf litter to increase the nutrition pool and organic matter, thus transforming organic waste into valuable vermicompost by the grinding and digesting process. The activity of earthworms is believed to increase beneficial micro flora and inhibit harmful pathogenic microbes. The vermicompost is rich in micro- and macronutrients along with microbial enzymes (Lavelle and Martin 1992).

25.4 Earthworms as Nutrient Boosters for Plants

Soil fertility is the inherent capacity of soil to supply nutrients in adequate amounts and in suitable proportions. An increase in plant growth has been observed in response to earthworms in various pot experiments. Different mechanisms by which earthworms can increase plant growth have been demonstrated and suggested. These include increased incorporation of organic matter that accelerates mineralization, the impact of metabolism products on plant growth, increased aeration, and improved water relations and permeability in poorly structured soils; these have been well elaborated by various workers.

Earthworms contribute to the formation of soil aggregates, improvement in soil aeration, and porosity (Edwards and Bohlen 1996) by ingestion of soil and partial breakdown of organic matter along with intimate mixing of these components and ejection of this material as surface or subsurface casts. Earthworm casts contain more water-stable aggregates than the surrounding soil and through their activity influence both the drainage of water from soil and the moisture-holding capacity of soil, both of which are important factors for plant productivity (Edwards and Bohlen 1996).

25.4.1 Role of Earthworms and Associated Microbes in Soil Fertility

Earthworms have the ability to mineralize organic matter and release the nutrients in available forms that are easily taken up by plants (Edwards and Bohlen 1996). The casts of earthworms have higher base exchangeable phosphorus, potassium, manganese, and total exchangeable calcium that act as boosters for soil productivity. Earthworms favor nitrogen-fixing bacteria and help in bacterial population and soil aeration. The lower horizons of the soil are enriched by the stimulation of microbial activity in casts that enhance the transformation of soluble nitrogen into microbial protein, thereby preventing their loss through leaching. Lee (1985) argued that nitrogenous products of earthworm metabolism are returned to the soil through casts, urine, mucoproteins, and dead tissues of earthworms. There is a complex interrelationship between earthworms and microorganisms. Most of the species of microorganisms that occur in the alimentary canal of earthworms are the same as those in the soil in which the earthworms live. Earthworm casts have higher densities of fungi, actinomycetes, bacteria, and higher enzyme activity than the surrounding soil, as reported by Lachnicht and Hendrix (2001). Earthworms are very important for inoculating soils with microorganisms as most of the microorganisms in soil are in a dormant stage with low metabolic activity, awaiting suitable conditions like the earthworm gut (Lachnicht and Hendrix 2001) or mucus (Lavelle et al. 1983) to become active. Macrofauna, namely Earthworms, are major compounds of soil and form a large proportion of the macrofauna biomass through the rapid incorporation of detritus into mineral soils.

Earthworms affect the nutrient supply by producing aggregates and causing pores in the soil that alter physical properties, cycling of nutrients, and growth of plants (Curry and Schmidt 2007; Flegel and Schrader 2000). The assemblage of organomineral aggregates, their stability, and organic matter concentration impacts the physical properties of soil and the dynamics of soil organic matter. Some ecological processes within the functional domain of earthworms are also affected. The casts of earthworms are rich in organic carbon and nitrogen and have been reported to increase these minerals by a factor of 1.5 and 1.3, respectively, as compared to nonintegrated soils.

25.4.2 Earthworm and Nutrient Enrichment

The chemical fertility of the soil is dependent on the availability of nutrients as they play an important role in the productivity of crops. Suitable management of organic materials and earthworms can encourage an appreciable increase in the productivity of crops as earthworms also speed up the weathering process (Carpenter et al. 2007) to release the essential elements that promote growth. The decomposition of organic matter to release nutrients and trace elements is one of the most important functions of soil biodiversity (Schinner et al. 2012), and is enhanced by fungi (saprotrophic and mutualistic) as they increase the rate of mineralization (Hoffland et al. 2004). Plants are unable to utilize gaseous nitrogen (Bernhard 2010), hence the task of fixing nitrogen is performed by free-living microorganisms which form root nodules in legumes. The best examples are cyanobacteria and other genera of bacteria and actinomycetes, or symbiotic bacteria, which are enhanced by the presence of earthworms (Gordon and Wheeler 2012; Llorens-Marès et al. 2015; Gresshoff et al. 2015).

Organic manure like vermicompost acts as a source of nutrients and organic matter, increases size, diversity, and activity of the microbes in soil, influences soil structure and nutrient turnover, and brings appreciable changes in physical, chemical, and biological parameters of the soil (Albiach et al. 2000). Fertilizer produced through the activity of earthworms contains higher levels of organic matter, organic carbon, total and available nitrogen, phosphorus, potassium, and micronutrients, and increased microbial and enzymatic activity (Edwards and Bohlen 1996; Parthasarathi et al. 2007). Orozco et al. (1996) and Parthasarathi (2004) also reported that vermicompost contains nutrients such as nitrates, exchangeable phosphorus and potassium, and calcium and magnesium soluble in forms that are easily taken up by plants. Tomati et al. (1990) and Parthasarathi et al. (2006) also confirmed that vermicompost contains higher concentrations of humic acid and biologically active substances like plant growth regulators.

25.4.3 Vermicompost, Microbes, and Crop Production

Esinea fetida can tolerate soils that are nearly half as salty as seawater, i.e. 15 g/kg of soil, and also improve their biology and chemistry. Live earthworms were applied by farmers at Phaltan in the Satara district of Maharashtra (India) to the sugarcane crops grown in saline soils as irrigated by saline ground water. The soil nitrogen, phosphate, and potash was increased by 37%, 66%, and 10%, respectively, and decreased the chloride content by 46% within a year. The yield was 125 tons/hectare of sugarcane with marked improvement in soil chemistry (Sinha et al. 2009). A better yield of potato (*Solanum tuberosum*) was reported by the application of vermicompost in reclaimed sodic soil due to a reduction in sodicity (ESP) from the initial 96.74 to 73.68 in almost 12 weeks. The average available nitrogen concentration of the soil was improved from an initial 336.00 to 829.33 kg/ha (Ansari 2008).

Earthworms and their vermicompost are exceptional for plant growth and support crop production without the application of chemical fertilizers. CSIRO Australia found that earthworms (*Aporrectodea trapezoids*) increased the growth of wheat crops (*Triticum aestivum*) by 39%, grain yield by 35%, increased the protein value of the grain by 12%, and also acted as a good repellent for different pests (Baker et al. 1997). The fertility of the soil has been improved by the introduction of earthworms in cherry fields and results were quite appreciable after a period of 3 years compared with chemical fertilizers (Webster 2005). Similarly, Ohio State University in Columbus, OH, USA indicated that vermicompost improves the growth rate of vegetables. Application of vermicompost to a transplant grown in vermicompost had the highest amount of red marketable fruit at harvest with no symptoms of early blight lesions on the fruit. The yield of pea (*Pisum sativum*) was also higher with the recommended nitrogen, phosphorus, and potash along with the application of vermicompost. Vadiraj et al. (1998) reported that application of vermicompost produced herbage yields of coriander cultivars more than when compared to those obtained with chemical fertilizers.

Mamta et al. (2012) indicated that the different treatments affected the seed germination of the test crop significantly. Plant height, number of leaves, and fruit weight were higher in the vermicompost-treated field as compared to the control field, and no incidence of disease was reported in the fruit of *Solanum melongena* treated with vermicompost. The fresh weight of *Chrysanthemum chinensis* flowers increased with the application of vermicompost. With the application of 10 t ha^{-1} of vermicompost along with 50% of the recommended dose of a nitrogen/phosphorus/potassium (NPK) fertilizer the number of flowers per plant (26), flower diameter (6 cm), and yield (0.5 t ha^{-1}) were maximum. However, the vase life of flowers (11 days) was high with the combined application of vermicompost at 15 t ha^{-1} and 50% of the recommended dose of NPK fertilizer (Nethra et al. 1999).

Rao et al. (2010) studied the effect of vermicompost on the growth and yield of onion (*Allium cepa*) and indicated that there is a synergistic relationship between the nutrients of the vermicompost and chemical fertilizers which boost the mobility of mineral nutrients for better production of the onions. The height of the plant increased significantly compared to the respective control by up to 51.60% after

30 days. The length of leaves increased up to 52.6% after 60 days, and after 90 days the leaf size increased by 71.4%. The increase was reported to be up to 56.65% after 120 days, which was significant compared to the respective control. The total yield of the onion harvest was enhanced significantly when subjected to the statistical application of analysis of variance (ANOVA).

Azarmi et al. (2008) determined the impact of vermicompost on growth, yield, and fruit quality of tomato (*Lycopersicum esculentum* var. Super Beta) under field conditions. Growth characteristics such as leaf number, leaf area, and shoot dry weights were determined, and the results indicated that the addition of vermicompost at a rate of 15 t ha^{-1} increased growth and yield compared to control tomatoes. Vermicompost at a rate of 15 t ha^{-1} increased electrical conductivity of fruit juice and percentage of fruit dry matter up to 30% and 24%, respectively. The content of potassium, phosphorus, iron, and zinc in the plant tissue increased 55%, 73%, 32%, and 36%, respectively, when compared to untreated plots. A significant increase in the mean values of total organic carbon, humus, nitrogen, phosphorous, and potassium was observed while converting garden waste, kitchen waste, and cow dung into vermicompost by using *Eisenia fetida* (Singh et al. 2000; Kumar and Singh 2001; Wani and Rao 2013).

A significant effect of using vermicompost as compared to humus manure has been observed. Potatoes and beets grown under the influence of vermicompost showed a yield increase of 1.5 and 1.2 times, respectively, as compared to the control. It was observed that flowering, ripening of crops, and development of the roots increased significantly.

The effects of vermicompost application in reclaimed sodic soils on the productivity of potato (*Solanum tuberosum*), spinach (*Spinacia oleracea*), and turnip (*Brassica campestris*) were studied by Ansari (Ansari 2008). The overall productivity of vegetable crops revealed that the requirement of vermicompost for leafy crops like spinach was lower (4 tons/ha), whereas that for tuber crops like potato and turnip was higher (6 tons/ha). Physical, chemical, and biological properties along with the growth, yield, and nutrient content of beans (*Phaseolus vulgaris*) was amended with vermicompost at 5 tons/ha. The pore space, water holding capacity, and cation exchange capacity of the soils in the field increased significantly. However, particle size, bulk density, pH, and electrical conductivity were reduced. An increase in organic carbon, calcium, magnesium, zinc, manganese, sodium, iron, copper, and microbial activity was also reported in this study.

Dhanalakshmi et al. (2014) reported an increase in root length, shoot length, and branch and leaf number in the seeds of okra, brinjal, and chili sown in vermicompost-containing soil. Joshi and Vig (2010) reported that branch and leaf number and almost all the growth, yield, and quality parameters were increased under the influence of vermicompost. The mean stem diameter, mean plant height, yield per plant, marketable yield per plant, mean leaf number, and total plant biomass increased significantly in a vermicompost medium compared to control plants in the case of tomato, *Lycopersicum esculentum* L. Vermicompost stimulates the growth of various horticultural crop plants such as strawberry (Arancon et al. 2004), garlic (Suthar 2009), sweet corn (Lazcano et al. 2011), and groundnut (Kumar et al. 2014).

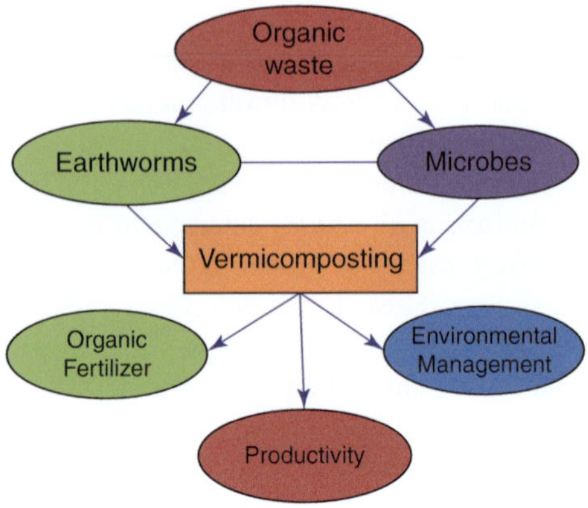

Fig. 25.2 Significance of vermicomposting

Vermicomposting shows positive effects on aromatic and medicinal plants (Prabha et al. 2007), fruit crops such as banana and papaya (Reddy et al. 2014), and ornamental plants such as chrysanthemum (Hidalgo and Harkess 2002), geranium (Chand et al. 2007), and marigolds (Shadanpour et al. 2011).

25.4.4 Vermicompost in Forest Trees

The positive effects of vermicompost on forestry species such as acacia, eucalyptus, and pine trees were reported by Lazcano et al. (2010a, b). Seed germination in several plant species such as green grams (Karmegam et al. 1999) and tomato plants (Zaller 2007) is triggered by vermicompost. The nutritional quality of vermicomposting in spinach (Peyvast et al. 2008), strawberries (Singh et al. 2008), Chinese cabbage (Wang et al. 2010) and sweet corn (Lazcano et al. 2011) has been well documented.

Earthworms change the environment for all other soil-inhabiting organisms including plants, whose roots can increase their uptake of various minerals due to the presence of earthworms. Some earthworm species selectively feed on plant residues at the soil surface (e.g., *Lumbricus terrestris* L.) whereas others feed on deeper residues (e.g., Octolasion). However, the mechanisms operate simultaneously and none of them has been experimented in isolation, taking into consideration all the different factors, namely initial chemical and physical nature, crop species, and the species of earthworm present. The significance of vermicomposting is shown in Fig. 25.2.

25.5 Earthworm and Eco-Biome

The presence of earthworms is very important for the agro-ecosystem and they are considered to be major invertebrates in terms of biomass and activity. Microorganisms are the natural diet for earthworms and the feeding behavior of earthworms is clearly associated with the ecological group. The epigiec earthworm species live near the soil surface and prefer litter layers of forest soil; they usually do not burrow into the soil and ingest litter rather than soil, and hence they are popularly known as litter transformers. Several research studies have shown an increase in microbial activities in the presence of epigeic earthworms that has been associated with an increase in surface area for decomposition, a reduction in immobilization by fungi, and a modification of the consumption of microorganism communities.

The anecic earthworms prefer to live in permanent and semi-permanent burrows in soil layers that have high concentrations of minerals. These types of earthworms feed on organic matter mixed with soil particles and mostly form middens. Earthworm food procurement is dependent on the quality of food. Anecic earthworms prefer nitrogen-rich food and do not like food with a high lignin content or that is colonized by *Fusarium laterium* or *Tricoderma* species (Cooke and Luxton 1980). Endogenic earthworms are found at a depth up to 15 cm in the mineral horizon. They have the ability to consume more soil than other species and are commonly known as geophageous or organic matter feeders.

The different microbial species including fungi, bacteria, yeast, actinomycetes, and protozoa were estimated in the gut casts of different earthworm species qualitatively and quantitatively, although the interaction between the two categories seems to be complex. The increase in the availability of plant nutrients is dependent on microflora present in the earthworm's gut (Petersen and Luxton 1982; Edwards and Bohlen 1996). Microorganisims like *Pseudomonas*, *Rhizobium*, *Bacillus*, *Azosprillium*, *Azotobacter*, etc. along with rhizospheric soil, are reported to get activated within the gut of earthworm under ideal conditions. (Sinha et al. 2010). During the earlier times, Edwards and Loft (1977) reported a higher number of the aerobes *L. terrestris*, *Allolobophora caliginosa* and *A. terrestris* compared to soil in the gut of earthworms. The increase in plate counts of total bacteria, proteolytic bacteria and actinomycetes was observed through the earthworms gut (Parle 1963; Pedersen and Hendriksen 1993; Devliegher and Verstraete 1995). *Pseudomonas oxalaticus* an oxalate degrading bacterium was isolated from the intestine of *Pheretima* species (Khambata and Bhat 1953) and *Streptomyces lipmanii*, an actinomycete was identified in the gut of *Eisenia lucens* (Contreras 1980). The endogenous microflora *L. terrestris* and *Octolasion cyaneum* was found in the gut of earthworms with the help of scanning electron micrographs (Jolly et al. 1993). N_2-fixing anaerobic bacteria viz., *Clostridium butyricum*, *C. beijerinckii* and *C. paraputrificum* were recorded in the gut of *E. foetida* (Citernesi et al. 1977). A higher concentration of aerobes and anaerobes were present in *Lumbricus rubellus* and *Octolasium lacteum* (Karsten and Drake 1995).

Earthworms in association with other microorganisms have the potential to mineralize and humify organic matter and help in the facilitation of metal ion chelation. It was observed that microorganisms aid the earthworms in their growth and reproduction (Pizl and Novakova 2003). *Eisenia fetida savigny* hatched from sterile cocoons were not able to reach sexual maturity in sterilized soil and appreciable results were reported with the inoculation of mobile protozoa in its food.

Currently, the interest towards microbial activation by earthworms in agricultural ecosystems has widened the scope of vermicomposting; however, the actual presence of symbionts in the earthworm gut is highly controversial (Curry and Schmidt 2007). Research has shown that there exists a small difference between bacterial communities in the soil, gut and fresh casts of *L. terrestis* which indicates that an indigenous microbial community is unlikely to exist (Egert and Horn 2004). This was further supported (McLean et al. 2006; Jayasinghe et al. 2009) during studies on the impact of radical diet shift on the gut microbiota of *Lumbricus rubella* . Byzov et al. (2009) reported gut symbiosis in earthworms and surprisingly they formed different microorganisms in the earthworm gut as compared to the surrounding soil. Despite the shift in food source and habitat changes, the development of different bacterial communities was strongly associated with ecological group. Moreover, due to the presence of all bacteria in both the earthworm gut and in soil it was not possible to determine symbiotic metabolic interaction with earthworms (Thakuria et al. 2010). The real presence of different symbionts in the earthworm gut and their actual role needs to be studied further by involving large diversity of earthworms in different geographical regions.

25.5.1 Microbiome of the Earthworm

The earthworm gut is a mobile micro-habitat for most of the dormant soil microorganisms as they have easy access to food, a free ride and shelter (Lavelle et al. 1995). The presence of suitable amounts of carbon in a soluble form and nutrient resources in the casts make the activities of microorganisms possible. The increase in microbial respiration rate in the casts is an indication that microbial activity exists there. Furthermore, effects of microbial activity and invertebrate activity have been seen in burrows (Graff 1971). *Lumbricus terrestris*, *Cellulomonas* sp., and *Promicromona* sp., are the dominant bacteria whereas *Bacillus* sp., was reported in the surrounding soils with no fungal species, (Tiunov and Dobrovolskaya 2002). The loss of carbon through the secretions of earthworms and nitrogen through nephridia may be the reason for the presence of different species. A higher number of actinomycetes was reported in earthworm casts than the surrounding soil which shows that casts may be a suitable microhabitat for this microorganism.

With the help of microscopic observations, earthworm species specific microbial symbionts in the ampulla of nephridia were described during the earlier times (Knop 1926). During the advancing stages of research, it was confirmed that the presence of symbionts in nephridia are members of a monophyletic branch of the genus *Acidovorax* (Schramm et al. 2003). Furthermore the result showed earthworms harbor distinct gene sequence types associated with *Acidovorax* sp., whereas, the same earthworm species from different continents have similar symbiont sequences. The discovery of symbionts in the nephridia of E. *fetida* has opened new directions in the field of research and several workers have reported bacterial colonization in the nephridia of earthworms (Davidson and Stahl 2006). Davidson and Stahl (2006)) advocated that nephridia symbionts are not acquired from the environment but are directly transferred from the adults to the capsules during mating. In *E. fetida* and *Acidovorax* sp., cells are very high in mating mucus and in egg capsules. In due course new species *verminephrobacte*r sp., and *verminephrosactor eiseniac* were defined based on the isolation, characterization and the unique ecology of genus *Acidovorax* and *E.fetida* (Pinel et al. 2008). Subsequently v*erminephrobacter* sp. ecology was investigated by Lund et al. (2010) and was reported from 19 earthworm species. The species were colonized by different types of organisms, namely *Lumbricus terrestris L.*, *Aporrectodea caliginosa* Savigny, *Aporrectodea longa* Ude, *Aporrectodea rosea* Savigny, *Dendrobaena veneta* Bouché, *Dendrobaena octaedra* Savigny, *and Dendrobaena attemsi* Michaelsen.

25.5.2 Interaction of Worms with Microbes

The interaction of different detritivorous earthworms with microorganisms enhances the stabilization of organic matter that modifies physical, chemical, and biological properties of soil. Earthworms are important players that stimulate the production of microorganisms through fragmentation and organic matter ingestion. Although earthworms decrease the overall biomass and activity during the vermicomposting process, this decrease is higher for fungi than for bacteria, as fungi are being selectively used as food by the earthworms. The shaping of microorganisms in organic waste is also carried out through the process of vermicomposting, which indicates that detritivorous earthworms directly modify the decomposer community composition in the short term and accelerate organic matter decomposition. Grazing on microorganisms increases microbial activity at first, which then decreases the availability of these resources for the microbial communities and consequently their activity (Dominguez 2010).

Earthworms play an important role in the reduction of total coliforms during the vermicomposting process as it has been found that 98% of the total coliforms are reduced in the gut of earthworm species *Eisenia andrei*, *E. fetida*, *and Eudrilus eugenia*. Monroy (2006) also found a drastic reduction in the population of total coliforms after 2 weeks of vermicomposting with *E. fetida*, which indicates that vermicomposting is effective in decreasing the levels of human pathogens during the stabilization of biosolids and other organic waste. The decease in *E. coli* may be due to the competitive environment in the gut of earthworms as it is home to a number of microflora. The characterization of microbial communities by PFLA profiles (Zelles 1999) revealed that earthworm activity greatly influenced microbial community structure and function. The activity of earthworms decreased the viable microbial biomass by approximately four to five times relative to control without earthworms.

The inoculation of different strains of *Azotobacter* in vermicompost has been found to increase *Azosprillum lipoferum*, and the inoculated phosphate-solubilizing bacterium *P. striata* caused a significant effect on the available phosphorus content in vermicompost. The addition of rock phosphate inoculated with *P. striata* led to more availability of phosphorus due to the production of organic acids by the bacteria that solubilized the rock phosphate (Premono et al. 1996). Kumar and Singh (2001) assessed the impact of inoculation of vermicompost with the nitrogen-fixing *Azotobacter chroococcum*, *Azospirillum lipoferum*, and the phosphate-solubilizing *Pseudomonas striata* on nitrogen and phosphorous contents of the vermicompost. It was found that the inoculation of nitrogen-fixing bacteria into vermicompost enhances the concentration of nitrogen and phosphorous. The inoculation of *Pseudomonas striata* in vermicompost significantly improved the available phosphorous with the addition of rock phosphate. The inoculated bacterial strains proliferated rapidly during the incubation period and fixed nitrogen and solubilized added and natural phosphate during the process.

The population of soil microorganisms (Binet et al. 1998), microbial numbers, and biomass is stimulated and accelerated by earthworms (Edwards and Bohlen 1996) by improving aeration through burrowing actions. The diversity and increase in the microbe population within the digestive tract of the earthworm is due to the hospitable conditions and the presence of nutrient-rich organic wastes in the gut that provide energy and act as a substrate for the growth of microorganisms (Tiwari et al. 1989). Actinobacteria and Gammaproteobacteria are plentiful in vermicompost (Vivas et al. 2009). *Nitrobacter*, *Azotobacter*, *Rhizobium*, phosphate solubilizers, and actinomycete counts exceeded 10^{-10}/g of vermicompost (Suhane 2007). A considerable increase in total viable counts of actinomycetes and bacteria in the earthworm-treated compost was reported by Haritha Devi et al. (2009). *Aeromonas hydrophila* in *E. foetida* (Toyota and Kimura 2000) and fluorescent pseudomonads in *L. terrestris* (Devliegher and Verstraete 1995) are indications that specific phylogenetic groups of bacteria exist in different species of earthworms.

The potential degraders (*Pseudomonas*, *Paenibacillus*, *Azoarcus*, *Burkholderia*, *Spiroplasm*, *Acaligenes*, and *Acidobacterium*) of several categories of organics have been shown to be associated with the earthworm intestine and vermicasts (Singleton et al. 2003). Vaz-Moreira et al. (2008) documented different species of microorganisms from vermicompost. The Firmicutes included *Bacillus benzoevorans*, *B. cereus*, *B. licheniformis*, *B. megaterium*, *B. pumilus*, *B. subtilis*, and *B. macrolides*. Actinobacteria that were present in the vermicompost are *Cellulosimicrobium cellulans*, *Microbacterium* spp., *M. oxydans*, proteobacteria such as *Pseudomonas* spp. and *P. Libaniensis*, and ungrouped genotypes included *Sphingomonas* sp. and *Kocuria palustris*. The yeasts, namely *Geotrichum* sp. and *Williopsis californica* have also been reported in vermicomposts (Nechitaylo et al. 2010). The presence of *Verminephrobacter eiseniae*, a novel nephridial symbiont isolated from *E. foetida*, *Ochrobactrum* sp., *Massilia* sp., and *Leifsonia* sp., was reported by Pinel et al. (2008), and bacteria belonging to the families Aeromonadaceae, Comamonadaceae, Enterobacteriaceae, Flavobacteriaceae, Moraxellaceae, Pseudomonadaceae, Sphingobacteriaceae, Actinobacteria, and Microbacteriaceae were extracted from the earthworm alimentary canal (Byzov et al. 2009).

Fungal species like *Saksenae vasiformis*, *Mucor plumbeus*, *Cladosporium carrionii*, *C. herbacium*, *Alternaria* sp., *Cunninghamella echinulata*, *Mycetia sterila*,

Syncephalostrum racemosum, Curvalaria lunata, C. geniculata, and Geotrichum candidum were found to be digested by earthworms. Bacteria species like *Pseudomonas aeruginosa, Bacterium antitratum, Mima polymorpha, Enterobacter aerogenes, E. cloacae, Proteus vulgaris, P. mirabilis, P. rettgeri, Escherichia coli, Staphylococus citreus, Bacillus subtilis, B. cereus, Enterococci*, and *Micrococci* were also completely digested.

Soils infested with soil-borne pathogens and augmented with earthworms (*Lumbricus terrestris*) have been found to reduce diseases of susceptible cultivars of asparagus (*Asparagus officinalis*), eggplant (*Solanum melongena*), and tomato (*Solanum lycopersicum*) in greenhouse studies. Earthworm activity has been found to reduce the incidence of diseases in soils planted with asparagus that was infested with *Fusarium oxysporum* f. sp., *Asparagi*, and *F. proliferatum*, eggplant with *Verticillium dahliae*, and tomato with *F. oxysporum* f. sp., lycopersici Race 1. When soils were augmented with earthworms, plant weights were increased 60–80% and disease infestations were reduced to 50–70% (Stephens et al. 1993).

Under microcosm-controlled conditions the stimulatory effect of earthworms (*Lumbricus terrestris* L.) on soil microbial activity was studied. The microbial stimulation observed in the presence of a soil invertebrate has been correlated to the utilization of additional nutritive substances provided by the activity of earthworms. The stimulation of microbial activity was established with the increased density of the protozoan population, which was 3–19 times greater in the presence of earthworms (Daane and Haggblom 1999; Senapati et al. 1999).

Conclusions

Vermicomposts are equipped with all the nutrients that are essential for plant growth and the associated microbes are better organic amendments that act as a panacea for soil reclamation, improve soil fertility, promote plant growth, and control pathogens, pests, and nematodes for sustainable agriculture. Hence, earthworms along with microbial diversity affect the fertility of soil, which is of considerable importance for agricultural ecosystems and can be considered to be sustainable for waste land reclamation as well. Earthworms, due to their burrowing activity, increase the soil porosity, ventilate the soil, mineralize the soil, and also recycle nutrients, and are capable of changing the microbiological properties of fresh organic matter during the active phase of vermicomposting. Synthetic fertilizers and pesticides in the field enhance productivity but also disturb the diversity of microbes and earthworms in the soil. The introduction of earthworms in agricultural ecosystems may not only reduce the burden of synthetic fertilizers but may also activate beneficial microbes and improve the quality of the environment. The understanding of microbial activity is of vital importance for the management of soil fertility as soil-disturbing activities reduce the nutrient pool in agricultural fields that will help in planning management strategies for better productivity and yield. This may also have important implications for the optimization of this process and contribute to better understanding the relationships between earthworms and microorganisms during the decomposition of organic matter.

The diverse and complex relationship observed between earthworms and microorganisms in earthworm casts and burrows is an indication that these media increase microorganism activities. The specific associations for some earthworm species have been revealed in this study and the choice of particular organisms by earthworms is still a mystery.

The gut and nephredia of earthworm species is a home to different microbes; however, it will be of interest to study the ecological behavior of different organisms in the gut and nephridia of earthworm species along with fluctuations in the nutrient supply.

References

Aira M, Monroy F, Domınguez J (2007) *Eisenia fetida* (Oligochaeta: Lumbricidae) modifies the structure and physiological capabilities of microbial communities improving carbon mineralization during vermicomposting of pig manure. Microb Ecol 54:662–671

Aira M, Sampedro L, Monroy F, Domínguez J (2008) Detritivorous earthworms directly modify the structure, thus altering the functioning of a micro decomposer food web. Soil Biol Biochem 40:2511–2516

Albiach R, Canet R, Pomares F, Ingelmo F (2000) Microbial biomass content and enzymatic activities after the application of organic amendments to a horticultural soil. Bioresour Technol 75:43–48

Ansari AA (2008) Effect of vermicompost on the productivity of potato (*Solanum tuberosum*), spinach (*Spinacia oleracea*) and turnip (*Brassica campestris*). World J Agric Sci 4(3):333–336

Arancon NQ, Edwards CA, Bierman P, Welch C, Metzger JD (2004) Influences of vermicomposts on field strawberries: effects on growth and yields. Bioresour Technol 93:145–153

Azarmi R, Ziveh PS, Satari MR (2008) Effect of vermicompost on growth, yield nd nutrational status of tomato (*Lycopersicum esculentum*). Pak J Biol Sci 11(14):1797–1802

Baker GH, Williams PM, Carter PJ, Long NR (1997) Influence of lumbricid earthworms on yield and quality of wheat and clover in glasshouse trials. J Soil Biol Biochem 29(3/4):599–602

Bardgett R (2005) The biology of soil: a community and ecosystem approach. Oxford University Press, Oxford, UK, p 242

Bardgett RD, Frankland JC, Whittaker JB (1993) The effects of agricultural practise on the soil biota of some upland grasslands. Agric Ecosyst Environ 45:25–45

Benitez E, Sainz H, Nogales R (2005) Hydrolytic enzyme activities of extracted humic substances during the vermicomposting of a lignocellulosic olive waste. Bioresour Technol 96:785–790

Bernhard A (2010) The nitrogen cycle: processes, players, and human impact. Nat Educ Knowl 2(2):12

Bhadauria T, Saxena, KG (2007) Influence of landscape modification on earthworm biodiversity in the Garhwal region of central Himalayas. Proceedings of Indo-US workshop on Vermitechnology in human welfare, Coimbatore. pp 80–95

Bianchina JN (2009) Development of a flow system for the determination of Cd in fiel alcohol using vermicompost as bioabsorbent. Talanta 78:333–336

Binet F, Fayolle L, Pussard M (1998) Significance of earthworms in stimulating soil microbial activity. Biol Fertil Soils 27:79–84

Bohlen PJ, Edwards CA (1995) Earthworm effects on N dynamics and soil respiration in microcosms receiving organic and inorganic nutrients. Soil Biol Biochem 27:341–348

Brown GG (1995) How do earthworms affect microfloral and faunal community diversity? Plant Soil 170:209–231

Byzov BA, Nechitaylo TY, Bumazhkin BK, Kurakov AV, Golyshin PN, Zvyagintsev DG (2009) Culturable microorganisms from the earthworm digestive tract. Microbiology 78:360–368

Carpenter D, Hodson ME, Eggleton P, Kirk C (2007) Earthworm induced mineral weathering: preliminary results. ISEE8: International Symposium on Earthworm Ecology 8:341 Krakov, Poland
Chand S, Pande P, Prasad A, Anwar M, Patra DD (2007) Influence of integrated supply of vermicompost and zinc-enriched compost with two graded levels of iron and zinc on the productivity of geranium. Commun Soil Sci Plant Anal 38:2581–2599
Citernesi U, Neglia R, Seritti A, Lepidi AA, Filippi C, Bagnoli G, Nuti MP, Galluzzi R (1977) Nitrogen fixation in the gastro-enteric cavity of soil animals. Soil Biol Biochem 9:71–72
Clapperton MJ, Lee NO, Binet F, Conner RL (2001) Earthworms indirectly reduce the effect of take-all (Gaeumannomyces graminis var. tritici) on soft white spring wheat (*Triticium aestivum* cv. Fielder). Soil Biol Biochem 33:1531–1538
Contreras E (1980) Studies on the intestinal actinomycete flora of *Eisenia lucens* (Annelida, Oligochaeta). Pedobiologia 20:411–416
Cooke A, Luxton M (1980) Effect of microbes on food selection by *Lumbricus terrestris* L. Rev Écol Biol Sol 17:365–370
Curry JP, Schmidt O (2007) The feeding ecology of earthworms – a review. Pedobiologia 50:463–477
Daane LL, Haggblom MM (1999) Earthworm egg capsules as vectors for the environmental introduction of biodegradative bacteria. Appl Environ Microbiol 65(6):2376–2381
Darwin C (1881) The formation of vegetable moulds through the action of worms. Murray Publications, London
Davidson SK, Stahl DA (2006) Transmission of nephridial bacteria of the earthworm *Eisenia fetida*. Appl Environ Microbiol 72:769–775
Devliegher W, Verstraete W (1995) *Lumbricus terrestris* in a soil core experiment: nutrient-enrichment processes (NEP) and gut-associated processes (GAP) and their effect on microbial biomass and microbial activity. Soil Biol Biochem 27:1573–1580
Dhanalakshmi V, Remia KM, Shanmugapriyan R, Shanthi K (2014) Impact of addition of vermicompost on vegetable plant growth. Int Res J Biol Sci 3(12):56–61
Doube BM, Schmidt O, Killham K, Correll R (1997) Influence of mineral soil on the palatability of organic matter for lumbricid earthworms: a simple food preference study. Soil Biol Biochem 29:569–575
Domínguez J, Aira M, Brandón G (2010) Vermicomposting: earthworms enhance the work of microbes. In: InsamI H, Franke-Whittle I, Goberna M (eds) Microbes at work: from wastes to resources. Springer, Berlin Heidelberg
Edwards CA, Bohlen PJ (1996) The role of earthworms in organic matter and nutrient cycles. In: Biology and ecology of earthworms. Chapman and Hall, New York, pp 155–180
Edwards CA, Lofty R (1977) The biology of earthworms. Chapmann and Hall, London
Egert M, Marhan S, Wagner B, Scheu S, Friedrich MW (2004) Molecular profiling of 16S rRNA genes reveals diet-related differences of microbial communities in soil, gut, and casts of Lumbricus terrestris L. (Oligochaeta: Lumbricidae). FEMS Microbiol Ecol 48:187-197
Enami Y, Okano S, Yada H, Nakamura Y (2001) Influence of earthworm activity and rice straw application on the soil microbial community structure analyzed by PLFA pattern. Eur J Soil Biol 37:269–272
Farrell FC, Jaffee BA, Strong DR (2006) The nematode-trapping fungus Arthrybotrys oligospora in soil of the bodega marine reserve: distribution and dependence on nematode-parasitized larvae. Soil Biol Biochem 38:1422–1429
Flegel M, Schrader S (2000) Importance of food quality on selected enzyme activities in earthworm casts (*Dendrobaena octaedra*, Lumbricidae). Soil Biol Biochem 32:1191–1196
Gordon JC, Wheeler CT (2012) Biological nitrogen fixation in forest ecosystems: foundations and applications, vol 9. Springer Science & Business Media, Hauge/Boston/London
Graff O (1971) Beeinflussen Regenwurmröhren die flanzenernährung. Landbauforschung Volkenrode 21:303–320
Gresshoff PM, Hayashi S, Biswas B, Mirzaei S, Indrasumunar A, Reid D, Samuel S, Tollenaere A, Hameren BV, Hastwell A, Scott P, Ferguson BJ (2015) The value of biodiversity in legume symbiotic nitrogen fixation and nodulation for biofuel and food production. J Plant Physiol 172:128–136

Haritha Devi S, Vijayalakshmi K, Pavana Jyotsna K, Shaheen SK, Jyothi K, Surekha Rani M (2009) Comparative assessment in enzyme activities and microbial populations during normal and vermicomposting. J Environ Biol 30:1013–1017

Harley JL (1971) Fungi in ecosystems. J Ecol 59:653–668

Hidalgo PR, Harkess RL (2002) Earthworm casting as a substrate amendment for *chrysanthemum* production. Hortscience 37(7):1035–1039

Hildebrand JG (1995) Analysis of chemical signals by nervous systems. PNAS 92:67–74

Hoffland E, Kuyper TW, Wallander H, Plassard C, Gorbushina AA, Haselwandte K, Holmstrom S, Landeweert R, Lundstrom US, Rosling A, Sens R, Smit MM, VanHee PAW, VanBreemen N (2004) The role of fungi in weathering. Front Ecol Environ 2:258–264

Horn MA, Schramm A, Drake HL (2003) The earthworm gut: an ideal habitat for ingested N_2O-producing microorganisms. Appl Environ Microbiol 69:1662–1669

Ismail SA (1995) Earthworms in soil fertility management. In: Thampan PK (ed) Organic agriculture. Peekay Tree Crops Development Foundation, Cochin, pp 77–100

Ismail SA (2005) The earthworm book. Other India Press, Mapusa, Goa, 101pp

Ismail SA, Seshadri CV, Jeeji Bai N, Surya Kumar CR (1993) Composting through earthworms, Monograph series, vol 35. Shri AMM Murugappa Chettier Research Centre, Chennai, p 38

Jayasinghe BATD, Parkinson D (2009) Earthworms as the vectors of actinomycetes antagonistic to litter decomposer fungi. Appl Soil Ecol 43:1–10

Jolly JM, Lappin-Scott HM, Anderson JM, Clegg CD (1993) Scanning electron microscopy of the gut microflora of two earthworms: *Lumbricus terrestris* and *Octolasion cyaneum*. Microbial Ecol 26:235–245

Joshi R, Vig AP (2010) Effect of vermicompost on growth, yield and quality of tomato (*Lycopersicum esculentum* L). Afr J Basic Appl Sci 2(3–4):117–123

Kale RD, Bano K (1986) Field trials with vermicompost (vee comp. E. 8. UAS) on organic fertilizers. In: Dass MC, Senapati BK, Mishra PC (eds) Proceedings of the national seminar on organic waste utilization. Sri Artatrana Ront, Burla, pp 151–157

Karmegam N, Alagumalai K, Daniel T (1999) Effect of vermicompost on the growth and yield of green gram (*Phaseolus aureus* Roxb.) Trop Agric 76:143–146

Karsten GR, Drake HL (1995) Comparative assessment of the aerobic and anaerobic microfloras of earthworm guts and forest soils. Appl Environ Microbiol 61:1039–1044

Khambata SR, Bhat JV (1953) Studies on a new oxalate-decomposing bacterium, pseudomonas oxalaticus. J Bacteriol 66:505–507

Knapp BA, Seeber J, Podmirseg SM, Meyer E, Insam H (2008) Application of denaturing gradient gel electrophoresis (DGGE) for analysing the gut microflora of *Lumbricus rubellus* Hoffmeister under different feeding conditions. J Entomol Res 98:271–279

Knop (1926) Bakterien und Bakteroiden bei Oligochäten. *Z. Morphol.* Ökologie Tiere 6:588–624

Kumar V, Singh KP (2001) Enriching vermicompost by nitrogen fixing and phosphate solubilizing bacteria. Bioresour Technol 76:173–175

Kumar DS, Kumar PS, KumarVU AG (2014) Influence of biofertilizer mixed flower waste vermicompost on the growth, yield and quality of groundnut (*Arachis hypogea*). World Appl Sci J 31(10):1715–1721

Lachnicht SL, Hendrix PF (2001) Interaction of earthworm *Diplocardia mississippiensis* (Megascolecidae) with microbial and nutrient dynamics in subtropical Spodosol. Soil Biol Biochem 33:1411–1417

Lavelle P, Martin A (1992) Small-scale and large-scale effects of endogeic earthworms on soil organic matter dynamics in soils of the humid tropics. Soil Biol Biochem 24(12):1491–1498

Lavelle P, Spain AV (2001) Soil ecology. Kluwer Academic Publishers, Dordrecht

Lavelle P, Zaidi Z, Schaefer R (1983) Interactions between earthworms, soil organic matter and microflora in an African savanna soil. In: Lebrum P, Andre HM, de Medts A, Gregoire-Wibo C, Wauthy G (eds) New trends in soil biology. Louvain-la-Neuve, Dieu Brichart, pp 253–259

Lavelle P, Lattaud C, Trigo D, Barois I (1995) Mutualism and biodiversity in soils. Plant Soil 170:23–33

Lazcano C, Sampedro L, Zas R, Domínguez J (2010a) Vermicompost enhances germination of the maritime pine (*Pinus pinaster* ait.) New Forestry 39:387–400

Lazcano C, Sampedro L, Zas R, Dominguez J (2010b) Assessment of plant growth promotion by vermicompost in different progenies of maritime pine (*Pinus pinaster* ait.) Compost Sci Util 18:111–118
Lazcano C, Revilla P, Malvar RA, Domínguez J (2011) Yield and fruit quality of four sweet corn hybrids (*Zea mays*) under conventional and integrated fertilization with vermicompost. J Sci Food Agric 91(7):1244–1253
Lee KE (1985) Earthworms, their ecology and relationships with soil and land use. Academic Press, Sydney, p 411
Llorens-Marès T, Yooseph S, Goll J, Hoffman J, Vila-Costa M, Borrego CM, Dupont CL, Casamayor EO (2015) Connecting biodiversity and potential functional role in modern euxinic environments by microbial metagenomics. ISME J Nat 9(7):1648–1661
Lores M, Gómez-Brandón M, Pérez-Díaz D, Domínguez J (2006) Using FAME profiles for the characterization of animal wastes and vermicomposts. Soil Biol Biochem 38:2993–2996
Lund MB, Holmstrup M, Lomstein BA, Damgaard C, Schramm A (2010) Beneficial effect of Verminephrobacter nephridial symbionts on the fitness of the earthworm aporrectodea tuberculate. Appl Environ Microbiol 76(14):4738–4743
Maboeta MS, Van Rensburg L (2003) Vermicomposting of industrially produced wood chips and sewage sludge utilizing Eisenia Foetida. Ecotoxicol Environ Saf 56:265–270
Mamta, Wani KA, Rao RJ (2012) Effect of vermicompost on growth of brinjal plant (*Solanum melongena)* under field conditions. J New Biol Rep 1(1):25–28
Marhan S, Kandeler E, Scheu S (2007) Phospholipid fatty acid profiles and xylanase activity in particle size fractions of forest soil and casts of *Lumbricus terrestris* L. (Oligochaeta, Lumbricidae). Appl Soil Ecol 35:412–422
Martin A (1991) Shortand long term effects of the endogeic earthworm *Millsonia anomala* (Omodeo) (Megascolecidæ, Oligochæta) of tropical savannas, on soil organic matter. Biol Fertil Soils 11(3):234–238
McLean MA, Migge-Kleian S, Parkinson D (2006) Earthworm invasions of ecosystems devoid of earthworms: effect on soil microbes. Biol Invasions 8:1257–1273
Monroy F (2006) Efecto das miñocas (clase Oligochaeta) sobre a comunidade descompoñedora durante o proceso de vermicompostaxe. PhD diss., Universidade de Vigo, Spain
Mukherjee RN, Julka JM (1984) On the occurrence of the soil protozoa in the intestine of earthworm *Amynthas morrisi* (Beddard) in Himachal Pradesh. J Soil Biol Ecol 4(1):60–61
Nechitaylo TY, Yakimov MM, Godinho M, Timmis KN, Belogolova E, Byzov BA, Kurakov AV, Jones DL, Golyshin PN (2010) Effect of the earthworms Lumbricus terrestris and Aporrectodea caliginosa on bacterial diversity in soil. Microbial Ecol 59:574–587
Neilson R, Boag B (2003) Feeding preferences of some earthworm species common to upland pastures in Scotland. Pedobiologia 47:1–8
Nethra NN, Jayaprasad KV, Kale RD (1999) China aster (*Callistephus chinensis* (L)) cultivation using vermicompost as organic amendment. Crop Res 17(2):209–215
Orozco FMJ, Cegarra LM, Trujillo RA (1996) Vermicomposting of coffee pulp using the earthworm *Eisenia fetida*: effect on C and N contents and the availability of nutrient. Biol Fertil Soils 22:162–166
Parle JN (1963) A microbiological study of earthworm casts. J Gen Microbiol 31:13–22
Parthasarathi K (2004) Vermicomposts produced by four species of earthworms from sugar mill wastes (pressmud). Ind J Life Sci 1:41–46
Parthasarathi K, Gunasekaran G, Ranganathan LS (2006) Efficiency of mono and polycultured earthworms in humification of organic wastes. J Ann Uni Sci 42:127–134
Parthasarathi K, Ranganathan LS, Anandi V, Zeyer J (2007) Diversity of microflora in the gut and casts of tropical composting earthworms reared on different substrates. J Environ Biol 28:87–97
Pathma J, Rahul GR, Kamaraj Kennedy R, Subashri R, Sakthivel N (2011) Secondary metabolite production by bacterial antagonists. J Biol Cont 25:165–181
Pedersen JC, Hendriksen NB (1993) Effect of passage through the intestinal tract of detritivore earthworms (*Lumbricus spp.*) on the number of selected gram-negative and total bacteria. Biol Fertil Soils 16:227–232

Petersen H, Luxton MA (1982) A comparative analysis of soil fauna populations and their role in decomposition process. Oikos 39:287–388

Peyvast G, Olfati JA, Madeni S, Forghani A (2008) Effect of vermicompost on the growth and yield of spinach (*Spinacia oleracea* L.) J Food Agric Environ 6(1):110–113

Pinel N, Davidson SK, Stahl DA (2008) *Verminephrobacter eiseniae* gen. Nov., sp. nov., a nephridial symbiont of the earthworm *Eisenia foetida*(savigny). Int J Syst Evol Microbiol 58:2147–2157

Pizl V, Novakova A (2003) Interactions between microfungi and *Eisenia andrei* (Oligochaeta) during cattle manure vermicomposting. Pedobiologia 47:895–899

Prabha ML, Jayraaj IA, Jayaraj R, Rao DS (2007) Effect of vermicompost and compost on growth parameters of selected vegetable and medicinal plants. Asian J Microbiol Biotechnol Environ Sci 9(2):321–326

Premono EM, Moawad MA, Vlek PLG (1996) Effect of phosphate-solubilizing pseudomonas putida on the growth of maize and its survival in the rhizosphere. Indones J Crop Sci 11:13–23

Rao KR, Mulani AC, Parlekar GY, Shah NV (2010) Effect of vermicompost on the growth and yield of onion (*Allium cepa*). Karnataka J Agric Sci 23(2):361–363

Reddy YTN, Kurian RM, Ganeshamurthy AN, Pannersalvam P, Prasad SR (2014) Effect of organic practices on growth, fruit yield, quality and soil health of papaya cv. Arka Prabhat Indian Horticult J 4(1):9–13

Satchell JE (1967) Lumbricidae. In: Burges A, Raw F (eds) Soil biology. Academic Press, London, pp 259–322

Scheu S (1987) Microbial activity and nutrient dynamics in earthworm casts (Lumbricidae). Biol Fertil Soils 5(3):230–234

Schinner F, Ohlinger R, Kandeler E, Margesin R (eds) (2012) Methods in soil biology. Springer Science & Business Media

Schramm A, Davidson SK, Dodsworth JA, Drake HL, Stahl DA, Dubilier N (2003) *Acidovorax*-like symbionts in the nephridia of earthworms. Environ Microbiol 67:804–809

Senapati BK, Lavelle P, Giri S, Pashanasi B, Alegre J, Decaens T, Jiménez JJ, Albrecht A, Blanchart E, Mahieux M, Rousseaux L, Thomas R, Panigrahi PK, Venkatachalan M (1999) Soil earthworm technologies for tropical agro-ecosystems. In: Lavelle P, Brussaard L, Hendrix P (eds) Earthworm management in tropical agro-ecosystems. CABI, Wallingford, pp 199–237

Shadanpour F, Torkashvand AM, Majd KH (2011) The effect of cow manure vermicompost as the planting medium on the growth of marigold. Ann Biol Res 6:109–115

Singh KP, Kumar V, Hooda JS (2000) The effect of inoculation with *Eisenia foetida*, N-fixing and P-solubilizing microorganisms on decomposition of cattle dung and crop residues. Biol Hort Agril 18(2):103–112

Singh R, Sharma RR, Kumar S, Gupta RK, Patil RT (2008) Vermicompost substitution influences growth, physiological disorders, fruit yield and quality of strawberry (Fragaria x ananassa Duch.) Bioresour Technol 99:8507–8511

Singleton DR, Hendrixb PF, Colemanb DC, Whitmana WB (2003) Identification of uncultured bacteria tightly associated with the intestine of the earthworm Lumbricus rubellus (Lumbricidae; Oligochaeta). Soil Biol Biochem 35:1547–1555

Sinha RK, Herat S, Valani D, Chauhan K (2009) Vermiculture and sustainable agriculture. Am Euras J Agric Environ Sci 5:51–55

Sinha RK, Agarwal S, Chauhan K, Valani D (2010) The wonders of earthworms and its vermicompost in farm production: Charles Darwin's 'friends of farmers', with potential to replace destructive chemical fertilizers from agriculture. Agric Sci 1:76–94

Stephens PM, Davoren CW, Doube BM, Ryder MH (1993) Reduced superiority of Rhizoctonia solani disease on wheat seedlings associated with the presence of the earthworm Aporrectodea trapezoids. Soil Biol Biochem 11:1477–1484

Suhane RK (2007) Vermicompost. Publication of Rajendra Agriculture University, Pusa, p 88

Suthar S (2009) Vermicomposting of vegetable-market solid waste using *Eisenia fetida*: impact of bulking material on earthworm growth and decomposition rate. Ecol Eng 35(5):914–920

Thakuria D, Schmid O, Finan D, Egan D, Doohan FM (2010) Gut wall bacteria of earthworms: a natural selection process. ISME J 4:357–366
Tiunov AV, Dobrovolskaya TG (2002) Fungal and bacterial communities in *Lumbricus terrestris* burrow walls: a laboratory experiment. Pedobiologia 46:595–605
Tiwari SC, Tiwari BK, Mishra RR (1989) Microbial populations, enzyme activities and nitrogen, phosphorous, potassium enrichment in earthworm casts and in the surrounding soil of pine apple plantation. Biol Fertil Soils 8:178–182
Tomati U, Galli E, Grappelli A, Dihena G (1990) Effect of earthworm casts on protein synthesis in radish (Raphanus Sativum) and lettuce (*Lactuca sativa*) seedlings. Biol Fertil Soils 9:288–299
Toyota K, Kimura M (2000) Microbial community indigenous to the earthworm *Eisenia foetida*. Biol Fertil Soils 31:187–190
Vadiraj BA, Siddagangaiah D, Potty SN (1998) Response of coriander (*Coriandrum sativum* L.) cultivars to graded levels of vermicompost. J Spices Aromatic Crops 7:141–143
Vaz-Moreira I, Silva ME, Manaia CM, Nunes OC (2008) Diversity of bacterial isolates from commercial and homemade composts. Microbial Ecol 55:714–772
Vivas A, Moreno B, Garcia-Rodriguez S, Benitez E (2009) Assessing the impact of composting and vermicomposting on bacterial community size and structure, and functional diversity of an olive-mill waste. Bioresour Technol 100:1319–1326
Wang D, Shi Q, Wang X, WeiM HJ, Liu J, Yang F (2010) Influence of cow manure vermicompost on the growth, metabolite contents, and antioxidant activities of Chinese cabbage (*Brassica campestris ssp. chinensis*). Biol Fertil Soils 46:689–696
Wani KA, Mamta, Rao RJ (2013) Bioconversion of garden waste, kitchen waste and cow dung into value-added products using earthworm *Eisenia fetida*. Saudi J Biol Sci 20(2):149–1540
Webster KA (2005) Vermicompost increases yield of cherries for three years after a single application. EcoResearch, South Australia
Zaller JG (2007) Vermicompost as a substitute for peat in potting media: effects on germination, biomass allocation, yields and fruit quality of three tomato varieties. Sci Hortic 112:191–199
Zelles L (1999) Fatty acid patterns of phospholipids and lipopolysaccharides in the characterization of microbial communities in soil: a review. Biol Fertil Soils 29:111–129

Organic Farming, Food Quality, and Human Health: A Trisection of Sustainability and a Move from Pesticides to Eco-friendly Biofertilizers

26

Nitika Thakur

Abstract

The organic amendments that were witnessed in the "green phase" during the 1960s boosted food production, but at the expense of environmental sustainability. These methods increased food production but ultimately disturbed the physical, chemical, and biological properties of soil over years of use. The beneficial soil organisms were exploited and the power of "biological resistance" in crops was reduced, making them more prone to pests and diseases. As a result, no part of the world is left free of harmful pesticide residues today. Over time, it was realized that these residues are toxic for soil and society. Use of chemical fertilizers has not only led to sensational increases in the quality and quantity of crops, but has also resulted in the alteration of the total soil profile resulting in a reduction of beneficial microbes leading to an imbalance in ecology. This has ultimately devastated the resources of farmers, who are the building the path of our nation. Excessive use of non-renewable energy chemicals often tends to destroy the physiochemical properties of soil, reduce friendly predators, and enhance residual hazards in seeds and to human health and the environment. The use of beneficial microbial inoculants along with organic manures is considered to be an alternative requirement for crops. The technological approaches to the use of organic manures and biofertilizers in farming have proved to be effective means of upgrading soil structure, increasing water-holding capacity, enhancing soil fertility, and increasing crop yields. On the whole it can be deduced from the present studies that by integrating correct combinations of organic production technologies, production levels comparable to conventional practices can be achieved in tomato crops with improved soil-nutrient status and productivity.

N. Thakur
Shoolini University of Biotechnology and Management Sciences, Solan, HP, India
e-mail: nitikathakur45@gmail.com

V. Kumar et al. (eds.), *Probiotics in Agroecosystem*,
DOI 10.1007/978-981-10-4059-7_26

26.1 Introduction

The present agricultural practices mainly depend on high-priced inputs like mineral fertilizers to attain a high yield and also involve the application of chemical pesticides against relevant pathogens and pests. The application of chemical fertilizers not only extensively damages the helpful microbes in the soil but also causes detrimental effects on human health as well as environmental hazards, and reduces the soil fertility. It is now well established that application of nitrogen can result in nitrate leaching through the soil profile due to groundwater contamination. The issues and concerns about the destructive effects of using increasing amounts of chemical fertilizers have led to a strong move toward alternative strategies to ensure high yields coupled with crop safety and protection.

The indiscriminate use of hazardous pesticides and herbicides could result in diverse changes in the biological balance, increasing the incidence of lethal diseases like cancer, through undesirable harmful residues present in the produce. Industrialized production methods have clearly shown several limitations indicating global contamination of the food chain and water through toxic pesticide residues and a reduced nutritive value of food in cultivation practices carried out by farmers.

Because of these adventitious properties, tomato producers often use large amounts of chemical fertilizer, which is not sustainable due to the ill effects on the soil and environment through the high involvement of non-renewable energy in production input used to ultimately enhance the yield and quality of crops. The modern approach, often referred to as organic agriculture, seeks to introduce agricultural cultivation practices that are eco-friendly and maintain the sustainable ecological balance of the ecosystem. The growth pattern of organic agricultural land has increased tremendously from 1999 to the present time (Fig. 26.1).

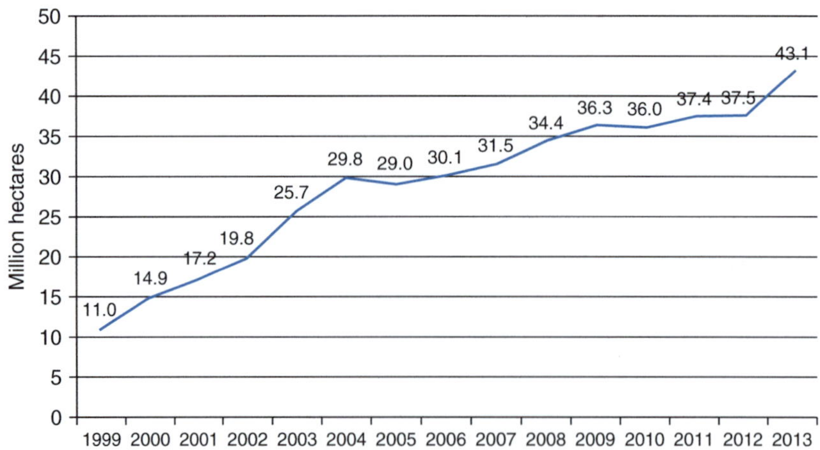

Fig. 26.1 Growth of organic agricultural land (1999–2013) (Source-ICCOA)

Organic matter is an excellent source of available nutrients, and their incorporation in soil can maintain high activity of microbial populations with increased values of biomass content, basal respiration, and total organic carbon (Tonfack et al. 2009).

The poor health of soil due to disease contamination after repeated use and the desire to implement optimal conditions for plant growth have led to the modern trend of growing plants in soil-less media. While soil-less media incur additional costs for growing systems and chemical fertilizers, they offer earlier growth and higher yields.

The biopriming of seeds with beneficial microorganisms provides long-term protection from yield-threatening fungal or bacterial diseases by creating a protection shell around the seed- root system, which provides a stronger and healthier root system leading to increased crop productivity and gradually to better yields. *Trichoderma harzianum* and *Trichoderma viride* are two widely used species that have been used for about 87 different crops, 70 and 18 soil and foliar pathogens, respectively (Sharma et al. 2014).

Organic agriculture is a multidirectional management system that furnishes the health of the agrological ecosystem. Sustainable farming, quality of food, and human health argue that environmental agents should be directed toward organic products. The use of management practices via the off-farm inputs require locally adapted systems taking into account the prevailing regional conditions. This is accomplished by using diverse methods such as agronomic, biological, and mechanical methods as opposed to using synthetic materials to fulfill any specific function. The focus is on maintaining soil fertility for generations, producing poison-free food for consumers, securing productivity, meeting competition from likely cheaper imports, achieving high water percolation, recharging groundwater, developing nitrogen- and phosphate-fixing microbes involved in transferring atmospheric moisture, soil enrichment through transfer of biomass from agro-waste, emergence of mixed farming systems, new marketing channels, premium prices, and higher product demand, on a worldwide basis (Figs. 26.2 and 26.3).

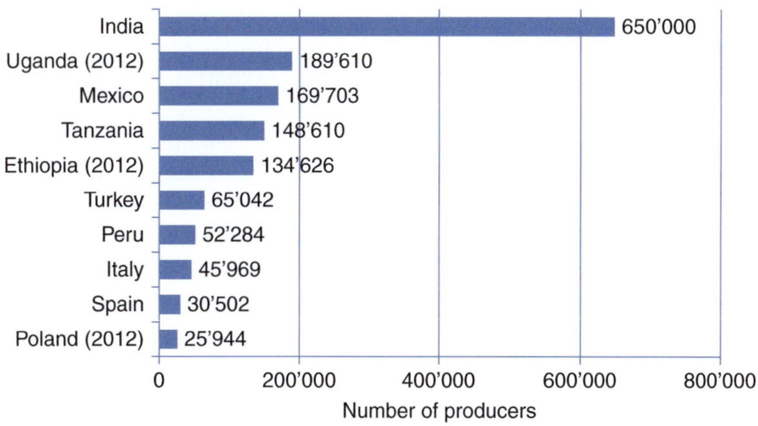

Fig. 26.2 The ten countries with the largest number of organic producers (2013) (Source: ICCOA)

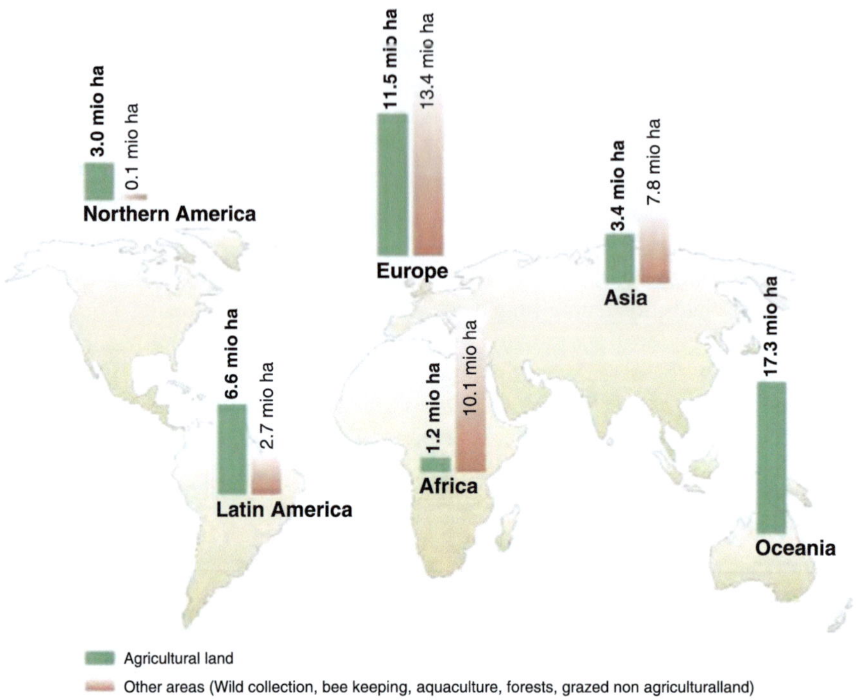

Fig. 26.3 Organic agriculture worldwide: statistics (Source: ICCOA)

Global markets for organic products are growing, hence satisfying the criteria for food safety (less incidence of diseases like mad cow disease and cancer, etc.), health aspects (over 20% more vitamins and minerals), price premiums, environmental concerns, and sustainability.

Parameters highlighting food security and safety are of primary concern to each individual. Thus, quality can be defined as a strong characteristic of food that determines the value of acceptability to a consumer. The increasing consumer awareness about the relationship between food, health, and environmental concerns has led to the increasing demand for organically cultivated food.

Organic food commodities contain lower pesticide and nitrate levels than conventionally grown fruit and vegetables. This may be considered beneficial in relation to antioxidants (polyphenolic compounds). Organic foods do not involve the use of synthetic fertilizers; they possess efficient biochemical energy to synthesize the important secondary plant metabolites as well as naturally occurring toxins. The present scenario of tomato production by the farmers is confined to the conventional open-field cultivation system with varied agro-climatic conditions in the Solan district of Himachal Pradesh. The favorable positioning of the State in the Himalayan region provides a great scope for the implementation of organic farming. The policies framed by the State government on organic farming in 2010 relate to 30,110 farmers with an area of 17,848 ha with a future vision in mind of converting around

200 villages to complete bio-villages. However, the government has already initiated the process of registration and certification for using organic fertilizers to organically cultivate tomatoes, but the farmers are still unaware of the incorporation of organic recommendations.

The economic parameter in organic farming stands out as an important issue to the farmers. Beneficial economics with good incentives will be the greatest boost for the adoption of organic practices in crop husbandry. Achieving circumstances that direct favorable economic conditions for organic farming becomes a priority. The history of organic tomato in HP is only 3–4 years. Although farmers have gradually been shifting to organic practices, the switch-over is not complete in the majority of cases. They have not been able to develop the mindset required for organic cultivation of tomatoes. Many farmers are not ready to put in the labor required for the preparation of inputs under the organic system; they look to markets for input supply. The shift has been gradual and the change in mindset even slower. However, many inorganic farmers are presently realizing the deleterious effects of using hazardous products on human health and the environment. Hence, they are in the process of reducing the doses and frequency of chemical products and relying more on farmyard manure (FYM), vermicompost (VC), and biofertilizers.

26.2 Organic Farming Systems: Quantity Coupled with Quality

Considering the ill effects on natural resources and global marketing demand for quality products, there is a need to switch from intensive chemical cropping systems to organic farming systems that will not only help in yield sustainability (Fig. 26.4) (Barbier 1987) but also earn higher foreign exchange from export (Jangir et al. 2008). On other hand, the small farmers with less than 1 ha of land are facing many problems related to debt, vagaries of rainfall and nature, lack of investment, and soil fatigue. The food quality concept can be defined in many different ways, such as through the quality of the fresh produce, which is often judged by examining the

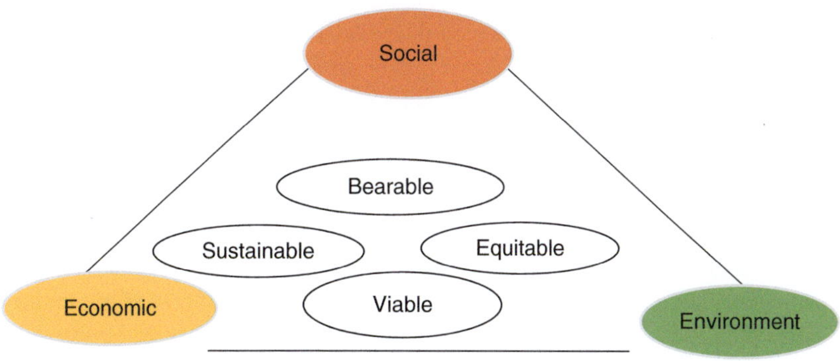

Fig. 26.4 Scheme of sustainable development (Source: Barbier 1987)

visual characteristics such as size, shape, and color. The present review deals with a combinatorial study and breakthrough aimed at differentiating between conventional and organic farming systems, thus ensuring the realization of the criteria of good soil fertility, nutrition, quality, productivity, yield, economics, and food security and safety of vegetable crops and fruit. This review highlights those aspects of nursery management, soil fertility, crop yield, productivity, economics, and food quality, which are the most crucial for the promotion of good health, specifically focusing on the available evidence for four criteria:

- *Safety and security of food*
  The first question relates to the extent of organic and non-organic foods that contain potentially harmful chemicals and pathogens.
- *Primary nutrients*
  The second question deals with the contribution of organic and non-organic foods toward a balanced diet.
- *Secondary nutrients*
  The third question focuses on the effect of different farming practices on the concentration and range of secondary plant compounds.
- *Microbiological hazards*
  The fourth question focuses on the final test of nutritional quality for a food to support growth and development thus reducing the microbiological hazards.

26.3 Effect of Microbial and Inorganic Fertilizers

26.3.1 Disease-Free Nursery Management

The necessity of healthy seed selection integrated with nursery raising (free from disease) is the most important requirement for achieving better crop performance in addition to lower abiotic and biotic yield constraints and lower incidence of insect-pest-disease (IPD), in order to achieve attractive economic returns for the farmers. The reduction witnessed in potential yield in hills by the farmers is due to the increasing incidence of pre- and post-emergence insect pests and diseases from the initial nursery raising to the final period of harvesting, where the incidence of serious diseases like damping off (*Pythium aphanidermatum*), fusarial wilt (*Fusarium oxysporum*), bacterial wilt (*Ralstonia solani*), etc., can be seen to be drastically ruining the crop diversity and quality. This review thus discusses healthy nursery management through various organic approaches.

The history of biological control can be traced back to the era of 1965, where the interactive ideas of Baker and Synder (Baker and Snyder 1965) emphasized the significance of the biocontrol formulations. Expanding these valuable ideas, biocontrol agents have gained momentum in nursery management, seedling germination, seedling vigor, and disease control in recent times as these technologies not only minimize the hazardous aspects of chemicals but are also found to be cheap and efficient in disease-control strategies. With the advent of biological control practices, several successful uses of fungal biocontrol agents like *Trichoderma* spp. have been investigated for controlling the soil-borne diseases caused by pathogens like *Sclerotium*, spp., *Fusarium*, *Pythium*, and *Phytophthora* (Cook and Baker 1983). The increased concern for environmental awareness, safety, and security of chemicals has evoked an interest in inbuilt microbial control of pathogens. In this regard, many microorganisms have been exploited as significant biocontrol agents for successful nursery raising and management. *Trichoderma* spp., (Raguchander et al. 1997), *Bacillus* spp., (Copper and Campbell 1986), and *Pseudomonas* spp. (Vidyasekaran et al. 1997) are specially incorporated in terms of the formulation and delivery system for research because of their abundant natural occurrence, biocontrol potential against fungal and nematode diseases, and host defense. *Trichoderma* has gained maximum attention as a biocontrol agent due to the fact that it is effective against a large number of soil-borne plant pathogenic fungi, has suppressive effects on some root nematodes without adversely affecting beneficial microbes like *Rhizobium*, and is capable of promoting growth of certain crops.

Various types of compost have been employed advantageously in nursery management, generally prepared by combining carbonaceous wastes such as sawdust with nitrogen and other nutrients contained in the manure (Galler et al. 1978) and also sewage sludges (Anonymous 1982). The pathogens are difficult to manage by chemical methods and there is a current interest to promote their biological management. The suppression of disease caused by soil-borne pathogens on application of vermicompost has been reported (Jack 2010). In the earlier *in vitro* studies, it was found that certain soil-borne fungal and bacterial plant pathogens are suppressed by earthworm exudates (Reddy et al. 2012).

The results of *in vitro* antimicrobial assay and pot culture studies conducted earlier (Agbenin and Marley 2006) formed the basis on which to evaluate the effects of seed treatment with aqueous extracts (10%) of vermicompost prepared from different substrates (agricultural wastes, leaves of *Azadiracta indica* (neem), *Parthenium hysterophorous* (parthenium) and *Lantana camara* (lantana)), and soil application of vermicomposted neem. The response of susceptible crops, e.g., tomato (*Lycopersicon esculentum*) and egg plant (*Solanum melongena*) in the infected fields were evaluated for developing effective biological management. Application of vermicompost alone was not enough for protecting the plants against disease, but combining the same with aqueous seed treatment is necessary to achieve complete disease uprooting and an increase in yields (Reddy et al. 2012).

Various studies have focused on the use of botanicals in combination with arbuscular mycorrhizal fungi (AMF) recording higher crop uniformity, better mineral uptake (Bethlenfalvay et al. 1988) and improved tolerance to soil-borne pathogens (Pozo and Azcon Aguilar 2007). Evidence from several studies highlights the activities of soil-borne pathogens and their antagonists, which are greatly influenced by the presence of various plant products present in the soil. These beneficial products not only alter the physiochemical characteristics of the soil, but also increase the density of antagonist inoculums by serving as a substrate medium for their growth (Champawat and Sharma, 2003; Neelamegam and Govindarajalu, 2002), which further results in overall suppression of diseases. The most devastating fungal disease is the damping off caused by *Pythium phytopthora, Fusarium* spp., and *Sclerotium* spp., which results in 50–60% losses in nursery plants (Srivastava and Singh 2000). Damping off is a serious disease of tomato nurseries that results in high seedling mortality. The findings of Kabdal et al. (2010) incorporated single and combined formulations of biocontrol agents, namely *Trichodema herzianum* 0.4% and *Pseudomonas fluorescens* 1.0% for the healthy management of capsicum nursery. All biological formulations were found to be significantly effective in increasing seedling emergence (71.9%) and vigor (135.2%) with marked decreases in pre-emergence rot (52.5%).

To tackle the increasing incidence of damping off of tomato under mid-hill conditions of the north western zone of the Himalayas, Hooda et al. (2011) highlighted the combinatorial selection of 17 locally available plant extracts (10% w/v) (*Lantana camara* (@10%), neem-based commercial formulations (10% w/v), botanical fungicide, raw neem oil (0.05% v/v), neem cake extract , cow urine (20% v/v), and cow dung ash. Among the selected extracts, i.e. *Thuja compacta, Azadirachtin* (Achook), neem cake extracts, cow urine and dung formulations were found to be the most promising for the management of pre-damping off and also significantly increased the mean seedling emergence (60.1%) and vigor (81.3%) in tomatoes treated with *Lantana camara* extracts as compared to the uninoculated controls. The study clearly indicated that the treatment of tomato seeds with extracts of *Lantana camara*, neem cake, and cow urine can be effectively utilized as a cost-effective, eco-friendly, and suitable alternative method for hilly areas where pesticide availability is scarce. The incorporation of plant growth–promoting rhizobacteria (PGPRs) with BCAs has been successfully integrated in tomatoes (Muthuraju et al. 2002) and green grams (Thilagavathi et al. 2007). Srinivasan and Mathivanan (2011) recorded a gradual increase in tomato seedling growth and a reduction in disease incidence both in nurseries and under field conditions with the use of a consortium of antagonist fungi alone and in combination with PGPR and biocontrol agents. The studies conducted by Kumar et al. (2010) focused on the development of modern strategies for the management of damping off on tomatoes caused by *Pythium aphanidermatum* with maha panch gavya (MPG), in combination with biocontrol agents (neem products) in nursery beds. Soil application of MPG with neem cakes in nursery beds improved seedling stand (63%), seedling height (27.09 cm), and decreased seedling mortality with high disease control (100%) in inoculated soil, thus ensuring one of the major components in managing plant diseases, especially soil-borne diseases, in organic farming.

Recent research carried out in Bangalore (Sudharani et al. 2014) highlights the effectiveness of selected bio-control agents in combination against damping off and wilt pathogens of cabbage crop. The results revealed that the treatment combination recorded highest germination percentages (91% and 95%, respectively) and took a minimum of days for 50% germination over pathogen-inoculated treatment. A maximum reduction of pre- (9.09%) and post- (6.14%) damping off was reported in Treatment T_{10} on a par with T_{12}.

The combination of *Azotobacter chroococcum* + *Bacillus megaterium* + *Pseudomonas fluorescens* + *Bacillus subtilis* + *Trichoderma harzianum* showed enhanced seedling vigor, total biomass, least disease incidence, and more biocontrol efficiency.

26.3.2 Soil Fertility, Health, and Quality

The improvement in soil fertility in organic farming through the use of composts and on-farm input relies on improved understanding about the effects of application methods on soil fertility along with the improved technology transfer of research results into practice.

The application of soil amendments has been associated with desirable soil properties including a water-holding capacity, lower bulk density, and beneficial microorganisms (Doran 1995). A similar correlated study highlights the fact that microbial activity and biomass is recorded higher in fields with organic amendments than in conventional fields (Drinkwater et al. 1995).

Bulluck over two successive years (1996 and 1997) conducted a field experiment emphasizing examining the effects of organic and chemical soil fertility amendments on soil microbial communities along with soil physical and chemical properties at three organic and three conventional farms in Virginia and Maryland, respectively. Two treatments including use of composted yard waste or cattle manure and synthetic soil amendment were applied to three replicated plots. A canonical correlation was figured out, which showed more negativity in fields with conventional history and synthetic fertilizers in comparison with the positive relationship in fields with organic production. Propagated densities of *Trichoderma* spp., thermophilic microorganisms, and enteric bacteria were higher in soils of organic fields. The concentration of major elements (calcium (Ca), potassium (K), magnesium (Mg), and manganese (Mn)) was also reported to be higher. On the whole, organic application resulted in increased beneficial soil microorganisms, reduced pest-pathogen population, and increased soil organic matter, thus improving soil health and fertility. The application of compost amendments results in providing benefits including pH stabilization and faster water infiltration rate (Stamatiadis et al. 1995). The integration of manures and composts tend to positively increase soil organic matter content thus reducing bulk density and increasing porosity, which in turn will have a significant impact on the protection of soil against erosion. This increase in the soil organic matter content is directly related to the increase in cation exchange capacity.

Sanwal et al. (2007) advocated an integrated approach toward organic manure application to assess the effects on residual soil fertility, quality, and yield parameters in turmeric. A significantly increased rhizome yield in the range of 16–103% was recorded with the application of FYM at18 t/ha, on a par with 10 t/ha poultry manure. The outcome of this approach resulted in not only the highest crop yield but also improved soil fertility and productivity.

26.3.3 Nutrient Uptake, Growth, and Yield Status

Yan et al. (2002) in his 3-month study analyzed compost maturity, which affects crop nutrient uptake, and recorded that after application of dairy cattle, swine, and poultry manure pellets released 31.5%, 41.6%, and 51.3% of nitrogen (N), respectively. A 2-year field trial was conducted by Trivedi et al. (2012) to assess the response of guava varieties to the integrated application of organic manures, inorganic fertilizers, and bio-fertilizers (2005–07). The sardar variety was recorded to have greater plant height when compared to Allahabad safeda, which registered a higher total soluble sugar (TSS), available nitrate, P_2O_5, and K_2O content in the soil. The application of castor cake proved best for attaining maximum plant height and nitrate uptake. The incorporation of vermicompost and FYM revealed big enhancements in N, K, and carbon content. The approach led to a greatly increased high fruit yield and available P_2O_5 content in soil with the addition of biofertilizer.

Patel et al. (2009) conducted successive experiments over a period of 3 years to work out the influence of microbial and inorganic fertilizers in combination with 0.4% micronutrients on different growth parameters along with attributes like yield, leaf nutrient status, and resultant changes in the rhizosphere in sweet oranges. The treatment consisted of application of full dosea of N (300 g), P (250 g), K (300 g), *Azospirillum* (5 g), AMF (5 g), and micronutrients (Cu + Fe + B + Zn 0.4%). This integrated application of macro- and micronutrients showed a great increase in plant height, canopy spread, fruit yield, quality, and juice content. The use of AMF along with *Azospirillum* proved beneficial in improving biological properties of soil.

Prativa and Bhattari (2011) carried out a field experiment at the Integrated Research Farm (Himalayan College of Agricultural Sciences and Technology (HICAST)) located in Nepal during 2009, to scrutinize and study the effect of integrated nutrient management (INM) on growth parameters of tomatoes with a randomized complete block design with nine treatments replicated three times. The outcome of the study clearly indicated that the combination of organic manures with inorganic fertilizers was found to be better in improving the overall growth and soil macro-micronutrient status than the sole application of either of these nutrients. The maximum plant height and number of leaves per plant were recorded with treatment combining 16.66 m/ha FYM + 8.33 m/ha vermicompost + NPK. The highest number of clusters and maximum fruit weight and yield (25.74 mt/ha) were recorded with the treatment with 16.66 mt/ha FYM + 8.33 mt/ha vermicompost + NPK. Similarly, the maximum organic matter percentage was also observed with the treatment with an application of 10 m/ha vermicompost.

Similar types of experiments were conducted (2012) on strawberry and consisted of a combination of five successive nutrient source treatments; T_1-FYM with *Azotobacter*, PSB and oil cake, T_2-poultry manure with *Azotobacter*, wood ash, PSB, and oil cake, T_3-FYM with *Azospirillum*, PSB, and oil cake, T_4-poultry manure with *Azospirillum*, wood ash, PSB, and oil cake, and T_5 comprised of the recommended dose of NPK (340:150:340 kg/ha). The results showed maximum plant growth and fruit yield (132.75 q/ha) with T_2 treatment closely followed by T_4. On the other hand, maximum available N (370.29 kg/ha) and phosphorus (P) (22.11 kg/ ha) were recorded with treatment T_4 with a gain of 36.29 and 4.61 kg/ ha, respectively. Maximum potassium (331.79 kg/ha) was obtained with treatment T_2 with a gain of 12.29 kg/ha. The varying degrees of difference were postulated in a population of bio-fertilizers that showed a maximum increase in the case of *Azotobacter*, PSB in treatment T_2 and *Azospirillum*, which recorded the maximum increase with treatment T_4. The highest yield and sustainability was found in T_2 and T_4 treatments.

Sepat Naval et al. (2012) conducted a field experiment during the kharif season of 2008 and 2009 at the Defence Institute of High Altitude Research, Leh and Ladakh, to evaluate the effect of biofertilizer, fertility levels and cow manure on growth, yield and quality of tomato var. Sultan in Trans Himalayan. Results revealed that the treatments with 100% NPK either in combination with each other or with *Azotobacter* had a significant effect on plant growth and economic attributes over control. However, application of 50% NPK + FYM + *Azotobacter* gave values of plant height (79 cm), branches (7.5), clusters of fruit (11), fruits cluster (4.8), fruit size (6.3 cm), weight of fruit (113.3 g), and fruit yield: plant (1.48 kg) and (12.3 q ha^{-1}) that were on a par with the values obtained with 100% NPK+FYM+*Azotobacter* and were significantly higher over other treatments.

Nutrient management affects both productivity and quality of produce and also contributes to input costs of production, and was fully justified by Singh et al. (2012) in their findings when they evaluated the influence of various levels of organic and synthetic nutrient sources on morphomatrix productivity and soil quality attributes of NA-7 Aonla trees during 2007–08 in a hot semi-arid ecosystem. A different application of cakes and FYM resulted in higher yield and quality. Maximum yield per plant (32.15 kg) was recorded with the plants treated with FYM and standard doses of NPK. A similar hike was observed in quality parameters like total sugars, vitamin C, and phenols with considerable improvement in soil properties.

An investigation carried out by Ramakrishnan and Selvakumar (2012) aimed at evaluating the effect of biofertilizer application on the growth and yield of tomato plants. After the transplanting process the tomato seedlings were treated with different formulations: T_0-Control, T_1-*Azotobacter*, T_2-*Azospirillum*, and T_3-*Azotobacter* with *Azospirillum*. The observations were based on recording significantly high performance in plant dry weight (g plant-1), height (cm), number of leaves per plant, number of fruits/plant, yield/plant (g), average fruit weight/plant (g), and protein content. On the whole, the treatments comprising *Azotobacter* with *Azospirillum* showed a significantly ($P < 0.05$) maximum yield when compared with single inoculations and control (428.41 g). The overall results suggested that inoculation combinations improve plant mineral concentration through nitrogen fixation, thereby altering fruit production in tomato plants.

The studies carried out by Chatterjee (2013) aimed to assess the influence of the integrated use of FYM, vermicompost, and inorganic fertilizers on plant nutrient uptake and the post-harvest status of tomato cultivation with 14 designed treatments. The pooled data revealed that the treatment T_3 was recorded with 17% higher potassium over treatment T_1. The results further pointed toward treatment T_{13}, which was recorded as having the maximum Kcontent in fruit as well as in plant residues (2.37%). The finding also emphasized that the application in which the vermicompost was supplemented with Azophos and 75% of inorganic fertilizer resulted in a maximum uptake of macronutrients by the tomato plants. Thus, the integrated intervention of diverse sources of nutrients not only increases the plant nutrient uptake, but also improves the soil fertility post-harvest and subsequently helps to attain the much desired crop production with sustainable soil health.

Haque et al. (2013) evaluated the effect of bio slurry on the performance of vegetable crops (cabbage, spinach, and brinjal) and two oil seed mustard plants. The trial was carried out in 25 locations with 90 farmers in the country. Four nutrient management packages, namely inorganic fertilizer IPNS with poultry manure/cow dung/poultry slurry/cow dung slurry along with farmers' practice were trialled on different vegetable crops. The workers reached the conclusion that there was increased plant height, yield, and weight per plant with all aspects, namely 0.23 g per plant using organic fertilizers like cow dung (CD), poultry manure (PM), and poultry manure (PM) slurry treatments. The yield from the poultry manure (PM) showed a reasonable increase in the growth of cabbage and showed increasing percentages from 6.9% to 11% per plant. A higher yield in brinjal was recorded in the treatment of plants in the field using organic fertilizer with bio-slurry. The yield surprisingly extended up to 90–118% over inorganic fertilizer and farmers' practice, thus affording a good financial turnover per ha in cost return. The highest growth of spinach in the field using IPNS + bio-slurry combination was recorded with the highest economic turnover compared to T_1 and T_3 treatments.

26.3.4 Crop (Fruit) Quality, Yield, and Economics (Benefit-to-Cost Ratio)

The aim of the study conducted by Yanar et al. (2011) was to evaluate the effects of different organic manures on yields and fruit quality of tomato compared during growing periods under field conditions. During the initial growing period (2006), the organic fertilizers used were Ormin K, N (40 kg/ha every week), composted poultry manure (CPM), and composted cattle manure (CCM) after first flowering and 5 t/ha after first harvest (liquid form).

Based on the initial year results, organic fertilizers used in year 2007 growing periods were F_1 (20 ton/ha CCM; 1 t/ha CPM used before planting; 40 kg/ha Coplex and 20 kg/ha N every week) and F_2 (20 t/ha CCM before planting; 500 kg/ha Ormin K before planting; 30 kg/ha Coplex and 30 kg/ha N every week). Inorganic fertilizers used as control (N: 450, P_2O_5: 350, K_2O: 600, CaO: 50, S: 200, and Mg: 50 kg/ha) were tested too. The tomato cultivars used in this study were Alida Fı in the 2006 growing period and Alida Fı, Yankı Fı, and Maya Fı during the 2007 growing period, respectively. During 2006, the highest yields were obtained from CPM, CCM, and control treatments. During 2007, marketable yields were the same for F_1 fertilizer treatment and the control application.

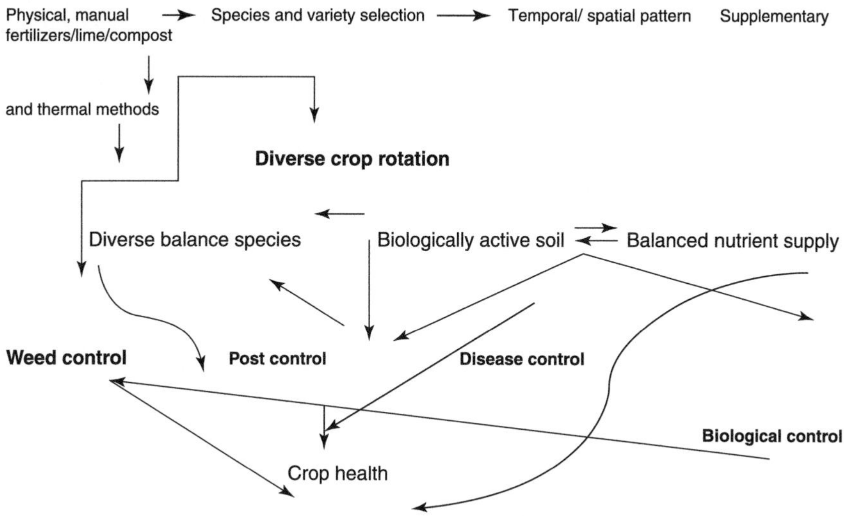

Fig. 26.5 Diagrammatic representation of structural and tactical components of organic production systems (Stockdale et al. 2000)

There was no significant difference among the treatments. However, it was observed that fruit cracking rates were higher in organic fertilizer treatments than the inorganic fertilizer treatment.

The results indicated that parameters like firmness and color value decreased significantly in all treatments during storage. However, TSS and reducing sugar values significantly increased in fruit analysis during storage. Application of microbial fertilizers and their combination significantly affected the quality parameters. The data suggested that organically produced fruit maintain their quality during storage for a period that is comparable to that of conventionally grown fruits. Similarly, the linkage of different components of organic crop production systems (Fig. 26.5) (Stockdale et al. 2000) also plays an important role in upgrading crop quality and health.

Kapoulas et al. (2011) focused on comparing the fruit quality parameters in different tomato cultivars (Robin-F_1, Amati $F_{1,}$ and Elfida F_1) obtained from organic and conventional greenhouse production in North-Eastern parts of Greece. Higher levels of sugar and vitamin C were recorded in conventional systems, while those grown organically contained increased amounts of carotenoids. Elphida cultivar was seen with the highest content of TSS (5.08%), sugar content (4.10 mg/100 g) and lycopene (37.5 mg/kg) in three varieties. The fruit flavor in organic production was much better than the tomatoes from conventional production because of the favorable ratios of total sugar and acid. Organically grown tomatoes had a softer texture and were preferred because of their better taste and juiciness, whereas the conventional tomatoes were described as dry and having less aroma.

Six fresh market tomatoes and three processing varieties that were harvested at the "mature green" stage were evaluated for total reducing and non-reducing sugar trends as well as marketability during a period study of 32 days storage under ambient conditions (Tadesse et al. 2012). The studies were undertaken using a randomized complete block design replicated three times. These tomato varieties experienced significant effects on overall quality and maintenance. At harvest, the highest sugar content was seen in Marglobe, whereas the processing tomato variety Roma VF showed a higher sugar content than the other two processing varieties, but the two processing varieties were better in their chemical quality characteristics than the former varieties.

Kachari and Korla (2012) reported a considerable increase in yield (33.94 t/ha) with a benefit:cost ratio amounting to 1:2.58 over the uninoculated control (FYM) during 2006–2007, with objectives to evaluate the influence of biofertilizers (*Azotobacter*, *Azospirillum*, AMF, PSB-1) and inorganic fertilizers on the quality and economics of cauliflower. The randomized block design consisted of 21 treatments. The formulation of bio-fertilizers with inorganic fertilizers showed the best results as compared to control (FYM).

Choudhary et al. (2012) reported the effect of different organic sources on subsequent parameters of sprouting broccoli under semi-arid conditions of Rajasthan. A significant increase in plant height, number of leaves, leaf area, diameter of head, total head yield, and chlorophyll content in head was recorded under various organic fertility levels.

Chatterjee et al. (2014), in order to study the adequate tomato plant nutrient needs for optimum growth and yield, incorporated field work that aimed at working out the effect of different combinations (15) of organic and inorganic nutrient sources on soil and crop profile. The results revealed that the tomato parameters responding to nitrogen use were greatly influenced by the application of different nutrient sources. It was observed successively that vermicompost was found to be the best organic nutrient source over farmyard manure. Inoculation with biofertilizer showed more positive results than the uninoculated treatments.

26.3.5 Microbiological Food Quality and Safety

Food safety can be defined as the assurance of the food quality that it will not cause any harm to the consumer when it is in the preparation or eating process according to its intended use (FAO/WHO 1997). It has been estimated that 82 food-borne illness outbreaks were associated with the consumption of fresh produce during 1996–2008. This time period was linked with tomato-associated outbreaks accounting for 1,927 illnesses and three deaths. These tomato-associated outbreaks were considered fatal. There are many factors that may play a role in the increased incidence of food-borne illness outbreaks that implicate fresh produce, such as a population consisting of aged plants, more complex supply and global trade, improved surveillance participating in the detection of food-borne illness, improvement in the advances of epidemiological investigation, and upgrading to the latest methods to detect pathogens.

The European Food Safety Authority (EFSA), Parma, Italy, 2014, advised that tomatoes may be minimally processed for obtaining ready-to-eat products. These steps include initial selection to final storage. The epidemiological source from the EU has identified one outbreak of Salmonella and one of Norovirus associated with tomato consumption from 2007 to 2012, which were considered in the context of the whole food chain. Available estimates of the Salmonella and Norovirus occurrence in tomatoes were evaluated together for prevention of contamination with relevant microbiological criteria.

It was concluded that each farm habitat represents a unique collection of risk factors that can influence persistence of pathogens in tomato production. The implementation of appropriate food safety management systems including good agricultural practices and good manufacturing practices should be the most important objectives of a tomato producer. According to Lairon (2009), food nutritional value, quality, and safety vary widely around the world. Attaining these three goals is one of the major priorities for the near future. In line with several published literature reviews, the French Agency for Food Safety (AFSSA) performed a critical evaluation on the nutritional and hygienic quality of organic food. The review generally underlines the following major points:

1. The organic plant products are known to contain more dry matter and minerals containing more antioxidants and micronutrients such as phenols and salicylic acid.
2. The organic animal products contain more fatty acids (polyunsaturated).
3. It is believed that 94–100% of organic foods do not contain any pesticide residues.
4. Organic vegetables contain far less nitrates. Thus, organic agricultural systems have been shown to be able to produce food with high quality standards.

26.3.6 Risk Level of Microbial Contamination

Machado et al. (2006) attempted to evaluate organically grown horticultural crops for their microbiological safety and quality. The study outlines six different treatments that were applied to the three species of vegetables (lettuce, (*Lactuva sativa*), radish (*Raphanus sativus*), and spinach (*Tetragonia expansa*)), which consisted of a mineral fertilizer in combination with liquid biofertilizers.

The samples were examined for most probable number to detect the presence of *Escherichia coli* and *Salmonella* spp. They were considered acceptable if they did not contain *Salmonella* spp. However, most samples of vegetables like lettuce contained $>10^2$ total coliforms/g of product, whereas none of the samples of spinach or radish presented >10^2 fecal coliforms/g and only a smaller amount (6.6%) of lettuce samples contained >10^2 fecal coliforms/g.

26.3.7 Nutritional Quality of Food (Vitamins, Nutrients, Toxins, Antioxidant Activity, and Pesticide Residue)

26.3.7.1 Flavor: Sugar-to-Acid Ratio

Tomato flavor is judged by the amount of acid and sugar present in it because sugar-acid interaction is directly correlated to overall flavor intensity including sourness and sweetness in tomatoes (DeBruyn et al. 1971; Stevens et al. 1997). As a result, relatively high sugars and acids are generally required for the best flavor (Kader 1986). The main component responsible for flavor is soluble solid content and titratable acidity (Kader 1986), which is believed most likely to match the consumer perception with best internal quality (Artes et al. 1999).

26.3.7.2 Antioxidant Activity, Vitamin C, and Nitrate Levels

It is known that consumers now look for safe foods produced in a local environment that is eco-friendly. These consumer demands are believed to be satisfied by organic food as organic crops have less nitrates and reduced pesticide residues and more nutritional elements than conventional crops. In the majority of cases higher levels of phenols and polyphenols have been reported in organic food stuffs such as apples (Lucarini et al. 1999), peaches (Carbonaro et al. 2002), potatoes (Hamouz et al. 1999), onions (Ren et al. 2001), tomatoes (Mitchell et al. 2007), peppers (Perez-Lopez et al. 2007), oranges (Tarozzi et al. 2006), and olive oil (Gutierrez et al. 1999). It has been reported in a recent review (Rembiałkowska et al. 2005) that organically cultivated foods contain higher amounts of phenolic compounds. Benbrook et al. (2008) in their study indicated higher levels of polyphenols in organically produced foods in comparison to ones produced conventionally. Polyphenols represent a varied class of secondary metabolites with increasing antioxidative properties as well as preventative properties such as being neuroprotective and cardioprotective (Ortuno et al. 2007). The important group of polyphenols identified as diminishing the incidence of various diseases are found to have higher contents of flavanols (Caris-Veynard et al. 2004; Shankar et al. 2007). According to Ren et al. (2001), extracted juices from the organic spinach, onion, and cabbage had 50–120% higher antioxidant activity than the juices extracted from conventionally produced vegetables. Similarly, antioxidant activity of currants grown organically was also recorded as 30% higher than through conventional methods (Kazimierczak et al. 2008. The meta-analysis carried out by Benbrook et al. (2008) on organic crops revealed that organic food contains more beneficial substances such as quercetin, kaempferol (55%), essential vitamins, and phosphorus. On the other hand, Magkos et al. (2003) in his reports describe protein quality in some organic cereal crops and vegetables that was higher than reported in conventionally produced ones. The presence of harmful substances such as nitrates was lower in organic crops (Abu et al. 2007).

Barrett et al. (2007) emphasized the importance of their findings regarding agricultural cultivation systems, which are crucial factors in determining the food quality. The studies were carried out by comparing four different growers of tomatoes under both commercial organic and conventional systems of farming. The goal of the study was to map out the comparison between the quality and nutritional value of tomatoes under both systems of cultivation. The sole analysis of variance results indicated that tomato juice prepared organically was higher in total soluble solids (5.96^0 Brix) for Terranova Ranch Growers as compared to conventional tomato juice (5.56^0 Brix), while the grower Romenger and sons reported higher ascorbic acid (1153 μg/g), lycopene (1345 μg/g), and total phenolics (1811 μg/g) with organic cultivation.

The aim of the study conducted by Ragab et al. (2010) was to evaluate the antioxidant profile of organically and conventionally cultivated tomato and carrot samples purchased from local markets of the Al-Qassim region, (Saudi Arabia), over six successive months.

The antioxidant activity components of both systems varied throughout the period of study. It was observed that the antioxidant activity coupled with antioxidant components was in a higher range than in the conventional tomatoes. In contrast, a smaller antioxidant capacity and vitamin C content were observed in organic carrots.

The nutritional constituents present in tomato and carrot showed their response to this production method. The organic tomatoes had higher values of nutritional contents (dry matter, soluble sugars, and oils) than conventional ones, whereas with organic carrots, higher levels of protein and minerals and lower sugar content were recorded as compared with conventional cultivation. Tomatoes were analyzed for ascorbic acid, phenolic compounds, lycopene content, and antioxidant activity.

Oliveira et al. (2013) in their recent study showed that tomatoes that were grown organically contained more vitamin C and sugars than conventional tomatoes. In the present research, the weights and biochemical properties of tomatoes from both systems of cultivation were compared. The outcome showed that tomatoes grown organically were approximately smaller (about 40%), accumulated more compounds, and thus developed more stress-linked conditions than under conventional techniques. This increased stress may be the reason organic tomatoes have higher sugar levels, vitamins, and pigment molecules. Based on these findings, the researchers suggested that strategies developed for fruit and vegetable cultivation should focus on plant stress management with efforts to increase yield and stabilize fruit size.

The development of nutritional management techniques plays a significant role in improving the overall quality of tomatoes. This fact was truly justified by the field studies conducted in New Delhi (India) with chemical fertilizers and control treatments (effective microorganisms (EM) with compost alone and in combination) on the evaluation of compost on the antioxidant activities and defense enzyme activities of tomatoes. The results revealed an increase of 31.83% tomato yield with the combined use of compost and half the recommended doses of chemical fertilizers (N50 + P30 + K25 + EM compost @ 5 t/ha). A significant increase in fruit quality in terms of lycopene content (35–63%), antioxidant activity (24–63%), and defense activity (11–54%) traced a positive correlation among fruit quality parameters with beneficial soil microbiological activities. This ultimately led to the conclusion that the positive impact exerted by EM compost could be adopted as an eco-friendly method for high quality product production.

Studies carried out in Greece compared ten types of olives, 11 types of tomatoes, and 18 types of legumes from conventional and organic farming for important nutritional properties. Natural black olives exhibited higher TAC (44.15) and total phenolic content. Natural green olives showed higher TAC (44.15 μmol $FeSO_4$/g) and total phenolic content (0.79 mg/GAE/g) than Spanish style green olives. Organic lentils exhibited lower predicted iron bioavailability than conventional lentils (% dialyzable iron 3.07 and 8.9, respectively, and % ferrous dialyzable iron 2.76 and 7.04 mg/GAE/g, respectively). In legumes, differences in total or ferrous iron dialyzability were observed. Giant and elephant beans exhibited the highest total iron dialyzability, while lentils the lowest. Likewise, in the organic and conventional type of "formula" tomatoes a small difference was also observed (p¼0.04). However, no differences were observed between conventional and organic types of tomato pulp (p¼0.31). The highest amount of dialyzable iron was found in "cherry" and "santorini" tomatoes. The organic tomato cherry cherellino yielded 6.8215 μmol $Feso_4$/g total antioxidant activity, 1.31 mg/GAE/g total phenolics, 0.23 mg/g ascorbic acid and percent dialyzable Fe (50.90%), and Zn (52.30%) as compared to conventional ones.

26.3.8 Pesticide and Toxin Residue Levels

Pesticides are used in controlling serious threats caused by insects, diseases, and weeds in agriculture. Increasing pesticide concentration affects the ongoing microbial activity and ultimately leads to destruction of soil fertility and productivity. Similar efforts for studying the deleterious effects of pesticides were highlighted by Diallo (1986), revealing that the toxicity of insecticides is mainly attributed to distortion of foliage necrosis and yellowing, thus causing a reduction in yield.

A pesticide residue analysis was performed by Baker (2002) to quantify differences between organic and chemically grown fruit and vegetable (fresh). The data collected on food residues from different markets that were conventionally grown, integrated pest management (IPM)-grown (NDR), and organically grown were compared using data reported from three different test programs. It was recorded that multiple pesticides are highest in conventionally grown and IPM samples when compared to organic cultivation.

Glover and Tetteh (2008) studied the increasing rates of pesticide application of lindane, undane, ditahne, and karate, (156.0, 244.0, and 312.0 g ha −1; (125.0, 187.5, and 250.0 g ha^{-1}; 166.6, 209.8 , and 333.3 g ha^{-1}), respectively on, okro, eggs, and tomatoes, to find out the advantages that are offered to a farmer by their use. It was seen that yields of garden eggs were reduced by the application of lindane. Doses higher than L20 were observed (i.e. 244 and 312 g ha^{-1}) to affect the yield drastically, whereas the yields of okra were higher than the control at all levels. The application of unden had the optimum effect on garden egg yields followed by tomatoes and okro. Increasing rates of unden on okro did not have any significant effect. The results indicated that the pesticide application greatly influenced the fungal population (50–70% reduction) compared to the bacterial population in the soil (23.0–38.4% reduction). Okro yields were higher than the control at every lindane application. It can be concluded that pesticide application had a greater effect on fungal populations than on bacterial populations in the soil.

Abreu et al. in (2007) worked on anti-nutritional and toxic components of potato tubers. For this, a comparison of glycoalkaloids was carried out for organic and conventional potato tubers in Portugal. Although differences were observed for one potato variety, the other varieties' glycoalkaloid levels were much higher in conventionally grown crops (79.5 mg/kg) than the those grown organically (44.6 mg/kg).

June 2009 (Source: The Organic Center, AAAS Session) saw the advent of a report about chemicals affecting the endocrine system, stating that there is evidence for altered health systems resulting in increased infertility, cancers, obesity, etc. (Source-Pediatrics, 2010).

In brief, the meta-analysis included 240 reports including 17 human studies, comparing organically and conventionally grown food, and reported that organic foods are considered safer and healthier than conventional foods (Beyond Pesticides, 2015).

26.3.9 Marketability, Consumer's Perceptions, and Preferences

The CONDOR project was the first to examine attitudes and behavior in relation to both fresh and processed organic foods and to do so across a number of the EU member states. It involved the development of a theoretically based consumer decision-making model for the purchase of organic food and the testing of this model in eight EU member states. The study highlights consumer perceptions on theory of planned behavior, which was modified slightly for this project. According to this theory, consumer behavior is co-determined by: (a) the individual's decisions as reflected in behavioral intentions; and (b) situational constraints and facilitators. Consumer intentions are co-determined by: (a) attitudes toward the behavior; (b) perceived social pressure; and (c) perceived control or self-efficacy. Finally, these three constructs are based on the person's relevant beliefs and evaluations.

Consumer surveys have indicated that people believe organically grown foods are better in terms of safety, nutritional quality, and taste than their conventional counterparts (Hammitt 1990; Davies et al. 1995). Organic buyers display a different lifestyle pattern than do conventional buyers.

The taste factor of crops has not been taken into consideration, which generally affects consumer perception. During the 1990s, tomatoes with no taste were termed "water bombe" by the unsatisfied German consumer (Baldwin et al. 1988). The reason for these tasteless tomatoes was accredited to more attention being paid to parameters like yields, pest resistance, product stability, durability, and price, while the taste attribute was fully ignored.

A consumer with a preference toward "organic" products in supermarkets is usually viewing these products as organic commodities, without considering brands and packaging. "Organic commodities" are considered as products with better quality as these are produced in local and safe environments. In contrast to safe organic food, conventional foods are generally linked with negative properties.

The commercial production of fresh tomatoes has been seen in about 20 states on a large scale. The USDA 2012 Census of Agriculture recorded an approximately 10% reduction in tomato production with a 20% rise in the number of farms. The census data showed that the highest growth was in farms with 5-acre dimensions or less.

Comparing the era of 2011, the per capita consumption of tomatoes in the USA slightly decreased to 17.3 pounds per person from 17.9 pounds/person. The USDA Economic Research Service estimates depict the largest use of processed tomatoes to be in sauces (35%), followed by paste (18%), canned whole tomato products (17%), and lastly catsup and juice (15% each).

Sanjuan et al. (2003) investigated willingness of consumers to adopt an organic lifestyle, which generally includes various factors pertaining to natural food choice, a balanced life, a positive attitude toward health, and social improvement. The studies showed that the preference of consumers and their willingness to go for organic food ranged from 22% to 37% for vegetables.

Carroll et al. (2013) carried out an experiment regarding consumer choice and willingness to pay for a locally grown, safe organic food from five mid-Atlantic states. This study relied on a mail survey of consumers to determine preferences for organic food in markets and groceries.

Order of preferences in local and state program versions showed various differences. For the three largest states, Virginia, Pennsylvania, and Maryland, a local product was preferred. Overall, these findings form a basis for the increased interest in food products produced locally within the region and the expansion of such programs.

To maintain the good growth of a farmer, good markets promoting locally produced organic food with beneficial incentives need to be upgraded to provide novel ways of attracting customers.

According to recent research in Greece (Anastasiadis and Van Dam 2014), purchasers are basically attracted by sustainability concerns that include ecofriendly and natural modes of production and cultivation. The study also confirms initial findings (Essoussi and Zahaf 2008; Zanoli and Naspetti 2002) that emphasise the importance of health, which presents a real motivation for purchasing organic produce.

26.4 Recommendations and Future Directions

Steadily growing public concerns about soil health, crop quality, productivity, pesticide residues, food safety, security, environmental quality, and ground-water contamination call for an initial objective of comparing conventional and organic farming systems in accordance with the following factors:

1. The large prevalence of weed and unwanted pathogen populations.
2. The differences in soil properties
3. Yield parameters, growth, and health attributes.
4. Economically viable methods

Below are some of the researchable issues for future consideration:

The organic sector is moving forward to upgrade organic farming systems by developing seed and planting materials organically to eradicate the biggest problems regarding soil fertility. Pest and diseases can be accurately controlled in organic systems, but some other points need to be taken in to consideration:

- The production of healthy seeds and nurseries, which is dependent on the various breeding programs.
- The maintenance of lower disease pressure via the improvement and strengthening of organic cultural practices.
- The development of varieties with increased tolerance against wide varieties of diseases and variable pathogens.
- Increased focus on improving methods for a good seed and plant stock production.
- The seed producers should have knowledge about selecting the best locations with lower disease pressure.
- The seed health standards should be improved for fighting against a high risk of seed-borne diseases.
- For the successful establishment of organic seed production all communication gaps should be resolved and a mutual commitment between the various groups from cultivation to production (farmers, traders, breeders, and government) needs to be made.

Conclusion

A major conclusion can be drawn in favor of going organic by highlighting various parameters including increasing soil health and protecting soil fertility from erosion via improved drainage systems. The new technological interventions in organic farming used to protect soil fertility include barrier crops, crop rotations, mulching, green manuring, application of on and off-farm inputs, and use of biodynamic preparations for insect pest and disease management. Organic farming methods rely on a high level of microbial activity in the soil and at the same time contribute increased abundance and diversity of those same beneficial soil microorganisms. Some of the benefits of this include: better uptake of minerals which enhances the nutrient supply, improvement in crop vigor, and reduced nutrient run-off.

Soil communities feed on organic matter in the soil while nutrients are made available to plants by those soil organisms that rely on them for their food and survival. The increased level of soil organic matter is also an indicator for rating a good soil, as it furnishes many important functions that are important to organic agricultural systems. Therefore, efforts should be made in the direction of developing methods to increase soil organic matter, which will serve to accomplish an important goal for organic farmers.

References

Abreu P, Relva A, Matthew S, Gomes Z, Morais Z (2007) High performance liquid chromatographic determination of glycoalkaloids in potatoes from conventional, integrated and organic crop systems. Food Control 18:40–44

Abu ZTR, Al Ismail K, Shatat F (2007) Effect of organic and conventional systems on fruit quality of strawberry (fragaria x ananassa duch) grown under plastic house conditions in the Jordan Valley. Acta Horticulture (ISHS) 741:159–171

Agbenin ON, Marley PS (2006) In-vitro assay of some plant extracts against *Fusarium oxysporum* sp. Lycopersici causal agent of tomato wilt. J Plant Prot Res 46(3):215–220

Anastasiadis F, Van Dam YK (2014) Consumer driven supply chains: the case of Dutch organic tomato. Agric Eng Int CIGR J Special Issue 2014: Agri-food and biomass supply chain. pp 11–20.

Anonymous (1982) Sludge may help nurseries. J For 80:34

Artes F, Conesa MA, Hernandez S, Gill MI (1999) Keeping quality of fresh cut tomato. Post Harvest Biol Technol 17:153–162

Baker KF, Snyder WC (1965) Ecology of soil borne plant pathogens, prelude to biological control. University of California Press, Berkeley, Los Angeles, p 571

Baldwin EA, Scott JW, Einstein MA, Malundo TMM, Carr BT, Shewfelt RL (1988) Relationship between sensory and instrumental analysis for tomato flavor. J Am Soc Hortic Sci 123:906–915

Barbier EB (1987) The concept of sustainable economic development. Environ Conserv 14(2):101–110

Barrett DM, Weakley C, Diaz JV, Watnik M (2007) Qualitative and nutritional differences in processing tomatoes grown under commercial organic and conventional production systems. J Food Sci 72:9

Benbrook CH, Zhao X, Yanez J, Davies N, Andrews P (2008) New evidence confirms the nutritional superiority of plant based organic foods. State of Science Review. 2008. http://www.organic-center.org/science.nutri.php?action=view&report_id=126

Bethlenfalvay GJ, Brown MS, Ames RN, Thomas RS (1988) Effects of drought on host and endophyte development in mycorrhizal soybeans in relation to water use and phosphate uptake. J Plant Physiol 72:565–571

Carbonaro M, Mattera M, Nicoli S, Bergamo P, Cappelloni M (2002) Modulation of antioxidant compounds in organic vs. conventional fruit (peach *Prunus persica* L., and pear *Pyrus communis* L.) J Agric Food Chem 50(19):9–11

Caris-Veynard C, Amiot MJ, Tyssandier V, Grasselly D, Buret M, Mikolajczak M (2004) Influence of organic versus conventional agricultural practice on the antioxidant micro constituent content of tomato and derived purees consequence on antioxidant plasma status in humans. J Agric Food Chem 52:6503–6509

Carroll KA, Bernard JC, Pesek JD (2013) Consumer preferences for tomatoes: the influence of local, organic, and state program promotions by purchasing venue. J Agric Resour Econ 38(3):379–396

Champawat RS, Sharma RS (2003) Integrated management of nursery diseases in Brinjal, Chilli and Onion. J Mycol Plant Pathol 33:290–291

Chatterjee R, Bandyopadhyay S, Chandra JJ (2014a) Impact of organic amendments and inorganic fertilizers on production potential, nitrogen use efficiency and nitrogen balance in tomato (*Solanum lycopersicum* Mill.) Int J Sci Res Knowl 2(5):233–240

Choudhary S, Soni AK, Jat NK (2012) Effect of organic and inorganic sources of nutrients on growth, yield and quality of sprouting broccoli cv. CBH-1. Indian J Hortic 69(4):550–554

Chowdhury MTI, Razzaque MA, Khan MSI (2011) Chlorinated pesticide residue status in tomato, potato and carrot. J Exp Sci 2(1):01–05

Cook RJ, Baker KF (1983) The nature and Practice of Biological control of plant pathogens. Journal of American Phytopathology Society, St. Paul. p 539.

Copper AL, Campbell R (1986) The effect of artificially inoculated antagonistic bacteria on the prevalence of take all disease of wheat in field experiments. J Appl Bacteriol 60:155–160

Davies A, Titterington AJ, Cochrane C (1995) Who buys organic food? Br Food J 97:17–23
DeBruyn J, Garretsen F, Kooistra E (1971) Variation in taste and chemical composition of the tomato. Euphytica 20:214–227
Diallo A (1986) Insecticides and the environment (References to tropical environment). Ecosystem management in developing countries. 1986; UNEP Postgraduate Training Course on Eecosystem Management. Technical University of Dresden GDR
Doran J (1995) Building soil quality. In: Proceedings of the 1995 conservation workshop on opportunities and challenges in sustainable agriculture. Alberta Conservation Tillage Society and Alberta Agriculture Conservation, Development Branch, Red Deer. pp 151–158
Drinkwater LE, Letourneau DK, Workneh F, Van Bruggen AHC, Shennan C (1995) Fundamental differences between conventional and organic tomato agro ecosystems in California. Appl Ecol 5:1098–1112
Essoussi LH, Zahaf M (2008) Decision making process of community organic food consumers: an exploratory study. J Consum Mark 25(2):95–104
FAO/WHO (1997). Codex alimentarius food hygiene basic texts. Joint FAO/WHO food standards programme, Codex Alimentarius Commission. Pub. # M-83
Galler WS, Davey CB, Meyer WL, Airan DS (1978). Animal waste composting with carbonaceous material. U.S. Environmental Protection Agency, Washington, D.C. EPA-600/2, 78:154–196
Glover AM, Tetteh FM (2008) Effect of pesticide application rate on yield of vegetables and soil microbial communities. West Afr J App Ecol 12
Gutierrez F, Arnaud T, Albi MA (1999) Influence of ecological cultivation on virgin olive oil quality. J Am Oil Chem Soc 76:617–621
Hammitt JK (1990) Risk perceptions and food choice: an exploratory analysis of organic versus conventional produce buyers. Risk Anal 10:367–374
Hamouz K, Lachman J, Vokal B, Pivec V (1999) Influence of environmental conditions and way of cultivation on the polyphenol and ascorbic acid content in potato tubers. Hortic Sci 45:293–298
Haque MA, Jahiruddin M, Rahman MM, Saleque MA (2013) Usability of bioslurry to improve system productivity and economic return under potato-rice cropping system. Res Agric Livest Fish 2(1):27–33
Hooda KS, Joshi D, Dhar S, Bhatt JC (2011) Management of damping off of tomato with botanicals and bio-products in North Western Himalayas. Indian J Hortic 68(2):219–223
Jack A (2010) In: Edward CA, Arancon NQ, Sherman R (eds). Earthworms, organic wastes and Enviornmental management. Vermiculture Technology. CRC Press, Boca Raton. p 623
Jangir RP, Rathore MS, Bisnoi HR, Sundria MM (2008) National workshop on spices and aromatic plants. Agricultural Research Station, Mandor, 6–7 Feb 2008.
Kabdal P, Hooda KS, Joshi D, Hedau NK, Pandey KN (2010) Biocontrol agents in the health management of capsicum nursery. Indian J Hortic 67(1):70–72
Kachari M, Korla BN (2012) Studies on influence of bio-fertilizers on quality and economics of cauliflower cv. PSB K-1 production. Indian J Hortic 69(2):215–220
Kader AA (1986) Effect of post harvest handling procedures on tomato quality. Acta Hortic 190:209–221
Kapoulas N, Zoran SI, Mihal Đ, Radmila T, Lidija M (2011) Effect of organic and conventional production practices on nutritional value and antioxidant activity of tomatoes. Afr J Biotechnol 10(71):15938–15945
Kazimierczak R, Hallman E, Rusaczonek A, Rembialkowska E (2008) Antioxidant content in black currants from organic and conventional cultivation. Food Sci Technol Res 2(11):57–61
Kumar R, Hooda I, Karwasra SS (2010) Efficacy of Mahapancha Gavya (mpg) in controlling damping-off in tomato caused by *Pythium aphanidermatum*. J Agric Res 35(1):11–16
Lairon D (2009) Nutritional quality and safety of organic food. A review. Agronomical Sustainable Dev Biochem 48(26):6157–6165
Lucarini M, Carbonaro M, Nicoli S, Aguzzi A, Cappelloni M, Ruggeri S (1999) Endogenous markers for organic versus conventional plant products. In: Agri-food quality II: quality management of fruits and vegetables. The Royal Society of Chemistry, Cambridge, UK, pp 306–310

Machado DC, Carla MM, Isabel DC, Natan F, Maria C, Dantas P (2006) Microbiological quality of organic vegetables produced in soil treated with different types of manure and mineral fertilizer. Braz J Microbiol 37:538–544

Magkos F, Arvaniti F, Zampelas A (2003) Organic food: nutritious food or food for thought: a review of the evidence. Int J Food Sci Nutr 54(5):357–371

Mitchell AE, Hong Y, Koh E, Barrett DM, Bryant DE, Denison RF (2007) Ten-year comparison of the influence of organic and conventional crop management practices on the content of flavonoids in tomatoes. J Agric Food Chem 55:6154–6159

Muthuraju R, Boby VU, Suvarna VC, Jayasheela N (2002) Interactive effects of *Glomusmosseae*, *Pseudomonas fluorescens*and *Azospirillum brasilense* on growth and yield of tomato. J Soil Biol Ecol 22:8–15

Neelamegam R, Govindarajalu T (2002) Integrated application of *Trichoderma viride* and farm yard manure to control damping off of tomato. J Biol Control 16:65–69

Oliveira AB, Moura CF, Gomes-Filho E, Marco CA, Urban L (2013) The impact of organic farming on quality of tomatoes is associated to increased oxidative stress during fruit development. PLoS One 8(2):e56354

Ortuno A, Benavente-Garcia O, Castillo J, Alcaraz M, Vicente V, Del Rio JA (2007) Beneficial action of citrus flavonoids on multiple cancer related biological pathways. Curr Cancer Drug Targets 7:795–809

Patel VB, Singh SK, Asrey R, Nain L, Singh AK, Singh L (2009) Microbial and inorganic fertilizers application influenced vegetative growth, yield, leaf nutrient status and soil microbial biomass in sweet orange cv. Mosambi. Indian J Hortic 66(2):163–168

Perez Lopez AJ, Lopez Nicolas JM, Nunez Delicado E, Del Amor FM, Carbonell Barrachina AA (2007a) Effects of agricultural practices on color, carotenoids composition and minerals contents of sweet peppers, cv. Almuden. J Agric Food Chem 55:8158–8164

Pozo MJ, Azcon Aguilar C (2007) Unravelling mycorrhiza-induced resistance. Curr Opin Plant Biol 4:393–398

Prativa KC, Bhattari BP (2011) Effect of Integrated Nutrient Management on the growth, yield and soil nutrient status in tomato. Nepal J Sci Technol 12:23–28

Ragab A, Mergawi E, Redhaiman KA (2010) Effect of organic and conventional production practices on antioxidant activity, antioxidant constituents and nutritional value of tomatoes and carrots in Saudi Arabia markets. Food Agric Environ 8(3, 4):253–258

Raguchander T, Rajappan K, Samlappan R (1997) Evaluating methods of application of biocontrol agent in the control of mungbean root rot. Indian Phytopathol 50(2):229–234

Ramakrishnan K, Selvakumar G (2012) Effect of biofertilizers on enhancement of growth and yield on Tomato (*Solanum lycopersicum* Mill.) Int J Res Bot 2(4):20–23

Reddy SA, Davis Joseph B, Radha Dinakar K (2012) Vermicompost as a biocontrol agent in suppression of two soil borne plant pathogens in the field. Acta Biol Indica 1(2):137–142

Rembiałkowska E, Hallmann E, Szafirowska A (2005) Nutritive quality of tomato fruits from organic and conventional cultivation. In: Edwards JSA, Kowrygo B, Rejman K, (eds). Culinary arts and sciences V. Global and national perspectives. Worshipful Comapany of Cooks Research Centre: Poole. pp 193–202.

Ren H, Endo H, Hayashi T (2001) Antioxidative and antimutagenic activities and polyphenol content of pesticide free and organically cultivated green vegetables using water soluble chitosan as a soil modifier and leaf surface spray. J Sci Food Agric 81:1426–1432

Sanjuan AI, Sanchez M, Gil JM, Gracia A, Soler F (2003) Brakes to organic market enlargement in Spain: consumers' and retailers' attitudes and willingness to pay. Int J Consum Stud 27:134–144

Sanwal SK, Laxminarayana K, Yadav RK, Rai N, Yadav DS, Bhuyan M (2007) Effect of organic manures on soil fertility, growth, physiology, yield and quality of turmeric. Indian J Hortic 64(4):444–449

Sepat Naval K, Kumar A, Yadav J, Srivastava RB (2012) Effect of Integrated Nutrient Management on growth, yield and quality of tomato in Trans Himalayan. Ann Plant Soil Res 14(2):120–123

Shankar S, Ganapathy S, Srivastava RK (2007) Green tea polyphenols: biology and therapeutic implications in cancer. Front Biosci 12:4881

Sharma P, Sharma M, Raja M, Shanmugam V (2014) Status of *Trichoderma* research in India: a review. Indian Phytopathol 67(1):1–19

Singh AK, Singh S, Appa Rao VV (2012) Influence of organic and inorganic nutrient sources on soil properties and quality of aonla in hot semi-arid ecosystem. Indian J Hortic 69(1):50–54

Srinivasan K, Mathivanan N (2011) Plant growth promoting microbial consortia mediated classical biocontrol of sunflower necrosis virus disease. J Biopest 4(1):65–72

Srivastava AK, Singh RK (2000) Extent of lysis of *Rhizoctonia solani* cell wall preparation by different hyperparasites. J Mycopathol Res 38:129–131

Stamatiadis S, Werner M, Buchanan M (1995) Field assessment of soil quality as affected by compost and fertilizer application in a broccoli field (San Benito County, California). Appl Soil Ecol 12:217–225

Stevens MA, Kader AA, Albright Holton M, Algazi M (1997) Genotypic variation for flavor and composition in fresh tomatoes. J Am Soc Hortic Sci 102:880–689

Stockdale EA, Lampkin N, Hovi M, Keatinge R, Lennartsson EKM, Macdonald DW et al (2000) Agronomic and environmental implications of organic farming systems. Adv Agron 70:261–327

Sudharani M, Shivaprakash MK, Prabhavathi MK (2014) Role of consortia of biocontrol agents and PGPR's in the production of cabbage under nursery condition. Int J Curr Microbiol Appl Sci 3(6):1055–1064

Tadesse T, Seyoum W, Woldetsadik K (2012) Effect of varieties on changes in sugar content and marketability of tomato stored under ambient conditions. Afr J Agric Res 7(14):2124–2130

Tarozzi A, Hrelia S, Angeloni C, Morroni F, Biagi P, Guardigli M (2006) Antioxidant effectiveness of organically and non-organically grown red oranges in cell culture systems. Eur J Nutr 45:152–158

The Organic Center, AAAS Session 2009 – "Living soil, food quality and the future of food", February 2009.

Thilagavathi R, Saravana Kumar D, Ragupathi N, Samiyappan RA (2007) Combination of biocontrol agents improves the management of dry root. Phytopathol Mediterr 46:157–167

Tonfack LB, Bernadac A, Youmbi E, Mbouapouognigni VP, Ngueguim M, Akoa A (2009) Impact of organic and inorganic fertilizers on tomato vigor, yield and fruit composition under tropical and soil conditions. Fruits 64:167–177

Trivedi YV, Patel NL, Ahlawat TR, Gaikwad SS, Bhalerao PP (2012) Impact of organic manures and inorganic fertilizers on growth, yield, nutrient uptake and soil nutrient status in guava. Indian J Hortic 69(4):501–506

Vidyasekaran P, Rabidran R, Muthamila M, Rajappan K, Subramanian N, Vasumathi K (1997) Development of a powder formulation of *Pseudomonas fluorescens* for control of rice blast. Indian Phytopathol 46:291–297

Yan W, Yamamoto K, Yakushido K (2002) Changes in nitrate content in different soil layers after the application of livestock waste compost pellets in a sweet corn field. Soil Sci Plant Nutr 48:165–170

Yanar D, Naif G, Yusuf Y, Mine A, Perihan C (2011) Effect of different organic fertilizers on yield and fruit quality of ındeterminate tomato. Sci Res Essays 6(17):3623–3628

Zanoli R, Naspetti S (2002) Consumer motivations in the purchase of organic food: A means-end approach. Br Food J 104(8):643–653

Role of Bioremediation Agents (Bacteria, Fungi, and Algae) in Alleviating Heavy Metal Toxicity

27

Zaid ul Hassan, Shafaqat Ali, Muhammad Rizwan, Muhammad Ibrahim, Muhammad Nafees, and Muhammad Waseem

Abstract

Heavy metals are environmental contaminants globally. They have polluted agricultural soils and caused detrimental effects on our ecosystem. Toxic effects of heavy metals have been reported in plants, animals, humans, and microorganisms. Heavy metal remediation is essential to preserve the health of agricultural soils and would lead to enhanced crop growth and yield. Various techniques and strategies have been used in recent years to remediate contaminated soils, but most of them were costly, environmentally unfriendly, and negatively affect soil properties. However, use of microbes to remediate heavy metals has been found to be cost effective and environmentally clean. Microbes enhance stability in agricultural soil health, which leads to sustained plant growth and development under stressful conditions. Particular agents used for bioremediation are bacteria, fungi, and algae. Bacterially-mediated processes have been used to alleviate heavy metal toxicity. Endophytic bacteria have greater potential to tolerate and remediate heavy metals stress. Bacterial strains showed potential to alleviate heavy metals from the rhizosphere of target plant species and improve their growth. Arbuscular mycorrhizal fungi alleviate heavy metal toxicity by inhibiting their uptake and translocation in plant parts. In addition, many morphological and physiological changes are induced by

Z.u. Hassan • S. Ali (✉) • M. Rizwan • M. Ibrahim
Department of Environmental Sciences and Engineering, Government College University, Allama Iqbal Road, 38000 Faisalabad, Pakistan
e-mail: shafaqataligill@yahoo.com

M. Nafees
Institute of Soil & Environmental Sciences, University of Agriculture Faisalabad, Faisalabad, Pakistan

M. Waseem
Department of Microbiology, Government College University, Allama Iqbal Road, 38000 Faisalabad, Pakistan

V. Kumar et al. (eds.), *Probiotics in Agroecosystem*,
DOI 10.1007/978-981-10-4059-7_27

fungi. Macro- and micro-algae have been reported to alleviate heavy metal toxicity mostly in marine systems. Reports suggested that applications of the above bioremediation agents alleviate heavy metal stress, enhance phytoremediation capacity in combination with plant growth–promoting bacteria, and ultimately improve plant growth attributes.

27.1 Introduction

Soil is an important resource on which the lives of plants, animals, and microorganisms are heavily dependent. Addition of soil pollutants largely disturbs soil microbial function (Swain and Abhijita 2013). Heavy metals are among the most problematic pollutants; they extensively take part in soil contamination through various anthropogenic activities (Panagos et al. 2013; Liu et al. 2013; Waterlot et al. 2013; Chodak et al. 2013). Toxic effects of heavy metals have been observed in plants, humans, and animals. Heavy metal stress poses severe threats to agricultural crops by inhibiting plant growth parameters and yield as documented by many researchers (Hu et al. 2013; Liu et al. 2015; Khan et al. 2016, Dheeba et al. 2015; Gill et al. 2015). The non-biodegradable nature of metals enhances their availability and longevity in soils. The longer persistence of metals in soils causes carcinogenic and mutagenic effects, and becomes part of our food chain (Ali et al. 2013; Ahemad and Kibret 2013). Heavy metal concentrations above the threshold limit causes disturbances in microbial activity and soil health (Huang et al. 2009). Thus, metal remediation is essential in unsuitable agricultural applications. Various techniques and strategies have been used to clean up heavy metals (Hashim et al. 2011). Conventional techniques used for the cleanup of metals are usually too costly to be used as they adversely affect soil health indicators (Rajkumar et al. 2010). One emerging, cheap, and most economically feasible technique is the use of microbes to remediate heavy metals. Bioremediation uses the metabolic potential of microorganisms to remediate heavy metals biochemically (Huang et al. 2013). Bioremediation of heavy metals has been investigated using both bacterial (Nath et al. 2012; Poornima et al. 2014) and fungal species (Andrade et al. 2010; Medina et al. 2010).

Microbes have gained attraction with regard to their roles in improving plant growth and resisting metal accumulation in contaminated soils (Glick 2010). Similarly, Khan et al. (2013) reported that addition of microbes not only improves plant growth traits but also remediates contaminated metals present in soil. Microbes help plants to overcome toxic effects of metals and ultimately enhance plant growth parameters and phytoremediation activity, as described by Weyens et al. (2010).

The aim of the present study was to evaluate the potential role of bioremediation agents (bacteria, fungi, and algae) in alleviating heavy metal toxicity in agricultural soils and improving plants growth attributes.

Table 27.1 Anthropogenic sources of selected heavy metals

Heavy metals	Sources	References
Cr	Tanneries, steel industries, flying ash from the burning of coal	Khan et al. (2007)
Pb	Herbicides, batteries, insecticides, aerial emissions from petrol	Wuana and Okieimen (2011)
Hg	Medical waste, coal burning, and Au-Ag mining	Wuana and Okieimen (2011)
Ni	Battery manufacturing, steel alloys, kitchen appliances, surgical instruments, industrial effluents	Tariq et al. (2006)
Cu	Pesticides and fertilizers usage	Khan et al. (2007)
Cd	Electroplating, plastic burning, phosphate fertilizer, paints and pigments	Pulford and Watson (2003)
As	Wood storage and pesticides	Thangavel and Subbhuraam (2004)

27.2 Sources of Heavy Metals in Environment

Heavy metals enter our environment through both natural (erosion, volcanic activity, minerals weathering) and anthropogenic (mining, pesticides, smelting, electroplating, sludge waste, industrial discharge) sources (Wuana and Okieimen 2011). Anthropogenic sources of some heavy metals are given below in Table 27.1.

27.3 Heavy Metal Toxicity in Plants

Plants are sensitive to heavy metal stress. Heavy metals are toxic to plants at lower (Ahmad et al. 2012) and higher concentrations as mentioned by Wuana and Okieimen (2011)). Toxic effects of heavy metals in plants have been observed by many researchers. Heavy metals disturb plant physiological parameters (Villiers et al. 2011), causing ultrastructural and biochemical changes (Gamalero et al. 2009). Cadmium (Cd) is one of the most toxic heavy metals. It inhibits plant growth, photosynthetic parameters, and nutrient uptake, and causes visible injury (Guo et al. 2008; Mohanpuria et al. 2007), root injury, and ultimately plant death (Mohanpuria et al. 2007). Cd negative effects on plant physiological, morphological, and biochemical traits have been investigated by Farid et al. (2013)). Cd toxic effects were also noted in oilseed crops as well as in declines in seed germination and seedling growth in *Brassica napus* (Irfan et al. 2014) and inhibition of enzyme activities in Indian mustard (Bashir et al. 2015). Chromium (Cr) also negatively disturbs various physiological parameters (Ali et al. 2011). Seed germination is inhibited by Cr stress in *Triticum aestivum* (Datta et al. 2011), *Glycine max*, *Vigna radiata*, *V. angularis* (Jun et al. 2009), *Brassica oleracea var. acephala* (Ozdener et al. 2011), celery seedlings (Scoccianti et al. 2006), and *T. aestivum* (Vajpayee et al. 2011). Inhibition in plant growth by Cr stress has been reported in *G. Americana* (Barbosa et al. 2007) and celery seedlings (Scoccianti et al. 2006). Cr stress decreases chlorophyll content, carotenoids (Redondo-Gomez et al. 2011), photosynthetic traits, and transpiration rate in *A. viridis* (Liu et al. 2008), celery seedlings (Scoccianti et al. 2006) and *O. sativa* (Ahmad et al. 2011). Furthermore, inhibitions in enzyme activities were reported by Cr toxicity (Subrahmanyam 2008). Cobalt (Co) uptake and accumulation in plants depends on target species (Bakkaus et al. 2005). Plant growth and biomass were disturbed by Co toxicity in barley (*Hordeum vulgare L.*), oilseed rape (*Brassica napus L.*), and tomato (*Lycopersicon esculentum L.*) (Li et al. 2009). Chlorophyll content, nutrient uptake, enzyme activity, and transpiration rate were badly affected by Co stress in cauliflower (Chatterjee and Chatterjee 2000), and caused slowed plant growth and loss of biomass in other plants (Li et al. 2009). Lead (Pb) is another of the toxic heavy metals that reduces plant growth, biomass, and enzyme activity, and increases oxidative stress, and water imbalance (Reddy et al. 2005). Pb accumulates in plant growth parts and negatively affects chlorophyll content (Ahmad et al. 2012), photosynthesis (Najeeb et al. 2014; Reddy et al. 2005), and photosynthetic biosynthesis in *Brassica rape* L (Cenkci et al. 2010). Pb stress caused inhibited plant growth traits in *B. oleracea var. Botrytis* (Theriappan et al. 2011) and disturbed metabolic functions (Ashraf et al. 2011). Mercury (Hg) application disturbs metabolic function, water imbalance, and oxidative stress (Zhou et al. 2007). Silicon (Si) is one of the abundant metals in the earth's crust that is beneficial to plants (Kamenidou et al. 2009). It has been used for detoxification purposes in the peanut (Shi et al. 2010) and cucumber (Feng et al. 2010). Copper (Cu) is also a micronutrient that is usually beneficial to plant growth parameters (Yruela 2009). However, it shows toxic effects at higher concentrations. It was reported that Cu inhibits plant growth parameters and induces oxidative stress (Adrees et al. 2015a, 2015b; Zaheer et al. 2015). Gas exchange parameters, chlorophyll, and protein content were decreased in *Brassica napus L* under Cu stress (Zaheer et al. 2015). Nutrient imbalance, disturbed cell membrane function, and chlorosis are caused by nickel (Ni) stress (Gajewska et al. 2006).

27.4 Mechanisms of Heavy Metal Detoxification

Plants have developed various mechanisms at the cellular level to detoxify heavy metals and thus tolerate heavy metal stress. Mechanisms involved are immobilization, plasma membrane exclusion, chelation, and sequestration of heavy metals by ligands, restriction of heavy metal uptake and transport, ROS effects like upregulation of antioxidants, synthesis of heavy metal transporters, and biosynthesis of proline and signal molecules like salicylic acid and nitric oxide (Sharma and Dietz 2009). Moreover, symbiotic association between plants and rhizobacteria efficiently improve plant growth by increasing mineral nutrition and alleviating heavy metal toxicity on target plants (Titah et al. 2013). However, many mechanisms are adopted by plants to combat heavy metals.

27.5 Cleanup of Heavy Metals in Contaminated Soils

Heavy metal contamination has been increased in the surrounding environment over time ages (Govindasamy et al. 2011). Clean-up of heavy metals is necessary to alleviate their impact on ecosystems, which is a severe challenge with regard to cost effectiveness and technique (Barceló and Poschenrieder 2003). Many physical, chemical, and biological techniques have been used for the remediation of heavy metals with deleterious effects on cost effectiveness and technique (Sheoran et al. 2011). Physiochemical and conventional techniques used for remediation purposes are soil incineration, landfill, excavation, soil washing, leaching, solidification, and soil flushing (Wuana and Okieimen 2011). These approaches negatively affect the soil's physical and chemical structure. In addition, these techniques are costly; they require extra labour, damage soil-living organisms, and cause pollution problems (Ali et al. 2013). These techniques do not completely remove heavy metals but transform them from one form to another (Lambert et al. 2000). Thus, advanced research is needed regarding cost and environmentally safe methods to clean heavy metals in contaminated soils. Biological techniques using plants and microorganisms have been used to remove toxic pollutants from the environment (Singh et al. 2009). Biological approaches are cost effective and environmentally friendly methods for the removal of toxic pollutants (Doble and Kumar 2005). Some biological techniques used are bioremediation, phytoremediation, bioventing, bioleaching, bioaugmentation, biostimulation, etc. Among these techniques, bioremediation and phytoremediation are the most useful. These techniques also maintain soil physical status unlike physiochemical approaches as documented by Beskoski et al. (2011)). Bioremediation is a biological method used for the remediation of heavy metals (Boopathy 2000). For instance, *Bacillus spp.* and *Pseudomonas aeruginosa* have been used to alleviate Zn and Cu stress (Kumar et al. 2011). Phytoremediation is another biological technique that can be enhanced if we use microbes as compared to phytoremediation alone. Therefore, plant-microbe interactions have gained importance due to the potential of microbes to bioaccumulate heavy metals as described by Hadi and Bano (2010) (Fig. 27.1).

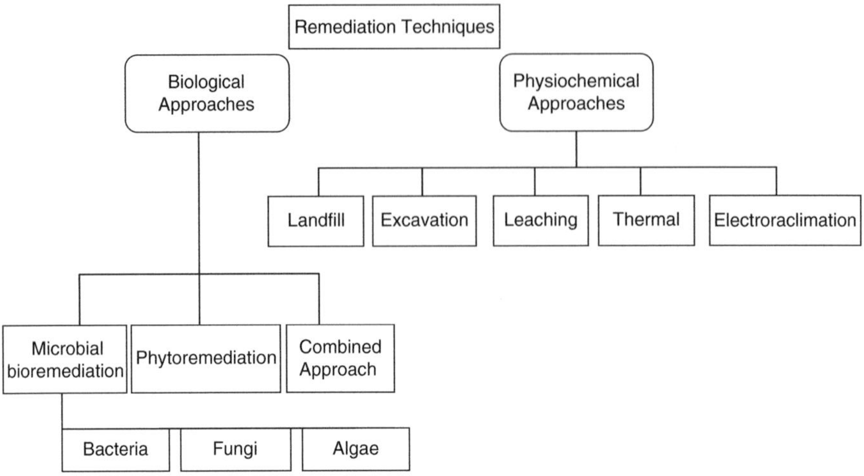

Fig. 27.1 Remediation strategies for heavy metals (Adapted from Ullah et al. (2015))

27.6 Phytoremediation of Heavy Metals (Techniques and Strategies)

Various phytoremediation techniques have been used to remediate heavy metals.

27.6.1 Phytoextraction

Phytoextraction is the uptake or extraction of heavy metals from soils or water using plant roots and their translocation to above ground plant parts like shoots (Rafati et al. 2011). Generally, harvesting of root biomass is not viable, which leads to the need for phytoextraction as metal translocation to above ground plant parts is central for biochemical processes (Tangahu et al. 2011).

27.6.2 Phytofiltration

Phytofiltration is the removal of soil contaminants from the surface and waste water by plants (Mukhopadhyay and Maiti 2010). In phytofiltration, absorption or adsorption of contaminants and their movement in underground water is minimized.

27.6.3 Phytostabilization

Phytostabilization is the stabilization of pollutants in contaminated soils (Singh 2012). Phytostabilization inhibits mobilization and bioavailability of contaminants in the environment and their entry into groundwater (Erakhrumen 2007). Plants prevent heavy metal entry into soils through sorption, precipitation, and complexation (Wuana and Okieimen 2011). Plants have the ability to reduce the toxic effects of heavy metals and their detrimental effects.

27.6.4 Phytovolatilization

In phytovolatilization, plants take up pollutants from soil and convert them into volatile forms so that they easily release into the atmosphere. This technique is mostly used for organic pollutants. Phytovolatilization does not remove organic pollutants completely but changes them from soil to atmosphere from where they can be redeposited (Padmavathiamma and Li 2007).

27.6.5 Phytodegradation

Phytodegradation is the degradation of organic pollutants by plants with the help of enzymes (Vishnoi and Srivastava 2008). Plants do not completely remove accumulated organic pollutants but detoxify them by metabolic activities. That is why green plants (green liver) are beneficial for the biosphere. Phytodegradation is limited to organic pollutants only because other heavy metals are nonbiodegradable. Different transgenic plants have been used for this purpose (Doty et al. 2007).

27.6.6 Rhizodegradation

Rhizodegradation is the degradation of organic pollutants by rhizobial microorganisms (Mukhopadhyay and Maiti 2010). Enhanced microbial activities of microbes improved degradation of organic pollutants. Plant roots release exudates in the form of nutrients to soil microbes and increase microbial activity due to the wealthy nutrient environment. In addition to this, plants also release enzymes that have the ability to degrade soil organic pollutants (Yadav et al. 2010).

27.6.7 Phytodesalination

Phytodesalination is one of the emerging techniques of phytoremediation (Zorrig et al. 2012). Halophytic plants are usually used to remove salts from salt disturbed soils (Sakai et al. 2010) because halophytic plants have a greater ability to cope with heavy metals as compared to glycophytic plants (Manousaki and Kalogerakis 2011).

27.7 Microbial Bioremediation as an Emerging Technique

Various traditional physiochemical methods have been used in economically developed countries to remediate heavy metals like filtration, evaporation, electrochemical application, reverse osmosis, ion exchange, chemical precipitation, oxidation, and reduction. However, these methods are very expensive in addition to causing other harmful and toxic effects leading to environmental pollution (Ahluwalia and Goyal 2007). Therefore, new cost-effective and environmental friendly techniques are required to remediate heavy metals. Bioremediation has gained the interest of researchers to remediate particular metals of interest. Microorganisms have the ability to remediate toxic pollutants without generating toxic byproducts (Kothe et al. 2005). Soil bioremediation is a great challenge because of soil heterogeneity and sediments that require well adapted microbes for their remediation (Tabak et al. 2005). There is a need for novel technologies to understand the mechanisms of heavy metal toxicity on living cells. Microbial models are helpful to study oxidative stress at molecular, biochemical, and cellular levels because oxidative damage caused by heavy metals is similar in all cell organizations. Antioxidant enzymes reduce the production of reactive oxygen species and repair damaged macromolecules (Poljsak et al. 2010). The role of microorganisms has been reported to remediate environmental pollutants (Ray and Ray 2009; Ruta et al. 2010; Nath et al. 2012; Poornima et al. 2014). Agents used for bioremediation purposes are bacteria (Glick 2010), fungi (Meier et al. 2012), macroalgae, and microalgae (Brinza et al. 2007).

27.8 Role of Microbes in Plant Growth and Metal Extraction

Microorganisms can enhance plant growth under stress conditions. Plant growth–promoting bacteria (PGPR) and arbuscular mycorrhizal fungi (AMF) have been found to have the most beneficial effects on plant growth. Many researchers have found that these microbes not only stimulate the plant growth and biomass that leads to the production of phytohormones such as indoleacetic acid (IAA) and ethylene, but they also improve plant nutritive status. Bacteria enhance plant growth and stress tolerance by using enzyme 1-aminocyclopropane-1-carboxylic acid (ACC) deaminase, which leads to reduced ethylene production (Glick 2004). Microbes interact with other microbes in the soil environment, which leads to stable agriculture, which depends mainly on biological processes to sustain plant growth, development, and soil health under stress conditions (Nadeem et al. 2014). Enhanced plant growth and biomass by endophytic bacteria have been observed mostly in hyperaccumulator plants (Ghosh et al. 2011). It was reported that the *Bacillus subtilis* strain enhanced *Brassica juncea* growth by producing IAA and also Ni accumulation (Zaidi et al. 2006). Similarly, *Pseudomonas* strain enhanced the growth of *Brassica juncea* by producing IAA and enhanced trace element extraction. On the other hand, *Bacillus* sp. did not produce IAA. Therefore, enhanced plant growth and Cr extraction were not seen to a great extent (Rajkumar et al. 2005).

27.9 Mechanisms to Alleviate Heavy Metal Stress by Bacteria

Microbial cell metabolism requires metal cations to carry out various biochemical processes. Higher concentrations of these metal ions are toxic as they form complexes in microbial cells and ultimately inhibit bacterial growth (Ahemad and Malik

2011). Bacteria undergo resistance mechanisms, by which they immobilize, mobilize, uptake, and transform heavy metals. In this way, bacteria alleviate heavy metal ion toxicity. Major mechanisms employed by bacteria are physical sequestration, complexation, exclusion, and detoxification. Intracellular and extracellular materials bind heavy metals and prevent their entry in bacterial cells (Ahemad and Kibret 2013). These substances and the production of siderophores secreted by bacteria remove toxic effects of heavy metals and limit bioavailability of metals by forming complexes (Rajkumar et al. 2010). The production of particular metabolites (Rajkumar et al. 2010) and bacterial exhibit transporters have the ability to detoxify heavy metals (Ahemad 2012). Metal ions are compartmentalized after entering bacterial cells that use the sequestration detoxification method (Ahemad 2012). Bacteria also use methylation as a detoxification mechanism (Rajkumar et al. 2010).

27.10 Role of Bacteria in Alleviating Heavy Metal Toxicity

Overall plant associated microbes enhance phytoremediation processes directly by altering metal accumulation through biogeochemical processes and indirectly increase plant growth parameters (Rajkumar et al. 2012). Yang et al. (2012) investigated in field experiments that rhizobial microflora enhanced the ability of plants to uptake metals from soils. Addition of arsenate-reducing bacteria enhanced arsenic (As) accumulation, *P. vittata* growth, and reduced As leaching. Increased plant biomass by 53%, As uptake by 44%, and decreased As leaching from 21% to 71% were observed. Arsenate-reducing bacteria enhanced plant ability to remove As from contaminated soils. Bacterial strain OSM29 showed observable potential in removing heavy metals such as Cd, Cr, Cu, and Ni from the rhizosphere of cauliflower (Oves et al. 2013). Endophyte bacteria have a greater metal adaptation potential to heavy metals stress (Idris et al. 2006). It has been reported that endophytic bacteria enhanced phytoremediation by improving plant growth and inhibited metal stress by suppressing their translocation in plant parts (Ma et al. 2011). Bacteria-mediated processes have been used to alleviate heavy metal toxicity (Ma et al. 2015). Inoculation of endophytic bacteria (*Sphingomonas SaMR12*) enhanced Cd accumulation, plant biomass, root surface, area, and root tips. Increased root secretion of oxalic, succinic, and citric acid caused by bacterial inoculation alleviated Cd toxic effects and reduced oxidative stress. In addition, *SaMR12* enhanced Cd tolerance by inhibiting oxidative stress and finally improved Cd extraction by the target plant (Chen et al. 2014; Chen et al. 2014). Rahmanian et al. (2011) investigated entry of HM-resistant microbes in target plant species. A reduction in plant biomass and decreased Cd uptake in plant parts were observed. Combinations of bacteria and mushrooms have been used to remediate heavy metals. Recently, Li et al. (2016) investigated the combinations of *Agrocybe aegerita* and *Serratia* spp. on Ni and Cd stress. Increased bacteria number, soil enzymatic activities, and a great number of heavy metals in *A. aegerita* suggested that combined application of *Agrocybe aegerita* and *Serratia* spp. energetically alleviated heavy metal stress as compared to sole treatments. In addition, the impacts of treatment on bacterial community structure and composition highlighted the fact that integration of bacteria and mushrooms was a beneficial method for the bioremediation of soils containing heavy metals.

Several mechanisms have been adapted by bacteria for metal resistance such as siderophore production (Schalk et al. 2011) and compartmentalization inside the cell (Ahemad et al. 2012) (Table 27.2).

Table 27.2 Endophyte bacteria enhanced phytoremediation of contaminated heavy metals

Endophyte bacteria	Target plant	Heavy metal	Mechanism	References
Pseudomonas koreensis AGB-1	*Miscanthus sinensis*	Cd, Pb, Cu, Zn and As	Increased plant biomass, chlorophyll content, enzyme activities, protein content	Babu et al. (2015)
Bacillus thuringiensis GDB-1	*Alnus firma*	Cd, Pb, Cu, Zn and As	Increased biomass, chlorophyll content, metal accumulation	Babu et al. (2013)
Pseudomonas sp. Lk9	*Solanum nigrum*	Cd, Zn, Cu and Cr	Enhanced plant dry biomass, nutrients and metals uptake	Chen et al. (2014) and Chen et al. (2014)
Rahnella sp. JN6	*Polygonum pubescens*	Cd, Pb and Zn	Improve plant growth, metal tolerance and accumulation	He et al. (2013)
Bacillus sp. SLS18	*Sorghum bicolor L.*	Cd and Mn	Enhanced plant biomass and metal uptake	Luo et al. (2012)
B. pumilus E2S2, Bacillus sp. E1S2, Bacillussp. E4S1, Achromobacter sp. E4L5 and Stenotrophomonas sp. E1L	*Sedumplumbizincicola*	Cd, Zn and Pb	Enhanced plant growth, biomass, leaf chlorophyll content and metals uptake	Ma et al. (2015)
Endophytes belonged to Firmicutes, Proteobacteria and Actinobacteria	*Pteris vittata and P. multifida*	As	Reduce and oxidized As forms	Zhu et al. (2014)
Rahnella sp. JN27	*Zea mays*	Cd	Improve metal uptake and target plant growth	Yuan et al. (2014)
Exiguobacterium sp	*Vigna radiata*	As	Enhanced plant growth, biomass, chlorophyll contents, reduce oxidative stress,	Pandey and Bhatt (2016)

Recent research suggests that endophytic bacteria alleviate metal toxicity by precipitation (Babu et al. 2015), buildup of metals and sequestration (Shin et al. 2012), biotransformation to less toxic forms (Zhu et al. 2014), and adsorption (Luo et al. 2011).

27.11 Metal Detoxification Mechanisms in Fungi

Detoxification mechanisms involved in fungi are different from eukaryotes (Bellion et al. 2006). Extracellular mechanisms involved are chelation, precipitation, and cell wall binding. Intercellular mechanisms include binding to sulfur compounds, organic acids, peptides, polyphosphates, and transport into intracellular compartments. These substances play a major role in metal detoxification (Bellion et al. 2006). In addition, antioxidant defense systems that cope with heavy metal toxic effects are directly or indirectly involved in detoxification mechanisms (Bellion et al. 2006).

27.12 Role of Fungi in Alleviating Heavy Metal Toxicity

Plants adapt themselves to stress conditions; their growth and yield are adversely affected by heavy metal stress. It was observed that arbuscular mycorrhizal (AM) fungi establish a symbiotic relationship with target plants that leads to plant heavy metal tolerance. AM fungi have developed various strategies to alleviate heavy metal toxicity and reduce threats to the food chain. Strategies used by AM fungi are chelation of heavy metals inside fungal cells, adsorption to chitin in fungal cells, and immobilization of metals (Upadhyaya et al. 2010). AM fungi have the ability to fasten heavy metals far away from the rhizosphere with the help of a glycoprotein also called glomalin (Gohre and Paszkowski 2006). A report suggested that 1 g of glomalin extracted 0.08 mg Cd, 1.12 mg Pb, and 4.3 mg Cu from contaminated soils (Gonzalez-Chavez et al. 2004). Physiological and morphological changes caused by AM fungi lead to plant protection against metal stress (Miransari 2016). It has been observed that AM fungi inoculation alleviates heavy metal stress (Hildebrandt et al. 2007) and enhances growth of plants used for phytoremediation purposes (Carrasco et al. 2011). Similarly, AM fungi appreciably inhibit Cd and Pb uptake into target plant roots and translocation in above-ground plant parts. A decrease in electrolyte leakage and lipid peroxidation, and enhanced enzyme activities were noted by AM fungi stressed pigeonpea (Garg and Aggarwal 2012). Interaction between PGPR and AM fungi has also been evaluated regarding their ability to enhance plant growth under stressful conditions (Nadeem et al. 2014). Gharemaleki et al. (2010a, b) investigated that inoculation of plant growth–promoting rhizobacteria and fungi enhanced Cd and Zn uptake and translocation in corn (*Zea mays*). Combined application of PGPR and fungi were found to be effective regarding plant growth and remediation of Cd and Zn. Similarly, Zhang et al. (2010) noted the effects of inoculation with AM fungi on lead uptake, translocation, and stress alleviation in corn seedlings. Results showed that AM fungi inoculation alleviated oxidative stress induced by lead in mycorrhizal *Zea mays L* seedlings. Improvement in height, biomass, and basal diameter of *Zea mays L* seedlings was observed with clear alleviation in lead stress. In addition, application of the fungal isolates *Rhizophagus irregularis* and *Funneliformis mosseae* enhanced growth and biomass in sunflower plants by alleviating Cd, Zn, and Cu phytotoxicity, as mentioned by Hassan et al. (2013).

Other findings showed that AM fungi *Glomus intraradices* improved the growth of *Helianthus annuus* (Ker and Charest 2010) and *B. coddii* (Orlowska et al. 2011) and enhanced Ni extraction. AM fungi enhanced the dry mass of *Pteris vittata* and improved As uptake (Al Agely et al. 2005). Biró et al. (2005) documented the buffer effects of AM fungi on trace elements and other metals. A clear reduction in metals (Cr, Zn, Cu, Cd, Pb, Al, As, and Se) was noted in Barley (*Hordeum vulgare L.*). In addition, AM fungi significantly alleviate Zn phytotoxicity and enhance plant tolerance to Zn stress (Navarro et al. 2008; Gonzalez-Guerrero et al. 2009; Cavagnaro et al. 2010; Watts-Williams et al. 2013). The above reports suggest that fungi not only alleviate heavy metal toxicity but also enhance plant growth parameters.

27.13 Mechanisms of Metal Removal by Algae

Usually microorganisms remove heavy metals from solutions by precipitation, intra- and extra-cellular accumulation, cell surface biosorption, and complexation facilitated microorganisms (Cossich et al. 2002). Metal ions are captured in the cellular structure and adsorbed on binding sites in cellular structure, which is called biosorption (Malik 2004). Remediation of heavy metals by algae is carried out through bioaccumulation (using living cells) and biosorption (using non-living products). Biosorption was found to be quite rapid and effective (Aksu 1998).

27.14 Role of Algae in Alleviating Heavy Metal Toxicity

Detoxification of heavy metals involves various mechanisms such as precipitation, oxidation/reduction, sedimentation, complexation, microbial activity and uptake, adsorption, and cation anion exchange (Matagi et al. 1998). The use of aquatic plants particularly micro- and macro-algae has gained importance due to their metal absorption ability and ability to take up toxic metals from the surrounding environment (Mitra et al. 2012). At present, the role of microalgae in bioremediation is being extensively used (Abdel Hameed and Ebrahim 2007). Microalgae have been used for the removal of metal ions from the contaminated environment (Bitton 2011) and also take part in biosorption and bioaccumulation of toxic metals. Moreover, the ability of algae to remove metals from wastewater has also been appreciably documented (Zeraatkar et al. 2016). Microalgae remove heavy metals from contaminated water through two mechanisms, i.e. metabolism dependent,

Table 27.3 Heavy metal removal by specific microalgae

Microalgae	Heavy metals	Initial conc. (mg/L)	Final removal conc. (mg/g)	References
C. vulgaris	Cd	25–150	58.4	Aksu and Donmez (2006)
Desmodesmus pleiomorphus	Zn	1–30	360.2	Monteiro et al. (2009)
D. pleiomorphus	Cd	0.5–5	61.2	Monteiro et al. (2010)
Chlorella vulgaris	Ni	100	28.6 (immobilized)	Al-Rub et al. (2004)
Chlamydomonas reinhardtii	Cd	20–400	77.62	Tuzun et al. (2005)

Table 27.4 Heavy metals removal by specific macroalgae

Macroalgae	Heavy metals	Initial conc. (mg/L)	Final removal conc. (mg/g)	References
Ulva reticulata	Zn	1500	125.5	Senthilkumar et al. (2006)
Cladophora fascicularis	Cu	12.7–254.2	70.54	Deng et al. (2007)
Spirogyra insignis	Cd	10–150	87.7	Romera et al. (2007)
Sargassum wightii	Cu	100–1000	115	Vijayaraghavan and Prabu (2006)
Fucus spiralis	Cd	10–150	114.9	Romera et al. (2007)
	Zn		53.2	
Asparagopsis armata	Cd	10–150	32.3	Romera et al. (2007)
	Zn		21.6	
Chondrus crispus	Cd	10–150	75.2	Romera et al. (2007)
	Zn		45.7	

lower concentration uptake into cells, and biosorption (Matagi et al. 1998). Shanab et al. (2012) investigated the tolerance and removal of Cd, Pb, and Hg metals by using microalgae isolates *Phormidium ambiguum*, *Pseudochlorococcum typicum*, and *Scenedesmus quadricauda var. quadrispina* (chlorophyta) from fresh water. Macroalgae have been used for the biomonitoring and removal of heavy metals in the marine system (Gosavi et al. 2004) (Tables 27.3 and 27.4).

27.15 Conclusion and Future Perspectives

The toxic effects of heavy metals are well known worldwide. Many techniques and strategies have been used to remediate their toxicity in living organisms including plants, animals, and microorganisms. One of the emerging, cost-effective, and environmentally friendly techniques is bioremediation. The most specific bioremediation mediators are bacteria, fungi, and algae. Phytoremediation is an emerging technology that uses plants and microbes to clean polluted soils. Reports showed that bacteria alone and in combination with particular PGPRs alleviated heavy metal toxicity and enhanced plant growth attributes. AM fungi bind heavy metals far away from the rhizosphere by using glycoprotein. AM fungi inoculation alleviates heavy metal stress and enhances plant tolerance to metals. An improvement in plant growth parameters was observed by the interaction between fungi and PGPRs and has also been reported with algae. Both micro-and macro-algae take part in biosorption of toxic metals and alleviate heavy metal-induced phytotoxic effects on marine systems. Increased tolerance and removal of heavy metals by algae strains was also investigated.

However, further research is needed to investigate other beneficial biological methods and mechanisms involved in remediating heavy metal stress and improving plant growth. Biological fertilizers could be applied by using bacteria and fungi to provide essential nutrients to plant growth. However, more study is needed regarding exact and clear mechanisms involved in the removal of heavy metals by bacteria, fungi, and algae. Further development of innovative techniques is required to overcome existing ones.

References

Abdel Hameed MSA, Ebrahim OH (2007) Biotechnological potentials uses of immobilized algae. Int J Agric Biol 9:183–192

Adrees M, Ali S, Rizwan M, Ibrahim M, Abbas F, Farid M, Zia-ur-Rehman M, Irshad MK, Bharwana SA (2015a) The effect of excess copper on growth and physiology of important food crops: a review. Environ Sci Pollut Res. http://dx.doi.org/10.1007/s11356-015-4496-5

Adrees M, Ali S, Rizwan M, Rehman M, Ibrahim M, Abbas F, Farid M, Qayyum MF, Irshad MK (2015b) Mechanisms of silicon-mediated alleviation of heavy metal toxicity in plants: a review. Ecotoxicol Environ Saf 119:186–197

Ahemad M (2012) Implication of bacterial resistance against heavy metals in bioremediation: a review. IIOAB J 3:39–46

Ahemad M, Kibret M (2013) Recent trends in microbial biosorption of heavy metals: a review. Biochem Mol Biol 1:19–26

Ahemad M, Malik A (2011) Bioaccumulation of heavy metals by zinc resistant bacteria isolated from agricultural soils irrigated with wastewater. Bacteriol J 2:12–21

Ahluwalia SS, Goyal D (2007) Microbial and plant derived biomass for removal of heavy metals from wastewater. Bioresour Technol 98:2243–2257

Ahmad M, Wahid A, Ahmad SS, Butt ZA, Tariq M (2011) Ecophysiological responses of rice (*Oryza sativa* L.) to hexavalent chromium. Pak J Bot 43:2853–2859

Ahmad P, Ozturk M, Gucel S (2012) Oxidative damage and antioxidants induced by heavy metal stress in two cultivars of mustard (L) plants. Fres Environ Bull 21:2953–2961

Aksu Z (1998) Biosorption of heavy metals by microalgae in batch and continous systems. In: Wong YS, NFY T (eds) Algae for wastewater treatment. Springer, Berlin, pp 37–53

Aksu Z, Donmez G (2006) Binary biosorption of cadmium (II) and nickel (II) onto dried *Chlorella vulgaris*: co-ion effect on monocomponent isotherm parameters. Process Biochem 41:860–868

Al Agely A, Sylvia DM, Ma LQ (2005) Mycorrhizae increase arsenic uptake by the hyperaccumulator Chinese brake fern (*Pteris vittata* L). J Environ Qual 6:2181–2186

Ali S, Zeng F, Qiu L, Zhang GP (2011) The effect of chromium and aluminum on growth, root morphology, photosynthetic parameters and transpiration of the two barley cultivars differing in Al tolerance. Biol Plant 55:291–296

Ali H, Khan E, Sajad MA (2013) Phytoremediation of heavy metals-concepts and applications. Chemosphere 91:869–881

Al-Rub FAA, El-Naas MH, Benyahia F, Ashour I (2004) Biosorption of nickel on blank alginate beads, free and immobilized algal cells. Process Biochem 39:1767–1773

Andrade SAL, Gratão PL, Azevedo RA, Silveira APD, Schiavinato MA, Mazzafera P (2010) Biochemical and physiological changes in jack bean under mycorrhizal symbiosis growing in soil with increasing Cu concentrations. Environ Exp Bot 68:198–207. doi:10.1016/j.envexpbot.2009.11.009

Ashraf MY, Azhar N, Ashraf M, Hussain M (2011) Influence of lead on growth and nutrient accumulation in canola (*Brassica napus* L.) cultivars. J Environ Biol 32:659–666

Babu AG, Kim GD, Oh BT (2013) Enhancement of heavy metal phytoremediation by *Alnus firma* with endophytic *Bacillus thuringiensis* GDB-1. J Hazard Mater 250-251:477–483

Babu AG, Shea PJ, Sudhakar D, Jung IB, Oh BT (2015) Potential use of *Pseudomonas koreensis* AGB-1 in associated with *Miscanthus sinensis* to remediate heavy metal (loid)-contaminated mining site soil. J Enviorn Manage 151:160–166

Bakkaus E, Gouget B, Gallien JP, Khodja H, Carrot H, Morel JL, Collins R (2005) Concentration and distribution of cobalt in higher plants: the use of micro-PIXE spectroscopy. Nucl Instr Meth B 231:350–356

Barbosa RMT, de Almeida AF, Mielke MS, Loguercio LL, Mangabeira PAO, Gomes FP (2007) A physiological analysis of *Genipa americana* L.: a potential phytoremediator tree for chromium polluted watersheds. Environ Exp Bot 61:264–271

Barceló J, Poschenrieder C (2003) Phytoremediation: principles and perspectives. Contrib Sci 2:333–344

Bashir H, Ibrahim MM, Bagheri R, Ahmad J, Arif IA, Baig MA, Qureshi MI (2015) Influence of sulfur and cadmium on antioxidants, phytochelatins and growth in Indian mustard. J Plant Sci. doi:10.1093/aobpla/plv001

Bellion M, Courbot M, Jacob C, Blaudez D, Chalot M (2006) Extracellular and cellular mechanisms sustaining metal tolerance in ectomycorrhizal fungi. FEMS Microbiol Lett 254:173–181

Beskoski VP, Gojgic-Cvijovic G, Milic J, Ilic M, Miletic S, Solevic T, Vrvic MM (2011) Ex situ bioremediation of a soil contaminated by mazut (heavy residual fuel oil)- a field experiment. Chemosphere 83:34–40

Biró B, Posta K, Füzy A, Kadar I, Németh T (2005) Mycorrhizal functioning as part of the survival mechanisms of barley (*Hordeum vulgare* L.) at long-term heavy metal stress. Acta Biol Szegedien 49:65–67

Bitton G (2011) Wastewater microbiology, 4th edn. A John Witney & Sons Inc, Hoboken, pp 482–485

Boopathy R (2000) Factors limiting bioremediation technologies. Bioresour Technol 74:63–67

Brinza L, Dring MJ, Gavrilescu M (2007) Marine micro and macro algal species as biosorbents for heavy metals. Environ Eng Manag J 6:237–251

Carrasco L, Azcón R, Kohler J, Roldán A, Caravaca F (2011) Comparative effects of native filamentous and arbuscular mycorrhizal fungi in the establishment of an autochthonous, leguminous shrub growing in a metal-contaminated soil. Sci Total Environ 409:1205–1209

Cavagnaro TR, Dickson S, Smith FA (2010) Arbuscular mycorrhizas modify plant response to soil zinc addition. Plant Soil 329:307–313

Cenkci S, Cigerci IH, Yildiz M, Özay C, Bozdag A, Terzi H (2010) Lead contamination reduces chlorophyll biosynthesis and genomic template stability in *Brassica rapa* L. Environ Exp Bot 67:467–473

Chen L, Luo SL, Li XJ, Wan Y, Chen JL, Liu CB (2014) Interaction of Cd hyperaccumulator Solanum nigrum L. and functional endophyte Pseudomonas sp. Lk9 on soil heavy metals uptake. Soil Biol. Biochemist 68:300–308

Chen B, Zhang Y, Rafiq MT, Khan KY, Pan F, Yang X, Feng Y (2014) Improvement of cadmium uptake and accumulation in *Sedum alfredii* by endophytic bacteria *Sphingomonas SaMR12*: Effects on plant growth and root exudates. Chemosphere 117:367–373

Chodak M, Gołębiewski M, Morawska-Płoskonka J, Kuduk K, Niklińska M (2013) Diversity of microorganisms from forest soils differently polluted with heavy metals. Appl Soil Ecol 64:7–14

Cossich ES, Tavares CRG, Ravagnani TMK (2002) Biosorption of chromium (III) by *sargassium* sp. Biomass. Electron J Biotechnol 5:133–140
Datta JK, Bandhyopadhyay A, Banerjee A, Mondal NK (2011) Phytotoxic effect of chromium on the germination, seedling growth of some wheat (*Triticum aestivum* L.) cultivars under laboratory condition. J Agric Technol 7:395–402
Deng L, Zhu X, Wang X, Su Y, Su H (2007) Biosorption of copper (II) from aqueous solutions by green alga *Cladophora fascicularis*. Biodegradation 18:393–402
Dheeba B, Sampathkumar P, Kannan K (2015) Fertilizers and mixed crop cultivation of chromium tolerant and sensitive plants under chromium toxicity. J Toxicol. doi:10.1155/2015/367217
Doble M, Kumar A (2005) Biotreatment of industrial effluents. Elsvier Butterworth-Heinemann, Boston
Doty SL, Shang QT, Wilson AM, Moore AL, Newman LA, Strand SE (2007) Enhanced metabolism of halogenated hydrocarbons in transgenic plants containing mammalian P450 2E1. Proc Natl Acad Sci U S A 97:6287–6291
Erakhrumen AA (2007) Phytoremediation: an environmentally sound technology for pollution prevention, control and remediation in developing countries. Educ Res Rev 2:151–156
Farid M, Shakoor MB, Ehsan S, Ali S, Zubair M, Hanif MA (2013) Morphological, physiological and biochemical responses of different plant species to Cd stress. IJCBS 3:53–60
Feng JP, Shi QH, Wang XF, Wei M, Yang FJ, Xu HN (2010) Silicon supplementation ameliorated the inhibition of photosynthesis and nitrate metabolism by cadmium (Cd) toxicity in *Cucumis sativus* L. Sci Hortic 123:521–530
Gajewska E, Sklodowska M, Slaba M, Mazur J (2006) Effect of nickel on antioxidative enzyme activities, proline and chlorophyll contents in wheat shoots. Biol Plant 50:653–659
Gamalero E, Lingua G, Berta G, Glick BR (2009) Beneficial role of plant growth promoting bacteria and arbuscular mycorrhizal fungi on plant responses to heavy metal stress. Can J Microbiol 55:501–514
Garg N, Aggarwal N (2012) Effect of mycorrhizal inoculations on heavy metal uptake and stress alleviation of *Cajanus cajan* (L.) Millsp. Genotypes grown in cadmium and lead contaminated soils. Plant Growth Regul 66:9–26
Gharemaleki T, Besharati H, Rasouli-Sadaghiani MH, Tavasoli A (2010a) Effect of soil microbial activity in phytoremediation of Zn. International Soil Science Congress on "Management of natural resources to sustain soil health and quality", Samsun, Turkey. p 326
Gharemaleki T, Rasouli-Sadaghiani MH, Besharati H, Tavasoli A (2010b) Plant growth-promoting microorganisms effect on Cd uptake by *Zea mays* in a contaminated soil. International Soil Science Congress on "Management of natural resources to sustain soil health and quality", Samsun, Turkey. pp 1135–1140
Ghosh P, Rathinasabapathi B, Ma LQ (2011) Arsenic-resistant bacteria solubilized arsenic in the growth media and increased growth of arsenic hyperaccumulator *Pteris vittata* L. Bioresour Technol 102:8756–8761
Gill RA, Zang L, Ali B, Farooq MA, Cui P, Yang S, Zhou W (2015) Chromium-induced physiochemical and ultrastructural changes in four cultivars of *Brassica napus* L. Chemosphere 120:154–164
Glick BR (2004) Teamwork in phytoremediation. Nat Biotechnol 22:526–527
Glick BR (2010) Using soil bacteria to facilitate phytoremediation. Biotechnol Adv 28:367–374
Gohre V, Paszkowski U (2006) Contribution of arbuscular mycorrhizal symbiosis to heavy metal phytoremediation. Plant 223:1115–1122
Gonzalez-Chavez MC, Carrillo-Gonzalez R, Wright SF, Nichols KA (2004) The role of glomalin, a protein produced by arbuscular mycorrhizal fungi, in sequestering potentially toxic elements. Environ Pollut 130:317–323
Gonzalez-Guerrero M, Benabdellah K, Ferrol N, Azcon-Aguilar C (2009) Mechanisms underlying heavy metal tolerance in arbuscular mycorrhizas. In: Azcon-Aguilar C, Barea JM, Gianinazzi S, Gianinazzi-Pearson V (eds) Mycorrhizas functional processes and ecological impact. Springer, Berlin, pp 107–122

Gosavi K, Sammut J, Jankowski J (2004) Macro algal biomanitors of trace metal contamination in acid sulfate soil aquaculture ponds. Sci Total Environ 324:25–39

Govindasamy C, Arulpriya M, Ruban P, Francisce LJ, Ilayaraja A (2011) Concentration of heavy metals in seagrasses tissue of the palk strait, Bay of Bangal. Int J Enviorn Sci 2:145–153

Guo J, Dai X, Xu W, Ma M (2008) Over expressing GSHI and AsPCSI simultaneously increases the tolerance and accumulation of cadmium and arsenic in *Arabidopsis thaliana*. Chemosphere 72:1020–1026

Hadi F, Bano A (2010) Effect of diazotrophs (Rhizobium and Azobactor) on the growth of maize (*Zea mays* L) and accumulation of Lead (Pb) in different plant parts. Pak J Bot 42:4363–4370

Hashim MA, Mukhopadhyay S, Sahu JN, Sengupta B (2011) Remediation technologies for heavy metal contaminated groundwater. J Environ Manag 92:2355–2388

Hassan SE, Hijri M, St-Arnaud M (2013) Effect of arbuscular mycorrhizal fungi on trace metal uptake by sunflower plants grown on cadmium contaminated soil. New Biotechnol 30:780–787

He H, Ye Z, Yang D, Yan J, Xiao L, Zhong T, Yuan M, Cai X, Fang Z, Jing Y (2013) Characterization of endophytic *Rahnella* sp. JN6 from *Polygonum pubescens* and its potential in promoting growth and Cd, Pb, Zn uptake by *Brassica napus*. Chemosphere 90:1960–1965

Hildebrandt U, Regvar M, Bothe H (2007) Arbuscular mycorrhiza and heavy metal tolerance. Phytochemistry 68:139–146

Hu YA, Liu XP, Bai JM, Shih KM, Zeng EY, Cheng HF (2013) Assessing heavy metal pollution in the surface soils of a region that had undergone three decades of intense industrialization and urbanization. Environ Sci Pollut Res Int 20:6150–6159

Huang S, Peng B, Yang Z, Chai L, Zhou L (2009) Chromium accumulation, microorganism population and enzyme activities in soils around chromium-containing slag heap of steel alloy factory. Trans Nonferrous Metals Soc China 19:241–248

Huang L, Xie J, Lv B, Shi X, Li G, Liang F, Lian J (2013) Optimization of nutrient component for diesel oil degradation by *Acinetobacter beijerinckii* ZRS. Mar Pollut Bull 76(1–2):325–332. doi:10.1016/j marpolbul.2013.03.037

Idris R, Kuffner M, Bodrossy L, Puschenreiter M, Monchy S, Wenzel WW, Sessitsch A (2006) Characterization of Ni-tolerant methylobacteria associated with the hyperaccumulating plant *Thlaspi goesingense* and description of *Mathylobacterium goesingense* sp. nov. Syst Appl Microbiol 29:634–644

Irfan M, Ahmad A, Hayat S (2014) Effect of cadmium on the growth and antioxidant enzymes in two varieties of *Brassica juncea*. Saudi J Biol Sci 21:125–131

Jun R, Ling T, Guanghua Z (2009) Effects of chromium on seed germination, root elongation and coleoptile growth in six pulses. Int J Environ Sci Tech 6:571–578

Kamenidou S, Cavins TJ, Marek S (2009) Evaluation of silicon as a nutritional supplement for greenhouse zinnia production. Sci Hortic 119:297–301

Ker K, Charest C (2010) Nickel remediation by AM-colonized sunflower. Mycorrhiza 20:399–406

Khan MA, Ahmad I, Ur Rahman I (2007) Effect of environmental pollution onheavy metals content of Withania somnifera. J Chin Chem Soc 54:339343

Khan S, Afzal M, Iqbal S, Khan QM (2013) Plant-bacteria partnerships for the remediation of hydrocarbon contaminated soils. Chemosphere 90:1317–1332

Khan MU, Shahbaz N, Waheed S, Mahmood A, Shinwari ZK, Malik RN (2016) Comparative health risk surveillance of heavy metals via dietary foodstuff consumption in different land use types of Pakistan. Hum Ecol Risk Assess An Int J 22:168–186

Kothe E, Bergmann H, Buchel G (2005) Molecular mechanisms in biogeo-interactions: from a case study to general mechanisms. Chemie der Erde Geochem 65:7–27

Kumar A, Bisht BS, Joshi VD, Dhewa T (2011) Review on bioremediation of polluted environment: a management tool. Int J Environ Sci 1:1079–1093

Lambert M, Leven B, Green R (2000) New methods of cleaning up heavy metals in soils and water. In: Environmental science and Technology Briefs for citizens. Kansas State University, Manhattan

Li HF, Gray C, Mico C, Zhao FJ, McGrath SP (2009) Phytotoxicity and bioavailability of cobalt to plants in a range of soils. Chemosphere 75:979–986

Li X, Dong S, Yao Y, Shi W, Wu M, Xu H (2016) Inoculation of bacteria for the bioremediation of heavy metals contaminated soil by *Agrocybe aegerita*. RSC Adv 6:65816–65824

Liu D, Zou J, Wang M, Jiang W (2008) Hexavalent chromium uptake and its effects on mineral uptake, antioxidant defence system and photosynthesis in *Amaranthus viridis* L. Bioresour Technol 99:2628–2636

Liu M, Huang B, Bi X, Ren Z, Shenga G, Fu J (2013) Heavy metals and organic compounds contamination in soil from an e-waste region in South China. Environ Sci Process Impacts 15:919–929

Liu L, Zhang X, Zhong T (2015) Pollution and health risk assessment of heavy metals in urban soil in China. Hum Ecol Risk Assess An Int. doi:10.1080/10807039.2015.1078226

Luo SL, Wan Y, Xiao X, Guo H, Chen L, Xi Q, Zeng G, Liu C, Chen J (2011) Isolation and characterization of endophytic bacterium LRE07 from cadmium hyperaccumulator *Solanum nigrum* L. and its potential for remediation. Appl Microbiol Biotechnol 89:1637–1644

Luo S, Xu T, Chen L, Chen J, Rao C, Xiao X, Wan Y, Zeng G, Long F, Liu C, Liu Y (2012) Endophyte-assisted promotion of biomass production and metal uptake of energy crop sweet sorghum by plant-growth-promoting endophyte *Bacillus* sp. SLS18. Appl Microbiol Biotechnol 93:1745–1753

Ma Y, Rajkumar M, Luo YM, Freitas H (2011) Inoculation of endophytic bacteria on host and non-host plants- effects on plant growth and Ni uptake. J Hazard Mater 195:230–237

Ma Y, Oliveira RS, Nai FJ, Rajkumar M, Luo YM, Rocha I, Freitas H (2015) The hyperaccumulator Sedum Plumbizincicola harbors metal-resistant endophytic bacteria that improve its phytoextraction capacity in multi-metal contaminated soil. J Environ Manag 156:62–69

Malik A (2004) Metal bioremediation through growing cells. Environ Int 30:261–278

Manousaki E, Kalogerakis N (2011) Halophytes present new opportunities in phytoremediation of heavy metals and saline soils. Ind Eng Chem Res 50:656–660

Matagi S, Swaiand D, Mugabe R (1998) A review of heavy metal removal mechanisms in wetlands. Afr J Trop Hydrobiol Fish 8:23–35

Medina A, Roldán A, Azcón R (2010) The effectiveness of arbuscular-mycorrhizal fungi and *Aspergillus Niger* or *Phanerochaete chrysosporium* treated organic amendments from olive residues upon plant growth in a semi-arid degraded soil. J Environ Manag 91(2547):2553. doi:10.1016/j. jenvman.2010.07.008

Meier S, Borie F, Bolan N, Cornejo P (2012) Phytoremediation of metal-polluted soils by Arbuscular Mycorrhizal fungi. Crit Rev Environ Sci Technol 42:741–775

Miransari, M (2016). Stress and Mycorrhizal Plant. Recent Advances on Mycorrhizal Fungi. Pagano, MC, Cham, Springer International Publishing Switzerland, pp 63–79

Mitra N, Rezvan Z, Seyed Ahmad M, Gharaie M, Hosein M (2012) Studies of water arsenic and boron pollutants and algae phytoremediation in three springs, Iran. Int J Ecosys 2:32

Mohanpuria P, Rana NK, Yadav SK (2007) Cadmium induced oxidative stress influence on glutathione metabolic genes of *Camellia sinensis* (L.) O Kuntze. Environ Toxicol 22:368–374

Monteiro CM, Marques APGC, Castro PML, Malcata FX (2009) Characterization of *Desmodesmus pleiomorphus* isolated from a heavy metal-contaminated site: biosorption of zinc. Biodegradation 20:629–641

Monteiro CM, Castro PML, Malcata FX (2010) Cadmium removal by two strains of *Desmodesmus pleiomorphus* cells. Water Air Soil Pollut 208:17–27

Mukhopadhyay S, Maiti SK (2010) Phytoremediation of metal enriched mine waste: a review. Global J Environ Res 4:135–150

Nadeem SM, Ahmad M, Zahir ZH, Javaid A, Ashraf M (2014) The role of mycorrhizae and plant growth promoting rhizobacteria (PGPR) in improving crop productivity under stressful environments. Biotechnol Adv 32:429–448

Najeeb U, Ahmad W, Zia MH, Malik Z, Zhou W (2014) Enhancing the lead phytostabilization in wetland plant *Juncus ef usus* L. through somaclonal manipulation and EDTA enrichment. Arab J Chem (2017) 10:S3310–S3317. http://dx.doi.org/10.1016/j.arabjc.2014.01.009

Nath S, Deb B, Sharma I (2012) Isolation and characterization of cadmium and lead resistant bacteria. Glob Adv Res J Microbiol 1(11):194–198

Navarro E, Baun A, Behra R, Hartmann NB, Filser J, Miao AJ, Quigg A, Santschi PH, Sigg L (2008) Environmental behavior and ecotoxicity of engineered nanoparticles to algae, plants, and fungi. Ecotoxicol 17:372–386

Orlowska E, Przybylowicz W, Orlowski D, Turnau K, Mesjasz-Przybylowicz J (2011) The effect of mycorrhiza on the growth and elemental composition of Ni-hyperaccumulating plant *Berkheya coddii* Roessler. Environ Pollut 159:3730–3738

Oves M, Khan MS, Zaidi A (2013) Biosorption of heavy metals by *Bacillus thuringiensis* strain OSM29 originating from industrial effluent contaminated north Indian soil. Saudi J Biol Sci 20:121–129

Ozdener Y, Aydin BK, Aygun SF, Yurekli F (2011) Effect of hexavalent chromium on the growth and physiological and biochemical parameters on *Brassica oleracea* L. var. acephala DC. Acta Biol Hung 62:463–476

Padmavathiamma PK, Li LY (2007) Phytoremediation technology: hyperaccumulation metals in plants. Water Air Soil Pollut 184:105–126

Panagos P, Van Liedekerke M, Yigini Y, Montanarella L (2013) Contaminated sites in Europe: review of the current situation based on data collected through a European network. J Environ Pub Health 1–11

Pandey N, Bhatt N (2016) Role of soil associated *Exiguobacterium* in reducing arsenic toxicity and promoting plant growth in *Vigna radiate*. Eur J Soil Biol 75:142–150

Poljsak B, Pócsi I, Raspor P, Pesti M (2010) Interference of chromium with biological systems in yeast and fungi: a review. J Basic Microbiol 50:21–36

Poornima M, Kumar RS, Thomas PD (2014) Isolation and molecular characterization of bacterial Strains from tannery effluent and reduction of chromium. Int J Curr Microbiol Appl Sci 3:530–538

Pulford I, Watson C (2003) Phytoremediation of heavy metal-contaminated land by trees-a review. Environ Int 29:529–540

Rafati M, Khorasani N, Moattar F, Shirvany A, Moraghebi F, Hosseinzadeh S (2011) Phytoremediation potential of *Populus alba* and *Morus alba* for cadmium, chromium and nickel absorption from polluted soil. Int J Environ Res 5:961–970

Rahmanian M, Khodaverdiloo H, Rezaee Danesh Y, Rasouli Sadaghiani MH (2011) Effects of heavy metal resistant soil microbes inoculation and soil Cd concentration on growth and metal uptake of millet, couch grass and Alfalfa. Afr J Microbiol Res 5:403–410

Rajkumar M, Lee KJ, Lee WH, Banu JR (2005) Growth of *Brassica juncea* under chromium stress: influence of siderophores and indole-3-acetic acid-producing rhizosphere bacteria. J Environ Biol 26:693–699

Rajkumar M, Ae N, Prasad MNV, Freitas H (2010) Potential of siderophore-producing bacteria for improving heavy metal phytoextraction. Trends Biotechnol 28:142–149

Ray S, Ray MK (2009) Bioremediation of heavy metal toxicity-with special Reference to chromium Al Ameen. J Med Sci 2:57–63

Reddy AM, Kumar SG, Jyonthsnakumari G, Thimmanaik S, Sudhakar C (2005) Lead induced changes in antioxidant metabolism of horsegram (*Macrotylomauniflorum (Lam.)Verdc.*) and Bengal gram (*Cicerarietinum L.*) Chemosphere 60:97–104

Redondo-Gomez S, Mateos-Naranjo E, Vecino-Bueno I, Feldman SR (2011) Accumulation and tolerance characteristics of chromium in a cordgrass Cr hyperaccumulator, *Spartina argentinensis*. J Hazard Mater 185:862–869

Romera E, Gonzalez F, Ballester A, Blazquez ML, Munoz JA (2007) Comparative study of biosorption of heavy metals using different types of algae. Bioresour Technol 98:3344–3353

Ruta L, Paraschivescu C, Matache M, Avramescu S, Farcasanu IC (2010) Removing heavy metals from synthetic effluents using "kamikaze" *Saccharomyces cerevisiae* cells. Appl Microbiol Biotechnol 85:763–771

Sakai Y, Ma Y, Xu C, Wu H, Zhu W, Yang J (2010) Phytodesalination of a salt affected soil with four halophytes in China. J Arid Land Stud 22:17–20

Schalk IJ, Hannauer M, Braud A (2011) New roles of bacterial siderophores in metal transport and tolerance. Environ Microbiol 13:2844–2854

Scoccianti V, Crinelli R, Tirillini B, Mancinelli V, Speranza A (2006) Uptake and toxicity of Cr (Cr^{3+}) in celery seedlings. Chemosphere 64:1695–1703

Senthilkumar R, Vijayaraghavan K, Thilakavathi M, Iyer PVR, Velan M (2006) Seaweeds for the remediation of wastewaters contaminated with zinc (II) ions. J Hazard Mater B 136:791–799

Shanab S, Essa A, Shalaby E (2012) Bioremoval capacity of three heavy metals by some microalgae species (Egyptian Isolates). Plant Signal Behav 7:392–399

Sharma SS, Dietz KJ (2009) The relationship between metal toxicity and cellular redox imbalance. Trends Plant Sci 14:43–50

Sheoran V, Sheoran A, Poonia P (2011) Role of hyperaccumulators in phytoextraction of metals from contaminated mining sites: a review. Crit Rev Environ Sci Technol 41:168–214

Shi GR, Cai QS, Liu CF, Wu L (2010) Silicon alleviates cadmium toxicity in peanut plants in relation to cadmium distribution and stimulation of antioxidative enzymes. Plant Growth Regul 61:45–52

Shin M, Shim J, You Y, Myung H, Bang KS, Cho M, Kamala-Kannan S, Oh BT (2012) Characterization of lead resistant endophytic Bacillus sp. MN3-4 and its potential for promoting lead accumulation in metal hyperaccumulator *Alnus firma*. J Hazard Mater 199-200:314–320

Singh S (2012) Phytoremediation: a sustainable alternative for environmental challenges. Int J Gr Herb Chem 1:133–139

Singh A, Kuhad RC, Ward OP (2009) Biological remediation of soil: an overview of global market and available technologies. In: Advances in applied bioremediation. Springer, Berlin/Heidelberg

Subrahmanyam D (2008) Effects of chromium toxicity on leaf photosynthetic characteristics and oxidative changes in wheat *Triticum aestivum L.* Photosynthetica 46:339–345

Tabak HH, Lens P, Van Hullebusch ED, Dejonghe W (2005) Developments in bioremediation of soils and sediments polluted with metals and radionuclides e1. Microbial processes and mechanisms affecting bioremediation of metal contamination and influencing metal toxicity and transport. Rev Environ Sci Biotechnol 4:115–156

Tangahu BV, Abdullah SRS, Basri H, Idris M, Anuar N, Mukhlisin M (2011) A review on heavy metals (As, Pb, and Hg) uptake by plants through phytoremediation. Int J Chem Eng

Tariq M, Ali M, Shah Z (2006) Characteristics of industrial effluents and their possible impacts on quality of underground water. Soil Environ 25:64–69

Thangavel P, Subbhuraam C (2004) Phytoextraction: role of hyperaccumulators in metal contaminated soils. Proc Indian Natl Sci Acad Part B 70:109–130

Theriappan P, Gupta AK, Dasarathan P (2011) Accumulation of proline under salinity and heavy metal stress in cauliflower seedlings. J Appl Sci Environ Manage 15:251–255

Titah HS, Abdullaha SRS, Mushrifah I, Anuar N, Basri H, Mukhlisin M (2013) Effect of applying rhizobacteria and fertilizer on the growth of *Ludwigiaoctovalvis* for arsenic uptake and accumulation in phytoremediation. Ecol Eng 58:303–313

Tuzun I, Bayramoglu G, Yalcin E, Basaran G, Celik G, Arica MY (2005) Equilibrium and kinetic studies on biosorption of Hg(II), Cd(II) and Pb(II) ions onto microalgae *Chlamydomonas* reinhardtii. J Environ Manag 77:85–92

Ullah A, Heng S, Munis MFH, Fahad S (2015) Phytoremediation of heavy metals assisted by plant growth promoting (PGP) bacteria: a review. Environ Exp Bot 117:28–40

Upadhyaya H, Panda SK, Bhattacharjee MK, Dutta S (2010) Role of Arbuscular Mycorrhiza in heavy metal tolerance in plants: Prospects for phytoremediation. J Phytol 2:16–27

Vajpayee P, Khatoon I, Patel CB, Singh G, Gupta KC, Shanker R (2011) Adverse effects of chromium oxide nano-particles on seed germination and growth in *Triticum aestivum L.* J Biomed Nanotechnol 7:205–206

Vijayaraghavan K, Prabu D (2006) Potential of Sargassum wightii biomass for copper (II) removal from aqueous solutions: application of different mathematical models to batch and continuous biosorption data. J Hazard Mater B 137:558–564

Villiers F, Ducruix C, Hugouvieux V (2011) Investigating the plant response to cadmium exposure by proteomic and metabolomic approaches. Proteomics 11:1650–1663

Vishnoi SR, Srivastava PN (2008) Phytoremediation-green for environmental clean. In: The 12th world Lake Conference. Jaipur, Rajasthan, India, pp 1016–1021

Waterlot C, Bidar G, Pelfrêne A, Roussel H, Fourrier H, Douay F (2013) Contamination, fractionation and availability of metals in urban soils in the vicinity of former lead and zinc smelters, France. Pedosphere 23:143–159

Watts-Williams SJ, Patti AF, Cavagnaro TR (2013) Arbuscular mycorrhizas are beneficial under both deficient and toxic soil zinc conditions. Plant Soil 371:299–312

Weyens N, Truyens S, Dupae J, Newman L, Taghavi S, van der Lelie D, Carleer R, Vangronsveld J (2010) Potential of the TCE-Degrading endophyte *Pseudomonas putida* W619-TCE to improve plant growth and reduce TCE phytotoxicity and evapotranspiration in poplar cuttings.S. Environ Pollut 158:2915–S2919

Wuana RA, Okieimen FE (2011) Heavy metals in contaminated soils: a review of sources, chemistry, risks and best available strategies for remediation. ISRN Ecol 2011:1–20

Yadav R, Arora P, Kumar S, Chaudhury A (2010) Perspectives for genetic engineering of poplars for enhanced phytoremediation abilities. Ecotoxicolgy 19:1574–1588

Yang Q, Tu S, Wang G, Liao X, Yan X (2012) Effectiveness of applying arsenate reducing Bacteria to enhance arsenic removal from polluted soils by *Pteris Vittata* L. Int J Phytoremediation 14:89–99

Yruela I (2009) Copper in plants: acquisition transport and interactions. Funct Plant Biol 36:409–430

Yuan M, He H, Xiao L, Zhong T, Liu L, Li S, Deng P, Ye Z, Jing Y (2014) Enhancement of Cd phytoextraction by two *Amaranthus* species with endophytic *Rahnella* sp. JN27. Chemosphere 103:99–104

Zaheer IE, Ali S, Rizwan M, Farid M, Shakoor MB, Gill RA, Najeeb U, Iqbal N, Ahmad R (2015) Citric acid assisted phytoremediation of copper by *Brassica napus* L. Ecotoxicol Environ Saf 120:310–317

Zaidi S, Usmani S, Singh BR, Musarrat J (2006) Significance of *Bacillus subtilis* strain SJ-101 as a bioinoculant for concurrent plant growth promotion and nickel accumulation in *Brassica juncea*. Chemosphere 64:991–997

Zeraatkar AK, Ahmadzadeh H, Talebi AF, Moheimani NR, McHenry MP (2016) Potential use of algae for heavy metal bioremediation, a critical review. J Environ Manag. doi:10.1016/j.jenvman.2016.06.059

Zhang HH, Tang M, Chen H, Zheng CL, Niu ZC (2010) Effect of inoculation with AM fungi on lead uptake, translocation and stress alleviation of *Zea mays* L. seedlings planting in soil with increasing lead concentrations Eur. J Biol 46:306–311

Zhou ZS, Huang SQ, Guo K, Mehta SK, Zhang PC, Yang ZM (2007) Metabolic adaptations to mercury-induced oxidative stress in roots of *Medicago sativa L.* J Inorganic Biochem 101:1–9

Zhu LJ, Guan DX, Luo J, Rathinasabapathi B, Ma LQ (2014) Characterization of arsenic-resistant endophytic bacteria from hyperaccumuators *Pteris vittata* and *Pteris multifida*. Chemosphere 113:9–16

Zorrig W, Rabhi M, Ferchichi S, Smaoui A, Abdelly C (2012) Phytodesalination: a solution for salt-affected soils in arid and semi-arid regions. J Arid Land Stud 22:299–302